For the Instructor

Test Bank
0-534-37460-3

The test bank includes 3 tests per chapter as well as 3 final exams. The tests comprise multiple-choice, free-response, true/false, and fill-in-the-blank questions.

Complete Solutions Manual
0-534-37459-X

The complete solutions manual provides worked out solutions to all of the problems in the text.

Text-Specific Video Series
0-534-37644-4

This set of videotapes is available free upon adoption of the text. There is one tape for each chapter of the text and each is broken down into 10-20 minute problem-solving lessons that cover every section of the chapter.

BCA Testing
0-534-38964-3

Brooks/Cole Assessment is a revolutionary, Internet-ready text-specific testing suite that allows instructors to customize exams and track student progress in an accessible, browser-based format. *BCA* offers full algorithmic generation of problems and free-response mathematics. No longer are you limited to multiple-choice or true/false test questions. *BCA* completely integrates testing and course-management, simplifying your routine tasks. Test results flow automatically to your gradebook and you can communicate easily to individuals, sections, or entire courses.

For the Student

Explorations in Beginning and Intermediate Algebra Using the TI-82/83 with Integrated Appendix Notes for the TI-85/86, 2/e

Deborah J. Cochener and Bonnie M. Hodge, both of Austin Peay State University
0-534-36149-8

This user-friendly workbook improves student understanding and retention of algebra concepts through a series of activities and guided explorations using the graphing calculator. An ideal supplement for any beginning or intermediate algebra course, *Explorations in Beginning and Intermediate Algebra, 2/e* is an ideal tool for integrating technology without sacrificing course content. It clearly and succinctly teaches keystrokes so your class time can be devoted to investigations rather than to how to use a graphing calculator.

Mastering Mathematics: How to Be a Great Math Student, 3/e

Richard Manning Smith, Bryant College
0-534-34947-1

Providing solid tips for every stage of study, *Mastering Mathematics* stresses the importance of a positive attitude and gives students the tools to succeed in their math courses. This practical guide will help students avoid mental blocks during math exams, identify and improve areas of weakness, get the most out of class time, study more effectively, overcome a perceived "low math ability," be successful on math tests, get back on track when feeling "lost," and much more!

Beginning and Intermediate Algebra Activities Manual

Debbie Garrison, Judy Jones, and Jolene Rhodes, all of Valencia Community College
Instructor Edition: 0-534-35356-8;
Student Edition: 0-534-35355-X

Designed as a stand-alone supplement for any beginning or intermediate algebra text, *Activities in Beginning and Intermediate Algebra* is a collection of activities that incorporates the recommendations from the NCTM and from AMATYC's *Crossroads*. Activities can be used during class or in a laboratory setting to introduce, teach, or reinforce a topic. This set of activities facilitates discovery learning, collaborative learning, use of graphing technology, connections with other areas of mathematics and other disciplines, oral and written communication, real data collection, and active learning.

Student Resources continued on next page

Additional Resources

Conquering Math Anxiety: A Self-Help Workbook

Cynthia Arem, Pima Community College

0-534-18876-1

This comprehensive workbook provides a variety of exercises and worksheets along with detailed explanations of methods to help "math-anxious" students deal with and overcome math fears.

Active Arithmetic and Algebra: Activities for Prealgebra and Beginning Algebra

0-534-36771-2

Judy Jones, Valencia Community College

This activities manual includes a variety of approaches to learning mathematical concepts. Sixteen activities, including puzzles, games, data collection, graphing, and writing activities are included.

Student Solutions Manual

0-534-37458-1

The student solutions manual provides worked out solutions to the odd-numbered problems in the text.

BCA Tutorial Instructor and Student Versions

This text-specific, interactive tutorial software is delivered via the Web (at **http://bca.brookscole.com**) and is offered in both student and instructor versions. Like *BCA Testing*, it is browser-based, making it an intuitive mathematical guide, even for students with little technological proficiency. *BCA Tutorial* allows students to work with real math notation in real time and provides instant analysis and feedback. The tracking program built into the instructor version of the software enables instructors to monitor student progress carefully.

Interactive Video Skillbuilder CD-ROM

Packaged with the book, this CD-ROM contains more than 8 hours of video instruction. The problems worked during each video lesson are listed next to the viewing screen, so students can work them ahead of time, if they choose. In order to help students evaluate their progress, each section contains a 10-question Web quiz and each chapter contains a chapter test, with answers to each problem provided.

The Math Student's Guide to the TI-83 Graphing Calculator

0-534-37802-1

The Math Student's Guide to the TI-86 Graphing Calculator

0-534-37801-3

Trish Cabral, Butte College

This instructive video is designed for students who are new to the graphing calculator or for those who would like to brush up on their skills. Topics covered include basic calculations, the custom menu, graphing, advanced graphing, matrix operations, trigonometry, parametric equations, polar coordinates, calculus, Statistics I and one variable data, and Statistics II with linear regression. This valuable tool is 105 minutes in length and covers all of the important functions of a graphing calculator.

Web site www.brookscole.com/ mathematics

The Brooks/Cole Mathematics Resource Center offers book-specific student and instructor resources, discipline specific links, and a complete catalog of Brooks/Cole mathematics products.

If you are interested in value-added bundling of any of these additional resources with the text please contact your local Brooks/Cole representative.

Chapter 1 – An Introduction to Algebra

Ideas for Instruction	Print Resources	Media Resources for Instructors	Media Resources for Students

Explorations in Beginning and Intermediate Algebra
• Activities 1-4
 	
Activities for Beginning and Intermediate Algebra
• Activities 22-25
 	
Active Arithmetic and Algebra
• Bar Graphs and Line Graphs
• The Algebraic Match-Up
• Understanding Fractions
• The Prealgebra Pizza Party
	Test Bank
	• Chapter 1
	  Test Form A
	  Test Form B
	  Test Form C
	
	Complete Solutions Manual
	• Chapter 1
	  Pages 1-38
	
	Student Solutions Manual
	• Chapter 1
	  Pages 1-21
	
	How to Be a Great Math Student
	
	Conquering Math Anxiety
		BCA Testing
		http://bca.brookscole.com
		
		BCA Tutorial
		http://bca.brookscole.com
		
		Web Site
		www.brookscole.com/ mathematics
		
		Text-Specific Videos
		• Chapter 1
		
		TI-83 and TI-86 Graphing Calculator Videos
			Interactive Video SkillBuilder
			• Chapter 1
			
			BCA Tutorial
			http://bca.brookscole.com
			
			Web Site
			www.brookscole.com/ mathematics
			
			Text-Specific Videos
			• Chapter 1
			
			TI-83 and TI-86 Graphing Calculator Videos

Chapter 2 – Real Numbers, Equations, and Inequalities

Ideas for Instruction	Print Resources	Media Resources for Instructors	Media Resources for Students

Explorations in Beginning and Intermediate Algebra
• Activities 4-5, 8
 	
Activities for Beginning and Intermediate Algebra
• Activities 1-2, 4
 	
Active Arithmetic and Algebra
• Addition and Subtraction with Integers
• A Polynomial Puzzle
• Equation Triangles
• Writing Percent Problems
• Verifying the Formula for the Sum of Angles of a Polygon
• Volume of Food Containers
	Test Bank
	• Chapter 2
	  Test Form A
	  Test Form B
	  Test Form C
	
	Complete Solutions Manual
	• Chapter 2
	  Pages 39-100
	
	Student Solutions Manual
	• Chapter 2
	  Pages 22-53
	
	How to Be a Great Math Student
	
	Conquering Math Anxiety
		BCA Testing
		http://bca.brookscole.com
		
		BCA Tutorial
		http://bca.brookscole.com
		
		Web Site
		www.brookscole.com/ mathematics
		
		Text-Specific Videos
		• Chapter 2
		
		TI-83 and TI-86 Graphing Calculator Videos
			Interactive Video SkillBuilder
			• Chapter 2
			
			BCA Tutorial
			http://bca.brookscole.com
			
			Web Site
			www.brookscole.com/ mathematics
			
			Text-Specific Videos
			• Chapter 2
			
			TI-83 and TI-86 Graphing Calculator Videos

Chapter 3 – Graphs, Linear Equations, and Functions

Ideas for Instruction	Print Resources	Media Resources for Instructors	Media Resources for Students
Explorations in Beginning and Intermediate Algebra • Activities 10-11, 21-26, 28 **Activities for Beginning and Intermediate Algebra** • Activities 3, 7-10, 12, and 16 **Active Arithmetic and Algebra** • Describing a Linear Relationship • Bouncing Ball • Painters' Pay	**Test Bank** • Chapter 3 Test Form A Test Form B Test Form C **Complete Solutions Manual** • Chapter 3 Pages 101-150 **Student Solutions Manual** • Chapter 3 Pages 56-83 **How to Be a Great Math Student** **Conquering Math Anxiety**	**BCA Testing** http://bca.brookscole.com **BCA Tutorial** http://bca.brookscole.com **Web Site** www.brookscole.com/mathematics **Text-Specific Videos** • Chapter 3 **TI-83 and TI-86 Graphing Calculator Videos**	**Interactive Video SkillBuilder** • Chapter 3 **BCA Tutorial** http://bca.brookscole.com **Web Site** www.brookscole.com/mathematics **Text-Specific Videos** • Chapter 3 **TI-83 and TI-86 Graphing Calculator Videos** (

Chapter 4 – Exponents and Polynomials

Ideas for Instruction	Print Resources	Media Resources for Instructors	Media Resources for Students
Explorations in Beginning and Intermediate Algebra • Activities 6-7	**Test Bank** • Chapter 4 Test Form A Test Form B Test Form C **Complete Solutions Manual** • Chapter 4 Pages 151-203 **Student Solutions Manual** • Chapter 4 Pages 84-114 **How to Be a Great Math Student** **Conquering Math Anxiety**	**BCA Testing** http://bca.brookscole.com **BCA Tutorial** http://bca.brookscole.com **Web Site** www.brookscole.com/mathematics **Text-Specific Videos** • Chapter 4 **TI-83 and TI-86 Graphing Calculator Videos**	**Interactive Video SkillBuilder** • Chapter 4 **BCA Tutorial** http://bca.brookscole.com **Web Site** www.brookscole.com/mathematics **Text-Specific Videos** • Chapter 4 **TI-83 and TI-86 Graphing Calculator Videos**

Chapter 5 – Roots and Radicals

Ideas for Instruction	Print Resources	Media Resources for Instructors	Media Resources for Students
Explorations in Beginning and Intermediate Algebra • Activities 9, 20, 29 **Activities for Beginning and Intermediate Algebra** • Activity 22	**Test Bank** • Chapter 5 Test Form A Test Form B Test Form C **Complete Solutions Manual** • Chapter 5 Pages 204-250 **Student Solutions Manual** • Chapter 5 Pages 115-141 **How to Be a Great Math Student** **Conquering Math Anxiety**	**BCA Testing** **http://bca.brookscole.com** **BCA Tutorial** **http://bca.brookscole.com** **Web Site** **www.brookscole.com/ mathematics** **Text-Specific Videos** • Chapter 5 **TI-83 and TI-86 Graphing Calculator Videos**	**Interactive Video SkillBuilder** • Chapter 5 **BCA Tutorial** **http://bca.brookscole.com** **Web Site** **www.brookscole.com/ mathematics** **Text-Specific Videos** • Chapter 5 **TI-83 and TI-86 Graphing Calculator Videos**

Chapter 6 – Factoring and Quadratic Equations

Ideas for Instruction	Print Resources	Media Resources for Instructors	Media Resources for Students
Explorations in Beginning and Intermediate Algebra • Activities 15-17, 30 **Activities for Beginning and Intermediate Algebra** • Activities 5-6, 49	**Test Bank** • Chapter 6 Test Form A Test Form B Test Form C **Complete Solutions Manual** • Chapter 6 Pages 251-313 **Student Solutions Manual** • Chapter 6 Pages 142-175 **How to Be a Great Math Student** **Conquering Math Anxiety**	**BCA Testing** **http://bca.brookscole.com** **BCA Tutorial** **http://bca.brookscole.com** **Web Site** **www.brookscole.com/ mathematics** **Text-Specific Videos** • Chapter 6 **TI-83 and TI-86 Graphing Calculator Videos**	**Interactive Video SkillBuilder** • Chapter 6 **BCA Tutorial** **http://bca.brookscole.com** **Web Site** **www.brookscole.com/ mathematics** **Text-Specific Videos** • Chapter 6 **TI-83 and TI-86 Graphing Calculator Videos**

Resource Integration Guide—Make the Connection!

Chapter 7 – Proportion and Rational Expressions

Ideas for Instruction	Print Resources	Media Resources for Instructors	Media Resources for Students

Activities for Beginning and Intermediate Algebra
• Activities 18-19
 	
Active Arithmetic and Algebra
• Mathematics of Drawing the Human Figure
• Circular Relationships
• Painting the Classroom
	Test Bank
	• Chapter 7
	  Test Form A
	  Test Form B
	  Test Form C
	
	Complete Solutions Manual
	• Chapter 7
	  Pages 314–364
	
	Student Solutions Manual
	• Chapter 7
	  Pages 176–204
	
	How to Be a Great Math Student
	
	Conquering Math Anxiety
		BCA Testing
		http://bca.brookscole.com
		
		BCA Tutorial
		http://bca.brookscole.com
		
		Web Site
		www.brookscole.com/mathematics
		
		Text-Specific Videos
		• Chapter 7
		
		TI-83 and TI-86 Graphing Calculator Videos
			Interactive Video SkillBuilder
			• Chapter 7
			
			BCA Tutorial
			http://bca.brookscole.com
			
			Web Site
			www.brookscole.com/mathematics
			
			Text-Specific Videos
			• Chapter 7
			
			TI-83 and TI-86 Graphing Calculator Videos

Chapter 8 – Solving Systems of Equations and Inequalities

Ideas for Instruction	Print Resources	Media Resources for Instructors	Media Resources for Students

Explorations in Beginning and Intermediate Algebra
• Activities 12, 27
 	
Activities for Beginning and Intermediate Algebra
• Activities 11, 13-15
	Test Bank
	• Chapter 8
	  Test Form A
	  Test Form B
	  Test Form C
	
	Complete Solutions Manual
	• Chapter 8
	  Pages 365–422
	
	Student Solutions Manual
	• Chapter 8
	  Pages 205–236
	
	How to Be a Great Math Student
	
	Conquering Math Anxiety
		BCA Testing
		http://bca.brookscole.com
		
		BCA Tutorial
		http://bca.brookscole.com
		
		Web Site
		www.brookscole.com/mathematics
		
		Text-Specific Videos
		• Chapter 8
		
		TI-83 and TI-86 Graphing Calculator Videos
			Interactive Video SkillBuilder
			• Chapter 8
			
			BCA Tutorial
			http://bca.brookscole.com
			
			Web Site
			www.brookscole.com/mathematics
			
			Text-Specific Videos
			• Chapter 8
			
			TI-83 and TI-86 Graphing Calculator Videos

Elementary Algebra

Books in the Tussy and Gustafson Series

In paperback:

Basic Mathematics for College Students, Second Edition
Student edition: ISBN 0-534-37643-6
Instructor's edition: ISBN 0-534-38585-0

Developmental Mathematics for College Students
Student edition: ISBN 0-534-38031-X
Instructor's edition: ISBN 0-534-38584-2

Prealgebra, Second Edition
Student edition: ISBN 0-534-37642-8
Instructor's edition: ISBN 0-534-38309-2

Introductory Algebra, Second Edition
Student edition: ISBN 0-534-37641-X
Instructor's edition: ISBN 0-534-38571-0

Intermediate Algebra, Second Edition
Student edition: ISBN 0-534-37640-1
Instructor's edition: ISBN 0-534-38586-9

In hardcover:

Elementary Algebra, Second Edition
Student edition: ISBN 0-534-38629-6
Instructor's edition: ISBN 0-534-39119-2

Intermediate Algebra, Second Edition
Student edition: ISBN 0-534-38628-8
Instructor's edition: ISBN 0-534-39120-6

Elementary and Intermediate Algebra, Second Edition
Student edition: ISBN 0-534-38627-X
Instructor's edition: ISBN 0-534-39118-4

Elementary Algebra

Second Edition

Alan S. Tussy
Citrus College

R. David Gustafson
Rock Valley College

BROOKS/COLE

THOMSON LEARNING

• Australia • Canada • Mexico • Singapore • Spain • United Kingdom • United States

BROOKS/COLE

THOMSON LEARNING

Sponsoring Editor: Jennifer Huber/Robert W. Pirtle

Assistant Editor: Rachael Sturgeon

Marketing Team: Leah Thomson

Marketing Communications: Samantha Cabaluna

Marketing Assistant: Maria Salinas

Production Editors: Ellen Brownstein/Scott Brearton

Manuscript Editor: David Hoyt

Permissions Editor: Sue Ewing

Interior Design: Carolyn Deacy

Cover Design: Vernon T. Boes

Cover Illustration: George Abe

Interior Illustration: Lori Heckelman

Print Buyer: Kristine Waller

Typesetting: The Clarinda Company

Cover Printing: Phoenix Color Corp.

Printing and Binding: R.R. Donnelley & Sons, Willard

For more information about this or any other Brooks/Cole product, contact:
BROOKS/COLE
511 Forest Lodge Road
Pacific Grove, CA 93950 USA
www.brookscole.com
1-800-423-0563 (Thomson Learning Academic Resource Center)

For permission to use material from this work, contact us by
www.thomsonrights.com
fax: 1-800-730-2215
phone: 1-800-730-2214

Printed in the United States of America

10   9   8   7   6   5   4   3   2   1

Library of Congress Cataloging-in-Publication Data
Tussy, Alan S., [date]
    Elementary algebra / Alan S. Tussy, R. David Gustafson.—2nd ed.
      p. cm.
    Includes index.
    ISBN 0-534-38629-6—ISBN 0-534-39119-2
    1. Algebra I. Gustafson, R. David (Roy David), [date] II. Title.

QA152.3.T85 2001
512.9—dc21   2001037432

To Margaret Kavelaar
and
Gordon Guillaume,
two fine mathematics instructors
and outstanding
department chairpersons.

AST

To Dr. John A. Schumaker,
my teacher, mentor, colleague, and friend.

RDG

Contents

M101
Review M102
M102
Optional

Optional

Preface

For the Instructor

Algebra is used to describe numerical relationships. It is a language in its own right. The purpose of this textbook is to teach students how to read, write, speak, and think mathematically using the language of algebra. It presents all the topics associated with a first course in algebra. We have used a blend of the traditional and the reform instructional approaches to do this. In this book, you will find the vocabulary, practice, and well-defined pedagogy of a traditional approach. You will also find that we emphasize the reasoning, modeling, communicating, and technological skills that are such a big part of today's reform movement.

This textbook expands the students' mathematical reasoning abilities and gives them a set of mathematical survival skills that will help them succeed in a world that increasingly requires that every person become a better analytical thinker.

Features of the Text

Chapter 1   An Innovative Introduction to Algebra

The best way to learn a new language is to be immediately immersed in it. Therefore, Chapter 1 begins with an introduction of the fundamental algebraic concepts of variable, equation, function, and graphing. We show the students how to translate English phrases to mathematical symbols, and we introduce a problem-solving strategy that is used throughout the book. From the start, students see how algebra is a powerful tool that they can use to solve problems.

Interactivity

Most worked examples in the text are accompanied by Self Checks. This feature allows students to practice skills discussed in the example by working a similar problem. Because the Self Check problems are adjacent to the worked examples, students can easily refer to the solution and author's notes of the example as they solve the Self Check. Author's notes are used to explain the steps in the solutions of examples. The notes are extensive so as to increase the student's ability to read and write mathematics.

Example titles highlight the ▶
concept being discussed.

Author's notes explain the steps ▶
in the solution process.

Most examples have Self ▶
Checks. The answers are
provided.

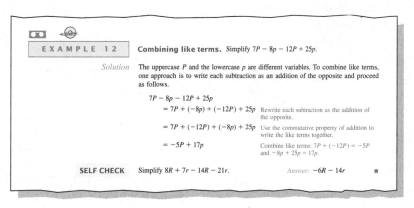

In-Depth Coverage of Geometry

Perimeter, area, and volume, as well as many other geometric concepts, are used in a variety of contexts throughout the book. We have included many drawings to help students improve their ability to spot visual patterns in their everyday lives.

◄ Geometric topics are presented in a practical setting.

Geometric topics appear ▶ throughout the text and are presented in a way that reinforces important algebra skills.

2.6 Formulas 149

ILLUSTRATION 16

56. NATIVE AMERICAN DWELLING The teepees constructed by the Blackfoot Indians were cone-shaped tents made of long poles and animal hide, about 10 feet high and about 15 feet across at the ground. (See Illustration 13.) Estimate the volume of a teepee with these dimensions, to the nearest cubic foot.

ILLUSTRATION 13

57. IGLOO During long journeys, some Canadian Inuit (Eskimos) built winter houses of snow blocks piled in the dome shape shown in Illustration 14. Estimate the volume of an igloo to the nearest cubic foot having an interior height of 5.5 feet.

ILLUSTRATION 14

58. PYRAMID The Great Pyramid at Giza in northern Egypt is one of the most famous works of architecture in the world. Use the information in Illustration 15 to find the volume to the nearest cubic foot.

ILLUSTRATION 15

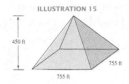
450 ft
755 ft
755 ft

59. BARBECUING See Illustration 16. Use the fact that the fish is 18 inches long to find the area of the bar-

60. SKATEBOARDING A "half-pipe" ramp used for skateboarding is in the shape of a semicircle with a radius of 8 fee... nearest tenth o... that the skatcb...

61. OHM'S LAW ... tronics. Solve ... peres) when th... tance R is 12 o...

62. GROWTH OF ... P grows to $... $A = P(1 + rt)... years, a woma... that pays 6%.

63. POWER LOSS ... current I passe... $P = I^2R$. Solv... amperes, find ...

64. FORCE OF G... jects in Illustra... gravitation, F,... where G is a c... them. Solve fo...

65. THERMODY... tion is given b... mula for the p...

---

148 Chapter 2 / Real Numbers, Equations, and Inequalities

43. $C = \dfrac{5F - 160}{9}$; for F **44.** $F = \dfrac{GMm}{d^2}$; for d^2

APPLICATIONS

In Exercises 45–66, a calculator will be helpful with some problems.

45. CARPENTRY Find the perimeter and area of the truss shown in Illustration 3.

ILLUSTRATION 3

6 ft   10 ft   10 ft
16 ft

46. CAMPERS Find the area of the window of the camper shell shown in Illustration 4.

ILLUSTRATION 4

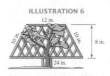
56 in.
14 in.

47. ARCHERY To the nearest tenth, find the circumference and area of the target shown in Illustration 5.

ILLUSTRATION 5

8 in.

48. GEOGRAPHY The circumference of the earth is about 25,000 miles. Find its diameter to the nearest mile.

49. LANDSCAPING Find the perimeter and the area of the redwood trellis in Illustration 6.

ILLUSTRATION 6

12 in.
10 in.   10 in.   8 in.
24 in.

50. VOLUME To the nearest hundredth, find the volume of the soup can shown in Illustration 7.

ILLUSTRATION 7

6 in.
SOUP
3 in.

51. "THE WALL" The Vietnam Veterans Memorial is a black granite wall recognizing the more than 58,000 Americans who lost their lives or remain missing. A diagram of the wall is shown in Illustration 8. Find the total area of the two triangular-shaped surfaces on which the names are inscribed.

ILLUSTRATION 8

10 ft
245 ft   245 ft

52. SIGNAGE Find the perimeter and area of the service station sign shown in Illustration 9.

ILLUSTRATION 9

2.75 ft
Chevron
2.75 ft

53. RUBBER MEETS THE ROAD A sport truck tire has the road surface "footprint" shown in Illustration 10. Estimate the perimeter and area of the tire's footprint. (*Hint:* First change the dimensions to decimals.)

ILLUSTRATION 10

$6\frac{3}{8}$ in.
$7\frac{1}{2}$ in.

54. SOFTBALL The strike zone in fastpitch softball is between the batter's armpit and top of her knees, as shown in Illustration 11. Find the area of the strike zone.

ILLUSTRATION 11

26 in.
17 in.

55. FIREWOOD The dimensions of a cord of firewood are shown in Illustration 12. Find the area on which the wood is stacked and the volume the cord of firewood occupies.

ILLUSTRATION 12

4 ft
4 ft   8 ft

Coordinate Graphing Appears Early

The foundation for coordinate graphing is laid in Chapters 1 and 2, where the students graph many different types of real numbers on the number line. In Chapter 3, students learn how to graph lines. They quickly learn that the graph of an equation in two variables is not always a straight line.

Problem-Solving Strategy

One of the major objectives of this textbook is to make students better problem solvers. To this end, we use a five-step problem-solving strategy throughout the book. The five steps are: *Analyze the problem, Form an equation, Solve the equation, State the conclusion,* and *Check the result.*

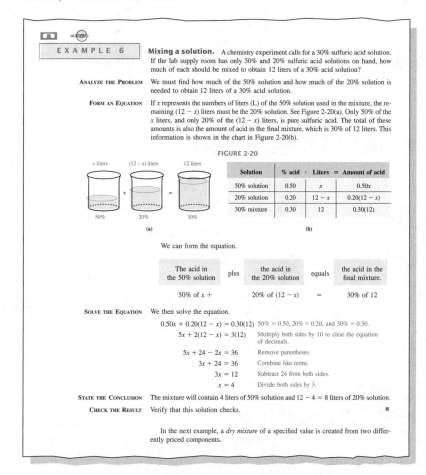

Applications and Connections to Other Disciplines

A distinguishing feature of this book is its wealth of application problems. We have included numerous applications from disciplines such as science, economics, business, manufacturing, history, and entertainment, as well as mathematics.

Every application problem ▶
has a title

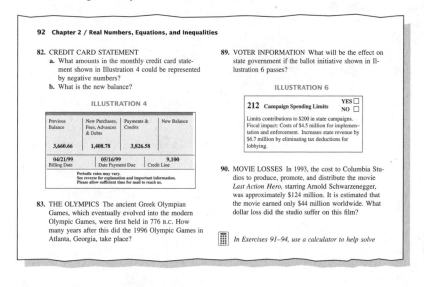

Study Sets—More Than Just Exercises

The problems at the end of each section are called Study Sets. Each Study Set includes Vocabulary, Notation, and Writing problems designed to help students improve their ability to read, write, and communicate mathematical ideas. The problems in the Concepts section of the Study Sets encourage students to engage in independent thinking and reinforce major ideas through exploration. In the Practice section of the Study Sets, students get the drill necessary to master the material. In the Applications section, students deal with real-life situations that involve the topics being studied. Each Study Set concludes with a Review section consisting of problems selected from previous sections.

◀ Each Study Set contains Vocabulary, Concepts, Notation, Practice, Applications, Writing, and Review sections.

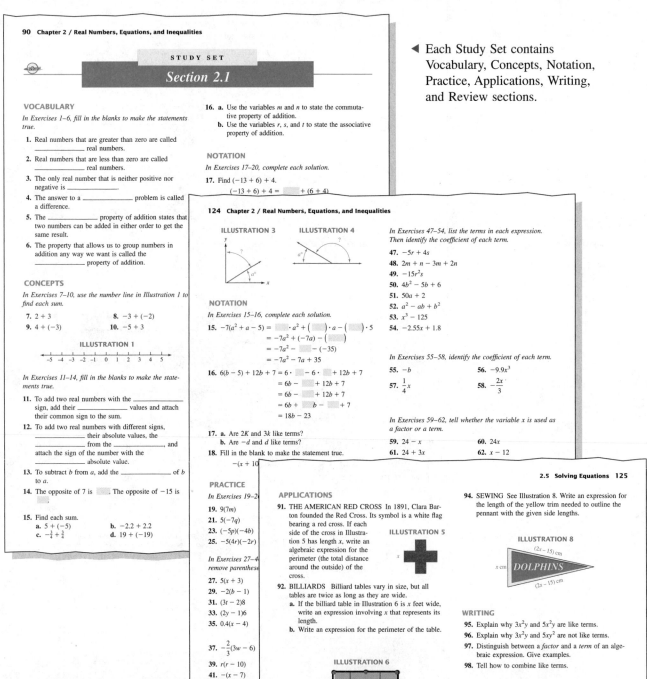

90 Chapter 2 / Real Numbers, Equations, and Inequalities

STUDY SET
Section 2.1

VOCABULARY

In Exercises 1–6, fill in the blanks to make the statements true.

1. Real numbers that are greater than zero are called _____ real numbers.
2. Real numbers that are less than zero are called _____ real numbers.
3. The only real number that is neither positive nor negative is _____.
4. The answer to a _____ problem is called a difference.
5. The _____ property of addition states that two numbers can be added in either order to get the same result.
6. The property that allows us to group numbers in addition any way we want is called the _____ property of addition.

CONCEPTS

In Exercises 7–10, use the number line in Illustration 1 to find each sum.

7. $2 + 3$
8. $-3 + (-2)$
9. $4 + (-3)$
10. $-5 + 3$

ILLUSTRATION 1

$-5 \ -4 \ -3 \ -2 \ -1 \ 0 \ 1 \ 2 \ 3 \ 4 \ 5$

In Exercises 11–14, fill in the blanks to make the statements true.

11. To add two real numbers with the _____ sign, add their _____ values and attach their common sign to the sum.
12. To add two real numbers with different signs, _____ their absolute values, the _____ from the _____, and attach the sign of the number with the _____ absolute value.
13. To subtract b from a, add the _____ of b to a.
14. The opposite of 7 is _____. The opposite of -15 is _____.
15. Find each sum.
   a. $5 + (-5)$ b. $-2.2 + 2.2$
   c. $-\frac{3}{4} + \frac{3}{4}$ d. $19 + (-19)$

16. a. Use the variables m and n to state the commutative property of addition.
   b. Use the variables r, s, and t to state the associative property of addition.

NOTATION

In Exercises 17–20, complete each solution.

17. Find $(-13 + 6) + 4$.
   $(-13 + 6) + 4 = \boxed{\ } + (6 + 4)$

124 Chapter 2 / Real Numbers, Equations, and Inequalities

ILLUSTRATION 3   ILLUSTRATION 4

NOTATION

In Exercises 15–16, complete each solution.

15. $-7(a^2 + a - 5) = \boxed{\ } \cdot a^2 + \left(\boxed{\ }\right) \cdot a - \left(\boxed{\ }\right) \cdot 5$
   $= -7a^2 + (-7a) - \left(\boxed{\ }\right)$
   $= -7a^2 - \boxed{\ } - (-35)$
   $= -7a^2 - 7a + 35$

16. $6(b - 5) + 12b + 7 = 6 \cdot \boxed{\ } - 6 \cdot \boxed{\ } + 12b + 7$
   $= 6b - \boxed{\ } + 12b + 7$
   $= 6b - \boxed{\ } + 12b + 7$
   $= 6b + \boxed{\ } b - \boxed{\ } + 7$
   $= 18b - 23$

17. a. Are $2K$ and $3k$ like terms?
   b. Are $-d$ and d like terms?
18. Fill in the blank to make the statement true.
   $-(x + 10$

PRACTICE

In Exercises 19–2

19. $9(7m)$
21. $5(-7q)$
23. $(-5p)(-4b)$
25. $-5(4r)(-2r)$

In Exercises 27–4 remove parentheses

27. $5(x + 3)$
29. $-2(b - 1)$
31. $(3t - 2)8$
33. $(2y - 1)6$
35. $0.4(x - 4)$

37. $-\frac{2}{3}(3w - 6)$
39. $r(r - 10)$
41. $-(x - 7)$
43. $17(2x - y + 2$
45. $-(-14 + 3p$

In Exercises 47–54, list the terms in each expression. Then identify the coefficient of each term.

47. $-5r + 4s$
48. $2m + n - 3m + 2n$
49. $-15r^2s$
50. $4b^2 - 5b + 6$
51. $50a + 2$
52. $a^2 - ab + b^2$
53. $x^3 - 125$
54. $-2.55x + 1.8$

In Exercises 55–58, identify the coefficient of each term.

55. $-b$
56. $-9.9x^3$
57. $\frac{1}{4}x$
58. $-\frac{2x}{3}$

In Exercises 59–62, tell whether the variable x is used as a factor or a term.

59. $24 - x$
60. $24x$
61. $24 + 3x$
62. $x - 12$

2.5 Solving Equations 125

APPLICATIONS

91. THE AMERICAN RED CROSS In 1891, Clara Barton founded the Red Cross. Its symbol is a white flag bearing a red cross. If each side of the cross in Illustration 5 has length x, write an algebraic expression for the perimeter (the total distance around the outside) of the cross.

ILLUSTRATION 5

92. BILLIARDS Billiard tables vary in size, but all tables are twice as long as they are wide.
   a. If the billiard table in Illustration 6 is x feet wide, write an expression involving x that represents its length.
   b. Write an expression for the perimeter of the table.

ILLUSTRATION 6

93. PING-PONG Write an expression for the perimeter of the ping-pong table shown in Illustration 7.

94. SEWING See Illustration 8. Write an expression for the length of the yellow trim needed to outline the pennant with the given side lengths.

ILLUSTRATION 8

$(2x - 15)$ cm
x cm DOLPHINS
$(2x - 15)$ cm

WRITING

95. Explain why $3x^2y$ and $5x^2y$ are like terms.
96. Explain why $3x^2y$ and $5xy^2$ are not like terms.
97. Distinguish between a *factor* and a *term* of an algebraic expression. Give examples.
98. Tell how to combine like terms.

REVIEW

In Exercises 99–102, evaluate each expression given that $x = -3$, $y = -5$, and $z = 0$.

99. $x^2z(y^3 - z)$
100. $z - y^3$
101. $\dfrac{x - y^2}{2y - 1 + x}$
102. $\dfrac{2y + 1}{x} - x$

Group Work

A one-page feature called Accent on Teamwork appears near the end of each chapter. It gives the instructor a set of problems that can be assigned as group work or to individual students as outside-of-class projects.

Key Concepts

Eight key algebraic concepts are highlighted in one-page Key Concept features, appearing near the end of each chapter. Each Key Concept page summarizes a concept and gives students an opportunity to review the role it plays in the overall picture.

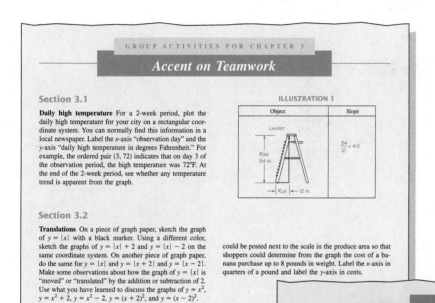

GROUP ACTIVITIES FOR CHAPTER 3

Accent on Teamwork

Section 3.1

Daily high temperature For a 2-week period, plot the daily high temperature for your city on a rectangular coordinate system. You can normally find this information in a local newspaper. Label the x-axis "observation day" and the y-axis "daily high temperature in degrees Fahrenheit." For example, the ordered pair $(3, 72)$ indicates that on day 3 of the observation period, the high temperature was $72°F$. At the end of the 2-week period, see whether any temperature trend is apparent from the graph.

Section 3.2

Translations On a piece of graph paper, sketch the graph of $y = |x|$ with a black marker. Using a different color, sketch the graphs of $y = |x| + 2$ and $y = |x| - 2$ on the same coordinate system. On another piece of graph paper, do the same for $y = |x|$ and $y = |x + 2|$ and $y = |x - 2|$. Make some observations about how the graph of $y = |x|$ is "moved" or "translated" by the addition or subtraction of 2. Use what you have learned to discuss the graphs of $y = x^2$, $y = x^2 + 2$, $y = x^2 - 2$, $y = (x + 2)^2$, and $y = (x - 2)^2$.

ILLUSTRATION 1

Object	Slope
Ladder — Rise 54 in., Run 12 in.	$\frac{54}{12} = 4.5$

could be posted next to the scale in the produce area so that shoppers could determine from the graph the cost of a banana purchase up to 8 pounds in weight. Label the x-axis in quarters of a pound and label the y-axis in cents.

KEY CONCEPT

Describing Linear Relationships

In Chapter 3, we discussed four ways to mathematically describe linear relationships between two quantities.

Equations in Two Variables

The general form of the equation of a line is $Ax + By = C$. Two very useful forms of the equation of a line are the slope–intercept form and the point–slope form.

1. Write the equation of a line with a slope of -3 and a y-intercept of $(0, -4)$.

2. Write the equation of the line that passes through $(5, 2)$ and $(-5, 0)$. Answer in slope–intercept form.

Rectangular Coordinate Graphs

The graph of an equation is a "picture" of all of its solutions (x, y). Important information can be obtained from a graph.

3. Complete the table of solutions for $2x - 4y = 8$. Then graph the equation.

4. See Illustration 1.
   a. What information does the y-intercept of the graph give us?
   b. What is the slope of the line and what does it tell us?

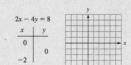

$2x - 4y = 8$

x	y
0	
	0
-2	

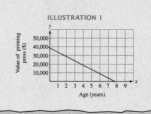

ILLUSTRATION 1

Value of printing press ($): 50,000, 40,000, 30,000, 20,000, 10,000

Age (years): 1 2 3 4 5 6 7 8 9

Functions and Modeling

The concept of function is introduced in Chapter 3 and is stressed throughout the text. Students learn to use function notation, graph functions, and write functions that mathematically model many interesting real-life situations. By the end of the course, students will recognize families of functions, their graphs, and areas of application.

Real data is integrated ▶ throughout the text.

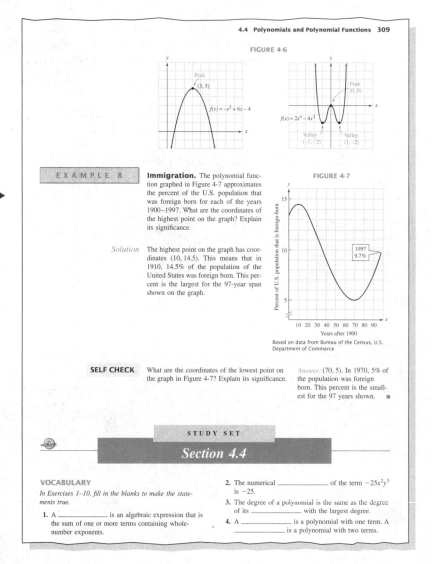

Systematic Review

Each Study Set ends with a Review section that contains problems similar to those in previous sections. Each chapter ends with a Chapter Review and a Chapter Test. The chapter reviews have been designed to be "user friendly." In a unique format, the reviews list the important concepts of each section of the chapter in one column, with appropriate review problems running parallel in a second column. In addition, Cumulative Review Exercises appear after Chapters 2, 4, 6, and 8.

CHAPTER REVIEW

Section 1.1

CONCEPTS

Tables, bar graphs, and *line graphs* are used to describe numerical relationships.

Describing Numerical Relationships

REVIEW EXERCISES

1. Illustration 1 lists the worldwide production of wide-screen TVs. Use the data to construct a bar graph. Describe the trend in the production in words.

ILLUSTRATION 1

Year	Production (millions of units)
'95	3
'96	5
'97	7
'98	11

Source: Electronic Industries Association of Japan

2. Consider the line graph in Illustra-

The result of an addition is called the *sum;* of a subtraction, the *difference;* of a multiplication, the *product;* and of a division, the *quotient.*

Variables are letters used to stand for numbers.

An *equation* is a mathematical sentence that contains an = sign. Variables and/or numbers can be combined with the operations of addition, subtraction, multiplication, and division to create *algebraic expressions.*

CHAPTER 2

Test

1. Add $(-6) + 8 + (-4)$.

2. Subtract $1.4 - (-0.8)$.

3. Multiply $(-2)(-3)(-5)$.

4. Divide $\dfrac{-22}{-11}$.

5. Evaluate $-7[(-5)^2 - 2(3 - 5^2)]$.

6. Evaluate $\dfrac{3(20 - 4^2)}{-2(6 - 2^2)}$.

In Problems 7–8, let $x = -2$, $y = 3$, and $z = 4$. Evaluate each expression.

7. $xy + z$

8. $\dfrac{z + 4y}{2x}$

9. What is the coefficient of the term $6x$?

10. How many terms are in the expression $3x^2 + 5x - 7$?

In Problems 11–14, simplify each expression.

11. $5(-4x)$

12. $-8(-7t)(4t)$

13. $3(x + 2) - 3(4 - x)$

14. $-1.1d^2 - 3.8d^2$

In Problems 15–22, solve each equa

15. $12x = -144$

16.

17. $\dfrac{c}{7} = -1$

18.

19. $2(x - 7) = -15$

20.

21. $23 - 5(x + 10) = -12$

22.

In Problems 23–24, solve each equa indicated.

23. $d = rt$; for t

24. $A = P + Prt$; for r

25. COMMERCIAL REAL ESTATE *Net absorption* is a term used to indicate how much office space in a city is being purchased. Use the information from the graph in Illustration 1 to determine the eight-quarter average net absorption figure for Long Beach, California.

ILLUSTRATION 1

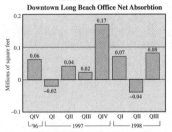

Downtown Long Beach Office Net Absorbtion

Based on information from *Los Angeles Times* (Oct. 13, 1998) Section C10.

CHAPTERS 1–2

Cumulative Review Exercises

In Exercises 1–4, tell whether the expression is an equation.

1. $m - 25$

2. $t = 25r$

3. $\dfrac{p + 5}{2}$

4. $x + 1 = 24$

In Exercises 5–6, classify each number as a natural number, a whole number, an integer, a rational number, an irrational number, and a real number. Each number may be in several classifications.

5. 3

6. -0.25

In Exercises 7–8, graph each set of numbers on the number line.

7. The natural numbers between 2 and 7

8. The real numbers between 2 and 7

In Exercises 9–12, evaluate each expression.

9. $|-12|$

10. $|15|$

11. $-|15|$

12. $-|-12|$

 In Exercises 13–16, use a calculator to find each square root to the nearest hundredth.

13. $\sqrt{77}$

14. $\sqrt{0.26}$

15. $\sqrt{\pi}$

16. $\sqrt{\dfrac{7}{5}}$

In Exercises 17–20, write each product using exponents.

17. $3 \cdot 3 \cdot 3$

18. $8 \cdot \pi \cdot r \cdot r$

19. $4 \cdot x \cdot x \cdot y \cdot y$

20. $m \cdot m \cdot m \cdot n$

In Exercises 21–24, evaluate each expression.

21. $13 - 3 \cdot 4$

22. $-2 \cdot 7^2$

23. $6 + 3\left(\dfrac{5}{2}\right) - \dfrac{1}{2}$

24. $\dfrac{12^2 - 4^2 - 2}{2(7 - 4)}$

In Exercises 25–26, write each phrase as an algebraic expression.

25. The sum of the width w and 12

26. Four less than a number n

In Exercises 27–30, complete each table of values.

27. t	$t^2 - 4$
0	
1	
3	

28. t	$(t - 4)^2$
0	
1	
3	

Calculators

For instructors who wish to use calculators as part of the instruction in this course, the text includes an Accent on Technology feature that introduces keystrokes and shows how scientific calculators and graphing calculators can be used to solve problems. In the Study Sets, logos are used to denote problems that require a scientific calculator or a graphing calculator .

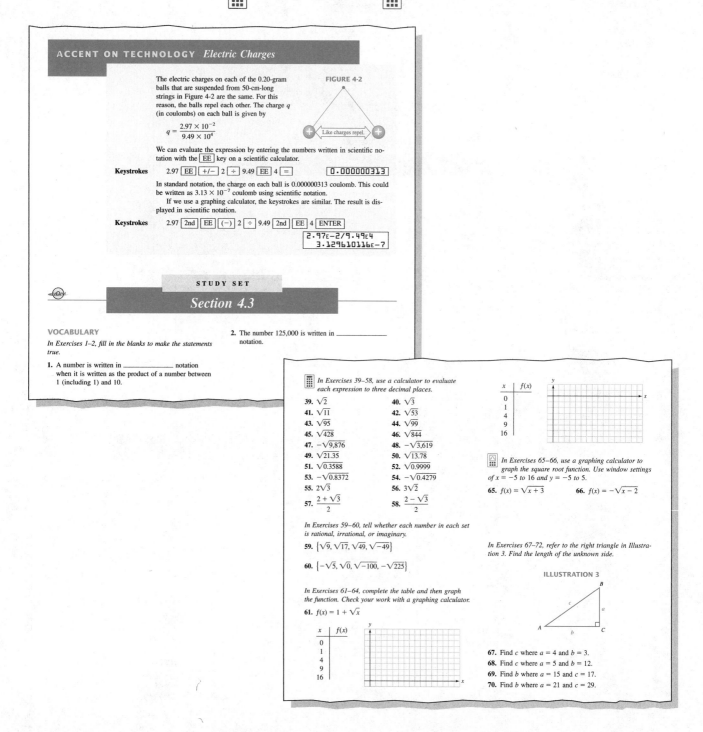

ACCENT ON TECHNOLOGY *Electric Charges*

The electric charges on each of the 0.20-gram balls that are suspended from 50-cm-long strings in Figure 4-2 are the same. For this reason, the balls repel each other. The charge q (in coulombs) on each ball is given by

$$q = \frac{2.97 \times 10^{-2}}{9.49 \times 10^{4}}$$

FIGURE 4-2

Like charges repel.

We can evaluate the expression by entering the numbers written in scientific notation with the $\boxed{\text{EE}}$ key on a scientific calculator.

Keystrokes   2.97 $\boxed{\text{EE}}$ $\boxed{+/-}$ 2 $\boxed{\div}$ 9.49 $\boxed{\text{EE}}$ 4 $\boxed{=}$ $\boxed{\text{0.000000313}}$

In standard notation, the charge on each ball is 0.000000313 coulomb. This could be written as 3.13×10^{-7} coulomb using scientific notation.

If we use a graphing calculator, the keystrokes are similar. The result is displayed in scientific notation.

Keystrokes   2.97 $\boxed{\text{2nd}}$ $\boxed{\text{EE}}$ $\boxed{(-)}$ 2 $\boxed{\div}$ 9.49 $\boxed{\text{2nd}}$ $\boxed{\text{EE}}$ 4 $\boxed{\text{ENTER}}$

$\boxed{\begin{array}{l}\text{2.97e-2/9.49e4}\\ \text{3.129610116e-7}\end{array}}$

STUDY SET

Section 4.3

VOCABULARY

In Exercises 1–2, fill in the blanks to make the statements true.

1. A number is written in _____ notation when it is written as the product of a number between 1 (including 1) and 10.

2. The number 125,000 is written in _____ notation.

In Exercises 39–58, use a calculator to evaluate each expression to three decimal places.

39. $\sqrt{2}$ **40.** $\sqrt{3}$
41. $\sqrt{11}$ **42.** $\sqrt{53}$
43. $\sqrt{95}$ **44.** $\sqrt{99}$
45. $\sqrt{428}$ **46.** $\sqrt{844}$
47. $-\sqrt{9,876}$ **48.** $-\sqrt{3,619}$
49. $\sqrt{21.35}$ **50.** $\sqrt{13.78}$
51. $\sqrt{0.3588}$ **52.** $\sqrt{0.9999}$
53. $-\sqrt{0.8372}$ **54.** $-\sqrt{0.4279}$
55. $2\sqrt{3}$ **56.** $3\sqrt{2}$
57. $\dfrac{2 + \sqrt{3}}{2}$ **58.** $\dfrac{2 - \sqrt{3}}{2}$

x	$f(x)$
0	
1	
4	
9	
16	

In Exercises 59–60, tell whether each number in each set is rational, irrational, or imaginary.

59. $\left\{\sqrt{9}, \sqrt{17}, \sqrt{49}, \sqrt{-49}\right\}$

60. $\left\{-\sqrt{5}, \sqrt{0}, \sqrt{-100}, -\sqrt{225}\right\}$

In Exercises 61–64, complete the table and then graph the function. Check your work with a graphing calculator.

61. $f(x) = 1 + \sqrt{x}$

x	$f(x)$
0	
1	
4	
9	
16	

In Exercises 65–66, use a graphing calculator to graph the square root function. Use window settings of $x = -5$ to 16 and $y = -5$ to 5.

65. $f(x) = \sqrt{x + 3}$ **66.** $f(x) = -\sqrt{x - 2}$

In Exercises 67–72, refer to the right triangle in Illustration 3. Find the length of the unknown side.

ILLUSTRATION 3

67. Find c where $a = 4$ and $b = 3$.
68. Find c where $a = 5$ and $b = 12$.
69. Find b where $a = 15$ and $c = 17$.
70. Find b where $a = 21$ and $c = 29$.

Appendixes

A review of arithmetic fractions and decimal fractions is included in Appendix I. Appendix II covers the mean, median, and mode. For each, problem sets are included for student practice. Appendix III gives a table of roots and powers.

Student Support

We have included many features that make *Elementary Algebra* very accessible to students. (See the examples starting on page xi.)

Worked Examples

The text contains over 425 worked examples, many with several parts. Explanatory notes make the examples easy to follow.

Author's Notes

Author's notes, printed in red, are used to explain the steps in the solutions of examples. The notes are extensive; complete sentences are used so as to increase the students' ability to read and write mathematics.

A special logo shows which ▶ examples are included in the videotape series.

Each step is explained using ▶ detailed author's notes.

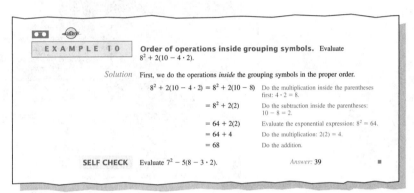

EXAMPLE 10 **Order of operations inside grouping symbols.** Evaluate $8^2 + 2(10 - 4 \cdot 2)$.

Solution First, we do the operations *inside* the grouping symbols in the proper order.

$$8^2 + 2(10 - 4 \cdot 2) = 8^2 + 2(10 - 8) \quad \text{Do the multiplication inside the parentheses first: } 4 \cdot 2 = 8.$$
$$= 8^2 + 2(2) \quad \text{Do the subtraction inside the parentheses: } 10 - 8 = 2.$$
$$= 64 + 2(2) \quad \text{Evaluate the exponential expression: } 8^2 = 64.$$
$$= 64 + 4 \quad \text{Do the multiplication: } 2(2) = 4.$$
$$= 68 \quad \text{Do the addition.}$$

SELF CHECK Evaluate $7^2 - 5(8 - 3 \cdot 2)$. *Answer:* 39

Self Checks

There are more than 365 Self Check problems that allow students to practice the skills demonstrated in the worked examples.

Warnings

Throughout the text, students are warned about common mistakes and how to avoid them.

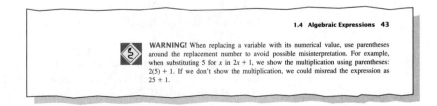

1.4 Algebraic Expressions 43

WARNING! When replacing a variable with its numerical value, use parentheses around the replacement number to avoid possible misinterpretation. For example, when substituting 5 for x in $2x + 1$, we show the multiplication using parentheses: $2(5) + 1$. If we don't show the multiplication, we could misread the expression as $25 + 1$.

Videotapes

The videotape series that accompanies this book uses eye-catching computer graphics to show students the steps in solving many examples in the text. A video logo 👓 placed next to an example indicates that the example is taught on tape. In addition, the tapes present the solutions of two Study Set problems from each secti on.

Functional Use of Color

For easy reference, definition boxes (light blue with a green title), strategy boxes (light yellow with a red title), and rule or property boxes (bright yellow with royal blue title), are color-coded. In addition, the book uses color to highlight terms and expressions that you would point to in a classroom discussion.

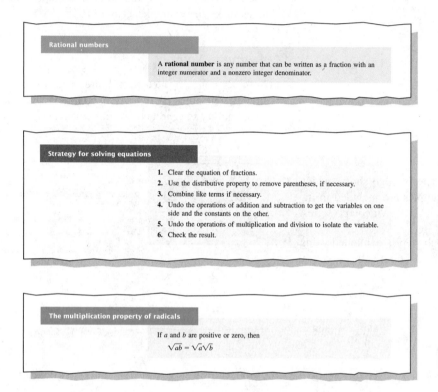

Rational numbers

A **rational number** is any number that can be written as a fraction with an integer numerator and a nonzero integer denominator.

Strategy for solving equations

1. Clear the equation of fractions.
2. Use the distributive property to remove parentheses, if necessary.
3. Combine like terms if necessary.
4. Undo the operations of addition and subtraction to get the variables on one side and the constants on the other.
5. Undo the operations of multiplication and division to isolate the variable.
6. Check the result.

The multiplication property of radicals

If a and b are positive or zero, then

$$\sqrt{ab} = \sqrt{a}\sqrt{b}$$

Problems and Answers

The book includes more than 5,600 carefully graded exercises. In the Student Edition, Appendix IV provides the answers to the odd-numbered exercises in the Study Sets, as well as all the answers to the Key Concept, Chapter Review, Chapter Test, and Cumulative Review problems.

Reading and Writing Mathematics

Also included (on pages xxv–xxvi) are two features to help students improve their ability to read and write mathematics. "Reading Mathematics" helps students get the most out of the examples in this book by showing them how to read the solutions properly. "Writing Mathematics" highlights the characteristics of a well-written solution.

Study Skills and Math Anxiety

These two topics are discussed in detail in the section entitled "For the Student" at the end of this preface. In "Success in Elementary Algebra," students are asked to design a personal strategy for studying and learning the material. "Taking a Math Test" helps students prepare for a test and then gives them suggestions for improving their performance.

Ancillaries for the Instructor

Complete Solutions Manual

The *Complete Solutions Manual* provides worked-out solutions to all of the problems in the text.

Test Bank

The *Test Bank* includes three tests per chapter as well as three final exams. The tests contain a combination of multiple-choice, free-response, true/false, and fill-in-the-blank questions.

BCA Testing

Brooks/Cole Assessment is an internet-ready, text-specific testing suite that allows instructors to customize exams and track student progress in an accessible, browser-based format. *BCA* offers full algorithmic generation of problems and free-response mathematics, and completely integrates the testing and course management components. Test results flow automatically to your grade book and you can easily communicate to individuals, sections, or entire courses.

Text-Specific Video Series

The tapes work through examples from each section of the text, with additional solutions to two Study Set problems from each section.

Ancillaries for the Student

Student Solutions Manual

The *Student Solutions Manual* contains worked-out solutions to the odd-numbered problems in the text.

BCA Tutorial Student and Instructor Versions

This text-specific, interactive software is delivered via the Web (http://bca.brookscole.com) and is offered in both student and instructor versions. Like *BCA Testing,* it is browser based, making it an intuitive mathematical guide, even for students with little technological proficiency. *BCA Tutorial* allows students to work with real math notation in real time, and provides instant analysis and feedback. The tracking program built into the instructor version of the software enables instructors to monitor student progress carefully.

Interactive Video Skillbuilder CD

Packaged with the book, this single CD-ROM contains more than 8 hours of video instruction. The problems worked during each video lesson are listed next to the viewing screen, so students can work them ahead of time if they choose. In order to help students evaluate their progress, each section contains a 10-question Web quiz and each chapter contains a chapter test with answers provided.

Acknowledgments

We are grateful to the instructors who have reviewed the text at various stages of its development. Their comments and suggestions have proven invaluable in making this a better book. We sincerely thank all of them for lending their time and talent to this project.

Julia Brown
Atlantic Community College

John Coburn
*Saint Louis Community College–
Florissant Valley*

Sally Copeland
Johnson County Community College

Ben Cornelius
Oregon Institute of Technology

James Edmondson
Santa Barbara Community College

Judith Jones
Valencia Community College

Therese Jones
Amarillo College

Elizabeth Morrison
Valencia Community College

Angelo Segalla
Orange Coast College

June Strohm
*Pennsylvania State Community College–
DuBois*

Rita Sturgeon
San Bernardino Valley College

Jo Anne Temple
Texas Technical University

Sharon Testone
Onondaga Community College

Marilyn Treder
Rochester Community College

We want to express our gratitude to Karl Hunsicker, Cathy Gong, Dave Ryba, Terry Damron, Marion Hammond, Lin Humphrey, Doug Keebaugh, Robin Carter, Tanja Rinkel, Bob Billups, Liz Tussy, and the Citrus College Library staff for their help with some of the application problems in the textbook.

We would also like to thank Bob Pirtle, Ellen Brownstein, David Hoyt, Lori Heckelman, Vernon Boes, Melissa Henderson, Erin Wickersham, Michelle Paolucci, and The Clarinda Company for their help in creating this book.

Alan S. Tussy
R. David Gustafson

▶

For the Student

Success in Elementary Algebra

To be successful in mathematics, you need to know how to study it. The following checklist will help you develop your own personal strategy to study and learn the material. The suggestions listed below require some time and self-discipline on your part, but it will be worth the effort. This will help you get the most out of this course.

As you read each of the following statements, place a check mark in the box if you can truthfully answer Yes. If you can't answer Yes, think of what you might do to make the suggestion part of your personal study plan. You should go over this checklist several times during the term to be sure you are following it.

Preparing for the Class

☐   I have made a commitment to myself to give this course my best effort.

☐   I have the proper materials: a pencil with an eraser, paper, a notebook, a ruler, a calculator, and a calendar or day planner.

☐   I am willing to spend a minimum of two hours doing homework for every hour of class.

☐   I will try to work on this subject every day.

☐   I have a copy of the class syllabus. I understand the requirements of the course and how I will be graded.

☐ I have scheduled a free hour after the class to give me time to review my notes and begin the homework assignment.

Class Participation

☐ I will regularly attend the class sessions and be on time.

☐ When I am absent, I will find out what the class studied, get a copy of any notes or handouts, and make up the work that was assigned when I was gone.

☐ I will sit where I can hear the instructor and see the chalkboard.

☐ I will pay attention in class and take careful notes.

☐ I will ask the instructor questions when I don't understand the material.

☐ When tests, quizzes, or homework papers are passed back and discussed in class, I will write down the correct solutions for the problems I missed so that I can learn from my mistakes.

Study Sessions

☐ I will find a comfortable and quiet place to study.

☐ I realize that reading a math book is different from reading a newspaper or a novel. Quite often, it will take more than one reading to understand the material.

☐ After studying an example in the textbook, I will work the accompanying Self Check.

☐ I will begin the homework assignment only after reading the assigned section.

☐ I will try to use the mathematical vocabulary mentioned in the book and used by my instructor when I am writing or talking about the topics studied in the course.

☐ I will look for opportunities to explain the material to others.

☐ I will check all of my answers to the problems with those provided in the back of the book (or with the *Student Solutions Manual*) and reconcile any differences.

☐ My homework will be organized and neat. My solutions will show all the necessary steps.

☐ I will work some review problems every day.

☐ After completing the homework assignment, I will read the next section to prepare for the coming class session.

☐ I will keep a notebook containing my class notes, homework papers, quizzes, tests, and any handouts—all in order by date.

Special Help

☐ I know my instructor's office hours and am willing to go in to ask for help.

☐ I have formed a study group with classmates that meets regularly to discuss the material and work on problems.

☐ When I need additional explanation of a topic, I view the video and check the web site.

☐ I take advantage of extra tutorial assistance that my school offers for mathematics courses.

☐ I have purchased the *Student Solutions Manual* that accompanies this text, and I use it.

To follow each of these suggestions will take time. It takes a lot of practice to learn mathematics, just as with any other skill.

No doubt, you will sometimes become frustrated along the way. This is natural. When it occurs, take a break and come back to the material after you have had time to clear your thoughts. Keep in mind that the skills and discipline you learn in this course will help make for a brighter future. Good luck!

Taking a Math Test

The best way to relieve anxiety about taking a mathematics test is to know that you are well-prepared for it and that you have a plan. Before any test, ask yourself three questions. When? What? How?

When Will I Study?

1. When is the test?

2. When will I begin to review for the test?

3. What are the dates and times that I will reserve for studying for the test?

What Will I Study?

1. What sections will the test cover?
2. Has the instructor indicated any types of problems that are guaranteed to be on the test?

How Will I Prepare for the Test?

Put a check mark by each method you will use to prepare for the test.

☐ Review the class notes.

☐ Outline the chapter(s) on a piece of poster board to see the big picture and to see how the topics relate to one another.

☐ Recite the important formulas, definitions, vocabulary, and rules into a tape recorder.

☐ Make flash cards for the important formulas, definitions, vocabulary, and rules.

☐ Rework problems from the homework assignments.

☐ Rework each of the Self Check problems in the text.

☐ Form a study group to discuss and practice the topics to be tested.

☐ Complete the appropriate Chapter Review(s) and the Chapter Test(s).

☐ Review the Warnings given in the text.

☐ Work on improving my speed in answering questions.

☐ Review the methods that can be used to check my answers.

☐ Write a sample test, trying to think of the questions the instructor will ask.

☐ Complete the appropriate Cumulative Review Exercises.

☐ Get organized the night before the test. Have materials ready to go so that the trip to school will not be hurried.

☐ Take some time to relax immediately before the test. Don't study right up to the last minute.

Taking the Test

Here are some tips that can help improve your performance on a mathematics test.

- When you receive the test, scan it, looking for the types of problems you had expected to see. Do them first.
- Read the instructions carefully.
- Write down any formulas or rules as soon as you receive the test.
- Don't spend too much time on any one problem until you have attempted all the problems.
- If your instructor gives partial credit, at least try to begin a solution.
- Save the most difficult problems for last.
- Don't be afraid to skip a problem and come back to it later.
- If you finish early, go back over your work and look for mistakes.

Reading Mathematics

To get the most out of this book, you need to learn how to read it correctly. A mathematics textbook must be read differently than a novel or a newspaper. For one thing, you need to read it slowly and carefully. At times, you will have to reread a section to understand its content. You should also have pencil and paper with you when reading a mathematics book, so that you can work along with the text to understand the concepts presented.

Perhaps the most informative parts of a mathematics book are its examples. Each example in this textbook consists of a problem and its corresponding solution. One form of solution that is used many times in this book is shown in the diagram below. It is important that you follow the "flow" of its steps if you are to understand the mathematics involved. For this solution form, the basic idea is this:

- A property, rule, or procedure is applied to the original expression to obtain an equivalent expression. We show that the two expressions are equivalent by writing an equals sign between them. The property, rule, or procedure that was used is then listed next to the equivalent expression in the form of an author's note, printed in red.
- The process of writing equivalent expressions and explaining the reasons behind them continues, step by step, until the final result is obtained.

The solution in the following diagram consists of three steps, but solutions have varying lengths.

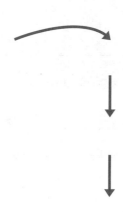

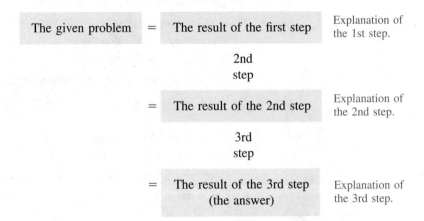

A solution (one of the basic forms)

1st step

| The given problem | = | The result of the first step | Explanation of the 1st step. |

2nd step

| | = | The result of the 2nd step | Explanation of the 2nd step. |

3rd step

| | = | The result of the 3rd step (the answer) | Explanation of the 3rd step. |

Writing Mathematics

One of the major objectives of this course is for you to learn how to write solutions to problems properly. A written solution to a problem should explain your thinking in a series of neat and organized mathematical steps. Think of a solution as a mathematical essay—one that your instructor and other students should be able to read and understand. Some solutions will be longer than others, but they must all be in the proper format and use the correct notation. To learn how to do this will take time and practice.

To give you an idea of what will be expected, let's look at two samples of student work. In the first, we have highlighted some important characteristics of a well-written solution. The second sample is poorly done and would not be acceptable.

$$\text{Evaluate: } 35 - 2^2 \cdot 3.$$

A well-written solution:

The problem was copied ▶ from the textbook.

$$35 - 2^2 \cdot 3 = 35 - 4 \cdot 3$$
$$= 35 - 12$$
$$= 23$$

◀ The first step of the solution is written here.
◀ The steps are written under each other in a neat, organized manner.

▲ The equals signs are lined up vertically.

A poorly written solution:

The problem wasn't ▶ copied from the text.

$$2^2 = 4 = 35 - 4 \cdot 3$$
$$\underbrace{}_{12}$$

SUB: 35
$$\underline{-12}$$
$$23 \rightarrow \boxed{23}$$

◀ An equals sign is improperly used.

◀ The work is disorganized and difficult to follow.

Elementary Algebra

1

An Introduction to Algebra

CAMPUS CONNECTION

The Accounting Department

In accounting, students study some of the basic billing practices used in the health care industry. The students need to know how to work with percents, because many health care plans pay only a percentage of a patient's medical bill. For example, one popular coverage pays 80% of the costs and requires the patient to pay the remaining 20% (called the *copayment*). In this chapter, we discuss how the algebraic concepts of *variable* and *equation* can be used to solve percent problems from many different disciplines, including accounting.

> *Algebra is a mathematical language that can be used to solve many types of real-world problems.*

▶ 1.1   Describing Numerical Relationships

In this section, you will learn about

> Tables ■ Bar graphs ■ Line graphs ■ Vocabulary ■ Symbols and notation ■ Variables, algebraic expressions, and equations ■ Constructing tables

Introduction   Using the vocabulary, symbols, and notation of algebra, we can mathematically describe (or **model**) the real world. From an algebraic model, we can make observations and predict outcomes. To solve problems using algebra, you will need to learn how to read it, write it, and speak it. In this section, we begin to explore the language of algebra by introducing some algebraic methods that are used to describe numerical relationships.

Tables

A production planner at a bicycle manufacturing plant must order parts for upcoming production runs. To order the correct number of tires, she uses the **table** in Figure 1-1.

After locating the number of bicycles to be manufactured in the left-hand column, she scans across the table to the corresponding entry in the right-hand column to find the number of tires to order. For example, if 400 bikes are to be manufactured, the table shows that 800 tires should be ordered.

FIGURE 1-1

Bicycles to be manufactured	Tires to order
100	200
200	400
300	600
400	800

Bar Graphs

The information in the table in Figure 1-1 can also be presented in a **bar graph.** The bar graph in Figure 1-2 has a **horizontal axis** labeled "Number of bicycles to be manufactured" and has been scaled in units of 100 bicycles. The **vertical axis** of the graph, labeled "Number of tires to be ordered," is scaled in units of 100 tires. The bars directly over each of the production amounts (100, 200, 300, and 400 bicycles) extend to a height indicating the corresponding number of tires to order. For example, if 300 bikes are to be manufactured, the height of the bar indicates that 600 tires should be ordered.

FIGURE 1-2

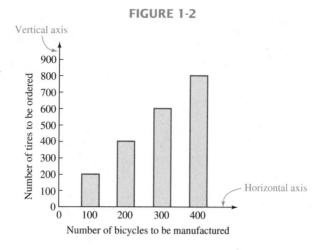

Line Graphs

A third way to present the information shown in the table in Figure 1-1 is with a **line graph.** Instead of using a bar to denote the number of tires to order for a given size of production run, we use a heavy dot drawn at the correct "height." See Figure 1-3(a). After drawing the four data points for 100, 200, 300, and 400 bicycles, we connect them with line segments to create the line graph. See Figure 1-3(b).

FIGURE 1-3

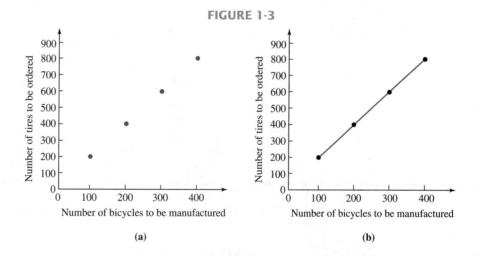

(a)   (b)

The line graph not only presents all the information contained in the table and the bar graph, but it also provides additional information that they do not. We can use the line graph to find the number of tires to order for a production run of a size that is not shown in the table or the bar graph.

EXAMPLE 1

Reading a line graph. Use the line graph in Figure 1-3(b) to determine the number of tires needed if 250 bicycles are manufactured.

Solution First, locate the number 250 (between 200 and 300) on the horizontal axis. Then draw a line straight up to intersect the graph. (See Figure 1-4.) From the point of intersection, draw a horizontal line to the left that intersects the vertical axis. We see that the number of tires to order is 500.

FIGURE 1-4

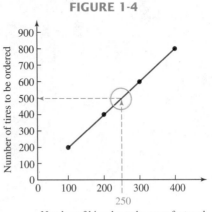

Number of bicycles to be manufactured

SELF CHECK   Use the line graph in Figure 1–4 to find the number
of tires needed if 350 bicycles are manufactured.   *Answer:* 700   ■

Vocabulary

After working with a table, a bar graph, and a line graph, it is evident that there is a relationship between the number of tires to order and the number of bicycles to be manufactured. Using words, we can express this relationship as follows:

"The number of tires to order is two times the number of bicycles to be manufactured."

In algebra, the word **product** is used to indicate the answer to a multiplication. Therefore, we can restate the relationship this way:

"The number of tires to order is the product of two and the number of bicycles to be manufactured."

To indicate other arithmetic operations, we will use the following words:

■ **sum** to indicate the answer to an addition: The sum of 5 and 6 is 11.

■ **difference** to indicate the answer to a subtraction: The difference of 6 and 2 is 4.

■ **quotient** to indicate the answer to a division: The quotient of 6 and 3 is 2.

EXAMPLE 2   **Vocabulary.** Express each statement in words, using one of the words *sum, product, difference,* or *quotient:*   **a.** $22 \div 11 = 2$ and   **b.** $22 + 11 = 33$.

Solution   **a.** The quotient of 22 and 11 is 2.
b. The sum of 22 and 11 is 33.

SELF CHECK   Express each statement in words: *Answers:*   **a.** The difference of
a. $22 - 11 = 11$ and   **b.** $22 \times 11 = 242$.   22 and 11 is 11. **b.** The product of 22 and 11 is 242.   ■

Construct a tbl for all
4 operations showing:
Operat'n, Vocab., Symbols

Symbols and Notation

In algebra, we will use many symbols and notations. Because the letter *x* is often used in algebra and could be confused with the multiplication sign $\times$, we usually write multiplication in another form.

Symbols that are used for multiplication

Symbol	Name	Example
$\times$	times sign	$6 \times 4 = 24$ or $\begin{array}{r} 451 \\ \times\ 53 \\ \hline 23{,}903 \end{array}$
$\cdot$	raised dot	$6 \cdot 4 = 24$ or $451 \cdot 53 = 23{,}903$
( )	parentheses	$(6)4 = 24$ or $6(4) = 24$ or $(6)(4) = 24$

There are several ways to indicate division. In algebra, the form most often involves a fraction bar.

Symbols that are used for division

Symbol	Name	Example
$\div$	division sign	$16 \div 4 = 4$ or $3{,}400 \div 20 = 170$
$\overline{)\ }$	long division	$4\overline{)16}\ \ \overset{4}{}$ or $20\overline{)3{,}400}\ \ \overset{170}{}$
—	fraction bar	$\dfrac{16}{4} = 4$ or $\dfrac{3{,}400}{20} = 170$

Variables, Algebraic Expressions, and Equations

Another way to describe the relationship between the number of tires to order and the number of bicycles being manufactured uses *variables*. **Variables** are letters that stand for numbers. If we let the letter t stand for the number of tires to be ordered and b for the number of bicycles to be manufactured, we can translate the relationship from words to mathematical symbols.

The number of tires to order	is	two	times	the number of bicycles to be manufactured.
t	$=$	2	$\cdot$	b

The statement $t = 2 \cdot b$ is called an **equation.** Equations are mathematical sentences that contain an $=$ sign, which indicates that two quantities are equal. Some examples of equations are

$$3 + 5 = 8 \qquad x + 5 = 20 \qquad 17 - t = 14 - t$$

In the equation $t = 2 \cdot b$, the variable b is multiplied by 2. When we multiply a variable by another number or multiply a variable by another variable, we don't need to use a multiplication symbol.

$2b$ means $2 \cdot b$ xy means $x \cdot y$ abc means $a \cdot b \cdot c$

Using this form, we can write the equation $t = 2 \cdot b$ as $t = 2b$. The notation $2b$ on the right-hand side of the equation is called an **algebraic expression** or, more simply, an **expression.** Algebraic expressions are the building blocks of equations.

Algebraic expressions

Variables and/or numbers can be combined with the operations of addition, subtraction, multiplication, and division to create **algebraic expressions.**

[Handwritten margin notes:]

$\frac{xy}{}$

Alg Exp. or Eq.? Alg.

How many var? 2

What operations? mult.

What would make it
an eq.? $x = y$
  or $xy = 6$
  or $xy = z$

$2 + 3 = 5$
eq. or alg. express.?

Here are some examples of algebraic expressions.

$2a + 7$ This algebraic expression is a combination of the numbers 2 and 7, the variable a, and the operations of multiplication and addition.

$\dfrac{10 - y}{3}$ This algebraic expression is a combination of the numbers 10 and 3, the variable y, and the operations of subtraction and division.

$15mn(2m)$ This algebraic expression is a combination of the numbers 15 and 2, the variables m and n, and the operation of multiplication.

Using the equation $t = 2b$ to express the relationship between the number of tires to order and the number of bicycles to be manufactured has an advantage over the other methods we have discussed. It can be used to determine the exact number of tires to order for a production run of *any* size.

EXAMPLE 3

Using an equation. Find the number of tires needed for a production run of 178 bicycles.

Solution

To find the number of tires needed, we use the equation that describes this numerical relationship.

$t = 2b$ The describing equation.

$t = 2(178)$ Replace b, which stands for the number of bicycles, with 178. Use parentheses to show the multiplication.

$t = 356$ Do the multiplication: $2(178) = 356$.

Therefore, 356 tires will be needed.

SELF CHECK

Find the number of tires needed if 604 bicycles are to be manufactured.   *Answer:* 1,208   ■

Constructing Tables

Equations such as $t = 2b$, which express a known relationship between two or more variables, are called **formulas.** Formulas are used in many fields, such as economics, biology, nursing, and construction. In the next example, we will see that the results found using the formula $t = 2b$ can be presented in table form.

[Handwritten margin note:] Use Ex. 26.

EXAMPLE 4

Constructing a table. Find the number of tires to order for production runs of 233 and 852 bicycles. Present the results in a table.

Solution

We begin by constructing a table with appropriate column headings. The size of each production run (233 and 852) is entered in the left-hand column of the table.

Bicycles to be manufactured	Tires to order
233	
852	

Next, we use the equation $t = 2b$ to find the number of tires needed if 233 and 852 bikes are to be manufactured.

$t = 2b$	$t = 2b$
$t = 2(233)$ Replace b with 233.	$t = 2(852)$ Replace b with 852.
$t = 466$	$t = 1,704$

Finally, we enter these results in the right-hand column of the table: 466 tires for 233 bicycles manufactured and 1,704 tires for 852 bicycles manufactured.

Bicycles to be manufactured	Tires to order
233	466
852	1,704

SELF CHECK

Find the number of tires to order for production runs of 87 and 487 bicycles. Present the results in a table.

Answer:

Bicycles to be manufactured	Tires to order
87	174
487	974

STUDY SET

Section 1.1

1-25 , 27-71 odd

VOCABULARY

In Exercises 1–12, fill in the blanks to make the statements true.

1. The answer to an addition problem is called the ____sum____.

2. The answer to a subtraction problem is called the ____difference____.

3. The answer to a multiplication problem is called the ____product____.

4. The answer to a division problem is called the ____quotient____.

5. ____Variables____ are letters that stand for numbers.

6. The symbols ( ) are called ____parentheses____.

7. Variables and/or numbers can be combined with the operations of addition, subtraction, multiplication, and division to create algebraic ____expressions____.

8. An ____equation____ is a mathematical sentence that contains an = sign.

9. An equation such as $t = 2b$, which expresses a known relationship between two or more variables, is called a ____formula____.

10. In Illustration 1, a ____line____ graph is shown.

11. In Illustration 1, the ____horizontal____ axis of the graph has been scaled in units of _1_ second.

12. In Illustration 1, the ____vertical____ axis of the graph has been scaled in units of 50 ____feet____.

ILLUSTRATION 1

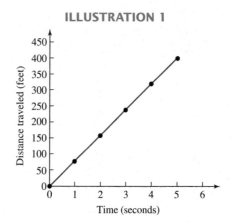

CONCEPTS

In Exercises 13–20, classify each item as either an algebraic expression or an equation.

13. $18 + m = 23$ equation

14. $18 + m$
algebraic expression

15. $y - 1$
algebraic expression

16. $y - 1 = 2$ equation

17. $30x$
algebraic expression

18. $t = 16b$ equation

19. $r = \dfrac{2}{3}$ equation

20. $\dfrac{c - 7}{5}$

algebraic expression

21. a. What operations does the expression $5x - 16$ contain?   multiplication, subtraction
   b. What variable(s) does it contain? x

22. a. What operations does the expression $\frac{12 + t}{25}$ contain?
   addition, division
   b. What variable(s) does it contain? t

23. a. What operations does the equation
   $4 + 1 = 20 - m$ contain?   addition, subtraction
   b. What variable(s) does it contain? m

24. a. What operations does the equation $y + 14 = 5(6)$
   contain?   addition, multiplication
   b. What variable(s) does it contain? y

25. See Illustration 2. As the railroad crossing guard drops, the measure of angle 1 increases, while the measure of angle 2 decreases. At any instant, the sum of the measures of the two angles is 90°.
   a. Complete the table.

ILLUSTRATION 2

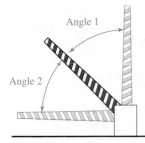

Angle 1	Angle 2
0°	90°
30°	60°
45°	45°
60°	30°
90°	0°

   b. Use the data in the table to construct a line graph for values of angle 1 from 0° to 90°.

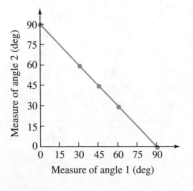

26. See Illustration 3. As the legs on the keyboard stand are widened, the measure of angle 1 will increase, and in turn, the measure of angle 2 will decrease. For any position, the sum of the measures of the angles is 180°.
   a. Complete the table.

ILLUSTRATION 3

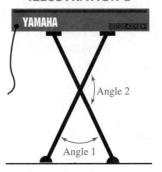

Angle 1	Angle 2
50°	130°
60°	120°
70°	110°
80°	100°
90°	90°

   b. Use the data in the table to construct a line graph for values of angle 1 from 50° to 90°.

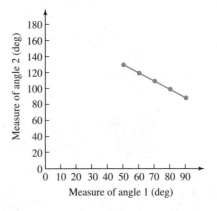

27. a. Explain what the dotted lines help us find in the graph in Illustration 4.
   They help us determine that 15-year-old machinery is worth $35,000.
   b. According to the graph, as the machinery ages, what happens to its value?   It decreases.

ILLUSTRATION 4

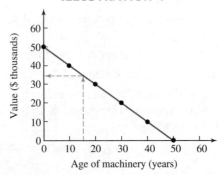

28. a. Use the line graph in Illustration 5 to find the income received from 30, 50, and 70 customers.
   $250, $350, $450

b. According to the graph, as the number of customers increases, what happens to the income?

It increases.

ILLUSTRATION 5

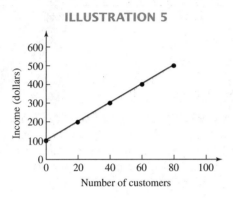

NOTATION

In Exercises 29–32, write each multiplication using a raised dot · and then using parentheses ( ).

29. 5×6 $5 \cdot 6, 5(6)$ **30.** 4×7 $4 \cdot 7, 4(7)$

31. 34×75 **32.** 90×12
$34 \cdot 75, 34(75)$ $90 \cdot 12, 90(12)$

In Exercises 33–40, write each multiplication without using a multiplication symbol.

33. $4 \cdot x$ $4x$ **34.** $5 \cdot y$ $5y$

35. $3 \cdot r \cdot t$ $3rt$ **36.** $22 \cdot q \cdot s$ $22qs$

37. $l \cdot w$ lw **38.** $b \cdot h$ bh

39. $P \cdot r \cdot t$ Prt **40.** $l \cdot w \cdot h$ lwh

In Exercises 41–44, write each division using a fraction bar.

41. $32 \div x$ $\frac{32}{x}$ **42.** $y \div 15$ $\frac{y}{15}$

43. $30\overline{)90}$ $\frac{90}{30}$ **44.** $20\overline{)80}$ $\frac{80}{20}$

PRACTICE

In Exercises 45–52, express each statement using one of the words sum, difference, product, or quotient.

45. $18(24)$
the product of 18 and 24

46. $45 \cdot 12$
the product of 45 and 12

47. $11 - 9$
the difference of 11 and 9

48. $65 + 89$
the sum of 65 and 89

49. $2x$
the product of 2 and x

50. $16t$
the product of 16 and t

51. $\dfrac{66}{11}$
the quotient of 66 and 11

52. $12 \div 3$
the quotient of 12 and 3

In Exercises 53–56, translate the word model into an equation. (Hint: You will need to use variables.)

53.

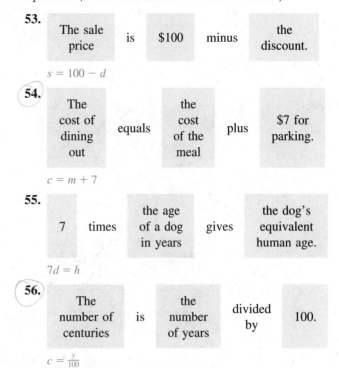

| The sale price | is | $100 | minus | the discount. |

$s = 100 - d$

54. | The cost of dining out | equals | the cost of the meal | plus | $7 for parking. |

$c = m + 7$

55. | 7 | times | the age of a dog in years | gives | the dog's equivalent human age. |

$7d = h$

56. | The number of centuries | is | the number of years | divided by | 100. |

$c = \frac{y}{100}$

In Exercises 57–64, translate the word model into an equation. (Hint: You will need to use variables.)

57. The amount of sand that should be used is the product of 3 and the amount of cement used. $s = 3c$

58. The number of waiters needed is the quotient of the number of customers and 10. $w = \frac{c}{10}$

59. The weight of the truck is the sum of the weight of the engine and 1,200. $w = e + 1{,}200$

60. The number of classes that are still open is the difference of 150 and the number of classes that are closed. $o = 150 - c$

61. The profit is the difference of the revenue and 600.
$p = r - 600$

62. The distance is the product of the rate and 3. $d = 3r$

63. The quotient of the number of laps run and 4 is the number of miles run. $\frac{l}{4} = m$

64. The sum of the tax and 35 is the total cost.
$t + 35 = c$

In Exercises 65–68, use the given equation (formula) to complete each table.

65. $d = 360 + l$

Lunch time (minutes)	School day (minutes)
30	390
40	400
45	405

66. $b = 1,024k$

Kilobytes	Bytes
1	1,024
5	5,120
10	10,240

67. $t = 1,500 - d$

Deductions	Take-home pay
200	1,300
300	1,200
400	1,100

68. $w = \dfrac{s}{12}$

Inches of snow	Inches of water
12	1
24	2
72	6

In Exercises 69–72, use the data to find an equation that describes the relationship between the two quantities. Then state the relationship in words.

69.

Eggs	Dozen
24	2
36	3
48	4

$d = \frac{e}{12}$; the number of dozen eggs is the quotient of the number of eggs and 12.

70.

Input voltage	Output voltage
60	50
110	100
220	210

$o = i - 10$; the output voltage is the difference of the input voltage and 10.

71.

Couples	Individuals
20	40
100	200
200	400

$i = 2c$; the number of individuals is the product of 2 and the number of couples.

72.

Benefit package ($)	Total compensation ($)
4,000	39,000
5,000	40,000
6,000	41,000

$T = b + 35,000$; the total compensation is the sum of the benefit package and 35,000.

APPLICATIONS

73. CHAIR PRODUCTION Use the diagram shown in Illustration 6 to write six formulas that could be used by planners to order the necessary number of legs, arms, seats, backs, arm pads, and screws for a production run of c chairs.

$l = 4c$, $a = 2c$, $S = c$, $b = c$, $p = 2c$, $s = 20c$

ILLUSTRATION 6

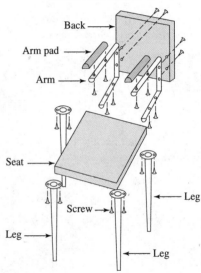

74. STAIRCASE PRODUCTION Write four formulas that could be used by the job superintendent to order the necessary number of staircase parts for a tract of h homes, each of which will have a staircase as shown in Illustration 7.

$b = 16h$, $p = 3h$, $r = 2h$, $t = 8h$

ILLUSTRATION 7

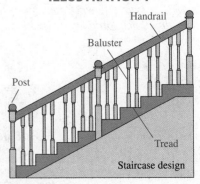

Staircase design

75. SPARE PARTS When an automobile company has parts manufactured for the assembly-line production of a car, it orders 500 more than necessary, to stock dealership service departments. Write an equation to find the number of left-side front doors that should be manufactured for a production run of *x* cars. $d = x + 500$

76. NO-SHOWS Park rangers have noticed that each weekend in the summer, on average, 15 people who had made campground reservations fail to show up. Write an equation the rangers could use to predict the weekend occupancy of the campground if *r* reservations have been made.
$o = r - 15$

77. RELIGIOUS BOOKS Illustration 8 shows the annual sales of books on religion, sprituality, and inspiration for the years 1991–1997. Graph the data using a bar graph. Then describe any trend that is apparent. On the horizontal axis, use the label 1 to represent the year 1991, 2 to represent 1992, and so on.
Sales have steadily increased.

ILLUSTRATION 8

Year	Book sales
1991	36,651,000
1992	50,104,000
1993	60,449,000
1994	70,541,000
1995	74,794,000
1996	78,022,000
1997	91,627,000

Based on data from the
Book Industry Study Group

Sales of Religious Books

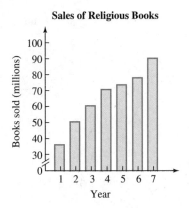

78. U.S. CRIME STATISTICS Property crimes include burglary, theft, and motor vehicle theft. Graph the property crime rates listed in Illustration 9 using a bar graph. Is an overall trend apparent?
Property crime rates steadily declined.

ILLUSTRATION 9

Year	Victimizations per 1,000 households
1991	354
1992	325
1993	319
1994	310
1995	291
1996	266

Based on data from the Bureau of Justice Statistics

U.S. Property Crime Rates

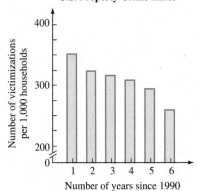

WRITING

79. Many people misuse the word *equation* when discussing mathematics. What is an equation? Give an example.

80. Explain the difference between an algebraic expression and an equation. Give an example of each.

81. Which do you think is more informative, a bar graph or a line graph? Explain your reasoning.

82. Create a bar graph that shows, on average, how many hours of television you watch each day of the week. Let Sunday be day 1, Monday be day 2, and so on.

▶ 1.2 The Real Numbers

In this section, you will learn about

Sets of numbers ■ Order on the number line ■ Rational numbers (fractions and mixed numbers) ■ Decimals ■ Irrational numbers ■ The real numbers ■ Opposites ■ Absolute value

Introduction   In this course, we will work with many types of numbers. We will solve problems involving temperatures that are negative, see units of time expressed as fractions, and express amounts of money using decimals. Some of the geometric figures we will encounter will have dimensions that are expressed as square roots. In this section, we define each type of number we will use in the course and show that they are part of a larger collection of numbers called **real numbers.**

Sets of Numbers

Table 1-1 shows the low temperatures for Rockford, IL during the first week of January. In the left column, we have used the numbers 1, 2, 3, 4, 5, 6, and 7 to denote the calendar days of the month. This collection of numbers is called a **set,** and the members (or **elements**) of the set can be listed within **braces** {   }.

{1, 2, 3, 4, 5, 6, 7}

TABLE 1-1

Day of the month	Low temperature (°F)
1	4
2	−5
3	−7
4	0
5	3
6	6
7	6

Each of the numbers 1, 2, 3, 4, 5, 6, and 7 is a member of a basic set of numbers called the **natural numbers.** The natural numbers are the numbers that we count with.

Natural numbers

The set of **natural numbers** is {1, 2, 3, 4, 5, 6, 7, 8, 9, 10, . . .}.

The three dots used in this definition indicate that the list of natural numbers continues forever.

The natural numbers together with 0 make up another set of numbers called the **whole numbers.**

Whole numbers

The set of **whole numbers** is {0, 1, 2, 3, 4, 5, 6, 7, 8, 9, 10, . . .}.

Since every natural number is also a whole number, we say that the set of natural numbers is a **subset** of the whole numbers.

Table 1-1 contains positive and negative temperatures. For example, on the 2nd day of the month, the low temperature was $-5°$ ($5°$ below zero). On the 5th day, the low temperature was $3°$ ($3°$ above zero). The numbers used to represent the temperatures listed in the table are members of a set of numbers called the **integers.**

Integers

> The set of **integers** is $\{. \ . \ . \ , -4, -3, -2, -1, 0, 1, 2, 3, 4, . \ . \ .\}$.

Since every whole number is also an integer, we say that the set of whole numbers is a subset of the set of integers. The natural numbers are also a subset of the integers.

Order on the Number Line

We can illustrate sets of numbers with a **number line.** Like a ruler, a number line is straight and has uniform markings, as in Figure 1-5. The arrowheads indicate that the number line continues forever to the left and to the right. Numbers to the left of 0 have values that are less than 0; they are called **negative numbers.** Numbers to the right of 0 have values that are greater than 0; they are called **positive numbers.** The number 0 is neither positive nor negative.

FIGURE 1-5

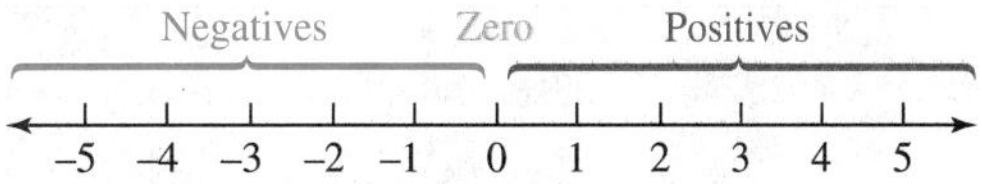

Negative numbers can be used to describe amounts that are less than 0, such as a checking account that is $75 overdrawn ($-$75), an elevation of 200 feet below sea level (-200 ft), and 5 seconds before liftoff (-5 sec).

Using a process known as **graphing,** a single number or a set of numbers can be represented on a number line. The **graph of a number** is the point on the number line that corresponds to that number. *To graph a number* means to locate its position on the number line and then to highlight it by using a heavy dot.

E X A M P L E   1   **Graphing on the number line.** Graph the integers between -4 and 5.

Solution   To graph these integers, we locate their positions on the number line and highlight each position by drawing a dot.

SELF CHECK   Graph the integers between -2 and 2.   *Answer:*

As we move to the right on a number line, the values of the numbers increase. As we move to the left, the values of the numbers decrease. In Figure 1-6, we know that 5 is greater than -3, because the graph of 5 lies to the right of the graph of -3. We also know that -3 is less than 5, because its graph lies to the left of the graph of 5.

FIGURE 1-6

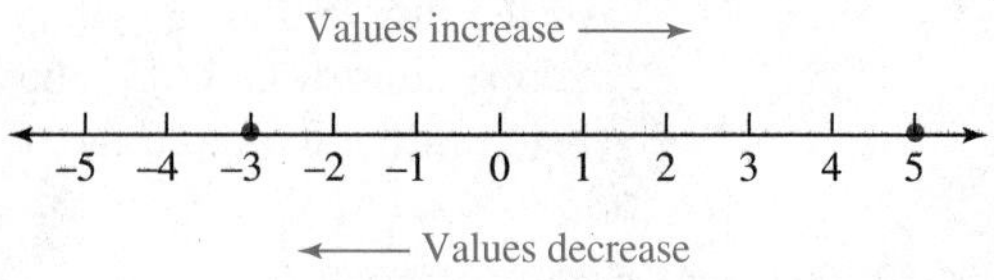

Values increase
−5 −4 −3 −2 −1 0 1 2 3 4 5
Values decrease

The **inequality symbol** $>$ ("is greater than") can be used to show that 5 is greater than -3, and the inequality symbol $<$ ("is less than") can be used to show that -3 is less than 5.

$5 > -3$ Read as "5 is greater than -3."

$-3 < 5$ Read as "-3 is less than 5."

To distinguish between these two inequality symbols, remember that each one points to the smaller of the two numbers involved.

$5 > -3$ $\qquad$ $-3 < 5$

Points to the smaller number.

EXAMPLE 2 **Inequality symbols.** Use one of the symbols $>$ or $<$ to make each statement true: **a.** -4 ___ 4 and **b.** -2 ___ -3.

Solution **a.** Since -4 is to the left of 4 on the number line, $-4 < 4$.

b. Since -2 is to the right of -3 on the number line, $-2 > -3$.

SELF CHECK Use one of the symbols $>$ or $<$ to make each statement true: **a.** 1 ___ -1 and **b.** -5 ___ -4.

Answers: **a.** $>$, **b.** $<$ ■

By extending the number line to include negative numbers, we can represent more situations graphically. In Figure 1-7, the line graph illustrates the low temperatures listed in Table 1-1. The vertical axis is scaled in units of degrees Fahrenheit, and temperatures below zero (negative temperatures) are graphed. For example, for the second day of the month, the low was -5°F.

FIGURE 1-7

Low temperature (°F)
8
7
6
5
4
3
2
1
0
−1
−2
−3
−4
−5
−6
−7
−8
1 2 3 4 5 6 7
Day of the month

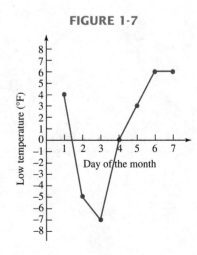

Rational Numbers (Fractions and Mixed Numbers)

In this course, we will work with positive and negative fractions. For example, the time it takes for a motorist to complete the commute home might be $\frac{3}{4}$ of an hour, or a surveyor might indicate that a building's foundation has fallen below finished grade by expressing its elevation as $-\frac{7}{8}$ of an inch. In a fraction, the number above the fraction bar is called the **numerator**. The number below is called the **denominator**. If the numerator is less than the denominator, the fraction is called a **proper fraction**. If the numerator is greater than or equal to the denominator, the fraction is called an **improper fraction**.

$$\text{Proper fraction} \quad \frac{3}{4} \quad \begin{array}{l} \longleftarrow \text{Numerator} \longrightarrow \\ \longleftarrow \text{Denominator} \longrightarrow \end{array} \quad \frac{25}{12} \quad \text{Improper fraction}$$

We will also work with **mixed numbers**—numbers that are the sum of a whole number and a proper fraction. For example, a piece of fabric could be $5\frac{1}{2}$ yards long, or a job could take $3\frac{2}{5}$ days to complete. Fractions and mixed numbers are part of the set of numbers called the **rational numbers**.

Rational numbers

> A **rational number** is any number that can be written as a fraction with an integer numerator and a nonzero integer denominator.

Fractions such as $\frac{3}{4}$ and $\frac{25}{12}$ are rational numbers, because they have an integer numerator and a nonzero integer denominator. The fraction $-\frac{7}{8}$ is a rational number, because it can be written in the form $\frac{-7}{8}$. Mixed numbers such as $5\frac{1}{2}$ and $3\frac{2}{5}$ are also rational numbers, because they can be written as fractions with an integer numerator and a nonzero integer denominator:

$$5\frac{1}{2} = \frac{11}{2} \qquad \text{and} \qquad 3\frac{2}{5} = \frac{17}{5}$$

All integers are rational numbers, because they can be written as fractions with a denominator of 1. For example, $-4 = \frac{-4}{1}$ and $0 = \frac{0}{1}$. Therefore, the set of integers is a subset of the rational numbers.

EXAMPLE 3

Graphing fractions and mixed numbers. Graph $-4\frac{1}{4}, 3\frac{7}{8}, -\frac{5}{3}$, and $\frac{1}{10}$.

Solution It is easier to locate the graph of $-\frac{5}{3}$ if we express it as $-1\frac{2}{3}$.

$$\begin{array}{c} -4\frac{1}{4} \qquad\qquad -\frac{5}{3} \qquad \frac{1}{10} \qquad\qquad 3\frac{7}{8} \\ \xleftarrow{\hspace{1cm}} \bullet \;|\; \;|\; \bullet\; |\; \bullet\; |\; \;|\; \;|\; \bullet\; |\; \xrightarrow{\hspace{1cm}} \\ -5 \;\; -4 \;\; -3 \;\; -2 \;\; -1 \;\; 0 \;\; 1 \;\; 2 \;\; 3 \;\; 4 \;\; 5 \end{array}$$

Since the graph of $-\frac{5}{3}$ lies to the left of the graph of $\frac{1}{10}$, we can write $-\frac{5}{3} < \frac{1}{10}$.

SELF CHECK Graph $-\frac{2}{3}, -\frac{7}{4}$, and $2\frac{1}{8}$.

Answer:

$$\begin{array}{c} -\frac{7}{4} \quad -\frac{2}{3} \qquad\qquad 2\frac{1}{8} \\ \xleftarrow{\hspace{0.5cm}} \bullet \;|\; \bullet\; |\; \;|\; \;|\; \bullet\; |\; \xrightarrow{\hspace{0.5cm}} \\ -2 \;\; -1 \;\; 0 \;\; 1 \;\; 2 \;\; 3 \end{array}$$ ∎

Decimals

We will also work with decimals. Here are some examples of decimals.

- The price of apples is $0.89 a pound.
- A dragster was clocked at 203.156 miles per hour.
- The first-quarter loss for a business was $-\$2.7$ million.

EXAMPLE 4

Graphing decimals. Graph −3.75, 2.5, −0.9, and 1.25.

Solution   Recall that $0.75 = \frac{3}{4}$, so $-3.75 = -3\frac{3}{4}$, and $0.25 = \frac{1}{4}$, so $1.25 = 1\frac{1}{4}$.

$$
\begin{array}{c}
 -3.75 \qquad -0.9 \quad\quad 1.25\ \ 2.5 \\
\overset{\displaystyle \bullet}{\underset{-4}{|}}\ \overset{}{\underset{-3}{|}}\ \overset{}{\underset{-2}{|}}\ \overset{\bullet}{\underset{-1}{|}}\ \overset{}{\underset{0}{|}}\ \overset{\bullet}{\underset{1}{|}}\ \overset{\bullet}{\underset{2}{|}}\ \overset{}{\underset{3}{|}}\ \overset{}{\underset{4}{|}}
\end{array}
$$

Notice that the graph of −0.9 lies to the right of the graph of −3.75. Therefore, we can write −0.9 > −3.75.

SELF CHECK

Graph −1.25, −2.5, and 1.75.

Answer:

$$
\begin{array}{c}
-2.5\ -1.25 \qquad\quad 1.75 \\
\overset{}{\underset{-3}{|}}\ \overset{\bullet}{\underset{-2}{|}}\ \overset{\bullet}{\underset{-1}{|}}\ \overset{}{\underset{0}{|}}\ \overset{}{\underset{1}{|}}\ \overset{\bullet}{\underset{2}{|}}
\end{array}
$$

Decimals such as 0.89, 203.156, and −2.7 are called **terminating decimals.** Since they can be written as fractions with integer numerators and nonzero integer denominators, they are rational numbers.

$$0.89 = \frac{89}{100} \qquad 203.156 = 203\frac{156}{1,000} = \frac{203{,}156}{1,000} \qquad -2.7 = -2\frac{7}{10} = \frac{-27}{10}$$

We will also work with **repeating decimals** such as 0.33333. . . and 2.161616. . . . (The three dots indicate that the digits continue in the pattern shown forever.) Any repeating decimal can be expressed as a fraction with an integer numerator and a nonzero integer denominator. For example, $0.333\ldots = \frac{1}{3}$, and $2.161616\ldots = 2\frac{16}{99} = \frac{214}{99}$. Since every repeating decimal can be written as a fraction, repeating decimals are also rational numbers.

Irrational Numbers

Some decimals cannot be written as fractions. In Figure 1-8, the length of each wire used to anchor the volleyball net is $\sqrt{2}$ (read as "the **square root** of 2") yards. Expressed in decimal form,

$$\sqrt{2} = 1.414213562\ldots$$

FIGURE 1-8

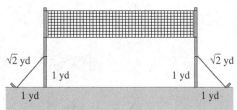

This **nonterminating, nonrepeating decimal** cannot be written as a fraction with an integer for its numerator and a nonzero integer for its denominator. Therefore, $\sqrt{2}$ is not a rational number. It is called an **irrational number.** When working with irrational numbers, it is often beneficial to approximate them. We can use a scientific calculator to approximate square roots.

ACCENT ON TECHNOLOGY *Approximating Irrational Numbers*

We can use the $\boxed{\sqrt{}}$ key (the square root key) on a calculator to find square roots. For example, to find $\sqrt{2}$ using a scientific calculator, we enter 2 and press the $\boxed{\sqrt{}}$ key.

Keystrokes   2 $\boxed{\sqrt{}}$ $\boxed{1.414213562}$

We see that $\sqrt{2} \approx 1.414213562$. The symbol $\approx$ means "is approximately equal to." If the problem calls for an approximate answer, we can round this decimal. For example, to the nearest hundredth, $\sqrt{2} \approx 1.41$.

To approximate $\sqrt{2}$ using a graphing calculator, we use the following keystrokes:

Keystrokes $\boxed{\text{2nd}}$ $\boxed{\sqrt{}}$ $\boxed{2}$ $\boxed{)}$ $\boxed{\text{ENTER}}$ $\boxed{\begin{array}{l}\sqrt{(2)} \\ \qquad 1.414213562\end{array}}$

An irrational number often used in geometry is π (read as "pi"). To find the approximate value of π using a scientific calculator, we simply press the $\boxed{\pi}$ key.

Keystrokes $\boxed{\pi}$ (you may have to use a $\boxed{\text{2nd}}$ or $\boxed{\text{Shift}}$ key first) $\boxed{3.141592654}$

Rounded to the nearest thousandth, $\pi \approx 3.142$.

To approximate π with a graphing calculator, we press the following keys:

Keystrokes $\boxed{\text{2nd}}$ $\boxed{\pi}$ $\boxed{\text{ENTER}}$ $\boxed{\begin{array}{l}\pi \\ \qquad 3.141592654\end{array}}$

Some other examples of irrational numbers are 2π (this means $2 \cdot \pi$), $\sqrt{3}$, $3\sqrt{2}$ (this means $3 \cdot \sqrt{2}$), $-\sqrt{7}$, and $\sqrt{27}$.

EXAMPLE 5

Graphing irrational numbers. Graph $2\sqrt{2}$, $-\sqrt{7}$, and π.

Solution   To locate these numbers on a number line, we can use a calculator to approximate them. To the nearest tenth, $2\sqrt{2} = 2 \cdot \sqrt{2} \approx 2.8$, $-\sqrt{7} \approx -2.6$, and $\pi \approx 3.1$.

SELF CHECK   Graph $-\sqrt{3}$ and $\frac{\pi}{2}$.   *Answer:*

The Real Numbers

The set of rational numbers together with the set of irrational numbers form the set of **real numbers.** This means that every real number can be written as either a terminating; a repeating; or a nonterminating, nonrepeating decimal. Thus, the set of real numbers is the set of all decimals. All of the points on a number line represent the set of real numbers.

The real numbers

A **real number** is any number that is either a rational or an irrational number.

Figure 1-9 shows how the sets of numbers discussed in this section are related; it also gives some specific examples of each type of number.

FIGURE 1-9

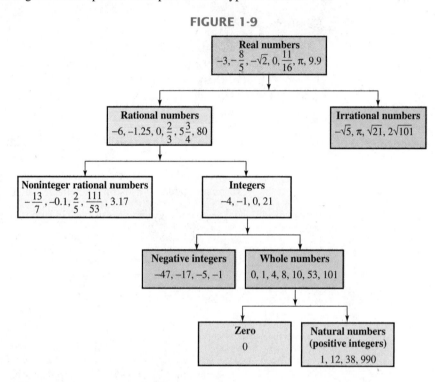

EXAMPLE 6

Classifying real numbers. Tell which numbers in the following set are natural numbers, whole numbers, integers, rational numbers, irrational numbers, and real numbers: $\left\{-3.4, \frac{12}{5}, 0, -6, 1\frac{3}{4}, -\pi, 16\right\}$.

Solution

Natural numbers: 16

Whole numbers: 0, 16

Integers: 0, −6, 16

Rational numbers: $-3.4, \frac{12}{5}, 0, -6, 1\frac{3}{4}, 16$

Irrational numbers: $-\pi$

Real numbers: $-3.4, \frac{12}{5}, 0, -6, 1\frac{3}{4}, -\pi, 16$

SELF CHECK

Use the instructions for Example 6 with the set $\left\{0.4, -\sqrt{2}, -\frac{2}{7}, 45, -2, \frac{13}{4}\right\}$

Answers: natural numbers: 45; whole numbers: 45; integers: 45, −2; rational numbers: 0.4, $-\frac{2}{7}$, 45, −2, $\frac{13}{4}$; irrational numbers: $-\sqrt{2}$; real numbers: 0.4, $-\sqrt{2}$, $-\frac{2}{7}$, 45, −2, $\frac{13}{4}$ ■

Opposites

In Figure 1-10, we see that −4 and 4 are both a distance of 4 away from 0. Because of this, we say that −4 and 4 are **opposites** or **additive inverses**.

FIGURE 1-10

Opposites

Two numbers represented by points on a number line that are the same distance away from the origin, but on opposite sides of it, are called **opposites** or **additive inverses.**

To write the opposite of a number, a − symbol is used. For example, the opposite, or additive inverse, of 6 can be written as −6. The opposite of 0 is 0, so −0 = 0. Since the opposite of −6 is 6, we have −(−6) = 6. In general, if a represents any real number; then

$$-(-a) = a$$

Absolute Value

The **absolute value** of a number gives the distance between the number and 0 on a number line. To indicate absolute value, a number is inserted between two vertical bars. For the example shown in Figure 1-10, we would write $|-4| = 4$. This notation is read as "The absolute value of negative 4 is 4," and it tells us that the distance between −4 and 0 is 4 units. In Figure 1-10, we also see that $|4| = 4$.

Absolute value

The **absolute value** of a number is the distance on a number line between the number and 0.

 WARNING! Absolute value expresses distance. The absolute value of a number is always positive or zero—never negative.

EXAMPLE 7

Evaluating absolute values. Evaluate each absolute value: **a.** $|-18|$, **b.** $\left|-\frac{7}{8}\right|$, **c.** $-|-\pi|$.

Solution

a. Since -18 is a distance of 18 from 0 on the number line,

$$|-18| = 18$$

b. Since $-\frac{7}{8}$ is a distance of $\frac{7}{8}$ from 0 on the number line,

$$\left|-\tfrac{7}{8}\right| = \tfrac{7}{8}$$

c. Since $-\pi$ is a distance of π from 0 on the number line, $|-\pi| = \pi$. Therefore, $-|-\pi| = -(\pi) = -\pi$.

SELF CHECK

Evaluate each absolute value: **a.** $|100|$, **b.** $\left|-\frac{1}{2}\right|$, and **c.** $-|-\sqrt{2}|$.

Answers: **a.** 100, **b.** $\frac{1}{2}$, **c.** $-\sqrt{2}$ ∎

EXAMPLE 8

Comparing real numbers. Insert one of the symbols $>$, $<$, or $=$ in the blank to make a true statement: **a.** $-(-3.9)$ ____ 3, **b.** $-\left|-\frac{4}{5}\right|$ ____ $\left|\sqrt{5}\right|$.

Solution **a.** $-(-3.9) > 3$, because $-(-3.9) = 3.9$ and $3.9 > 3$.

b. $-\left|-\frac{4}{5}\right| < \left|\sqrt{5}\right|$, because $-\left|-\frac{4}{5}\right| = -\frac{4}{5}$, $\left|\sqrt{5}\right| = \sqrt{5} \approx 2.2$, and $-\frac{4}{5} < 2.2$.

SELF CHECK

Insert one of the symbols $>$, $<$, or $=$ in the blank to make a true statement:

a. $-(-7)$ ____ 12 and **b.** $3\frac{3}{4}$ ____ $\left|-\frac{5}{4}\right|$. *Answers:* **a.** $<$, **b.** $>$ ∎

STUDY SET

Section 1.2

VOCABULARY

In Exercises 1–12, fill in the blanks to make the statements true.

1. The set of ____whole____ numbers is $\{0, 1, 2, 3, 4, 5, \dots\}$.

2. The set of ____natural____ numbers is $\{1, 2, 3, 4, 5, \dots\}$.

3. Numbers less than zero are ____negative____, and numbers greater than zero are ____positive____.

4. The set of ____real____ numbers is the set of all decimals.

5. A ____rational____ number can be written as a fraction with an ____integer____ numerator and a non-zero integer denominator.

6. A decimal such as 0.25 is called a ____terminating____ decimal, while 0.333 . . . is called a ____repeating____ decimal.

7. The set of ____integers____ is $\{\dots, -2, -1, 0, 1, 2, \dots\}$.

8. The symbols $<$ and $>$ are ____inequality____ symbols.

9. An ____irrational____ number cannot be written as a fraction with an integer for its numerator and denominator.

10. If the numerator of a fraction is less than the denominator, it is called a ____proper____ fraction.

11. The ____absolute value____ of a number is the distance on a number line between the number and 0.

12. Two numbers represented by points on a number line that are the same distance away from the origin, but on opposite sides of it, are called ____opposites____ or ____additive inverses____.

CONCEPTS

13. Show that each of the following numbers is a rational number by expressing it as a fraction with an integer in its numerator and a nonzero integer in its denominator: 6, -9, $-\frac{7}{8}$, $3\frac{1}{2}$, -0.3, 2.83. $\frac{6}{1}, \frac{-9}{1}, \frac{-7}{8}, \frac{7}{2}, \frac{-3}{10}, \frac{283}{100}$

14. Represent each situation using a signed number.
    a. A trade deficit of $15 million $-\$15$
    b. A rainfall total 0.75 inch below average -0.75 in.
    c. A score $12\frac{1}{2}$ points under the standard $-12\frac{1}{2}$ pts
    d. A building foundation $\frac{5}{16}$ inch above grade $+\frac{5}{16}$ in.

15. What two numbers are a distance of 8 away from 5 on the number line? 13 and -3

16. Suppose the variable m stands for a negative number. Use an inequality to state this fact. $m < 0$

17. The variables a and b represent real numbers. Use an inequality symbol, $<$ or $>$, to make each statement true.

    a.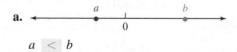

    a $<$ b

    b.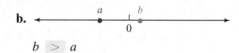

    b $>$ a

    c.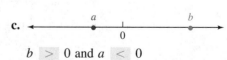

    b $>$ 0 and a $<$ 0

18. Each year from 1990–1997, the United States imported more goods and services from Japan than it exported to Japan. This caused trade *deficits*, which can be represented by negative numbers. See Illustration 1.

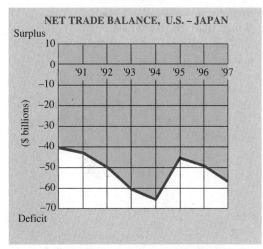

ILLUSTRATION 1

NET TRADE BALANCE, U.S. – JAPAN

Based on data from the Department of Commerce and the Bureau of Economic Analysis

a. In what year was the deficit the worst? Estimate the deficit then. '94; −$65 billion
b. In what year was the deficit the smallest? Estimate the deficit then. '90; −$40 billion

19. The diagram in Illustration 2 can be used to show how the natural numbers, whole numbers, integers, rational numbers, and irrational numbers make up the set of real numbers. If the natural numbers are represented as shown, label each of the other sets.

ILLUSTRATION 2

Real Numbers

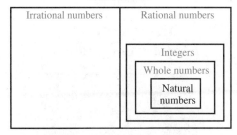

| Irrational numbers | Rational numbers |
| Integers |
| Whole numbers |
| Natural numbers |

20. Which number graphed below has the largest absolute value? r

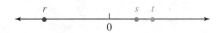

NOTATION

21. $\sqrt{5}$ is read "the ___square root___ of 5."
22. $|-15|$ is read "the ___absolute value___ of −15."
23. The symbol $\approx$ means ___is approximately equal to___.
24. The symbols { }, called ___braces___, are used when writing a set.
25. In $\frac{3}{4}$, 3 is the ___numerator___, and 4 is the ___denominator___ of the fraction.

26. The grouping symbols ( ) are called ___parentheses___.

27. Explain what 4π means, and then use a calculator to approximate it to the nearest tenth.
4π means $4 \cdot \pi$; $4\pi \approx 12.6$

28. Explain what $2\sqrt{3}$ means, and then use a calculator to approximate it to the nearest tenth.
$2\sqrt{3}$ means $2 \cdot \sqrt{3}$; $2\sqrt{3} \approx 3.5$

In Exercises 29–44, insert one of the symbols $>$, $<$, or $=$ in the blank to make each statement true.

29. $-2 \;>\; -3$
30. $0 \;<\; 32$
31. $|3.4| \;>\; \sqrt{10}$
32. $0.08 \;>\; 0.079$
33. $-|-1.1| \;<\; -1$
34. $-(-5.5) \;=\; -\left(-5\frac{1}{2}\right)$
35. $-\left(-\frac{5}{8}\right) \;>\; -\left(-\frac{3}{8}\right)$
36. $-19\frac{2}{3} \;<\; -19\frac{1}{3}$
37. $\left|-\frac{15}{2}\right| \;=\; 7.5$
38. $\sqrt{39} \;<\; 3\pi$
39. $\frac{99}{100} \;=\; 0.99$
40. $|2| \;>\; -|-2|$
41. $0.333\ldots \;>\; 0.3$
42. $\left|-2\frac{2}{3}\right| \;>\; -\left(-\frac{3}{2}\right)$
43. $-(-1) \;>\; \left|-\frac{15}{16}\right|$
44. $-0.666\ldots \;<\; 0$

PRACTICE

In Exercises 45–46, tell which numbers in the given set are natural numbers, whole numbers, integers, rational numbers, irrational numbers, and real numbers.

45. $\left\{ -\frac{5}{6}, 35.99, 0, 4\frac{3}{8}, \sqrt{2}, -50, \frac{17}{5} \right\}$
natural: none; whole: 0; integers: 0, −50; rational: $-\frac{5}{6}$, 35.99, 0, $4\frac{3}{8}$, −50, $\frac{17}{5}$; irrational: $\sqrt{2}$, real: all

46. $\left\{ -0.001, 10\frac{1}{2}, 6, 3\pi, \sqrt{7}, -23, -5.6 \right\}$
natural: 6; whole: 6; integers: 6, −23; rational: −0.001, $-10\frac{1}{2}$, 6, −23, −5.6; irrational: 3π, $\sqrt{7}$; real: all

In Exercises 47–48, tell whether each statement is true or false.

47. a. Every whole number is an integer. true
b. Every integer is a natural number. false
c. Every integer is a whole number. false
d. Irrational numbers are nonterminating, nonrepeating decimals. true

48. a. Irrational numbers are real numbers. true
b. Every whole number is a rational number. true
c. Every rational number can be written as a fraction. true
d. Every rational number is a whole number. false

49. a. Write the statement $-6 < -5$ using an inequality symbol that points in the other direction.
$-5 > -6$
b. Write the statement $16 > -25$ using an inequality symbol that points in the other direction.
$-25 < 16$

50. If we begin with the number −4 and find its opposite, and then find the opposite of that result, what number do we obtain?   −4

In Exercises 51–52, graph each set of numbers on the number line.

51. $\left\{-\pi,\ 4.25,\ -1\frac{1}{2},\ -0.333\ldots,\ \sqrt{2},\ -\frac{35}{8}\right\}$

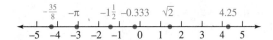

52. $\left\{\pi,\ -2\frac{1}{8},\ 2.75,\ -\sqrt{17},\ \frac{17}{4},\ -0.666\ldots\right\}$

In Exercises 53–60, use a calculator to approximate each irrational number to the nearest thousandth.

53. $\sqrt{5}$ 2.236   **54.** $\sqrt{19}$ 4.359

55. $\sqrt{99}$ 9.950   **56.** $\sqrt{42}$ 6.481

57. $2\sqrt{5}$ 4.472   **58.** $\dfrac{\sqrt{3}}{2}$ 0.866

59. 5π 15.708   **60.** $\dfrac{\pi}{4}$ 0.785

In Exercises 61–68, write each expression in simpler form.

61. The opposite of 5   −5   **62.** The opposite of −9   9

63. The opposite of $-\dfrac{7}{8}$ $\frac{7}{8}$ **64.** The opposite of 6.56   −6.56

65. $-(-10)$ 10   **66.** $-(-1)$ 1

67. $-(-2.3)$ 2.3   **68.** $-\left(-\frac{3}{4}\right)$ $\frac{3}{4}$

APPLICATIONS

69. BANKING Later in this course, we will use a table such as the one in Illustration 3 to solve banking problems. Which numbers shown here are natural numbers, whole numbers, integers, rational numbers, irrational numbers, and real numbers?

ILLUSTRATION 3

Type of account	Principal	Rate	Time (years)	Interest
Checking	$135.75	0.0275	$\frac{31}{365}$	$0.32
Savings	$5,000	0.06	$2\frac{1}{2}$	$750

natural, whole, integers: 750, 5,000; rational: all; irrational: none; real: all

70. DRAFTING The drawing in Illustration 4 shows the dimensions of an aluminum bracket.
   a. Which numbers shown are natural numbers, whole numbers, integers, rational numbers, irrational numbers, and real numbers?
   natural, whole, integers: 9; rational: 9, $\frac{15}{16}$, $3\frac{1}{8}$, 1.765; irrational: 2π, 3π, $\sqrt{89}$; real: all
   b. Use a calculator to approximate all the irrational numbers in the drawing to the nearest thousandth. $3\pi \approx 9.425$, $2\pi \approx 6.283$, $\sqrt{89} \approx 9.434$

ILLUSTRATION 4

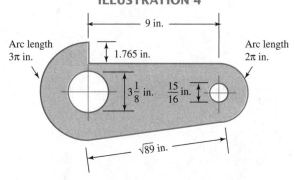

71. AUTOMOBILE INDUSTRY See Illustration 5.
   a. Estimate the net income ($ billions) for the Chrysler Corporation for each of the years from 1990 to 1997.
   '90: $0.3; '91: −$0.8; '92: $0.9; '93: −$2.3; '94: $3.8; '95: $2.0; '96: $3.8; '97: $2.8
   b. What does a negative net income indicate?
   The company lost money.

ILLUSTRATION 5
Chrysler Corporation

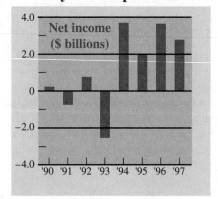

Based on data from *Business Week* and the American Automobile Manufacturers Association

72. GOVERNMENT DEBT A budget *deficit* indicates that the government's outlays (expenditures) were more than the receipts (revenue) it took in that year. See Illustration 6.
   a. For the years 1980–1998, when was the federal budget deficit the worst? Estimate the size of the deficit?   1992, −$290 billion

b. The 1998 budget *surplus* was the first in nearly three decades. Estimate it. Explain what it means to have a budget surplus.

$70 billion; the government takes in more money than it spends.

ILLUSTRATION 6
Federal Budget Deficit/Surplus

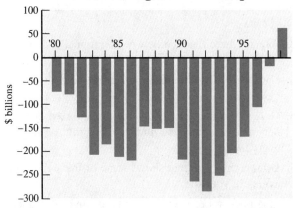

Based on data from U.S. Census Bureau and the *Los Angeles Times* (Oct. 1, 1998)

73. 🖩 TIRES The distance a tire rolls in one revolution can be found by computing the circumference of the circular tire using the formula $C = \pi d$, where d is the diameter of the tire. How far will the tire shown in Illustration 7 roll in one revolution? Answer to the nearest tenth of an inch.   81.7 in.

ILLUSTRATION 7

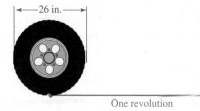

One revolution

74. 🖩 HULA HOOP The length of plastic pipe needed to form a hula hoop can be found by computing the circumference of the circular hula hoop using the formula $C = \pi d$, where d is its diameter. Find the length of pipe needed to form the hula hoop shown in Illustration 8. Answer to the nearest tenth of an inch.   106.8 in.

ILLUSTRATION 8

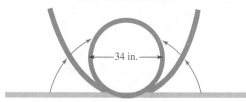

75. FROZEN FOODS In freezing some baked goods for cold storage, the temperature of the product is

lowered, as shown in Illustration 9. Use the data to draw a line graph.

ILLUSTRATION 9

Time since removed from oven (min)	Temperature of product (° Celsius)
5	50
10	10
15	0
20	−5
25	−10

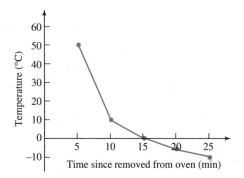

76. APPLE COMPUTER Use the data given in Illustration 10 to draw a bar graph of the company's net income for each quarter of 1997 and the first three quarters of 1998.

ILLUSTRATION 10

🍎 **Press Release—July 15, 1998** 🍎

Cupertino, California— Apple Computer, Inc. today announced a profit of $101 million for the fiscal 1998 third quarter that ended June 26. With a 1998 first-quarter profit of $47 million and a second-quarter profit of $55 million, Apple has now rebounded from a difficult 1997 in which the company experienced four consecutive quarters with losses of $120 million, $708 million, $56 million, and $161 million, respectively.

Apple Computer, Inc.

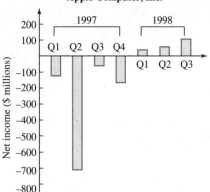

77. VELOCITY AND SPEED In science, a positive velocity normally indicates movement to the right, and a negative velocity indicates movement to the left. The speed of an object is the absolute value of its velocity. Find the speed of the car and the motorcycle in Illustration 11.   55 mph, 30 mph

ILLUSTRATION 11

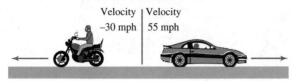

78. TARGET PRACTICE In Illustration 12, which artillery shell landed farther from the target? How can the concept of absolute value be applied to answer this question?

The shell that landed on the point marked −18—it's farther away from the target, which is located at 0.

ILLUSTRATION 12

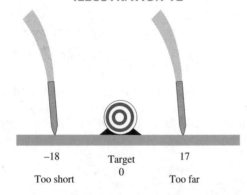

In Exercises 79–82, refer to the historical time line in Illustration 13.

ILLUSTRATION 13
MAYA CIVILIZATION

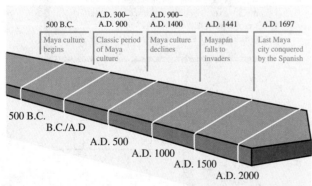

Based on data from *People in Time and Place, Western Hemisphere* (Silver Burdett & Ginn Inc., 1991), p. 129

79. What basic unit was used to scale the time line?
500 years

80. On the time line, what symbolism was used to represent zero?   B.C./A.D.

81. On the time line, what could be thought of as positive and what could be thought of as negative numbers?   pos: A.D.; neg: B.C.

82. Express the dates for the Maya civilization using positive and negative numbers.   −500 to 1697

WRITING

83. Explain the difference between a rational and an irrational number.

84. Can two different numbers have the same absolute value? Explain.

85. Give two examples each of fractions, mixed numbers, decimals, and negative numbers that you use in your everyday life.

86. Draw a time line that shows some of the significant events in your life.

87. In some parts of the country, there is negative unemployment in the computer information systems field. Explain what is meant by a −5% unemployment rate.

88. In writing courses, students are warned not to use double negatives in their compositions. Identify the double negative in the following sentence. Then rewrite the sentence so that it conveys the same idea without using a double negative. "No one didn't turn in the homework."

REVIEW

In Exercises 89–92, express each statement in words, using one of the words sum, difference, product, *and* quotient.

89. $7 - 5 = 2$ The difference of 7 and 5 is 2.

90. $5(6) = 30$ The product of 5 and 6 is 30.

91. $30 \div 15 = 2$ The quotient of 30 and 15 is 2.

92. $12 + 12 = 24$ The sum of 12 and 12 is 24.

In Exercises 93–94, use the given equation to complete the table.

93. $T = 15g$

Number of gears	Number of teeth
10	150
12	180
15	225

94. $p = r - 200$

Revenue	Profit
1,000	800
5,000	4,800
10,500	10,300

▶ 1.3

Exponents and Order of Operations

In this section, you will learn about

Exponents ■ Order of operations ■ Grouping symbols
■ Applications

Introduction   In this course, we will perform six operations with real numbers: addition, subtraction, multiplication, division, raising to a power, and finding a root. Quite often, we will have to **evaluate** (find the value of) expressions containing more than one operation. In that case, we need to know the order in which the operations are to be performed. That is the focus of this section.

Exponents

The multiplication statement $3 \cdot 5 = 15$ has two parts: the two numbers that are being multiplied, and the answer. The answer (15) is called the **product,** and the numbers that are being multiplied (3 and 5) are called **factors.**

In the expression $3 \cdot 3 \cdot 3 \cdot 3 \cdot 3$, the number 3 is used as a factor five times. We call 3 a *repeated factor*. To express a repeated factor, we can use an **exponent.**

Exponent and base

An **exponent** is used to indicate repeated multiplication. It tells how many times the **base** is used as a factor.

The exponent is 5.
↓

$$\underbrace{3 \cdot 3 \cdot 3 \cdot 3 \cdot 3}_{} = \qquad 3^5$$

↑   ↑
Five repeated factors of 3.   The base is 3.

In the **exponential expression** a^n, a is the base, and n is the exponent. The expression a^n is called a **power of a.** Some examples of powers are

5^2 Read as "5 to the second power" or "5 squared." Here, $a = 5$ and $n = 2$.

9^3 Read as "9 to the third power" or "9 cubed." Here, $a = 9$ and $n = 3$.

$(-2)^5$ Read as "-2 to the fifth power." Here, $a = -2$ and $n = 5$.

EXAMPLE 1

Repeated factors. Write each expression using exponents.

a. $4 \cdot 4 \cdot 4$

b. $8 \cdot 8 \cdot 15 \cdot 15 \cdot 15 \cdot 15$

c. Sixteen cubed

Solution   **a.** In $4 \cdot 4 \cdot 4$, the 4 is repeated as a factor 3 times. So $4 \cdot 4 \cdot 4 = 4^3$.

b. $8 \cdot 8 \cdot 15 \cdot 15 \cdot 15 \cdot 15 = 8^2 \cdot 15^4$

c. 16^3

SELF CHECK

Write each expression using exponents:
a. (12)(12)(12)(12)(12)(12),   **b.** $2 \cdot 9 \cdot 9 \cdot 9$,   and   **c.** fifty squared.

Answers: **a.** 12^6,   **b.** $2 \cdot 9^3$,   **c.** 50^2 ∎

In the next example, we use exponents to rewrite expressions involving repeated variable factors.

E X A M P L E   2

Using exponents with variables. Write each product using exponents.

a. $(a)(a)(a)(a)(a)(a)$

b. $4 \cdot \pi \cdot r \cdot r$

Solution   **a.** $(a)(a)(a)(a)(a)(a) = a^6$ *a* is repeated as a factor 6 times.

b. $4 \cdot \pi \cdot r \cdot r = 4\pi r^2$ *r* is repeated as a factor 2 times.

SELF CHECK

Write each product using exponents:
a. $y \cdot y \cdot y \cdot y$ and   **b.** $12 \cdot b \cdot b \cdot b \cdot c$.

Answers: **a.** y^4,   **b.** $12b^3c$ ∎

E X A M P L E   3

Evaluating exponential expressions. Find the value of each of the following: **a.** 3^2,   **b.** 5^3,   **c.** 10^1,   **d.** 2^5.

Solution   We write the base as a factor the number of times indicated by the exponent and then do the multiplication.

a. $3^2 = 3 \cdot 3 = 9$ The base is 3, the exponent is 2.

b. $5^3 = 5 \cdot 5 \cdot 5 = 125$ The base is 5, the exponent is 3.

c. $10^1 = 10$ The base is 10, the exponent is 1.

d. $2^5 = 2 \cdot 2 \cdot 2 \cdot 2 \cdot 2 = 32$ The base is 2, the exponent is 5.

SELF CHECK

Which of the numbers 3^4, 4^3, and 5^2 is the largest?   *Answer:* $3^4 = 81$ ∎

ACCENT ON TECHNOLOGY *The Squaring Key*

A plastic tarp, used to cover the infield of a baseball field when it rains, is in the shape of a square with sides 125 feet long. To find the area covered by the tarp, we can use the formula for the area of a square.

$A = s^2$ The formula for the area of a square with side length *s*.

$A = \mathbf{125}^2$ Substitute 125 for *s*.

We can use the squaring key $\boxed{x^2}$ on a scientific calculator to find the square of a number. To find 125^2, we enter these numbers and press these keys:

Keystrokes   125 $\boxed{x^2}$ $\boxed{15625}$

Using a graphing calculator, we can find 125^2 with the following keystrokes:

Keystrokes   125 $\boxed{x^2}$ $\boxed{\text{ENTER}}$ $\boxed{\begin{array}{r} 125^2 \\ 15625 \end{array}}$

The plastic tarp covers an area of 15,625 square feet (ft^2).

ACCENT ON TECHNOLOGY *The Exponential Key*

A store owner sent two friends a letter advertising her store's low prices. The ad closed with the following request: "Please send a copy of this letter to two of your friends." If all those receiving letters respond, how many letters will be circulated in the 10th level of the mailing?

Table 1-2 shows how a pattern develops. On the first level, 2 letters are mailed by the owner to her two friends. On the second level, the two friends mail out 2 letters each, for a total of 4 (or 2^2) letters. On the third level, those four people mail two letters each, for a total of 8 (or 2^3) letters. Therefore, the tenth level will mail out 2^{10} letters. We can use the exponential key $\boxed{y^x}$

TABLE 1-2

Level	Number of letters circulated
1st	$2 = 2^1$
2nd	$4 = 2^2$
3rd	$8 = 2^3$
10th	$? = 2^{10}$

(on some calculators, it is labeled x^y) to raise a number to a power. To find 2^{10} using a scientific calculator, we enter these numbers and press these keys:

Keystrokes   2 $\boxed{y^x}$ 10 $\boxed{=}$ $\boxed{ 1024}$

To evaluate 2^{10} using a graphing calculator, we press the following keys:

Keystrokes   2 $\boxed{\wedge}$ 10 $\boxed{\text{ENTER}}$ $\boxed{\begin{array}{l} 2\text{^}10 \\ 1024 \end{array}}$

On the 10th level of this chain letter mailing, $2^{10} = 1,024$ letters will be circulated.

Order of Operations

Suppose you have been asked to contact a friend if you see a certain type of oriental rug for sale while you are traveling in Turkey. While in Turkey, you spot the rug and send the following E-mail message.

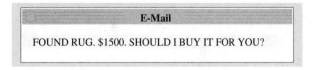

E-Mail
FOUND RUG. $1500. SHOULD I BUY IT FOR YOU?

The next day, you get this response from your friend.

E-Mail
NO PRICE TOO HIGH! REPEAT...NO! PRICE TOO HIGH.

Something is wrong. One part of the response says to buy the rug at any price. The other part of the response says not to buy it, because it's too expensive. The placement of the exclamation point makes us read the two parts of the response differently, resulting in different interpretations. When reading a mathematical statement, the same kind of confusion is possible. For example, consider the expression

$$2 + 3 \cdot 6$$

This expression contains two operations: addition and multiplication. We can do the calculations in two ways. We can do the addition first and then do the multiplication. Or we can do the multiplication first and then do the addition. However, we get different results.

Method 1: Add first

$2 + 3 \cdot 6 = 5 \cdot 6$ Add 2 and 3 first.

$ = 30$ Multiply 5 and 6.

Method 2: Multiply first

$2 + 3 \cdot 6 = 2 + 18$ Multiply 3 and 6 first.

$ = 20$ Add 2 and 18.

If we don't establish a uniform order of operations, the expression $2 + 3 \cdot 6$ has two different answers. To avoid this possibility, we always use the following set of priority rules.

Order of operations

1. Evaluate all exponential expressions.
2. Do all multiplications and divisions, working from left to right.
3. Do all additions and subtractions, working from left to right.

It may not be necessary to apply all of these steps in every problem. For example, the expression $2 + 3 \cdot 6$ does not contain any exponential expressions. So we next look for multiplications and divisions to perform. To correctly evaluate $2 + 3 \cdot 6$, we apply the rules for the order of operations:

$$2 + 3 \cdot 6 = 2 + 18$$ Do the multiplication first: $3 \cdot 6 = 18$.

$$ = 20$$ Do the addition.

Therefore, the correct result when evaluating $2 + 3 \cdot 6$ is 20.

EXAMPLE 4

Order of operations. Evaluate $3 \cdot 2^3 - 4$.

Solution   To find the value of this expression, we must do the operations of multiplication, raising to a power, and subtraction. The rules for the order of operations tell us to begin by evaluating the exponential expression.

$$3 \cdot 2^3 - 4 = 3 \cdot 8 - 4$$ Evaluate the exponential expression: $2^3 = 8$.

$$ = 24 - 4$$ Do the multiplication: $3 \cdot 8 = 24$.

$$ = 20$$ Do the subtraction.

SELF CHECK   Evaluate $2 \cdot 3^2 + 17$.   *Answer:* 35   ■

EXAMPLE 5

Order of operations. Evaluate $30 - 4 \cdot 5 + 9$.

Solution   To evaluate this expression, we must do the operations of subtraction, multiplication, and addition. The rules for the order of operations tell us to begin with the multiplication.

$$30 - 4 \cdot 5 + 9 = 30 - 20 + 9$$ Do the multiplication: $4 \cdot 5 = 20$.

$$ = 10 + 9$$ Working from left to right, do the subtraction: $30 - 20 = 10$.

$$ = 19$$ Do the addition.

SELF CHECK   Evaluate $40 - 9 \cdot 4 + 10$.   *Answer:* 14   ■

EXAMPLE 6	**Order of operations.** Evaluate $\dfrac{160}{4} - 6(2)3$.

Solution   This expression contains the operations of division, subtraction, and multiplication. Working from left to right, we begin with the division.

$$\frac{160}{4} - 6(2)3 = \mathbf{40} - 6(2)3 \quad \text{Do the division: } \tfrac{160}{4} = 40.$$

$$= 40 - (12)3 \quad \text{Do the multiplication: } 6(2) = 12.$$

$$= 40 - 36 \quad \text{Do the multiplication: } (12)3 = 36.$$

$$= 4 \quad \text{Do the subtraction.}$$

SELF CHECK   Evaluate $\dfrac{240}{8} - 3(2)4$.   *Answer:* 6   ■

EXAMPLE 7	**Order of operations.** Evaluate $5^5 - 3 \cdot 2^4$.

Solution   This expression contains the operations of raising to a power, subtraction, and multiplication. We begin by evaluating the exponential expressions.

$$5^5 - 3 \cdot 2^4 = \mathbf{3{,}125} - 3 \cdot \mathbf{16} \quad \text{Use a calculator to evaluate the exponential}$$
$$\text{expressions: } 5^5 = 3{,}125 \text{ and } 2^4 = 16.$$

$$= 3{,}125 - 48 \quad \text{Do the multiplication: } 3 \cdot 16 = 48.$$

$$= 3{,}077 \quad \text{Do the subtraction.}$$

SELF CHECK   Evaluate $6^4 - 4 \cdot 3^4$.   *Answer:* 972   ■

Grouping Symbols

Grouping symbols serve as mathematical punctuation marks. They help determine the order in which an expression is to be evaluated. Examples of grouping symbols are parentheses ( ), brackets [ ], braces { }, and the fraction bar —.

Order of operations when grouping symbols are present

If the expression contains grouping symbols, do all calculations within each pair of grouping symbols, working from the innermost pair to the outermost pair, in the following order:

1. Evalute all exponential expressions.

2. Do all multiplications and divisions, working from left to right.

3. Do all additions and subtractions, working from left to right.

If the expression *does not* contain grouping symbols, begin with step 1.

In a fraction, first simplify the numerator and denominator separately. Then simplify the fraction, whenever possible.

In the next example, there are two similar-looking expressions. However, because of the parentheses, we evaluate them in a different order.

E X A M P L E 8	**Working with grouping symbols.** Evaluate each expression: **a.** $15 - 6 + 4$ and **b.** $15 - (6 + 4)$.

Solution   **a.** This expression does not contain any grouping symbols.

$$15 - 6 + 4 = 9 + 4 \quad \text{Working from left to right, we first do the subtraction:}$$
$$15 - 6 = 9.$$
$$= 13 \quad \text{Do the addition.}$$

b. Since the expression contains grouping symbols, we must do the operation inside the parentheses first.

$$15 - (6 + 4) = 15 - 10 \quad \text{Do the addition inside the parentheses: } 6 + 4 = 10.$$
$$= 5 \quad \text{Do the subtraction.}$$

SELF CHECK   Evaluate each expression: **a.** $30 - 9 + 3$ and
b. $30 - (9 + 3)$. *Answers:*   **a.** 24,   **b.** 18   ∎

E X A M P L E 9	**Evaluating expressions containing grouping symbols.** Evaluate $(6 - 3)^2$.

Solution   This expression contains parentheses. By the rules for the order of operations, we must do the operation within the parentheses first.

$$(6 - 3)^2 = 3^2 \quad \text{Do the subtraction inside the parentheses: } 6 - 3 = 3.$$
$$= 9 \quad \text{Evaluate the exponential expression.}$$

SELF CHECK   Evaluate $(12 - 6)^3$. *Answer:* 216   ∎

E X A M P L E 1 0	**Order of operations inside grouping symbols.** Evaluate $8^2 + 2(10 - 4 \cdot 2)$.

Solution   First, we do the operations *inside* the grouping symbols in the proper order.

$$8^2 + 2(10 - 4 \cdot 2) = 8^2 + 2(10 - 8) \quad \text{Do the multiplication inside the parentheses first: } 4 \cdot 2 = 8.$$
$$= 8^2 + 2(2) \quad \text{Do the subtraction inside the parentheses: } 10 - 8 = 2.$$
$$= 64 + 2(2) \quad \text{Evaluate the exponential expression: } 8^2 = 64.$$
$$= 64 + 4 \quad \text{Do the multiplication: } 2(2) = 4.$$
$$= 68 \quad \text{Do the addition.}$$

SELF CHECK   Evaluate $7^2 - 5(8 - 3 \cdot 2)$. *Answer:* 39   ∎

E X A M P L E 1 1	**Working with a fraction bar.** Evaluate $\dfrac{3(15) - 12}{2 + 3^2}$.

Solution   We evaluate the numerator and denominator separately. Then we do the division indicated by the fraction bar.

$$\frac{3(15) - 12}{2 + 3^2} = \frac{45 - 12}{2 + 9} \quad \begin{array}{l} \text{In the numerator, do the multiplication: } 3(15) = 45. \\ \text{In the denominator, evaluate } 3^2 = 9. \end{array}$$
$$= \frac{33}{11} \quad \begin{array}{l} \text{In the numerator, do the subtraction.} \\ \text{In the denominator, do the addition.} \end{array}$$
$$= 3 \quad \text{Do the division.}$$

SELF CHECK   Evaluate $\dfrac{5(10) + 15}{4^2 - 3}$.   *Answer:* 5

ACCENT ON TECHNOLOGY *Order of Operations and Parentheses*

Calculators have the rules for the order of operations built in. A left parenthesis key $\boxed{(}$ and a right parenthesis key $\boxed{)}$ should be used when grouping symbols are needed. To evaluate $\frac{320}{20 - 16}$ with a scientific calculator, we enter these numbers and press these keys:

Keystrokes $320\ \boxed{\div}\ \boxed{(}\ 20\ \boxed{-}\ 16\ \boxed{)}\ \boxed{=}$ $\boxed{\qquad\qquad\qquad\qquad 80}$

To evaluate $\frac{320}{20 - 16}$ with a graphing calculator, we enter these numbers and press these keys:

Keystrokes $320\ \boxed{\div}\ \boxed{(}\ 20\ \boxed{-}\ 16\ \boxed{)}\ \boxed{\text{ENTER}}$ $\boxed{\begin{array}{l} 320/(20-16) \\ \qquad\qquad\quad 80 \end{array}}$

The answer is 80.

If an expression contains more than one pair of grouping symbols, we begin by working inside the innermost pair and then work to the outermost pair.

$$\begin{array}{c} \text{Innermost parentheses} \\ \downarrow \qquad \downarrow \\ 25 + 2[13 - 3(4 - 2)] \\ \uparrow \qquad\qquad \uparrow \\ \text{Outermost brackets} \end{array}$$

EXAMPLE 12   **Grouping symbols inside grouping symbols.** Evaluate $25 + 2[13 - 3(4 - 2)]$.

Solution   We begin by working inside the innermost grouping symbols, the parentheses.

$$
\begin{aligned}
25 + 2[13 - 3(4 - 2)] &= 25 + 2[13 - 3(2)] && \text{Do the subtraction inside the parentheses: } 4 - 2 = 2. \\
&= 25 + 2[13 - 6] && \text{Do the multiplication inside the brackets: } 3(2) = 6. \\
&= 25 + 2(7) && \text{Do the subtraction inside the brackets: } 13 - 6 = 7. \\
&= 25 + 14 && \text{Do the multiplication.} \\
&= 39 && \text{Do the addition.}
\end{aligned}
$$

SELF CHECK   Evaluate $9 + 4[16 - 2(9 - 2)]$.   *Answer:* 17

Applications

EXAMPLE 13   **Travelers' checks.** The following table shows the number of each denomination of travelers' check contained in a booklet of 25 checks. Find the total value of the booklet.

Denomination	$20	$50	$100	$500
Number	15	5	3	2

Solution We can find the total value by adding the values of each of the denominations. First, we describe this in words; then we translate to mathematical symbols.

Total value	=	value of all the 20's	+	value of all the 50's	+	value of all the 100's	+	value of all the 500's.

Total value = 15(20) + 5(50) + 3(100) + 2(500)

Multiply each denomination by the number of checks.

Total value = 300 + 250 + 300 + 1,000   Do the multiplications.

Total value = 1,850   Do the additions.

The total value of the booklet is $1,850. ∎

The **arithmetic mean** (or **average**) of a set of numbers is a value around which the values of the numbers are grouped. When finding the mean, we usually need to apply the rules for the order of operations.

Finding an arithmetic mean

To find the **mean** of a set of values, divide the sum of the values by the number of values.

EXAMPLE 14

Customer service. To measure its effectiveness in serving customers, a store had the telephone company electronically record the number of times the telephone rang before an employee answered it. The results of the week-long survey are shown in Table 1-3. Find the average number of times the phone rang before an employee answered it that week.

TABLE 1-3

Number of rings	Occurrences
1	11
2	46
3	45
4	28
5	20

Solution To find the total number of rings, we multiply each of the *number of rings* (1, 2, 3, 4, and 5 rings) by the respective number of occurrences and add those subtotals.

Total number of rings = 11(1) + 46(2) + 45(3) + 28(4) + 20(5)

To find the total number of calls received, we add the number of occurrences in the right-hand column of the table.

Total number of calls received = 11 + 46 + 45 + 28 + 20

To find the average, we divide the total number of rings by the total number of calls and apply the rules for the order of operations to evaluate the expression.

$$\text{Average} = \frac{11(1) + 46(2) + 45(3) + 28(4) + 20(5)}{11 + 46 + 45 + 28 + 20}$$

$$\text{Average} = \frac{11 + 92 + 135 + 112 + 100}{150}$$
In the numerator, do the multiplication. In the denominator, do the addition.

$$\text{Average} = \frac{450}{150}$$
Do the addition.

$$\text{Average} = 3$$
Do the division.

The average number of times the phone rang before it was answered was 3. ∎

STUDY SET

Section 1.3

VOCABULARY

In Exercises 1–6, fill in the blanks to make the statements true.

1. In the multiplication statement $6 \cdot 7 = 42$, the numbers being multiplied are called _____factors_____.

2. In the exponential expression x^2, x is the _____base_____, and 2 is the _____exponent_____.

3. 10^2 can be read as ten _____squared_____, and 10^3 can be read as ten _____cubed_____.

4. 7^5 is the fifth _____power_____ of seven.

5. The arithmetic _____mean_____ or _____average_____ of a set of numbers is a value around which the values of the numbers are grouped.

6. An _____exponent_____ is used to represent repeated multiplication.

CONCEPTS

7. Given: $4 + 5 \cdot 6$.
   a. What operations does this expression contain?
      addition and multiplication
   b. Evaluate the expression in two different ways and state the two possible results. 54, 34
   c. Which result from part b is correct, and why?
      34; multiplication is to be done before addition

8. a. What repeated multiplication does 5^3 represent?
      $5 \cdot 5 \cdot 5$
   b. Write a multiplication statement where the factor x is repeated 4 times. Then write the expression in simpler form using an exponent. $x \cdot x \cdot x \cdot x = x^4$
   c. How can we represent the repeated *addition* $3 + 3 + 3 + 3 + 3$ in a simpler form? 5(3)

9. a. How is the mean (or average) of a set of scores found?
      Divide the sum of the scores by the number of scores.
   b. Find the average of 75, 81, 47, and 53. 64

10. In the expression $8 + 2[15 - (6 + 1)]$, which grouping symbols are *innermost* and which are *outermost*?
    innermost: parentheses; outermost: brackets

11. a. What operations does the expression $12 + 5^2 \cdot 3$ contain? addition, power, multiplication
    b. In what order should they be performed?
       power, multiplication, addition

12. a. What operations does the expression $20 - (2)^2 + 3(1)$ contain?
       subtraction, power, addition, multiplication
    b. In what order should they be performed?
       power, multiplication, subtraction, addition

13. Consider the expression $\frac{36 - 4(7)}{2(10 - 8)}$. In the numerator, what operation should be done first? In the denominator, what operation should be done first?
    multiplication; subtraction

14. Explain the differences in evaluating $4 \cdot 2^2$ and $(4 \cdot 2)^2$.
    In $4 \cdot 2^2$, find the power, then multiply. In $(4 \cdot 2)^2$ multiply, then find the power.

15. To evaluate each expression, what operation should be performed first?
    a. $80 - 3 + 5 - 2^2$ power
    b. $80 - (3 + 5) - 2^2$ addition
    c. $80 - 3 + (5 - 2)^2$ subtraction

16. To evaluate each expression, what operation should be performed first?
    a. $(65 - 3)^3$ subtraction
    b. $65 - 3^3$ power
    c. $6(5) - (3)^3$ power

NOTATION

17. Write an exponential expression with a base of 12 and an exponent of 6. 12^6

18. Tell the name of each grouping symbol: ( ), [ ], ——. parentheses, brackets, fraction bar

In Exercises 19–22, complete each solution.

19. $50 + 6 \cdot 3^2 = 50 + 6(\boxed{9})$
    $= 50 + \boxed{54}$
    $= 104$

20. $100 - (25 - 8 \cdot 2) = 100 - (25 - \boxed{16})$
    $= 100 - \boxed{9}$
    $= 91$

21. $19 - 2[(1 + 2) \cdot 3] = 19 - 2[(\boxed{3}) \cdot 3]$
    $= 19 - 2(\boxed{9})$
    $= 19 - \boxed{18}$
    $= 1$

22. $\dfrac{46 - 2^3}{3(5) + 4} = \dfrac{46 - \boxed{8}}{\boxed{15} + 4}$
    $= \dfrac{\boxed{38}}{\boxed{19}}$
    $= 2$

PRACTICE

In Exercises 23–24, translate each word model to an equation. (Hint: You will need to use variables.)

23.

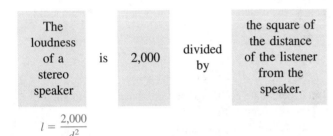

| The distance fallen by a skydiver | is | 16 | times | the square of the time he has been falling. |

$$d = 16t^2$$

24.

| The loudness of a stereo speaker | is | 2,000 | divided by | the square of the distance of the listener from the speaker. |

$$l = \frac{2,000}{d^2}$$

25. See Illustration 1. The formula for the volume of a cube is $V = s^3$. Complete the table.

ILLUSTRATION 1

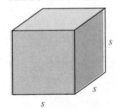

Side length of cube (in.)	Volume of cube (in.³)
1	1
2	8
3	27
4	64

26. See Illustration 2. Examine the data in the table in the right column and then write an equation that mathematically describes the relationship. $A = s^2$

ILLUSTRATION 2

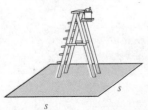

Length of a side of a painter's tarp (ft)	Area covered by tarp (ft²)
5	25
6	36
7	49
8	64

In Exercises 27–34, write each product using exponents.

27. $b \cdot b \cdot b$ b^3

28. $m \cdot m \cdot m \cdot m \cdot m$ m^5

29. $10 \cdot 10 \cdot k \cdot k \cdot k$ 10^2k^3

30. $5(5)(5)(i)(i)$ 5^3i^2

31. $8 \cdot \pi \cdot r \cdot r \cdot r$ $8\pi r^3$

32. $8 \cdot \pi \cdot r \cdot r \cdot r$ $8\pi r^3$

33. $6 \cdot x \cdot x \cdot y \cdot y \cdot y$ $6x^2y^3$

34. $76 \cdot s \cdot s \cdot s \cdot s \cdot t$ $76s^4t$

In Exercises 35–84, evaluate each expression.

35. $12 - 2 \cdot 3$ 6

36. $9 + 5 \cdot 3$ 24

37. $15 + (30 - 4)$ 41

38. $(15 + 30) - 4$ 41

39. $100 - 8(10) + 60$ 80

40. $50 - 2(5) - 7$ 33

41. $22 - (15 - 3)$ 10

42. $(33 - 8) - 10$ 15

43. $2(9) - 2(5)$ 8

44. $75 - 7^2$ 26

45. $5^2 + 13^2$ 194

46. $3^3 - 2^3$ 19

47. $3 \cdot 8^2$ 192

48. $(3 \cdot 4)^2$ 144

49. $8 \cdot 5 - 4 \div 2$ 38

50. $9 \cdot 5 - 6 \div 3$ 43

51. $14 + 3(7 - 5)$ 20

52. $5(10 + 2) - 1$ 59

53. $4 + 2[26 - 5(3)]$ 26

54. $64 - 6[15 - (3)3]$ 28

55. $(10 - 3)^2$ 49

56. $(12 - 2)^3$ 1,000

57. $10 - 3^2$ 1

58. $12 - 2^3$ 4

59. $19 - (45 - 41)^2$ 3

60. $200 - (6 - 5)^3$ 199

61. $2 + 3 \cdot 2^2 \cdot 4$ 50

62. $5 \cdot 2^2 \cdot 4 - 30$ 50

63. $3(4)(5)(6)$ 360

64. $1(2)(3)(4)$ 24

65. $5[9(2) - 2(8)]$ 10

66. $[6(5) - 5(5)]4$ 20

67. $75 - 3 \cdot 1^2$ 72

68. $175 - 2 \cdot 3^4$ 13

69. $5(150 - 3^3)$ 615

70. $6(130 - 4^3)$ 396

71. $(4 + 2 \cdot 3)^4$ 10,000

72. $(17 - 5 \cdot 2)^3$ 343

73. $3(2)^5(2)^2$ 384

74. $5(2)^3(3)^2$ 360

75. $6\left(\dfrac{25}{5}\right) - \dfrac{36}{9} + 1$ 27

76. $2\left(\dfrac{15}{5}\right) - \dfrac{6}{2} + 9$ 12

77. $\dfrac{5(68 - 32)}{9}$ 20

78. $\dfrac{5 \cdot 50 - 160}{9}$ 10

79. $\dfrac{(6 - 5)^4 + 21}{27 - 4^2}$ 2

80. $\dfrac{(4^3 - 10) - 4}{5^2 - 4(5)}$ 10

81. $\dfrac{13^2 - 5^2}{3(9 - 5)}$ 12

82. $\dfrac{72 - (2 - 2 \cdot 1)}{10^2 - (90 + 2^2)}$ 12

83. $\dfrac{8^2 - 10}{2(3)(4) - 5(3)}$ 6

84. $\dfrac{40 - 1^3 - 2^4}{3(2 + 5) + 2}$ 1

In Exercises 85–88, evaluate each expression using a calculator.

85. $5^7 - (45 \cdot 489)$ 56,120

86. $\dfrac{3(3,246 - 1,111)}{561 - 546}$ 427

87. $54^3 - 16^4 + 19(3)$
91,985

88. $\dfrac{36^2 - 2(48)}{25^2 - 325}$ 4

ILLUSTRATION 5

Allstate	$2,672	Mercury	$1,370
Auto Club	$1,680	State Farm	$2,737
Farmers	$2,485	20th Century	$1,692

Criteria: Six-month premium. Husband, 45, drives a 1995 Explorer, 12,000 annual miles. Wife, 43, drives a 1996 Dodge Caravan, 12,000 annual miles. Son, 17, is an occasional operator. All have clean driving records.

APPLICATIONS

Write a numerical expression describing each situation and evaluate it using the rules for the order of operations. A calculator may be helpful for some problems.

89. CASH AWARDS A contest is to be part of a promotional kickoff for a new children's cereal. The prizes to be awarded are shown in Illustration 3.
   a. How much money will be awarded in the promotion? $11,875
   b. What is the average cash prize? $95

ILLUSTRATION 3

Coloring Contest

Grand prize: Disney World vacation plus $2,500
Four 1st place prizes of $500
Thirty-five 2nd place prizes of $150
Eighty-five 3rd place prizes of $25

90. DISCOUNT COUPONS An entertainment book contains 50 coupons good for discounts on dining, lodging, and attractions. (See Illustration 4.)
   a. Overall, how much money can a person save if all of the coupons are used? $700
   b. What is the average savings per coupon? $14

ILLUSTRATION 4

Type of coupon	Amount of discount	Number of coupons
Dining	$10	32
Lodging	$25	11
Attractions	$15	7

91. STAKES RACES One weekend, a thoroughbred race track held eight stakes races, with cash awards to the race winners of $300,000, $200,000, $175,000, $300,000, $200,000, $175,000, $150,000, and $150,000. What was the average stakes award that weekend? $206,250

92. AUTO INSURANCE See the premium comparison in Illustration 5. What is the average six-month insurance premium? $2,106

93. SPREADSHEETS The spreadsheet in Illustration 6 contains data collected by a chemist. For each row, the sum of the values in columns A and B is to be subtracted from the product of 6 and the value in column C. That result is then to be divided by 12 and entered in column D. Use this information to complete the spreadsheet.

ILLUSTRATION 6

	A	B	C	D
1	20	4	8	2
2	9	3	16	7
3	1	5	11	5

94. DOG SHOWS The final score for each dog competing in a "toy breeds" competition is computed by dividing the sum of the judges' marks, after the highest and lowest have been dropped, by 6. (See Illustration 7.)
   a. What was their order of finish?
      Pomeranian, Terrier, Pekinese
   b. Did any judge rate all the dogs the same?
      yes, judge 6

ILLUSTRATION 7

Judge	1	2	3	4	5	6	7	8
Terrier	14	11	11	10	12	12	13	13
Pekinese	10	9	8	11	11	12	9	10
Pomeranian	15	14	13	11	14	12	10	14

95. CABLE TELEVISION The number of new households that subscribed each month to a newly installed cable television system are shown in the line graph in Illustration 8.
   a. Express the number of new subscribers each month as a power of 5.
      $5 = 5^1$, $25 = 5^2$, $125 = 5^3$, $625 = 5^4$
   b. If the trend continues, predict the number of new subscribers in the 5th month. 3,125

ILLUSTRATION 8

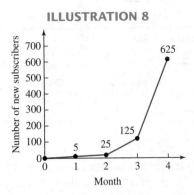

96. TURNTABLES The number of record turntables sold by a large music store declined over a six-year period, as shown in Illustration 9. Express the number of turntables sold in each of the years as a power.
$729 = 3^6, 243 = 3^5, 81 = 3^4, 27 = 3^3, 9 = 3^2, 3 = 3^1$

ILLUSTRATION 9

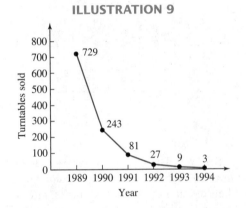

WRITING

97. Explain the difference between 2^3 and 3^2.

98. Explain why rules for the order of operations are necessary.

99. Consider the process of evaluating $25 - 3 \cdot 5$. Explain how the insertion of grouping symbols, $(25 - 3) \cdot 5$, causes the expression to be evaluated in a different order.

100. In what settings do you encounter or use the concept of arithmetic mean (average) in your everyday life?

REVIEW

101. Find $|-5|$.   5

102. List the set of integers. $\{. . . , -2, -1, 0, 1, 2, . . .\}$

103. True or false: Every real number can be expressed as a decimal.   true

104. True or false: Irrational numbers are nonterminating, nonrepeating decimals.   true

105. What two numbers are a distance of 6 away from -3 on the number line? -9 and 3

106. Graph $\left\{-2.5, \sqrt{5}, \frac{11}{3}, -0.333. . . , 0.75\right\}$ on the number line.

▶ 1.4 Algebraic Expressions

In this section, you will learn about

Translating from English to mathematical symbols ■ Representing two unknown quantities with a variable ■ Looking for hidden operations ■ Algebraic expressions involving more than one operation ■ Number and value ■ Evaluating algebraic expressions

---

Introduction   When solving problems, we must often write an equation that mathematically describes the situation in question. In this section, you will see how the facts of a problem can be *translated* to mathematical symbols, which are then assembled to form algebraic expressions—the building blocks of equations.

Translating from English to Mathematical Symbols

To describe numerical relationships, we can translate the words of a problem into mathematical symbols. Below, we list some key words and phrases that are used to represent the operations of addition, subtraction, multiplication, and division and show how they can be translated to form algebraic expressions.

Addition

The phrase	translates to the algebraic expression
the sum of a and 8	$a + 8$
4 plus c	$4 + c$
16 added to m	$m + 16$
4 more than t	$t + 4$
20 greater than F	$F + 20$
T increased by r	$T + r$
Exceeds y by 35	$y + 35$

Subtraction

The phrase	translates to the algebraic expression
the difference of 23 and P	$23 - P$
550 minus h	$550 - h$
w less than 108	$108 - w$
7 decreased by j	$7 - j$
M reduced by x	$M - x$
12 subtracted from L	$L - 12$
5 less f	$5 - f$

Multiplication

The phrase	translates to the algebraic expression
the product of 4 and x	$4x$
20 times B	$20B$
twice r	$2r$
$\frac{3}{4}$ of m	$\frac{3}{4}m$

Division

The phrase	translates to the algebraic expression
the quotient of R and 19	$\dfrac{R}{19}$
s divided by d	$\dfrac{s}{d}$
the ratio of c to d	$\dfrac{c}{d}$
k split into 4 equal parts	$\dfrac{k}{4}$

E X A M P L E 1	**Writing algebraic expressions.** Write each phrase as an algebraic expression.

    a. The sum of the length *l* and the width 20

    b. *b* less than the capacity *c*

    c. The product of the weight *w* and 2,000

Solution   **a.** **Key word:** *sum*   **Translation:** add
The phrase translates to $l + 20$.

    b. **Key phrase:** *less than*   **Translation:** subtract
The capacity *c* is to be made less, so we subtract *b* from it: $c - b$.

    c. **Key word:** *product*   **Translation:** multiply
The weight *w* is to be multiplied by 2,000: $2{,}000w$.

SELF CHECK   Write each phrase as an algebraic expression:
a. The distance *d* divided by *r*, **b.** 80 cents less than *t* cents, and **c.** $\frac{2}{3}$ of the time *T*.   *Answers:* **a.** $\frac{d}{r}$, **b.** $t - 80$, **c.** $\frac{2}{3}T$ ∎

Representing Two Unknown Quantities with a Variable

In the next two examples, we will use a variable to describe two unknown quantities.

E X A M P L E 2	**Writing an algebraic expression.** A butcher trims 4 ounces of fat from a roast that originally weighed *x* ounces. Write an algebraic expression that represents the weight of the roast after it is trimmed.

Solution   We let $x =$ the original weight of the roast (in ounces).

    Key word: *trimmed*   **Translation:** subtract

After 4 ounces of fat have been trimmed, the weight of the roast is $(x - 4)$ ounces.

SELF CHECK   When a secretary rides the bus to work, it takes her *m* minutes. If she drives her own car, her travel time exceeds this by 15 minutes. How can we represent the time it takes her to get to work by car?   *Answer:* $(m + 15)$ minutes ∎

E X A M P L E 3	**Writing an algebraic expression.** The swimming pool in Figure 1-11 is *x* feet wide. If it is to be sectioned into 8 equally wide swimming lanes, write an algebraic expression that represents the width of each lane.

FIGURE 1-11

x

Solution   Let $x =$ the width of the swimming pool.

    Key phrase: *sectioned into 8 equally wide lanes*   **Translation:** divide

The width of each lane is $\dfrac{x}{8}$ feet.

SELF CHECK   A handyman estimates that it will take the same amount of time to sand as to paint some kitchen cabinets. If the entire job takes x hours, how can we express the time it will take him to do the painting?

Answer: $\dfrac{x}{2}$ hours ∎

When solving problems, the variable to be used is rarely specified. You must decide what the unknown quantities are and how they will be represented using variables. The next example illustrates how to approach these situations.

E X A M P L E   4

Naming two unknown quantities. The value of a collectible doll is three times that of an antique toy truck. Express the value of each, using a variable.

Solution   There are two unknown quantities. Since the doll's value is related to the truck's value, we will let $x =$ the value of the toy truck.

> **Key phrase:** *3 times*   **Translation:** multiply by 3

The value of the doll is $3x$.

SELF CHECK   The McDonald's Chicken Deluxe sandwich has 5 less grams of fat than the Quarter-Pounder hamburger. Express the grams of fat in each sandwich using a variable.

Answers: $x =$ the number of grams of fat in the hamburger, $x - 5 =$ the number of grams of fat in the chicken sandwich ∎

WARNING! A variable is used to represent an unknown number. Therefore, in Example 4, it would be incorrect to write, "Let $x =$ toy truck," because the truck is not a number. We need to write, "Let $x =$ the *value* of the toy truck."

Looking for Hidden Operations

When analyzing problems, we aren't always given key words or key phrases to help establish what mathematical operation to use. Sometimes a careful reading of the problem is needed to determine the hidden operations.

E X A M P L E   5

Hidden operations. Disneyland, located in Anaheim, California, was in operation 16 years before the opening of Walt Disney World, in Orlando, Florida. Euro Disney, in Paris, France, was constructed 21 years after Disney World. Use algebraic expressions to express the ages (in years) of each of these Disney attractions.

Solution   The ages of Disneyland and Euro Disney are both related to the age of Walt Disney World. Therefore, we will let $x =$ the age of Walt Disney World.

In carefully reading the problem, we find that Disneyland was built 16 years *before* Disney World, so its age is more than that of Disney World.

> **Key phrase:** *more than*   **Translation:** add

In years, the age of Disneyland is $x + 16$. Euro Disney was built 21 years *after* Disney World, so its age is less than that of Disney World.

> **Key phrase:** *less than*   **Translation:** subtract

In years, the age of Euro Disney is $x - 21$. The results are summarized in Table 1-4.

TABLE 1-4

Attraction	Age
Disneyland	$x + 16$
Disney World	x
Euro Disney	$x - 21$

∎

EXAMPLE 6	**Hidden operations.** **a.** How many months are in x years? **b.** How many yards are in i inches?

Solution **a.** Since there are no key words, we must decide what operation is needed to find the number of months. Each of the x years contains 12 months; this indicates multiplication. Therefore, the number of months is $12 \cdot x$, or $12x$.

b. Since there are no key words, we must decide what operation is called for. 36 inches make up one yard. We need to see how many groups of 36 inches are in i inches. This indicates division; therefore, the number of yards is $\frac{i}{36}$.

SELF CHECK **a.** h hours is how many days? **b.** How many dimes equal the value of x dollars?   *Answers:* **a.** $\dfrac{h}{24}$, **b.** $10x$ ■

Algebraic Expressions Involving More Than One Operation

In all of the preceding examples, each algebraic expression contained only one operation. We now examine expressions involving two operations.

EXAMPLE 7	**An expression involving two operations.** In the second semester, student enrollment in a retraining program at a college was 32 more than twice that of the first semester. Use a variable to express the student enrollment in the program each semester.

Solution Since the second-semester enrollment is expressed in terms of the first-semester enrollment, we let $x =$ the enrollment in the first semester.

> **Key phrase:** *more than*   **Translation:** add
> **Key word:** *twice*   **Translation:** multiply by 2

The enrollment for the second semester is $2x + 32$.

SELF CHECK The number of votes received by the incumbent in a congressional election was 55 less than three times the challenger's vote. Use a variable to express the number of votes received by each candidate.

Answers: $x =$ the number of votes received by the challenger, $3x - 55 =$ the number of votes received by the incumbent   ■

Number and Value

Some problems deal with quantities that have value. In these problems, we must distinguish between *the number of* and *the value of* the unknown quantity. For example, to find the value of 3 quarters, we multiply the number of quarters by the value (in cents) of one quarter. Therefore, the value of 3 quarters is $3 \cdot 25¢ = 75¢$.

The same distinction must be made if the number is unknown. For example, the value of n nickels is not $n¢$. The value of n nickels is $n \cdot 5¢ = (5n)¢$. For problems of this type, we will use the relationship

Number $\cdot$ value $=$ total value

EXAMPLE 8	**Number–value problems.** Suppose a roll of paper towels sells for 79¢. Find the cost of **a.** five rolls of paper towels, **b.** x rolls of paper towels, and **c.** $x + 1$ rolls of paper towels.

Solution In each case, we will multiply the *number* of rolls of paper towels by the *value* of one roll (79¢) to find the total cost.

a. The cost of 5 rolls of paper towels is $5 \cdot 79¢ = 395¢$, or $3.95.

b. The cost of x rolls of paper towels is $x \cdot 79¢ = (79 \cdot x)¢ = (79x)¢$.

c. The cost of $x + 1$ rolls of paper towels is
$(x + 1) \cdot 79¢ = 79 \cdot (x + 1)¢ = 79(x + 1)¢$.

SELF CHECK Find the value of **a.** six $50 savings bonds,
b. t $100 savings bonds, and **c.** $(x - 4)$
$1,000 savings bonds.

Answers: **a.** $300, **b.** $100t,
c. $1,000(x - 4) ∎

Evaluating Algebraic Expressions

To **evaluate** an algebraic expression, we replace its variable or variables with specific numbers and then apply the rules for the order of operations. Consider the algebraic expression $x^2 - 2x$. If we are told that $x = 3$, we can substitute 3 for x in $x^2 - 2x$ and evaluate the resulting expression.

$$x^2 - 2x = (3)^2 - 2(3) \quad \text{Replace } x \text{ with } 3.$$
$$= 9 - 2(3) \quad \text{Find the power: } (3)^2 = 9.$$
$$= 9 - 6 \quad \text{Do the multiplication: } 2(3) = 6.$$
$$= 3 \quad \text{Do the subtraction.}$$

It is often necessary to evaluate an algebraic expression for *several* values of its variable. When doing that, we can show the results in a **table of values.** Suppose we want to evaluate the expression $x^2 - 2x$ for $x = 2$, $x = 4$, and $x = 5$. (See Table 1-5.)

In the column headed "x," we list each value of the variable to be used in the evaluations. In the column headed "$x^2 - 2x$," we write the result of each evaluation.

TABLE 1-5

x	$x^2 - 2x$
2	0
4	8
5	15

Evaluate for $x = 2$:
$$x^2 - 2x = (2)^2 - 2(2)$$
$$= 4 - 2(2)$$
$$= 4 - 4$$
$$= 0$$

Evaluate for $x = 4$:
$$x^2 - 2x = (4)^2 - 2(4)$$
$$= 16 - 2(4)$$
$$= 16 - 8$$
$$= 8$$

Evaluate for $x = 5$:
$$x^2 - 2x = (5)^2 - 2(5)$$
$$= 25 - 2(5)$$
$$= 25 - 10$$
$$= 15$$

The two columns of a table of values are sometimes headed with the terms **input** and **output,** as shown in Table 1-6. The x-values are the "inputs" into the expression $x^2 - 2x$, and the resulting values are thought of as the "outputs."

TABLE 1-6

Input x	Output $x^2 - 2x$
2	0
4	8
5	15

E X A M P L E   9

Ballistics. If a toy rocket is shot into the air with an initial velocity of 80 feet per second, its height (in feet) after t seconds in flight is given by the algebraic expression

$$80t - 16t^2$$

How many seconds after the launch will it hit the ground?

Solution   We can substitute positive values for t, the time in flight, until we find the one that gives a height of 0. At that time, the rocket will be on the ground. We will begin by finding the height after the rocket has been in flight for 1 second ($t = 1$) and record the result in a table.

t	$80t - 16t^2$
1	64

Evaluate for $t = 1$:
$$80t - 16t^2 = 80(1) - 16(1)^2$$
$$= 64$$

After 1 second in flight, the height of the rocket is 64 feet. We continue to pick more values of t until we find out when the height is 0.

t	$80t - 16t^2$
2	96
3	96
4	64
5	0

Evaluate for $t = 2$:
$$80t - 16t^2 = 80(2) - 16(2)^2$$
$$= 96$$

Evaluate for $t = 3$:
$$80t - 16t^2 = 80(3) - 16(3)^2$$
$$= 96$$

Evaluate for $t = 4$:
$$80t - 16t^2 = 80(4) - 16(4)^2$$
$$= 64$$

Evaluate for $t = 5$:
$$80t - 16t^2 = 80(5) - 16(5)^2$$
$$= 0$$

We see that for $t = 5$, the height of the rocket is 0. Therefore, the rocket will hit the ground in 5 seconds.

SELF CHECK   In Example 9, suppose the height of the rocket was given by $112t - 16t^2$. Complete the table to find out how many seconds after launch it would hit the ground.

t	$112t - 16t^2$
1	
3	
5	
7	

Answer: 7 (the heights are 96, 192, 160, and 0)

WARNING! When replacing a variable with its numerical value, use parentheses around the replacement number to avoid possible misinterpretation. For example, when substituting 5 for x in $2x + 1$, we show the multiplication using parentheses: $2(5) + 1$. If we don't show the multiplication, we could misread the expression as $25 + 1$.

EXAMPLE 10

Evaluating algebraic expressions. If $a = 3$, $b = 3$, $c = 2$, and $d = 9$, evaluate $\dfrac{6b - a}{d - c^2}$.

Solution   The given expression contains four variables. We substitute the values for a, b, c, and d into the expression and apply the rules for the order of operations.

$$\frac{6b - a}{d - c^2} = \frac{6(3) - 3}{9 - (2)^2} \qquad \text{In the numerator, replace } b \text{ with 3 and } a \text{ with 3.}$$
$$\text{In the denominator, replace } d \text{ with 9 and } c \text{ with 2.}$$

$$= \frac{18 - 3}{9 - 4} \qquad \text{In the numerator, do the multiplication: } 6(3) = 18.$$
$$\text{In the denominator, find the power: } (2)^2 = 4.$$

$$= \frac{15}{5} \qquad \text{In the numerator, do the subtraction.}$$
$$\text{In the denominator, do the subtraction.}$$

$$= 3 \qquad \text{Do the division.}$$

SELF CHECK   If $e = 4$, $f = 10$, $g = 6$, and $h = 2$, evaluate $\dfrac{f + 4e}{h^3 - g}$.

Answer: 13   ■

ACCENT ON TECHNOLOGY *Evaluating Algebraic Expressions*

The rotating drum of a clothes dryer is a cylinder. (See Figure 1-12.) To find the capacity of the dryer, we can find its volume by evaluating the algebraic expression $\pi r^2 h$, where r represents the radius and h represents the height (although the cylinder is lying on its side) of the drum. If we substitute 13.5 for r and 20 for h, we obtain $\pi(13.5)^2(20)$. Using a scientific calculator, we can evaluate the expression by entering these numbers and pressing these keys:

FIGURE 1-12

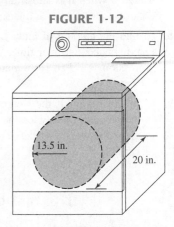

13.5 in.   20 in.

Keystrokes   13.5 $\boxed{x^2}$ $\boxed{\times}$ $\boxed{\pi}$ $\boxed{\times}$ 20 $\boxed{=}$ $\boxed{\text{11451.10522}}$

Using a graphing calculator, we can evaluate the expression by entering these numbers and pressing these keys:

Keystrokes   13.5 $\boxed{x^2}$ $\boxed{\times}$ $\boxed{\text{2nd}}$ $\boxed{\pi}$ $\boxed{\times}$ 20 $\boxed{\text{ENTER}}$

$\boxed{\begin{array}{l} 13.5^2 * \pi * 20 \\ \qquad\qquad 11451.10522 \end{array}}$

To the nearest cubic inch, the capacity of the dryer is 11,451 in.³.

STUDY SET

Section 1.4

VOCABULARY

In Exercises 1–4, fill in the blanks to make the statements true.

1. To _____evaluate_____ an algebraic expression, we substitute the values for the variables and then apply the rules for the order of operations.

2. Variables and/or numbers can be combined with the operation symbols of addition, subtraction, multiplication, and division to create algebraic _____expressions_____.

3. When translated to mathematical symbols, words such as *decreased* and *reduced* indicate the operation of _____subtraction_____.

4. A variable is a letter that stands for a _____number_____.

CONCEPTS

5. Write two algebraic expressions that contain the variable x and the numbers 6 and 20.
   $6 + 20x$; $\frac{6-x}{20}$ (answers may vary)

6. **a.** How many days are in w weeks? $7w$
   b. D days is how many weeks? $\frac{D}{7}$

7. When evaluating $3x - 6$ for $x = 4$, what misunderstanding can occur if we don't write parentheses around 4 when it is substituted for the variable?
   We would obtain $34 - 6$; it looks like 34, not 3(4).

8. To find the amount of money that will accumulate in a savings account over time, we can use the algebraic expression $P + Prt$. How many variables does it contain? 3

9. **a.** In Illustration 1, the weight of the van is 500 pounds less than twice the weight of the car. Express the weight of the van and the car using the variable x.
   $x =$ weight of the car; $2x - 500 =$ weight of the van

ILLUSTRATION 1

   b. Suppose you learn that the actual weight of the car is 2,000 pounds. What is the weight of the van? 3,500 lb

10. See Illustration 2.
   a. If we let b represent the length of the beam, write an algebraic expression for the length of the pipe.
   $b - 15$

   b. If we let p represent the length of the pipe, write an algebraic expression for the length of the beam.
   $p + 15$

ILLUSTRATION 2

15 ft

11. **a.** In Illustration 3, what algebraic expression was evaluated? $8x - x^2$
   b. For what values of x was it evaluated? 3, 4, 5
   c. What was the value of the expression when it was evaluated for $x = 4$? 16

ILLUSTRATION 3

x	$8x - x^2$
3	15
4	16
5	15

12. Complete the table in Illustration 4.

ILLUSTRATION 4

Type of coin	Number	Value in cents	Total value in cents
Penny	12	1	12
Nickel	n	5	$5n$
Dime	d	10	$10d$
Quarter	5	25	125
Half dollar	$x + 5$	50	$50(x + 5)$

NOTATION

In Exercises 13–14, complete each solution.

13. Evaluate the expression $9a - a^2$ for $a = 5$.
$$9a - a^2 = 9(\,5\,) - (\,5\,)^2$$
$$= 9(5) - 25$$
$$= 45 - 25$$
$$= 20$$

14. Evaluate the expression $b^2 - 4ac$ for $a = 1$, $b = 6$, and $c = 5$.
$$b^2 - 4ac = (\,6\,)^2 - 4(\,1\,)(5)$$
$$= 36 - 4(1)(\,5\,)$$
$$= 36 - 20$$
$$= 16$$

PRACTICE

In Exercises 15–40, write each phrase as an algebraic expression. If no variable is given, use x as the variable.

15. The sum of the length l and 15 $l + 15$

16. The difference of a number and 10 $x - 10$

17. The product of a number and 50 $50x$

18. Three-fourths of the population p $\frac{3}{4}p$

19. The ratio of the amount won w and lost l $\frac{w}{l}$

20. The tax t added to c $c + t$

21. P increased by p $P + p$

22. 21 less than the total height h $h - 21$

23. The square of k minus 2,005 $k^2 - 2,005$

24. s subtracted from S $S - s$

25. J reduced by 500 $J - 500$

26. Twice the attendance a $2a$

27. 1,000 split n equal ways $\frac{1,000}{n}$

28. Exceeds the cost c by 25,000 $c + 25,000$

29. 90 more than the current price p $p + 90$

30. 64 divided by the cube of y $\frac{64}{y^3}$

31. The total of 35, h, and 300 $35 + h + 300$

32. x decreased by 17 $x - 17$

33. 680 fewer than the entire population p $p - 680$

34. Triple the number of expected participants $3x$

35. The product of d and 4, decreased by 15 $4d - 15$

36. Forty-five more than the quotient of y and 6 $\frac{y}{6} + 45$

37. Twice the sum of 200 and t $2(200 + t)$

38. The square of the quantity 14 less than x $(x - 14)^2$

39. The absolute value of the difference of a and 2
   $|a - 2|$

40. The absolute value of a, decreased by 2 $|a| - 2$

In Exercises 41–44, if n represents a number, write a word description of each algebraic expression. (Answers may vary.)

41. $n - 7$
   7 less than a number

42. $n^2 + 7$
   the square of a number, increased by 7

43. $7n + 4$
   the product of 7 and a number, increased by 4

44. $3(n + 1)$
   three times the sum of a number and 1

45. Express how many minutes there are in   **a.** 5 hours and   **b.** h hours. $300; 60h$

46. A woman watches television x hours a day. Express the number of hours she watches TV   **a.** in a week and   **b.** in a year. $7x; 365x$

47. **a.** Express how many feet are in y yards. $3y$
   b. Express how many yards are in f feet. $\frac{f}{3}$

48. If a car rental agency charges 29¢ a mile, express the rental fee if a car is driven x miles. $29x¢$

49. A model's skirt is x inches long. The designer then lets the hem down 2 inches. How can we express the length (in inches) of the altered skirt? $x + 2$

50. A soft drink manufacturer produced c cans of cola during the morning shift. Write an expression for how many six-packs of cola can be assembled from the morning shift's production. $\frac{c}{6}$

51. The tag on a new pair of 36-inch-long jeans warns that after washing, they will shrink x inches in length. Express the length (in inches) of the jeans after they are washed. $36 - x$

52. A caravan of b cars, each carrying 5 people, traveled to the state capital for a political rally. Express how many people were in the car caravan. $5b$

53. A sales clerk earns \$$x$ an hour. Express how much he will earn in   **a.** an 8-hour day and   **b.** a 40-hour week. $\$8x; \$40x$

54. A caterer always prepares food for 10 more people than the order specifies. If p people are to attend a reception, write an expression for the number of people she should prepare for. $p + 10$

55. Tickets to a circus cost \$5 each. Express how much tickets will cost for a family of x people if they also pay for two of their neighbors. $\$5(x + 2)$

56. If each egg is worth e¢, express the value (in cents) of a dozen eggs. $12e¢$

In Exercises 57–60, evaluate each algebraic expression for the given value of the variable.

57. $6x$ for $x = 7$ 42

58. $\frac{p - 15}{30}$ for $p = 75$ 2

59. $3(t - 6)$ for $t = 8$ 6

60. $y^3 - 2y^2 - 2$ for $y = 3$
   7

In Exercises 61–68, complete each table of values.

61.
g	$g^2 - 7g + 1$
0	1
7	1
10	31

62.
f	$5(16 - f)^2$
7	405
8	320
9	245

63.
s	$\frac{5s + 36}{s}$
1	41
6	11
12	8

64.
a	$2,500a + a^3$
2	5,008
4	10,064
5	12,625

65.
Input x	Output $2x - \frac{x}{2}$
100	150
300	450

66.
Input x	Output $\frac{x}{3} + \frac{x}{4}$
12	7
36	21

67.

Input a	Output $3a^2 + 1$
4	49
8	193

68.

Input b	Output $50 - 2b^2$
3	32
5	0

In Exercises 69–74, evaluate each algebraic expression.

69. $a^2 + b^2$ for $a = 5$ and $b = 12$ 169

70. $\dfrac{x + y}{2}$ for $x = 8$ and $y = 10$ 9

71. $\dfrac{s + t}{s - t}$ for $s = 23$ and $t = 21$ 22

72. $3r^2h$ for $r = 4$ and $h = 10$ 480

73. $\dfrac{h(b + c)}{2}$ for $h = 5$, $b = 7$, and $c = 9$ 40

74. $b^2 - 4ac$ for $a = 2$, $b = 8$, and $c = 4$ 32

In Exercises 75–76, use a calculator to evaluate each expression. Round to the nearest tenth.

75. Find the volume of a basketball having a radius r of 4.5 inches by evaluating

$$\frac{4\pi r^3}{3}$$ 381.7 in.^{3}

76. Find the volume of a cone with a radius r of 3 inches and a height h of 8 inches by evaluating

$$\frac{\pi r^2 h}{3}$$ 75.4 in.^{3}

APPLICATIONS

In Exercises 77–80, use a variable to represent one unknown. Then write an algebraic expression to describe the second unknown.

77. TRAINING PROGRAM Of the original group that entered a lifeguard training program, six candidates dropped out. After graduation, those remaining were divided into eight equal-size squads for beach patrol duty. Write an expression for the number of lifeguards in each patrol squad.
Let n = original number entering program; $\frac{n-6}{8}$ = number in each squad.

78. INVESTMENTS The value of a gold coin is $350 more than twice that of a silver coin. Write an expression for the value of the gold coin.
Let s = value of the silver coin; $2s + 350$ = value of the gold coin.

79. OCCUPANCY RATE An apartment owner lowered the monthly rent, and soon the number of apartments rented doubled. Then water damage forced three units

to be vacated. Write an expression for the number of apartments that are now occupied.
Let a = number of apts occupied before rent was lowered; $2a - 3$ = number now occupied.

80. SCHEDULING The manager of a restaurant schedules 4 times as many workers per weekend shift as per weekday shift. If a holiday falls on a weekend, 6 additional workers are scheduled. Write an expression for the number of employees needed for a holiday weekend shift.
Let e = number of employees for a weekday shift; $4e + 6$ = number of employees for a holiday weekend shift.

In Exercises 81–86, solve each problem.

81. ROCKETRY The algebraic expression $64t - 16t^2$ gives the height of a toy rocket (in feet) t seconds after being launched. Find the height of the rocket for each of the times shown in Illustration 5. Present your results in an input/output table.

ILLUSTRATION 5

t	h
0	0
0.5	28
1	48
1.5	60
2	64
2.5	60
3	48
3.5	28
4	0

82. FREE FALL A tennis ball is dropped from the 1,024-foot-tall Chrysler Building in New York. The algebraic expression $1,024 - 16t^2$ gives the height of the ball (in feet) t seconds after it is dropped. Complete the input/output table in Illustration 6; then approximate *when* the ball would have fallen half of the height of the building.
about $5\frac{1}{2}$ seconds after it is dropped

ILLUSTRATION 6

t	$1,024 - 16t^2$
2	960
3	880
4	768
5	624
6	448
7	240
8	0

83. PACKAGING The lengths, widths, and heights (in inches) of three cardboard shipping boxes are recorded in columns B, C, and D (respectively) of the spreadsheet in Illustration 7. To find the *surface area* for the box described in row 1, the computer uses the formula

SUM(2*B1*C1, 2*B1*D1, 2*C1*D1)

where 2*B1*C1 means 2 *times* the number in cell B1 *times* the number in cell C1. The result of the computation is recorded in cell E1.

a. Use this information to complete the spreadsheet.
b. What are the units of your answers? in.²

ILLUSTRATION 7

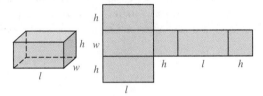

	B	C	D	E
1	12	6	6	360
2	18	12	12	1,152
3	18	24	18	2,376

84. GRAPHIC ARTS An artist's design of a logo for a company used words to describe the relative sizes of each of the elements. See Illustration 8.

ILLUSTRATION 8

Base is 2 inches longer than side.

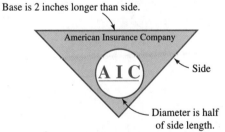

American Insurance Company

A I C

Side

Diameter is half of side length.

a. Describe the lengths of the base and sides, and the diameter, using a variable.

x = length of side, $x + 2$ = length of base, $\frac{x}{2}$ = diameter

b. If the company decides to have the sides of the logo be 8 inches long, what would be the dimensions of the other elements of the design?

base: 10 in.; diameter: 4 in.

85. ENERGY CONSERVATION A fiberglass blanket wrapped around a water heater helps prevent heat loss. See Illustration 9. Find the number of square feet of heater surface the blanket covers by evaluating the algebraic expression $2\pi rh$. Round to the nearest square foot. 69 ft²

ILLUSTRATION 9

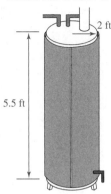

2 ft

5.5 ft

86. LANDSCAPING A grass strip is to be planted around a tree, as shown in Illustration 10. Find the number of square feet of sod to order by evaluating the expression $\pi(R^2 - r^2)$. Round to the nearest square foot. 235 ft²

ILLUSTRATION 10

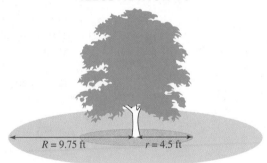

$R = 9.75$ ft $r = 4.5$ ft

WRITING

87. What is an algebraic expression? Give some examples.

88. What is a variable? How are variables used in this section?

REVIEW

89. Simplify -0.   0

90. Is the statement $-5 > -4$ true or false?   false

91. Evaluate $\left|-\frac{2}{3}\right|$. $\frac{2}{3}$

92. Evaluate $2^3 \cdot 3^2$.   72

93. Write $c \cdot c \cdot c \cdot c$ in exponential form. c^4

94. Evaluate $15 + 2[15 - (12 - 10)]$.   41

95. Find the mean (average) of the three test scores 84, 93, and 72.   83

96. Fill in the blanks to make the statement true: In the multiplication statement $5 \cdot x = 5x$, 5 and x are called __factors__, and $5x$ is called the __product__.

▶ **1.5**   # Solving Equations

In this section, you will learn about

> Equations ■ Checking solutions ■ The subtraction property of
> equality ■ The addition property of equality ■ The division
> property of equality ■ The multiplication property of equality
> ■ Using algebra to solve percent problems

Introduction   The idea of an equation is one of the most useful concepts in all of algebra. Equations are mathematical sentences that can be used to describe real-life situations. In this section, we will introduce some basic types of equations and discuss four fundamental properties that are used to solve them. Then we will show how some of the algebraic concepts discussed in this chapter can be used to solve problems involving percent.

Equations

Recall that an **equation** is a statement indicating that two expressions are equal. In the equation $x + 5 = 15$, the expression $x + 5$ is called the **left-hand side,** and 15 is called the **right-hand side.**

An equation can be true or false. For example, $10 + 5 = 15$ is a true equation, whereas $11 + 5 = 15$ is a false equation. An equation containing a variable can be true or false, depending upon the value of the variable. If $x = 10$, the equation $x + 5 = 15$ is true, because

> $\mathbf{10} + 5 = 15$ Substitute 10 for x.

However, this equation is false for all other values of x.

Any number that makes an equation true when substituted for its variable is said to **satisfy** the equation. Such numbers are called **solutions** or **roots** of the equation. Because 10 is the only number that satisfies $x + 5 = 15$, it is the only solution of the equation.

Checking Solutions

EXAMPLE 1

Checking a solution. Is 9 a solution of $3y - 1 = 2y + 7$?

Solution   We substitute 9 for y in the equation and simplify each side. If 9 is a solution, we will obtain a true statement.

$$3y - 1 = 2y + 7 \qquad \text{The original equation.}$$
$$3(9) - 1 \stackrel{?}{=} 2(9) + 7 \qquad \text{Substitute 9 for } y.$$
$$27 - 1 \stackrel{?}{=} 18 + 7 \qquad \text{Do the multiplication.}$$
$$26 = 25 \qquad \text{Do the subtraction and the addition.}$$

Since the resulting equation is not true, 9 is *not* a solution.

SELF CHECK   Is 25 a solution of $2(46 - x) = 41$?   *Answer:* no   ■

EXAMPLE 2

Checking a solution. Verify that 6 is a solution of the equation $x^2 - 5x - 6 = 0$.

Solution   We substitute 6 for x in the equation and simplify.

$$x^2 - 5x - 6 = 0 \quad \text{The original equation.}$$
$$(6)^2 - 5(6) - 6 \stackrel{?}{=} 0 \quad \text{Substitute 6 for } x.$$
$$36 - 30 - 6 \stackrel{?}{=} 0 \quad \text{Evaluate the power: } (6)^2 = 36. \text{ Do the multiplication: } 5(6) = 30.$$
$$0 = 0 \quad \text{Do the subtractions on the left-hand side.}$$

Since the resulting equation is true, 6 is a solution.

SELF CHECK   Is 8 a solution of the equation $\frac{m-4}{4} = \frac{m+4}{12}$?   *Answer:* yes   ■

The Subtraction Property of Equality

In practice, we will not be told the solutions of an equation. We will need to find the solutions (that is, solve the equation) ourselves. To develop an understanding of the procedures used to solve an equation, we will first examine $x + 2 = 5$ and make some observations as we solve it.

We can think of the scales shown in Figure 1-13(a) as representing the equation $x + 2 = 5$. The weight (in grams) on the left-hand side of the scale is $x + 2$, and the weight (in grams) on the right-hand side is 5. Because these weights are equal, the scale is in balance. To find x, we need to isolate it. That can be accomplished by removing 2 grams from the left-hand side of the scale. Common sense tells us that we must also remove 2 grams from the right-hand side if the scales are to remain in balance. In Figure 1-13(b), we can see that x grams will be balanced by 3 grams. We say that we have *solved* the equation and that the *solution* is 3.

FIGURE 1-13

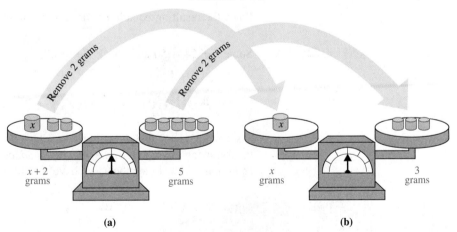

(a)   (b)

These observations suggest a property of equality: *If the same quantity is subtracted from equal quantities, the results will be equal quantities.* We can express this property, called the **subtraction property of equality,** in symbols.

Subtraction property of equality

Let a, b, and c be any real numbers.

If $a = b$, then $a - c = b - c$.

When we use this property, the resulting equation will be equivalent to the original equation.

Equivalent equations

Two equations are **equivalent** when they have the same solutions.

In the previous example, we found that $x + 2 = 5$ is equivalent to $x = 3$. This is true because both of these equations have a solution of $x = 3$.

We now show how to solve the equation using an algebraic approach.

EXAMPLE 3

Solving an equation. Solve $x + 2 = 5$ and check the result.

Solution   To isolate x on the left-hand side of the equation, we use the subtraction property of equality. We can undo the addition of 2 by subtracting 2 from both sides.

$$x + 2 = 5$$
$$x + 2 - 2 = 5 - 2 \quad \text{Subtract 2 from both sides.}$$
$$x = 3 \quad \text{Do the subtractions: } 2 - 2 = 0 \text{ and } 5 - 2 = 3.$$

We check by substituting 3 for x in the original equation and simplifying. If 3 is the solution, we will obtain a true statement.

$$x + 2 = 5 \quad \text{The original equation.}$$
$$3 + 2 \overset{?}{=} 5 \quad \text{Substitute 3 for } x.$$
$$5 = 5 \quad \text{Do the addition: } 3 + 2 = 5.$$

Since the resulting equation is true, 3 is a solution.

SELF CHECK   Solve $x + 24 = 50$ and check the result.   *Answer:* 26   ■

The Addition Property of Equality

A second property that we will use to solve equations involves addition. It is based on the following idea: *If the same quantity is added to equal quantities, the results will be equal quantities.* In symbols, we have the following property.

Addition property of equality

Let a, b, and c be any real numbers.

If $a = b$, then $a + c = b + c$.

To illustrate the addition property of equality, we can think of the scales shown in Figure 1-14(a) as representing the equation $x - 2 = 3$. To find x, we need to add 2 grams of weight to each side. The scales will remain in balance. From the scales in Figure 1-14(b), we can see that x grams will be balanced by 5 grams. Thus, $x = 5$.

We now show how to solve a similar equation using an algebraic approach.

FIGURE 1-14

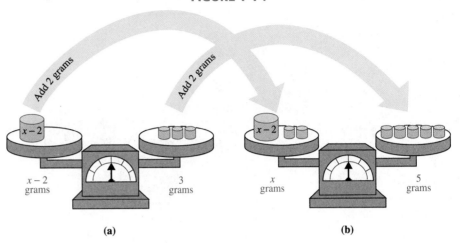

(a)   (b)

EXAMPLE 4

Solving an equation. Solve $19 = y - 7$ and check the result.

Solution   To isolate the variable y on the right-hand side, we use the addition property of equality. We can undo the subtraction of 7 by adding 7 to both sides.

$$19 = y - 7$$
$$19 + 7 = y + 7 - 7 \quad \text{Add 7 to both sides.}$$
$$26 = y \qquad\qquad \text{Do the operations: } 19 + 7 = 26 \text{ and } 7 - 7 = 0.$$
$$y = 26 \qquad\qquad \text{If } 26 = y, \text{ then } y = 26.$$

We check by substituting 26 for y in the original equation and simplifying.

$$19 = y - 7 \qquad \text{The original equation.}$$
$$19 \overset{?}{=} 26 - 7 \quad \text{Substitute 26 for } y.$$
$$19 = 19 \qquad\; \text{Do the subtraction: } 26 - 7 = 19.$$

This is a true statement, so 26 is a solution.

SELF CHECK   Solve $75 = b - 38$ and check the result.   *Answer:* 113   ■

The Division Property of Equality

We can think of the scales in Figure 1-15(a) as representing the equation $2x = 8$. Since $2x$ means $2 \cdot x$, the equation can be written as $2 \cdot x = 8$. The weight (in grams) on the left-hand side of the scale is $2 \cdot x$, and the weight (in grams) on the right-hand side is 8. Because these weights are equal, the scale is in balance.

To find x, we need to isolate it on the left-hand side of the scale. That can be accomplished by removing half of the weight on the left-hand side. Common sense tells us that if the scale is to remain in balance, we need to remove half of the weight from the right-hand side. We can think of the process of removing half of the weight as dividing the weight by 2. In Figure 1-15(b), we can see that x grams will be balanced by 4 grams. Thus, $x = 4$.

These observations suggest a property of equality: *If equal quantities are divided by the same nonzero quantity, the results will be equal quantities.* We can express this property, called the **division property of equality,** in symbols.

FIGURE 1-15

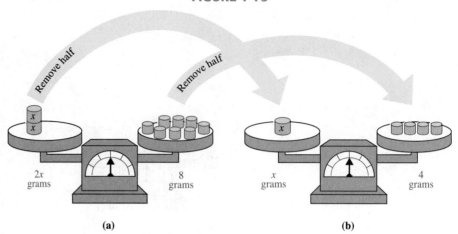

(a)   (b)

Division property of equality

Let a, b, and c be any real numbers.

If $a = b$, then $\dfrac{a}{c} = \dfrac{b}{c}$ $(c \neq 0)$

EXAMPLE 5   **Solving an equation.** Solve the equation $2x = 8$ and check the result.

Solution   Recall that $2x = 8$ means $2 \cdot x = 8$. To isolate x on the left-hand side of the equation, we use the division property of equality: We undo the multiplication by 2 by dividing both sides of the equation by 2.

$$2x = 8$$
$$\frac{2x}{2} = \frac{8}{2} \quad \text{To undo the multiplication by 2, divide both sides by 2.}$$
$$x = 4 \quad \text{Do the divisions: } \tfrac{2}{2} = 1 \text{ and } \tfrac{8}{2} = 4.$$

The solution is 4. Check it as follows:

$$2x = 8 \quad \text{The original equation.}$$
$$2 \cdot 4 \overset{?}{=} 8 \quad \text{Substitute 4 for } x.$$
$$8 = 8 \quad \text{Do the multiplication: } 2 \cdot 4 = 8.$$

SELF CHECK   Solve the equation $16x = 176$ and check the result.   *Answer:* 11   ■

The Multiplication Property of Equality

Sometimes we need to multiply both sides of an equation by the same nonzero number to solve it. This concept is summed up in the multiplication property of equality. It is based on the following idea: *If equal quantities are multiplied by the same nonzero quantity, the results will be equal quantities.* In symbols, we have the following property.

Multiplication property of equality

Let a, b, and c be any real numbers.

If $a = b$, then $ca = cb$ $\quad (c \neq 0)$

To illustrate the multiplication property of equality, we can think of the scales shown in Figure 1-16(a) as representing the equation $\frac{x}{3} = 25$. The weight on the left-hand side of the scale is $\frac{x}{3}$ grams, and the weight on the right-hand side is 25 grams. Because these weights are equal, the scale is in balance. To find x, we triple (or multiply by 3) the weight on each side. The scales will remain in balance. From the scales shown in Figure 1-16(b), we can see that x grams will be balanced by 75 grams. Thus, $x = 75$.

FIGURE 1-16

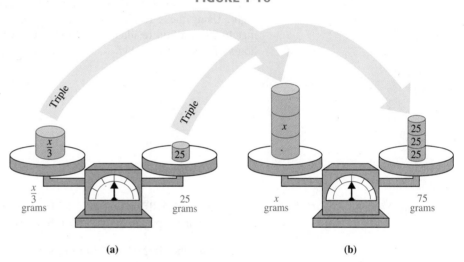

(a)   (b)

We now show how to solve a similar equation using an algebraic approach.

EXAMPLE 6

Solving an equation. Solve the equation $\dfrac{s}{5} = 15$ and check the result.

Solution

To isolate s on the left-hand side, we use the multiplication property of equality. We can undo the division of the variable by 5 by multiplying both sides by 5.

$$\frac{s}{5} = 15$$

$$5 \cdot \frac{s}{5} = 5 \cdot 15 \quad \text{Multiply both sides by 5.}$$

$$s = 75 \quad \text{Do the multiplication: } 5 \cdot 15 = 75.$$

Check: $\dfrac{s}{5} = 15 \quad$ The original equation.

$$\frac{75}{5} \stackrel{?}{=} 15 \quad \text{Substitute 75 for } s.$$

$$15 = 15 \quad \text{Do the division: } \tfrac{75}{5} = 15.$$

SELF CHECK

Solve the equation $\dfrac{t}{24} = 3$ and check the result.

Answer: 72

Using Algebra to Solve Percent Problems

Percents are often used to present numeric information. Stores use them to advertise discounts; manufacturers use them to describe the content of their products; and banks use them to list interest rates for loans and savings accounts. Percent problems occur in three types. Examples of these are shown below.

- What is 28% of 270?
- 14 is what percent of 52?
- 80 is 20% of what number?

Using the concept of a variable, along with the translating skills studied in Section 1.4 and the equation-solving skills of this section, we will now solve these problems.

EXAMPLE 7

Hours of sleep. Figure 1-17 shows the average number of hours United States residents sleep each night. Using these results, how many of the approximately 270 million residents would we expect to get 8 hours of sleep each night?

FIGURE 1-17

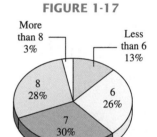

Based on data from *The Macmillan Visual Almanac* (Blackbirch Press, 1996), p. 121

Solution The circle graph tells us that 28% of U.S. residents get 8 hours of sleep each night. Since there are approximately 270 million residents, we need to find:

What number is 28% of 270 million?

To do this, we let a variable represent this unknown number.

Let x = the number of residents who get 8 hours of sleep.

Then we translate the words of the problem into an equation. In this case, the word *of* indicates multiplication, and the word *is* means equals.

What number	is	28%	of	270?	270 represents 270 million.
x	=	28%	$\cdot$	270	Translate from words to mathematical symbols.

To find the number of people who get 8 hours of sleep, we need to find x.

$x = 28\% \cdot 270$ The original equation.

$x = 0.28 \cdot 270$ Write 28% as a decimal: 28% = 0.28.

$x = 75.6$ Use a calculator to do the multiplication on the right-hand side.

According to the survey, 75.6 million U.S. residents would get, on average, 8 hours of sleep each night.

We can check this result using estimation: 28% is approximately 25% $\left(\text{or } \frac{1}{4}\right)$, and 270 million is approximately 280 million. If we find $\frac{1}{4}$ of 280 $\left(\frac{1}{4} \cdot 280\right)$, the result is 70. This is close to 75.6, so our answer seems reasonable. ■

In the statement "28% of 270 is 75.6," the number 75.6 is called the **amount,** 28% is the **percent,** and 270 is the **base.** The relationship between the amount, the percent, and the base is shown in the **percent formula.**

The percent formula

$$\text{Amount} = \text{percent} \cdot \text{base}$$

EXAMPLE 8

Best-selling song. In 1993, Whitney Houston's "I Will Always Love You" was at the top of Billboard's music charts for 14 weeks. What percent of the year did she have the #1 song? (Round to the nearest percent.)

Solution   First, we write the problem in a form that can be translated to an equation. We are asked to find the percent, knowing that for 14 out of the 52 weeks in a year, she had the #1 song. Therefore, we let x = the unknown percent and translate the words of the problem into an equation.

14	is	what percent	of	52?
14	=	x	·	52

14 is the amount, x is the percent, and 52 is the base.

To find x, we solve the equation.

$14 = x \cdot 52$ The original equation.

$14 = 52x$ Write $x \cdot 52$ as $52x$.

$\dfrac{14}{52} = \dfrac{52x}{52}$ To isolate x, undo the multiplication by 52 by dividing both sides by 52.

$0.2692307 \approx x$ Use a calculator to do the division.

$26.92307\% \approx x$ To write the decimal as a percent, multiply by 100 and write a % sign.

$x \approx 27\%$ Round to the nearest percent.

To the nearest percent, Whitney Houston had the best-selling song 27% of the year. We can check this result using estimation. 14 weeks out of 52 weeks is approximately $\frac{14}{50}$ or $\frac{28}{100}$, which is 28%. The answer of 27% seems reasonable.

SELF CHECK   In 1956, Elvis Presley's hit "Hound Dog" was the best-selling song for 11 weeks. To the nearest percent, what percent of the year was this?   *Answer:* 21%   ∎

EXAMPLE 9

Aging population. By the year 2050, the U.S. Census Bureau estimates that about 20%, or 80 million, of the U.S. population will be over 65 years of age. If this is true, what is the population of the country predicted to be at that time?

Solution   First, we write the problem in words. Then we let x = the predicted population in 2050 and translate to form an equation.

80	is	20%	of	what number?
80	=	20%	·	x

80 is the amount, 20 is the percent, and x is the base.

Now we solve the equation to find x.

$$80 = 20\% \cdot x \qquad \text{The original equation.}$$

$$80 = 0.20 \cdot x \qquad \text{Write 20\% as a decimal by dividing by 100 and dropping the \% sign.}$$

$$80 = 0.20x \qquad \text{Write } 0.20 \cdot x \text{ as } 0.20x.$$

$$\frac{80}{0.20} = \frac{0.20x}{0.20} \qquad \text{To undo the multiplication by 0.20, divide both sides by 0.20.}$$

$$400 = x \qquad \text{Use a calculator to do the division.}$$

The Census Bureau is predicting a population of 400 million, of which 80 million are expected to be over the age of 65. We can check the reasonableness of this result using estimation. 80 million out of a population of 400 million could be expressed as $\frac{80}{400}$ or $\frac{40}{200}$ or $\frac{20}{100}$, which is 20%. The answer checks.

SELF CHECK
In 1997, Census Bureau records indicated that about 1.49%, or 4 million, of the U.S. population were over 85 years old. To the nearest million, what was the bureau using as the country's estimated population?   *Answer:* 268 million   ∎

Percents are often used to describe how a quantity has changed. For example, a health care provider might increase the cost of medical insurance by 3%, or a police department might decrease the number of officers assigned to street patrols by 10%. To describe such changes, we use **percent increase** or **percent decrease.**

E X A M P L E   1 0

Minimum wage. In August of 1996, Congress passed, and President Clinton signed, a bill that raised the federal hourly minimum wage to $5.15 on September 1, 1997. See Table 1-7. Find the percent increase in the minimum wage that this legislation provided.

TABLE 1-7

The Hourly Federal Minimum Wage	
1996	**September 1, 1997**
$4.75	$5.15

Source: U.S. Bureau of Labor Statistics

Solution   The minimum wage increased from 1996 to 1997. To find the percent increase, we first find the *amount* of increase by subtracting the smaller number from the larger.

$$\$5.15 - \$4.75 = \$0.40 \qquad \text{Subtract the 1996 minimum wage from the 1997 minimum wage.}$$

Next, we find what percent of the *original* wage ($4.75) the $0.40 increase is. We let $x =$ the unknown percent and translate the words to an equation.

What percent	of	$4.75	is	$0.40?
x	$\cdot$	4.75	$=$	0.40

x is the percent, 4.75 is the base, and 0.40 is the amount.

$$x \cdot 4.75 = 0.40 \qquad \text{The original equation.}$$

$$4.75x = 0.40 \qquad \text{Rewrite } x \cdot 4.75 \text{ as } 4.75x.$$

$$\frac{4.75x}{4.75} = \frac{0.40}{4.75} \qquad \text{To isolate } x, \text{ divide both sides by 4.75.}$$

$$x \approx 0.0842105263 \qquad \text{Use a calculator to do the division.}$$

$$x \approx 8.4\% \qquad \text{To write the decimal as a percent, multiply by 100 and insert a \% sign. Round to the nearest tenth of a percent.}$$

The hourly federal minimum wage increased 8.4% from 1996 to 1997.   ∎

STUDY SET

Section 1.5

VOCABULARY

In Exercises 1–8, fill in the blanks to make the statements true.

1. An _____equation_____ is a statement indicating that two expressions are equal.

2. Any number that makes an equation true when substituted for its variable is said to _____satisfy_____ the equation. Such numbers are called _____solutions_____.

3. To _____check_____ the solution of an equation, we substitute the value for the variable in the original equation and see whether the result is a _____true_____ statement.

4. In the statement "10 is 50% of 20," 10 is called the _____amount_____, 50% is the _____percent_____, and 20 is the _____base_____.

5. Two equations are _____equivalent_____ when they have the same solutions.

6. In mathematics, the word *of* often indicates _____multiplication_____, and _____is_____ means equals.

7. To solve an equation, we _____isolate_____ the variable on one side of the equals sign.

8. When solving an equation, the objective is to find all the values for the _____variable_____ that will make the equation true.

CONCEPTS

9. For each equation, tell what operation is performed on the variable. Then tell how to undo that operation to isolate the variable.
   a. $x - 8 = 24$ subtraction of 8, addition of 8
   b. $x + 8 = 24$ addition of 8, subtraction of 8
   c. $\dfrac{x}{8} = 24$ division by 8, multiplication by 8
   d. $8x = 24$ multiplication by 8, division by 8

10.
ILLUSTRATION 1

   a. What equation is represented by the scale in Illustration 1? $x + 5 = 7$
   b. How will the scale react if 5 pounds is removed from the left side?   It will not be in balance.
   c. How will the scale react if 5 pounds is then removed from the right side?   It will be in balance.

   d. What algebraic property do the steps listed in parts b and c illustrate?
   subtraction property of equality
   e. After parts b and c, what equation will the scale represent, and what is its significance?
   $x = 2$; it is the solution of the original equation.

11. Given $x + 6 = 12$,
   a. What forms the left-hand side of the equation?
   $x + 6$
   b. Is this equation true or false?   neither
   c. Is $x = 5$ a solution?   no
   d. Does $x = 6$ satisfy the equation?   yes

12. Complete the following properties, and then give their names.
   a. If $x = y$ and c is any number, then $x + c = $ _____$y + c$_____.   addition property of equality
   b. If $x = y$ and c is any nonzero number, then $cx = $ _____cy_____.
   multiplication property of equality

NOTATION

In Exercises 13–14, complete the solution of each equation.

13. Solve $x + 15 = 45$.
$$x + 15 = 45$$
$$x + 15 - \boxed{15} = 45 - \boxed{15}$$
$$x = 30$$

14. Solve $8x = 40$.
$$8x = 40$$
$$\frac{8x}{\boxed{8}} = \frac{40}{\boxed{8}}$$
$$x = 5$$

In Exercises 15–16, translate each sentence into an equation.

15. 12 is 40% of what number? $12 = 0.40 \cdot x$
16. 99 is what percent of 200? $99 = x \cdot 200$

17. When computing with percents, the percent must be changed to a decimal or a fraction. Change each percent to a decimal.
   a. 35%   0.35   **b.** 3.5%   0.035
   c. 350%   3.5   **d.** $\frac{1}{2}$%   0.005

18. Change each decimal to a percent.
   a. 0.9   90%   **b.** 0.09   9%
   c. 9   900%   **d.** 0.999   99.9%

PRACTICE

In Exercises 19–36, tell whether the given number is a solution of the equation. (Exercises 35 and 36 require a calculator.)

19. $x + 12 = 18$; $x = 6$ yes

20. $x - 50 = 60$; $x = 110$ yes

21. $2b + 3 = 15$; $b = 5$ no

22. $5t - 4 = 16$; $t = 4$ yes

23. $0.5x = 2.9$; $x = 5$ no

24. $1.2 + x = 4.7$; $x = 3.5$ yes

25. $33 - \dfrac{x}{2} = 30$; $x = 6$ yes

26. $\dfrac{x}{4} + 98 = 100$; $x = 8$ yes

27. $|c - 8| = 10$; $c = 20$ no

28. $|30 - r| = 15$; $r = 20$ no

29. $3x - 2 = 4x - 5$; $x = 12$ no

30. $5y + 8 = 3y - 2$; $y = 5$ no

31. $x^2 - x - 6 = 0$; $x = 3$ yes

32. $y^2 + 5y - 3 = 0$; $y = 2$ no

33. $\dfrac{2}{a + 1} + 5 = \dfrac{12}{a + 1}$; $a = 1$ yes

34. $\dfrac{2t}{t - 2} - \dfrac{4}{t - 2} = 1$; $t = 4$ no

35. $\sqrt{x - 5} + 1 = 15$; $x = 201$ yes

36. $\sqrt{15 + y} - 3 = 20$; $y = 514$ yes

In Exercises 37–64, use a property of equality to solve each equation. Check all solutions.

37. $x + 7 = 10$ 3

38. $15 + y = 24$ 9

39. $a - 5 = 66$ 71

40. $x - 34 = 19$ 53

41. $0 = n - 9$ 9

42. $3 = m - 20$ 23

43. $9 + p = 90$ 81

44. $16 + k = 71$ 55

45. $9 + p = 9$ 0

46. $88 + j = 88$ 0

47. $203 + f = 442$ 239

48. $y - 34 = 601$ 635

49. $4x = 16$ 4

50. $5y = 45$ 9

51. $369 = 9c$ 41

52. $840 = 105t$ 8

53. $4f = 0$ 0

54. $0 = 60k$ 0

55. $23b = 23$ 1

56. $16 = 16h$ 1

57. $\dfrac{x}{15} = 3$ 45

58. $\dfrac{y}{7} = 12$ 84

59. $\dfrac{l}{24} = 2$ 48

60. $\dfrac{k}{17} = 8$ 136

61. $35 = \dfrac{y}{4}$ 140

62. $550 = \dfrac{w}{3}$ 1,650

63. $0 = \dfrac{v}{11}$ 0

64. $\dfrac{d}{49} = 0$ 0

In Exercises 65–76, translate each problem from words to an equation, and then solve the equation.

65. What number is 48% of 650?   312

66. What number is 60% of 200?   120

67. What percent of 300 is 78?   26%

68. What percent of 325 is 143?   44%

69. 75 is 25% of what number?   300

70. 78 is 6% of what number?   1,300

71. What number is 92.4% of 50?   46.2

72. What number is 2.8% of 220?   6.16

73. What percent of 16.8 is 0.42?   2.5%

74. What percent of 2,352 is 199.92?   8.5%

75. 128.1 is 8.75% of what number?   1,464

76. 1.12 is 140% of what number?   0.8

APPLICATIONS

77. LAND OF THE RISING SUN The flag of Japan is a red disc (representing sincerity and passion) on a white background (representing honesty and purity).
    a. What is the area of the rectangular-shaped flag in Illustration 2?   6 ft²
    b. To the nearest tenth of a square foot, what is the area of the red disc? (*Hint:* $A = \pi r^2$)   1.2 ft²
    c. Use the results from parts a and b to find what percent of the area of the Japanese flag is occupied by the red disc.   20%

ILLUSTRATION 2

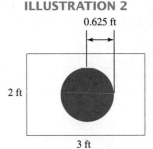

78. COLLEGE ENTRANCE EXAMS On the Scholastic Aptitude Test, or SAT, a high school senior scored 550 on the mathematics portion and 700 on the verbal portion. What percent of the maximum 1,600 points did this student receive?   78.125%

79. GENEALOGY Through an extensive computer search, a genealogist determined that worldwide, 180 out of every 10 million people had his last name. What percent is this?   0.0018%

80. AREA The total area of the 50 states and the District of Columbia is 3,618,770 square miles. If Alaska covers 591,004 square miles, what percent is this of the U.S. total (to the nearest percent)?   16%

81. FEDERAL OUTLAYS Illustration 3 shows the breakdown of U.S. federal budget for fiscal year 1997. If total spending was approximately $1,601 billion, how much was paid as interest on the national debt (to the nearest billion dollars)? $240 billion

ILLUSTRATION 3

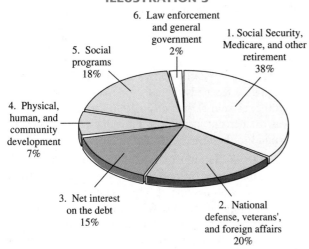

6. Law enforcement and general government 2%
5. Social programs 18%
1. Social Security, Medicare, and other retirement 38%
4. Physical, human, and community development 7%
3. Net interest on the debt 15%
2. National defense, veterans', and foreign affairs 20%

Based on 1998 Federal Income Tax Form 1040

82. DENTAL RECORDS The dental chart for an adult patient is shown in Illustration 4. The dentist marks each tooth that has had a filling. To the nearest percent, what percent of this patient's teeth have fillings? 19%

ILLUSTRATION 4

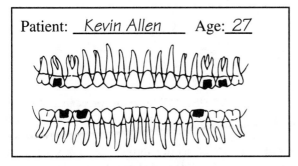

Patient: _Kevin Allen_   Age: _27_

83. TIPPING When paying with a Visa Card, the user must fill in the amount of the gratuity (tip) and then compute the total. Complete the sales draft in Illustration 5 if a 15% tip, rounded up to the nearest dollar, is to be left for the waiter.

ILLUSTRATION 5

STEAK STAMPEDE	
Bloomington, MN	
Server #12\ AT	
VISA	67463777288
NAME	DALTON/ LIZ
AMOUNT	$75.18
GRATUITY $	12.00
TOTAL $	87.18

84. INCOME TAX Use the Schedule X Tax Table shown in Illustration 6 to compute the amount of federal income tax if the amount of taxable income entered on Form 1040, line 39, is $39,909. $7,879.02

ILLUSTRATION 6

If the amount on Form 1040, line 39, is Over—	But not over—	Enter on Form 1040, line 40		of the amount over—
$0	$25,350	.... 15%		$0
25,350	61,400	$3,802.50 +	28%	25,350
61,400	128,100	13,896.50 +	31%	61,400
128,100	278,450	34,573.50 +	36%	128,100
278,450	....	88,699.50 +	39.6%	278,450

85. INSURANCE COST A college student's good grades earned her a student discount on her car insurance premium. What was the percent decrease, to the nearest percent, if her annual premium was lowered from $1,050 to $925? 12%

86. POPULATION The magazine *American Demographics* estimated the population of the United States as 270,539,805 on August 31, 1998. For the month of September, 1998, it estimated that the number of births was 427,810, the number of deaths was 219,341, and net immigration was 75,279.
  a. What was the magazine's estimate of the U.S. population as of September 30, 1998? 270,823,553
  b. What was the percent increase in the U.S. population for the month of September, 1998, according to this magazine? about 0.1048822%

87. CHARITABLE GIVING Nonprofit organizations receive contributions from individuals, corporations, foundations, and bequests. In 1997, individuals contributed $109 billion to nonprofit organizations. If this was 76% of all charitable giving for that year, what was the total amount given to nonprofit organizations in 1997 (to the nearest billion dollars)? $143 billion

88. NUTRITION The Nutrition Facts label from a can of New England Clam Chowder is shown in Illustration 7.
  a. Use the information on the label to determine the number of grams of fat and the number of grams of saturated fat that should be consumed daily. Round to the nearest gram. fat: 65 g; saturated fat: 20 g
  b. To the nearest percent, what percent of the calories in a serving of clam chowder come from fat? 58%

89. COPAYMENT "80/20 health care coverage" means that the plan pays 80% of the medical bill, and the patient pays the remaining 20% (called the *copayment*). Under this plan, what was the cost of an office visit to a dermatologist if the patient's copayment was $11.30? $56.50

ILLUSTRATION 7

Nutrition Facts

Serving Size 1 cup (240mL)
Servings Per Container about 2

Amount per serving	
Calories 240 Calories from Fat 140	
	% Daily Value*
Total Fat 15 g	**23%**
Saturated Fat 5 g	**25%**
Cholesterol 10 mg	**3%**
Sodium 980 mg	**41%**
Total Carbohydrate 21 g	**7%**
Dietary Fiber 2 g	**8%**
Sugars 1 g	
Protein 7 g	

90. EARTHQUAKE INSURANCE A homeowner received a settlement check of $21,568.50 from her insurance company to cover damages caused by an earthquake. If her policy required her to pay a 10% deductible, what was the total dollar amount of the damages to the home? $23,965.00

91. EXPORTS The bar graph in Illustration 8 shows United States exports to Mexico for the years 1992 to 1997.
  a. Between what two years was there a decline in U.S. exports? Find the percent decrease, to the nearest percent.   1994–1995; 10%
  b. Between what two years was there the most dramatic increase in exports? Find the percent increase, to the nearest percent.
  1996–1997; 25%

ILLUSTRATION 8

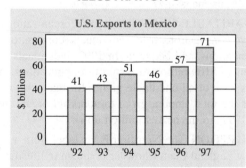

U.S. Exports to Mexico

Based on data from *Los Angeles Times* (May 5, 1997), Business Section

92. PROJECTIONS A market research company was hired to project the demand for professional house-cleaning services in the future. A survey was taken, and projections were made. Complete the last two columns in Illustration 9, rounding to the nearest tenth. Would you classify the projected growth in this industry as small, moderate, or large?
moderate

ILLUSTRATION 9

Households using a maid, housekeeper, or professional cleaning service				
Year	Year	Year	% change	% change
1996	2000	2006	1996–2000	2000–2006
9,436,000	9,999,000	10,740,000	6.0%	7.4%

Based on data from *Demographics Magazine* (Nov. 1996), p. 4

93. AUCTION A pearl necklace of former First Lady Jacqueline Kennedy Onassis, originally valued at $700, was sold at auction in 1996 for $211,500. What was the percent increase in the value of the necklace? (Round to the nearest percent.)   30,114%

94. GEOLOGY See Illustration 10. Geologists have found that a pile of loose sand always takes the form of a cone with a slope of $33\frac{1}{3}$%. This is called the natural **angle of repose.** When more sand is added to the top of the pile, some will trickle down the sides until the $33\frac{1}{3}$% slope is restored. Find the slope of this cone of sand by finding what percent of the horizontal length of 45 feet the 15-foot vertical rise represents.   33.333. . .% = $33\frac{1}{3}$%

ILLUSTRATION 10

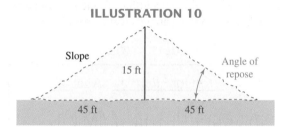

Slope   15 ft   Angle of repose   45 ft   45 ft

95. SPORTS STATISTICS In sports, percentages are most often expressed as three-place decimals instead of percents. For example, if a basketball player makes 75.8% of his free throws, the sports page will list this as .758. Use this format to complete Illustration 11.

ILLUSTRATION 11

All-Time Best Regular-Season Winning Percentages		
Team	**Won–lost record**	**Winning percentage**
1996 Chicago Bulls Basketball	72–10	.878
1972 Miami Dolphins Football	14–0	1.000
1906 Chicago Cubs Baseball	116–36	.763

96. TAXES AND PENALTIES
   a. What is the hotel room tax rate if a $6.84 tax is charged on a room costing $72 a night? 9.5%
   b. What is the late penalty, expressed as a percent, if a charge of $70.75 is applied to a car registration fee of $283 that was not paid to the Motor Vehicle Department on time? 25%

WRITING

97. What does it mean to solve an equation?

98. Write a real-life situation that could be described by "45 is what percent of 50?"

99. After solving an equation, how do we check the solution?

100. Explain what it would mean if the attendance at an amusement park reached 105% of capacity.

REVIEW

101. What is the output of the expression $9 - 3x$ if 3 is the input? 0

102. Write a formula that would give the number of eggs in d dozen. $e = 12d$

103. Translate to symbols: the difference of 45 and x. $45 - x$

104. Evaluate $\dfrac{2^3 + 3(5 - 3)}{15 - 4 \cdot 2}$. 2

105. Approximate 3π to the nearest tenth. 9.4

106. True or false? $-23 > -24$ true

▶ **1.6**   # Problem Solving

In this section, you will learn about

     Writing equations ■ A problem-solving strategy ■ Drawing diagrams ■ Constructing tables

Introduction   One of the objectives of this course is for you to become a better problem solver. The key to problem solving is to understand the problem and then to devise a plan for solving it. In this section, we will introduce a five-step problem-solving strategy. We will also show how drawing a diagram or constructing a table is often helpful in visualizing the facts of a given problem.

Writing Equations

To solve a problem, it is often necessary to write an equation that describes the given situation. To write an equation, we must analyze the facts of the problem, looking for two different ways to describe the same quantity. As an introduction to this procedure, let's consider the following statement.

> If a charity fundraising drive can raise $13 million more, the goal of $87 million will be reached.

Since we are not told the amount already raised, we will let the variable m represent this unknown amount. See Figure 1-18. The key word *more* tells us that if we add $13 million to the m dollars already raised, the goal will be reached. Therefore, the goal can be represented by the algebraic expression $m + 13$. We now have two ways to represent the fundraising goal: $m + 13$ and 87. We can use an equation to state this.

$$m + 13 = 87$$

↑   ↑

This describes the fundraising goal in one way.   This describes the fundraising goal in another way.

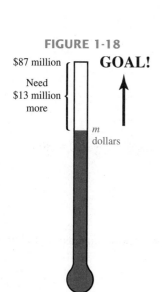

FIGURE 1-18

$87 million   **GOAL!**

Need $13 million more

m dollars

EXAMPLE 1

Writing equations. Using the facts in the following statement, write an equation that describes the same quantity in two different ways.

If Nina had twice as much money in the bank as she has now, she would have enough to pay her tuition bill of $1,500.

Solution   We don't know how much money Nina has in the bank, so we let x = the amount currently in her account. The key word *twice* tells us to *multiply* the amount currently in her account by 2 to find an expression for the cost of tuition. Therefore, $2 \cdot x$, or $2x$, is the cost of tuition in dollars. We now have two ways to represent the tuition: $2x$ and 1,500. Thus,

$$2x = 1,500$$

SELF CHECK

Based on the following statement, write an equation that describes the same quantity in two different ways. "After having 5 feet trimmed off the top, the height of a pine tree is 46 feet."
(*Hint:* Let x = the original height of the tree.)   *Answer:* $x - 5 = 46$ ■

EXAMPLE 2

Writing equations. Using the facts in the following situation, write an equation that describes the same quantity in two different ways.

Three friends pooled their money to purchase some Lotto tickets. When one of the tickets won, they split the cash prize evenly. Each of them received $95.

Solution   We don't know how much money the winning Lotto ticket was worth, so we let x = the amount of money won. The key phrase *split evenly* tells us to *divide* the amount of money won by 3 to find an expression for each person's share. Therefore, each person receives $\frac{x}{3}$ dollars. We are told that each person won $95, so we now have two ways to represent the amount won by each person: $\frac{x}{3}$ and 95. Thus,

$$\frac{x}{3} = 95$$

SELF CHECK

Based on the following facts, write an equation that describes the same amount in two different ways. "The attendance at a political rally was 250 people more than the organizers had anticipated. 800 people were present." (*Hint:* Let x = the number of people that were anticipated.)   *Answer:* $x + 250 = 800$ ■

A Problem-Solving Strategy

To become a good problem solver, you need a plan to follow, such as the following five-step problem-solving strategy.

Strategy for problem solving

1. **Analyze the problem** by reading it carefully to understand the given facts. What information is given? What are you asked to find? What vocabulary is given? Often, a diagram or table will help you visualize the facts of the problem.

2. **Form an equation** by picking a variable to represent the quantity to be found. Then express all other unknown quantities as expressions involving that variable. Key words or phrases can be helpful. Finally, write an equation expressing a quantity in two different ways.

3. **Solve the equation.**

4. **State the conclusion.**

5. **Check the result** in the words of the problem.

EXAMPLE 3

Systems analysis. An engineer found that a company's telephone use would have to increase by 350 calls per hour before the system would reach the maximum capacity of 1,500 calls per hour. Currently, how many calls are being made each hour on the system?

ANALYZE THE PROBLEM   We are asked to find the number of calls currently being made each hour. We are given two facts:

■ The maximum capacity of the system is 1,500 calls per hour.

■ If the number of calls increases by 350, the system will reach capacity.

FORM AN EQUATION   Let n = the number of calls currently being made each hour. To form an equation involving n, we look for a key word or phrase in the problem.

Key phrase: *increase by 350*   **Translation:** addition

The key phrase tells us to add 350 to the current number of calls to obtain an expression for the maximum capacity of the system. Therefore, we can write the maximum capacity of the system in two ways.

The current number of calls per hour	increased by	350	is	the maximum capacity of the system.
n	$+$	350	$=$	$1{,}500$

SOLVE THE EQUATION
$$n + 350 = 1{,}500$$
$$n + 350 - 350 = 1{,}500 - 350 \quad \text{To undo the addition of 350, subtract 350 from both sides.}$$
$$n = 1{,}150 \quad \text{Do the subtractions.}$$

STATE THE CONCLUSION   Currently, 1,150 calls per hour are being made.

CHECK THE RESULT   If 1,150 calls are currently being made each hour and an increase of 350 calls per hour occurs, then $1{,}150 + 350 = 1{,}500$ calls will be made each hour. This is the capacity of the system. The solution checks.   ■

| EXAMPLE 4 | **Marine recruitment.** The annual number of Marine recruits from a certain county tripled after an intense recruiting program was conducted at the area's high schools. If 384 students from the county decided to join the Marines, what was the previous year's county recruitment total? |

ANALYZE THE PROBLEM   We are asked to find the county recruitment total for last year. We are given two facts about the situation:

■ 384 recruits were signed this year.

■ The number of recruits this year is triple that of last year.

FORM AN EQUATION   Let r = the number of recruits last year. To form an equation involving r, we look for a key word or phrase in the problem.

Key word: *tripled*   **Translation:** multiplication by 3

The key word tells us that we can multiply last year's number of recruits by 3 to obtain an expression for the number of recruits this year. Therefore, we can write this year's number of recruits in two ways.

3	times	the number of recruits last year	is	384.
3	·	r	=	384

SOLVE THE EQUATION

$$3r = 384$$

$$\frac{3r}{3} = \frac{384}{3}$$ To undo the multiplication by 3, divide both sides by 3.

$$r = 128$$ Do the divisions.

STATE THE CONCLUSION   The number of recruits last year was 128.

CHECK THE RESULT   If we multiply last year's number of recruits (128) by 3, we get $3 \cdot 128 = 384$. This is the number of recruits for this year. The solution checks. ■

Drawing Diagrams

When solving problems, diagrams are often helpful; they allow us to visualize the facts of the problem.

| EXAMPLE 5 | **Airline travel.** On a book tour that took her from New York City to Chicago to Los Angeles and back to New York City, an author flew a total of 4,910 miles. The flight from New York to Chicago was 714 miles, and the flight from Chicago to L.A. was 1,745 miles. How long was the direct flight back to New York City? |

ANALYZE THE PROBLEM   We are asked to find the length of the flight from L.A. to New York City. We are given the following three facts.

■ The total miles flown on the tour was 4,910.

■ The flight from New York to Chicago was 714 miles.

■ The flight from Chicago to L.A. was 1,745 miles.

In the diagram in Figure 1-19, we see that the three parts of the tour form a triangle. We know the lengths of two of the sides of the triangle (714 and 1,745) and the perimeter of the triangle (4,910).

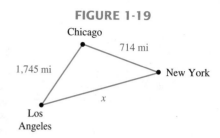

FIGURE 1-19

Chicago

714 mi

1,745 mi

New York

x

Los Angeles

FORM AN EQUATION We will let *x* = the length (in miles) of the flight from L.A. to New York and label the appropriate side of the triangle in Figure 1-19. There are two ways to describe the total number of miles traveled by the author on the book tour.

The miles from New York to Chicago	plus	the miles from Chicago to L.A.	plus	the miles from L.A. to New York	is	4,910.
714	+	1,745	+	*x*	=	4,910

SOLVE THE EQUATION

$$714 + 1{,}745 + x = 4{,}910$$

$$2{,}459 + x = 4{,}910 \qquad \text{Simplify the left-hand side of the equation:}$$
$$714 + 1{,}745 = 2{,}459.$$

$$2{,}459 - \mathbf{2{,}459} + x = 4{,}910 - \mathbf{2{,}459} \qquad \text{Subtract 2,459 from both sides to isolate } x.$$

$$x = 2{,}451 \qquad \text{Do the subtractions.}$$

STATE THE CONCLUSION The flight from L.A. to New York was 2,451 miles.

CHECK THE RESULT If we add the three flight lengths, we get 714 + 1,745 + 2,451 = 4,910. This was the total number of miles flown on the book tour. The solution checks. ■

EXAMPLE 6

Eye surgery. A surgical technique called **radial keratotomy** is sometimes used to correct nearsightedness. This procedure involves equally spaced incisions in the cornea, as shown in Figure 1-20. Find the angle between each incision.

FIGURE 1-20

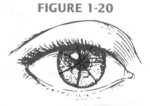

Solution

ANALYZE THE PROBLEM From the diagram in Figure 1-21, we see that there are 7 angles of equal measure. We also know that 1 complete revolution is 360°.

FIGURE 1-21

FORM AN EQUATION We will let *x* = the measure of one of the angles. We then have:

7	times	the measure of one of the angles	is	360°.
7	·	*x*	=	360

SOLVE THE EQUATION

$$7x = 360$$

$$\frac{7x}{7} = \frac{360}{7} \qquad \text{Divide both sides by 7.}$$

$$x \approx 51.42857143 \qquad \text{Use a calculator to do the division.}$$

$$x \approx 51.4° \qquad \text{Round to the nearest tenth of a degree.}$$

STATE THE CONCLUSION The incisions are approximately 51.4° apart.

CHECK THE RESULT If we multiply 51 by 7, we obtain 357. This is close to 360, so the result of 51.4° seems reasonable.

Constructing Tables

Sometimes it is helpful to organize the given facts of the problem in a table.

EXAMPLE 7

Labor statistics. The number of women in the U.S. labor force has grown steadily over the past 100 years. From 1900 to 1940, the number grew by 8 million. By 1980, it had increased by an additional 32 million. By 1997, the number rose another 15 million; by the end of that year, 60 million women were in the labor force. How many women were in the labor force in 1900?

ANALYZE THE PROBLEM   We are to find the number of women in the U.S. labor force in 1900. We know the following:

- The number grew by 8 million, increased by 32 million, and rose another 15 million.
- The number of women in the labor force in 1997 was 60 million.

We will let x represent the number of women (in millions) in the labor force in 1900. We can write algebraic expressions to represent the number of women in the work force in 1940, 1980, and 1997 by translating key words. See Table 1-8.

TABLE 1-8

Year	Women in the labor force (millions)	
1900	x	
1940	$x + 8$	**Key word:** *grew* **Translation:** addition
1980	$x + 8 + 32$	**Key word:** *increased* **Translation:** addition
1997	$x + 8 + 32 + 15$	**Key word:** *rose* **Translation:** addition

FORM AN EQUATION   There are two ways to represent the number of women (in millions) in the 1997 labor force: $x + 8 + 32 + 15$ and 60. Therefore,

$$x + 8 + 32 + 15 = 60$$

SOLVE THE EQUATION

$$x + 8 + 32 + 15 = 60$$
$$x + 55 = 60 \qquad \text{Simplify: } 8 + 32 + 15 = 55.$$
$$x + 55 - 55 = 60 - 55 \qquad \text{To undo the addition of 55, subtract 55 from both sides.}$$
$$x = 5 \qquad \text{Do the subtractions.}$$

STATE THE CONCLUSION   There were 5 million women in the U.S. labor force in 1900.

CHECK THE RESULT   Adding the number of women (in millions) in the labor force in 1900 and the increases, we get $5 + 8 + 32 + 15 = 60$. In 1997, there were 60 million, so the solution checks.

STUDY SET

Section 1.6

VOCABULARY

In Exercises 1–4, fill in the blanks to make the statements true.

1. A letter that is used to represent a number is called a ___variable___.

2. To ___solve___ an equation means to find all the values of the variable that make the equation true.

3. An ___equation___ is a mathematical statement that two quantities are equal.

4. Phrases such as *increased by* and *more than* indicate the operation of ___addition___.

CONCEPTS

5. Put the steps of the five-step problem-solving strategy listed below in the proper order.

State the result. Solve the equation. Check the result. Analyze the problem. Form an equation.

analyze, form, solve, state, check

6. Fill in the blanks to make the statements true. When solving a real-world problem, we let a _____variable_____ represent the unknown quantity. Then we write an _____equation_____ that describes the same quantity in two different ways. Finally, we _____solve_____ the equation for the variable to find the unknown.

In Exercises 7–8, a diagram is given. Tell what the variable represents, and then write an equation that describes the same quantity in two ways.

7.

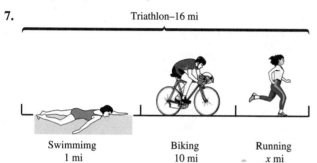

Triathlon–16 mi

Swimmimg 1 mi   Biking 10 mi   Running x mi

x = the length of the running portion of the triathlon;
$1 + 10 + x = 16$

8.

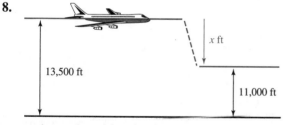

13,500 ft   x ft   11,000 ft

x = the number of feet the plane descended;
$13{,}500 - x = 11{,}000$

In Exercises 9–12, a statement about a situation and a table containing facts about the situation are given. Tell what the variable represents, and then write an equation that describes the same quantity in two ways.

9. The sections of a 430-page book are being assembled.

Section	Number of pages
Table of Contents	4
Preface	x
Text	400
Index	12

x = the number of pages in the preface;
$4 + x + 400 + 12 = 430$

10. A hamburger chain sold a total of 31 million hamburgers in its first four years of business.

Years in business	Cumulative number of hamburgers sold (millions)
1	x
2	x + 5
3	x + 5 + 8
4	x + 5 + 8 + 16

x = number of hamburgers sold the first year;
$x + 5 + 8 + 16 = 31$

11. Two accounts earned a total of $678 in interest for the year.

Account	Interest earned ($)
Savings	x
Checking	73

x = the amount of interest earned on the savings account;
$x + 73 = 678$

12. A motorist traveled from Toledo, OH to Columbus, OH at 55 mph and from Columbus to Cincinnati, OH at 50 mph. The entire trip covered 253 miles.

	Speed (mph)	Distance (mi)
Toledo to Columbus	55	145
Columbus to Cincinnati	50	x

x = the distance from Columbus to Cincinnati;
$145 + x = 253$

PRACTICE

In Exercises 13–26, an occupation is listed along with a statement that someone with that occupation might make. Identify the key word or phrase in the sentence and the operation it indicates.

13. *Financial planner:* The profits from the sale were disbursed equally among the investors.
disbursed equally; division

14. *Auditor:* The cost of the project skyrocketed by a factor of 10. factor of 10; multiplication

15. *Park ranger:* Two feet of the snow pack had melted before February 1. melted; subtraction

16. *Lawyer:* The city annexed 15 uninhabited houses that were under county jurisdiction. annexed; addition

17. *Ecologist:* After the flood, 6 acres of the marshy land were reclaimed. reclaimed; addition

18. *Race car driver:* The new tires helped shave off 2.5 seconds from each lap. shave off; subtraction

19. *Landscaper:* Four feet of the embankment had eroded.   eroded; subtraction

20. *Developer:* Plans were drawn up to bisect the lot with a one-way street.   bisect; division

21. *Archaeologist:* The area to be excavated was sectioned off uniformly.   sectioned off uniformly; division

22. *Carpenter:* The rough-cut lumber needed to be sanded down a sixteenth of an inch.
    sanded down; subtraction

23. *School administrator:* Two of the cheerleaders had to be cut from the squad for disciplinary reasons.
    cut; subtraction

24. *Producer:* Because of overflow crowds, the play's run was extended two weeks.   extended; addition

25. *Optometrist:* The patient's pupils dilated to twice their size.   dilated to twice their size; multiplication

26. *Nurse:* The amount of required paperwork has quadrupled over the last few years.
    quadrupled; multiplication

In Exercises 27–34, write an equation that describes the same quantity in two ways.

27. An existing 1,000-foot water line had to be extended to a length of 1,525 feet to reach a new restroom facility. Let x = the length of the extension.
    $1,000 + x = 1,525$

28. Because of overgrazing, state agriculture officials determined that the 4,500 head of cattle currently on the ranch had to be reduced to 2,750. Let x = the number of head of cattle to be removed. $4,500 - x = 2,750$

29. Because 4 people didn't show up for their appointments, the dentist saw only 8 patients on Wednesday. Let x = the number of scheduled appointments.
    $x - 4 = 8$

30. After the management of a preschool decided to open up an additional 15 spaces, the total number of children attending the school reached 260. Let x = the number of children who attended before the enrollment was increased. $x + 15 = 260$

31. A length of gold chain, cut into 12-inch-long pieces, makes five bracelets. Let x = the length of the chain. $\frac{x}{12} = 5$

32. The 24 ounces of walnuts used to make a fruitcake was twice what was called for in the recipe. Let x = the number of ounces of walnuts called for in the recipe. $24 = 2x$

33. When the value of 15 postage stamps was calculated, the total was found to be 495 cents. Let x = the value of one stamp. $15x = 495$

34. The overtime hours available for 8 employees were distributed equally, so that each of them got to work an additional 11 hours. Let x = the total number of overtime hours. $\frac{x}{8} = 11$

In Exercises 35–36, use the outline of the five-step problem-solving strategy to help answer each question.

35. **MAJOR REQUIREMENTS** The business department of a college reduced by 6 the number of units of course work needed to obtain a degree. The department now requires the completion of 28 units. What was the old unit requirement?

ANALYZE THE PROBLEM
What are you asked to find? the old unit requirement

We know that

■   The unit requirement was reduced by 6 .

■   The new unit requirement is 28 .

FORM AN EQUATION
Let x = the old unit requirement

   Key word: *reduced*   **Translation:** subtract

We can express the new requirement in two ways.

The old unit requirement	reduced by	6	is	the new unit requirement.
x	−	6	=	28

SOLVE THE EQUATION
$$x - 6 = 28$$
$$x + 6 - 6 = 28 + 6$$
$$x = 34$$

STATE THE CONCLUSION
The old unit requirement was 34 units.

CHECK THE RESULT
If we reduce the old unit requirement of 34 by 6, we have $34 - 6 = 28$. The answer checks.

36. **BUSINESS LOSSES** After the membership fee to a health spa was raised, the number of new members joining each week was half of what it used to be. If, on average, 27 people per week are now joining, how many used to join each week?

ANALYZE THE PROBLEM
What are you asked to find? the number that used to join each week

We know that

■   27 people are now joining each week.

■   The number of people now joining is
    half of what it used to be.

FORM AN EQUATION

Let $x = $ the number that used to join each week

Key phrase: *half of*   **Translation:** _____divide by 2_____

We can express the number now joining in two ways.

The number that used to join each week	divided by	2	is	the number that are now joining each week.
x	$\div$	2	$=$	27

SOLVE THE EQUATION

$$\frac{x}{2} = 27$$

$$2\left(\frac{x}{2}\right) = 2\,(27)$$

$$x = 54$$

STATE THE CONCLUSION

The number of people that used to join each week was 54 .

CHECK THE RESULT

If we divide the number of people that used to join each week by 2 , we have $\frac{54}{2} = 27$. The answer checks.

APPLICATIONS

You can probably solve Exercises 37–46 without using algebra. Nevertheless, you should use the methods discussed in this section to solve the problems, so that you can gain experience in applying these concepts and procedures. This will prepare you for more challenging problems later in the course.

37. ICE CREAM   People in the United States consume more ice cream per capita than any other people in the world—47 pints per year. If this is 20 more pints than Canadians eat, what is the yearly per-capita consumption of ice cream in Canada?   27 pints

38. ENTERTAINMENT   According to *Forbes* magazine, Oprah Winfrey made an estimated $125 million in 1998. This was $67 million more than Harrison Ford's estimated earnings for that year. How much did Harrison Ford make in 1998? $58 million

39. TENNIS   Billie Jean King won 40 Grand Slam tennis titles in her career. This is 14 less than the all-time leader, Martina Navratilova. How many Grand Slam titles did Navratilova win?   54 .

40. MONARCHY   George III reigned as king of Great Britain for 59 years. This is four years less than the longest-reigning British monarch, Queen Victoria. For how many years did Queen Victoria rule?   63

41. COST OVERRUN   At completion, the construction cost of a subway system totaled $564 million. This was larger than the original estimate by a factor of 3. What was the original estimate for the subway system? $188 million

42. FLOODING   Torrential rains caused the width of a river to swell to 84 feet. If this was twice its normal size, how wide was the river before the flooding?   42 ft

43. ATM RECEIPT   Use the information on the automatic-teller receipt in Illustration 1 to find the balance in the account before the withdrawal. $322.00

ILLUSTRATION 1

HOME SAVINGS OF AMERICA			
Thank you for letting us serve all your financial needs.			
TRAN.	DATE	TIME	TERM
0286.	1/16/99	11:46 AM	HSOA822
CARD NO.			6125 8
WITHDRAWAL OF			$35.00
FROM CHECKING ACCT.			3325256-612
CHECKING BAL.			$287.00

44. EDUCATION IN THE UNITED STATES   One category in the circle graph shown in Illustration 2 is not labeled with a percent. What percent should be written there?   24%

ILLUSTRATION 2

Highest level of education attained by persons 25 years and older

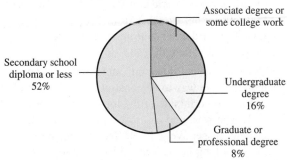

Based on data from *Digest of Education Statistics*

45. TV NEWS   An interview with a world leader was edited into equally long segments and broadcast in parts over a three-day period on a TV news program. If each daily segment of the interview lasted 9 minutes, how long was the original interview?   27 min

46. MISSING CHILD   Six search and rescue squads were called to a mountain wilderness park to look for a lost child. If each squad was able to cover 16 acres, over how large an area was the search conducted?   96 acres

In Exercises 47–50, use a table to help organize the facts of the problem; then find the solution.

47. **STATEHOOD** From 1800 to 1850, 15 states joined the Union. From 1851 to 1900, an additional 14 states entered. Three states joined from 1901 to 1950. Since then, Alaska and Hawaii are the only others to enter the Union. How many states were part of the Union prior to 1800?   16

48. **STUDIO TOUR** Over a four-year span, improvements in a Hollywood movie studio tour caused it to take longer. The first year, 10 minutes were added to the tour length. The second, third, and fourth years, 5 minutes were added. If the tour now lasts 135 minutes, how long was it originally?   110 min

49. **ANATOMY** A premed student has to know the names of all 206 bones that make up the human skeleton. So far, she has memorized the names of the 60 bones in the feet and legs, the 31 bones in the torso, and the 55 bones in the neck and head. How many more names does she have to memorize?   60

50. **ORCHESTRA** A 98-member orchestra is made up of a woodwind section with 19 musicians, a brass section with 23 players, a two-person percussion section, and a large string section. How many musicians make up the string section of the orchestra?   54

In Exercises 51–54, draw a diagram to help organize the facts of the problem, and then find the solution.

51. **BERMUDA TRIANGLE** The Bermuda Triangle is a triangular region in the Atlantic Ocean where many ships and airplanes have disappeared. The perimeter of the triangle is about 3,075 miles. It is formed by three imaginary lines. The first, 1,100 miles long, is from Melbourne, Florida, to Puerto Rico. The second, 1,000 miles long, stretches from Puerto Rico to Bermuda. The third extends from Bermuda back to Florida. Find its length.   975 mi

52. **FENCING** To cut down on vandalism, a lot on which a house was to be constructed was completely fenced. The north side of the lot was 205 feet in length. The west and east sides were 275 and 210 feet long, respectively. If 945 feet of fencing was used, how long is the south side of the lot?   255 ft

53. **SPACE TRAVEL** The 364-foot-tall Saturn V rocket carried the first astronauts to the moon. Its first, second, and third stages were 138, 98, and 46 feet tall, respectively. Atop the third stage was the Lunar Module, and from it extended a 28-foot escape tower. How tall was the Lunar Module?   54 ft

54. **PLANETS** Mercury, Venus, and the earth have approximately circular orbits about the sun. Earth is the farthest from the sun at 93 million miles, and Mercury is the closest, at 36 million miles. The orbit of Venus is about 31 million miles from that of Mer-

cury. How far is the earth's orbit from that of Venus?   26 million mi

In Exercises 55–56, solve each problem.

55. **STOP SIGN** Find the measure of one angle of the octagonal stop sign shown in Illustration 3. (*Hint:* The sum of the measures of the angles of an octagon is 1080°.)   135°

ILLUSTRATION 3

56. **FERRIS WHEEL** What is the measure of the angle between each of the "spokes" of the Ferris wheel shown in Illustration 4?   30°

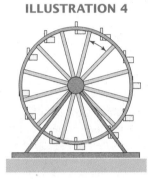

ILLUSTRATION 4

WRITING

57. How can a variable and an equation be used to find an unknown quantity?

58. How can a diagram or a table be helpful when solving problems?

59. From what you have learned in this section, what is the key to problem solving?

60. Briefly explain what should occur in each step of the five-step problem-solving strategy.

REVIEW

61. If an 85-seat restaurant added an additional 25 seats, what is the percent increase in seating (to the nearest percent)?   29%

62. What two numbers are a distance of 8 away from 4 on the number line?   −4 and 12

63. What is 35% of 7,800?   2,730

64. Solve: $\dfrac{t}{25} = 6$.   150

65. Is $x = 34$ a solution of $x - 12 = 20$?   no

66. Write 0.898 as a percent.   89.8%

67. Evaluate $2x^2 - 3x$ for $x = 4$.   20

68. Evaluate $2 + 3[24 - 2(5 - 2)]$.   56

Variables

One of the major objectives of this course is for you to become comfortable working with **variables**. In Chapter 1, we have used the concept of variable in four ways.

Stating Mathematical Properties

Variables have been used to state properties of mathematics in a concise, "shorthand" notation.

1. Complete the statement of the subtraction property of equality: Let a, b, and c be any three numbers. If $a = b$, then $a - c = $ _____ $b - c$ _____.

2. Complete the statement of the multiplication property of equality: Let a, b, and c be any three numbers ($c \neq 0$). If $a = b$, then $ca = $ _____ cb _____.

Stating Relationships between Quantities

Variables are letters that stand for numbers. We have used variables to express known relationships between two or more quantities. These written relationships are called **formulas.**

3. Translate the word model to an equation (formula) that mathematically describes the situation.

The total cost is the sum of the purchase price of the item and the sales tax. $C = p + t$

4. Use the data in the table to state the relationship between the quantities using a formula. $b = 2t$

Picnic tables	Benches needed
2	4
3	6
4	8

Writing Algebraic Expressions

Variables and numbers (called constants) have been combined with the operations of addition, subtraction, multiplication, and division to create **algebraic expressions.**

5. One year, a cruise company did x million dollars worth of business. After a television celebrity was signed as a spokeswoman for the company, its business increased by $4 million the next year. Use an algebraic expression to indicate the amount of business the cruise company had in the year the celebrity was the spokeswoman.

$x + 4 = $ amount of business ($ millions) the year with the celebrity

6. Evaluate the algebraic expression for the given values of the variable, and enter the results in the table.

x	$3x^2 - 2x + 1$
0	1
4	41
6	97

Writing Equations to Solve Problems

To solve problems, we have let a variable represent an unknown quantity, written an equation containing the variable, and then solved the equation to find the unknown. In Exercises 7 and 8, complete the statement: Let $x = $ _____.

7. AREA The total area covered by the United States is 3,618,770 square miles. If Maine covers 33,265 square miles, what percent of the U.S. total is this?

Let $x = $ the percent of the U.S. total that Maine's area represents.

8. CANVASSING Eight volunteers were hired to conduct a survey of the residents of a city. If each volunteer was able to canvass 4 city blocks, how many city blocks were canvassed by the volunteers?

Let $x = $ the number of city blocks canvassed.

Accent on Teamwork

Section 1.1

Production planning In the Study Set for Section 1.1, Exercise 73 on page 10 asks you to write a series of equations that would assist a production planner in ordering the correct number of parts for a production run of *c* chairs. See Illustration 1.

a. Decide on a product for which you will be the production planner. Make a detailed drawing of it, like the one in Illustration 1.

b. For one component of your product, create a table giving the number that should be ordered for production runs of 50, 100, 150, and 200 units. Do the same for a second component of your product using a bar graph, and for a third component using a line graph.

c. For each of the three components in part b, write an equation describing the number of components that need to be ordered for a production run of *u* units.

ILLUSTRATION 1

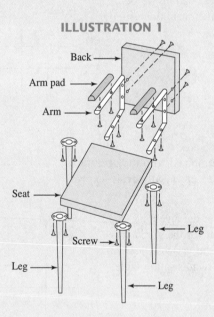

Section 1.2

Real numbers Give some examples of situations in your everyday life where you encounter the types of numbers listed below.

- Whole numbers
- Zero
- Negative numbers
- Fractions
- Decimals

Section 1.3

Order of operations To make a cake from a mix, the instructions must be followed carefully. Otherwise, the results can be disastrous. Think of two other multistep processes and explain why the steps must be performed in the proper order, or the outcome is adversely affected. Think of two processes where the order in which the steps are performed does not affect the outcome.

Section 1.4

Evaluating algebraic expressions Find five examples of cylinders. Measure and record the diameter *d* of their bases and their heights *h*. Express the measurements as decimals. Find the radius *r* of each base by dividing the diameter by 2. Then find each volume by evaluating the expression $\pi r^2 h$. Round to the nearest tenth of a cubic unit. See the Accent on Technology on page 43 for an example. Present your results in a table of the form shown in Illustration 2.

ILLUSTRATION 2

Cylinder	*d*	*r*	*h*	Volume
Container of salt	$3\frac{1}{4}$ in. (3.25 in.)	$1\frac{5}{8}$ in. (1.625 in.)	$5\frac{3}{8}$ in. (5.375 in.)	44.6 in.^{3}

Section 1.5

Subtraction property of equality Check out a scale and some weights from your school's science department. Use them as part of a class presentation to explain how the subtraction property of equality is used to solve the equation $x + 2 = 5$. See the discussion and Figure 1-13 on page 49 for some suggestions on how to do this.

Section 1.6

Translation Study Exercises 13–26 of the Study Set for Section 1.6 on page 67. These problems contain key words and phrases used in occupations. When translated to mathematical symbols, the words and phrases suggest addition, subtraction, multiplication, and division. Give two new examples for each operation.

Section 1.1

Describing Numerical Relationships

CONCEPTS

Tables, bar graphs, and *line graphs* are used to describe numerical relationships.

REVIEW EXERCISES

1. Illustration 1 lists the worldwide production of wide-screen TVs. Use the data to construct a bar graph. Describe the trend in the production in words.

The production of wide-screen TVs is increasing.

ILLUSTRATION 1

Year	Production (millions of units)
'95	3
'96	5
'97	7
'98	11

Source: Electronic Industries Association of Japan

2. Consider the line graph in Illustration 2 that shows the number of cars parked in a mall parking structure from 6 P.M. to 12 midnight on a Saturday.

   a. What units are used to scale the horizontal and vertical axes?

   1 hr; 100 cars

   b. How many cars were in the parking structure at 11 P.M.?

   100

   c. At what time did the parking structure have 500 cars in it?

   7 P.M.

ILLUSTRATION 2

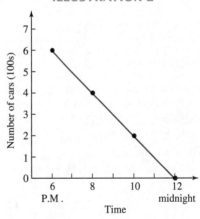

The result of an addition is called the *sum;* of a subtraction, the *difference;* of a multiplication, the *product;* and of a division, the *quotient.*

3. Express each statement in words.

   a. $15 - 3 = 12$

   The difference of 15 and 3 is 12.

   c. $15 \div 3 = 5$

   The quotient of 15 and 3 is 5.

   b. $15 + 3 = 18$

   The sum of 15 and 3 is 18.

   d. $15 \cdot 3 = 45$

   The product of 15 and 3 is 45.

4. a. Write the multiplication 4×9 in two ways: first with a raised dot $\cdot$, and then using parentheses. $4 \cdot 9$; $4(9)$

   b. Write the division $9 \div 3$ without using the symbols $\div$ or $)\overline{}$. $\frac{9}{3}$

Variables are letters used to stand for numbers.

5. Write each multiplication without a multiplication symbol.

   a. $8 \cdot b$ $8b$ **b.** $x \cdot y$ xy **c.** $2 \cdot l \cdot w$ $2lw$ **d.** $P \cdot r \cdot t$ Prt

An *equation* is a mathematical sentence that contains an = sign. Variables and/or numbers can be combined with the operations of addition, subtraction, multiplication, and division to create *algebraic expressions.*

6. Classify each item as either an algebraic expression or an equation.

   a. $5 = 2x + 3$ equation   **b.** $2x + 3$ algebraic expression

   c. $\dfrac{t + 6}{12}$ algebraic expression   **d.** $P = 2l + 2w$ equation

Equations that express a known relationship between two or more variables are called *formulas.*

7. Translate the word model to an equation that mathematically describes this situation: The number of tellers needed by a bank is the quotient of the number of bank accounts and 350. $T = \frac{a}{350}$

8. Use the equation (formula) $n = b + 5$ to complete the table in Illustration 3.

9. Use the data in Illustration 4 to write an equation (formula) that mathematically describes the relationship between the two quantities; then state the relationship in words.

$f = 50c$; the total fees are the product of 50 and the number of children.

ILLUSTRATION 3

Number of brackets	Number of nails
5	10
10	15
20	25

ILLUSTRATION 4

Number of children	Total fees (dollars)
1	50
2	100
4	200

Section 1.2

The Real Numbers

The *natural numbers:*
$\{1, 2, 3, 4, 5, 6, \ldots\}$
The *whole numbers:*
$\{0, 1, 2, 3, 4, 5, 6, \ldots\}$
The *integers:*
$\{\ldots, -3, -2, -1, 0, 1, 2, 3, \ldots\}$

10. Which number is a whole number but not a natural number?   0

11. Graph each member of the set $\{-3, 5, 0, -1\}$ on the number line.

12. Represent each of these situations with a signed number.
   a. A budget deficit of $65 billion   **b.** 206 feet below sea level -206
      $-\$65$

Two *inequality symbols* are
$>$ "is greater than"
$<$ "is less than"

13. Use one of the symbols $>$ or $<$ to make each statement true.
   a. $0 \boxed{<} 5$ **b.** $6 \boxed{>} 4$

   c. $-12 \boxed{>} -13$ **d.** $-3 \boxed{<} 2$

A *rational number* is any number that can be written as a fraction with an integer in its numerator and a nonzero integer in its denominator.

14. Show that each of the following numbers is a rational number by expressing it as a fraction.
   a. 5 $\frac{5}{1}$ **b.** -12 $\frac{-12}{1}$

   c. 0.7 $\frac{7}{10}$ **d.** $4\frac{2}{3}$ $\frac{14}{3}$

Rational numbers are either *terminating* or *repeating* decimals.

An *irrational number* is a nonterminating, nonrepeating decimal.

15. Graph each member of the set $\left\{-\pi, 0.333\ldots, 3.75, -\frac{17}{4}, \frac{7}{8}\right\}$ on the number line.

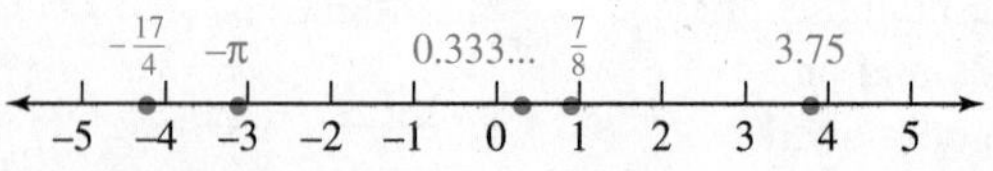

$$-\frac{17}{4} \quad -\pi \qquad \quad 0.333\ldots \quad \frac{7}{8} \qquad \quad 3.75$$

16. Use a calculator to approximate 3π and $2\sqrt{2}$ to the nearest hundredth.
   $3\pi \approx 9.42$; $2\sqrt{2} \approx 2.83$

A *real number* is any number that is either a rational or an irrational number.

The natural numbers are a *subset* of the whole numbers. The whole numbers are a subset of the integers. The integers are a subset of the rational numbers.

17. Tell whether each statement is true or false.
  a. All integers are whole numbers. false
  b. π is an irrational number. true
  c. The real numbers are the set of all decimals. true
  d. A real number is either rational or irrational. true

18. Tell which numbers in the given set are natural numbers, whole numbers, integers, rational numbers, irrational numbers, and real numbers.
  $\left\{-\frac{4}{5}, 99.99, 0, \sqrt{2}, -12, 4\frac{1}{2}, 0.666. . . , 8\right\}$
  natural: 8; whole: 0, 8; integers: 0, -12, 8; rational: $-\frac{4}{5}$, 99.99, 0, -12, $4\frac{1}{2}$, 0.666. . . , 8; irrat: $\sqrt{2}$; real: all

Two numbers represented by points on a number line that are the same distance away from the origin, but on opposite sides of it, are called *opposites* or *additive inverses*.

19. Write the expression in simpler form.
  a. The opposite of 10 -10
  b. The opposite of -3 3
  c. $-\left(-\frac{9}{16}\right)$ $\frac{9}{16}$
  d. -0 0

The *absolute value* of a number is the distance on the number line between the number and 0.

20. Insert one of the symbols $>$, $<$, or $=$ in the blank to make each statement true.
  a. $|-6|$ $\boxed{>}$ $|5|$
  b. $|-39|$ $\boxed{=}$ 39
  c. -9 $\boxed{>}$ $-|-10|$
  d. $\left|-\frac{1}{4}\right|$ $\boxed{<}$ 0.333. . .

Section 1.3

Exponents and Order of Operations

An *exponent* is used to represent repeated multiplication. In the *exponential expression* a^n, a is the base, and n is the exponent.

21. Write each expression using exponents.
  a. $8 \cdot 8 \cdot 8 \cdot 8 \cdot 8$ 8^5
  b. $(2)(2)(2)$ 2^3
  c. $5 \cdot 5 \cdot 5 \cdot 9 \cdot 9$ $5^3 \cdot 9^2$
  d. $a \cdot a \cdot a \cdot a$ a^4
  e. $9 \cdot \pi \cdot r \cdot r$ $9\pi r^2$
  f. $x \cdot x \cdot x \cdot y \cdot y \cdot y \cdot y$ $x^3 y^4$
  g. one hundred squared 100^2
  h. the sixth power of one 1^6

22. Evaluate each expression.
  a. 9^2 81
  b. 2 cubed 8
  c. 2^5 32
  d. 15^1 15

23. CHECKERBOARD Use the formula $A = s^2$ to find the area of a checkerboard that has sides of length 15 inches. 225 in.^{2}

24. STORAGE At a preschool, each child's jacket, sack lunch, and papers are kept in individual cubicles called "cubbyholes" by the kids. The cubbyholes are plywood cubes with an open front and have sides 1.75 feet long. Draw a picture of a cubicle and then find its volume using the formula $V = s^3$. Round the result to the nearest tenth of a cubic foot. 5.4 ft^3

Order of operations
If an expression contains grouping symbols, do all calculations working from the innermost pair to the outermost pair, in the following order:
1. Evaluate all exponential expressions.

25. How many operations does the expression $5 \cdot 4 - 3^2 + 1$ contain, and in what order should they be performed?
  4; power, multiplication, subtraction, addition

26. Evaluate each expression.
  a. $24 - 3 \cdot 6$ 6
  b. $3^2 + 10^3$ 1,009
  c. $6 \cdot 5 - 18 \div 9$ 28
  d. $2(3)(4)(5)$ 120
  e. $800 - 3 \cdot 4^4$ 32
  f. $8\left(\frac{12}{3}\right) - \frac{30}{6} + 2$ 29
  g. $3(5)^2\left(\frac{24}{6}\right)$ 300
  h. $60 \div 20 + 10$ 13

2. Do all multiplications and divisions, working from left to right.

3. Do all additions and subtractions, working from left to right.

If the expression does not contain grouping symbols, begin with step 1.

In a fraction, simplify the numerator and the denominator separately. Then simplify the fraction, whenever possible.

The *arithmetic mean* (or *average*) is a value around which number values are grouped.

$$\text{Mean} = \frac{\text{sum of values}}{\text{number of values}}$$

27. Tell the name of each type of grouping symbol: ( ), [ ], —.
parentheses, brackets, fraction bar

28. Evaluate each expression.

a. $(3 + 4 \cdot 2) - 3^2$ 2

b. $2 + 3[2 \cdot 8 - (3^2 + 2 \cdot 3)]$ 5

c. $(200 - 3^3 + 1) - 64$ 110

d. $2(5 - 4)^2$ 2

e. $\dfrac{8^2 - 10}{2(3)(4) - 2 \cdot 3^2}$ 9

f. $\dfrac{5 + (80 - 8 + 4)}{3 + (6 - 2 \cdot 3)}$ 27

29. COMPARISON SHOPPING Using the prices listed in Illustration 5, find the average (mean) cost of a new Ford Explorer. $21,134

30. WALK-A-THON Use the data in Illustration 6 to find the average (mean) donation to a charity walk-a-thon. $20

ILLUSTRATION 5

Dealer	Price of Explorer
Inland Motors	$19,981
Ford City	$21,456
Central Ford	$20,599
C.J. Smith Ford	$22,500

ILLUSTRATION 6

Donation	Number received
$5	20
$10	65
$20	25
$50	5
$100	10

Section 1.4   Algebraic Expressions

In order to describe numerical relationships, we need to translate the words of a problem into mathematical symbols.

Key words and *key phrases* are used to represent the operations of addition, subtraction, multiplication, and division.

31. Write each phrase as an algebraic expression.

a. 25 more than the height h
$h + 25$

b. 15 less than the cutoff score s
$s - 15$

c. $\frac{1}{2}$ of the time t $\frac{1}{2}t$

d. the product of 6 and x $6x$

32. An advertisement stated that the regular price p of a living room set was slashed $500 during a spring clearance sale. Express the sale price of the furniture, in dollars, using an algebraic expression. $p - 500$

33. The time a man spent standing in line to pay his fees was three times longer than it took to complete his driving test. Express each length of time using a variable. x = time to complete test; $3x$ = time spent standing in line

34. See Illustration 7.

a. If we let *n* represent the length of the nail, write an algebraic expression for the length of the bolt (in cm). $n + 4$

b. If we let *b* represent the length of the bolt, write an algebraic expression for the length of the nail (in cm). $b - 4$

ILLUSTRATION 7

4 cm

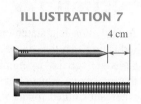

Sometimes you must rely on common sense and insight to find *hidden operations.*

35. a. How many years are in *d* decades? $10d$

b. If you have *x* donuts, how many dozen donuts do you have? $\frac{x}{12}$

c. Five years after a house was constructed, a patio was added. How old, in years, is the patio if the house is *x* years old? $x - 5$

Number · value = total value

36. Complete the table in Illustration 8.

ILLUSTRATION 8

Type of coin	Number	Value in cents	Total value in cents
Nickel	6	5	30
Dime	*d*	10	10*d*

When we replace the variable, or variables, in an algebraic expression with specific numbers and then apply the rules for the order of operations, we are *evaluating* the algebraic expression.

37. Complete the table of values in Illustration 9.

ILLUSTRATION 9

x	$20x - x^3$
0	0
1	19
4	16

ILLUSTRATION 10

input	output
0	0
15	2
60	8

38. Complete the input/output table in Illustration 10 for the algebraic expression $\frac{x}{3} - \frac{x}{5}$.

39. Evaluate each algebraic expression for the given value(s) of the variable(s).

a. $6x$ for $x = 6$ 36

b. $7x^2 - \frac{x}{2}$ for $x = 4$ 110

c. $b^2 - 4ac$ for $b = 10$, $a = 3$, and $c = 5$ 40

d. $2(24 - 2c)^3$ for $c = 9$ 432

e. $3r^2h + 1$ for $r = 7$ and $h = 5$ 736

f. $\frac{x + y}{x - y}$ for $x = 19$ and $y = 17$ 18

40. Use a calculator to find the volume, to the nearest tenth of a cubic inch, of the ice cream waffle cone in Illustration 11 by evaluating the algebraic expression $\frac{\pi r^2 h}{3}$. 17.7 in.³

ILLUSTRATION 11

1.5-inch radius

7.5-inch height

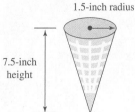

Section 1.5

An *equation* is a statement indicating that two expressions are equal. Any number that makes an equation true when substituted for its variable is said to *satisfy* the equation. Such numbers are called *solutions* or *roots*.

To *solve an equation,* isolate the variable on one side of the equation by undoing the operations performed on it.

If the same number is added to, or subtracted from, both sides of an equation, an equivalent equation results.
If $a = b$, then $a + c = b + c$.
If $a = b$, then $a - c = b - c$.

If both sides of an equation are multiplied, or divided, by the same nonzero number, an *equivalent* equation results.
If $a = b$, then $ca = cb \ (c \neq 0)$.
If $a = b$, then $\dfrac{a}{c} = \dfrac{b}{c} \ (c \neq 0)$.

We can translate a percent problem from words into an equation. A variable is used to stand for the unknown number; *is* can be translated to an $=$ sign; and *of* means multiply.

The percent formula:
   Amount = percent · base

Solving Equations

41. Tell whether the given number is a solution of the equation.

   a. $x - 34 = 50$; $x = 84$ yes   **b.** $5y + 2 = 12$; $y = 3$ no

   c. $\frac{x}{5} = 6$; $x = 30$ yes   **d.** $a^2 - a - 1 = 0$; $a = 2$ no

   e. $5b - 2 = 3b + 3$; $b = 3$ no   **f.** $\dfrac{2}{y+1} = \dfrac{12}{y+1} - 5$; $y = 1$ yes

42. Fill in the blanks to make the statement true: When solving the equation $x + 8 = 10$, we are to find all the values of the ___variable___ that make the equation a ___true___ statement.

43. Solve each equation. Check all solutions.

   a. $x - 9 = 12$ 21   **b.** $y + 15 = 32$ 17

   c. $4 = v - 1$ 5   **d.** $100 = 7 + x$ 93

   e. $2x = 40$ 20   **f.** $120 = 15c$ 8

   g. $x - 11 = 0$ 11   **h.** $p + 3 = 3$ 0

   i. $\dfrac{t}{8} = 12$ 96   **j.** $3 = \dfrac{q}{26}$ 78

   k. $6b = 0$ 0   **l.** $\dfrac{x}{14} = 0$ 0

44. Translate "16 is 5% of an unknown number" into an equation. $16 = 0.05x$

45. COST OF LIVING A retired trucker receives a monthly Social Security check of $764. If she is to receive a 3.5% cost-of-living increase soon, how much larger will her check be? $26.74

46. 4.81 is 2.5% of what number?   192.4

47. FAMILY BUDGET It is recommended that a family pay no more than 30% of its monthly income (after taxes) on housing. If a family has an after-tax income of $1,890 per month and pays $625 in housing costs each month, are they within the recommended range?   no

48. AUTOMOBILE SALES In 1997, 8,272,074 passenger cars were sold in the United States. Use the circle graph in Illustration 12 to determine the number sold by General Motors.   2,688,424

ILLUSTRATION 12

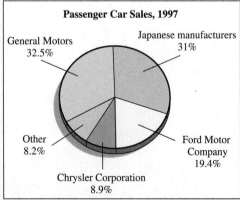

Based on data from the American Automobile Manufacturers Association

49. COLLECTIBLES A collector of football trading cards paid $6 for a 1984 Dan Marino rookie card several years ago. If the card is now worth $100, what is the percent of increase in the card's value? (Round to the nearest percent.)   1,567%

Section 1.6

An *equation* is a mathematical statement that two quantities are equal.

To solve a problem, follow these steps:
1. Analyze the problem.
2. Form an equation.
3. Solve the equation.
4. State the conclusion.
5. Check the result.

Drawing a diagram or creating a table is often helpful in problem solving.

Problem Solving

50. Write an equation that describes the same quantity in two ways: At the end of the holiday season, a bakery had sold 165 cherry cheesecakes. This was 28 cheesecakes less than the projected sales total. Let x = the number of cheesecakes the bakery had hoped to sell. $x - 28 = 165$

51. SOCIAL WORK A human services program assigns each of its social workers a caseload of 75 clients. How many clients are served by this program if it employs 20 social workers?   1,500

52. HISTORIC TOUR A driving tour of three historic cities is an 858-mile round trip. Beginning in Boston, the drive to Philadelphia is 296 miles. From Philadelphia to Washington, DC is another 133 miles. How long would the return trip to Boston be?   429 mi

53. CARDS A standard deck of 54 playing cards contains 2 jokers, 4 aces, and 12 face cards; the remainder are numbered cards. How many numbered cards are there in a standard deck?   36

54. CHROME WHEELS Find the measure of the angle between each of the spokes on the wheel shown in Illustration 13.   60°

ILLUSTRATION 13

CHAPTER 1

Test

The graph in Illustration 1 shows the cost to hire a security guard. Use the graph to answer Problems 1 and 2.

ILLUSTRATION 1

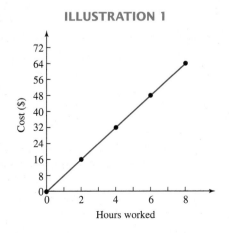

1. What will it cost to hire a security guard for 3 hours? $24

2. If a school was billed $40 for hiring a security guard for a dance, for how long did the guard work?   5 hr

3. Use the formula $f = \frac{a}{5}$ to complete the table.

Area in square miles	Number of fire stations
15	3
100	20
350	70

4. Graph each member of the set on the number line.
$$\left\{-1\tfrac{1}{4}, \sqrt{2}, -3.75, \tfrac{7}{2}, 0.5\right\}$$

5. Tell whether each statement is true or false.
   a. Every integer is a rational number. true
   b. Every rational number is an integer. false
   c. π is an irrational number. true
   d. 0 is a whole number. true

6. Insert the proper symbol, $>$ or $<$, in the blanks to make each statement true.
   a. $-2 \;>\; -3$
   b. $-|-7| \;<\; 8$
   c. $|-4| \;<\; -(-5)$
   d. $\left|-\tfrac{7}{8}\right| \;>\; 0.5$

7. Find the volume of a styrofoam ice chest that is a cube with sides of length 10 inches. 1,000 in.^{3}

8. Rewrite each product using exponents:
   a. $9 \cdot 9 \cdot 9 \cdot 9 \cdot 9$ 9^5
   b. $3 \cdot x \cdot x \cdot z \cdot z \cdot z.$ $3x^2z^3$

9. Evaluate 5^6. 15,625

10. Evaluate $8 + 2 \cdot 3^4$. 170

11. Evaluate $9^2 - 3[45 - 3(6 + 4)]$. 36

12. Evaluate $\dfrac{4^3 - 2 \cdot 13}{4\left(\tfrac{25}{5}\right) - 1^4}$. 2

13. What is a real number?
A real number is any number that can be written as a decimal.

14. Explain the difference between an expression and an equation.
An equation is a mathematical sentence that contains an $=$ sign. An expression does not contain an $=$ sign.

15. SICK DAYS Use the data in Illustration 2 to find the average (mean) number of sick days used by this group of employees this year. 4

ILLUSTRATION 2

Name	Sick days	Name	Sick days
Chung	4	Ryba	0
Cruz	8	Nguyen	5
Damron	3	Tomaka	4
Hammond	2	Young	6

16. Complete the table in Illustration 3.

ILLUSTRATION 3

x	$2x - \dfrac{30}{x}$
5	4
10	17
30	59

17. Evaluate the expression $2lw + w^2$ for $l = 4$ and $w = 8$. 128

18. A rock band recorded x songs for a CD. Technicians had to delete two songs from the album, because of poor sound quality. Express the number of songs on the CD using an algebraic expression.
$x - 2 =$ number of songs on the CD

19. What is the value of q quarters in cents? $25q$

20. Is $x = 3$ a solution of the equation $2x + 3 = 4x - 6$?
no

21. Solve $x + 11 = 24$. 13 **22.** Solve $22t = 110$. 5

23. Solve $5 = x - 25$. 30 **24.** Solve $\tfrac{c}{10} = 55$. 550

25. DOWN PAYMENT To buy a house, a woman was required to make a down payment of \$11,400. What did the house sell for if this was 15% of the purchase price? \$76,000

26. VIDEO SALES Illustration 4 shows the number of videos sold nationwide during the fourth quarter of 1997 and the first quarter of 1998. Find the percent of decrease in the number of videos sold (to the nearest percent). What do you think is the reason for such a dramatic decrease?
37%. There is an increase in the number of videos sold during the holiday season (4th quarter). Then sales fall back to normal levels in the 1st quarter of the next year.

ILLUSTRATION 4

Quarterly Video Sales (in millions of units)	
4th qtr, 1997	1st qtr, 1998
97	61

27. MULTIPLE BIRTHS IN THE UNITED STATES In 1994, about 4,500 women gave birth to three or more babies at one time. This is quadruple (4 times) the number of such births in 1974, 20 years earlier. How many multiple births occurred in the United States in 1974? 1,125

28. GREAT LAKES The Great Lakes have a total surface area of about 94,000 square miles. In square miles, Lake Huron covers 23,000, Lake Michigan 22,000, Lake Erie 10,000, and Lake Ontario 7,000. Find the surface area of Lake Superior. 32,000 mi^2

2

Real Numbers, Equations, and Inequalities

CAMPUS CONNECTION

The Automotive Technology Department

In automotive courses, instructors stress that mechanics need a solid understanding of mathematics to service today's technologically advanced cars. During their training, automotive students will use different types of *formulas* dealing with such things as carburetors, brakes, and electronic ignition systems. One basic formula they will use is $d = rt$, where d is the distance traveled by a car, t is the time traveled, and r is the rate or speed. This chapter will help you build the mathematical foundation necessary to be able to work formulas from many different disciplines, including automotive technology.

In this chapter, we will learn how to add, subtract, multiply, and divide real numbers, and we will use these skills to solve equations and inequalities.

▶ 2.1   Adding and Subtracting Real Numbers

In this section, you will learn about

Adding two real numbers with the same sign ■ Adding two real numbers with different signs ■ Properties of addition ■ Subtracting real numbers ■ Solving equations

Introduction   Recall that all of the points on a number line represent the set of real numbers. Real numbers that are greater than zero are *positive real numbers*. (See Figure 2-1.) Positive numbers can be written with or without a + sign. For example, 2 = +2 and 4.75 = +4.75. Real numbers that are less than zero are *negative real numbers*. They are always written with a − sign. For example, negative 2 = −2 and negative 4.75 = −4.75.

FIGURE 2-1

Negatives   Zero   Positives

$$-5 \quad -4 \quad -3 \quad -2 \quad -1 \quad 0 \quad 1 \quad 2 \quad 3 \quad 4 \quad 5$$

 WARNING! Zero is neither positive nor negative.

We use real numbers to describe many situations. Words such as *gain, above, up, to the right,* and *in the future* indicate positive numbers. Words such as *loss, below, down, to the left,* and *in the past* indicate negative numbers.

In words	*In symbols*	*Meaning*
16 degrees above 0	$+16°$	positive sixteen degrees
5 degrees below 0	$-5°$	negative five degrees
a balance of $590.80	$590.80	positive five hundred ninety dollars and eighty cents
$104.93 overdrawn	$-$104.93	negative one hundred four dollars and ninety-three cents
750 feet above sea level	750	positive seven hundred fifty
$25\frac{1}{2}$ feet below sea level	$-25\frac{1}{2}$	negative twenty-five and one-half

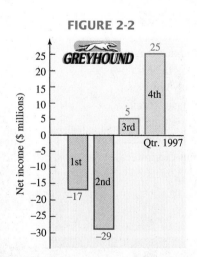

FIGURE 2-2

The bar graph in Figure 2-2 shows the 1997 quarterly profits and losses of Greyhound Bus Lines. In the figure, the first-quarter loss of $17 million and the second-quarter loss of $29 million are represented by the negative numbers −17 and −29. The third-quarter profit of $5 million and a fourth-quarter profit of $25 million are represented by 5 and 25. To find Greyhound's 1997 annual net income, we must add these positive and negative numbers:

Annual income = $-17 + (-29) + 5 + 25$

In this section, we will discuss how to add and subtract positive and negative real numbers and find the annual profit (or loss) of Greyhound.

Adding Two Real Numbers with the Same Sign

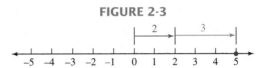

$2 + 3$
both positive

To explain the addition of signed numbers, we can use a number line, as shown in Figure 2-3. To compute $2 + 3$, we begin at the **origin** (the zero point) and draw an arrow 2 units long, pointing to the right. This represents 2. From that point, we draw an arrow 3 units long, also pointing to the right. This represents 3. We end up at 5; therefore, $2 + 3 = 5$.

FIGURE 2-3

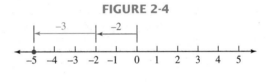

$-2 + (-3)$
both negative

To compute $-2 + (-3)$ on a number line, we begin at the origin and draw an arrow 2 units long, pointing to the left. This represents -2. From there, we draw an arrow 3 units long, also pointing to the left. This represents -3. We end up at -5, as shown in Figure 2-4; therefore, $-2 + (-3) = -5$.

FIGURE 2-4

As a check, think of this problem in terms of money. If you had a debt of \$2 (-2) and incurred another debt of \$3 (-3), you would have a debt of \$5 (-5).

From the first two examples, we observe that both arrows point in the same direction and build on each other. The result has the same sign as the two real numbers that are being added.

$$2 \ + \ 3 \ = \ 5 \qquad \text{and} \qquad -2 \ + \ (-3) \ = \ -5$$

positive + positive = positive   negative + negative = negative

These observations suggest the following rule.

Adding two real numbers with the same sign

To add two real numbers with the same sign, add their absolute values and attach their common sign to the sum.

If both real numbers are positive, the sum is positive. If both real numbers are negative, the sum is negative.

E X A M P L E   1

Adding real numbers with the same sign. Find the sum: $-25 + (-18)$.

Solution   Since both real numbers are negative, the answer will be negative.

$$-25 + (-18) = -43 \quad \text{Add their absolute values, 25 and 18, to get 43. Use their common sign.}$$

SELF CHECK   Find the sum: $-45 + (-12)$.   *Answer:* -57 ∎

Adding Two Real Numbers with Different Signs

$5 + (-2)$
one positive, one negative

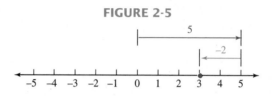

To compute $5 + (-2)$ on a number line, we start at the origin and draw an arrow 5 units long, pointing to the right; this represents 5. From there, we draw an arrow 2 units long, pointing to the left; this represents -2. We end up at 3, as shown in Figure 2-5. Therefore, $5 + (-2) = 3$. In terms of money, if you had \$5 (+5) and lost \$2 (−2), you would have \$3 (+3) left.

FIGURE 2-5

$-5 + 3$
one negative, one positive

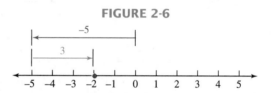

To compute $-5 + 3$ on a number line, we start at the origin and draw an arrow 5 units long, pointing to the left; this represents -5. From there, we draw an arrow 3 units long, pointing to the right; this represents 3. We end up at -2, as shown in Figure 2-6. Therefore, $-5 + 3 = -2$. In terms of money, if you owed a friend \$5 (−5) and paid back \$3 (+3), you would still owe your friend \$2 (−2).

FIGURE 2-6

From the previous examples, we observe that the arrows point in opposite directions and that the longer arrow determines the sign of the result.

$5 + (-2) = 3$	and	$-5 + 3 = -2$
positive + negative = positive		negative + positive = negative

These observations suggest the following rule.

Adding two real numbers with different signs

To add two real numbers with different signs, subtract their absolute values (the smaller from the larger). To this result, attach the sign of the number with the larger absolute value.

EXAMPLE 2

Adding real numbers with different signs. Find each sum: **a.** $-17 + 32$, **b.** $5.4 + (-7.7)$.

Solution **a.** Since 32 has the larger absolute value, the answer will be positive.

$-17 + 32 = 15$ Subtract their absolute values, 17 from 32, to get 15.

b. Since -7.7 has the larger absolute value, the answer will be negative.

$5.4 + (-7.7) = -2.3$ Subtract their absolute values, 5.4 from 7.7, to get 2.3. Use a − sign.

SELF CHECK   Find each sum: **a.** $63 + (-87)$ and
b. $-\frac{1}{5} + \frac{3}{5}$.

Answers: **a.** -24,   **b.** $\frac{2}{5}$ ∎

E X A M P L E   3   **Adding several real numbers.** Find the 1997 annual net income of Greyhound Bus Lines from the data given in the graph in Figure 2-2.

Solution   To find the annual net income, add the 1997 quarterly profits and losses. We use the rules for order of operations and do the additions as they occur from left to right.

$$-17 + (-29) + 5 + 25 = -46 + 5 + 25 \quad \text{Add: } -17 + (-29) = -46.$$
$$= -41 + 25 \quad \text{Add: } -46 + 5 = -41.$$
$$= -16$$

In 1997, Greyhound lost $16 million.

SELF CHECK   Add $-7 + 13 + (-5) + 10$.   *Answer:* 11   ∎

ACCENT ON TECHNOLOGY *The Sign Change Key*

A scientific calculator can add positive and negative numbers.

■ You don't have to do anything special to enter positive numbers. When you press 5, for example, a positive 5 is entered.

■ To enter a negative 17 on a scientific calculator, you must press the $\boxed{+/-}$ key after entering 17. This key is called the *opposite* or *sign change* key.

To find the annual net income of Greyhound Bus Lines, we must find the sum $-17 + (-29) + 5 + 25$. To do so, we enter these numbers and press these keys:

Keystrokes $17 \boxed{+/-} \boxed{+} 29 \boxed{+/-} \boxed{+} 5 \boxed{+} 25 \boxed{=}$ $\boxed{\qquad\qquad -16}$

Using a graphing calculator, we enter a negative value by first pressing the negation key $\boxed{(-)}$. To find the sum, we press these keys:

Keystrokes $\boxed{(-)} 17 \boxed{+} \boxed{(-)} 29 \boxed{+} 5 \boxed{+} 25 \boxed{\text{ENTER}}$

$$\boxed{\begin{array}{r} {}^-17 + {}^-29 + 5 + 25 \\ -16 \end{array}}$$

The sum is -16. Greyhound's 1997 net loss was $16 million.

Properties of Addition

The operation of addition has some special properties. The first property, called the **commutative property,** states that two real numbers can be added in either order to get the same result. For example, when adding the numbers 10 and -25, we see that

$$10 + (-25) = -15 \quad \text{and} \quad -25 + 10 = -15$$

To state the **commutative property of addition** concisely, we use variables.

The commutative property of addition

If a and b represent any real numbers, then

$$a + b = b + a$$

To find the sum of three numbers, we first add two of them and then add the third to that result. For example, we can add $-3 + 7 + 5$ in two ways.

Method 1: Group -3 and 7

$(-3 + 7) + 5 = 4 + 5$ Because of the parentheses, add -3 and 7 first to get 4.

$= 9$ Then add 4 and 5.

Method 2: Group 7 and 5

$-3 + (7 + 5) = -3 + 12$ Because of the parentheses, add 7 and 5 first to get 12.

$= 9$ Then add -3 and 12.

Either way, the sum is 9, which suggests that it doesn't matter how we group or "associate" numbers in addition. This property is called the **associative property of addition.**

The associative property of addition

If a, b, and c represent any real numbers, then

$$(a + b) + c = a + (b + c)$$

E X A M P L E   4

"Jeopardy." A contestant on the game show "Jeopardy" answered the first question correctly to win $100, missed the second question to lose $200, answered the third question correctly to win $300, and answered the fourth question incorrectly to lose $400. Find her net gain or loss after four questions.

Solution   "To win $100" can be represented by 100. "To lose $200" can be represented by -200. "To win $300" can be represented by 300, and "to lose $400" can be represented by -400. Her net gain or loss is the sum of these four numbers. We can find the sum by doing the additions from left to right. An alternate method, which uses the commutative and associative properties of addition, is to add the positives, then add the negatives, and finally add those results.

$100 + (-200) + 300 + (-400) = 100 + 300 + (-200) + (-400)$ Reorder the terms.

$= (100 + 300) + [(-200) + (-400)]$ Group the positives together. Group the negatives together.

$= 400 + (-600)$ Add the positives. Add the negatives.

$= -200$

After four questions, she had a net loss of $200.   ∎

Whenever we add zero to a number, the number remains the same. For example,

$$0 + 8 = 8, \qquad 2.3 + 0 = 2.3, \qquad \text{and} \qquad -16 + 0 = -16$$

These examples suggest the **addition property of zero.**

Addition property of zero

If a represents any real number, then

$$a + 0 = a \qquad \text{and} \qquad 0 + a = a$$

Two numbers that are the same distance away from the origin, but on opposite sides of it, are called **opposites** or **additive inverses**. For example, 10 is the additive inverse of -10, and -10 is the additive inverse of 10. Whenever we add opposites or additive inverses, the result is 0.

$$10 + (-10) = 0, \qquad -\tfrac{4}{5} + \tfrac{4}{5} = 0 \qquad 56.8 + (-56.8) = 0$$

Adding opposites (additive inverses)

If a represents any number, then

$$a + (-a) = 0$$

Subtracting Real Numbers

The subtraction $5 - 2$ can be thought of as taking 2 away from 5. We can use the number line shown in Figure 2-7 to illustrate this. Beginning at the origin, we draw an arrow of length 5 units pointing to the right. From that point, we move back 2 units to the left. The result, 3, is called the **difference.**

FIGURE 2-7

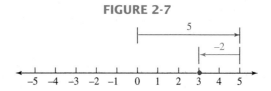

Figure 2-7 looks like the illustration for the addition problem $5 + (-2)$ shown in Figure 2-5. In the problem $5 - 2$, we subtracted 2 from 5. In the problem $5 + (-2)$, we added -2 (which is the opposite of 2) to 5. In each case, the result is 3.

Subtracting 2.   Adding the opposite of 2.

$$5 - 2 = 3 \qquad\qquad 5 + (-2) = 3$$

The results are the same.

This observation suggests the following rule.

Subtracting real numbers

If a and b represent any real numbers, then

$$a - b = a + (-b)$$

This rule indicates that *subtraction is the same as adding the opposite of the number to be subtracted.* We won't need this rule for every subtraction problem. For example, $5 - 2$ is obviously 3. However, for more complicated problems such as $-8 - (-3)$, where the result is not obvious, the subtraction rule will be helpful.

$$-8 - (-3) = -8 + 3 \quad \text{To subtract } -3, \text{ add the opposite of } -3, \text{ which is } 3.$$
$$= -5 \qquad \text{Do the addition.}$$

EXAMPLE 5

Adding the opposite. Find **a.** $-13 - 18$, **b.** $-45 - (-27)$, and **c.** $\frac{1}{4} - \left(-\frac{1}{8}\right)$.

Solution

a. To subtract 18 from -13, we use the subtraction rule.

$$-13 - 18 = -13 + (-18)$$

To subtract 18, add the opposite of 18, which is -18.

$$= -31$$

Add their absolute values, 13 and 18, to get 31. Keep their common sign.

b. To subtract -27 from -45, we use the subtraction rule.

$$-45 - (-27) = -45 + 27$$

To subtract -27, add the opposite of -27, which is 27.

$$= -18$$

Subtract their absolute values, 27 from 45, to get 18. Use the sign of the number with the greater absolute value, which is -45.

c. The lowest common denominator (LCD) for the fractions is 8.

$$\frac{1}{4} - \left(-\frac{1}{8}\right) = \frac{2}{8} - \left(-\frac{1}{8}\right)$$

Express $\frac{1}{4}$ in terms of eighths: $\frac{1}{4} = \frac{2}{8}$.

$$= \frac{2}{8} + \frac{1}{8}$$

Add the opposite of $-\frac{1}{8}$, which is $\frac{1}{8}$.

$$= \frac{3}{8}$$

Add the numerators: $2 + 1 = 3$. Write the sum over the common denominator 8.

SELF CHECK

Find: **a.** $-32 - 25$, **b.** $1.7 - (-1.2)$, and **c.** $-\frac{1}{2} - \frac{1}{8}$.

Answers: **a.** -57, **b.** 2.9, **c.** $-\frac{5}{8}$ ∎

ACCENT ON TECHNOLOGY *U.S. Temperature Extremes*

The record high temperature in the United States was 134°F in Death Valley, California, on July 10, 1913. The record low was -80°F at Prospect Creek, Alaska, on January 23, 1971. See Figure 2-8. To find the difference between these two temperatures, we subtract:

$$134 - (-80)$$

We can subtract positive and negative real numbers using a scientific calculator. To find $134 - (-80)$, we enter these numbers and press these keys:

FIGURE 2-8

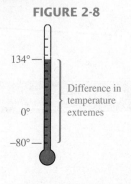

Keystrokes   134 $\boxed{-}$ 80 $\boxed{+/-}$ $\boxed{=}$ $\boxed{ 214}$

If we use a graphing calculator, we enter these numbers and press these keys:

Keystrokes   134 $\boxed{-}$ $\boxed{(-)}$ 80 $\boxed{\text{ENTER}}$ $\boxed{\begin{array}{l}134--80 \\ 214\end{array}}$

The difference in the record high and low temperatures is 214°F.

EXAMPLE 6

Constructing a table of values. Evaluate the expression $5 - x$ for $x = -10$ and $x = 7$. Show the results in an input/output table of values.

Solution   To evaluate $5 - x$ for $x = -10$ and $x = 7$, we substitute these numbers for x and simplify.

x	$5 - x$
-10	15
7	-2

Evaluate for x = −10:
$$5 - x = 5 - (-10)$$
$$= 15$$

Evaluate for x = 7:
$$5 - x = 5 - 7$$
$$= -2$$

SELF CHECK   Complete the following input/output table of values.

x	$7 + x$
-8	
-4	

Answers: $-1, 3$ ■

Solving Equations

The following examples will show that solutions of equations can be negative numbers.

EXAMPLE 7

Solving equations. Solve $x + 5 = -13$ and check the result.

Solution   To isolate x on the left-hand side of the equation, we use the subtraction property of equality. To undo the addition of 5, we subtract 5 from both sides.

$$x + 5 = -13$$
$$x + 5 - 5 = -13 - 5 \quad \text{Subtract 5 from both sides.}$$
$$x = -18 \quad \text{Do the subtractions: } 5 - 5 = 0 \text{ and } -13 - 5 = -18.$$

We check by substituting -18 for x in the original equation.

$$x + 5 = -13 \quad \text{The original equation.}$$
$$-18 + 5 \stackrel{?}{=} -13 \quad \text{Substitute } -18 \text{ for } x.$$
$$-13 = -13 \quad \text{Do the addition: } -18 + 5 = -13.$$

Since the result is true, -18 is a solution.

SELF CHECK   Solve $-22 = 7 + y$ and check the result.   *Answer:* -29 ■

EXAMPLE 8

Solving equations. Solve $-23 = y - 14$.

Solution   To isolate y on the right-hand side of the equation, we use the addition property of equality. To undo the subtraction of 14, we add 14 to both sides.

$$-23 = y - 14$$
$$-23 + 14 = y - 14 + 14 \quad \text{Add 14 to both sides.}$$
$$-9 = y \quad \text{Do the additions: } -23 + 14 = -9 \text{ and } -14 + 14 = 0.$$
$$y = -9 \quad \text{If } -9 = y, \text{ then } y = -9.$$

Check the result.

SELF CHECK   Solve $-43 = -7 + p$ and check the result.   *Answer:* -36 ■

Section 2.1

VOCABULARY

In Exercises 1–6, fill in the blanks to make the statements true.

1. Real numbers that are greater than zero are called ____positive____ real numbers.

2. Real numbers that are less than zero are called ____negative____ real numbers.

3. The only real number that is neither positive nor negative is ____zero____.

4. The answer to a ____subtraction____ problem is called a difference.

5. The ____commutative____ property of addition states that two numbers can be added in either order to get the same result.

6. The property that allows us to group numbers in addition any way we want is called the ____associative____ property of addition.

CONCEPTS

In Exercises 7–10, use the number line in Illustration 1 to find each sum.

7. $2 + 3$ 5
8. $-3 + (-2)$ -5
9. $4 + (-3)$ 1
10. $-5 + 3$ -2

ILLUSTRATION 1

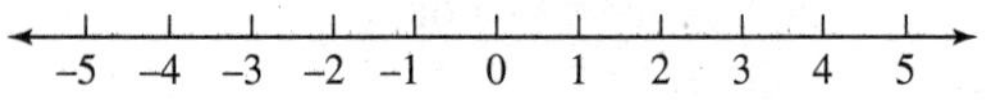

$$\begin{array}{ccccccccccc} & & & & & & & & & & \\ -5 & -4 & -3 & -2 & -1 & 0 & 1 & 2 & 3 & 4 & 5 \end{array}$$

In Exercises 11–14, fill in the blanks to make the statements true.

11. To add two real numbers with the ____same____ sign, add their ____absolute____ values and attach their common sign to the sum.

12. To add two real numbers with different signs, ____subtract____ their absolute values, the ____smaller____ from the ____larger____, and attach the sign of the number with the ____larger____ absolute value.

13. To subtract b from a, add the ____opposite____ of b to a.

14. The opposite of 7 is -7. The opposite of -15 is 15.

15. Find each sum.
   a. $5 + (-5)$ 0
   b. $-2.2 + 2.2$ 0
   c. $-\frac{3}{4} + \frac{3}{4}$ 0
   d. $19 + (-19)$ 0

16. a. Use the variables m and n to state the commutative property of addition. $m + n = n + m$
   b. Use the variables r, s, and t to state the associative property of addition. $r + (s + t) = (r + s) + t$

NOTATION

In Exercises 17–20, complete each solution.

17. Find $(-13 + 6) + 4$.
$$(-13 + 6) + 4 = \boxed{-13} + (6 + 4)$$
$$= -13 + \boxed{10}$$
$$= -3$$

18. Find $-9 + (9 + 43)$.
$$-9 + (9 + 43) = \left(\boxed{-9} + 9\right) + 43$$
$$= \boxed{0} + 43$$
$$= 43$$

19. Solve $x + 23 = -12$.
$$x + 23 = -12$$
$$x + 23 - \boxed{23} = -12 - \boxed{23}$$
$$x = -35$$

20. Solve $-17 = b - 9$.
$$-17 = b - 9$$
$$-17 + \boxed{9} = b - 9 + \boxed{9}$$
$$-8 = b$$
$$b = -8$$

PRACTICE

In Exercises 21–38, find each sum.

21. $6 + (-8)$ -2
22. $4 + (-3)$ 1
23. $-6 + 8$ 2
24. $-21 + (-12)$ -33
25. $-65 + (-12)$ -77
26. $75 + (-13)$ 62
27. $-10.5 + 2.3$ -8.2
28. $-2.1 + 0.4$ -1.7
29. $-\dfrac{9}{16} + \dfrac{7}{16}$ $-\frac{1}{8}$
30. $-\dfrac{3}{4} + \dfrac{1}{4}$ $-\frac{1}{2}$
31. $-\dfrac{1}{4} + \dfrac{2}{3}$ $\frac{5}{12}$
32. $\dfrac{3}{16} + \left(-\dfrac{1}{2}\right)$ $-\frac{5}{16}$
33. $8 + (-5) + 13$ 16
34. $17 + (-12) + (-23)$ -18
35. $21 + (-27) + (-9)$ -15
36. $-32 + 12 + 17$ -3
37. $-27 + (-3) + (-13) + 22$ -21
38. $53 + (-27) + (-32) + (-7)$ -13

In Exercises 39–42, use a calculator to find each sum.

39. $3,718 + (-5,237)$ $-1,519$

40. $-5,235 + (-17,235)$ $-22,470$

41. $-237.37 + (-315.07) + (-27.4)$ -579.84

42. $-587.77 + (-1,732.13) + 687.39$ $-1,632.51$

In Exercises 43–58, find each difference.

43. $8 - (-3)$ 11 **44.** $17 - (-21)$ 38

45. $-12 - 9$ -21 **46.** $-25 - 17$ -42

47. $-19 - (-17)$ -2 **48.** $-30 - (-11)$ -19

49. $-1.5 - 0.8$ -2.3 **50.** $-1.5 - (-0.8)$ -0.7

51. $-25 - (-25)$ 0 **52.** $13 - (-13)$ 26

53. $0 - 4$ -4 **54.** $0 - (-3)$ 3

55. $-\dfrac{1}{8} - \dfrac{3}{8}$ $-\frac{1}{2}$ **56.** $-\dfrac{3}{4} - \dfrac{1}{4}$ -1

57. $-\dfrac{9}{16} - \left(-\dfrac{1}{4}\right)$ $-\frac{5}{16}$ **58.** $-\dfrac{1}{2} - \left(-\dfrac{1}{4}\right)$ $-\frac{1}{4}$

In Exercises 59–62, use a calculator to find each difference.

59. $8,713 - (-3,753)$ $12,466$

60. $-2,727 - 1,208$ $-3,935$

61. $-27,357.875 - 17,213.376$ $-44,571.251$

62. $-45,307.039 - (-27,592.47)$ $-17,714.569$

In Exercises 63–66, complete each input/output table.

63.

x	$x + 4$
-8	-4
-4	0
20	24

64.

x	$x - 5$
4	-1
0	-5
-8	-13

65.

x	$8 + x$
-8	0
2	10
-13	-5

66.

x	$-5 - x$
3	-8
-5	0
-15	10

In Exercises 67–70, tell whether the given number is a solution of the equation.

67. $57 + x = 12$; -45 yes

68. $p + 37 = 65$; -26 no

69. $-23 + t = -12$; -11 no

70. $-51 = y + 6$; -57 yes

In Exercises 71–78, solve each equation.

71. $x + 7 = -12$ -19 **72.** $5 + x = -11$ -16

73. $20 = -31 + x$ 51 **74.** $-5 = -12 + x$ 7

75. $x - 9 = -23$ -14 **76.** $-9 = y - 5$ -4

77. $a - 7 = -3$ 4 **78.** $-5 = b - 12$ 7

APPLICATIONS

In Exercises 79–90, solve each problem.

79. MILITARY SCIENCE During a battle, an army retreated 1,500 meters, regrouped, and advanced 2,400 meters. The next day, it advanced another 1,250 meters. Find the army's net gain. $2,150$ m

80. MEDICAL QUESTIONNAIRE Determine the risk of contracting heart disease for the woman whose responses are shown in Illustration 2. 4%

ILLUSTRATION 2

Age		Total Cholesterol	
Age	Points	Reading	Points
35	-4	280	3

Cholesterol		Blood Pressure	
HDL	Points	Systolic/Diastolic	Points
62	-3	124/100	3

Diabetic		Smoker	
	Points		Points
Yes	4	Yes	2

10-Year Heart Disease Risk			
Total Points	**Risk**	Total Points	**Risk**
-2 or less	1%	5	4%
-1 to 1	2%	6	6%
2 to 3	3%	7	6%
4	4%	8	7%

Source: National Heart, Lung, and Blood Institute

81. GOLF Illustration 3 shows the top four finishers from the 1997 Masters Golf Tournament. The scores for each round are related to *par*, the standard number of strokes deemed necessary to complete the course. A score of -2, for example, indicates that the golfer used 2 strokes less than par to complete the course. A score of $+5$ indicates the golfer used five strokes more than par.

a. Determine the tournament total for each golfer.

b. Tiger Woods won by the largest margin in the history of the Masters. What was the margin?   12 strokes

ILLUSTRATION 3

Leaderboard

	Round				
	1	2	3	4	Total
Tiger Woods	-2	-6	-7	-3	-18
Tom Kite	$+5$	-3	-6	-2	-6
Tommy Tolles	0	0	0	-5	-5
Tom Watson	$+3$	-4	-3	0	-4

82. CREDIT CARD STATEMENT
  a. What amounts in the monthly credit card state-ment shown in Illustration 4 could be represented by negative numbers? 3,660.66, 1,408.78
  b. What is the new balance? 1,242.86

ILLUSTRATION 4

Previous Balance	New Purchases, Fees, Advances & Debts	Payments & Credits	New Balance
3,660.66	**1,408.78**	**3,826.58**	

04/21/99 Billing Date	05/16/99 Date Payment Due	9,100 Credit Line

Periodic rates may vary.
See reverse for explanation and important information.
Please allow sufficient time for mail to reach us.

83. THE OLYMPICS The ancient Greek Olympian Games, which eventually evolved into the modern Olympic Games, were first held in 776 B.C. How many years after this did the 1996 Olympic Games in Atlanta, Georgia, take place? 2,772

84. SUBMARINE A submarine was cruising at a depth of 1,250 feet. The captain gave the order to climb 550 feet. Relative to sea level, find the new depth of the sub. −700 ft

85. TEMPERATURE RECORDS Find the difference between the record high temperature of 108°F set in 1926 and the record low of −52°F set in 1979 for New York State. 160°F

86. LIE DETECTOR TEST A burglar scored −18 on a lie detector test, a score that indicates deception. However, on a second test, he scored +3, a score that is inconclusive. Find the difference in the scores. 21

87. LAND ELEVATIONS The elevation of Death Valley, California, is 282 feet below sea level. The elevation of the Dead Sea in Israel is 1,312 feet below sea level. Find the difference in their elevations. 1,030 ft

88. STOCK EXCHANGE Many newspapers publish daily summaries of the stock market's activity. (See Illustration 5.) The last entry on the line for October 5 indicates that one share of Walt Disney Co. stock lost $\$\frac{3}{16}$ in value that day. How much did the value of a share of Disney stock rise or fall over the five-day period shown? lost $\$1\frac{1}{2}$

ILLUSTRATION 5

Oct. 5	42³/₄	23⁷/₈	Disney	.21	0.8	26	42172	25¹/₁₆	−³/₁₆
Oct. 6	42³/₄	23⁷/₈	Disney	.21	0.8	26	46600	25³/₈	+⁵/₁₆
Oct. 7	42³/₄	23⁷/₈	Disney	.21	0.8	26	46404	24¹⁵/₁₆	−³/₈
Oct. 8	42³/₄	23⁷/₈	Disney	.21	0.9	24	97098	23¹/₂	−1⁷/₁₆
Oct 9	42³/₄	23⁷/₈	Disney	.21			61333	23¹¹/₁₆	+³/₁₆

Based on data from *Los Angeles Times*

89. VOTER INFORMATION What will be the effect on state government if the ballot initiative shown in Il-lustration 6 passes? a gain of $2.2 million

ILLUSTRATION 6

212 Campaign Spending Limits	YES ☐ NO ☐

Limits contributions to $200 in state campaigns. Fiscal impact: Costs of $4.5 million for implemen–tation and enforcement. Increases state revenue by $6.7 million by eliminating tax deductions for lobbying.

90. MOVIE LOSSES In 1993, the cost to Columbia Stu-dios to produce, promote, and distribute the movie *Last Action Hero,* starring Arnold Schwarzenegger, was approximately $124 million. It is estimated that the movie earned only $44 million worldwide. What dollar loss did the studio suffer on this film? $80 million

In Exercises 91–94, use a calculator to help solve each problem.

91. SAHARA DESERT From 1980 to 1990, a satellite was used to trace the expansion and contraction of the southern boundary of the Sahara Desert in Africa (see Illustration 7). If movement southward is repre-sented with a negative number and movement north-ward with a positive number, use the data in the table to determine the net movement of the Sahara Desert boundary over the 10-year period. southward, 132 km

ILLUSTRATION 7

Years	Distance/Direction
1980–1984	240 km/South
1984–1985	110 km/North
1985–1986	30 km/North
1986–1987	55 km/South
1987–1988	100 km/North
1988–1990	77 km/South

Based on data from A. Dolgoff, *Physical Ge-ology* (D. C. Heath, 1996), p. 496

92. BANKING On February 1, Marta had $1,704.29 in a checking account. During the month, she made deposits of $713.87 and $1,245.57, wrote checks for $813.45, $937.49, and $1,532.79, and had a total of $500 in ATM withdrawals. Find her checking account balance at the end of the month.
−$120 (overdrawn $120)

93. CARD GAME In the second hand of a card game, Gonzalo was the winner and earned 50 points. Matt and Hydecki had to deduct the value of each of the cards left in their hands from their running point total. Use the information in Illustration 8 to update the score sheet. (Face cards are counted as 10 points and aces as 1 point.)

ILLUSTRATION 8

Matt   Hydecki

Running point total	Hand 1	Hand 2
Matt	+50	+29
Gonzalo	−15	+35
Hydecki	−2	−23

94. PROFITS AND LOSSES Odwalla®, a juice maker, reported a large first-quarter loss in 1997 because of a juice recall. Approximate the net loss during the nine quarters shown in Illustration 9.
≈$3.3 million

ILLUSTRATION 9

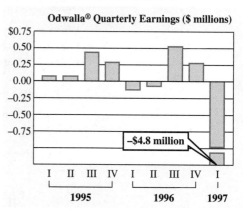

Based on company reports

WRITING

95. Explain why the sum of two positive numbers is always positive and why the sum of two negative numbers is always negative.

96. Is subtracting 2 from 10 the same as subtracting 10 from 2? Explain.

97. Explain why we need to subtract when we add two real numbers with different signs.

98. Explain why we can subtract by adding the opposite.

REVIEW

In Exercises 99–102, write each expression using exponents.

99. $c \cdot c \cdot c \cdot c$ c^4

100. $3 \cdot d \cdot d \cdot d$ $3d^3$

101. $a \cdot a \cdot b \cdot b \cdot b$ $a^2 b^3$

102. $3 \cdot x \cdot x \cdot x \cdot y \cdot y$
$3x^3 y^2$

▶ **2.2**

Multiplying and Dividing Real Numbers

In this section, you will learn about

  Multiplication of real numbers ■ Properties of multiplication
  ■ Division of real numbers ■ Properties of division ■ Division and
  zero ■ Solving equations

Introduction   In this course, we will often need to multiply or divide positive and negative numbers. For example,

  If the temperature drops 4° per hour for 5 hours, we can find the total drop in temperature by doing the multiplication $5(-4)$.

  If the temperature uniformly drops 30° over a 5-hour period, we can find the number of degrees it drops each hour by doing the division $\frac{-30}{5}$.

In this section, we will show how to do such multiplications and divisions.

Multiplication of Real Numbers

When multiplying two nonzero real numbers, the first factor can be positive or negative, and the second factor can be positive or negative. This means there are four possible combinations to consider.

Positive · positive

Positive · negative

Negative · positive

Negative · negative

$4(3)$
like signs, both positive

We begin by considering the product 4(3). Since both factors are positive, they have *like signs*. Because multiplication represents repeated addition, 4(3) equals the sum of four 3's.

$4(3) = 3 + 3 + 3 + 3$ Multiplication is repeated addition. Write 3 four times.

$\quad\ = 12$ The result is +12.

This example suggests that *the product of two positive numbers is positive.*

As a check, let's think of this problem in terms of money. If someone gave you $3 four times, you would have $12.

Multiplying two positive real numbers

To multiply two positive real numbers, multiply their absolute values. The product is positive.

$4(-3)$
unlike signs
one positive, one negative

Next, we consider 4(−3). The signs of these factors are *unlike*. According to the definition of multiplication, 4(−3) means that we are to add −3 four times.

$4(-3) = (-3) + (-3) + (-3) + (-3)$ Multiplication is repeated addition. Write −3 four times.

$\quad = \quad (-6) + (-3) + (-3)$ Add: −3 + (−3) = −6.

$\quad = \quad (-9) + (-3)$ Add: −6 + (−3) = −9.

$\quad = \quad -12$ Add: −9 + (−3) = −12. The result is negative.

This example suggests that *the product of a positive number and a negative number is negative.*

In terms of money, if you lost $3 four times in the lottery, you would lose a total of $12, which is denoted as −$12.

$-3(4)$
unlike signs
one negative, one positive

Next, consider −3(4). The signs of these factors are *unlike*. Because changing the order when multiplying does not change the result, −3(4) = 4(−3). Since 4(−3) = −12, we know that −3(4) = −12. This suggests that *the product of a negative number and a positive number is negative.*

Multiplying two real numbers with unlike signs

To multiply two real numbers with unlike signs, multiply their absolute values. Then make the product negative.

EXAMPLE 1

Multiplying two numbers with unlike signs. Multiply: **a.** $8(-12)$ and **b.** $(-15)(25)$.

Solution **a.** $8(-12) = -96$ Multiply the absolute values, 8 and 12, to get 96. Since the numbers have unlike signs, make the answer negative.

b. $(-15)(25) = -375$ Multiply the absolute values, 15 and 25, to get 375. Make the answer negative.

SELF CHECK   Multiply **a.** $20(-30)$ and **b.** $(-0.4)(2)$ *Answers:* **a.** -600, **b.** -0.8

$$-4(-3)$$
like signs, both negative

Finally, consider the product $(-4)(-3)$. To develop a rule for multiplying two negative numbers, we examine the following pattern, in which we multiply -4 and a series of factors that decrease by 1. After finding the first four products, we graph them on a number line, as shown in Figure 2-9.

This factor decreases
by 1 as you read down
the column.
↓

Look for a
pattern here.
↓

$$-4(3) = -12$$
$$-4(2) = -8$$
$$-4(1) = -4$$
$$-4(0) = 0$$
$$-4(-1) = ?$$
$$-4(-2) = ?$$
$$-4(-3) = ?$$

FIGURE 2-9

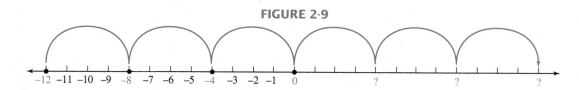

From the pattern, we see that the product increases by 4 each time. Thus,

$$-4(-1) = 4, \qquad -4(-2) = 8, \qquad \text{and} \qquad -4(-3) = 12$$

These results suggest that *the product of two negative numbers is positive.*

Multiplying two negative real numbers

To multiply two negative real numbers, multiply their absolute values. The product is positive.

EXAMPLE 2	**Multiplying two negative numbers.** Multiply **a.** $(-5)(-6)$, and **b.** $\left(-\frac{1}{2}\right)\left(-\frac{5}{8}\right)$.

Solution   **a.** $(-5)(-6) = 30$ Multiply the absolute values, 5 and 6, to get 30. Since both numbers are negative, the answer is positive.

b. $\left(-\frac{1}{2}\right)\left(-\frac{5}{8}\right) = \frac{5}{16}$ Multiply the absolute values, $\frac{1}{2}$ and $\frac{5}{8}$, to get $\frac{5}{16}$. The product is positive.

SELF CHECK   Multiply: **a.** $(-15)(-8)$ and   **b.** $-\frac{1}{4}\left(-\frac{1}{3}\right)$ *Answers:*   **a.** 120,   **b.** $\frac{1}{12}$ ■

We now summarize the rules for multiplying two real numbers.

Multiplying two real numbers

To multiply two real numbers, multiply their absolute values.

1. The product of two numbers with *like signs* is positive.
2. The product of two numbers with *unlike signs* is negative.

ACCENT ON TECHNOLOGY   *Bank Promotion*

To attract business, a bank gave a clock radio to each customer who opened a checking account. The radios cost the bank $12.75 each, and 230 new accounts were opened. Each of the 230 radios was given away at a cost of $12.75, which can be expressed as -12.75. To find how much money the promotion cost the bank, we need to find the product of 230 and -12.75.

We can multiply positive and negative numbers with a scientific calculator. To find the product $(230)(-12.75)$, we enter these numbers and press these keys:

Keystrokes   230 ☒ 12.75 ☒ ☒   ☒ -2932.5 ☒

Using a graphing calculator, we enter the following sequence:

Keystrokes   230 ☒ ☒ 12.75 ☒ ENTER ☒

```
230*⁻12.75
            -2932.5
```

The promotion cost the bank $2,932.50.

Properties of Multiplication

A special property of multiplication is that two real numbers can be multiplied in either order to get the same result. For example, when multiplying -6 and 5, we see that

$$-6(5) = -30 \quad \text{and} \quad 5(-6) = -30$$

This property is called the **commutative property of multiplication.**

The commutative property of multiplication

If a and b represent any real numbers, then

$$ab = ba$$

To find the product of three numbers, we multiply two of them and then multiply the third number by that result. For example, we can multiply $-3 \cdot 7 \cdot 5$ in two ways.

Method 1: Group -3 and 7

$(-3 \cdot 7)5 = (-21)5$ Because of the parentheses, multiply -3 and 7 first.

$\qquad\qquad = -105$ Then multiply -21 and 5.

Method 2: Group 7 and 5

$-3(7 \cdot 5) = -3(35)$ Because of the parentheses, multiply 7 and 5 first.

$\qquad\qquad = -105$ Then multiply -3 and 35.

Either way, the product is -105, which suggests that it doesn't matter how we group or "associate" numbers in multiplication. This property is called the **associative property of multiplication.**

The associative property of multiplication

If a, b, and c represent any real numbers, then

$$(ab)c = a(bc)$$

EXAMPLE 3

Multiplying more than two numbers. Multiply   **a.** $-5(-37)(2)$ and   **b.** $2(-3)(-2)(-3)$

Solution

Using the commutative and associative properties of multiplication, we can reorder and regroup the factors to simplify the computations.

a. $-5(-37)(2) = -10(-37)$ Think of the problem as $-5(2)(-37)$ and then multiply -5 and 2.

$\qquad\qquad\quad = 370$

b. $2(-3)(-2)(-3) = -6(6)$ Multiply the first two factors and multiply the last two factors.

$\qquad\qquad\qquad = -36$

SELF CHECK

Multiply: **a.** $-25(-3)(-4)$ and
b. $-1(-2)(-3)(-3)$.   *Answers:*   **a.** -300,   **b.** 18 ■

Whenever we multiply a number and 0, the product is 0. For example,

$$0 \cdot 8 = 0, \qquad 6.5(0) = 0, \qquad \text{and} \qquad 0(-12) = 0$$

We also see that whenever we multiply a number by 1, the number remains the same. For example,

$$6 \cdot 1 = 6, \qquad 4.53(1) = 4.53, \qquad \text{and} \qquad 1(-9) = -9$$

These examples suggest the **multiplication properties of 0 and 1.**

Multiplication properties of 0 and 1

If a represents any real number, then

$$a \cdot 0 = 0 \quad \text{and} \quad 0 \cdot a = 0$$
$$a \cdot 1 = a \quad \text{and} \quad 1 \cdot a = a$$

Division of Real Numbers

Every division fact containing three numbers can be written as an equivalent multiplication fact containing the same three numbers. For example,

$$\frac{15}{5} = 3 \quad \text{because} \quad 5(3) = 15$$

We will use this relationship between multiplication and division to develop the rules for dividing signed numbers. There are four cases to consider.

$$\frac{15}{5}$$

like signs, both positive

From the previous example, $\frac{15}{5} = 3$, we see that *the quotient of two positive numbers is positive.*

$$\frac{-15}{-5}$$

like signs, both negative

To determine the quotient of two negative numbers, we consider the division $\frac{-15}{-5} = ?$. We can do the division by examining its related multiplication fact: $-5(?) = -15$. To find the integer that should replace the question mark, we use the rules for multiplying signed numbers discussed earlier in this section.

Multiplication fact

$$-5(?) = -15$$

This must be *positive* 3 if the product is to be *negative* 15.

Division fact

$$\frac{-15}{-5} = 3$$

So the quotient is *positive* 3.

From this example, we see that *the quotient of two negative numbers is positive.*

$$\frac{15}{-5}$$

unlike signs
one positive, one negative

To determine the quotient of a positive number and a negative number, we consider $\frac{15}{-5} = ?$ and its equivalent multiplication fact $-5(?) = 15$.

Multiplication fact

$$-5(?) = 15$$

This must be *negative* 3 if the product is to be *positive* 15.

Division fact

$$\frac{15}{-5} = -3$$

So the quotient is *negative* 3.

From this example, we see that *the quotient of a positive number and a negative number is negative.*

$$\frac{-15}{5}$$

unlike signs
one negative, one positive

To determine the quotient of a negative number and a positive number, we consider $\frac{-15}{5} = ?$ and its equivalent multiplication fact $5(?) = -15$.

Multiplication fact	*Division fact*
$5(?) = -15$	$\dfrac{-15}{5} = -3$
This must be *negative* 3 if the product is to be *negative* 15.	So the quotient is *negative* 3.

From this example, we see that *the quotient of a negative number and a positive number is negative.*

We can now summarize the results from the previous discussion. Note that the rules for division are similar to those for multiplication.

Dividing two real numbers

To divide two real numbers, divide their absolute values.

1. The quotient of two numbers with *like signs* is positive.
2. The quotient of two numbers with *unlike signs* is negative.

EXAMPLE 4

Dividing two real numbers. Find each quotient: **a.** $\dfrac{66}{11}$, **b.** $\dfrac{-81}{-9}$, **c.** $\dfrac{-45}{9}$, and **d.** $\dfrac{28}{-7}$.

Solution   To divide numbers with like signs, we find the quotient of their absolute values and make the quotient positive.

a. $\dfrac{66}{11} = 6$ Dividing the absolute values, 66 by 11, we get 6. The answer is positive.

b. $\dfrac{-81}{-9} = 9$ Dividing the absolute values, 81 by 9, we get 9. The answer is positive.

To divide numbers with unlike signs, we find the quotient of their absolute values and make the quotient negative.

c. $\dfrac{-45}{9} = -5$ Dividing the absolute values, 45 by 9, we get 5. The answer is negative.

d. $\dfrac{28}{-7} = -4$ Dividing the absolute values, 28 by 7, we get 4. The answer is negative.

SELF CHECK   Find each quotient: **a.** $\dfrac{48}{12}$, **b.** $\dfrac{-63}{-9}$, **c.** $\dfrac{40}{-8}$, and **d.** $\dfrac{-49}{7}$.

Answers: **a.** 4, **b.** 7, **c.** −5, **d.** −7

Properties of Division

The examples

$$\frac{12}{1} = 12, \qquad \frac{-80}{1} = -80 \qquad \text{and} \qquad \frac{7.75}{1} = 7.75$$

illustrate that any number divided by 1 *is the number itself.* The examples

$$\frac{35}{35} = 1, \qquad \frac{-4}{-4} = 1, \qquad \text{and} \qquad \frac{0.9}{0.9} = 1$$

illustrate that *any number (except 0) divided by itself is 1.*

Division properties

If a represents any real number, then

$$\frac{a}{1} = a \qquad \text{and} \qquad \frac{a}{a} = 1 \quad (a \neq 0)$$

Division and Zero

We will now consider two types of division that involve zero. In the first case, we will examine division *of* zero; in the second case, division *by* zero.

To help explain the concept of division of zero, we consider the division $\frac{0}{5} = ?$ and its equivalent multiplication fact $5(?) = 0$.

Multiplication fact	*Division fact*
$5(?) = 0$	$\dfrac{0}{5} = 0$
↑	↑
This must be 0 if the product is to be 0.	So the quotient is 0.

This example suggests that the *quotient of zero divided by any nonzero number is zero.*

To illustrate that division by zero is not permitted, we consider the division $\frac{5}{0} = ?$ and its equivalent multiplication fact $0(?) = 5$.

Multiplication fact	*Division fact*
$0(?) = 5$	$\dfrac{5}{0} = \text{undefined}$
↑	↑
There is no number that gives 5 when multiplied by 0.	There is no quotient.

This example illustrates that the *quotient of any nonzero number divided by zero is undefined.*

EXAMPLE 5   **Division involving zero.** Find each quotient, if possible: **a.** $\dfrac{0}{13}$ and

b. $\dfrac{-13}{0}$.

Solution   **a.** $\frac{0}{13} = 0$ Because $13(0) = 0$.

b. Since $\frac{-13}{0}$ involves division by zero, the division is undefined.

SELF CHECK   Find each quotient, if possible: **a.** $\frac{4}{0}$ and   *Answers:* **a.** undefined, **b.** 0
b. $\frac{0}{17}$. ■

ACCENT ON TECHNOLOGY *Depreciation of a House*

During a period of 17.5 years, the value of a $124,930 house fell at a uniform rate to $97,105. To find how much the house depreciated per year, we must first find the change in its value by subtracting $124,930 from $97,105. To calculate this difference, we enter these numbers and press these keys on a scientific calculator:

Keystrokes   97105 $\boxed{-}$ 124930 $\boxed{=}$ $\boxed{-27825}$

-27825 represents a drop in value of $27,825. Since this depreciation occurred in 17.5 years, we divide $-27,825$ by 17.5 to find the amount of depreciation per year. With $-27,825$ already displayed, we only need to enter these numbers and press these keys:

Keystrokes $\boxed{\div}$ 17.5 $\boxed{=}$ $\boxed{-1590}$

If we use a graphing calculator to compute the amount of depreciation per year, we enter these numbers and press these keys:

Keystrokes   97105 $\boxed{-}$ 124930 $\boxed{\text{ENTER}}$ $\boxed{\div}$ 17.5 $\boxed{\text{ENTER}}$

```
97105-124930
            ‾27825
Ans/17.5
            ‾1590
```

The amount of depreciation per year was $1,590.

Solving Equations

In the following examples, we will use the division and multiplication properties of equality to solve equations involving negative numbers.

EXAMPLE 6   **The division property of equality.** Solve $-4x = 48$ and check the result.

Solution   Recall that $-4x$ indicates multiplication: $-4 \cdot x$. To undo the multiplication of x by -4, we divide both sides by -4.

$$-4x = 48$$

$$\frac{-4x}{-4} = \frac{48}{-4} \qquad \text{Divide both sides by } -4$$

$$x = -12$$

Check: $-4x = 48$

$-4(-12) \stackrel{?}{=} 48$ Substitute -12 for x.

$48 = 48$ Do the multiplication: $-4(-12) = 48$.

SELF CHECK

Solve $-45 = 9x$ and check the result.   *Answer:* -5 ■

EXAMPLE 7

The multiplication property of equality. Solve $\dfrac{b}{-15} = 3$ and check the result.

Solution   Recall that $\frac{b}{-15}$ indicates that b is to be divided by -15. To undo the division of b by -15, we multiply by -15.

$$\frac{b}{-15} = 3$$

$$-15\left(\frac{b}{-15}\right) = -15(3) \quad \text{Multiply both sides by } -15.$$

$$b = -45 \quad \text{Simplify each side.}$$

Check: $\dfrac{b}{-15} = 3$

$$\frac{-45}{-15} \stackrel{?}{=} 3 \quad \text{Substitute } -45 \text{ for } b.$$

$$3 = 3 \quad \text{Do the division.}$$

SELF CHECK   Solve $\frac{y}{23} = -12$ and check the result.   *Answer:* -276 ■

STUDY SET

Section 2.2

VOCABULARY

In Exercises 1–8, fill in the blanks to make the statements true.

1. The answer to a multiplication problem is called a _____product_____.

2. The answer to a division problem is called a _____quotient_____.

3. The numbers -4 and -6 are said to have _____like_____ signs.

4. The numbers -10 and $+12$ are said to have _____unlike_____ signs.

5. The _____commutative_____ property of multiplication states that two numbers can be multiplied in either order to get the same result.

6. _____Positive_____ numbers are greater than zero, and _____negative_____ numbers are less than zero.

7. Division by zero is _____undefined_____.

8. The statement $(ab)c = a(bc)$ expresses the _____associative_____ property of _____multiplication_____.

CONCEPTS

In Exercises 9–14, fill in the blanks to make the statements true.

9. The division fact $\frac{25}{-5} = -5$ is related to the multiplication fact $-5(-5) = 25$.

10. The expression $-5 + (-5) + (-5) + (-5)$ can be represented by the multiplication statement $4(-5)$.

11. The quotient of two numbers with _____unlike_____ signs is negative.

12. The product of two negative numbers is _____positive_____.

13. The product of zero and any number is 0.

14. The product of 1 and any number is that number.

15. Draw a number line from -6 to 6. Graph each of these products on the number line. What is the distance between each product? 3

$$-3(2), \quad -3(1), \quad -3(0), \quad -3(-1), \quad -3(-2)$$

16. a. Find $-1(8)$. In general, what is the result when a number is multiplied by -1?

-8, the opposite of that number

b. Find $\frac{8}{-1}$. In general, what is the result when a number is divided by -1?

-8, the opposite of that number

In Exercises 17–18, POS stands for a positive number and NEG stands for a negative number. Determine the sign of each result, if possible.

17. a. POS · NEG   NEG   **b.** POS + NEG

not possible to tell

c. POS − NEG   POS   **d.** $\dfrac{\text{POS}}{\text{NEG}}$ NEG

18. a. NEG · NEG   POS   **b.** NEG + NEG   NEG

c. NEG − NEG   **d.** $\dfrac{\text{NEG}}{\text{NEG}}$ POS

not possible to tell

19. Is -6 a solution of $\dfrac{x}{2} = -3$?   yes

20. Is -6 a solution of $-2x = -12$?   no

NOTATION

In Exercises 21–24, complete the solution.

21. Find $(-37 \cdot 5)2$.

$(-37 \cdot 5)2 = -37\left(\boxed{5} \cdot 2\right)$

$= -37\left(\boxed{10}\right)$

$= -370$

22. Find $-20[5(-79)]$.

$-20[5(-79)] = (-20 \cdot 5)\left(\boxed{-79}\right)$

$= \boxed{-100}\,(-79)$

$= 7{,}900$

23. Solve $-3x = 36$.

$-3x = 36$

$\dfrac{-3x}{\boxed{-3}} = \dfrac{36}{\boxed{-3}}$

$x = \boxed{-12}$

24. Solve $\dfrac{x}{-7} = -5$.

$\dfrac{x}{-7} = -5$

$\boxed{-7}\left(\dfrac{x}{-7}\right) = \boxed{-7}\,(-5)$

$x = \boxed{35}$

PRACTICE

In Exercises 25–70, find each product or quotient, if possible.

25. $(-6)(-9)$ 54   **26.** $(-8)(-7)$ 56

27. $12(-5)$ -60 **28.** $(-9)(11)$ -99

29. $-6 \cdot 4$ -24 **30.** $-8 \cdot 9$ -72

31. $-20(40)$ -800 **32.** $-10(10)$ -100

33. $-0.6(-4)$ 2.4   **34.** $-0.7(-8)$ 5.6

35. $1.2(-0.4)$ -0.48 **36.** $0(-0.2)$ 0

37. $\dfrac{1}{2}\left(-\dfrac{3}{4}\right)$ $-\frac{3}{8}$ **38.** $\dfrac{1}{3}\left(-\dfrac{5}{16}\right)$ $-\frac{5}{48}$

39. $-1\dfrac{1}{4}\left(-\dfrac{3}{4}\right)$ $\frac{15}{16}$ **40.** $-1\dfrac{1}{8}\left(-\dfrac{3}{8}\right)$ $\frac{27}{64}$

41. $0(-22)$ 0   **42.** $-8 \cdot 0$ 0

43. $-3(-4)(0)$ 0   **44.** $15(0)(-22)$ 0

45. $3(-4)(-5)$ 60   **46.** $(-2)(-4)(-5)$ -40

47. $(-4)(3)(-7)$ 84   **48.** $5(-3)(-4)$ 60

49. $(-2)(-3)(-4)(-5)$ 120   **50.** $(-3)(-4)(5)(-6)$ -360

51. $\dfrac{-6}{-2}$ 3   **52.** $\dfrac{-36}{9}$ -4

53. $\dfrac{4}{-2}$ -2 **54.** $\dfrac{-9}{3}$ -3

55. $\dfrac{80}{-20}$ -4 **56.** $\dfrac{-66}{33}$ -2

57. $\dfrac{-110}{-110}$ 1   **58.** $\dfrac{-200}{-200}$ 1

59. $\dfrac{-160}{40}$ -4 **60.** $\dfrac{-250}{-50}$ 5

61. $\dfrac{320}{-16}$ -20 **62.** $\dfrac{-180}{36}$ 5

63. $\dfrac{0}{150}$ 0   **64.** $\dfrac{225}{0}$ undefined

65. $\dfrac{-17}{0}$ undefined   **66.** $\dfrac{0}{-12}$ 0

67. $-\dfrac{1}{3} \div \dfrac{4}{5}$ $-\frac{5}{12}$ **68.** $-\dfrac{1}{8} \div \dfrac{2}{3}$ $-\frac{3}{16}$

69. $-\dfrac{3}{16} \div \left(-\dfrac{2}{3}\right)$ $\frac{9}{32}$ **70.** $-\dfrac{3}{25} \div \left(-\dfrac{2}{3}\right)$ $\frac{9}{50}$

In Exercises 71–76, use a calculator to do each operation.

71. $(-23.5)(47.2)$ $-1{,}109.2$

72. $(-435.7)(-37.8)$ 16,469.46

73. $(-6.37)(-7.2)(-9.1)$ -417.3624

74. $(5.2)(-8.2)(7.75)$ -330.46

75. $\dfrac{204.6}{-37.2}$ -5.5

76. $\dfrac{-30.56625}{-4.875}$ 6.27

In Exercises 77–80, complete each input/output table.

77.

x	$3x$
-1	-3
-5	-15
-10	-30

78.

x	$-3x$
4	-12
0	0
-4	12

79.

x	$\dfrac{x}{2}$
-2	-1
-6	-3
-8	-4

80.

x	$\dfrac{x}{4}$
8	2
0	0
-12	-3

In Exercises 81–88, solve each equation. Check each solution.

81. $3x = -3$ -1

82. $-4x = 36$ -9

83. $-54 = -18z$ 3

84. $-57 = -19x$ 3

85. $\dfrac{b}{3} = -5$ -15

86. $-3 = \dfrac{s}{11}$ -33

87. $-6 = \dfrac{t}{-7}$ 42

88. $\dfrac{v}{12} = -7$ -84

APPLICATIONS

In Exercises 89–100, use signed numbers to solve each problem.

89. TEMPERATURE CHANGE In a lab, the temperature of a fluid was decreased 6° per hour for 12 hours. What signed number indicates the change in temperature? $-72°$

90. BACTERIAL GROWTH To slowly warm a bacterial culture, biologists programmed a heating pad under the culture to increase the temperature 4° every hour for 6 hours. What signed number indicates the change in the temperature of the pad? $+24°$

91. GAMBLING A gambler places a $40 bet and loses. He then decides to go "double or nothing," and loses again. Feeling that his luck has to change, he goes "double or nothing" once more and, for the third time, loses. What signed number indicates his gambling losses? $-\$160$

92. REAL ESTATE A house has depreciated $1,250 each year for 8 years. What signed number indicates its change in value over that time period? $-\$10,000$

93. PLANETS The temperature on Pluto gets as low as $-386°$ F. This is twice as low as the lowest temperature reached on Jupiter. What is the lowest temperature on Jupiter? $-193°$F

94. CAR RADIATOR The instructions on the back of a container of antifreeze state, "A 50/50 mixture of antifreeze and water protects against freeze-ups down to $-34°$ F, while a 60/40 mix protects against freeze-ups down to one and one-half times that temperature." To what temperature does the 60/40 mixture protect? $-51°$ F

95. AIRLINE INCOME For the fourth quarter of 1997, Trans World Airlines' total net income was $-\$31$ million. The company's losses for the first quarter of 1998 were even worse, by a factor of about 1.8. What signed number indicates the company's total net income that quarter? $-\$55.8$ million

96. ACCOUNTING Illustration 1 shows the income statement for Converse Inc., the sports shoe company. The numbers in parentheses indicate losses. What signed number describes the company's average net income per quarter for the four quarters shown? $-\$5.1$ million/qtr

ILLUSTRATION 1

Converse® Inc. **INCOME STATEMENT**				
All dollar amounts in millions	2nd Qtr **Jun 98**	1st Qtr **Mar 98**	4th Qtr **Dec 97**	3rd Qtr **Sep 97**
Total Net Income	(1.5)	(1.2)	(17.9)	0.2

Based on information from Hoover's Online

97. QUEEN MARY The ocean liner Queen Mary was commissioned in 1936 and cost $22,500,000 to build. In 1967, the ship was purchased by the city of Long Beach, California for $3,450,000 and now serves as a hotel and convention center. What signed number indicates the annual average depreciation of the Queen Mary over the 31-year period from 1936 to 1967? Round to the nearest dollar. $-\$614,516$

98. COMPUTER SPREADSHEET The "formula" = SUM(A1:C1)/3 in cell D1 of the spreadsheet shown in Illustration 2 instructs the computer to add the values in cells A1, B1, and C1, then to divide that sum by 3, and finally to print the result *in place of the formula* in cell D1. What values will the computer print in the cells D1, D2, and D3? $-6, 2, -38$

ILLUSTRATION 2

Microsoft Excel-Book 1			
File **Edit** **View** **Insert** **Format** **Tools**			
	A	**B**	**C**	**D**
1	4	−5	−17	= SUM(A1:C1)/3
2	22	−30	14	= SUM(A2:C2)/3
3	−60	−20	−34	= SUM(A3:C3)/3
4				

Sheet 1 / Sheet 2 / Sheet 3 / Sheet 4 / Sheet 5

99. PHYSICS An oscilloscope is an instrument that displays electrical signals, which appear as wavy lines on a fluorescent screen. (See Illustration 3.) By switching the magnification setting (MAGNIFN.) to × 2, for example, the "height" of the crest and the "depth" of the trough of a graph will be doubled. Use signed numbers to indicate the crest height and the trough depth for each setting of the magnification dial.

    a. normal 5, −10   **b.** × 0.5 2.5, −5
    c. × 1.5 7.5, −15   **d.** × 2 10, −20

ILLUSTRATION 4

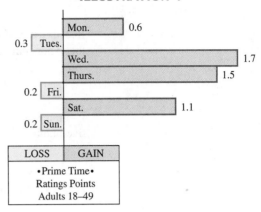

WRITING

101. Explain how you would decide whether the product of several numbers was positive or negative.

102. If the product of five numbers is negative, how many of them could be negative? Explain.

ILLUSTRATION 3

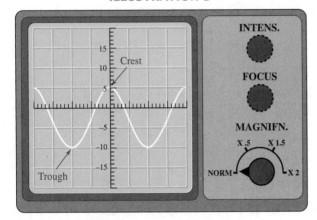

100. SWEEPS WEEK During "sweeps week," television networks make a special effort to gain viewers by showing unusually flashy programming. Use the information in Illustration 4 to determine the average daily gain (or loss) of ratings points by a network for the 7-day "sweeps period."

a gain of 0.6 of a rating point

REVIEW

103. Use the formula $V = s^3$ to find the volume of a cube with a side of length 4 inches. 64 in.^{3}

104. Write the subtraction statement $-3 - (-5)$ as addition of the opposite. $-3 + 5$

105. What is a real number?
  any number that can be written as a decimal

106. Describe the balance in a checking account that is overdrawn $65 using a signed number. $-\$65$

107. Find 34% of 612. 208.08

108. 18 is what percent of 90? 20%

▶ **2.3**

Order of Operations and Evaluating Algebraic Expressions

In this section, you will learn about

    Powers of real numbers ■ Order of operations ■ Evaluating algebraic expressions ■ Making tables

Introduction In Section 1.3, we saw that the order in which we do arithmetic operations is important. To guarantee that a simplification of a numerical expression has only one answer, we need to perform arithmetic operations in an agreed-upon order. In this section, we will review the rules for the order of operations as we evaluate expressions involving real numbers.

Powers of Real Numbers

Recall that 3 to the fourth power (or 3^4) is a shorthand way of writing $3 \cdot 3 \cdot 3 \cdot 3$. In the exponential expression 3^4, the base is 3 and the exponent is 4.

$$\text{Base} \longrightarrow 3^4 \longleftarrow \text{Exponent}$$

In the next example, we evaluate exponential expressions with bases that are negative numbers.

EXAMPLE 1

Powers of integers. Find each power: **a.** $(-3)^4$,   and   **b.** $(-3)^5$.

Solution   **a.** $(-3)^4 = (-3)(-3)(-3)(-3)$ Write -3 as a factor four times.

$\qquad\qquad = 9(-3)(-3)$ Work from left to right: $(-3)(-3) = 9$.

$\qquad\qquad = -27(-3)$ Work from left to right: $9(-3) = -27$.

$\qquad\qquad = 81$ Do the multiplication.

b. $(-3)^5 = (-3)(-3)(-3)(-3)(-3)$ Write -3 as a factor five times.

$\qquad\qquad = 9(-3)(-3)(-3)$ Work from left to right: $(-3)(-3) = 9$.

$\qquad\qquad = -27(-3)(-3)$ Work from left to right: $(9)(-3) = -27$.

$\qquad\qquad = 81(-3)$ Work from left to right: $(-27)(-3) = 81$.

$\qquad\qquad = -243$ Do the multiplication.

SELF CHECK   Find each power: **a.** $(-6)^2$ and   **b.** $(-5)^3$.   *Answers:*   **a.** 36,   **b.** -125 ∎

In part a of Example 1, -3 was raised to an even power, and the result was positive. In part b, -3 was raised to an odd power, and the result was negative. This suggests a general rule.

Even and odd powers of a negative number

When a negative number is raised to an even power, the result is positive.

When a negative number is raised to an odd power, the result is negative.

EXAMPLE 2

Powers of fractions and decimals. Find each power: **a.** $\left(-\dfrac{2}{3}\right)^3$ and **b.** $(0.6)^2$.

Solution   **a.**   In the expression $\left(-\frac{2}{3}\right)^3$, $-\frac{2}{3}$ is the base and 3 is the exponent.

$$\left(-\frac{2}{3}\right)^3 = \left(-\frac{2}{3}\right)\left(-\frac{2}{3}\right)\left(-\frac{2}{3}\right) \qquad \text{Write } -\tfrac{2}{3} \text{ as a factor three times.}$$

$$= \frac{4}{9}\left(-\frac{2}{3}\right) \qquad \text{Multiply: } \left(-\tfrac{2}{3}\right)\left(-\tfrac{2}{3}\right) = \tfrac{4}{9}.$$

$$= -\frac{8}{27} \qquad \text{Do the multiplication.}$$

b. In the expression $(0.6)^2$, 0.6 is the base and 2 is the exponent.

$$(0.6)^2 = (0.6)(0.6) \quad \text{Write 0.6 as a factor two times.}$$
$$= 0.36 \qquad \text{Do the multiplication.}$$

SELF CHECK   Find each power: **a.** $\left(-\frac{3}{4}\right)^3$ and   **b.** $(-0.3)^2$ *Answers:*   **a.** $-\frac{27}{64}$,   **b.** 0.09   ∎

WARNING! Although the expressions -4^2 and $(-4)^2$ look somewhat alike, they are not. In -4^2, the base is 4 and the exponent is 2. The $-$ sign in front of 4^2 means the opposite of 4^2. In $(-4)^2$, the base is -4 and the exponent is 2. When we find the value of each expression, it becomes clear that they are not equivalent.

$$-4^2 = -(4 \cdot 4) \quad \begin{array}{l}\text{Write 4 as a factor}\\\text{two times.}\end{array} \qquad\qquad (-4)^2 = (-4)(-4) \quad \begin{array}{l}\text{Write } -4 \text{ as a}\\\text{factor two times.}\end{array}$$
$$= -16 \qquad \begin{array}{l}\text{Multiply inside}\\\text{the parentheses.}\end{array} \qquad\qquad = 16 \qquad \begin{array}{l}\text{The product of two}\\\text{negative numbers}\\\text{is positive.}\end{array}$$

Different results

ACCENT ON TECHNOLOGY *Raising a Negative Number to a Power*

We can use a scientific calculator to raise negative numbers to powers. We need to use the change-sign key $\boxed{+/-}$ and the power key $\boxed{y^x}$. For example, to evaluate $(-7.36)^5$, we enter these numbers and press these keys:

Keystrokes   7.36 $\boxed{+/-}$ $\boxed{y^x}$ 5 $\boxed{=}$ $\boxed{-21596.78335}$

When using a graphing calculator to raise a negative number to a power, we must enter the parentheses that contain the base.

Keystrokes $\boxed{(}$ $\boxed{(-)}$ 7.36 $\boxed{)}$ $\boxed{\wedge}$ 5 $\boxed{\text{ENTER}}$ $\boxed{\begin{array}{l}(-7.36)\wedge5\\\qquad\quad-21596.78335\end{array}}$

Thus, $(-7.36)^5 = -21,596.78335$.

Order of Operations

To illustrate that the rules for the order of operations are necessary when working with negative numbers, let's evaluate $-2 + 3(-4)$. If we multiply first, we obtain -14. However, if we add first, we obtain -4.

Method 1: Multiply first

$$-2 + 3(-4) = -2 + (-12) \quad \begin{array}{l}\text{Multiply 3 and } -4\\\text{first: } 3(-4) = -12.\end{array}$$
$$= -14 \qquad \text{Add.}$$

Method 2: Add first

$$-2 + 3(-4) = 1(-4) \quad \begin{array}{l}\text{Add } -2 \text{ and 3}\\\text{first: } -2 + 3 = 1.\end{array}$$
$$= -4 \qquad \text{Multiply.}$$

Different results

To make sure that problems involving exponents, multiplication, division, addition, and subtraction have only one answer, we must apply the rules for the order of operations.

Order of operations

If the expression contains grouping symbols, do all calculations within each pair of grouping symbols, working from the innermost pair to the outermost pair, in the following order:

1. Evaluate all exponential expressions.

2. Do all multiplications and divisions, working from left to right.

3. Do all additions and subtractions, working from left to right.

If the expression does not contain grouping symbols, begin with step 1.

In a fraction, first simplify the numerator and denominator separately. Then simplify the fraction, whenever possible.

Because we do multiplications before additions, the correct calculation of $-2 + 3(-4)$ is

$$-2 + 3(-4) = -2 + (-12) \quad \text{Multiply first: } 3(-4) = -12.$$
$$= -14 \quad \text{Do the addition.}$$

E X A M P L E   3

Order of operations. Evaluate $-5 + 4(-3)^2$.

Solution   To evaluate (find the value of) this expression, we must perform three operations: addition, multiplication, and raising a number to a power. By the rules for the order of operations, we find the power first.

$$-5 + 4(-3)^2 = -5 + 4(9) \quad \text{Evaluate the exponential expression: } (-3)^2 = 9.$$
$$= -5 + 36 \quad \text{Do the multiplication: } 4(9) = 36.$$
$$= 31 \quad \text{Do the addition.}$$

SELF CHECK   Evaluate $-9 + 2(-4)^2$.   *Answer:* 23   ■

E X A M P L E   4

An expression containing grouping symbols. Evaluate $5^3 + 2(-8 - 3 \cdot 2)$.

Solution   We do the work within the parentheses first.

$$5^3 + 2(-8 - 3 \cdot 2) = 5^3 + 2(-8 - 6) \quad \text{Do the multiplication within the parentheses: } 3 \cdot 2 = 6.$$
$$= 5^3 + 2(-14) \quad \text{Do the subtraction within the parentheses: } -8 - 6 = -14.$$
$$= 125 + 2(-14) \quad \text{Evaluate the exponential expression: } 5^3 = 125.$$
$$= 125 + (-28) \quad \text{Do the multiplication: } 2(-14) = -28.$$
$$= 97 \quad \text{Do the addition.}$$

SELF CHECK   Evaluate $-3[5 + 3(-2)] + 3^2$.   *Answer:* 12   ■

EXAMPLE 5

An expression containing two pairs of grouping symbols. Evaluate $-4[-2 - 3(4 - 8^2)] - 2$.

Solution   We do the work within the innermost grouping symbols (the parentheses) first.

$$-4[-2 - 3(4 - 8^2)] - 2$$

$= -4[-2 - 3(4 - \mathbf{64})] - 2$ Evaluate the exponential expression within the parentheses: $8^2 = 64$.

$= -4[-2 - 3(-60)] - 2$ Do the subtraction within the parentheses: $4 - 64 = -60$.

$= -4[-2 - (-180)] - 2$ Do the multiplication within the brackets: $3(-60) = -180$.

$= -4(178) - 2$ Do the subtraction within the brackets by adding the opposite of -180, which is 180: $-2 + 180 = 178$.

$= -712 - 2$ Do the multiplication: $-4(178) = -712$.

$= -714$ Do the subtraction by adding the opposite of 2, which is -2: $-712 + (-2) = -714$.

SELF CHECK   Evaluate $-5[2(5^2 - 15) + 4] - 10$.   *Answer:* -130 ∎

EXAMPLE 6

Simplifying a fractional expression. Evaluate $\dfrac{-3(3 + 2) + 5}{17 - 3(-4)}$.

Solution   We simplify the numerator and the denominator separately.

$$\frac{-3(3 + 2) + 5}{17 - 3(-4)} = \frac{-3(5) + 5}{17 - (-12)}$$ In the numerator, do the addition within the parentheses. In the denominator, do the multiplication.

$$= \frac{-15 + 5}{17 + 12}$$ In the numerator, do the multiplication. In the denominator, write the subtraction as addition of the opposite of -12, which is 12.

$$= \frac{-10}{29}$$ Do the additions.

$$= -\frac{10}{29}$$ $\frac{-10}{29} = -\frac{10}{29}$.

SELF CHECK   Evaluate $\dfrac{-4(-2 + 8) + 6}{8 - 5(-2)}$.   *Answer:* -1 ∎

Evaluating Algebraic Expressions

We will often need to evaluate algebraic expressions that contain variables. To do so, we substitute the values for the variables and simplify the resulting expression.

EXAMPLE 7

Surface area of a swim fin. Divers use swim fins because they provide a much larger surface area to push against the water than do bare feet. Consequently, the diver can swim faster wearing them. In Figure 2-10, we see that the fin is in the shape of a

FIGURE 2-10

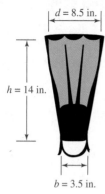

$d = 8.5$ in.

$h = 14$ in.

$b = 3.5$ in.

trapezoid. The algebraic expression $\frac{1}{2}h(b + d)$ gives the area of a trapezoid, where h is the height and b and d are the lengths of the lower and upper bases, respectively. To find the area of the top portion of the fin shown here, we evaluate the algebraic expression for $h = 14$, $b = 3.5$, and $d = 8.5$.

$$\frac{1}{2}h(b + d) = \frac{1}{2}(14)(3.5 + 8.5) \quad \text{Substitute 14 for } h, \text{ 3.5 for } b, \text{ and 8.5 for } d.$$

$$= \frac{1}{2}(14)(12) \qquad \text{Do the addition within the parentheses.}$$

$$= 7(12) \qquad \text{Work from left to right: } \frac{1}{2}(14) = 7.$$

$$= 84 \qquad \text{Do the multiplication.}$$

The fin has an area of 84 square inches.   ■

EXAMPLE 8

Evaluating algebraic expressions. Evaluate   **a.** $-y$ and **b.** $-3(y + x^2)$ when $x = 3$ and $y = -4$.

Solution   **a.** $-y = -(-4)$ Substitute -4 for y.

            $= 4$ The opposite of -4 is 4.

b. $-3(y + x^2) = -3(-4 + 3^2)$ Substitute 3 for x and -4 for y.

                $= -3(-4 + 9)$ Do the work within the parentheses first. Evaluate the exponential expression.

                $= -3(5)$ Do the addition within the parentheses.

                $= -15$ Do the multiplication.

SELF CHECK   Evaluate   **a.** $-x$ and   **b.** $5(x - y)$ when $x = -2$ and $y = 3$.   *Answers:*   **a.** 2,   **b.** -25 ■

EXAMPLE 9

Evaluating an algebraic expression containing an absolute value. Evaluate $\dfrac{3a^2 + 2b}{-2|a + b|}$ when $a = -4$ and $b = -3$.

Solution   In the denominator, the absolute value bars serve as grouping symbols.

$$\frac{3a^2 + 2b}{-2|a + b|} = \frac{3(-4)^2 + 2(-3)}{-2|-4 + (-3)|} \qquad \text{Substitute } -4 \text{ for } a \text{ and } -3 \text{ for } b.$$

$$= \frac{3(16) + 2(-3)}{-2|-7|} \qquad \begin{array}{l}\text{Find the value of } (-4)^2 \text{ in the numerator. In the}\\\text{denominator, do the addition inside the absolute}\\\text{value bars.}\end{array}$$

$$= \frac{48 + (-6)}{-2(7)} \qquad \begin{array}{l}\text{Do the multiplications in the numerator. Find the}\\\text{absolute value in the denominator.}\end{array}$$

$$= \frac{42}{-14} \qquad \begin{array}{l}\text{In the numerator, do the addition. Do the multipli-}\\\text{cation in the denominator.}\end{array}$$

$$= -3 \qquad \text{Do the division.}$$

SELF CHECK   Evaluate $\dfrac{4s^2 + 3t - 1}{-3|s - t|}$ when $s = -5$ and $t = 2$.   *Answer:* -5 ■

Making Tables

When we evaluate an algebraic expression in one variable for several values of the variable, we can keep track of the results in an input/output table.

EXAMPLE 10

Temperature conversion. The equation

$$F = \frac{9C + 160}{5}$$

is used to change temperatures given in degrees Celsius to temperatures in degrees Fahrenheit. Change each temperature to degrees Fahrenheit and give the results in a table.

a. The coldest temperature on the moon: $-170°C$.

b. The coldest recorded temperature on earth; Vostok, Antarctica, July 21, 1983: $-89.2°C$.

Solution

C	$F = \frac{9C + 160}{5}$
-170	-274
-89.2	-128.56

Evaluate for C = −170:

$$\frac{9C + 160}{5} = \frac{9(-170) + 160}{5}$$
$$= \frac{-1{,}530 + 160}{5}$$
$$= \frac{-1{,}370}{5}$$
$$= -274$$

Evaluate for C = −89.2:

$$\frac{9C + 160}{5} = \frac{9(-89.2) + 160}{5}$$
$$= \frac{-802.8 + 160}{5}$$
$$= \frac{-642.8}{5}$$
$$= -128.56$$

The corresponding temperatures in degrees Fahrenheit are $-274°F$ and approximately $-128.6°F$.

SELF CHECK

On January 22, 1943, the temperature in Spearfish, South Dakota, changed from $-20°C$ to $7.2°C$ in 2 minutes. Change each temperature to degrees Fahrenheit.

C	$F = \frac{9C + 160}{5}$
-20	
7.2	

Answers: $-4°F$, approximately $45°F$

Section 2.3

VOCABULARY

In Exercises 1–8, fill in the blanks to make the statements true.

1. The rules for the _____order_____ of operations guarantee that a simplification of a numerical expression results in a single answer.

2. $2x + 5$ is an example of an algebraic _____expression_____, whereas $2x + 5 = 7$ is an example of an _____equation_____.

3. To _____evaluate_____ an algebraic expression means to substitute the values for the variables and then apply the rules for the order of operations.

4. When we evaluate an algebraic expression in one variable for several values of the variable, we can keep track of the results in an input/output _____table_____.

5. In the expression $(-9)^3$, -9 is the _____base_____ and ⟨3⟩ is the exponent.

6. In the expression -8^4, ⟨8⟩ is the base and 4 is the _____exponent_____.

7. Some examples of grouping symbols are ( ) _____parentheses_____, [ ] _____brackets_____, and | | _____absolute value_____ bars.

8. The expression $(-4.5)^6$ is called a _____power_____ of -4.5.

CONCEPTS

In Exercises 9–10, fill in the blanks to make the statements true.

9. To simplify an expression with no grouping symbols, evaluate all _____exponential_____ expressions before doing any multiplications.

10. When simplifying an expression containing grouping symbols, do all calculations within each pair of grouping symbols, working from the _____innermost_____ pair to the _____outermost_____ pair.

11. Complete each input/output table and make an observation about the signs of the outputs.

a.
x	x^2
-3	9
-4	16
-5	25

b.
x	x^3
-3	-27
-4	-64
-5	-125

When a negative number is raised to an even power, the result is a positive number.

When a negative number is raised to an odd power, the result is a negative number.

12. Complete the table and make an observation about the outputs.

x	x^2	x^3	x^4	x^5	x^6	x^7
-1	1	-1	1	-1	1	-1

The signs of the outputs are alternating.

13. **a.** How many operations need to be performed to evaluate the expression $-3[5^2 - 6(3 - 2)]$? ⟨5⟩

b. List the operations in the order in which they should be performed.

subtraction, power, multiplication, subtraction, multiplication

14. If $x = -9$, find the value of

a. $-x$ ⟨9⟩   **b.** $-(-x)$ ⟨-9⟩

c. $-x^2$ ⟨-81⟩   **d.** $(-x)^2$ ⟨81⟩

NOTATION

In Exercises 15–16, complete each solution.

15. Evaluate $(-6)^2 - 2(5 - 4 \cdot 2)$.

$$(-6)^2 - 2(5 - 4 \cdot 2) = (-6)^2 - 2(5 - \boxed{8})$$
$$= (-6)^2 - 2(\boxed{-3})$$
$$= \boxed{36} - 2(-3)$$
$$= 36 - (\boxed{-6})$$
$$= 42$$

16. Evaluate $\dfrac{4x^2 - 3y}{9(x - y)}$ when $x = 4$ and $y = -3$.

$$\frac{4x^2 - 3y}{9(x - y)} = \frac{4(4)^2 - 3(-3)}{9[4 - (-3)]}$$
$$= \frac{4(\boxed{16}) - 3(\boxed{-3})}{9[\boxed{7}]}$$
$$= \frac{\boxed{64} - (\boxed{-9})}{\boxed{63}}$$
$$= \frac{73}{63}$$

PRACTICE

In Exercises 17–28, evaluate each expression.

17. $(-6)^2$ ⟨36⟩   18. -6^2 ⟨-36⟩

19. -4^4 ⟨-256⟩   20. $(-4)^4$ ⟨256⟩

21. $(-5)^3$ ⟨-125⟩   22. -5^3 ⟨-125⟩

23. $-(-6)^4$ ⟨$-1,296$⟩   24. $-(-7)^2$ ⟨-49⟩

25. $(-0.4)^2$ 0.16

26. $(-0.5)^2$ 0.25

27. $\left(-\dfrac{2}{5}\right)^3$ $-\frac{8}{125}$

28. $\left(-\dfrac{1}{4}\right)^3$ $-\frac{1}{64}$

In Exercises 29–58, evaluate each expression.

29. $3 - 5 \cdot 4$ -17

30. $-4 \cdot 6 + 5$ -19

31. $-3(5 - 4)$ -3

32. $-4(6 + 5)$ -44

33. $3 + (-5)^2$ 28

34. $4^2 - (-2)^2$ 12

35. $(-3 - 5)^2$ 64

36. $(-5 - 2)^2$ 49

37. $2 + 3\left(-\dfrac{25}{5}\right) - (-4)$
  -9

38. $12 + 2\left(-\dfrac{9}{3}\right) - (-2)$
  8

39. $(-2)^3\left(\dfrac{-6}{-2}\right)(-1)$ 24

40. $(-3)^3\left(\dfrac{-4}{-2}\right)(-1)$ 54

41. $\dfrac{-7 - 3^2}{2 \cdot 4}$ -2

42. $\dfrac{-5 - 3^3}{2^3}$ -4

43. $\dfrac{1}{2}\left(\dfrac{1}{8}\right) + \left(-\dfrac{1}{4}\right)^2$ $\frac{1}{8}$

44. $-\dfrac{1}{9}\left(\dfrac{1}{4}\right) + \left(-\dfrac{1}{6}\right)^2$ 0

45. $-2|4 - 8|$ -8

46. $-5|1 - 8|$ -35

47. $|7 - 8(4 - 7)|$ 31

48. $|9 - 5(1 - 8)|$ 44

49. $3 + 2[-1 - 4(5)]$
  -39

50. $4 + 2[-7 - 3(9)]$
  -64

51. $-3[5^2 - (7 - 3)^2]$
  -27

52. $3 - [3^3 + (3 - 1)^3]$
  -32

53. $-(2 \cdot 3 - 4)^3$ -8

54. $-(3 \cdot 5 - 2 \cdot 6)^2$ -9

55. $\dfrac{(3 + 5)^2 + |-2|}{-2(5 - 8)}$ 11

56. $\dfrac{|-25| - 8(-5)}{2^4 - 29}$ -5

57. $\dfrac{2[-4 - 2(3 - 1)]}{3[(3)(2)]}$ $-\frac{8}{9}$

58. $\dfrac{3[-9 + 2(7 - 3)]}{(8 - 5)(9 - 7)}$ $-\frac{1}{2}$

In Exercises 59–70, evaluate each expression given that $x = 3$, $y = -2$, and $z = -4$.

59. $2x - y$ 8

60. $2z - y$ -6

61. $-y + yz$ 10

62. $-z + x - 2y$ 11

63. $(3 + x)y$ -12

64. $(4 + z)y$ 0

65. $(x + y)^2(x - y)$ 5

66. $[(z - 1)(z + 1)]^2$ 225

67. $(4x)^2 + 3y^2$ 156

68. $4x^2 + (3y)^2$ 72

69. $\dfrac{2x + y^3}{y + 2z}$ $\frac{1}{5}$

70. $\dfrac{2z^2 - y}{2x - y^2}$ 17

In Exercises 71–76, evaluate each expression for the given values of the variables.

71. $b^2 - 4ac$; $a = -1$, $b = 5$, and $c = -2$ 17

72. $(x - a)^2 + (y - b)^2$; $x = -2$, $y = 1$, $a = 5$, and $b = -3$ 65

73. $a^2 + 2ab + b^2$; $a = -5$ and $b = -1$ 36

74. $\dfrac{x - a}{y - b}$; $x = -2$, $y = 1$, $a = 5$, and $b = 2$ 7

75. $\dfrac{n}{2}[2a + (n - 1)d]$; $n = 10$, $a = -4$, and $d = 6$ 230

76. $\dfrac{a(1 - r^n)}{1 - r}$; $a = -5$, $r = 2$, and $n = 3$ -35

 In Exercises 77–82, use a calculator to find the value of each expression.

77. $(-5.6)^4$ 983.4496

78. $(-8.3)^4$ 4,745.8321

79. $(-1.12)^5$
  -1.762341683

80. $(-4.07)^5$
  $-1,116.791362$

81. $(23.1)^2 - (14.7)(-61.9)$
  1,443.54

82. $12 - 7\left(-\dfrac{85.684}{34.55}\right)^3$
  118.770944

APPLICATIONS

In Exercises 83–88, a calculator may be helpful.

83. TRANSLATION When an American company exports its Advanced Formula Antifreeze to China, the product description on the back of the container, seen in Illustration 1, must be translated into Chinese. Since China uses the metric system of measurement, the temperatures in the description must be converted to degrees Celsius. Use the formula

$$C = \dfrac{5(F - 32)}{9}$$

to make the conversions to the nearest degree Celsius.
$-37°C$, $-64°C$

ILLUSTRATION 1

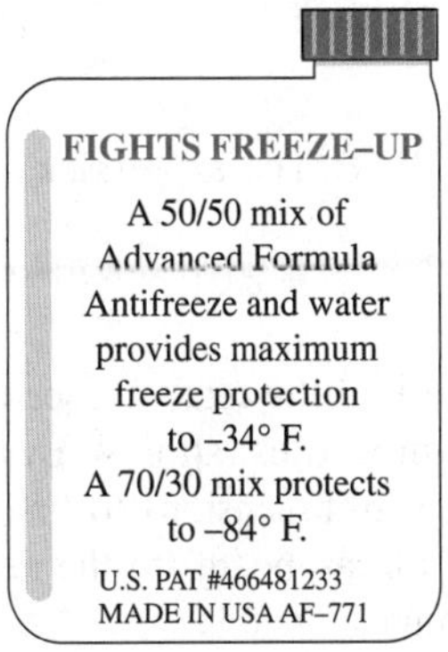

FIGHTS FREEZE–UP

A 50/50 mix of
Advanced Formula
Antifreeze and water
provides maximum
freeze protection
to –34° F.
A 70/30 mix protects
to –84° F.

U.S. PAT #466481233
MADE IN USA AF–771

84. TEMPERATURE ON MARS On Mars, maximum summer temperatures can reach 20°C. However, daily temperatures average $-33°C$. Convert each of these temperatures to degrees Fahrenheit. See Example 10. Round to the nearest degree. 68°F, $-27°F$

85. TIDES After reaching a high tide mark of 80 centimeters (cm) on the graduated pole shown in Illustration 2, the water level fell for 5 hours at an average rate of 21 centimeters per hour to reach the low tide mark. Write an expression that gives the low tide reading on the pole, then evaluate it.
$80 - 5(21) = -25$

ILLUSTRATION 2

Graduated pole for
determining tidal range

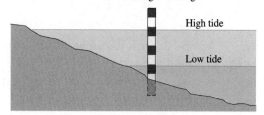

High tide

Low tide

86. GROWING SOD To determine the number of square feet of sod *remaining* in a field after filling an order (see Illustration 3), the manager of a sod farm uses the expression $20{,}000 - 3s$ (where s is the number of 1-foot-by-3-foot strips the customer has ordered). To sod a soccer field, a city orders 7,000 strips of sod. Evaluate the expression for this value of s and explain the result.

$-1{,}000$; the sod farm is short 1,000 ft^2 needed to fill the city's order

ILLUSTRATION 3

1-ft-by-3-ft strips of sod, cut and ready to be loaded on a truck for delivery

87. TRUMPET MUTE The expression

$$\pi[b^2 + d^2 + (b + d)s]$$

can be used to find the total surface area of the trumpet mute shown in Illustration 4. Evaluate the expression for the given dimensions to find the number of square inches of cardboard (to the nearest tenth) used to make the mute.   77.8 in.^{2}

ILLUSTRATION 4

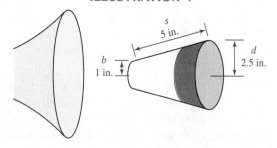

88. PET SUPPLIES The volume of the dog's water bowl shown in Illustration 5 can be found using the expression

$$\frac{\pi h(3a^2 + 3b^2 + h^2)}{6}$$

Evaluate the expression for the given dimensions to find the capacity of the bowl in cubic inches. Round to the nearest tenth.   75.4 in.^{3}

ILLUSTRATION 5

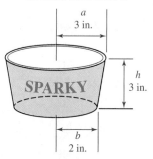

a
3 in.

h
3 in.

SPARKY

b
2 in.

WRITING

89. Explain why $3 + 5 \cdot 4$ can have two answers if we don't use the rules for the order of operations.

90. Explain how to evaluate $x^2 - y$ when $x = 2$ and $y = -3$.

91. 🖩 Does your calculator have the rules for the order of operations "built in"? Explain how you can tell.

92. Explain the difference in how the two expressions -3^2 and $(-3)^2$ are evaluated.

REVIEW

93. On the number line, graph the integers from -4 to 2.

$-5 \quad -4 \quad -3 \quad -2 \quad -1 \quad 0 \quad 1 \quad 2 \quad 3$

94. Write the inequality $7 < 12$ as an inequality that uses the symbol $>$. $12 > 7$

95. Write the expression $a \cdot a \cdot b \cdot b \cdot b$ using exponents. a^2b^3

96. Is 7 a solution of the equation $3x + 4 = 25$?   yes

97. Solve $12 + x = 17$. $x = 5$

98. Solve $x - 14 = -25$. $x = -11$

▶ **2.4**

Simplifying Algebraic Expressions

In this section, you will learn about

Simplifying algebraic expressions involving multiplication ■ The distributive property ■ Extending the distributive property ■ Terms of an algebraic expression ■ Coefficients of a term ■ Terms and factors ■ Like terms ■ Combining like terms

Introduction In this section, we will simplify algebraic expressions by replacing them with equivalent expressions that are less complicated. This will often require that we use the distributive property to combine like terms.

Simplifying Algebraic Expressions Involving Multiplication

To **simplify algebraic expressions,** we use properties of algebra to write the expressions in a less complicated form. Two properties used to simplify algebraic expressions are the associative and commutative properties of multiplication. Recall that the associative property of multiplication enables us to change the grouping of factors involved in a multiplication. The commutative property of multiplication enables us to change the order of the factors.

As an example, let's consider the expression $8(4x)$ and simplify it as follows:

$$8(4x) = 8 \cdot (4 \cdot x) \quad 4x = 4 \cdot x.$$

$$= (8 \cdot 4) \cdot x \quad \text{Apply the associative property of multiplication to group 4 with 8, instead of with } x.$$

$$= 32x \quad \text{Do the multiplication inside the parentheses: } 8 \cdot 4 = 32.$$

Since $8(4x) = 32x$, we say that $8(4x)$ simplifies to $32x$.

EXAMPLE 1

Simplifying algebraic expressions involving multiplication. Simplify each expression: **a.** $5(8t)$, **b.** $15a(-7)$.

Solution   **a.** $5(8t) = (5 \cdot 8)t$ Use the associative property of multiplication to regroup the factors.

$\qquad\qquad = 40t$ Do the multiplication inside the parentheses: $5 \cdot 8 = 40$.

b. $15a(-7) = 15(-7)a$ Use the commutative property of multiplication to change the order of the factors.

$\qquad\qquad = [15(-7)]a$ Use the associative property of multiplication to group the numbers together.

$\qquad\qquad = -105a$ Do the multiplication within the brackets: $15(-7) = -105$.

SELF CHECK   Simplify each expression: **a.** $9 \cdot 6s$ and **b.** $-5b(13)$.   *Answers:* **a.** $54s$, **b.** $-65b$ ■

In the next example, we will work with expressions involving two variables.

EXAMPLE 2

Simplifying algebraic expressions involving multiplication. Simplify each expression: **a.** $-5r(-6s)$ and   **b.** $3(7p)(-5p)$.

Solution   **a.** $-5r(-6s) = [-5(-6)][r \cdot s]$ Use the commutative and associative properties to group the numbers and to group the variables.

$= 30rs$ Do the multiplications within the brackets: $-5(-6) = 30$ and $r \cdot s = rs$.

b. $3(7p)(-5p) = [3(7)(-5)][p \cdot p]$ Use the commutative and associative properties to change the order and to regroup the factors.

$= (-105)(p^2)$ Do the multiplications within the brackets: $3(7)(-5) = -105$ and $p \cdot p = p^2$.

$= -105p^2$ Write the multiplication without parentheses.

SELF CHECK   Simplify each expression: **a.** $7p(-3q)$ and   *Answers:*   **a.** $-21pq$,
b. $-4(6m)(-2m)$.   **b.** $48m^2$ ∎

The Distributive Property

To introduce the **distributive property,** we will examine the expression $4(5 + 3)$, which can be evaluated in two ways.

Method 1: Rules for the order of operations: In this method, we compute the sum inside the parentheses first.

$4(5 + 3) = 4(8)$ Do the addition inside the parentheses first: $5 + 3 = 8$.

$= 32$ Do the multiplication.

Method 2: The distributive property: In this method, we distribute the 4 across 5 and 3, find each product separately, and add the results.

Distribute the 4.

$4(5 + 3) = 4(5 + 3)$

To apply the distributive property, we multiply each term inside the parentheses by the factor outside the parentheses.

First product   Second product

$= \quad 4(5) \quad + \quad 4(3)$

$= \quad 20 \quad + \quad 12$ Do the multiplications first: $4(5) = 20$ and $4(3) = 12$.

$= \quad 32$ Do the addition.

Notice that each method gives a result of 32.

We can interpret the distributive property geometrically. Figure 2-11 shows three rectangles that are divided into squares. Since the area of the rectangle on the left-hand side of the equals sign can be found by multiplying its width by its length, its area is $4(5 + 3)$ square units. We can evaluate this expression or we can count squares; either way, we see that the area is 32 square units.

FIGURE 2-11

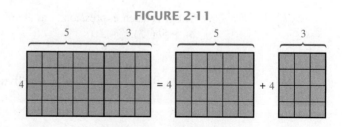

The area shown on the right-hand side is the sum of the areas of two rectangles: $4(5) + 4(3)$. Either by evaluating this expression or by counting squares, we see that this area is also 32 square units. Therefore,

$$4(5 + 3) = 4(5) + 4(3)$$

Figure 2-12 shows the general case where the width is a and the length is $b + c$.

FIGURE 2-12

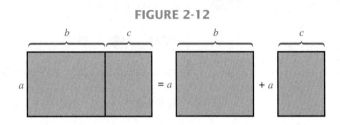

Using Figure 2-12 as a basis, we can state the distributive property in symbols.

The distributive property

If a, b, and c are real numbers, then

$$a(b + c) = ab + ac$$

Since subtraction is the same as adding the opposite, the distributive property also holds for subtraction.

The distributive property

If a, b, and c are real numbers, then

$$a(b - c) = ab - ac$$

We can use the distributive property to *remove the parentheses* when an expression is multiplied by a quantity. For example, to remove the parentheses in the expression $-5(x + 2)$, we proceed as follows:

$-5(x + 2) = -5(x) + (-5)(2)$	Distribute the multiplication by -5.
$= -5x + (-10)$	Do the multiplications.
$= -5x - 10$	Write the addition of -10 as subtraction of 10.

EXAMPLE 3

Applying the distributive property. Use the distributive property to remove parentheses: **a.** $3(x + 8)$, **b.** $-6(3y - 2)$, and **c.** $x(x + 2)$.

Solution

a.

$3(x + 8) = 3 \cdot x + 3 \cdot 8$	Distribute the 3.
$= 3x + 24$	Do the multiplications.

b.

$-6(3y - 2) = -6(3y) - (-6)(2)$	Distribute the -6.
$= -18y - (-12)$	Do the multiplications.
$= -18y + 12$	Add the opposite of -12, which is 12.

c.

$x(x + 2) = x \cdot x + x \cdot 2$	Distribute the x.
$= x^2 + 2x$	Do the multiplications. Recall that $x \cdot x = x^2$.

SELF CHECK   Remove parentheses: **a.** $5(p + 2)$,   *Answers:*   **a.** $5p + 10$,
b. $-8(2x - 4)$,   and   **c.** $p(p - 5)$.   **b.** $-16x + 32$,   **c.** $p^2 - 5p$ ■

WARNING!   If an expression contains parentheses, it does not necessarily mean that the distributive property can be applied. For example, the distributive property does not apply to the expressions

$$6(5x)   \quad \text{or} \quad   6(-7 \cdot y)   \qquad \text{Here a product is multiplied by 6.}$$

However, the distributive property does apply to the expressions

$$6(5 + x)   \quad \text{or} \quad   6(-7 - y)   \qquad \text{Here a sum or difference is multiplied by 6.}$$

Extending the Distributive Property

The distributive property can be extended to situations where there are more than two terms within parentheses.

The extended distributive property

If a, b, c, and d are real numbers, then

$$a(b + c + d) = ab + ac + ad   \qquad \text{and} \qquad   a(b - c - d) = ab - ac - ad$$

EXAMPLE 4   **Applying the extended distributive property.** Remove parentheses:
a. $5(x + y - z)$ and   **b.** $-0.3(3x - 4y + 7z)$.

Solution   **a.** $5(x + y - z) = 5 \cdot x + 5 \cdot y - 5 \cdot z$ Use the extended distributive property.

$$= 5x + 5y - 5z   \qquad \text{Write the multiplication without the multiplication dots.}$$

b. $-0.3(3x - 4y + 7z)$

$$= -0.3(3x) - (-0.3)(4y) + (-0.3)(7z)   \qquad \text{Distribute the } -0.3.$$
$$= -0.9x - (-1.2y) + (-2.1z)   \qquad \text{Do the multiplications.}$$
$$= -0.9x + 1.2y + (-2.1z)   \qquad \text{Do the subtraction of } -1.2y \text{ by adding its opposite, which is } 1.2y.$$

$$= -0.9x + 1.2y - 2.1z   \qquad \text{Write the addition of } -2.1z \text{ as a subtraction.}$$

SELF CHECK   Remove parentheses in the expression
$-7(2r + 5s - 8t)$.   *Answer:* $-14r - 35s + 56t$ ■

Since multiplication is commutative, we can write the distributive property in the following forms.

$$(b + c)a = ba + ca,   \qquad (b - c)a = ba - ca,   \qquad (b + c + d)a = ba + ca + da$$

EXAMPLE 5

Applying the distributive property. Multiply   **a.** $(6 + 4p)5$ and
b. $(-12 - 6x + 4y)\dfrac{1}{2}$.

Solution   **a.** $(6 + 4p)5 = 6 \cdot 5 + 4p \cdot 5$ Use the distributive property.

$= 30 + 20p$ Do the multiplications.

b. $(-12 - 6x + 4y)\dfrac{1}{2}$

$= -12 \cdot \dfrac{1}{2} - 6x \cdot \dfrac{1}{2} + 4y \cdot \dfrac{1}{2}$ Distribute the $\frac{1}{2}$.

$= -6 - 3x + 2y$ Do the multiplications.

SELF CHECK   Multiply $(-5x - 4y)8$.   *Answer:* $-40x - 32y$ ∎

To use the distributive property to simplify $-(x + 10)$, we note that the negative sign in front of the parentheses represents -1.

The $-$ sign represents -1.
↓   ↓
$-(x + 10) = -1(x + 10)$

$= -1(x) + (-1)(10)$ Use the distributive property to distribute -1.

$= -x + (-10)$ Do the multiplications.

$= -x - 10$ Write the addition of -10 as a subtraction.

EXAMPLE 6

Distributing a negative sign. Simplify $-(-12 - 3p)$.

Solution $-(-12 - 3p)$

$= -1(-12 - 3p)$ Change the $-$ sign in front of the parentheses to -1.

$= -1(-12) - (-1)(3p)$ Use the distributive property to distribute -1.

$= 12 - (-3p)$ Do the multiplications.

$= 12 + 3p$ To subtract $-3p$, add the opposite of $-3p$, which is $3p$.

SELF CHECK   Simplify $-(-5x + 18)$.   *Answer:* $5x - 18$ ∎

Terms of an Algebraic Expression

Addition signs separate algebraic expressions into parts called **terms.** The expression $5x + 8$ contains two terms, $5x$ and 8.

The $+$ signs separates the
expression into two terms.
↓
$5x$ $+$ 8
↑   ↑
First term   Second term

A term may be

- a number. Examples are 8, 98.6, and -45.

- a variable, or a product of variables (which may be raised to powers). Examples are x, s^3, rt, and a^2bc^4.

- a product of a number and one or more variables (which may be raised to powers). Examples are $-35x$, $\frac{1}{2}bh$, and πr^2h.

Since subtraction can be expressed as addition of the opposite, the expression $6x - 5$ can be written in the equivalent form $6x + (-5)$. We can then see that $6x - 5$ contains two terms, $6x$ and -5.

EXAMPLE 7

Identifying terms. List the terms in each expression: **a.** $-4p + 7 + 5p$, **b.** $-12r^2st$, and **c.** $y^3 + 8y^2 - 3y + 24$

Solution   **a.** $-4p + 7 + 5p$ has three terms: $-4p$, 7, and $5p$.

b. The expression $-12r^2st$ has one term: $-12r^2st$.

c. $y^3 + 8y^2 - 3y - 24$ can be written as $y^3 + 8y^2 + (-3y) + (-24)$. It contains four terms: y^3, $8y^2$, $-3y$, and 24.

SELF CHECK   List the terms in each expression: **a.** $\frac{1}{3}Bh$, **b.** $3q + 5q - 1.2$, and **c.** $b^2 - 4ac$.   *Answers:* **a.** $\frac{1}{3}Bh$, **b.** $3q$, $5q$, -1.2, **c.** b^2, $-4ac$ ■

Coefficients of a Term

In a term that is the product of a number and one or more variables, the number factor is called the **numerical coefficient,** or simply the **coefficient.** In the expression $5x$, 5 is the coefficient and x is the variable part. Other examples are shown in Table 2-1.

TABLE 2-1

Term	Coefficient	Variable part
$8y^2$	8	y^2
$-0.9pq$	-0.9	pq
$\frac{3}{4}b$	$\frac{3}{4}$	b
$-\frac{x}{6}$	$-\frac{1}{6}$	x
x	1	x
$-t$	-1	t
15	15	none

Notice that when there is no number in front of a variable, the coefficient is 1. For example, $x = 1x$. When there is only a negative sign in front of the variable, the coefficient is -1. For example, $-t = -1t$.

EXAMPLE 8

Identifying coefficients of terms. Identify the coefficient of each term in the expression $-7x^2 + 3x - 6$.

Solution

Term	Coefficient
$-7x^2$	-7
$3x$	3
-6	-6

SELF CHECK   Identify the coefficient of each term in the expression $p^3 - 12p^2 + 3p - 4$.   *Answers:* 1, -12, 3, -4 ■

Terms and Factors

It is important to distinguish between a *term* of an expression and a *factor* of a term. Terms are separated by an addition sign $+$. Factors are numbers or variables that are multiplied together.

EXAMPLE 9

Determine whether the variable x is a factor or a term: **a.** $18 + x$, **b.** $18x$, and **c.** $18 - 3x$.

Solution
a. x is a *term* of $18 + x$, since x and 18 are separated by a $+$ sign.

b. x is a *factor* of $18x$, since x and 18 are multiplied.

c. The expression $18 - 3x$ can be written $18 + (-3x)$. The $+$ sign separates the expression into two terms, 18 and $-3x$. The variable x is a *factor* of the second term.

SELF CHECK
Determine whether y is a factor or a term: **a.** $32y$, **b.** $-45 + y$, and **c.** $-45y - 25$.

Answers: **a.** factor, **b.** term, **c.** factor ■

Like Terms

The expression $5p + 7q - 3p + 12$, which can be written $5p + 7q + (-3p) + 12$, contains four terms, $5p$, $7q$, $-3p$, and 12. Since the variable of $5p$ and $-3p$ are the same, we say that these terms are **like** or **similar terms**.

Like terms (similar terms)

Like terms (or **similar terms**) are terms with exactly the same variables raised to exactly the same powers. Any numbers (called **constants**) in an expression are considered to be like terms.

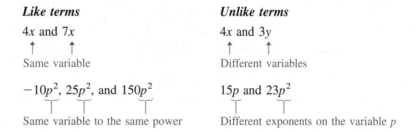

Like terms	*Unlike terms*
$4x$ and $7x$	$4x$ and $3y$
↑ ↑	↑ ↑
Same variable	Different variables
$-10p^2$, $25p^2$, and $150p^2$	$15p$ and $23p^2$
Same variable to the same power	Different exponents on the variable p

WARNING! When looking for like terms, don't look at the coefficients of the terms. Consider only their variable parts.

EXAMPLE 10

Identifying like terms. List like terms: **a.** $7r + 5 + 3r$, **b.** $x^4 - 6x^2 - 5$, and **c.** $-7m + 7 - 2 + m$.

Solution
a. $7r + 5 + 3r$ contains the like terms $7r$ and $3r$.

b. $x^4 - 6x^2 - 5$ contains no like terms.

c. $-7m + 7 - 2 + m$ contains two pairs of like terms: $-7m$ and m are like terms, and the constants, 7 and -2, are like terms.

SELF CHECK
List like terms: **a.** $5x - 2y + 7y$ and **b.** $-5pq + 17p - 12q - 2pq$.

Answers: **a.** $-2y$ and $7y$, **b.** $-5pq$ and $-2pq$ ■

Combining Like Terms

If we are to add (or subtract) objects, they must have the same units. For example, we can add dollars to dollars and inches to inches, but we cannot add dollars to inches. The same is true when we work with terms of an algebraic expression. They can be added or subtracted only when they are like terms.

This expression can be simplified, because it contains like terms.

$$3x + 4x$$

↑   ↑
Like terms
The variable parts are identical.

This expression cannot be simplified, because its terms are not like terms.

$$3x + 4y$$

↑   ↑
Unlike terms
The variable parts are not identical.

To simplify an expression containing like terms, we use the distributive property. For example, we can simplify $3x + 4x$ as follows:

$$3x + 4x = (3 + 4)x \quad \text{Apply the distributive property.}$$
$$= 7x \qquad \text{Do the addition in the parentheses: } 3 + 4 = 7.$$

We have simplified the expression $3x + 4x$ by **combining like terms**. The result is the equivalent expression $7x$. This example suggests the following general rule.

Combining like terms

To add or subtract like terms, combine their coefficients and keep the same variables with the same exponents.

EXAMPLE 11

Simplifying algebraic expressions. Simplify by combining like terms:
a. $-8p + (-12p)$, and **b.** $0.5s^2 - 0.3s^2$.

Solution   **a.** $-8p + (-12p) = -20p$ Add the coefficients of the like terms: $-8 + (-12) = -20$. Keep the variable p.

b. $0.5s^2 - 0.3s^2 = 0.2s^2$ Subtract: $0.5 - 0.3 = 0.2$. Keep the variable part s^2.

SELF CHECK   Simplify by combining like terms:
a. $5n + (-8n)$ and   **b.** $-1.2a^3 + (1.4a^3)$.   *Answers:*   **a.** $-3n$,   **b.** $0.2a^3$ ■

EXAMPLE 12

Combining like terms. Simplify $7P - 8p - 12P + 25p$.

Solution   The uppercase P and the lowercase p are different variables. To combine like terms, one approach is to write each subtraction as an addition of the opposite and proceed as follows.

$$7P - 8p - 12P + 25p$$
$$= 7P + (-8p) + (-12P) + 25p \quad \text{Rewrite each subtraction as the addition of the opposite.}$$
$$= 7P + (-12P) + (-8p) + 25p \quad \text{Use the commutative property of addition to write the like terms together.}$$
$$= -5P + 17p \qquad \text{Combine like terms: } 7P + (-12P) = -5P \text{ and } -8p + 25p = 17p.$$

SELF CHECK   Simplify $8R + 7r - 14R - 21r$.   *Answer:* $-6R - 14r$ ■

The expression in Example 12 contained two sets of like terms, and we rearranged the terms so that like terms were next to each other. With practice, you will be able to combine like terms without having to write them next to each other and without having to write each subtraction as addition of the opposite.

EXAMPLE 13

Combining like terms without rearranging terms. Simplify $4(x + 5) - 3(2x - 4)$.

Solution

$$4(x + 5) - 3(2x - 4)$$
$$= 4x + 20 - 6x + 12 \quad \text{Use the distributive property twice.}$$
$$= -2x + 32 \quad \text{Combine like terms: } 4x - 6x = -2x \text{ and } 20 + 12 = 32.$$

SELF CHECK Simplify $-5(y - 4) + 2(4y + 6)$. *Answer:* $3y + 32$ ∎

STUDY SET

Section 2.4

VOCABULARY

In Exercises 1–6, fill in the blanks to make the statements true.

1. To ___simplify___ an algebraic expression, we use properties of algebra to write the expression in a less complicated form.

2. A ___term___ is a number or a product of a number and one or more variables.

3. In the term $3x^2$, the number factor 3 is called the ___coefficient___ and x^2 is called the ___variable___ part.

4. Two terms with exactly the same variables and exponents are called ___like___ terms.

5. We can use the distributive property to ___remove___ the parentheses when an expression is multiplied by a quantity.

6. The ___commutative___ property of multiplication enables us to change the order of the factors involved in a multiplication.

CONCEPTS

7. What property does the statement $a(b + c) = ab + ac$ illustrate? the distributive property

8. Complete this statement:
$$a(b + c + d) = \underline{ab + ac + ad}$$

9. What application of the distributive property is demonstrated in Illustration 1? $2(3 + 4) = 2 \cdot 3 + 2 \cdot 4$

ILLUSTRATION 1

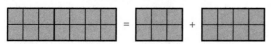

10. Complete this statement: To add or subtract like terms, combine their ___coefficients___ and keep the same variables and ___exponents___.

11. A board was cut into two pieces, as shown in Illustration 2. Add the lengths of the two pieces. How long was the original board? $x + 20 - x = 20$; 20 ft

ILLUSTRATION 2

x ft $(20 - x)$ ft

12. Let x equal the number of miles driven the first day of a 2-day driving trip. Translate the verbal model to mathematical symbols and simplify by combining like terms. $x + x + 100 = 2x + 100$

| the miles driven day 1 | plus | 100 miles more than the miles driven day 1 |

13. Two angles are **complementary angles** if the sum of their measures is 90°. If one of the angles has a measure of $a°$ (see Illustration 3), use an algebraic expression to represent the measure of its complement. $(90 - a)°$

14. Two angles are **supplementary angles** if the sum of their measures is 180°. If one of the angles has a measure of $a°$ (see Illustration 4), use an algebraic expression to represent the measure of its supplement. $(180 - a)°$

which are "like terms"

ILLUSTRATION 3   **ILLUSTRATION 4**

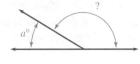

NOTATION

In Exercises 15–16, complete each solution.

15. $-7(a^2 + a - 5) = \boxed{-7} \cdot a^2 + \left(\boxed{-7}\right) \cdot a - \left(\boxed{-7}\right) \cdot 5$

$= -7a^2 + (-7a) - \left(\boxed{-35}\right)$

$= -7a^2 - \boxed{7a} - (-35)$

$= -7a^2 - 7a + 35$

16. $6(b - 5) + 12b + 7 = 6 \cdot \boxed{b} - 6 \cdot \boxed{5} + 12b + 7$

$= 6b - \boxed{30} + 12b + 7$

$= 6b + \boxed{12}\,b - \boxed{30} + 7$

$= 18b - 23$

17. a. Are $2K$ and $3k$ like terms?   no
   b. Are $-d$ and d like terms?   yes

18. Fill in the blank to make the statement true.

$-(x + 10) = - \boxed{1}\,(x + 10)$

PRACTICE

In Exercises 19–26, simplify each expression.

19. $9(7m)$ $63m$

20. $12n(8)$ $96n$

21. $5(-7q)$ $-35q$

22. $-7(5t)$ $-35t$

23. $(-5p)(-4b)$ $20bp$

24. $(-7d)(-7c)$ $49cd$

25. $-5(4r)(-2r)$ $40r^2$

26. $7t(-4t)(-2)$ $56t^2$

In Exercises 27–46, use the distributive property to remove parentheses.

27. $5(x + 3)$ $5x + 15$

28. $4(x + 2)$ $4x + 8$

29. $-2(b - 1)$ $-2b + 2$

30. $-7(p - 5)$ $-7p + 35$

31. $(3t - 2)8$ $24t - 16$

32. $(2q + 1)9$ $18q + 9$

33. $(2y - 1)6$ $12y - 6$

34. $(3w - 5)5$ $15w - 25$

35. $0.4(x - 4)$ $0.4x - 1.6$

36. $-2.2(2q + 1)$
   $-4.4q - 2.2$

37. $-\dfrac{2}{3}(3w - 6)$ $-2w + 4$

38. $\dfrac{1}{2}(2y - 8)$ $y - 4$

39. $r(r - 10)$ $r^2 - 10r$

40. $h(h + 4)$ $h^2 + 4h$

41. $-(x - 7)$ $-x + 7$

42. $-(y + 1)$ $-y - 1$

43. $17(2x - y + 2)$
   $34x - 17y + 34$

44. $-12(3a + 2b - 1)$
   $-36a - 24b + 12$

45. $-(-14 + 3p - t)$
   $14 - 3p + t$

46. $-(-x - y + 5)$
   $x + y - 5$

In Exercises 47–54, list the terms in each expression. Then identify the coefficient of each term.

47. $-5r + 4s$ $-5r, 4s;\ -5, 4$

48. $2m + n - 3m + 2n$ $2m, n, -3m, 2n;\ 2, 1, -3, 2$

49. $-15r^2s$ $-15r^2s;\ -15$

50. $4b^2 - 5b + 6$ $4b^2, -5b, 6;\ 4, -5, 6$

51. $50a + 2$ $50a, 2;\ 50, 2$

52. $a^2 - ab + b^2$ $a^2, -ab, b^2;\ 1, -1, 1$

53. $x^3 - 125$ $x^3, -125;\ 1, -125$

54. $-2.55x + 1.8$ $-2.55x, 1.8;\ -2.55, 1.8$

In Exercises 55–58, identify the coefficient of each term.

55. $-b$ -1

56. $-9.9x^3$ -9.9

57. $\dfrac{1}{4}x$ $\frac{1}{4}$

58. $-\dfrac{2x}{3}$ $-\frac{2}{3}$

In Exercises 59–62, tell whether the variable x is used as a factor or a term.

59. $24 - x$ term

60. $24x$ factor

61. $24 + 3x$ factor

62. $x - 12$ term

In Exercises 63–90, simplify each expression by combining like terms.

63. $3x + 17x$ $20x$

64. $12y - 15y$ $-3y$

65. $8x^2 - 5x^2$ $3x^2$

66. $17x^2 + 3x^2$ $20x^2$

67. $-4x + 4x$ 0

68. $-16y + 16y$ 0

69. $-7b^2 + 7b^2$ 0

70. $-2c^3 + 2c^3$ 0

71. $3x + 5x - 7x$ x

72. $-y + 3y + 2y$ $4y$

73. $13x^2 + 2x^2 - 5x^2$
   $10x^2$

74. $8x^3 - x^3 + 2x^3$ $9x^3$

75. $1.8h - 0.7h$ $1.1h$

76. $-5.7m + 4.3m$ $-1.4m$

77. $\dfrac{3}{5}t + \dfrac{1}{5}t$ $\frac{4}{5}t$

78. $\dfrac{3}{16}x - \dfrac{5}{16}x$ $-\frac{1}{8}x$

79. $4(y + 9) - 6y$
   $-2y + 36$

80. $-3(3 + z) + 2z$
   $-z - 9$

81. $2z + 5(z - 3)$ $7z - 15$

82. $12(m + 11) - 11$
   $12m + 121$

83. $8(c + 7) - 2(c - 3)$
   $6c + 62$

84. $9(z + 2) + 5(3 - z)$
   $4z + 33$

85. $2x + 4(X - x) + 3X$
   $7X - 2x$

86. $3p - 6(p + z) + p$
   $-2p - 6z$

87. $(a + 2) - (a - b)$
   $2 + b$

88. $3z + 2(Z - z) + Z$
   $3Z + z$

89. $x(x + 3) - 3x^2$
   $-2x^2 + 3x$

90. $2x + x(x - 3)$ $x^2 - x$

APPLICATIONS

91. THE AMERICAN RED CROSS In 1891, Clara Barton founded the Red Cross. Its symbol is a white flag bearing a red cross. If each side of the cross in Illustration 5 has length x, write an algebraic expression for the perimeter (the total distance around the outside) of the cross. 12x

ILLUSTRATION 5

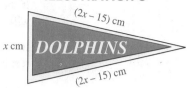

92. BILLIARDS Billiard tables vary in size, but all tables are twice as long as they are wide.
a. If the billiard table in Illustration 6 is x feet wide, write an expression involving x that represents its length. 2x ft
b. Write an expression for the perimeter of the table. 6x ft

ILLUSTRATION 6

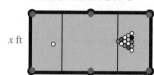

x ft

93. PING-PONG Write an expression for the perimeter of the ping-pong table shown in Illustration 7. $(4x + 8)$ ft

ILLUSTRATION 7

$(x + 4)$ ft — x ft

94. SEWING See Illustration 8. Write an expression for the length of the yellow trim needed to outline the pennant with the given side lengths. $(5x - 30)$ cm

ILLUSTRATION 8

$(2x - 15)$ cm

x cm *DOLPHINS*

$(2x - 15)$ cm

WRITING

95. Explain why $3x^2y$ and $5x^2y$ are like terms.

96. Explain why $3x^2y$ and $5xy^2$ are not like terms.

97. Distinguish between a *factor* and a *term* of an algebraic expression. Give examples.

98. Tell how to combine like terms.

REVIEW

In Exercises 99–102, evaluate each expression given that $x = -3$, $y = -5$, and $z = 0$.

99. $x^2z(y^3 - z)$ 0

100. $z - y^3$ 125

101. $\dfrac{x - y^2}{2y - 1 + x}$ 2

102. $\dfrac{2y + 1}{x} - x$ 6

▶ 2.5 Solving Equations

In this section, you will learn about

> Solving equations by using more than one property of equality ■ Simplifying expressions to solve equations ■ Identities and impossible equations ■ Problem solving

Introduction In Chapter 1, we solved some simple equations using the properties of equality. In this section, we will solve more complicated equations. Our objective is to develop a general strategy that can be used to solve any kind of linear equation.

Solving Equations by Using More Than One Property of Equality

Recall the following properties of equality:

- If the same quantity is added to (or subtracted from) equal quantities, the results will be equal quantities.

- If equal quantities are multiplied (or divided) by the same nonzero quantity, the results will be equal quantities.

We have already solved many simple equations using one of the properties listed above. For example, to solve $x + 6 = 10$, we isolate x by subtracting 6 from both sides.

$$x + 6 = 10$$
$$x + 6 - 6 = 10 - 6 \qquad \text{To undo the addition of 6, subtract 6 from both sides.}$$
$$x = 4 \qquad \text{Do the subtractions.}$$

To solve $2x = 10$, we isolate x by dividing both sides by 2.

$$2x = 10$$
$$\frac{2x}{2} = \frac{10}{2} \qquad \text{To undo the multiplication by 2, divide both sides by 2.}$$
$$x = 5 \qquad \text{Do the divisions: } \tfrac{2}{2} = 1 \text{ and } \tfrac{10}{2} = 5.$$

Sometimes several properties of equality must be applied in succession to solve an equation. For example, on the left-hand side of $2x + 6 = 10$, x is first multiplied by 2 and then 6 is added to that product. To isolate x, we use the rules for the order of operations in reverse. First, we undo the addition of 6, and then we undo the multiplication by 2.

$$2x + 6 = 10$$
$$2x + 6 - 6 = 10 - 6 \qquad \text{To undo the addition of 6, subtract 6 from both sides.}$$
$$2x = 4 \qquad \text{Do the subtractions.}$$
$$\frac{2x}{2} = \frac{4}{2} \qquad \text{To undo the multiplication by 2, divide both sides by 2.}$$
$$x = 2 \qquad \text{Do the divisions: } \tfrac{2}{2} = 1 \text{ and } \tfrac{4}{2} = 2.$$

EXAMPLE 1

Using two properties of equality. Solve and check: $-12x + 5 = 17$.

Solution   The left-hand side indicates that we must multiply x by -12 and then add 5. To isolate x, we undo these operations in the opposite order.

- To undo the addition of 5, we subtract 5 from both sides.

- To undo the multiplication by -12, we divide both sides by -12.

$$-12x + 5 = 17$$
$$-12x + 5 - 5 = 17 - 5 \qquad \text{Subtract 5 from both sides.}$$
$$-12x = 12 \qquad \text{Do the subtractions: } 5 - 5 = 0 \text{ and } 17 - 5 = 12.$$
$$\frac{-12x}{-12} = \frac{12}{-12} \qquad \text{Divide both sides by } -12.$$
$$x = -1 \qquad \text{Do the divisions: } \tfrac{-12}{-12} = 1 \text{ and } \tfrac{12}{-12} = -1.$$

Check: $-12x + 5 = 17$ The original equation.

$$-12(-1) + 5 \stackrel{?}{=} 17 \quad \text{Substitute } -1 \text{ for } x.$$

$$12 + 5 \stackrel{?}{=} 17 \quad \text{Do the multiplication: } -12(-1) = 12.$$

$$17 = 17 \quad \text{Do the addition.}$$

Since the statement $17 = 17$ is true, -1 satisfies the equation.

SELF CHECK   Solve and check: $8x - 13 = 43$.   *Answer:* 7   ■

EXAMPLE 2   **Using two properties of equality.** Solve and check: $\dfrac{2x}{3} = -6$.

Solution   The left-hand side indicates that we must multiply x by 2 and divide that product by 3. To solve this equation, we must undo these operations in the opposite order.

- To undo the division of 3, we multiply both sides by 3.
- To undo the multiplication by 2, we divide both sides by 2.

$$\frac{2x}{3} = -6$$

$$3\left(\frac{2x}{3}\right) = 3(-6) \quad \text{Multiply both sides by 3.}$$

$$2x = -18 \quad \text{Simplify: } \tfrac{3}{3} = 1 \text{ and } 3(-6) = -18.$$

$$\frac{2x}{2} = \frac{-18}{2} \quad \text{Divide both sides by 2.}$$

$$x = -9 \quad \text{Do the divisions: } \tfrac{2}{2} = 1 \text{ and } \tfrac{-18}{2} = -9.$$

Check: $\dfrac{2x}{3} = -6$ The original equation.

$$\frac{2(-9)}{3} \stackrel{?}{=} -6 \quad \text{Substitute } -9 \text{ for } x.$$

$$\frac{-18}{3} \stackrel{?}{=} -6 \quad \text{Do the multiplication: } 2(-9) = -18.$$

$$-6 = -6 \quad \text{Do the division.}$$

Since we obtain a true statement, -9 is a solution.

SELF CHECK   Solve and check: $\dfrac{7h}{16} = -14$.   *Answer:* -32 ■

Another approach can be used to solve the equation from Example 2, $\frac{2x}{3} = -6$. To isolate the variable, we will use the fact that the product of a number and its **reciprocal**, or **multiplicative inverse**, is 1. Since $\frac{2x}{3} = \frac{2}{3}x$, the equation can be rewritten as

$$\frac{2}{3}x = -6$$

To isolate x, we multiply both sides by $\frac{3}{2}$, the reciprocal (multiplicative inverse) of $\frac{2}{3}$.

$$\frac{3}{2}\left(\frac{2}{3}x\right) = \frac{3}{2}(-6) \quad \begin{array}{l}\text{The coefficient of } x \text{ is } \tfrac{2}{3}. \text{ Multiply both sides by the reciprocal of} \\ \tfrac{2}{3}, \text{ which is } \tfrac{3}{2}.\end{array}$$

$$\left(\frac{3}{2} \cdot \frac{2}{3}\right)x = \frac{3}{2}(-6) \quad \begin{array}{l}\text{Apply the associative property of multiplication to regroup} \\ \text{factors.}\end{array}$$

$$1x = -9 \quad \text{Do the multiplications: } \tfrac{3}{2} \cdot \tfrac{2}{3} = 1 \text{ and } \tfrac{3}{2}(-6) = -9.$$

$$x = -9 \quad 1x = x.$$

EXAMPLE 3

Isolating the variable. Solve $-0.2 = -0.8 - y$.

Solution To solve the equation, we begin by eliminating -0.8 from the right-hand side. We can do this by adding 0.8 to both sides.

$$-0.2 = -0.8 - y$$
$$-0.2 + \mathbf{0.8} = -0.8 - y + \mathbf{0.8} \quad \text{Add 0.8 to both sides.}$$
$$0.6 = -y \qquad \text{Do the additions: } -0.2 + 0.8 = 0.6 \text{ and}$$
$$-0.8 + 0.8 = 0.$$

Since $-y$ means $-1y$, the equation can be rewritten as $0.6 = -1y$. To isolate y, either multiply both sides or divide both sides by -1.

$$0.6 = -1y \quad \text{Write } -y \text{ as } -1y.$$
$$\frac{0.6}{-1} = \frac{-1y}{-1} \quad \text{Divide both sides by } -1.$$
$$-0.6 = y \qquad \text{Do the divisions: } \frac{0.6}{-1} = -0.6 \text{ and } \frac{-1}{-1} = 1.$$
$$y = -0.6$$

Verify that -0.6 satisfies the equation.

SELF CHECK   Solve $-6.6 - m = -2.7$.   *Answer:* -3.9 ∎

EXAMPLE 4

Using three properties of equality.   Solve $\dfrac{3x}{4} + 2 = -7$.

Solution The left-hand side indicates that we must multiply x by 3, divide that product by 4, and then add 2. To solve the equation, we must undo these operations in the opposite order.

- To undo the addition of 2, we subtract 2 from both sides.

- To undo the division by 4, we multiply both sides by 4.

- To undo the multiplication by 3, we divide both sides by 3.

$$\frac{3x}{4} + 2 = -7$$

$$\frac{3x}{4} + 2 - \mathbf{2} = -7 - \mathbf{2} \quad \text{Subtract 2 from both sides.}$$

$$\frac{3x}{4} = -9 \qquad \text{Do the subtractions: } 2 - 2 = 0 \text{ and } -7 - 2 = -9.$$

$$4\left(\frac{3x}{4}\right) = 4(-9) \quad \text{Multiply both sides by 4.}$$

$$3x = -36 \qquad \text{Simplify: } \tfrac{4}{4} = 1 \text{ and } 4(-9) = -36.$$

$$\frac{3x}{3} = \frac{-36}{3} \qquad \text{Divide both sides by 3.}$$

$$x = -12 \qquad \text{Do the divisions: } \tfrac{3}{3} = 1 \text{ and } \tfrac{-36}{3} = -12.$$

Verify that -12 satisfies the equation.

SELF CHECK   Solve $\dfrac{2}{3}b - 3 = -15$.

Answer: -18 ∎

Simplifying Expressions to Solve Equations

In the next example, we must use the distributive property to remove parentheses.

EXAMPLE 5

Combining like terms. Solve and check: $3(k + 1) - 5k = 0$.

Solution

$$3(k + 1) - 5k = 0$$

$$3k + 3(1) - 5k = 0 \qquad \text{Use the distributive property to remove parentheses.}$$

$$3k - 5k + 3 = 0 \qquad \text{Do the multiplication and rearrange terms.}$$

$$-2k + 3 = 0 \qquad \text{Simplify the left-hand side by combining like terms.}$$

$$-2k + 3 - 3 = 0 - 3 \qquad \text{To undo the addition of 3, subtract 3 from both sides.}$$

$$-2k = -3 \qquad \text{Do the subtractions: } 3 - 3 = 0 \text{ and } 0 - 3 = -3.$$

$$\frac{-2k}{-2} = \frac{-3}{-2} \qquad \text{To undo the multiplication by } -2, \text{ divide both sides by } -2.$$

$$k = \frac{3}{2} \qquad \text{Simplify: } \frac{-2}{-2} = 1 \text{ and } \frac{-3}{-2} = \frac{3}{2}.$$

Check:

$$3(k + 1) - 5k = 0 \qquad \text{The original equation.}$$

$$3\left(\frac{3}{2} + 1\right) - 5\left(\frac{3}{2}\right) \stackrel{?}{=} 0 \qquad \text{Substitute } \tfrac{3}{2} \text{ for } k.$$

$$3\left(\frac{3}{2} + \frac{2}{2}\right) - 5\left(\frac{3}{2}\right) \stackrel{?}{=} 0 \qquad 1 = \tfrac{2}{2}.$$

$$3\left(\frac{5}{2}\right) - 5\left(\frac{3}{2}\right) \stackrel{?}{=} 0 \qquad \text{Do the addition inside the parentheses.}$$

$$\frac{15}{2} - \frac{15}{2} \stackrel{?}{=} 0 \qquad \text{Do the multiplications.}$$

$$0 = 0 \qquad \text{Do the subtraction.}$$

SELF CHECK   Solve and check: $-5(x - 3) + 3x = 11$.   *Answer:* 2   ■

EXAMPLE 6

Variables on both sides of the equation. Solve and check: $3x - 15 = 4x + 36$.

Solution   To solve for x, all the terms containing x must be on the same side of the equation. We can eliminate $3x$ from the left-hand side by subtracting $3x$ from both sides.

$$3x - 15 = 4x + 36$$

$$3x - 15 - 3x = 4x + 36 - 3x \qquad \text{Subtract } 3x \text{ from both sides.}$$

$$-15 = x + 36 \qquad \text{Combine like terms: } 3x - 3x = 0 \text{ and } 4x - 3x = x.$$

$$-15 - 36 = x + 36 - 36 \qquad \text{To undo the addition of 36, subtract 36 from both sides.}$$

$$-51 = x \qquad \text{Do the subtractions: } -15 - 36 = -51 \text{ and } 36 - 36 = 0.$$

$$x = -51$$

Check: $3x - 15 = 4x + 36$ The original equation.

$3(-51) - 15 \stackrel{?}{=} 4(-51) + 36$ Substitute -51 for x.

$-153 - 15 \stackrel{?}{=} -204 + 36$ Do the multiplications.

$-168 = -168$ Simplify each side.

SELF CHECK   Solve and check: $3n + 48 = -4n - 8$.   *Answer:* -8 ■

 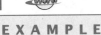

EXAMPLE 7   **Clearing an equation of fractions.** Solve $\dfrac{x}{6} - \dfrac{5}{2} = -\dfrac{1}{3}$.

Solution   Since integers are easier to work with, we will clear the equation of the fractions by multiplying both sides by the least common denominator (LCD), which in this case is 6.

$$\frac{x}{6} - \frac{5}{2} = -\frac{1}{3}$$

$6\left(\dfrac{x}{6} - \dfrac{5}{2}\right) = 6\left(-\dfrac{1}{3}\right)$ 6 is the smallest number that each denominator will divide exactly. Multiply both sides by 6 to clear the fractions.

$6\left(\dfrac{x}{6}\right) - 6\left(\dfrac{5}{2}\right) = 6\left(-\dfrac{1}{3}\right)$ On the left-hand side, remove parentheses.

$x - 15 = -2$ Do the multiplications.

$x - 15 + 15 = -2 + 15$ To undo the subtraction of 15, add 15 to both sides.

$x = 13$ Do the additions: $-15 + 15 = 0$ and $-2 + 15 = 13$.

Verify that 13 satisfies the equation.

SELF CHECK   Solve $\dfrac{x}{4} + \dfrac{1}{2} = -\dfrac{1}{8}$.   *Answer:* $-\dfrac{5}{2}$ ■

The preceding examples suggest the following strategy for solving equations.

Strategy for solving equations

1. Clear the equation of fractions.
2. Use the distributive property to remove parentheses, if necessary.
3. Combine like terms if necessary.
4. Undo the operations of addition and subtraction to get the variables on one side and the constants on the other.
5. Undo the operations of multiplication and division to isolate the variable.
6. Check the result.

EXAMPLE 8   **Applying the equation-solving strategy.** Solve and check: $\dfrac{3x + 11}{5} = x + 3$.

Solution

$$\frac{3x + 11}{5} = x + 3$$

$$5\left(\frac{3x + 11}{5}\right) = 5(x + 3)$$ Clear the equation of the fraction by multiplying both sides by 5.

$$3x + 11 = 5x + 15$$ On the left-hand side, simplify: $\frac{\overset{1}{\cancel{5}}}{1}\left(\frac{3x + 11}{\underset{1}{\cancel{5}}}\right)$. On the right-hand side, remove parentheses.

$$3x + 11 - 11 = 5x + 15 - 11$$ Subtract 11 from both sides.

$$3x = 5x + 4$$ Do the subtractions.

$$3x - 5x = 5x + 4 - 5x$$ To eliminate $5x$ from the right-hand side, subtract $5x$ from both sides.

$$-2x = 4$$ Combine like terms: $3x - 5x = -2x$ and $5x - 5x = 0$.

$$\frac{-2x}{-2} = \frac{4}{-2}$$ To undo the multiplication by -2, divide both sides by -2.

$$x = -2$$ Do the divisions.

Verify that -2 satisfies the equation.

SELF CHECK Solve and check: $\dfrac{3x + 23}{7} = x + 5$.

Answer: -3

 WARNING! Remember that when you multiply one side of an equation by a nonzero number, you must multiply the other side of the equation by the same number.

Identities and Impossible Equations

Equations in which some numbers satisfy the equation and others don't are called **conditional equations.** The equations in Examples 1–8 are conditional equations.

An equation that is true for all values of its variable is called an **identity.**

$x + x = 2x$ This is an identity, because it is true for all values of x.

An equation that is not true for any values of its variable is called an **impossible equation** or a **contradiction.** Such equations are said to have no solution.

$x = x + 1$ Because no number is 1 greater than itself, this is an impossible equation.

EXAMPLE 9 **Identities.** Solve $3(x + 8) + 5x = 2(12 + 4x)$.

Solution

$$3(x + 8) + 5x = 2(12 + 4x)$$
$$3x + 24 + 5x = 24 + 8x$$ Remove parentheses.
$$8x + 24 = 24 + 8x$$ Combine like terms.
$$8x + 24 - 8x = 24 + 8x - 8x$$ Subtract $8x$ from both sides.
$$24 = 24$$ Combine like terms: $8x - 8x = 0$.

Since the result $24 = 24$ is true for every number x, all values of x satisfy the original equation. This equation is an identity.

SELF CHECK Solve $3(x + 5) - 4(x + 4) = -x - 1$.

Answer: all values of x; this equation is an identity

EXAMPLE 10	**Impossible equations.** Solve $3(d + 7) - d = 2(d + 10)$.

Solution

$$3(d + 7) - d = 2(d + 10)$$
$$3d + 21 - d = 2d + 20 \qquad \text{Remove parentheses.}$$
$$2d + 21 = 2d + 20 \qquad \text{Combine like terms.}$$
$$2d + 21 - \mathbf{2d} = 2d + 20 - \mathbf{2d} \qquad \text{Subtract } 2d \text{ from both sides.}$$
$$21 = 20 \qquad \text{Combine like terms.}$$

Since the result $21 = 20$ is false, the original equation has no solution. It is an impossible equation.

SELF CHECK   Solve $-4(c - 3) + 2c = 2(10 - c)$.

Answer: No solution. This equation is an impossible equation. ■

Problem Solving

We can use equations to solve many types of problems. To set up and solve the equations, we will follow these steps:

Strategy for problem solving

1. **Analyze the problem** by reading it carefully to understand the given facts. What information is given? What vocabulary is given? What are you asked to find? Often, a diagram will help you visualize the facts of the problem.

2. **Form an equation** by picking a variable to represent the quantity to be found. Then express all other unknown quantities as expressions involving that variable. Finally, write an equation expressing a quantity in two different ways.

3. **Solve the equation.**

4. **State the conclusion.**

5. **Check the result** in the words of the problem.

EXAMPLE 11	**California coastline.** The first part of California's magnificent 17-Mile Drive scenic tour, shown in Figure 2-13, begins at the Pacific Grove entrance and travels to Seal Rock. It is 1 mile longer than the second part of the drive, which extends from Seal Rock to the Lone Cypress. The final part of the tour winds through the hills of the Monterey Peninsula, eventually returning to the entrance. This part of the drive is 1 mile longer than four times the length of the second part. How long is each of the three parts of 17-Mile Drive?

ANALYZE THE PROBLEM   In Figure 2-14, we "straighten out" the winding 17-Mile Drive so that it can be modeled with a line segment. The drive is composed of three parts. We need to find the length of each part.

FIGURE 2-13

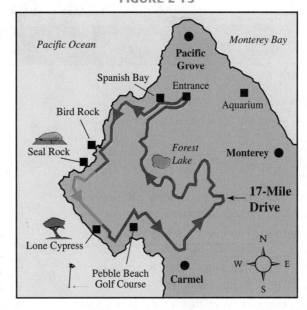

FORM AN EQUATION

Since the lengths of the first part and of the third part of the scenic drive are related to the length of the second part, we will let x represent the length of that part. We then express the other lengths in terms of that variable.

$x + 1$ represents the length of the first part of the drive.

$4x + 1$ represents the length of the third part of the drive.

FIGURE 2-14

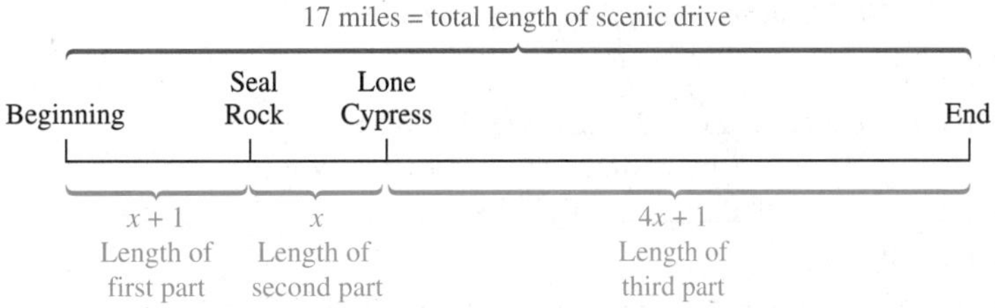

17 miles = total length of scenic drive

| Beginning | Seal Rock | Lone Cypress | | End |

| $x + 1$ | x | $4x + 1$ |
| Length of first part | Length of second part | Length of third part |

The sum of the lengths of the three parts of the drive must equal the total length of 17-Mile Drive.

The length of part 1	plus	the length of part 2	plus	the length of part 3	equals	the total length.
$x + 1$	$+$	x	$+$	$4x + 1$	$=$	17

SOLVE THE EQUATION

$x + 1 + x + 4x + 1 = 17$ The equation to solve.

$6x + 2 = 17$ Combine like terms: $x + x + 4x = 6x$.

$6x = 15$ To undo the addition of 2, subtract 2 from both sides.

$\dfrac{6x}{6} = \dfrac{15}{6}$ To undo the multiplication by 6, divide both sides by 6.

$x = 2.5$ Do the divisions.

STATE THE CONCLUSION

The second part of the scenic drive is 2.5 miles long. Because the first part of the drive is 1 mile longer than the second, it is 3.5 miles long. The length of the third part of the drive is 1 mile more than four times the second part, so it is $4(2.5) + 1$, or 11 miles long.

CHECK THE RESULT

Because the sum of 3.5 miles, 2.5 miles, and 11 miles is 17 miles, the solution checks.

STUDY SET

Section 2.5

VOCABULARY

In Exercises 1–6, fill in the blanks to make the statements true.

1. An _____equation_____ is a statement that two quantities are equal.
2. If a number is a solution of an equation, the number is said to _____satisfy_____ the equation.
3. In an equation that is an identity, all _____numbers_____ satisfy the equation.
4. If no numbers satisfy an equation, the equation is called an _____impossible_____ equation.
5. In $2(x - 7)$, "to remove parentheses" means to apply the _____distributive_____ property.
6. To solve an equation, we must _____isolate_____ the variable on one side of the equation.

CONCEPTS

In Exercises 7–10, fill in the blanks to make the statements true.

7. To solve the equation $2x - 7 = 21$, we first undo the _____subtraction_____ of 7 by adding 7 to both sides. We then undo the _____multiplication_____ by 2 by dividing both sides by 2.
8. To solve the equation $\frac{x}{2} + 3 = 5$, we first undo the _____addition_____ of 3 by subtracting 3 from both sides. We then undo the _____division_____ by 2 by multiplying both sides by 2.
9. To solve $3b - 8 = 2b + 1$, all the terms containing b should be on the same side. To do this, we can eliminate $2b$ on the right side by _____subtracting_____ $2b$ from both sides.
10. To solve $\frac{s}{3} + \frac{1}{4} = -\frac{1}{2}$, we can clear the equation of the fractions by _____multiplying_____ both sides of the equation by 12.

11. **a.** Simplify $3x + 5 - x$. $2x + 5$
    b. Solve $3x + 5 - x = 9$. 2
    c. Evaluate $3x + 5 - x$ for $x = 9$. 23
12. **a.** Simplify $3(x - 4) - 4x$. $-x - 12$
    b. Solve $3(x - 4) - 4x = 0$. -12
    c. Evaluate $3(x - 4) - 4x$ for $x = 0$. -12
13. The amount of seating in an auditorium can be described using algebraic expressions, as in Illustration 1.

ILLUSTRATION 1

Seating area	Number of seats
Main floor	$x + 200$
Box seats	$x - 50$
Balcony	x

a. The number of seats in the table is based on which seating area? balcony
b. How many more seats are there on the main floor than in the balcony? 200
c. What is the total seating in the auditorium? $3x + 150$

14. The circulation for a large newspaper can be described using algebraic expressions, as in Illustration 2.

ILLUSTRATION 2

Day	Circulation
Weekday	x
Saturday	$x - 75,000$
Sunday	$2x$

a. On which day are the circulation numbers based? weekdays
b. On which day is the circulation the smallest? Saturday
c. What is the total weekly circulation of the paper? $4x - 75,000$

15. What is the LCD for the fractions in the equation $\frac{x}{3} - \frac{4}{5} = \frac{1}{2}$? 30
16. One method of solving $-\frac{4}{5}x = 8$ is to multiply both sides of the equation by the reciprocal of $-\frac{4}{5}$. What is the reciprocal of $-\frac{4}{5}$? $-\frac{5}{4}$

NOTATION

In Exercises 17–18, complete each solution.

17. Solve the equation $2x - 7 = 21$.

$$2x - 7 = 21$$
$$2x - 7 + \boxed{7} = 21 + \boxed{7}$$
$$2x = \boxed{28}$$
$$\frac{2x}{\boxed{2}} = \frac{28}{\boxed{2}}$$
$$x = 14$$

18. Solve the equation $\frac{x}{2} + 3 = 5$.

$$\frac{x}{2} + 3 = 5$$

$$\frac{x}{2} + 3 - \boxed{3} = 5 - \boxed{3}$$

$$\frac{x}{2} = \boxed{2}$$

$$\boxed{2}\left(\frac{x}{2}\right) = \boxed{2}\,(2)$$

$$x = 4$$

19. Fill in the blanks to make the statements true.

a. $-x = \boxed{-1}\ x$.   **b.** $\frac{3x}{5} = \frac{3}{\boxed{5}}\ x$.

20. When checking a solution of an equation, the symbol $\overset{?}{=}$ is used. What does it mean?   is possibly equal to

PRACTICE

In Exercises 21–68, solve each equation. Check the result.

21. $2x + 5 = 17$ 6   **22.** $3x - 5 = 13$ 6

23. $-5q - 2 = 1$ $-\frac{3}{5}$ **24.** $4p + 3 = 2$ $-\frac{1}{4}$

25. $0.6 = 4.1 - x$ 3.5   **26.** $1.2 - x = -1.7$ 2.9

27. $-g = -4$ 4   **28.** $-u = -20$ 20

29. $-8 - 3c = 0$ $-\frac{8}{3}$ **30.** $-5 - 2d = 0$ $-\frac{5}{2}$

31. $-\frac{5}{6}k = 10$ -12 **32.** $\frac{2c}{5} = 2$ 5

33. $-\frac{t}{3} + 2 = 6$ -12 **34.** $\frac{x}{5} - 5 = -12$ -35

35. $\frac{2x}{3} - 2 = 4$ 9   **36.** $\frac{2}{5}y + 3 = 9$ 15

37. $\frac{x + 5}{3} = 11$ 28   **38.** $\frac{x + 2}{13} = 3$ 37

39. $\frac{y - 2}{7} = -3$ -19 **40.** $\frac{x - 7}{3} = -1$ 4

41. $2(-3) + 4y = 14$ 5   **42.** $4(-1) + 3y = 8$ 4

43. $-2x - 4(1) = -6$ 1   **44.** $-5x - 3(5) = 0$ -3

45. $3(x + 2) - x = 12$ 3   **46.** $2(x - 4) + x = 7$ 5

47. $-3(2y - 2) - y = 5$ $\frac{1}{7}$ **48.** $-(3a + 1) + a = 2$ $-\frac{3}{2}$

49. $3x + 2.5 = 2x$ -2.5 **50.** $5x + 7.2 = 4x$ -7.2

51. $9y - 3 = 6y$ 1   **52.** $8y + 4 = 4y$ -1

53. $\frac{1}{2} + \frac{x}{5} = \frac{3}{4}$ $\frac{5}{4}$ **54.** $\frac{1}{3} + \frac{c}{5} = -\frac{3}{2}$ $-\frac{55}{6}$

55. $\frac{2}{3} = -\frac{2x}{3} + \frac{3}{4}$ $\frac{1}{8}$ **56.** $-\frac{2}{9} = \frac{5x}{6} - \frac{1}{3}$ $\frac{2}{15}$

57. $3(a + 2) = 2(a - 7)$ -20 **58.** $9(t - 1) = 6(t + 2) - t$ $\frac{21}{4}$

59. $9(x + 11) + 5(13 - x) = 0$ -41

60. $3(x + 15) + 4(11 - x) = 0$ 89

61. $\frac{3t - 21}{2} = t - 6$ 9   **62.** $\frac{2t + 18}{3} = t - 8$ 42

63. $\frac{10 - 5s}{3} = s + 6$ -1 **64.** $\frac{40 - 8s}{5} = -2s$ -20

65. $2 - 3(x - 5) = 4(x - 1)$ 3

66. $2 - (4x + 7) = 3 + 2(x + 2)$ -2

67. $15x = x$ 0   **68.** $-7y = -8y$ 0

In Exercises 69–76, solve each equation. If it is an identity or an impossible equation, so indicate.

69. $8x + 3(2 - x) = 5(x + 2) - 4$ identity

70. $5(x + 2) = 5x - 2$ impossible equation

71. $-3(s + 2) = -2(s + 4) - s$ impossible equation

72. $21(b - 1) + 3 = 3(7b - 6)$ identity

73. $2(3z + 4) = 2(3z - 2) + 13$ impossible equation

74. $x + 7 = \frac{2x + 6}{2} + 4$ identity

75. $4(y - 3) - y = 3(y - 4)$ identity

76. $5(x + 3) - 3x = 2(x + 8)$ impossible equation

APPLICATIONS

77. CARPENTRY The 12-foot board in Illustration 3 has been cut into two sections, one twice as long as the other. How long is each section?   4 ft and 8 ft

ILLUSTRATION 3

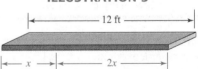

78. PLUMBING A 20-foot pipe has been cut into two sections, one 3 times as long as the other. How long is each section?   15 ft and 5 ft

79. ROBOTICS The robotic arm shown in Illustration 4 will extend a total distance of 18 feet. Find the length of each section.   5 ft, 9 ft, 4 ft

ILLUSTRATION 4

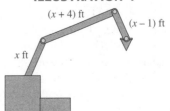

80. WINDOW DESIGN A window in the tank of an aquarium is shown in Illustration 5. If the total distance around the window is 36 feet, how long is each side? 4 ft, 10 ft, 4 ft, 4 ft, 10 ft, 4 ft

ILLUSTRATION 5

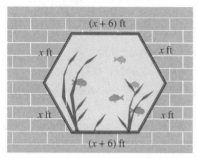

81. TOURING An American rock group plans to travel for a total of 38 weeks, making three major concert tours. They will be in Japan for 4 more weeks than they will be in Australia. Their stay in Sweden will be 2 weeks less than that in Australia. How many weeks will they be in each country?

Australia: 12 wk; Japan: 16 wk; Sweden: 10 wk

82. HOURLY PAY After an evaluation, an employee who was making $5 an hour was promised a raise. This week, the employee worked 54 hours. If he is paid time and a half for any hours over 40 and his paycheck was $332.45, did he receive the raise? If so, what is his new hourly rate? yes; $5.45/hr

83. COUNTING CALORIES A slice of pie with a scoop of ice cream has 850 calories. The calories in the pie alone are 100 more than twice the calories in the ice cream alone. How many calories are in the ice cream? 250

84. PUBLISHER'S INVENTORY A novel can be purchased in a hardcover edition for $15.95 or in paperback for $4.95. The publisher printed 11 times as many paperbacks as hardcover books, a total of 114,000 copies for both books. How many hardcover books were printed? 9,500

85. ATTORNEY'S FEES An attorney and her client each took half of the cash award that she negotiated in an out-of-court settlement. After paying her assistant $1,000, the attorney ended up making $12,000 from the case. What was the amount of the out-of-court settlement? $26,000

86. APARTMENT RENTAL In renting an apartment with two other friends, Jonathan agreed to pay the security deposit of $100. The three of them also agreed to contribute equally toward the monthly rent. Jonathan's first check to the apartment owner was for $225. What was the monthly rent for the apartment? $375

87. SOLAR HEATING One solar panel in Illustration 6 is 3.4 feet wider than the other. Find the width of each panel. 7.3 ft and 10.7 ft

ILLUSTRATION 6

18 ft

88. WASTE DISPOSAL Two tanks hold a total of 45 gallons of a toxic solvent. One tank holds 6 gallons more than twice the amount in the other. Can the smaller tank be emptied into a 10-gallon waste disposal canister? no

89. SHOPPING If you buy one bottle of vitamins, you can buy a second one for half price. If two bottles cost $2.25, find the regular price for one bottle. $1.50

90. BOTTLED WATER DELIVERY A driver left the plant with 300 bottles of drinking water on his truck. His route consisted of office buildings, each of which received 3 bottles of water. The driver returned to the plant at the end of the day with 117 bottles on the truck. To how many office buildings did he deliver? 61

91. CORPORATE DOWNSIZING In an effort to cut costs, a corporation has decided to lay off 5 employees every month until the number of employees totals 465. If 510 people are currently employed, how many months will it take to reach the employment goal? 9 mo

92. NET INCOME From the information given in Illustration 7, determine the net income of Sears, Roebuck and Co. for each quarter of 1997.

ILLUSTRATION 7

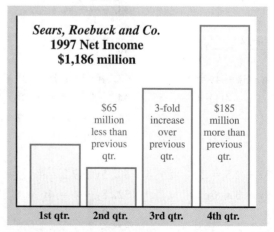

Based on data from Hoover's Online

in millions: $182, $117, $351, $536

WRITING

93. Explain the difference between *simplifying* an expression and *solving* an equation. Give some examples.

94. Briefly explain what should be accomplished in each of the five steps of the problem-solving strategy studied in this section.

REVIEW

 In Exercises 95–98, use a calculator to help solve each problem.

95. Find 82% of 168. 137.76

96. 29.05 is what percent of 415? 7%

97. What percent of 200 is 30? 15%

98. A woman bought a coat for $98.95 and some gloves for $7.95. If the sales tax was 6%, how much did the purchase cost her? $113.31

▶ 2.6 Formulas

In this section, you will learn about

Formulas from business ■ Formulas from science ■ Formulas from geometry ■ Solving formulas

Introduction A **formula** is an equation that is used to state a known relationship between two or more variables. Formulas are used in many fields: economics, physical education, anthropology, biology, automotive repair, and nursing, to name a few. In this section, we will consider formulas from business, science, and geometry.

N/A

Formulas from Business

A formula to find the sale price: To find the sale price of an item that has been discounted, we subtract the discount from the original price.

| Sale price | = | original price | − | discount |

Using variables to represent the sale price s, the original price p, and the discount d, this formula can be written as

$$s = p - d$$

EXAMPLE 1

Clearance sale. At a year-end clearance sale, a Jeep Wrangler sold for $13,450. If it was discounted $2,300, what was its original price?

Solution Since the sale price (s) was $13,450 and the discount (d) was $2,300, we substitute 13,450 for s and 2,300 for d in the formula $s = p - d$. Then we solve for p.

$$s = p - d$$
$$13{,}450 = p - 2{,}300 \qquad \text{Substitute 13,450 for } s \text{ and 2,300 for } d.$$
$$13{,}450 + 2{,}300 = p - 2{,}300 + 2{,}300 \qquad \text{To undo the subtraction of 2,300, add 2,300 to both sides.}$$
$$15{,}750 = p \qquad \text{Do the additions.}$$

The original price of the Jeep Wrangler was $15,750.

Because she maintained a 3.0 GPA, a high school senior received a "good student discount" on her auto insurance premium. The $124 discount lowered the six-month premium to $468. What would the premium be if she did not receive the discount?   *Answer:* $592 ■

A formula to find the retail price: To make a profit, a merchant must sell a product for more than he or she paid for it. The price at which the merchant sells the product, called the **retail price,** is the sum of what the item cost the merchant plus the **markup**.

$$\boxed{\text{Retail price}} \quad = \quad \boxed{\text{cost}} \quad + \quad \boxed{\text{markup}}$$

Using r to represent the retail price, c the wholesale cost, and m the markup, we can write this formula as

$$\boxed{r = c + m}$$

As an example, suppose a jeweler purchases a gold ring at a wholesale jewelry mart for $612.50. Then she sets the price of the ring at $837.95 for sale in her store. We can find the markup on the ring as follows:

$$r = c + m$$
$$837.95 = 612.50 + m \qquad \text{Substitute 837.95 for } r \text{ and 612.50 for } c.$$
$$837.95 - 612.50 = 612.50 + m - 612.50 \qquad \text{To undo the addition of 612.50, subtract 612.50 from both sides.}$$
$$225.45 = m \qquad \text{Do the subtractions.}$$

The markup on the ring is $225.45.

A formula for profit: The profit a business makes is the difference between the revenue (the money it takes in) and the costs.

$$\boxed{\text{Profit}} \quad = \quad \boxed{\text{revenue}} \quad - \quad \boxed{\text{costs}}$$

Using p to represent the profit, r the revenue, and c the costs, we can write this formula as

$$\boxed{p = r - c}$$

EXAMPLE 2

Charitable giving. In 1997, the Salvation Army collected $1.7 billion in donations. Of that amount, $1.4 billion went directly to the support of its programs. What were the 1997 administrative costs of the organization?

Solution   The charity collected $1.7 billion in revenue. We can think of the $1.4 billion that was spent on programs as profit. We need to find the administrative costs, c.

$$p = r - c$$ The formula for profit.

$$1.4 = 1.7 - c$$ Substitute 1.4 for p and 1.7 for r.

$$1.4 - 1.7 = 1.7 - c - 1.7$$ To eliminate 1.7, subtract 1.7 from both sides.

$$-0.3 = -c$$ Combine like terms on both sides.

$$\frac{-0.3}{-1} = \frac{-c}{-1}$$ Divide both sides by -1

$$0.3 = c$$ Do the divisions.

In 1997, the Salvation Army had administrative costs of $0.3 billion.

SELF CHECK   A PTA spaghetti dinner made a profit of $275.50. If the cost to host the dinner was $1,235, how much revenue did it generate?   *Answer:* $1,510.50   ∎

A formula for simple interest: When money is borrowed, the lender expects to be paid back the amount of the loan plus an additional charge for the use of the money. The additional charge is called **interest**. When money is deposited in a bank, the depositor is paid for the use of the money. The money the deposit earns is also called interest. In general, interest is the money that is paid for the use of money.

Interest is calculated in two ways: either as **simple interest** or as **compound interest.** To find simple interest, we use the formula

$$\boxed{\text{Interest}} \;=\; \boxed{\text{principal}} \;\cdot\; \boxed{\text{rate}} \;\cdot\; \boxed{\text{time}}$$

Using I to represent the simple interest, P the principal (the amount of money that is invested, deposited, or borrowed), r the annual interest rate, and t the length of time in years, we can write the formula as

$$\boxed{I = Prt}$$

EXAMPLE 3   **Retirement income.** One year after investing $15,000 in a mini-mall development, a retired couple received a check for $1,125 in interest. What interest rate did their money earn that year?

Solution   The couple invested $15,000 (the principal) for 1 year (the time) and made $1,125 (the interest). We need to find the annual interest rate.

$$I = Prt$$

$$1,125 = 15,000r(1)$$ Substitute 1,125 for I, 15,000 for P, and 1 for t.

$$1,125 = 15,000r$$ Simplify.

$$\frac{1,125}{15,000} = \frac{15,000r}{15,000}$$ To solve for r, undo the multiplication by 15,000 by dividing both sides by 15,000.

$$0.075 = r$$ Do the divisions.

$$7.5\% = r$$ To write 0.075 as a percent, multiply 0.075 by 100 and insert a % sign.

The couple received an annual rate of 7.5% that year.

SELF CHECK   A father loaned his daughter and son-in-law $12,200 at a 2% annual simple interest rate for a down payment on a house. If the interest on the loan amounted to $610, for how long was the loan?   *Answer:* 2.5 years ■

Formulas from Science

A formula for distance traveled: If we know the average rate (speed) at which we will be traveling and the time we will be traveling at that rate, we can find the distance traveled by using the formula

| Distance | = | rate | · | time |

Using d to represent the distance, r the average rate (speed), and t the time, we can write this formula as

$$d = rt$$

WARNING! When using this formula, the units must be the same. For example, if the rate is given in miles per hour, the time must be expressed in hours.

EXAMPLE 4   **Finding the rate.** As they migrate from the Bering Sea to Baja California, gray whales swim for about 20 hours each day, covering a distance of approximately 70 miles. Estimate their average swimming rate in miles per hour (mph).

Solution   Since the distance d is 70 miles and the time t is 20 hours, we substitute 70 for d and 20 for t in the formula $d = rt$, and then solve for r.

$$d = rt$$
$$70 = r(20) \quad \text{Substitute 70 for } d \text{ and 20 for } t.$$
$$\frac{70}{20} = \frac{20r}{20} \quad \text{To undo the multiplication by 20, divide both sides by 20.}$$
$$3.5 = r \quad \text{Do the divisions.}$$

The whales' average swimming rate is 3.5 mph.

SELF CHECK   An elevator in a building travels at an average rate of 288 feet per minute. How long will it take it to climb 30 stories, a distance of 360 feet?   *Answer:* 1.25 minutes ■

A formula for converting degrees Fahrenheit to degrees Celsius: Many marquees, like the one shown in Figure 2-15, flash two temperature readings, one in degrees Fahrenheit and one in degrees Celsius. The Fahrenheit scale is used in the American system of measurement. The Celsius scale is used in the metric system. As we noted in the Study Set in Section 2.3, there is a formula that shows how a Fahrenheit temperature F relates to a Celsius temperature C:

$$C = \frac{5(F - 32)}{9}$$

FIGURE 2-15

■ CITY SAVINGS
TEMP 30°C

EXAMPLE 5

Changing degrees Celsius to degrees Fahrenheit. Change the temperature reading on the sign in Figure 2-15 to degrees Fahrenheit.

Solution   Since the temperature C in degrees Celsius is 30°, we substitute 30 for C in the formula and solve for F.

$$C = \frac{5(F - 32)}{9}$$

$$30 = \frac{5(F - 32)}{9} \qquad \text{Substitute 30 for } C.$$

$$9(30) = 9\left[\frac{5(F - 32)}{9}\right] \qquad \text{To undo the division by 9, multiply both sides by 9.}$$

$$270 = 5(F - 32) \qquad \text{Simplify: } 9(30) = 270 \text{ and } \tfrac{9}{9} = 1.$$

$$270 = 5F - 5(32) \qquad \text{Distribute the 5.}$$

$$270 = 5F - 160 \qquad \text{Do the multiplication: } 5(32) = 160.$$

$$270 + 160 = 5F - 160 + 160 \qquad \text{To undo the subtraction of 160, add 160 to both sides.}$$

$$430 = 5F \qquad \text{Do the additions.}$$

$$\frac{430}{5} = \frac{5F}{5} \qquad \text{To undo the multiplication by 5, divide both sides by 5.}$$

$$86 = F \qquad \text{Do the divisions.}$$

Thus, 30°C is equivalent to 86°F.

SELF CHECK   Change −175°C, the temperature on Saturn, to degrees Fahrenheit.   *Answer:* −283°F   ■

Formulas from Geometry

The **perimeter** of a geometric figure is the distance around it. The **area** of the figure is the amount of surface that it encloses. Table 2-2 shows the formulas for the perimeter (P) and area (A) of several geometric figures.

TABLE 2-2

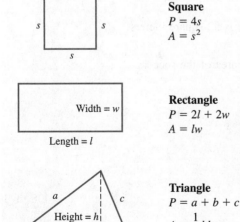

Square
$P = 4s$
$A = s^2$

Rectangle
$P = 2l + 2w$
$A = lw$

Triangle
$P = a + b + c$
$A = \dfrac{1}{2}bh$

Trapezoid
$P = a + b + c + d$
$A = \dfrac{1}{2}h(b + d)$

Circle
$C = 2\pi r$, where
   $\pi \approx 3.1416$
C is the circumference
   of the circle and
   r is its radius.
$A = \pi r^2$

FIGURE 2-16

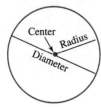

A **circle** (see Figure 2-16) is the set of all points in a plane that are a fixed distance from a point called its **center.** A segment drawn from the center of a circle to a point on the circle is called a **radius.** Since a **diameter** of a circle is a segment passing through the center that joins two points on the circle, the diameter D of a circle is twice as long as its radius r.

$$D = 2r$$

The perimeter of a circle is called its **circumference.** The formula for the circumference of a circle is

$$C = 2\pi r$$

E X A M P L E   6

Finding perimeters and areas. Find **a.** the perimeter of a square with sides 6 inches long   and   **b.** the area of a triangle with base 8 meters and height 13 meters.

Solution **a.** The perimeter of a square is given by the formula $P = 4s$, where P is the perimeter and s is the length of one side. Since the sides of the square are 6 inches long, we substitute 6 for s and simplify.

$$P = 4s$$
$$P = 4(6) \quad \text{Substitute 6 for } s.$$
$$= 24 \quad \text{Do the multiplication.}$$

The perimeter of the square is 24 inches.

b. The area of a triangle is given by the formula $A = \frac{1}{2}bh$. Since the base of the triangle is 8 meters and the height is 13 meters, we substitute 8 for b and 13 for h and simplify.

$$A = \frac{1}{2}bh$$

$$A = \frac{1}{2}(8)(13) \quad \text{Substitute 8 for } b \text{ and 13 for } h.$$

$$= 4(13) \quad \tfrac{1}{2}(8) = \tfrac{8}{2} = 4.$$

$$= 52 \quad \text{Do the multiplication.}$$

The area of the triangle is 52 square meters.

SELF CHECK

Find the perimeter and the area of the soccer field.

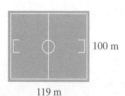

100 m

119 m

Answers: 438 m, 11,900 m^2 ■

E X A M P L E   7

Finding the area of a circle. To the nearest tenth, find the area of a circle with a diameter of 14 feet.

Solution   Since the radius of a circle is one-half its diameter, the radius of this circle is 7 feet. We can then substitute 7 for *r* in the formula for the area of a circle and simplify.

$$A = \pi r^2$$
$$A = \pi(7)^2$$
$$\quad = 49\pi \qquad \text{First, evaluate the exponential expression: } 7^2 = 49.$$
$$\quad \approx 153.93804 \qquad \text{Use a calculator to do the multiplication. Enter these numbers and press these keys on a scientific calculator: } 49 \;\boxed{\times}\; \boxed{\pi}\; \boxed{=}\;.$$

To the nearest tenth, the area is 153.9 square feet.

SELF CHECK   To the nearest hundredth, find the circumference of the circle of Example 7.   *Answer:* 43.98 ft   ■

The **volume** of a three-dimensional geometric solid is the amount of space it encloses. Table 2-3 shows the formula for the volume (*V*) of several solids.

TABLE 2-3

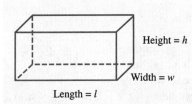

Height = *h*   **Rectangular solid**
Width = *w* $V = lwh$
Length = *l*

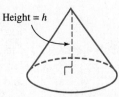

Height = *h*   **Cone** $V = \dfrac{1}{3}Bh*$

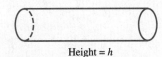

Height = *h*   **Cylinder** $V = Bh*$

Sphere $V = \dfrac{4}{3}\pi r^3$

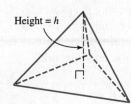

Height = *h*   **Pyramid** $V = \dfrac{1}{3}Bh*$

*B represents the area of the base.

EXAMPLE 8   **Finding volumes.** To the nearest tenth, find the volume of each figure.

a. 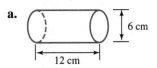 6 cm 12 cm   **b.**   6 m

Solution   **a.** The formula for the volume of a cylinder is $V = Bh$, where *B* is the area of the circular base and *h* is the height. We first find the area of the circular base.

$$A = \pi r^2 \qquad \text{The formula for the area of a circle.}$$
$$A = \pi(3)^2 \qquad \text{The radius is one-half the diameter: } \tfrac{1}{2}(6) = 3.$$
$$\quad = 9\pi$$

We then substitute 9π for B and 12 for h in the formula for the volume of a cylinder.

$V = Bh$	The formula for the volume of a cylinder.
$V = (9\pi)(12)$	Substitute 9π for B and 12 for h.
$= 108\pi$	Multiply: $9(12) = 108$.
≈ 339.2920066	Use a calculator.

To the nearest tenth, the volume is 339.3 cubic centimeters.

b. To find the volume of the sphere, we substitute 6 for r into the formula for the volume of a sphere.

$V = \dfrac{4}{3}\pi r^3$	The formula for the volume of a sphere.
$V = \dfrac{4}{3}\pi(6)^3$	Substitute 6 for r.
$= \dfrac{4}{3}\pi(216)$	$6^3 = 6 \cdot 6 \cdot 6 = 216$.
$= 288\pi$	$\frac{4}{3}(216) = \frac{4(216)}{3} = \frac{864}{3} = 288$.
≈ 904.7786842	Use a calculator.

To the nearest tenth, the volume is 904.8 cubic meters.

SELF CHECK   Find the volume of each figure: **a.**   a rectangular solid with length 7 inches, width 12 inches, and height 15 inches   and   **b.** a cone whose base has radius 12 meters and whose height is 9 meters. Give the answer to the nearest tenth.

Answers: **a.** 1,260 in.^{3},   **b.** 1,357.2 m^3 ∎

Solving Formulas

Suppose we wish to find the bases of several triangles whose areas and heights are known. It would be tedious to substitute values for A and h into the formula and then repeatedly solve the formula for b. A better way is to solve the formula $A = \frac{1}{2}bh$ for b first, and then substitute values for A and h and compute b directly.

To **solve an equation for a variable** means to isolate that variable on one side of the equation, with all other quantities on the opposite side.

EXAMPLE 9   **Solving formulas.** Solve $A = \frac{1}{2}bh$ for b.

Solution   To solve for b, we must isolate b on one side of the equation.

$A = \dfrac{1}{2}bh$	
$2A = 2 \cdot \dfrac{1}{2}bh$	To clear the equation of the fraction, multiply both sides by 2.
$2A = bh$	Simplify: $2 \cdot \frac{1}{2} = \frac{2}{2} = 1$.
$\dfrac{2A}{h} = \dfrac{bh}{h}$	To undo the multiplication by h, divide both sides by h.
$\dfrac{2A}{h} = b$	Simplify: $\frac{h}{h} = 1$.
$b = \dfrac{2A}{h}$	

SELF CHECK   Solve $A = \frac{1}{2}bh$ for h.   *Answer:* $h = \frac{2A}{b}$ ∎

EXAMPLE 10

Solving formulas. Solve $r = c + m$ for m.

Solution   To isolate m on the left-hand side, we undo the addition of c by subtracting c from both sides.

$$r = c + m$$
$$r - c = c + m - c \quad \text{Subtract } c \text{ from both sides.}$$
$$r - c = m \qquad\qquad \text{Combine like terms on the right-hand side: } c - c = 0.$$
$$m = r - c$$

SELF CHECK   Solve $p = r - c$ for r.   *Answer: $r = p + c$*   ■

EXAMPLE 11

Solving formulas. Solve $P = 2l + 2w$ for l.

Solution   To solve for l, we must isolate l on one side of the equation.

$$P = 2l + 2w$$
$$P - 2w = 2l + 2w - 2w \quad \text{To undo the addition of } 2w, \text{ subtract } 2w \text{ from both sides.}$$
$$P - 2w = 2l \qquad\qquad \text{Combine like terms.}$$
$$\frac{P - 2w}{2} = \frac{2l}{2} \qquad\qquad \text{To undo the multiplication by 2, divide both sides by 2.}$$
$$\frac{P - 2w}{2} = l \qquad\qquad \text{Simplify the right-hand side.}$$

If we solve the formula for l, we obtain $l = \dfrac{P - 2w}{2}$.

SELF CHECK   Solve $P = 2l + 2w$ for w.   *Answer: $w = \frac{P - 2l}{2}$*   ■

EXAMPLE 12

Solving formulas. Solve $C = \frac{5}{9}(F - 32)$ for F.

Solution   To solve for F, we must isolate F on one side of the equation.

$$C = \frac{5}{9}(F - 32)$$
$$9(C) = 9\left[\frac{5}{9}(F - 32)\right] \quad \text{Clear the fraction by multiplying both sides by 9.}$$
$$9C = 5(F - 32) \qquad\qquad \text{Simplify: } \frac{9}{9} = 1.$$
$$\frac{9C}{5} = \frac{5(F - 32)}{5} \qquad\qquad \text{To undo the multiplication by 5, divide both sides by 5.}$$
$$\frac{9}{5}C = F - 32 \qquad\qquad \text{Do the division: } \frac{5}{5} = 1.$$
$$\frac{9}{5}C + 32 = F - 32 + 32 \quad \text{To undo the subtraction of 32, add 32 to both sides.}$$
$$\frac{9}{5}C + 32 = F \qquad\qquad \text{Do the addition.}$$
$$F = \frac{9}{5}C + 32$$

SELF CHECK   Solve $F = \frac{9C + 160}{5}$ for C.   *Answer: $C = \frac{5F - 160}{9}$*   ■

STUDY SET

Section 2.6

VOCABULARY

In Exercises 1–8, fill in the blanks to make the statements true.

1. A ___formula___ is an equation that is used to state a known relationship between two or more variables.

2. The ___volume___ of a three-dimensional geometric solid is the amount of space it encloses.

3. The distance around a geometric figure is called its ___perimeter___.

4. A ___circle___ is the set of all points in a plane that are a fixed distance from a point called its center.

5. A segment drawn from the center of a circle to a point on the circle is called a ___radius___.

6. The amount of surface that is enclosed by a geometric figure is called its ___area___.

7. The perimeter of a circle is called its ___circumference___.

8. A segment passing through the center of a circle and connecting two points on the circle is called a ___diameter___.

CONCEPTS

9. Tell which geometric concept—perimeter, circumference, area, or volume—should be used to find the following:
   a. The amount of storage in a freezer   volume
   b. How far a bicycle tire rolls in one revolution   circumference
   c. The amount of land making up the Sahara Desert   area
   d. The distance around a Monopoly game board   perimeter

10. Tell which unit of measurement—ft, ft^2, or ft^3— would be appropriate when finding the following:
   a. The amount of storage inside a safe ft^3
   b. The ground covered by a sleeping bag lying on the floor ft^2
   c. The distance the tip of an airplane propeller travels in one revolution   ft
   d. The size of the trunk of a car ft^3

11. Write an expression for the area of the figure shown in Illustration 1. $(12x - 8)$ mm^2

ILLUSTRATION 1

4 mm

$(6x - 4)$ mm

12. Write an expression for the area of the figure shown in Illustration 2. $(2x + 6)$ cm^2

ILLUSTRATION 2

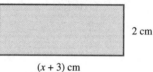

2 cm

$(x + 3)$ cm

13. a. Explain what it means to solve the equation $P = 2l + 2w$ for w.
      Isolate w on one side of the equation.
   b. Why can't we say that this equation is solved for a?
      $$a = 2b - a$$
      a is not isolated on one side of the equation.

14. What type of figure is the base of each of these three-dimensional solids, and, for each base, what is the formula used to find its area B?
   a. circle; πr^2 b. triangle; $\frac{1}{2}bh$

15. a. If the radius r of a circle is known, how can its diameter be found?   Multiply r by 2.
   b. If the diameter D of a circle is known, how can its radius r be found?   Divide D by 2.

16. Use variables to write the formula relating the following:
   a. Time, distance, rate $d = rt$
   b. Sale price, discount, original price $s = p - d$
   c. Markup, retail price, cost $r = c + m$
   d. Costs, revenue, profit $p = r - c$
   e. Interest rate, time, interest, principal $I = Prt$

17. Complete the table. (mi/sec means miles per second.)

	Rate	·	time	=	distance
Light	186,282 mi/sec		60 sec		11,176,920 mi
Sound	1,088 ft/sec		60 sec		65,280 mi

18. Complete the table.

Principal ·	rate ·	time =	interest
$2,500	5%	2 yr	$250
$15,000	4.8%	1 yr	$720

NOTATION

In Exercises 19–20, complete each solution.

19. Solve $V = \frac{1}{3}Bh$ for B.

$$V = \frac{1}{3}Bh$$

$$3\,(V) = 3\left(\frac{1}{3}\right)Bh$$

$$3V = Bh$$

$$\frac{3V}{h} = \frac{Bh}{h}$$

$$\frac{3V}{h} = B$$

$$B = \frac{3V}{h}$$

20. Solve $A = \frac{1}{2}h(B + b)$ for B.

$$A = \frac{1}{2}h(B + b)$$

$$2\,(A) = 2\left[\frac{1}{2}h(B + b)\right]$$

$$2A = h(B + b)$$

$$\frac{2A}{h} = \frac{h(B + b)}{h}$$

$$\frac{2A}{h} = B + b$$

$$\frac{2A}{h} - b = B + b - b$$

$$\frac{2A}{h} - b = B$$

PRACTICE   skip.

In Exercises 21–32, use a formula discussed in this section to solve each problem. A calculator will be helpful with some problems.

21. SENIOR CITIZEN DISCOUNT Anyone 65 years of age or older receives $1.75 off the regular price of admission at a certain zoo. If a senior pays $5.75, what is the regular price of admission? $7.50

22. REBATE A breadmaking machine that regularly sells for $119.99 costs only $99.99 if the customer takes advantage of a mail-in rebate. How much is the rebate offer worth? $20

23. VALENTINE'S DAY Find the markup on a dozen roses if a florist buys them wholesale for $12.95 and sells them for $37.50. $24.55

24. STICKER PRICE The factory invoice for a minivan shows that the dealer paid $16,264.55 for the vehicle. If the sticker price of the van is $18,202, how much over factory invoice is the sticker price? $1,937.45

25. HOLLYWOOD Figures for the summer of 1998 showed that the movie *Saving Private Ryan* had U.S. box-office receipts of $190 million. What were the production costs to make the movie, if, at that time, the studio had made a $125 million profit? $65 million

26. SERVICE CLUB After expenses of $55.15 were paid, a Rotary Club donated $875.85 in proceeds from a pancake breakfast to a local health clinic. How much did the pancake breakfast gross? $931

27. ENTREPRENEURS To start a mobile dog-grooming service, a woman borrowed $2,500. If the loan was for 2 years and the amount of interest was $175, what simple interest rate was she charged? 3.5%

28. BANKING Three years after opening an account that paid 6.45% annually, a depositor withdrew the $3,483 in interest earned. How much money was left in the account? 18,000

29. SWIMMING In 1930, a man swam down the Mississippi River from Minneapolis to New Orleans, a total of 1,826 miles. He was in the water for 742 hours. To the nearest tenth, what was his average rate swimming? 2.5 mph

30. ROSE PARADE Rose Parade floats travel down the 5.5-mile-long parade route at a rate of 2.5 mph. How long will it take a float to complete the parade if there are no delays? 2.2 hr

31. METALLURGY Change 2,212°C, the temperature at which silver boils, to degrees Fahrenheit. Round to the nearest degree. 4,014°F

32. LOW TEMPERATURES Cryobiologists freeze living matter to preserve it for future use. They can work with temperatures as low as −270°C. Change this to degrees Fahrenheit. −454°F

In Exercises 33–44, solve each formula for the given variable.

33. $E = IR$; for R $R = \frac{E}{I}$

34. $I = Prt$; for r $r = \frac{I}{Pt}$

35. $V = lwh$; for w $w = \frac{V}{lh}$

36. $d = rt$; for t $t = \frac{d}{r}$

37. $y = mx + b$; for x $x = \frac{y - b}{m}$

38. $P = 2l + 2w$; for l $l = \frac{P - 2w}{2}$

39. $A = P + Prt$; for t $t = \frac{A - P}{Pr}$

40. $V = \pi r^2 h$; for h $h = \frac{V}{\pi r^2}$

41. $V = \frac{1}{3}\pi r^2 h$; for h $h = \frac{3V}{\pi r^2}$

42. $K = \frac{wv^2}{2g}$; for w $w = \frac{2gK}{v^2}$

43. $C = \dfrac{5F - 160}{9}$; for F **44.** $F = \dfrac{GMm}{d^2}$; for d^2

$F = \frac{9C + 160}{5}$ $d^2 = \frac{GMm}{F}$

APPLICATIONS

In Exercises 45–66, a calculator will be helpful with some problems.

45. CARPENTRY Find the perimeter and area of the truss shown in Illustration 3.   36 ft, 48 ft²

ILLUSTRATION 3

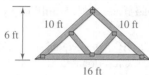

6 ft
10 ft   10 ft
16 ft

46. CAMPERS Find the area of the window of the camper shell shown in Illustration 4.   784 in.²

ILLUSTRATION 4

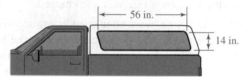

56 in.
14 in.

47. ARCHERY To the nearest tenth, find the circumference and area of the target shown in Illustration 5.
50.3 in., 201.1 in.²

ILLUSTRATION 5

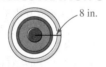

8 in.

48. GEOGRAPHY The circumference of the earth is about 25,000 miles. Find its diameter to the nearest mile.   7,958 mi

49. LANDSCAPING Find the perimeter and the area of the redwood trellis in Illustration 6.   56 in., 144 in.²

ILLUSTRATION 6

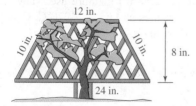

12 in.
10 in.   10 in.
8 in.
24 in.

50. VOLUME To the nearest hundredth, find the volume of the soup can shown in Illustration 7.   42.4 in.³

ILLUSTRATION 7

6 in.

SOUP
condensed
cream of chicken

3 in.

51. "THE WALL" The Vietnam Veterans Memorial is a black granite wall recognizing the more than 58,000 Americans who lost their lives or remain missing. A diagram of the wall is shown in Illustration 8. Find the total area of the two triangular-shaped surfaces on which the names are inscribed.   2,450 ft²

ILLUSTRATION 8

10 ft
245 ft   245 ft

52. SIGNAGE Find the perimeter and area of the service station sign shown in Illustration 9.
11 ft, 7.5625 ft²

ILLUSTRATION 9

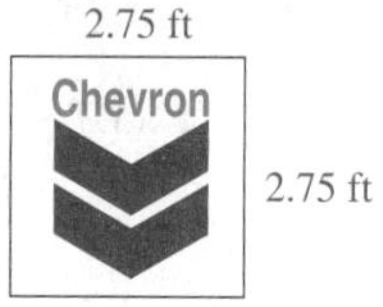

2.75 ft

Chevron

2.75 ft

53. RUBBER MEETS THE ROAD A sport truck tire has the road surface "footprint" shown in Illustration 10. Estimate the perimeter and area of the tire's footprint. (*Hint:* First change the dimensions to decimals.)   27.75 in., 47.8125 in.²

ILLUSTRATION 10

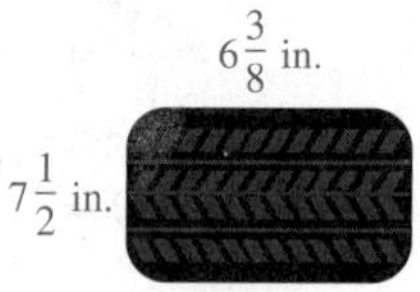

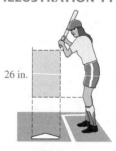

$6\frac{3}{8}$ in.

$7\frac{1}{2}$ in.

54. SOFTBALL The strike zone in fastpitch softball is between the batter's armpit and top of her knees, as shown in Illustration 11. Find the area of the strike zone.   442 in.²

ILLUSTRATION 11

26 in.

17 in.

55. FIREWOOD The dimensions of a cord of firewood are shown in Illustration 12. Find the area on which the wood is stacked and the volume the cord of firewood occupies.   32 ft², 128 ft³

ILLUSTRATION 12

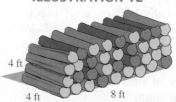

4 ft

4 ft   8 ft

56. NATIVE AMERICAN DWELLING The teepees constructed by the Blackfoot Indians were cone-shaped tents made of long poles and animal hide, about 10 feet high and about 15 feet across at the ground. (See Illustration 13.) Estimate the volume of a teepee with these dimensions, to the nearest cubic foot.   589 ft³

ILLUSTRATION 13

57. IGLOO During long journeys, some Canadian Inuit (Eskimos) built winter houses of snow blocks piled in the dome shape shown in Illustration 14. Estimate the volume of an igloo to the nearest cubic foot having an interior height of 5.5 feet.   348 ft³

ILLUSTRATION 14

58. PYRAMID The Great Pyramid at Giza in northern Egypt is one of the most famous works of architecture in the world. Use the information in Illustration 15 to find the volume to the nearest cubic foot. 85,503,750 ft³

ILLUSTRATION 15

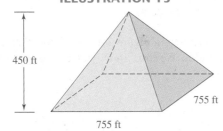
450 ft
755 ft
755 ft

59. BARBECUING See Illustration 16. Use the fact that the fish is 18 inches long to find the area of the barbecue grill to the nearest square inch.   254 in.²

ILLUSTRATION 16

60. SKATEBOARDING A "half-pipe" ramp used for skateboarding is in the shape of a semicircle with a radius of 8 feet, as shown in Illustration 17. To the nearest tenth of a foot, what is the length of the arc that the skateboarder travels on the ramp?   25.1 ft

ILLUSTRATION 17

8 ft
Plywood

61. OHM'S LAW The formula $E = IR$ is used in electronics. Solve it for I and find the current I (in amperes) when the voltage E is 48 volts and the resistance R is 12 ohms. $I = \dfrac{E}{R}$; $I = 4A$

62. GROWTH OF MONEY At a simple interest rate r, \$P grows to \$A in t years according to the formula $A = P(1 + rt)$. Solve the formula for P. After $t = 3$ years, a woman has \$4,357 on deposit in an account that pays 6%. What amount P did she start with?

$P = \dfrac{A}{1 + rt}$; $P = \$3,692.37$

63. POWER LOSS The power P lost when an electrical current I passes through a resistance R is given by $P = I^2R$. Solve for R. If P is 2,700 watts and I is 14 amperes, find R to the nearest hundredth of an ohm.

$R = \dfrac{P}{I^2}$; $R = 13.78\ \Omega$

64. FORCE OF GRAVITY The masses of the two objects in Illustration 18 are m and M. The force of gravitation, F, between the masses is given by

$$F = \frac{GmM}{d^2}$$

where G is a constant and d is the distance between them. Solve for m. $m = \dfrac{Fd^2}{GM}$

65. THERMODYNAMICS The Gibbs free-energy function is given by $G = U - TS + pV$. Solve this formula for the pressure p. $p = \dfrac{G - U + TS}{V}$

ILLUSTRATION 18

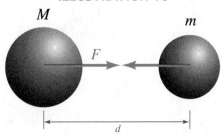

66. GEOMETRY The measure a of an interior angle of a regular polygon with n sides is given by the formula

$$a = 180°\left(1 - \frac{2}{n}\right)$$

See Illustration 19. Solve the formula for n. How many sides does a regular polygon have if an interior angle is 108°? (*Hint:* Distribute first.)

$n = \dfrac{360°}{180° - a}$; 5 sides

ILLUSTRATION 19

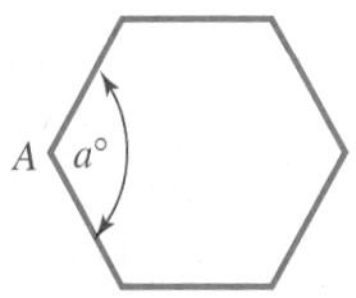

WRITING

67. The formula $P = 2l + 2w$ is also an equation, but an equation such as $2x + 3 = 5$ is not a formula. What equations do you think should be called formulas?

68. To solve the equation $s - A(s - 5) = r$ for the variable s, one student simply added $A(s - 5)$ to both sides to get $s = r + A(s - 5)$. Explain why this is not correct.

REVIEW

In Exercises 69–70, complete each input/output table.

69.

x	$x^2 - 3$
-2	1
0	-3
3	6

70.

x	$\frac{x}{3} + 2$
-6	0
0	2
12	6

In Exercises 71–74, classify each item as either an equation or an expression.

71. $18 + y = 7$ equation

72. $2y + 7$ expression

73. $\dfrac{3x - 2}{7}$ expression

74. $\dfrac{3x}{5} = 12$ equation

▶ **2.7**

Problem Solving

In this section, you will learn about

 Solving geometric problems ■ Solving number–value problems
 ■ Solving investment problems ■ Solving uniform motion problems
 ■ Solving mixture problems

Introduction   In this section, we will solve several different types of problems using the five-step problem-solving strategy.

Solving Geometric Problems

EXAMPLE 1

Dimensions of a garden. A gardener wants to use 62 feet of fencing bought at a garage sale to enclose a rectangular-shaped garden. Find the dimensions of the garden if its length is to be 4 feet longer than twice its width.

ANALYZE THE PROBLEM   We can make a sketch of the garden, as shown in Figure 2-17. We know that its length is to be 4 feet longer than twice its width. We also know that its perimeter is to be 62 feet.

FIGURE 2-17

$2w + 4$

FORM AN EQUATION   If we let w represent the width of the garden, then $2w + 4$ represents its length. Since the formula for the perimeter of a rectangle is $P = 2l + 2w$, the perimeter of the garden is $2(2w + 4) + 2w$, which is also 62. This fact enables us to form the equation.

2	times	the length	plus	2	times	the width	is	the perimeter.
2	$\cdot$	$(2w + 4)$	+	2	$\cdot$	w	=	62

SOLVE THE EQUATION   We then solve the equation.

$$2(2w + 4) + 2w = 62 \quad \text{The equation to solve.}$$
$$4w + 8 + 2w = 62 \quad \text{Use the distributive property to remove parentheses.}$$
$$6w + 8 = 62 \quad \text{Combine like terms.}$$
$$6w = 54 \quad \text{To undo the addition of 8, subtract 8 from both sides.}$$
$$w = 9 \quad \text{To undo the multiplication by 6, divide both sides by 6.}$$

STATE THE CONCLUSION   The width of the garden is 9 feet. Since $2w + 4 = 2 \cdot 9 + 4 = 22$, the length is 22 feet.

CHECK THE RESULT   If the garden has a width of 9 feet and a length of 22 feet, its length is 4 feet longer than twice the width ($2 \cdot 9 + 4 = 22$). Since its perimeter is ($2 \cdot 22 + 2 \cdot 9$) feet = 62 feet, the solution checks. ■

EXAMPLE 2

Isosceles triangles. If the vertex angle of an isosceles triangle is 56°, find the measure of each base angle.

ANALYZE THE PROBLEM   An **isosceles triangle** has two sides of equal length, which meet to form the **vertex angle.** In this case, the measurement of the vertex angle is 56°. We can sketch the triangle as shown in Figure 2-18. The **base angles** opposite the equal sides are also equal. We need to find their measure.

FIGURE 2-18

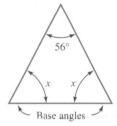

FORM AN EQUATION   If we let x represent the measure of one base angle, the measure of the other base angle is also x. Since the sum of the angles of any triangle is 180°, the sum of the base angles and the vertex angle is 180°. Therefore, we can form the equation.

One base angle	plus	the other base angle	plus	the vertex angle	is	180°.
x	+	x	+	56	=	180

SOLVE THE EQUATION   We then solve the equation.

$$x + x + 56 = 180 \quad \text{The equation to solve.}$$
$$2x + 56 = 180 \quad \text{Combine like terms.}$$
$$2x = 124 \quad \text{To undo the addition of 56, subtract 56 from both sides.}$$
$$x = 62 \quad \text{To undo the multiplication by 2, divide both sides by 2.}$$

STATE THE CONCLUSION   The measure of each base angle is 62°.

CHECK THE RESULT   The measure of each base angle is 62°, and the vertex angle measures 56°. Since $62° + 62° + 56° = 180°$, the solution checks. ■

Solving Number–Value Problems

Some problems deal with quantities that have a monetary value. In these problems, we must distinguish between the *number of* and the *value of* the unknown quantity. For problems of this type, we will use the relationship

Number · value = total value

EXAMPLE 3

Dining area improvements. A restaurant owner needs to purchase some new tables, chairs, and dinner plates for the dining area of her establishment. She plans to buy four chairs and four plates for each new table. She also needs 20 additional plates to keep in case of breakage. If a table costs $100, a chair $50, and a plate $5, how many of each can she buy if she takes out a small business loan for $6,500 to pay for the new items?

ANALYZE THE PROBLEM   We know the *value* of each item: Tables cost $100, chairs cost $50, and plates cost $5 each. We need to find the *number* of tables, chairs, and plates she can purchase for $6,500.

FORM AN EQUATION   The number of chairs and plates she needs depends on the number of tables she buys. So we let t be the number of tables to be purchased. Since every table requires four chairs and four plates, she needs to order $4t$ chairs. Because an additional 20 plates are needed, she should order $4t + 20$ plates. The total value of each purchase is the *product* of the number of items bought and the price, or value, of each item.

Item	Number purchased ·	Price per item =	Total value
Table	t	$100	$100t$
Chair	$4t$	$50	$50(4t)$
Plate	$4t + 20$	$5	$5(4t + 20)$

The total purchase can be expressed in two ways:

The value of the tables	+	the value of the chairs	+	the value of the plates	is	the total value of the purchase.
$100t$	+	$50(4t)$	+	$5(4t + 20)$	=	6,500

SOLVE THE EQUATION   We then solve the equation.

$$100t + 50(4t) + 5(4t + 20) = 6{,}500 \quad \text{The equation to solve.}$$
$$100t + 200t + 20t + 100 = 6{,}500 \quad \text{Do the multiplications.}$$
$$320t + 100 = 6{,}500 \quad \text{Combine like terms.}$$
$$320t = 6{,}400 \quad \text{Subtract 100 from both sides.}$$
$$t = 20 \quad \text{Divide both sides by 320.}$$

STATE THE CONCLUSION   The purchases are summarized as follows:

Item	Number purchased	Price per item	Total value
Table	$t = 20$	$100	$2,000
Chair	$4t = 80$	$50	$4,000
Plate	$4t + 20 = 100$	$5	$500
Total			$6,500

CHECK THE RESULT   Because the total purchase is $6,500, the solution checks.   ∎

Solving Investment Problems

To find the amount of interest I an investment earns, we use the formula

$I = Prt$

where P is the principal, r is the annual rate, and t is the time in years. When $t = 1$, the formula simplifies to $I = Pr$.

EXAMPLE 4

Paying tuition. A college student invested the $12,000 inheritance he received and decided to use the annual interest earned to pay his yearly tuition costs of $945. The highest rate offered by a savings and loan at that time was 6% annual simple interest. At this rate, he could not earn the needed $945, so he invested some of the money in a riskier, but more lucrative, investment offering a 9% return. How much did he invest at each rate?

ANALYZE THE PROBLEM We know $12,000 was invested for 1 year at two rates: 6% and 9%. We are asked to find the amount invested at each rate so that the total return would be $945.

FORM AN EQUATION Let x represent the amount invested at 6%. Then $12,000 - x$ represents the amount invested at 9%.

   If $\$x$ (the principal P) is invested at 6% (the rate r), the interest earned would be Pr or $\$0.06x$. At 9%, the rest of the inheritance money, $\$(12,000 - x)$, would earn $\$0.09(12,000 - x)$ interest. These facts are summarized in the following table.

	P	$\cdot$ r	$=$	I
Savings and loan	x	0.06		$0.06x$
Riskier investment	$12,000 - x$	0.09		$0.09(12,000 - x)$

The total interest earned can be expressed in two ways.

The interest earned at 6%	plus	the interest earned at 9%	is	the total interest.
$0.06x$	$+$	$0.09(12,000 - x)$	$=$	945

SOLVE THE EQUATION We then solve the equation.

$$0.06x + 0.09(12,000 - x) = 945 \qquad \text{The equation to solve.}$$

$$\mathbf{100}[0.06x + 0.09(12,000 - x)] = \mathbf{100}(945) \qquad \text{Multiply both sides by 100 to clear the equation of decimals.}$$

$$100(0.06x) + 100(0.09)(12,000 - x) = 100(945) \qquad \text{Distribute the 100.}$$

$$6x + 9(12,000 - x) = 94,500 \qquad \text{Do the multiplications by 100.}$$

$$6x + 108,000 - 9x = 94,500 \qquad \text{Remove parentheses.}$$

$$-3x + 108,000 = 94,500 \qquad \text{Combine like terms.}$$

$$-3x = -13,500 \qquad \text{Subtract 108,000 from both sides.}$$

$$x = 4,500 \qquad \text{Divide both sides by } -3.$$

STATE THE CONCLUSION The student invested $4,500 at 6% and $12,000 - $4,500 = $7,500 at 9%.

CHECK THE RESULT The first investment earned 6% of $4,500, or $270. The second earned 9% of $7,500, or $675. The total return was $270 + $675 = $945. The solution checks. ■

Solving Uniform Motion Problems

If we know the rate r at which we will be traveling and the time t we will be traveling at that rate, we can find the distance d traveled by using the formula

$$d = rt$$

EXAMPLE 5

Coast Guard rescue. A cargo ship, heading into port, radios the Coast Guard that it is experiencing engine trouble and that its speed has dropped to 3 knots. Immediately, a Coast Guard cutter leaves the port and speeds at a rate of 25 knots directly toward the disabled craft, which is 21 nautical miles away. How long will it take the Coast Guard cutter to reach the cargo ship?

ANALYZE THE PROBLEM   The diagram in Figure 2-19(a) shows the situation.

FIGURE 2-19

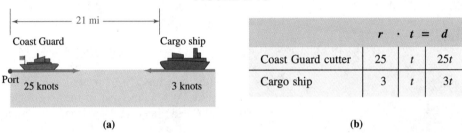

	r	$\cdot$ t	$=$ d
Coast Guard cutter	25	t	$25t$
Cargo ship	3	t	$3t$

(a)   (b)

We know the *rate* of each ship (25 knots and 3 knots), and we know that they must close a *distance* of 21 nautical miles between them. We don't know the *time* it will take them to do this.

FORM AN EQUATION   Let t represent the time it takes for the ships to meet. Using $d = rt$, we find that $25t$ represents the distance traveled by the Coast Guard cutter and $3t$ represents the distance traveled by the cargo ship. This information is recorded in the table in Figure 2-19(b). We can form the equation.

The distance the Coast Guard cutter travels	plus	the distance the cargo ship travels	is	the initial distance between the two ships.
$25t$	$+$	$3t$	$=$	21

SOLVE THE EQUATION   We then solve the equation.

$$25t + 3t = 21 \quad \text{The equation to solve.}$$
$$28t = 21 \quad \text{Combine like terms.}$$
$$t = \frac{21}{28} \quad \text{Divide both sides by 28.}$$
$$t = \frac{3}{4} \quad \text{Simplify the fraction: } \frac{21}{28} = \frac{\overset{1}{\cancel{7} \cdot 3}}{\underset{1}{\cancel{7} \cdot 4}}.$$

STATE THE CONCLUSION   The ships will meet in three-quarters of an hour, or 45 minutes.

CHECK THE RESULT   In three-quarters of an hour, the Coast Guard cutter travels $25 \cdot \frac{3}{4} = \frac{75}{4}$ nautical miles and the cargo ship travels $3 \cdot \frac{3}{4} = \frac{9}{4}$ nautical miles. Together, they travel $\frac{75}{4} + \frac{9}{4} = \frac{84}{4} = 21$ nautical miles. Since this is the initial distance between the ships, the solution checks. ∎

Solving Mixture Problems

We now discuss how to solve two types of mixture problems. In the first type, a *liquid mixture* of a desired strength is made from two solutions with different concentrations.

EXAMPLE 6

Mixing a solution. A chemistry experiment calls for a 30% sulfuric acid solution. If the lab supply room has only 50% and 20% sulfuric acid solutions on hand, how much of each should be mixed to obtain 12 liters of a 30% acid solution?

ANALYZE THE PROBLEM   We must find how much of the 50% solution and how much of the 20% solution is needed to obtain 12 liters of a 30% acid solution.

FORM AN EQUATION   If x represents the numbers of liters (L) of the 50% solution used in the mixture, the remaining $(12 - x)$ liters must be the 20% solution. See Figure 2-20(a). Only 50% of the x liters, and only 20% of the $(12 - x)$ liters, is pure sulfuric acid. The total of these amounts is also the amount of acid in the final mixture, which is 30% of 12 liters. This information is shown in the chart in Figure 2-20(b).

FIGURE 2-20

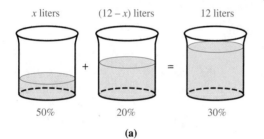

| x liters | $(12 - x)$ liters | 12 liters |

| 50% | 20% | 30% |

(a)

Solution	% acid	·	Liters	=	Amount of acid
50% solution	0.50		x		$0.50x$
20% solution	0.20		$12 - x$		$0.20(12 - x)$
30% mixture	0.30		12		$0.30(12)$

(b)

We can form the equation.

The acid in the 50% solution	plus	the acid in the 20% solution	equals	the acid in the final mixture.
50% of x	+	20% of $(12 - x)$	=	30% of 12

SOLVE THE EQUATION   We then solve the equation.

$$0.50x + 0.20(12 - x) = 0.30(12) \quad \text{50\% = 0.50, 20\% = 0.20, and 30\% = 0.30.}$$
$$5x + 2(12 - x) = 3(12) \quad \text{Multiply both sides by 10 to clear the equation of decimals.}$$
$$5x + 24 - 2x = 36 \quad \text{Remove parentheses.}$$
$$3x + 24 = 36 \quad \text{Combine like terms.}$$
$$3x = 12 \quad \text{Subtract 24 from both sides.}$$
$$x = 4 \quad \text{Divide both sides by 3.}$$

STATE THE CONCLUSION   The mixture will contain 4 liters of 50% solution and $12 - 4 = 8$ liters of 20% solution.

CHECK THE RESULT   Verify that this solution checks.   ∎

In the next example, a *dry mixture* of a specified value is created from two differently priced components.

EXAMPLE 7

Snack food. Because fancy cashews priced at $9 per pound were not selling, a market produce clerk decided to combine them with less expensive filberts and sell the mixture for $7 per pound. How many pounds of filberts, selling at $6 per pound, should be mixed with 50 pounds of cashews to obtain such a mixture?

ANALYZE THE PROBLEM

We know the value of the cashews ($9 per pound) and the filberts ($6 per pound). We also know that 50 pounds of cashews are to be mixed with an unknown number of pounds of filberts to obtain a mixture worth $7 per pound.

FORM THE EQUATION

To solve this problem, we use the formula $v = pn$, where v is value, p is the price per pound, and n is the number of pounds.

Suppose that x pounds of filberts are used in the mixture. At $6 per pound, they are worth $6x$. At $9 per pound, the 50 pounds of cashews are worth $9 \cdot 50 = \$450$. Their combined value will be $\$(6x + 450)$. We also know that the mixture weighs $(50 + x)$ pounds. At $7 per pound, that mixture will be worth $\$7(50 + x)$. This information is recorded in the table in Figure 2-21.

FIGURE 2-21

	p	$\cdot$ n	$=$ v
Filberts	6	x	$6x$
Cashews	9	50	450
Mixture	7	$50 + x$	$7(50 + x)$

We can form the equation.

The value of the filberts	plus	the value of the cashews	equals	the value of the mixture.
$6x$	$+$	450	$=$	$7(50 + x)$

SOLVE THE EQUATION

We then solve the equation.

$6x + 450 = 7(50 + x)$ The equation to solve.

$6x + 450 = 350 + 7x$ Remove parentheses.

$100 = x$ Subtract $6x$ and 350 from both sides.

STATE THE CONCLUSION

Thus, 100 pounds of filberts should be used in the mixture.

CHECK THE RESULT

The value of 100 pounds of filberts at $6 per pound is $600

The value of 50 pounds of cashews at $9 per pound is $450

The value of the mixture is $1,050

The value of 150 pounds of the mixture at $7 per pound is also $1,050. The solution checks. ∎

STUDY SET

Section 2.7

VOCABULARY

In Exercises 1–4, fill in the blanks to make the statements true.

1. The _____perimeter_____ of a triangle or a rectangle is the distance around it.

2. An _____isosceles_____ triangle is a triangle with two sides of the same length.

3. The equal sides of an isosceles triangle meet to form the _____vertex_____ angle.

4. The angles opposite the equal sides in an isosceles triangle are called _____base_____ angles, and they have _____equal_____ measures.

CONCEPTS

5. What is the sum of the measures of the angles of any triangle? 180°

6. Use a ruler to draw an isosceles triangle with sides 3 inches long and a base that is 2 inches long. Label the vertex and the base angles.

7. a. Complete Illustration 1, which shows the inventory of nylon brushes that a paint store carries.

ILLUSTRATION 1

Paint brush	Number ·	Value =	Total value
1 inch	$\frac{x}{2}$	$4	$2x
2 inch	x	$5	$5x
3 inch	$x + 10$	$7	$7(x + 10)

b. Which type of brush does the store have the largest number of? 3 in.

c. What is the least expensive brush? 1 in.

d. What is the total value of the inventory of nylon brushes? $(14x + 70)$

8. In the advertisement in Illustration 2, what are the principal, the rate, and the time for the investment opportunity shown? $30,000, 14%, 1 yr

ILLUSTRATION 2

> **Invest in Mini Malls!**
> Builder seeks daring people who want to earn big $$$$$$. In just 1 year, you will earn a gigantic 14% on an investment of only $30,000! Call now.

9. a. Complete Illustration 3, which gives the details about two investments that were made by a retired couple.

ILLUSTRATION 3

	P ·	r =	I
Certificate of deposit	x	0.04	0.04x
Brother-in-law's business	2x	0.06	(0.06)2x

b. How much more money was invested in the brother-in-law's business than in the certificate of deposit? twice as much

c. What is the total amount of interest the couple will make from these investments?
0.04x + 0.12x = 0.16x

10. a. Complete Illustration 4, which gives the details of each morning's commute by a husband and wife who travel in opposite directions.

ILLUSTRATION 4

	r ·	t =	d
Husband	35 mph	t hr	35t mi
Wife	45 mph	t hr	45t mi

b. Who is able to drive to work at the faster rate?
the wife

c. Which person is on the road longer?
They are both on the road for t hr.

d. Write an algebraic expression that represents the distance between their workplaces. 80t mi

11. Suppose the contents of the two barrels shown in Illustration 5 are poured into an empty third barrel.

ILLUSTRATION 5

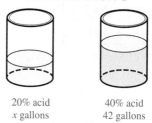

20% acid
x gallons

40% acid
42 gallons

a. How many gallons of liquid will the third barrel contain? $(x + 42)$ gal

b. What would be a *reasonable* estimate of the concentration of the solution in the third barrel— 19%, 32%, or 43% acid? 32%

12. Each bottle of dressing shown in Illustration 6 contains a mixture of oil and vinegar. After sitting overnight, the liquids separate completely, with the oil rising to the top. On each bottle, draw the line estimating where the separation would occur and shade the vinegar.

ILLUSTRATION 6

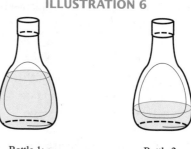

Bottle 1:
40% oil
60% vinegar

Bottle 2:
80% oil
20% vinegar

13. a. Complete Illustration 7, which gives the details about the ingredients of a box of breakfast cereal.

ILLUSTRATION 7

	Price ($/oz)	Amount (oz)	Value
Blueberries	$0.38	x	$0.38x
Bran Flakes	$0.08	14	$1.12
Blueberries & Bran Flakes Cereal	$0.21	$14 + x$	$0.21(14 + x)$

b. What does the cereal cost per ounce? $0.21

14. The ingredients of a weight-gain drink powder and their prices are shown in Illustration 8. Why can't the weight-gain powder's value be less than $3.50 a pound or more than $8.25 a pound?

ILLUSTRATION 8

Ingredients	Price (per lb)
Protein powder	$8.25
Carob powder	$3.50

PRACTICE

In Exercises 15–16, solve the equation by first clearing it of decimals.

15. $0.08x + 0.07(15,000 - x) = 1,110$ 6,000

16. $0.108x + 0.07(16,000 - x) = 1,500$ 10,000

17. Two angles are called **complementary angles** when the sum of their measures is 90°. Find the measures of the complementary angles shown in Illustration 9.   22°, 68°

ILLUSTRATION 9

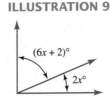

$(6x + 2)°$

$2x°$

18. Two angles are called **supplementary angles** when the sum of their measures is 180°. Find the measures of the supplementary angles shown in Illustration 10.   40°, 140°

ILLUSTRATION 10

$(4x + 40)°$

$(x + 15)°$

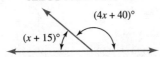

19. In Illustration 11, two lines intersect to form **vertical angles.** Use the fact that vertical angles have the same measure to find x.   15

ILLUSTRATION 11

$(2x + 5)°$ $(3x - 10)°$

20. Find the measures of the vertical angles shown in Illustration 12. (See Exercise 19.)   150°

ILLUSTRATION 12

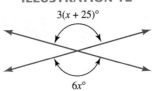

$3(x + 25)°$

$6x°$

APPLICATIONS

In Exercises 21–54, solve each problem; a diagram or table may be helpful in organizing the facts of the problem.

21. TRIANGULAR BRACING   The outside perimeter of the triangular brace in Illustration 13 is 57 feet. If all three sides are equal, find the length of each side.   19 ft

ILLUSTRATION 13

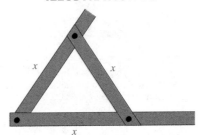

x x

x

22. CIRCUIT BOARD   The perimeter of the circuit board in Illustration 14 is 90 centimeters. Find its dimensions.   19 cm by 26 cm

ILLUSTRATION 14

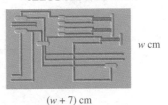

w cm

$(w + 7)$ cm

23. TRUSS   The truss in Illustration 15 is in the form of an isosceles triangle. Each of the two equal sides is 4 feet less than the third side. If the perimeter is 25 feet, find the lengths of the sides.   7 ft, 7 ft, and 11 ft

ILLUSTRATION 15

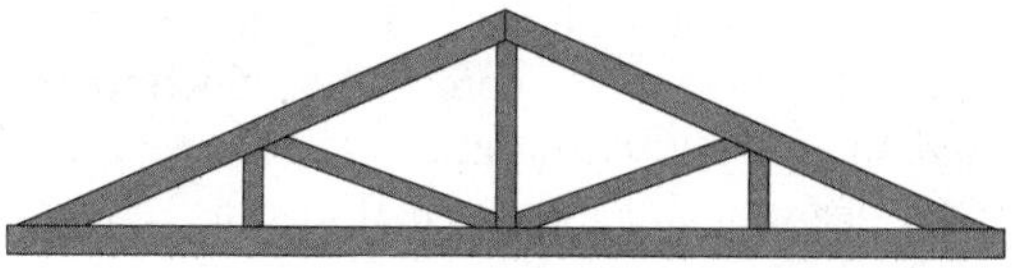

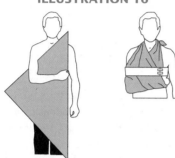

24. **FIRST AID** The sling shown in Illustration 16 is in the shape of an isosceles triangle with a perimeter of 144 inches. The longest side of the sling is 18 inches longer than either of the other two sides. Find the lengths of each side.   60 in., 42 in., and 42 in.

ILLUSTRATION 16

25. **SWIMMING POOL** The seawater Orthlieb Pool in Casablanca, Morocco is the largest swimming pool in the world. With a perimeter of 1,110 meters, this rectangular-shaped pool has a length that is 30 meters more than 6 times its width. Find its dimensions.
75 m by 480 m

26. **ART** The *Mona Lisa,* shown in Illustration 17, was completed by Leonardo da Vinci in 1506. The length of the picture is 11.75 inches less than twice the width. If the perimeter of the picture is 102.5 inches, find its dimensions.
21 in. by 30.25 in

ILLUSTRATION 17

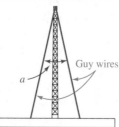

27. **GUY WIRES** The two guy wires shown in Illustration 18 form an isosceles triangle. One of the two equal angles of the triangle is 4 times the third angle (the vertex angle). Find the measure of the vertex angle.   20°

ILLUSTRATION 18

Guy wires

a

28. **MOUNTAIN BICYCLE** For the bicycle frame in Illustration 19, the angle that the horizontal crossbar makes with the seat support is 15° less than twice the angle at the steering column. The angle at the pedal

gear is 25° more than the angle at the steering column. Find these three angle measures.
42.5°, 70°, 67.5°

ILLUSTRATION 19

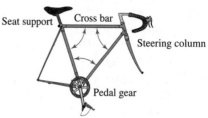

Seat support   Cross bar

Steering column

Pedal gear

29. **WAREHOUSING COSTS** A store warehouses 40 more portables than big-screen TV sets, and 25 fewer consoles than portables. Storage costs for the different TV sets are shown in Illustration 20. If storage costs $276 per month, how many big-screen sets are in stock?   12

ILLUSTRATION 20

Type of TV	Monthly cost
Portable	$1.50
Console	$4.00
Big screen	$7.50

30. **APARTMENT RENTAL** The owners of an apartment building rent 1-, 2-, and 3-bedroom units. They rent equal numbers of each, with the monthly rents given in Illustration 21. If the total monthly income is $36,550, how many of each type of unit are there?   17

ILLUSTRATION 21

Unit	Rent
One-bedroom	$550
Two-bedroom	$700
Three-bedroom	$900

31. **SOFTWARE SALES** Three software applications are priced as shown in Illustration 22. Spreadsheet and database programs sold in equal numbers, but 15 more word processing applications were sold than the other two combined. If the three applications generated sales of $72,000, how many spreadsheets were sold?   90

ILLUSTRATION 22

Software	Price
Spreadsheet	$150
Database	$195
Word processing	$210

32. INVENTORY With summer approaching, the number of air conditioners sold is expected to be double that of stoves and refrigerators combined. Stoves sell for $350, refrigerators for $450, and air conditioners for $500, and sales of $56,000 are expected. If stoves and refrigerators sell in equal numbers, how many of each appliance should be stocked?
20 stoves, 20 refrigerators, and 80 air conditioners

33. INTEREST INCOME On December 31, 1997, Terrell Washington opened two savings accounts. At the end of 1998, his bank mailed him the form shown in Illustration 23, for income tax purposes. If a total of $12,000 was initially deposited and if no further deposits or withdrawals were made, how much money was originally deposited in account number 721-94? $5,500

ILLUSTRATION 23

USA HOME SAVINGS	Copy B For Recipient Interest Income
This is important tax information and is being furnished to the Internal Revenue Service.	OMB No. 1545-0112 **1998**
RECIPIENT'S name **TERRELL WASHINGTON**	Form 1099–iNT

Acct. Number	Annual Percent Yield	Early Withdrawal Penalty
822–06	6%	.00
721–94	4.5%	.00
		Total Interest Income 637.50

34. MAKING A PRESENTATION A financial planner recommends a plan for a client who has $65,000 to invest. (See Illustration 24.) At the end of the presentation, the client asks, "How much will be invested at each rate?" Answer this question using the given information. $42,200 at 12%, $22,800 at 6.2%

ILLUSTRATION 24

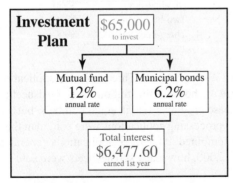

35. INVESTMENTS Equal amounts are invested in each of three accounts paying 7%, 8%, and 10.5%. If one year's combined interest income is $1,249.50, how much is invested in each account?
$4,900 in each account

36. RETIREMENT A professor wants to supplement her retirement income with investment interest. If she invests $15,000 at 6% interest, how much more would she have to invest at 7% to achieve a goal of $1,250 per year in supplemental income? $5,000

37. FINANCIAL PLANNING A plumber has a choice of two investment plans:

■ An insured fund that pays 11% interest

■ A risky investment that pays a 13% return

If the same amount invested at the higher rate would generate an extra $150 per year, how much does the plumber have to invest? $7,500

38. INVESTMENTS The amount of annual interest earned by $8,000 invested at a certain rate is $200 less than $12,000 would earn at a rate 1% lower. At what rate is the $8,000 invested?   8%

39. TORNADO During a storm, two teams of scientists leave a university at the same time in specially designed vans to search for tornadoes. The first team travels east at 20 mph and the second travels west at 25 mph, as shown in Illustration 25. If their radios have a range of up to 90 miles, how long will it be before they lose radio contact?   2 hr

ILLUSTRATION 25

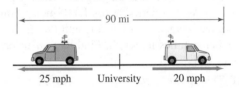

40. SEARCH AND RESCUE Two search-and-rescue teams leave base at the same time looking for a lost boy. The first team, on foot, heads north at 2 mph and the other, on horseback, south at 4 mph. How long will it take them to search a distance of 21 miles between them?   3.5 hr

41. SPEED OF TRAINS Two trains are 330 miles apart, and their speeds differ by 20 mph. Find the speed of each train if they are traveling toward each other and will meet in 3 hours.   65 mph and 45 mph

42. AVERAGE SPEED A car averaged 40 mph for part of a trip and 50 mph for the remainder. If the 5-hour trip covered 210 miles, for how long did the car average 40 mph?   4 hr

43. AIR TRAFFIC CONTROL An airliner leaves Berlin, Germany, headed for Montreal, Canada, flying at an average speed of 450 mph. At the same time, an airliner leaves Montreal headed for Berlin, averaging 500 mph. If the airports are 3,800 miles apart, when will the air traffic controllers have to make the pilots aware that the planes are passing each other?
4 hr into the flights

44. ROAD TRIP A bus, carrying the members of a marching band, and a truck, carrying their instruments, leave a high school at the same time. The bus travels at 65 mph and the truck at 55 mph. In how many hours will they be 75 miles apart? 7.5 hr

45. SALT SOLUTION How many gallons of a 3% salt solution must be mixed with 50 gallons of a 7% solution to obtain a 5% solution? 50

46. MAKING CHEESE To make low-fat cottage cheese, milk containing 4% butterfat is mixed with 10 gallons of milk containing 1% butterfat to obtain a mixture containing 2% butterfat. How many gallons of the richer milk must be used? 5

47. ANTISEPTIC SOLUTION A nurse wants to add water to 30 ounces of a 10% solution of benzalkonium chloride to dilute it to an 8% solution. How much water must she add? 7.5 oz

48. PHOTOGRAPHIC CHEMICALS A photographer wishes to mix 2 liters of a 5% acetic acid solution with a 10% solution to get a 7% solution. How many liters of 10% solution must be added? $1\frac{1}{3}$

49. MIXING FUELS How many gallons of fuel costing $1.15 per gallon must be mixed with 20 gallons of a fuel costing $0.85 per gallon to obtain a mixture costing $1 per gallon? See Illustration 26. 20

ILLUSTRATION 26

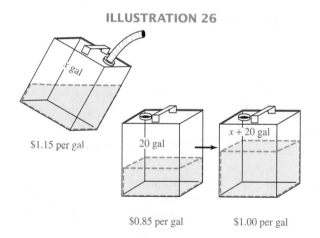

$1.15 per gal   20 gal $0.85 per gal   x + 20 gal $1.00 per gal

50. MIXING PAINT Paint costing $19 per gallon is to be mixed with 5 gallons of a $3-per-gallon thinner to make a paint that can be sold for $14 per gallon. How much paint will be produced? 16 gal

51. MIXING CANDY Lemon drops worth $1.90 per pound are to be mixed with jelly beans that cost $1.20 per pound to make 100 pounds of a mixture worth $1.48 per pound. How many pounds of each candy should be used?

40 lb lemon drops and 60 lb jelly beans

52. MIXING CANDY See Illustration 27. Twenty pounds of lemon drops are to be mixed with cherry chews to make a mixture that will sell for $1.80 per pound. How much of the more expensive candy should be used? 10 lb

ILLUSTRATION 27

Candy	Price per pound
Peppermint patties	$1.35
Lemon drops	$1.70
Licorice lumps	$1.95
Cherry chews	$2.00

53. BLENDING COFFEE A store sells regular coffee for $4 a pound and gourmet coffee for $7 a pound. To get rid of 40 pounds of the gourmet coffee, a shopkeeper makes a blend to put on sale for $5 a pound. How many pounds of regular coffee should he use? 80

54. BLENDING LAWN SEED A store sells bluegrass seed for $6 per pound and ryegrass seed for $3 per pound. How much ryegrass must be mixed with 100 pounds of bluegrass to obtain a blend that will sell for $5 per pound? 50 lb

WRITING

55. Create a mixture problem of your own, and solve it.

56. Use an example to explain the difference between the quantity and the value of the materials being combined in a mixture problem.

57. A car travels at 60 mph for 15 minutes. Why can't we multiply the rate, 60, and the time, 15, to find the distance traveled by the car?

58. Create a geometry problem that could be answered by solving the equation $2w + 2(w + 5) = 26$.

REVIEW

In Exercises 59–60, write each expression without using parentheses.

59. $-25(2x - 5)$ $-50x + 125$

60. $-12(3a + 4b - 32)$ $-36a - 48b + 384$

In Exercises 61–62, combine like terms.

61. $8p - 9q + 11p + 20q$ $19p + 11q$

62. $-5(t - 120) - 7(t + 5)$ $-12t + 565$

▶**2.8**

Inequalities

In this section, you will learn about

Inequality symbols ■ Graphing inequalities ■ Interval notation ■ Solving inequalities ■ Graphing compound inequalities ■ Solving compound inequalities ■ An application

Introduction   **Inequalities** are expressions indicating that two quantities are not necessarily equal. They appear in many situations:

■   An airplane is rated to fly at altitudes that are less than 36,000 feet.

■   To thaw ice, the temperature must be greater than 32°F.

■   To earn a B, I need a final exam score of at least 80%.

Inequality Symbols

We can use **inequality symbols** to show that two expressions are not equal.

Inequality symbols			
$\neq$	means	"is not equal to"
$<$	means	"is less than"
$>$	means	"is greater than"
$\leq$	means	"is less than or equal to"
$\geq$	means	"is greater than or equal to"

EXAMPLE 1

Reading inequalities.

a. $6 \neq 9$ is read as "6 is not equal to 9."

b. $x > 5$ is read as "x is greater than 5."

c. $5 \leq 5$ is read as "5 is less than or equal to 5." This is true, because $5 = 5$.

SELF CHECK   Write each inequality in words: **a.** $15 < 20$, **b.** $y \geq 9$,   and   **c.** $30 \leq 30$.   *Answers:*   **a.** 15 is less than 20.   **b.** y is greater than or equal to 9.   **c.** 30 is less than or equal to 30.   ■

If two numbers are graphed on a number line, the one to the right is the greater. For example, from Figure 2-22, we see that $-1 > -4$, because -1 lies to the right of -4.

FIGURE 2-22

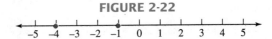

Inequalities can be written so that the inequality symbol points in the opposite direction. For example, the following statements both indicate that 27 is a smaller number than 32.

$27 < 32$ (27 is less than 32)   and $32 > 27$ (32 is greater than 27)

The following statements both indicate that 9 is greater than or equal to 6.

$9 \geq 6$ (9 is greater than or equal to 6)   and $6 \leq 9$ (6 is less than or equal to 9)

Variables can be used with inequality symbols to show mathematical relationships. For example, consider the statement, "You must be taller than 54 inches to ride the roller coaster." If we let h represent a person's height in inches, then to ride the roller coaster, $h > 54$ inches.

EXAMPLE 2

Writing inequalities. Express the following situation using an inequality symbol: "The occupancy of the dining room cannot exceed 200 people."

Solution   If p represents the number of people that can occupy the room, then p cannot be greater than (exceed) 200. Another way to state this is that p must be *less than or equal to* 200:

$$p \leq 200$$

SELF CHECK   Express the following statement using an inequality symbol: "The thermostat on the pool heater is set so that the water temperature t is at least 72°."   *Answer:* $t \geq 72$ ∎

Graphing Inequalities

Graphs of inequalities involving real numbers are **intervals** on the number line. For example, two versions of the graph of all real numbers x such that $x > -3$ are shown in Figure 2-23.

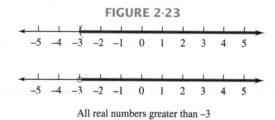

FIGURE 2-23

All real numbers greater than –3

The thick arrow pointing to the right shows that all numbers to the right of -3 are in the graph. The **parenthesis** (or open circle) at -3 indicates that -3 is not in the graph.

Interval Notation

The interval shown in Figure 2-23 can be expressed in **interval notation** as $(-3, \infty)$. Again, the first parenthesis indicates that -3 is not included in the interval. The infinity symbol ∞ does not represent a number. It indicates that the interval continues on forever to the right.

Figure 2-24 shows two versions of the graph of $x \leq 2$. The thick arrow pointing to the left shows that all numbers to the left of 2 are in the graph. The **bracket** (or closed circle) at 2 indicates that 2 is included in the graph. We can express this interval as $(-\infty, 2]$. Here the bracket indicates that 2 is included in the interval.

From now on, we will use parentheses or brackets when graphing intervals, because they are consistent with interval notation.

FIGURE 2-24

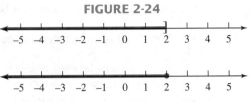

All real numbers less than or equal to 2

E X A M P L E   3

Writing an inequality from its graph. What inequality is represented by each graph?

a.

b.

Solution **a.** This is the interval $(-\infty, 3]$, which consists of all real numbers less than or equal to 3. The inequality is $x \leq 3$.

b. This is the interval $(-1, \infty)$, consisting of all real numbers greater than -1. The inequality is $x > -1$.

SELF CHECK What inequality is represented by each graph?

a.

b.

Answers:
a. $x \leq -1, (-\infty, -1]$
b. $x > -4, (-4, \infty)$ ■

Solving Inequalities

A **solution of an inequality** is any number that makes the inequality true. For example, 2 is a solution of $x \leq 3$, because $2 \leq 3$.

To solve more complicated inequalities, we will use the addition, subtraction, multiplication, and division properties of inequality. When we use one of these properties, the resulting inequality will always be equivalent to the original one.

Addition and subtraction properties of inequality

For real numbers a, b, and c,

    If $a < b$,   then $a + c < b + c$.

    If $a < b$,   then $a - c < b - c$.

Similar statements can be made for the symbols $>$, $\leq$, and $\geq$.

    The **addition property of inequality** can be stated this way: *If a quantity is added to both sides of an inequality, the resulting inequality will have the same direction as the original one.*

The **subtraction property of inequality** can be stated this way: *If a quantity is subtracted from both sides of an inequality, the resulting inequality will have the same direction as the original one.*

EXAMPLE 4

Solving inequalities. Solve $x + 3 > 2$ and graph its solution.

Solution

To isolate the x on the left-hand side of the $>$ sign, we proceed as we would when solving equations.

$$x + 3 > 2$$
$$x + 3 - 3 > 2 - 3 \quad \text{To undo the addition of 3, subtract 3 from both sides.}$$
$$x > -1 \quad \text{Do the subtractions: } 3 - 3 = 0 \text{ and } 2 - 3 = 2 + (-3) = -1.$$

The graph (see Figure 2-25) includes all points to the right of -1 but does not include -1. This represents all real numbers greater than -1. Expressed as an interval, we have $(-1, \infty)$.

FIGURE 2-25

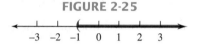

To check, we pick several numbers in the graph, such as 1 and 3, substitute each number for x in the inequality, and see whether it satisfies the inequality.

$x + 3 > 2$	$x + 3 > 2$
$1 + 3 \overset{?}{>} 2$ Substitute 1 for x.	$3 + 3 \overset{?}{>} 2$ Substitute 3 for x.
$4 > 2$ Do the addition.	$6 > 2$ Do the addition.

Since $4 > 2$, 1 satisfies the inequality. Since $6 > 2$, 2 satisfies the inequality. The solution appears to be correct.

SELF CHECK

Solve $x - 3 \leq -2$ and graph its solution. Then use interval notation to describe the solution.

Answer: $x \leq 1, (-\infty, 1]$

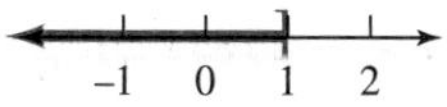

If both sides of the inequality $2 < 5$ are multiplied by a *positive* number, such as 3, another true inequality results.

$$2 < 5$$
$$3 \cdot 2 < 3 \cdot 5 \quad \text{Multiply both sides by 3.}$$
$$6 < 15 \quad \text{Do the multiplications: } 3 \cdot 2 = 6 \text{ and } 3 \cdot 5 = 15.$$

However, if we multiply both sides of $2 < 5$ by a *negative* number, such as -3, the direction of the inequality symbol must be reversed to produce another true inequality.

$$2 < 5$$
$$-3 \cdot 2 > -3 \cdot 5 \quad \text{Multiply both sides by the } negative \text{ number } -3 \text{ and reverse the direction of the inequality.}$$
$$-6 > -15 \quad \text{Do the multiplications: } -3 \cdot 2 = -6 \text{ and } -3 \cdot 5 = -15.$$

The inequality $-6 > -15$ is true, because -6 is to the right of -15 on the number line.

Multiplication and division properties of inequalities

For real numbers a, b, and c,

If $a < b$ and $c > 0$, then $ac < bc$.

If $a < b$ and $c < 0$, then $ac > bc$.

If $a < b$ and $c > 0$, then $\frac{a}{c} < \frac{b}{c}$.

If $a < b$ and $c < 0$, then $\frac{a}{c} > \frac{b}{c}$.

Similar statements can be made for the symbols $>$, $\leq$, and $\geq$.

The **multiplication property of inequality** can be stated this way:

If both sides of an inequality are multiplied by the same positive number, the resulting inequality will have the same direction as the original one.

If both sides of an inequality are multiplied by the same negative number, the resulting inequality will have the opposite direction from the original one.

The **division property of inequality** can be stated this way:

If both sides of an inequality are divided by the same positive number, the resulting inequality will have the same direction as the original one.

If both sides of an inequality are divided by the same negative number, the resulting inequality will have the opposite direction from the original one.

EXAMPLE 5

Solving inequalities. Solve $3x + 7 \leq -5$ and graph the solution.

Solution

$$3x + 7 \leq -5$$

$3x + 7 - 7 \leq -5 - 7$ To undo the addition of 7, subtract 7 from both sides.

$3x \leq -12$ Do the subtractions: $7 - 7 = 0$ and $-5 - 7 = -5 + (-7) = -12$.

$\dfrac{3x}{3} \leq \dfrac{-12}{3}$ To undo the multiplication by 3, divide both sides by 3.

$x \leq -4$ Do the divisions.

The graph (shown in Figure 2-26) consists of all real numbers less than or equal to -4. Using interval notation, we have $(-\infty, -4]$.

FIGURE 2-26

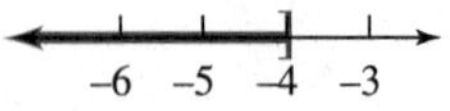

To check, we can pick a number in the graph, such as -6, and see whether it satisfies the inequality.

$$3x + 7 \leq -5$$

$3(-6) + 7 \overset{?}{\leq} -5$ Substitute -6 for x.

$-18 + 7 \overset{?}{\leq} -5$ Do the multiplication.

$-11 \leq -5$ Do the addition.

Since $-11 \leq -5$, -6 satisfies the inequality. The solution appears to be correct.

SELF CHECK
Solve $2x - 7 > -13$ and graph the solution. Then use interval notation to describe the solution.

Answer: $x > -3$, $(-3, \infty)$

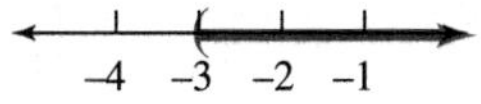

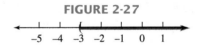

E X A M P L E   6

Reversing the inequality symbol. Solve $5 - 3x < 14$ and graph the solution.

Solution

$$5 - 3x < 14$$

$5 - 3x - 5 < 14 - 5$ To isolate $-3x$ on the left-hand side, subtract 5 from both sides.

$-3x < 9$ Do the subtractions: $5 - 5 = 0$ and $14 - 5 = 9$.

$\dfrac{-3x}{-3} > \dfrac{9}{-3}$ To undo the multiplication by -3, divide both sides by -3. Since we are dividing by -3, we reverse the direction of the $<$ symbol.

$x > -3$

The graph is shown in Figure 2-27. This is the interval $(-3, \infty)$, which consists of all real numbers greater than -3.

FIGURE 2-27

Check the result.

SELF CHECK
Solve $-2x - 5 \geq -7$ and graph the solution. Then use interval notation to describe the solution.

Answer: $x \leq 1$, $(-\infty, 1]$

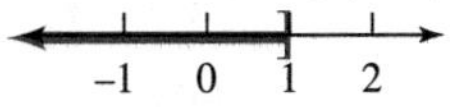

E X A M P L E   7

Solving inequalities. Solve $5(x + 1) \leq 2(x - 3)$ and graph the solution.

Solution

$$5(x + 1) \leq 2(x - 3)$$

$5x + 5 \leq 2x - 6$ Remove the parentheses on both sides.

$5x + 5 - 2x \leq 2x - 6 - 2x$ To eliminate $2x$ from the right side, subtract $2x$ from both sides.

$3x + 5 \leq -6$ Combine like terms on both sides.

$3x + 5 - 5 \leq -6 - 5$ To undo the addition of 5, subtract 5 from both sides.

$3x \leq -11$ Do the subtractions.

$\dfrac{3x}{3} \leq \dfrac{-11}{3}$ To undo the multiplication by 3, divide both sides by 3.

$x \leq -\dfrac{11}{3}$

The graph is shown in Figure 2-28. This is the interval $\left(-\infty, -\frac{11}{3}\right]$, which consists of all real numbers less than or equal to $-\frac{11}{3}$. We note that $-\frac{11}{3} = -3\frac{2}{3}$.

FIGURE 2-28

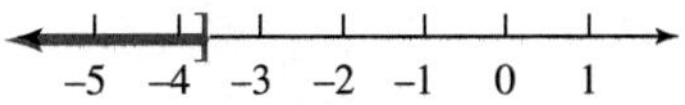

Check the result.

SELF CHECK   Solve $3(x - 2) > -(x + 1)$ and graph the solution. Then use interval notation to describe the solution.

Answer: $x > \dfrac{5}{4}, \left(\dfrac{5}{4}, \infty\right)$

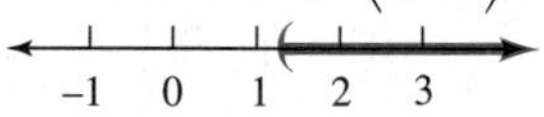

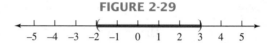

Graphing Compound Inequalities

NIA

Two inequalities can be combined into a **compound inequality** to indicate that numbers lie *between* two fixed values. For example, $-2 < x < 3$, is a combination of

$$-2 < x \quad \text{and} \quad x < 3$$

It indicates that x is greater than -2 and that x is also less than 3. The solution of $-2 < x < 3$ consists of all numbers that lie *between* -2 and 3. The graph of this interval appears in Figure 2-29. We can express this interval as $(-2, 3)$.

FIGURE 2-29

EXAMPLE 8   **Writing an inequality from its graph.** What inequality is represented by the graph below?

Solution $1 < x \leq 5$. This is the interval $(1, 5]$.

SELF CHECK   What inequality is represented by the graph below?

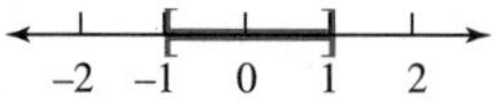

Answer: $-1 \leq x \leq 1$. This is the interval $[-1, 1]$.

EXAMPLE 9   **Graphing compound inequalities.** Graph the interval $-4 < x \leq 0$.

Solution   The interval $-4 < x \leq 0$ consists of all real numbers between -4 and 0, including 0. The graph appears in Figure 2-30. This is the interval $(-4, 0]$.

FIGURE 2-30

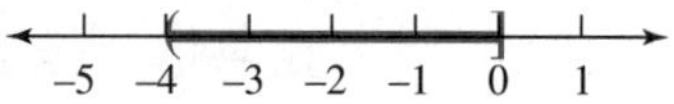

To check, we pick a number, such as -2, in the graph and see whether it satisfies the inequality. Since $-4 < -2 \leq 0$, the solution appears to be correct.

SELF CHECK   Graph the interval $-2 \leq x < 1$. Then use interval notation to describe the solution.

Answer: $[-2, 1)$

Solving Compound Inequalities

NIA

To solve compound inequalities, we use the same methods we used for solving equations. However, instead of applying the properties of equality to both sides of an equation, we will apply the properties of inequality to all three parts of the inequality.

EXAMPLE 10

Solving compound inequalities. Solve $-4 < 2(x - 1) \leq 4$ and graph the solution.

Solution

$$-4 < 2(x - 1) \leq 4$$

$-4 < 2x - 2 \leq 4$ Use the distributive property to remove parentheses.

$-2 < 2x \leq 6$ To undo the subtraction of 2, add 2 to all three parts.

$-1 < x \leq 3$ To undo the multiplication by 2, divide all three parts by 2.

The graph of the solution appears in Figure 2-31. This is the interval $(-1, 3]$.

FIGURE 2-31

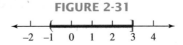

Check the solution.

SELF CHECK

Solve $-6 \leq 3(x + 2) \leq 6$ and graph the solution. Then use interval notation to describe the solution.   *Answer:* $-4 \leq x \leq 0, [-4, 0]$

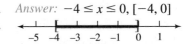

■

An Application

When solving problems, phrases such as "not more than," "at least," or "should exceed" suggest that an *inequality* should be written instead of an *equation*.

EXAMPLE 11

A student has scores of 72%, 74%, and 78% on three exams. What percent score does he need on the last exam to earn no less than a grade of B (80%)?

ANALYZE THE PROBLEM

We know three of the student's scores. We are to find what he must score on the last exam to earn at least a B grade.

FORM AN INEQUALITY

We can let x represent the score on the fourth (and last) exam. To find the average grade, we add the four scores and divide by 4. To earn no less than a grade of B, the student's average must be greater than or equal to 80%.

The average of the four grades	must be greater than or equal to	80.
$\dfrac{72 + 74 + 78 + x}{4}$	$\geq$	80

SOLVE THE INEQUALITY

We can solve this inequality for x.

$\dfrac{224 + x}{4} \geq 80$ $72 + 74 + 78 = 224$.

$224 + x \geq 320$ Multiply both sides by 4.

$x \geq 96$ Subtract 224 from both sides.

STATE THE CONCLUSION

To earn a B, the student must score 96% or better on the last exam. Of course, the student cannot score higher than 100%. The graph appears in Figure 2-32 on the next page. This is the interval $[96, 100]$.

FIGURE 2-32

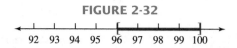

CHECK THE RESULT   Pick some numbers in the interval and verify that the average of the four scores will be 80% or greater.   ■

STUDY SET

Section 2.8

VOCABULARY

In Exercises 1–4, fill in the blanks to make the statements true.

1. An expression containing one of the symbols $>$, $<$, $\geq$, $\leq$, or $\neq$ is called an ____inequality____.

2. Graphs of inequalities involving real numbers are called ____intervals____ on the number line.

3. A ____solution____ of an inequality is any real number that makes the inequality true.

4. The inequality $-4 < x \leq 12$ is an example of a ____compound____ inequality.

CONCEPTS

In Exercises 5–8, fill in the blanks to make the statements true.

5. If a quantity is added to or subtracted from both sides of an inequality, the resulting inequality will have the ____same____ direction as the original one.

6. If both sides of an inequality are multiplied or divided by a positive number, the resulting inequality will have the ____same____ direction as the original one.

7. If both sides of an inequality are multiplied or divided by a negative number, the resulting inequality will have the ____opposite____ direction from the original one.

8. To solve compound inequalities, the properties of inequalities are applied to all ____three____ parts of the inequality.

9. The solution of an inequality is graphed below.

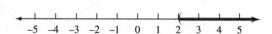

a. If 3 is substituted for the variable in the inequality, what type of statement will result?
   a true statement

b. If -3 is substituted for the variable in the inequality, what type of statement will result?
   a false statement

10. The solution of an inequality is graphed below.

a. If 3 is substituted for the variable in the inequality, what type of statement will result?
   a false statement

b. If -3 is substituted for the variable in the inequality, what type of statement will result?
   a true statement

11. Solve the inequality $2x - 4 > 12$ and give the solution:
   a. in words   all real numbers greater than 8
   b. using a graph   ◄———( )———►
                              8
   c. using interval notation $(8, \infty)$

12. Solve the compound inequality $-4 < 2x < 12$ and give the solution:
   a. in words   all real numbers between -2 and 6
   b. using a graph   ◄———( )———►
                          -2 6
   c. using interval notation $(-2, 6)$

NOTATION

In Exercises 13–18, fill in the blanks to make the statements true.

13. The symbol $<$ means "____is less than____."

14. the symbol $>$ means "____is greater than____."

15. The symbol $\geq$ means "____is greater than____ or equal to."

16. The symbol $\leq$ means "is less than ____or equal to____."

17. The symbol $\neq$ means "____is not equal to____."

18. In the interval [4, 8), the endpoint 4 is ____included____, but the endpoint 8 is not included.

19. Suppose you solve an inequality and obtain $-2 < x$. Rewrite this inequality so that x is on the left-hand side. $x > -2$

20. Explain what is wrong with the compound inequality $8 < x < -1$. 8 is not less than -1.

In Exercises 21–22, write each inequality so that the inequality symbol points in the opposite direction.

21. $17 \geq -2$ $-2 \leq 17$

22. $-32 < -10$ $-10 > -32$

In Exercises 23–24, complete each solution.

23. Solve $4x - 5 \geq 7$.

$$4x - 5 \geq 7$$
$$4x - 5 + \boxed{5} \geq 7 + \boxed{5}$$
$$4x \geq \boxed{12}$$
$$\frac{4x}{\boxed{4}} \geq \frac{12}{\boxed{4}}$$
$$x \geq 3$$

24. Solve $\dfrac{-x}{2} + 4 < 5$.

$$\frac{-x}{2} + 4 < 5$$
$$\frac{-x}{2} + 4 - \boxed{4} < 5 - \boxed{4}$$
$$\frac{-x}{2} < \boxed{1}$$
$$\boxed{2}\left(\frac{-x}{2}\right) < \boxed{2}\,(1)$$
$$\boxed{-x} < 2$$
$$\frac{-x}{\boxed{-1}} > \frac{2}{\boxed{-1}}$$
$$x > -2$$

PRACTICE

In Exercises 25–28, graph each inequality. Then describe the graph using interval notation.

25. $x < 5$

$(-\infty, 5)$

26. $x \geq -2$

$[-2, \infty)$

27. $-3 < x \leq 1$

$(-3, 1]$

28. $-1 \leq x \leq 3$

$[-1, 3]$

In Exercises 29–32, write the inequality that is represented by each graph. Then describe the graph using interval notation.

29. $x < -1,\ (-\infty, -1)$

30. $x \geq 2,\ [2, \infty)$

31. $-7 < x \leq 2,\ (-7, 2]$

32. $-3 \leq x \leq 1,\ [-3, 1]$

In Exercises 33–62, solve each inequality, graph the solution, and then use interval notation to describe the solution.

33. $x + 2 > 5$ $x > 3,\ (3, \infty)$

34. $x + 5 \geq 2$
$x \geq -3,\ [-3, \infty)$

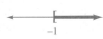

35. $-x - 3 \leq 7$
$x \geq -10,\ [-10, \infty)$

36. $-x - 9 > 3$
$x < -12,\ (-\infty, -12)$

37. $3 + x < 2$
$x < -1,\ (-\infty, -1)$

38. $5 + x \geq 3$
$x \geq -2,\ [-2, \infty)$

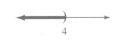

39. $2x - 0.3 \leq 0.5$
$x \leq 0.4,\ (-\infty, 0.4]$

40. $-3x - 0.5 < 0.4$
$x > -0.3,\ (-0.3, \infty)$

41. $-3x - 7 > -1$
$x < -2,\ (-\infty, -2)$

42. $-5x + 7 \leq 12$
$x \geq -1,\ [-1, \infty)$

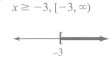

43. $-4x + 6 > 17$
$x < -\frac{11}{4},\ \left(-\infty, -\frac{11}{4}\right)$

44. $7x - 1 > 5$
$x > \frac{6}{7},\ \left(\frac{6}{7}, \infty\right)$

45. $\dfrac{2}{3}x \geq 2$ $x \geq 3,\ [3, \infty)$

46. $\dfrac{3}{4}x < 3$ $x < 4,\ (-\infty, 4)$

47. $-\dfrac{7}{8}x \le 21$

$x \ge -24, [-24, \infty)$

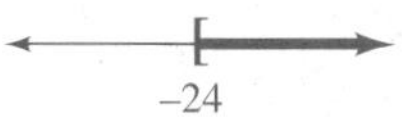

48. $-\dfrac{3}{16}x \ge -9$

$x \le 48, (-\infty, 48]$

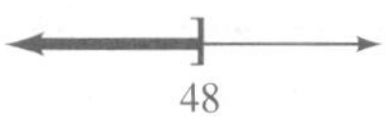

49. $2x + 9 \le x + 8$

$x \le -1, (-\infty, -1]$

50. $3x + 7 \le 4x - 2$

$x \ge 9, [9, \infty)$

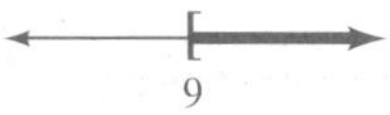

51. $9x + 13 \ge 8x$

$x \ge -13, [-13, \infty)$

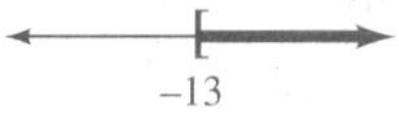

52. $7x - 16 < 6x$

$x < 16, (-\infty, 16)$

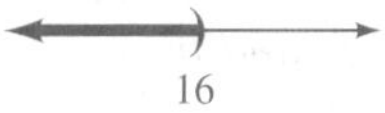

53. $8x + 4 > 3x + 4$

$x > 0, (0, \infty)$

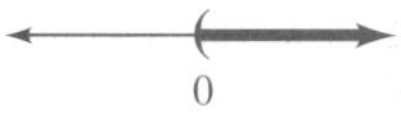

54. $7x + 6 \ge 4x + 6$

$x \ge 0, [0, \infty)$

55. $5x + 7 < 2x + 1$

$x < -2, (-\infty, -2)$

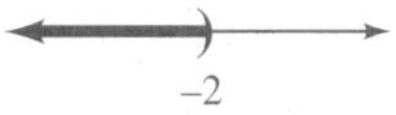

56. $7x + 2 \ge 4x - 1$

$x \ge -1, [-1, \infty)$

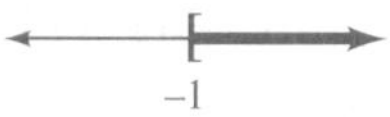

57. $7 - x \le 3x - 2$

$x \ge \dfrac{9}{4}, \left[\dfrac{9}{4}, \infty\right)$

58. $9 - 3x \ge 6 + x$

$x \le \dfrac{3}{4}, \left(-\infty, \dfrac{3}{4}\right]$

59. $3(x - 8) < 5x + 6$

$x > -15, (-15, \infty)$

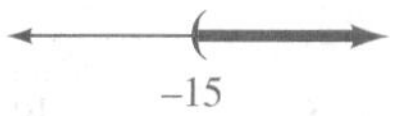

60. $9(x - 11) > 13 + 7x$

$x > 56, (56, \infty)$

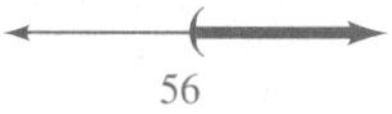

61. $8(5 - x) \le 10(8 - x)$

$x \le 20, (-\infty, 20]$

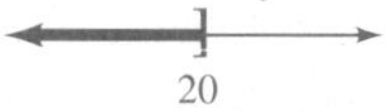

62. $17(3 - x) \ge 3 - 13x$

$x \le 12, (-\infty, 12]$

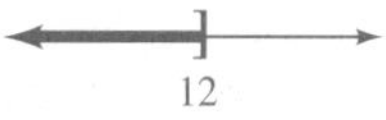

In Exercises 63–76, solve each inequality, graph the solution, and use interval notation to describe the solution.

63. $2 < x - 5 < 5$

$7 < x < 10, (7, 10)$

64. $3 < x - 2 < 7$

$5 < x < 9, (5, 9)$

65. $-5 < x + 4 \le 7$

$-9 < x \le 3, (-9, 3]$

66. $-9 \le x + 8 < 1$

$-17 \le x < -7, [-17, -7)$

67. $0 \le x + 10 \le 10$

$-10 \le x \le 0, [-10, 0]$

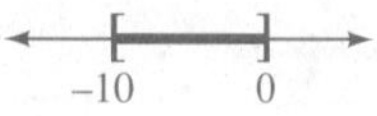

68. $-8 < x - 8 < 8$

$0 < x < 16, (0, 16)$

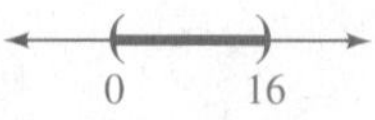

69. $4 < -2x < 10$

$-5 < x < -2, (-5, -2)$

70. $-4 \le -4x < 12$

$-3 < x \le 1 (-3, 1]$

71. $-3 \le \dfrac{x}{2} \le 5$

$-6 \le x \le 10, [-6, 10]$

72. $-12 < \dfrac{x}{3} < 0$

$-36 < x < 0, (-36, 0)$

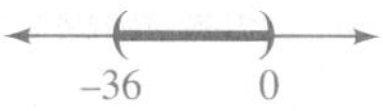

73. $3 \le 2x - 1 < 5$

$2 \le x < 3, [2, 3)$

74. $4 < 3x - 5 \le 7$

$3 < x \le 4, (3, 4]$

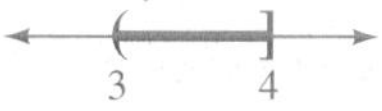

75. $0 < 10 - 5x \le 15$

$-1 \le x < 2, [-1, 2)$

76. $1 \le -7x + 8 \le 15$

$-1 \le x \le 1, [-1, 1)$

APPLICATIONS

In Exercises 77–88, express each solution as an inequality.

77. CALCULATING GRADES A student has test scores of 68%, 75%, and 79% in a government class. What must she score on the last exam to earn a B (80% or better) in the course? $s \ge 98\%$

78. OCCUPATIONAL TESTING Before taking on a client, an employment agency requires the applicant to average at least 70% on a battery of four job skills tests. If an applicant scored 70%, 74%, and 84% on the first three exams, what must he score on the fourth test to maintain a 70% or better average? $s \ge 52\%$

79. FLEET AVERAGES A car manufacturer produces three models in equal quantities. One model has an economy rating of 17 miles per gallon, and the second model is rated for 19 mpg. If governmental regulations require the manufacturer to have a fleet average of at least 21 mpg, what economy rating is required for the third model? $r \ge 27$ mpg

80. SERVICE CHARGES When the average daily balance of a customer's checking account falls below $500 in any week, the bank assesses a $5 service charge. Illustration 1 shows the daily balances of one customer. What must Friday's balance be to avoid the service charge? $b \ge \$869.20$

ILLUSTRATION 1

Day	Balance
Monday	$540.00
Tuesday	$435.50
Wednesday	$345.30
Thursday	$310.00

81. DOING HOMEWORK A Spanish teacher requires that students devote no less than 1 hour a day to their homework assignments. Write an inequality that describes the number of minutes a student should spend each week on Spanish homework. $t \geq 420$ min

82. CHILD LABOR A child labor law reads, "The number of hours a full-time student under 16 years of age can work on a weekday shall not exceed 4 hours." Write an inequality that describes the number of hours such a student can work Monday through Friday. $h \leq 20$ hr

83. SAFETY CODE Illustration 2 shows the acceptable and preferred angles of "pitch" or slope for ladders, stairs, and ramps. Use a compound inequality to describe each safe-angle range:
   a. Ramps or inclines $0° < a \leq 18°$
   b. Stairs $18° \leq a \leq 50°$
   c. Preferred range for stairs $30° \leq a \leq 37°$
   d. Ladders with cleats $75° \leq a < 90°$

ILLUSTRATION 2

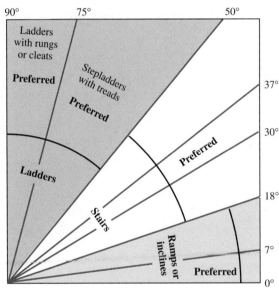

84. WEIGHT CHART Illustration 3 is used to classify the weight of a baby boy from birth to 1 year. Estimate the weight range w for boys in the following classifications, using a compound inequality:
   a. 10 months old, "heavy" 26 lb $\leq w \leq 31$ lb
   b. 5 months old, "light" 12 lb $\leq w \leq 14$ lb
   c. 8 months old, "average" 18.5 lb $\leq w \leq 20.5$ lb
   d. 3 months old, "moderately light"
       11 lb $\leq w \leq 13$ lb

85. LAND ELEVATIONS The land elevations in Nevada range from the 13,143-foot height of Boundary Peak to the Colorado River at 470 feet. Use a compound inequality to express the range of these elevations:
   a. in feet 470 ft $\leq x \leq 13{,}143$ ft
   b. in miles (round to the nearest tenth) (*Hint:* 1 mile is 5,280 feet.) 0.1 mi $\leq x \leq 2.5$ mi

ILLUSTRATION 3

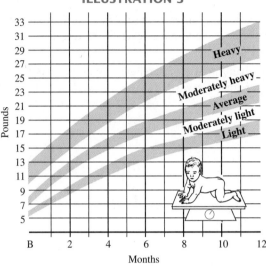

Based on data from *Better Homes and Gardens Baby Book* (Meredith Corp., 1969)

86. COMPARING TEMPERATURES To hold the temperature of a room between 19° C and 22°C, what Fahrenheit temperatures must be maintained? (*Hint:* Fahrenheit temperature F and Celsius temperature C are related by the formula $F = \frac{9C + 160}{5}$.)
   $66.2° < F < 71.6°$

87. DRAFTING In Illustration 4, the $\pm$ (read "plus or minus") symbol means that the width of a plug a manufacturer produces can range from $1.497 - 0.001$ inches to $1.497 + 0.001$ inches. Write the range of acceptable widths for the plug and the opening it fits into using compound inequalities.
   1.496 in. $\leq w \leq 1.498$ in.;
   1.5000 in. $\leq w \leq 1.5010$ in.

ILLUSTRATION 4

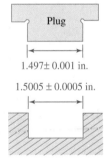

Plug

1.497 ± 0.001 in.

1.5005 ± 0.0005 in.

88. COUNTER SPACE In a large discount store, a rectangular counter is being built for the customer service department. If designers have determined that the outside perimeter of the counter (shown in red) needs to be at least 150 feet, use the plan in Illustration 5 to determine the acceptable values for x.
   $x \geq 35$ ft

ILLUSTRATION 5

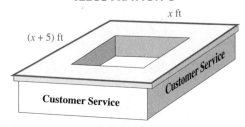

WRITING

89. Explain why multiplying both sides of an inequality by a negative number reverses the direction of the inequality.

90. Explain the use of parentheses and brackets for graphing intervals.

REVIEW

In Exercises 91–94, find each power.

91. -5^3 -125 **92.** $(-3)^4$ 81

93. -4^4 -256 **94.** $(-4)^4$ 256

Simplify and Solve

Two of the most often used instructions in this book are **simplify** and **solve.** In algebra, we *simplify expressions* and we *solve equations and inequalities.*

To simplify an expression, we write it in a less complicated form. To do so, we apply the rules of arithmetic, as well as algebraic concepts such as combining like terms, the distributive property, and the properties of 0 and 1.

To solve an equation or an inequality means to find the numbers that make the equation or inequality true when substituted for its variable. We use the addition, subtraction, multiplication, and division properties of equality or inequality to solve equations and inequalities. Quite often, we must simplify expressions on the left- or right-hand sides of an equation or inequality when solving it.

In Exercises 1–4, use the procedures and the properties that we have studied to simplify the expression in part a and to solve the equation or inequality in part b.

Simplify

1. a. $-3x + 2 + 5x - 10$ $2x - 8$

2. a. $4(y + 2) - 3(y + 1)$ $y + 5$

3. a. $\dfrac{1}{3}a + \dfrac{1}{3}a$ $\frac{2}{3}a$

4. a. $-(2x + 10)$ $-2x - 10$

Solve

b. $-3x + 2 + 5x - 10 = 4$ $x = 6$

b. $4(y + 2) = 3(y + 1)$ $y = -5$

b. $\dfrac{1}{3}a + \dfrac{1}{3} = \dfrac{1}{2}$ $a = \frac{1}{2}$

b. $-2x \geq -10$ $x \leq 5$

5. In the student's work on the right, where was the mistake made? Explain what the student did wrong.

The mistake is on the third line. The student made an equation out of the answer, which is $x - 6$, by writing "$0 =$" on the left and then solved that equation.

Simplify $2(x + 3) - x - 12$.

$2(x + 3) - x - 12 = 2x + 6 - x - 12$
$= x - 6$
$0 = x - 6$
$0 + 6 = x - 6 + 6$
$\boxed{6 = x}$

Accent on Teamwork

Section 2.1

Temperature extremes Check out a current world almanac from a library. Look up the record high and low temperatures for each state and calculate the difference in temperature extremes. Rank the states from largest to smallest differences in their record highs and lows.

Section 2.2

Operations with integers Prepare a presentation for the class in which you explain why the *sum* of -3 and -2 is negative and why the *product* of -3 and -2 is positive.

Section 2.3

Exponents Use a scientific calculator and the exponential key $\boxed{y^x}$ (on some calculators, it is labeled $\boxed{x^y}$) to decide whether each statement is true or false.

1. $7^5 = 5^7$ false

2. $2^3 + 7^3 = (2 + 7)^3$ false

3. $(-4)^4 = -4^4$ false

4. $\dfrac{10^3}{5^3} = 2^3$ true

5. $8^4 \cdot 9^4 = (8 \cdot 9)^4$ true

6. $2^3 \cdot 3^3 = 6^3$ true

7. $\dfrac{3^{10}}{3^2} = 3^5$ false

8. $[(1.2)^3]^2 = [(1.2)^2]^3$ true

Section 2.4

The distributive property Use colored paper to make models like those in Figures 2-11 and 2-12 on pages 116–117 to help you explain why $a(b + c) = ab + ac$ and $a(b - c) = ab - bc$.

Section 2.5

Solving equations Make a presentation to the class explaining how we "undo" operations to isolate the variable when solving an equation. As a visual aid, bring in a box, tied shut with string, that contains a toy wrapped in tissue paper. Compare the three-step process a person would use to get to the toy inside the box to the three-step process we could use to solve the equation $\frac{2x}{3} - 4 = 2$.

Section 2.6

Geometry gourmet Find snack foods that have the shapes of the geometric figures in Table 2-2 and Table 2-3 on pages 141 and 143. For example, tortilla chips can be triangular in shape, and malted milk balls are spheres. If you are unable to find a particular shape already available, decide on a way to make a snack in that shape. Make up several trays of the snacks to bring to class. Split the class up into groups, with each group having its own tray of snacks. Call out a geometric shape and have the students pick out the snack in that shape from the tray and eat it.

Section 2.7

Mixtures Get several cans of the same brand of orange juice concentrate and make up four pitchers of concentrate and water mixtures that are 10%, 30%, 50%, and 70% orange juice concentrate. For example, a 30% solution would consist of three small paper cups of concentrate and seven small paper cups of water. Pour small amounts of each mixture into clean cups. Have students not in your group taste each solution and see whether they can put the mixtures in order from least concentrated to most concentrated.

Section 2.8

Inequalities In most states, a person must be at least 16 years of age to have a driver's license. We can mathematically describe this situation with the inequality $a \geq 16$, where a represents a person's age in years. Think of other situations that can be described using an inequality. Try to come up with some examples that require compound inequalities.

Section 2.1

Adding and Subtracting Real Numbers

CONCEPTS

To add two real numbers with *like signs,* add their absolute values and attach their common sign to the sum.

To add two real numbers with *unlike signs,* subtract their absolute values, the smaller from the larger. To that result, attach the sign of the number with the larger absolute value.

Properties of the real numbers—the *commutative* and *associative* properties of addition:
$$a + b = b + a$$
$$(a + b) + c = a + (b + c)$$

To *subtract* real numbers, add the opposite:
$$a - b = a + (-b)$$

Solutions of equations can be negative numbers.

REVIEW EXERCISES

1. Add the numbers.
   a. $12 + 33$ 45
   b. $-45 + (-37)$ -82
   c. $-15 + 37$ 22
   d. $25 + (-13)$ 12
   e. $12 + (-8) + (-15)$ -11
   f. $-25 + (-14) + 35$ -4
   g. $-9.9 + (-2.4)$ -12.3
   h. $\dfrac{5}{16} + \left(-\dfrac{1}{2}\right)$ $-\frac{3}{16}$
   i. $35 + (-13) + (-17) + 6$ 11
   j. $-21 + (-11) + 32 + (-45)$ -45
   k. $0 + (-7)$ -7
   l. $-7 + 7$ 0

2. Tell what property of addition guarantees that the quantities are equal.
   a. $-2 + 5 = 5 + (-2)$
   commutative property of addition
   b. $(-2 + 5) + 1 = -2 + (5 + 1)$
   associative property of addition

3. Subtract the numbers.
   a. $45 - 64$ -19
   b. $-17 - 32$ -49
   c. $-27 - (-12)$ -15
   d. $3.6 - (-2.1)$ 5.7

4. Complete the following input/output tables.

   a.
x	$x + 5$
-2	3
0	5
-9	-4

   b.
x	$7 - x$
8	-1
0	7
-2	9

5. Solve each equation.
   a. $x + 12 = -17$ -29
   b. $-1.7 = y - 1.3$ -0.4
   c. $17 + p = -8$ -25
   d. $-8 + q = -5$ 3

6. ASTRONOMY *Magnitude* is a term used in astronomy to designate the brightness of celestial objects as viewed from the earth. Smaller magnitudes are associated with brighter objects, and larger magnitudes refer to fainter objects. See Illustration 1. For each of the following pairs of objects, by how many magnitudes do their brightnesses differ?
   a. A full moon and the sun   14
   b. The star Beta Crucis and a full moon   13.78

ILLUSTRATION 1

Object	Magnitude
Sun	-26.5
Full moon	-12.5
Beta Crucis	1.28

Based on data from *Exploration of the Universe* (Abell, Morrison, and Wolf; Saunders College Publishing, 1987)

7. **GEOGRAPHY** The tallest peak on earth is Mt. Everest at 29,028 feet. The greatest ocean depth is the Mariana Trench at $-36,205$ feet. Find the difference in the two elevations.   65,233 ft

---

Section 2.2

When multiplying two real numbers:
1. The product of two real numbers with *like signs* is positive.
2. The product of two real numbers with *unlike signs* is negative.

Multiplying and Dividing Real Numbers

8. Multiply the numbers.
   a. $-8 \cdot 7$ -56
   b. $(-9)(-6)$ 54
   c. $2(-3)(-2)$ 12
   d. $(-3)(4)(2)$ -24
   e. $(-3)(-4)(-2)$ -24
   f. $(-4)(-1)(-3)(-3)$ 36
   g. $-1.2(-5.3)$ 6.36
   h. $0.002(-1,000)$ -2
   i. $-\dfrac{2}{3}\left(\dfrac{1}{5}\right)$ $-\frac{2}{15}$
   j. $2\dfrac{1}{4}\left(-\dfrac{1}{3}\right)$ $-\frac{3}{4}$
   k. $-6 \cdot 0$ 0
   l. $(-3)(1)$ -3

Properties of the real numbers—the *commutative* and *associative* properties of multiplication:

$$ab = ba$$
$$(ab)c = a(bc)$$

9. Tell what property of multiplication guarantees that the quantities are equal.
   a. $(2 \cdot 3)5 = 2(3 \cdot 5)$
      associative property of multiplication
   b. $(-5)(-6) = (-6)(-5)$
      commutative property of multiplication

When dividing two real numbers:
1. The quotient of two real numbers with *like signs* is positive.
2. The quotient of two real numbers with *unlike signs* is negative.

Division *of zero* by a non-zero number is zero. Division *by zero* is undefined.

10. Do each division.
   a. $\dfrac{88}{44}$ 2
   b. $\dfrac{-100}{25}$ -4
   c. $\dfrac{-81}{-27}$ 3
   d. $\dfrac{0}{37}$ 0
   e. $-\dfrac{3}{5} \div \dfrac{1}{2}$ $-\frac{6}{5}$
   f. $\dfrac{-60}{0}$ undefined
   g. $\dfrac{-4.5}{1}$ -4.5
   h. $\dfrac{-5}{-5}$ 1

11. Solve each equation.
   a. $-12x = 24$ -2
   b. $36a = -108$ -3
   c. $\dfrac{b}{-5} = -4$ 20
   d. $\dfrac{f}{17} = -3$ -51

Section 2.3

An *exponent* is used to indicate repeated multiplication.

Order of Operations and Evaluating Algebraic Expressions

12. Find each power.
   a. 2^5 32
   b. $(-2)^5$ -32
   c. $(-3)^4$ 81
   d. $(-5)^3$ -125
   e. $(-0.8)^2$ 0.64
   f. $\left(-\dfrac{2}{3}\right)^3$ $-\frac{8}{27}$

Order of operations:
Work from the innermost pair of grouping symbols to the outermost pair in the following order:

1. Evaluate all exponential expressions.
2. Do all multiplications and divisions, working from left to right.
3. Do all additions and subtractions, working from left to right.

If the expression does not contain grouping symbols, begin with step 1. In a fraction, simplify the numerator and denominator separately. Then simplify the fraction, if possible.

13. Evaluate each expression.

   a. $4^3 + 2(-6 - 2 \cdot 2)$ 44
   b. $-5[-3 - 2(5 - 7^2)] - 5$ -430
   c. $\dfrac{-4(4 + 2) - 4}{18 - 4(-5)}$ $-\frac{14}{19}$
   d. $(-3)^3\left(\dfrac{-8}{2}\right) + 5$ 113
   e. $\dfrac{|-35| - 2(-7)}{2^4 - 23}$ -7
   f. $-9^2 + (-9)^2$ 0

14. Evaluate $3(x - y) - 5(x + y)$ when

   a. $x = 2$ and $y = -5$ 36
   b. $x = -3$ and $y = 3$ -18

15. Find the volume of a sphere with a radius r of 2 inches by evaluating the expression $\dfrac{4\pi r^3}{3}$. Round to the nearest tenth.   33.5 in.^{3}

16. Complete each input/output table.

a.

t	$50t - 16t^2$
1	34
-2	-164
3	6

b.

x	$x^3 - \frac{x+2}{2}$
-2	-8
-10	-996
0	-1

Section 2.4   Simplifying Algebraic Expressions

"To *simplify* an algebraic expression" means to write it in less complicated form.

The *distributive property:*
$a(b + c) = ab + ac$
$a(b - c) = ab - ac$

A *term* is a number or a product of a number and one or more variables. Addition signs separate algebraic expressions into terms.

In a term, the numerical factor is called the *coefficient.*

Like terms are terms with exactly the same variables raised to exactly the same powers.

17. Simplify each expression.

   a. $-4(7w)$ $-28w$
   b. $-3r(-5r)$ $15r^2$
   c. $3(-2x)(-4y)$ $24xy$
   d. $0.4(5.2f)$ $2.08f$

18. Write each expression without parentheses.

   a. $5(x + 3)$ $5x + 15$
   b. $-2(2x + 3 - y)$ $-4x - 6 + 2y$
   c. $-(a - 4)$ $-a + 4$
   d. $\dfrac{3}{4}(4c - 8)$ $3c - 6$

19. How many terms are in each expression?

   a. $3x^2 + 2x - 5$ 3
   b. $-12xyz$ 1

20. Identify the coefficient of each term.

   a. $2x - 5$ $2, -5$
   b. $16x^2 - 5x + 25$ $16, -5, 25$
   c. $\frac{1}{2}x + y$ $\frac{1}{2}, 1$
   d. $9.6t^2 - t$ $9.6, -1$

21. Simplify each expression by combining like terms.

   a. $8p + 5p - 4p$ $9p$
   b. $-5m + 2n - 2m - 2n$ $-7m$
   c. $6a + 2b - 8a - 12b$ $-2a - 10b$
   d. $5(p - 2q) - 2(3p + 4q)$ $-p - 18q$
   e. $x^2 - x(x - 1)$ x
   f. $8a^3 + 4a^3 - 20a^3$ $-8a^3$

22. Write an algebraic expression in simplified form for the perimeter of the triangle in Illustration 2. $(4x + 4)$ ft

ILLUSTRATION 2

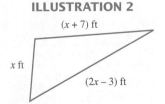

$(x + 7)$ ft

x ft

$(2x - 3)$ ft

Section 2.5

To solve an equation means to find all the values of the variable that, when substituted for the variable, make a true statement.

An equation that is true for all values of its variable is called an *identity*.

An equation that is not true for any values of its variable is called an *impossible equation*.

Solving Equations

23. Solve each equation.

   a. $5x + 4 = 14$ 2

   b. $-12y + 8 = 20$ -1

   c. $\dfrac{n}{5} - 2 = 4$ 30

   d. $\dfrac{b - 5}{4} = -6$ -19

   e. $5(2x - 4) - 5x = 0$ 4

   f. $-2(x - 5) = 5(-3x + 4) + 3$ 1

   g. $\dfrac{3}{4} = \dfrac{1}{2} + \dfrac{d}{5}$ $\frac{5}{4}$

   h. $-\dfrac{2}{3}f = 4$ -6

   i. $3(a + 8) = 6(a + 4) - 3a$
     identity, all values of a

   j. $2(y + 10) + y = 3(y + 8)$
     impossible equation, no solution

24. SOUND SYSTEM A 45-foot-long speaker wire is to be cut into three pieces. One piece is to be 15 feet long. Of the remaining pieces, one must be 2 feet less than 3 times the length of the other. Find the length of the shorter piece of wire.   8 ft

Section 2.6

A *formula* is an equation that is used to state a known relationship between two or more variables.

Sale price: $s = p - d$

Retail price: $r = c + m$

Profit: $p = r - c$

Distance: $d = rt$

Temperature: $C = \dfrac{5(F - 32)}{9}$

Formulas from geometry:

Square: $P = 4s$, $A = s^2$

Rectangle: $P = 2l + 2w$,
$A = lw$

Triangle: $P = a + b + c$
$A = \frac{1}{2}bh$

Trapezoid:
$P = a + b + c + d$
$A = \frac{1}{2}h(b + d)$

Formulas

25. A boat that is on sale for $13,998 has been discounted by $2,100. Find its original price. $16,098

26. Find the markup on a CD player whose wholesale cost is $219 and whose retail price is $395. $176

27. One month, a restaurant had sales of $13,500 and made a profit of $1,700. Find the expenses for the month. $11,800

28. INDY 500 In 1996, the winner of the Indianapolis 500-mile automobile race averaged 147.956 mph. To the nearest hundredth of an hour, how long did it take him to complete the race?   3.38 hr

29. JEWELRY MAKING Gold melts at about 1,065°C. Change this to degrees Fahrenheit.   1,949°F

30. CAMPING Find the perimeter of the air mattress in Illustration 3.   168 in.

31. CAMPING Find the amount of sleeping area on the top surface of the air mattress in Illustration 3.   1,440 in.²

ILLUSTRATION 3

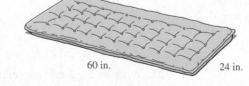

60 in.   24 in.

32. Find the area of a triangle with a base 17 meters long and a height of 9 meters.   76.5 m²

33. Find the area of a trapezoid with bases 11 inches and 13 inches long and a height of 12 inches.   144 in.²

Circle: $D = 2r$
$C = 2\pi r$
$A = \pi r^2$

34. To the nearest hundredth, find the circumference of a circle with a radius of 8 centimeters.   50.27 cm

35. To the nearest hundredth, find the area of the circle in Exercise 34.
201.06 cm^2

Rectangular solid: $V = lwh$

36. CAMPING Find the approximate volume of the air mattress in Illustration 3 if it is 3 inches thick.   4,320 in.^{3}

Cylinder: $V = Bh$

37. Find the volume of a 12-foot cylinder whose circular base has a radius of 0.5 feet. Give the result to the nearest tenth.   9.4 ft^3

Pyramid: $V = \frac{1}{3}Bh$
Cone: $V = \frac{1}{3}Bh$

38. Find the volume of a pyramid that has a square base, measuring 6 feet on a side, and a height of 10 feet.   120 ft^3

Sphere: $V = \frac{4}{3}\pi r^3$

39. HALLOWEEN After being cleaned out, a spherical-shaped pumpkin has an inside diameter of 9 inches. To the nearest hundredth, what is its volume?
381.70 in.^{3}

40. Solve each formula for the required variable.
   a. $A = 2\pi rh$ for h $h = \frac{A}{2\pi r}$ **b.** $P = 2l + 2w$ for l $l = \frac{P - 2w}{2}$

Section 2.7

To solve problems, use the five-step problem-solving strategy.
1. Analyze the problem.
2. Form an equation.
3. Solve the equation.
4. State the conclusion.
5. Check the result.

Problem Solving

41. UTILITY BILLS The electric company charges $17.50 per month, plus 18 cents for every kilowatt hour of energy used. One resident's bill was $43.96. How many kilowatt hours were used that month?   147

42. ART HISTORY *American Gothic,* shown in Illustration 4, was painted in 1930 by American artist Grant Wood. The length of the rectangular painting is 5 inches more than the width. Find the dimensions of the painting if it has a perimeter of $109\frac{1}{2}$ inches.
24.875 in. × 29.875 in. ($24\frac{7}{8}$ in. × $29\frac{7}{8}$ in.)

ILLUSTRATION 4

43. Find the missing angle measures of the triangle in Illustration 5.   76.5°, 76.5°

ILLUSTRATION 5

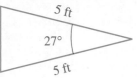
5 ft
27°
5 ft

Total value = number · value

44. What is the value of x video games each costing $45? $45x$

Interest = principal · rate · time
$I = Prt$

45. INVESTMENT INCOME A woman has $27,000. Part is invested for one year in a certificate of deposit paying 7% interest, and the remaining amount in a cash management fund paying 9%. After 1 year, the total interest on the two investments is $2,110. How much is invested at each rate?
$16,000 at 7%, $11,000 at 9%

Distance = rate · time
$$d = rt$$

46. WALKING AND BICYCLING A bicycle path is 5 miles long. A man walks from one end at the rate of 3 mph. At the same time, a friend bicycles from the other end, traveling at 12 mph. In how many minutes will they meet? 20

The value v of a commodity is its price per pound p times the number of pounds n:
$$v = pn.$$

47. MIXTURE A store manager mixes candy worth 90¢ per pound with gumdrops worth \$1.50 per pound to make 20 pounds of a mixture worth \$1.20 per pound. How many pounds of each kind of candy does he use? 10 lb of each

48. SOLUTION How much acetic acid is in x gallons of a solution that is 12% acetic acid? 0.12x gal

Section 2.8

Inequalities

An *inequality* is a mathematical expression that contains a $>$, $<$, $\geq$, $\leq$, or $\neq$ symbol.

49. Solve each inequality, graph the solution, and use interval notation to describe the solution.

a. $3x + 2 < 5$
  $x < 1, (-\infty, 1)$

b. $-5x - 8 > 7$
  $x < -3, (-\infty, -3)$

A *solution of an inequality* is any number that makes the inequality true.

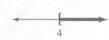

A *parenthesis* indicates that a number is not on the graph. A *bracket* indicates that a number is included in the graph.

c. $5x - 3 \geq 2x + 9$
  $x \geq 4, [4, \infty)$

d. $7x + 1 \leq 8x - 5$
  $x \geq 6, [6, \infty)$

Interval notation can be used to describe a set of real numbers.

e. $5(3 - x) \leq 3(x - 3)$
  $x \geq 3, [3, \infty)$

f. $-\dfrac{3}{4}x \geq -9$
  $x \leq 12, (-\infty, 12]$

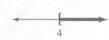

g. $8 < x + 2 < 13$
  $6 < x < 11, (6, 11)$

h. $0 \leq 2 - 2x < 6$
  $-2 < x \leq 1, (-2, 1]$

50. Graph the interval represented by $[-13, \infty)$.

51. SPORTS EQUIPMENT The acceptable weight of ping-pong balls used in competition can range between 2.40 and 2.53 grams. Express this range using a compound inequality. 2.40 g $< w <$ 2.53 g

CHAPTER 2

Test

1. Add $(-6) + 8 + (-4)$. -2

2. Subtract $1.4 - (-0.8)$. 2.2

3. Multiply $(-2)(-3)(-5)$. -30

4. Divide $\dfrac{-22}{-11}$. 2

5. Evaluate $-7[(-5)^2 - 2(3 - 5^2)]$. -483

6. Evaluate $\dfrac{3(20 - 4^2)}{-2(6 - 2^2)}$. -3

In Problems 7–8, let $x = -2$, $y = 3$, and $z = 4$. Evaluate each expression.

7. $xy + z$ -2

8. $\dfrac{z + 4y}{2x}$ -4

9. What is the coefficient of the term $6x$? 6

10. How many terms are in the expression $3x^2 + 5x - 7$? 3

In Problems 11–14, simplify each expression.

11. $5(-4x)$ $-20x$

12. $-8(-7t)(4t)$ $224t^2$

13. $3(x + 2) - 3(4 - x)$ $6x - 6$

14. $-1.1d^2 - 3.8d^2$ $-4.9d^2$

In Problems 15–22, solve each equation.

15. $12x = -144$ -12

16. $\dfrac{4}{5}t = -4$ -5

17. $\dfrac{c}{7} = -1$ -7

18. $3x = 5 - 2x$ 1

19. $2(x - 7) = -15$ $-\frac{1}{2}$

20. $\dfrac{m}{2} - \dfrac{1}{3} = \dfrac{1}{4}$ $\frac{7}{6}$

21. $23 - 5(x + 10) = -12$ -3

22. $5t + 7.2 = 12.7$ 1.1

In Problems 23–24, solve each equation for the variable indicated.

23. $d = rt$; for t $t = \frac{d}{r}$

24. $A = P + Prt$; for r $r = \frac{A - P}{Pt}$

25. COMMERCIAL REAL ESTATE *Net absorption* is a term used to indicate how much office space in a city is being purchased. Use the information from the graph in Illustration 1 to determine the eight-quarter average net absorption figure for Long Beach, California. 0.0475 million ft^2

ILLUSTRATION 1

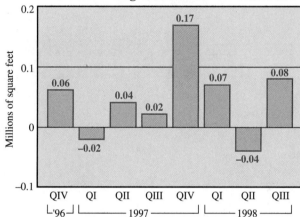

Based on information from *Los Angeles Times* (Oct. 13, 1998) Section C10.

26. PETS The spherical fishbowl shown in Illustration 2 is three-quarters full of water. To the nearest cubic inch, what is the volume of water in the bowl? 393 in.^{3}

ILLUSTRATION 2

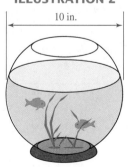

27. INVESTMENT PROBLEM Part of $13,750 is invested at 9% annual interest, and the rest is invested at 8%. After one year, the accounts paid $1,185 in interest. How much was invested at the lower rate? $5,250

28. TRAVEL TIMES A car leaves Rockford, Illinois at the rate of 65 mph, bound for Madison, Wisconsin. At the same time, a truck leaves Madison at the rate of 55 mph, bound for Rockford. If the cities are 72 miles apart, how long will it take for the car and the truck to meet? $\frac{3}{5}$ hr

29. MIXTURE PROBLEM How many liters of a 2% brine solution must be added to 30 liters of a 10% brine solution to dilute it to an 8% solution? 10

30. GEOMETRY If the vertex angle of an isosceles triangle is 44°, find the measure of each base angle. 68°

In Problems 31–32, solve each inequality, graph its solution, and use interval notation to describe the solution.

31. $-8x - 20 \leq 4$ ←———|━━━►
$x \geq -3, [-3, \infty)$ -3

32. $-4 \leq 2(x + 1) < 10$ ←———|━━━━)—►
$-3 \leq x < 4, [-3, 4)$ -3 4

33. Explain the difference between an equation and an inequality.

34. Explain this statement: "Subtraction is the same as addition of the opposite."

CHAPTERS 1–2
Cumulative Review Exercises

In Exercises 1–4, tell whether the expression is an equation.

1. $m - 25$ no

2. $t = 25r$ yes

3. $\dfrac{p + 5}{2}$ no

4. $x + 1 = 24$ yes

In Exercises 5–6, classify each number as a natural number, a whole number, an integer, a rational number, an irrational number, and a real number. Each number may be in several classifications.

5. 3 natural number, whole number, integer, rational, real

6. -0.25 rational, real

In Exercises 7–8, graph each set of numbers on the number line.

7. The natural numbers between 2 and 7

←—|——•——•——•——•——|—→
  2   3   4   5   6   7

8. The real numbers between 2 and 7

2   7

In Exercises 9–12, evaluate each expression.

9. $|-12|$ 12

10. $|15|$ 15

11. $-|15|$ -15

12. $-|-12|$ -12

In Exercises 13–16, use a calculator to find each square root to the nearest hundredth.

13. $\sqrt{77}$ 8.77

14. $\sqrt{0.26}$ 0.51

15. $\sqrt{\pi}$ 1.77

16. $\sqrt{\dfrac{7}{5}}$ 1.18

In Exercises 17–20, write each product using exponents.

17. $3 \cdot 3 \cdot 3$ 3^3

18. $8 \cdot \pi \cdot r \cdot r$ $8\pi r^2$

19. $4 \cdot x \cdot x \cdot y \cdot y$ $4x^2y^2$

20. $m \cdot m \cdot m \cdot n$ m^3n

In Exercises 21–24, evaluate each expression.

21. $13 - 3 \cdot 4$ 1

22. $-2 \cdot 7^2$ -98

23. $6 + 3\left(\dfrac{5}{2}\right) - \dfrac{1}{2}$ 13

24. $\dfrac{12^2 - 4^2 - 2}{2(7 - 4)}$ 21

In Exercises 25–26, write each phrase as an algebraic expression.

25. The sum of the width w and 12 $w + 12$

26. Four less than a number n $n - 4$

In Exercises 27–30, complete each table of values.

27. t	$t^2 - 4$
0	-4
1	-3
3	5

28. t	$(t - 4)^2$
0	16
1	9
3	1

29.

a	$\frac{a}{3} + 2$
0	2
3	3
6	4

30.

x	$\frac{x}{4} + \frac{x}{3}$
0	0
12	7
24	14

In Exercises 31–34, $a = 2$ and $b = 5$. Evaluate each expression.

31. $a^2 + b$ 9

32. $b^3 - 12a^2$ 77

33. $\dfrac{a + b}{b + 2}$ 1

34. $\dfrac{3(b^2 - 1)}{2a}$ 18

In Exercises 35–38, solve each equation.

35. $x + 9 = 13$ 4

36. $5x = 25$ 5

37. $\dfrac{y}{4} = 5$ 20

38. $t - 5 = 17$ 22

In Exercises 39–42, solve each problem.

39. Find 45% of 640.   288

40. What percent of 200 is 30?   15%

41. 45 is 15% of what number?   300

42. 20% of what number is 240?   1,200

In Exercises 43–46, let $x = -5$, $y = 3$, and $z = 0$. Evaluate each expression.

43. $(3x - 2y)z$ 0

44. $\dfrac{x - 3y + |z|}{2 - x}$ -2

45. $x^2 - y^2 + z^2$ 16

46. $\dfrac{x}{y} + \dfrac{y + 2}{3 - z}$ 0

In Exercises 47–50, find each power.

47. $(-6)^3$ -216

48. -6^2 -36

49. $-(-5)^2$ -25

50. $\left(-\dfrac{2}{3}\right)^3$ $-\frac{8}{27}$

In Exercises 51–54, evaluate each expression.

51. $2(5 - 3)^2$ 8

52. $5^2 - (8 - 4)^2$ 9

53. $\dfrac{2 + (2 + 6)^2}{2(9 - 6)}$ 11

54. $\dfrac{2[4 + 2(5 - 3)]}{3[3(2 \cdot 4 - 6)]}$ $\frac{8}{9}$

In Exercises 55–60, simplify each expression.

55. $-8(4d)$ $-32d$

56. $5(2x - 3y + 1)$
   $10x - 15y + 5$

57. $2x + 3x$ $5x$

58. $3a + 6a - 17a$ $-8a$

59. $q(q - 5) + 7q^2$
   $8q^2 - 5q$

60. $5(t - 4) + 3t$ $8t - 20$

61. What is the length of the longest side of the triangle in Illustration 1? $(x + 3)$ ft

62. Write an algebraic expression in simplest form for the perimeter of the triangle in Illustration 1. $3x$ ft

ILLUSTRATION 1

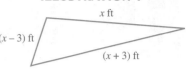

x ft

$(x - 3)$ ft

$(x + 3)$ ft

In Exercises 63–68, solve each equation.

63. $3x - 4 = 23$ 9

64. $\dfrac{x}{5} + 3 = 7$ 20

65. $-5p + 0.7 = 3.7$
   -0.6

66. $\dfrac{y - 4}{5} = 3$ 19

67. $-\dfrac{4}{5}x = 16$ -20

68. $-9(n + 2) - 2(n - 3) = 10$ -2

 *In Exercises 69–70, find the area of each figure.*

69. A rectangle with sides of 5 meters and 13 meters
   65 m^2

70. A circle with a radius of 5 centimeters (Give the answer to the nearest hundredth.)   78.54 cm^2

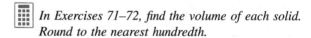 *In Exercises 71–72, find the volume of each solid. Round to the nearest hundredth.*

71. A 12-foot-long cylinder with a circular base with a radius of 0.5 feet   9.42 ft^3

72. A cone that is 10 centimeters tall and has a circular base whose diameter is 12 centimeters   376.99 cm^3

In Exercises 73–74, solve each formula for the given variable.

73. $P = 2l + 2w$ (for w) $w = \dfrac{P - 2l}{2}$

74. $A = P + Prt$ (for t) $t = \dfrac{A - P}{Pr}$

75. WORK Physicists say that *work* is done when an object is moved a distance d by a force F. To find the work done, we can use the formula

   $W = Fd$

   Find the work done in lifting the bundle of newspapers shown in Illustration 2 onto the workbench. (*Hint:* The force that must be applied to lift the papers is the weight of the newspapers.)   37.5 ft-lb

ILLUSTRATION 2

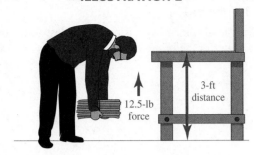

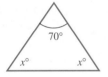

12.5-lb force

3-ft distance

76. WORK See Exercise 75. Find the weight of a 1-gallon can of paint if the amount of work done to lift it onto the workbench is 28.35 foot-pounds.
9.45 lb

In Exercises 77–78, find each unknown angle measure represented by a variable.

77.

55°, 55°

70°

$x°$ $x°$

78.

65°

115°

$y°$

In Exercises 79–82, solve each inequality, graph the solution, and use interval notation to describe the solution.

79. $x - 4 > -6$

$x > -2, (-2, \infty)$

–2

80. $-6x \geq -12$

$x \leq 2, (-\infty, 2]$

2

81. $8x + 4 \geq 5x + 1$

$x \geq -1, [-1, \infty)$

–1

82. $-1 \leq 2x + 1 < 5$

$-1 \leq x < 2, [-1, 2)$

–1   2

3

Graphs, Linear Equations, and Functions

CAMPUS CONNECTION

The Drafting Department

In drafting, students learn how to use AutoCAD, a computer-aided drafting program. As part of their coursework, they will draw the floor plan of a house on a grid, or *coordinate system,* displayed on a monitor. In this chapter, we will discuss the rectangular coordinate system. It allows us to represent mathematical relationships visually by graphing them. Learning how to use the rectangular coordinate system will give you the insight you'll need when you encounter coordinate systems in other disciplines such as drafting.

Relationships between two quantities can be described by a table, a graph, or an equation.

▶ 3.1 Graphing Using the Rectangular Coordinate System

In this section, you will learn about

The rectangular coordinate system ■ Graphing mathematical relationships ■ Reading graphs ■ Step graphs

Introduction It is often said, "A picture is worth a thousand words." In this section, we will show how numerical relationships can be described using mathematical pictures called **graphs.** We will also show how graphs are constructed and how we can obtain important information by reading graphs.

The Rectangular Coordinate System

When designing the Gateway Arch in St. Louis, shown in Figure 3-1(a), architects created a mathematical model called a **rectangular coordinate graph.** This graph, shown in Figure 3-1(b), is drawn on a grid called a **rectangular coordinate system.** This coordinate system is sometimes called a **Cartesian coordinate system,** after the 17th-century French mathematician René Descartes.

FIGURE 3-1

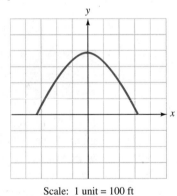

Scale: 1 unit = 100 ft

(a)   (b)

A rectangular coordinate system (see Figure 3-2) is formed by two perpendicular number lines. The horizontal number line is called the **x-axis,** and the vertical number line is called the **y-axis.** The positive direction on the *x*-axis is to the right, and the positive direction on the *y*-axis is upward. The scale on each axis should fit the data. For example, the axes of the graph of the arch shown in Figure 3-1(b) are scaled in units of 100 feet. If no scale is indicated on the axes, we assume that the axes are scaled in units of 1.

The point where the axes cross is called the **origin.** This is the zero point on each axis. The axes form a **coordinate plane** and divide it into four regions called **quadrants,** which are numbered as shown in Figure 3-2.

FIGURE 3-2

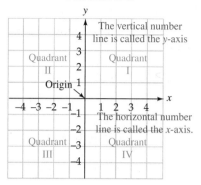

Each point in a coordinate plane can be identified by a pair of real numbers x and y written in the form (x, y). The first number x in the pair is called the **x-coordinate,** and the second number y is called the **y-coordinate.** The numbers in the pair are called the **coordinates** of the point. Some examples of such pairs are $(3, -4)$, $\left(-1, -\frac{3}{2}\right)$, and $(0, 2.5)$.

$$(3, -4)$$
$$\uparrow \qquad \uparrow$$

The x-coordinate The y-coordinate
is listed first.   is listed second.

The process of locating a point in the coordinate plane is called **graphing** or **plotting** the point. In Figure 3-3(a), we show how to graph the point with coordinates of $(3, -4)$. Since the **x-coordinate,** 3, is positive, we start at the origin and move 3 units to the *right* along the x-axis. Since the **y-coordinate,** -4, is negative, we then move *down* 4 units to locate point A. Point A is the **graph** of $(3, -4)$ and lies in quadrant IV.

To plot the point $(-4, 3)$, we start at the origin, move 4 units to the left along the x-axis, and then move up 3 units to locate point B. Point B lies in quadrant II.

FIGURE 3-3

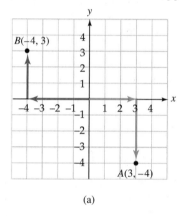

(a)

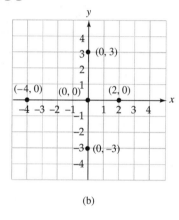

(b)

WARNING! Note that point A with coordinates $(3, -4)$ is not the same as point B with coordinates $(-4, 3)$. Since the order of the coordinates of a point is important, we call the pairs **ordered pairs.**

In Figure 3-3(b), we see that the points $(-4, 0)$, $(0, 0)$, and $(2, 0)$ lie on the x-axis. In fact, all points with a y-coordinate of zero will lie on the x-axis. We also see that the points $(0, -3)$, $(0, 0)$, and $(0, 3)$ lie on the y-axis. All points with an x-coordinate of zero lie on the y-axis. We can also see that the coordinates of the origin are $(0, 0)$.

E X A M P L E   1

Graphing points. Plot the points: **a.** $A(-2, 3)$,   **b.** $B\left(-1, -\frac{3}{2}\right)$,   **c.** $C(0, 2.5)$, and   **d.** $D(4, 2)$.

Solution   See Figure 3-4. (Note: If no scale is indicated on the axes, we assume that the axes are scaled in units of 1.)

a. To plot point A with coordinates $(-2, 3)$, we start at the origin, move 2 units to the *left* on the x-axis, and move 3 units *up*. Point A lies in quadrant II.

b. To plot point B with coordinates $\left(-1, -\frac{3}{2}\right)$, we start at the origin and move 1 unit to the *left* and $\frac{3}{2}$ (or $1\frac{1}{2}$) units *down*. Point B lies in quadrant III.

FIGURE 3-4

   c. To graph point C with coordinates $(0, 2.5)$, we start at the origin and move 0 units on the x-axis and 2.5 units *up*. Point C lies on the y-axis.

   d. To graph point D with coordinates $(4, 2)$, we start at the origin and move 4 units to the *right* and 2 units *up*. Point D lies in quadrant I.

SELF CHECK   Plot the points: **a.** $E(2, -2)$, **b.** $F(-4, 0)$,   *Answers:*
                c. $G\left(1.5, \frac{5}{2}\right)$, and **d.** $H(0, 5)$.

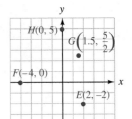

EXAMPLE 2

Orbit of the earth. The circle shown in Figure 3-5 is an approximate **graph** of the orbit of the earth. The graph is made up of infinitely many points, each with its own x- and y-coordinates. Use the graph to find the coordinates of the earth's position during the months of February, May, August, and December.

FIGURE 3-5

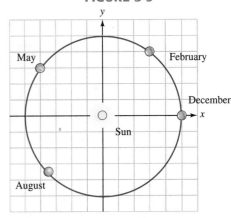

Scale: 1 unit = 18,600,000 mi

Solution   To find the coordinates of each position, we start at the origin and move left or right along the x-axis to find the x-coordinate and then up or down to find the y-coordinate.

Month	Position of earth on graph	Coordinates
February	3 units to the *right*, then 4 units *up*	$(3, 4)$
May	4 units to the *left*, then 3 units *up*	$(-4, 3)$
August	3.5 units to the *left*, then 3.5 units *down*.	$(-3.5, -3.5)$
December	5 units *right*, no units *up* or *down*	$(5, 0)$

Graphing Mathematical Relationships

Every day, we deal with quantities that are related:

- The distance that we travel depends on how fast we are going.
- Our weight depends on how much we eat.
- The amount of water in a tub depends on how long the water has been running.

   We often use graphs to visualize relationships between two quantities. For example, suppose we know the number of gallons of water that are in a tub at several time intervals after the water has been turned on. We can list that information in a **table of values** (see Figure 3-6).

   The information in the table can be used to construct a graph that shows the relationship between the amount of water in the tub and the time the water has been run-

FIGURE 3-6

Time (min)	Water in tub (gal)	
0	0	→ (0, 0)
1	8	→ (1, 8)
3	24	→ (3, 24)
4	32	→ (4, 32)

↑ *x*-coordinate   ↑ *y*-coordinate   ↑ The data in the table can be expressed as ordered pairs (*x*, *y*).

At various times, the amount of water in the tub was measured and recorded in the table of values.

(a)   **(b)**

ning. Since the amount of water in the tub depends on the time, we will associate *time* with the *x*-axis and *amount of water* with the *y*-axis.

To construct the graph in Figure 3-7, we plot the four ordered pairs and draw a line through the resulting data points. The *y*-axis is scaled in larger units (4 gallons) because the data range from 0 to 32 gallons.

FIGURE 3-7

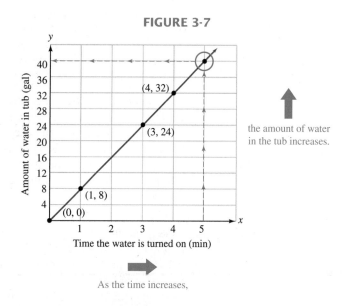

As the time increases,

From the graph, we can see that the amount of water in the tub steadily increases as the water is allowed to run. We can also use the graph to make observations about the amount of water in the tub at other times. For example, the dashed line on the graph shows that in 5 minutes, the tub will contain 40 gallons of water.

Reading Graphs

Valuable information can be obtained from a graph, as can be seen in the next example.

E X A M P L E   3

Reading a graph. The graph in Figure 3-8 shows the number of people in an audience before, during, and after the taping of a television show. On the *x*-axis, zero represents the time when taping began. Use the graph to answer the following questions, and record each result in a table of values.

a. How many people were in the audience when taping began?

b. What was the size of the audience 10 minutes before taping began?

c. At what times were there exactly 100 people in the audience?

FIGURE 3-8

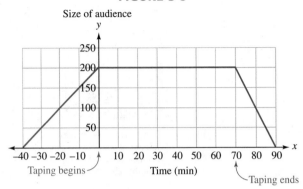

Solution

Time	Audience
0	200
−10	150
−20	100
80	100

a. The time when taping began is represented by zero on the *x*-axis. Since the point on the graph directly above zero has a *y*-coordinate of 200, the point (0, 200) is on the graph. The *y*-coordinate of this point indicates that 200 people were in the audience when the taping began. We enter this result in the table at the left.

b. Ten minutes before taping began is represented by −10 on the *x*-axis. Since the point on the graph directly above −10 has a *y*-coordinate of 150, the point (−10, 150) is on the graph. The *y*-coordinate of this point indicates that 150 people were in the audience 10 minutes before the taping began. We enter this result in the table.

c. We can draw a horizontal line passing through 100 on the *y*-axis. This line intersects the graph twice, at (−20, 100) and (80, 100). So there are two times when 100 people were in the audience. The first time was 20 minutes before taping began (−20), and the second time was 80 minutes after taping began (80). The *y*-coordinates of these points indicate that there were 100 people in the audience 20 minutes before and 80 minutes after taping began. We enter these results in the table.

SELF CHECK

Use the graph in Figure 3-8 to answer the following questions.   **a.** At what times were there exactly 50 people in the audience?   **b.** What was the size of the audience that watched the taping?   **c.** How long did it take for the audience to leave the studio after the taping ended?

Answers: **a.** 30 min before and 85 min after taping began,   **b.** 200,   **c.** 20 min   ■

Step Graphs

The graph in Figure 3-9 shows the cost of renting a trailer for different periods of time. For example, the cost of renting the trailer for 4 days is $60, which is the *y*-coordinate of the point (4, 60). The cost of renting the trailer for a period lasting over 4 and up to 5 days jumps to $70. Since the jumps in cost form steps in the graph, we call this graph a **step graph.**

E X A M P L E   4

Use the information in Figure 3-9 to answer the following questions. Write the results in a table of values.

a. Find the cost of renting the trailer for 2 days.

b. Find the cost of renting the trailer for $5\frac{1}{2}$ days.

c. How long can you rent the trailer if you have $50?

d. Is the rental cost per day the same?

FIGURE 3-9

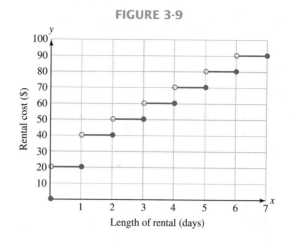

Length of rental (days)

Solution   **a.** The solid dot at the end of each step indicates the rental cost for 1, 2, 3, 4, 5, 6, or 7 days. Just as when we graphed inequalities, an open circle indicates that that point is not on the graph. We locate 2 days on the *x*-axis and move up to locate the point on the graph directly above the 2. Since the point has coordinates (2, 40), a 2-day rental would cost $40. We enter this ordered pair in the table at the left.

b. We locate $5\frac{1}{2}$ days on the *x*-axis and move straight up to locate the point with coordinates $\left(5\frac{1}{2}, 80\right)$, which indicates that a $5\frac{1}{2}$-day rental would cost $80. We then enter this ordered pair in the table.

c. We draw a horizontal line through the point labeled 50 on the *y*-axis. Since this line intersects one step in the graph, we can look down to the *x*-axis to find the *x*-values that correspond to a *y*-value of 50. From the graph, we see that the trailer can be rented for more than 2 and up to 3 days for $50. We write (3, 50) in the table.

d. No. If we look at the *y*-coordinates, we see that for the first day, the rental fee is $20. The second day, the cost jumps another $20. The third day, and all subsequent days, the cost jumps only $10. ∎

Length of rental (days)	Cost (dollars)
2	40
$5\frac{1}{2}$	80
3	50

STUDY SET

Section 3.1

VOCABULARY

In Exercises 1–6, fill in the blanks to make the statements true.

1. The pair of numbers $(-1, -5)$ is called an _____ordered_____ pair.

2. In the ordered pair $\left(-\frac{3}{2}, -5\right)$, the -5 is called the _____*y*-coordinate_____.

3. The point with coordinates (0, 0) is called the _____origin_____.

4. The *x*- and *y*-axes divide the coordinate plane into four regions called _____quadrants_____.

5. The point with coordinates (4, 2) can be graphed on a _____rectangular_____ coordinate system.

6. The process of locating the position of a point on a coordinate plane is called _graphing or plotting_ the point.

CONCEPTS

In Exercises 7–8, fill in the blanks to make the statements true.

7. To plot the point with coordinates $(-5, 4.5)$, we start at the _____origin_____ and move 5 units to the _____left_____ and then move 4.5 units _up_.

8. To plot the point with coordinates $\left(6, -\frac{3}{2}\right)$, we start at the _____origin_____ and move 6 units to the _____right_____ and then move $\frac{3}{2}$ units _____down_____.

9. Do (3, 2) and (2, 3) represent the same point?   no

10. In the ordered pair (4, 5), is the number 4 associated with the horizontal or the vertical axis?   horizontal

11. In which quadrant do points with a negative x-coordinate and a positive y-coordinate lie?
quadrant II

12. In which quadrant do points with a positive x-coordinate and a negative y-coordinate lie?
quadrant IV

13. Use the graph in Illustration 1 to complete the table.

ILLUSTRATION 1

Linear? (No)

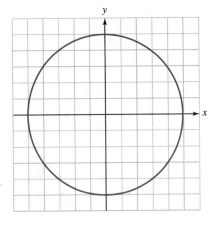

x	y
4	3
4	-3
0	5
0	-5
-3	4
-3	-4
-4	3
-4	-3
5	0

14. Use the graph in Illustration 2 to complete the table.

ILLUSTRATION 2

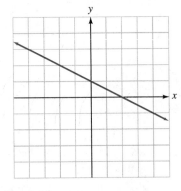

x	y
2	0
-2	2
4	-1
-4	3
0	1

The graph in Illustration 3 gives the heart rate of a woman before, during, and after an aerobic workout. In Exercises 15–22, use the graph to answer the questions.

15. What information does the point (−10, 60) give us?
10 min before the workout, her heart rate was 60 beats/min.

16. After beginning her workout, how long did it take the woman to reach her training-zone heart rate?   10 min

17. What was the woman's heart rate half an hour after beginning the workout?   150 beats/min

ILLUSTRATION 3

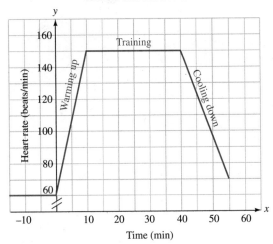

18. For how long did the woman work out at her training zone?   30 min

19. At what time was her heart rate 100 beats per minute?   approximately 5 min and 50 min after starting

20. How long was her cool-down period?   15 min

21. What was the difference in the woman's heart rate before the workout and after the cool-down period?
10 beats/min faster after cool-down

22. What was her approximate heart rate 8 minutes after beginning?   about 135 beats/min

NOTATION

23. Explain the difference between (3, 5), 3(5), and 5(3 + 5).
(3, 5) is an ordered pair, 3(5) indicates multiplication, and 5(3 + 5) is an expression containing grouping symbols.

24. In the table, which column contains values associated with the vertical axis of a graph?   the 2nd column

x	y
2	0
5	-2
-1	$-\frac{1}{2}$

25. Do these ordered pairs name the same point?
$\left(2.5, -\frac{7}{2}\right), \left(2\frac{1}{2}, -3.5\right), \left(2.5, -3\frac{1}{2}\right)$ yes

26. Do these ordered pairs name the same point?
$(-1.25, 4), \left(-1\frac{1}{4}, 4.0\right), \left(-\frac{5}{4}, 4\right)$ yes

PRACTICE

In Exercises 27–28, graph each point on the coordinate grid provided.

27. $A(-3, 4)$, $B(4, 3.5)$, $C\left(-2, -\frac{5}{2}\right)$, $D(0, -4)$, $E\left(\frac{3}{2}, 0\right)$, $F(3, -4)$

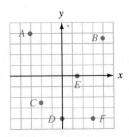

28. $G(4, 4)$, $H(0.5, -3)$, $I(-4, -4)$, $J(0, -1)$, $K(0, 0)$, $L(0, 3)$, $M(-2, 0)$

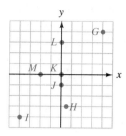

APPLICATIONS

29. CONSTRUCTION The graph in Illustration 4 shows a side view of a bridge design. Make a table with three columns; label them *rivets, welds,* and *anchors.* List the coordinates of the points at which each category is located.

rivets: $(-6, 0)$, $(-2, 0)$, $(2, 0)$, $(6, 0)$; welds: $(-4, 3)$, $(0, 3)$, $(4, 3)$; anchors: $(-6, -3)$, $(6, -3)$

ILLUSTRATION 4

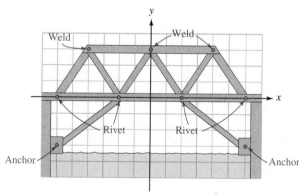

Scale: 1 unit = 8 ft

30. WATER PRESSURE The graph in Illustration 5 shows how the path of a stream of water changes when the hose is held at two different angles.
   a. At which angle does the stream of water shoot up higher? How much higher? 60°; 4 ft
   b. At which angle does the stream of water shoot out farther? How much farther? 30°; 4 ft

ILLUSTRATION 5

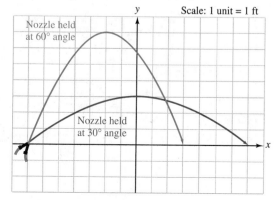

Scale: 1 unit = 1 ft

31. GOLF SWING To correct her swing, a golfer is videotaped and then has her image displayed on a computer monitor so that it can be analyzed by a golf pro. (See Illustration 6.) Give the coordinates of the points that are highlighted on the arc of her swing.
$(-3, 10)$, $(-2, 7)$, $(-1, 4.8)$, $(0, 3)$, $(1, 1.8)$, $(2.5, 0.5)$, $(4, 0)$

ILLUSTRATION 6

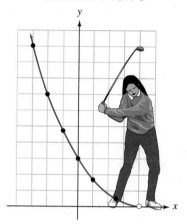

Scale: 1 unit = 6 in.

32. MEDICINE Scoliosis is a lateral curvature of the spine that can be more easily detected when a grid is superimposed over an X ray. In Illustration 7, find the coordinates of the "center points" of the indicated vertebrae. Note that T3 means the third thoracic vertebra, L4 means the fourth lumbar vertebra, and so on.

T3(2, 21), T6(3, 16), T9(3, 10), T11(2.5, 6.5), L1(1, 2.5), L2(0, 0), L4(−1, −5), L5(0, −8)

ILLUSTRATION 7

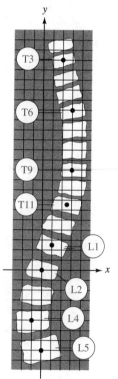

Scale: 1 unit = 0.5 in.

33. VIDEO RENTAL The charges for renting a video are shown in the graph in Illustration 8.
   a. Find the charge for a 1-day rental. $2
   b. Find the charge for a 2-day rental. $4
   c. What is the charge if a tape is kept for 5 days? $7
   d. What is the charge if a tape is kept for a week? $9

ILLUSTRATION 8

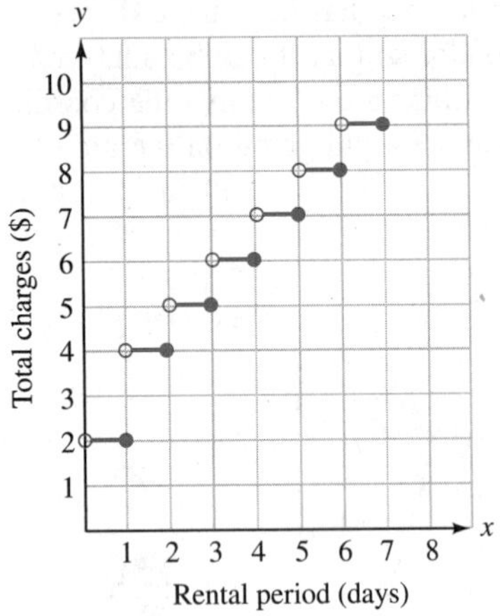

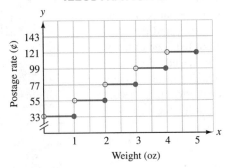

Total charges ($) vs. Rental period (days)

a. Estimate how far the truck can go on 7 gallons of gasoline. 35 mi
b. How many gallons of gas are needed to travel a distance of 20 miles? 4
c. How far can the truck go on 6.5 gallons of gasoline? 32.5 mi

36. VALUE OF A CAR The table in Illustration 11 shows the value y (in thousands of dollars) of a car that is x years old. Plot the ordered pairs and draw a line connecting the points.

ILLUSTRATION 11

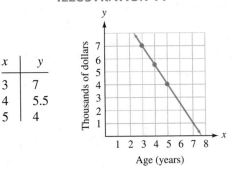

x	y
3	7
4	5.5
5	4

Thousands of dollars vs. Age (years)

34. POSTAGE RATES The graph shown in Illustration 9 gives the first-class postage rates in 1999 for mailing items weighing up to 5 ounces.

a. Find the postage costs for mailing each of the following letters first class: a 1-ounce letter, a 4-ounce letter, and a $2\frac{1}{2}$-ounce letter. 33¢, 99¢, 77¢
b. Find the difference in postage for a 3.75-ounce letter and a 4.75-ounce letter. 22¢
c. What is the heaviest letter that could be mailed for 55¢ first class? 2 oz

ILLUSTRATION 9

Postage rate (¢) vs. Weight (oz)

35. GAS MILEAGE The table in Illustration 10 gives the number of miles (y) that a truck can be driven on x gallons of gasoline. Plot the ordered pairs and draw a line connecting the points.

ILLUSTRATION 10

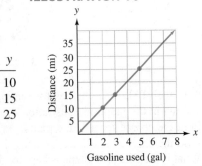

x	y
2	10
3	15
5	25

Distance (mi) vs. Gasoline used (gal)

a. What does the point (3, 7) on the graph tell you?
A 3-year-old car is worth $7,000.
b. Estimate the value of the car when it is 7 years old. $1,000
c. After how many years will the car be worth $2,500? 6

37. ROAD MAPS Road maps usually have a coordinate system to help locate cities. Use the map in Illustration 12 to locate Rockford, Mount Carroll, Harvard, and the intersection of state Highway 251 and U.S. Highway 30. Express each answer in the form (number, letter).
Rockford (5, B), Mount Carroll (1, C), Harvard (7, A), intersection (5, E)

ILLUSTRATION 12

38. BATTLESHIP In the game Battleship, the player uses coordinates to drop depth charges from a battleship to hit a hidden submarine. What coordinates should be used to make three hits on the exposed submarine shown in Illustration 13? Express each answer in the form (letter, number).
(E, 4), (F, 3), (G, 2)

ILLUSTRATION 13

WRITING

39. Explain why the point $(-3, 3)$ is not the same as the point $(3, -3)$.

40. Explain what is meant when we say that the rectangular coordinate graph of the St. Louis Gateway Arch is made up of *infinitely many* points.

41. Explain how to plot the point $(-2, 5)$.

42. Explain why the coordinates of the origin are $(0, 0)$.

REVIEW

43. Evaluate $-3 - 3(-5)$.   12

44. Evaluate $(-5)^2 + (-5)$.   20

45. What is the opposite of -8?   8

46. Simplify $|-1 - 9|$.   10

47. Solve $-4x + 7 = -21$.   7

48. Solve $P = 2l + 2w$ for w. $w = \frac{P - 2l}{2}$

49. Evaluate $(x + 1)(x + y)^2$ for $x = -2$ and $y = -5$.
   -49

50. Simplify $-6(x - 3) - 2(1 - x)$. $-4x + 16$

▶ **3.2**

Equations Containing Two Variables

In this section, you will learn about

Solving equations with two variables ■ Constructing tables of values ■ Graphing equations ■ Using different variables

Introduction In this section, we will discuss equations that contain two variables. Such equations are often used to describe relationships between two quantities. To see a mathematical picture of these relationships, we will construct graphs of their equations.

Solving Equations with Two Variables

We have previously solved equations containing one variable. For example, we can show that the solution of each of the following equations is $x = 3$.

$$2x + 3 = 9, \qquad -5x + 1 = 4 - 6x, \qquad \text{and} \qquad -3(x + 1) = 2x - 18$$

If we graph the solution $x = 3$ on a number line, we get the graph shown in Figure 3-10.

FIGURE 3-10

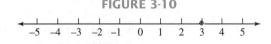

To describe relationships between two quantities mathematically, we use equations with two variables. Some examples of equations in two variables are

$$y = x - 1, \qquad y = x^2, \qquad y = |x|, \qquad \text{and} \qquad y = x^3$$

Solutions of equations in two variables are ordered pairs. For example, one solution of $y = x - 1$ is the ordered pair $(5, 4)$, because the equation is true when $x = 5$ and $y = 4$.

$y = x - 1$ The original equation.

$4 \overset{?}{=} 5 - 1$ Substitute 5 for x and 4 for y.

$4 = 4$ Do the subtraction on the right-hand side: $5 - 1 = 4$.

Since $4 = 4$ is a true statement, the ordered pair $(5, 4)$ is a solution, and we say that $(5, 4)$ **satisfies** the equation.

To see whether the ordered pair $(-1, -3)$ satisfies the equation, we substitute -1 for x and -3 for y.

$y = x - 1$ The original equation.

$-3 \overset{?}{=} -1 - 1$ Substitute -1 for x and -3 for y.

$-3 = -2$ Do the subtraction on the right-hand side: $-1 - 1 = -2$.

Since $-3 = -2$ is a false statement, $(-1, -3)$ does not satisfy the equation.

EXAMPLE 1

Verifying a solution. Is the ordered pair $(-2, 4)$ a solution of $y = 3x + 9$?

Solution   We substitute -2 for x and 4 for y and see if a true statement results.

$y = 3x + 9$ The original equation.

$4 \overset{?}{=} 3(-2) + 9$ Substitute -2 for x and 4 for y.

$4 \overset{?}{=} -6 + 9$ Do the multiplication: $3(-2) = -6$.

$4 = 3$ Do the addition: $-6 + 9 = 3$.

Since the equation $4 = 3$ is false, $x = -2$ and $y = 4$ is not a solution.

SELF CHECK   Is $(-1, -5)$ a solution of $y = 5x$?   *Answer:* yes   ■

EXAMPLE 2

Verifying a solution. Is the ordered pair $(3, -4)$ a solution of $y = 5 - x^2$?

Solution   We substitute 3 for x and -4 for y and see whether the resulting equation is a true statement.

$y = 5 - x^2$ The original equation.

$-4 \overset{?}{=} 5 - (3)^2$ Substitute 3 for x and -4 for y.

$-4 \overset{?}{=} 5 - 9$ Find the power: $(3)^2 = 9$.

$-4 = -4$ Do the subtraction: $5 - 9 = -4$.

Since the equation $-4 = -4$ is true, $x = 3$ and $y = -4$ is a solution.

SELF CHECK   Is the ordered pair $(-2, 0)$ a solution of $y = x^2 + 4$?   *Answer:* no   ■

Constructing Tables of Values

To find solutions of equations in x and y, we can pick numbers at random, substitute them for x, and find the corresponding values of y. For example, to find some ordered pairs that satisfy the equation $y = x - 1$, we can let $x = -4$ (called the **input value**), substitute -4 for x, and solve for y (called the **output value**).

$y = x - 1$		
x	y	(x, y)
-4	-5	$(-4, -5)$

$y = x - 1$ The original equation.

$y = -4 - 1$ Substitute the input -4 for x.

$y = -5$ The output is -5.

$y = x - 1$

x	y	(x, y)
−4	−5	(−4, −5)
−2	−3	(−2, −3)

$y = x - 1$

x	y	(x, y)
−4	−5	(−4, −5)
−2	−3	(−2, −3)
0	−1	(0, −1)

$y = x - 1$

x	y	(x, y)
−4	−5	(−4, −5)
−2	−3	(−2, −3)
0	−1	(0, −1)
2	1	(2, 1)

$y = x - 1$

x	y	(x, y)
−4	−5	(−4, −5)
−2	−3	(−2, −3)
0	−1	(0, −1)
2	1	(2, 1)
4	3	(4, 3)

The ordered pair $(-4, -5)$ is a solution. We list this ordered pair in the **table of values** (or **table of solutions**), shown on the previous page.

To find another ordered pair that satisfies $y = x - 1$, we let $x = -2$.

$y = x - 1$ The original equation.

$y = -2 - 1$ Substitute the input −2 for x.

$y = -3$ The output is −3.

A second solution is $(-2, -3)$, and we list it in the table of values.

If we let $x = 0$, we can find a third ordered pair that satisfies $y = x - 1$.

$y = x - 1$ The original equation.

$y = 0 - 1$ Substitute the input 0 for x.

$y = -1$ The output is −1.

A third solution is $(0, -1)$, which we also add to our table of values.

If we let $x = 2$, we can find a fourth solution.

$y = x - 1$ The original equation.

$y = 2 - 1$ Substitute the input 2 for x.

$y = 1$ The output is 1.

A fourth solution is $(2, 1)$, and we add it to our table of values.

If we let $x = 4$, we have

$y = x - 1$ The original equation.

$y = 4 - 1$ Substitute the input 4 for x.

$y = 3$ The output is 3.

A fifth solution is $(4, 3)$.

Since we can choose any real number for x, and since any choice of x will give a corresponding value of y, it is apparent that the equation $y = x - 1$ has *infinitely many solutions*. We have found five of them: $(-4, -5)$, $(-2, -3)$, $(0, -1)$, $(2, 1)$, and $(4, 3)$.

ACCENT ON TECHNOLOGY *Generating a Table of Values with a Graphing Calculator*

FIGURE 3-11

Constructing a table of values can be a tedious job. Instead of doing the work by hand (as we did above), we can use a graphing calculator to quickly generate solutions of an equation in two variables. Several brands of graphing calculators are available, and each has its own sequence of keystrokes to make a table of solutions. The instructions in this discussion are for a TI-83 graphing calculator, shown in Figure 3-11. For specific details about your calculator, please consult your owner's manual.

To construct a table of solutions for $y = x - 1$, we begin by entering the x-values that are to appear in the table. To do this, we press the keys

[2nd] [TBLSET] and enter the first value for x on the line labeled TblStart =.

Courtesy of Texas Instruments

(continued)

In Figure 3-12(a), the number -4 has been entered on that line. The other values for x that are to appear in the table are determined by entering an **increment** value on the line labeled ΔTbl $=$. Figure 3-12(a) shows that an increment of 2 was entered. This means that each x-value in the table will be 2 larger than the previous x-value.

To enter the equation $y = x - 1$, we press $\boxed{y =}$ and enter $x - 1$, as shown in Figure 3-12(b). (Ignore the subscript 1 on y_1; it is not relevant at this time.)

The final step is to the press the keys $\boxed{\text{2nd}}$ $\boxed{\text{TABLE}}$. This will display a table of solutions, as shown in Figure 3-12(c). The table contains all of the entries that we obtained by hand, as well as two additional solutions: $(6, 5)$ and $(8, 7)$.

FIGURE 3-12

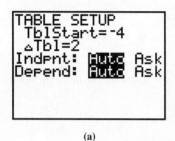
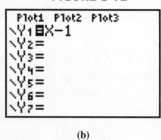
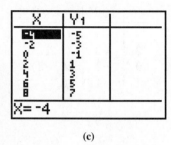

(a)   (b)   (c)

Graphing Equations

To graph the equation $y = x - 1$, we plot the ordered pairs listed in the table on a rectangular coordinate system, as shown in Figure 3-13(a). From the figure, we can see that the five points lie on a line.

In Figure 3-13(b), we draw a line through the points, because the graph of any solution of $y = x - 1$ will lie on this line. The arrowheads show that the line continues forever in both directions. The line is a picture of all the solutions of the equation $y = x - 1$. This line is called the **graph** of the equation.

FIGURE 3-13

$y = x - 1$

x	y	(x, y)
-4	-5	$(-4, -5)$
-2	-3	$(-2, -3)$
0	-1	$(0, -1)$
2	1	$(2, 1)$
4	3	$(4, 3)$

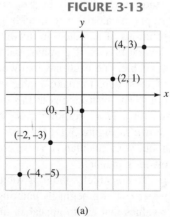
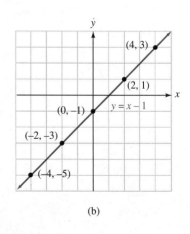

(a)   (b)

To graph an equation in x and y, we follow these steps.

Graphing an equation in *x* and *y*

1. Make a table of values containing several ordered pairs of numbers (x, y) that satisfy the equation. Do this by picking values for x and finding the corresponding values for y.

2. Plot each ordered pair on a rectangular coordinate system.

3. Carefully draw a line or smooth curve through the points.

Since we will usually choose a number for x and then find the corresponding value of y, the value of y depends on x. For this reason, we call y the **dependent variable** and x the **independent variable.** The value of the independent variable is the input value, and the value of the dependent variable is the output value.

EXAMPLE 3

Graphing equations. Graph $y = x^2$.

Solution   To make a table of values, we will choose numbers for x and find the corresponding values of y. If $x = -3$, we have

$$y = x^2 \qquad \text{The original equation.}$$
$$y = (-3)^2 \quad \text{Substitute the input } -3 \text{ for } x.$$
$$y = 9 \qquad \text{The output is 9.}$$

Thus, $x = -3$ and $y = 9$ is a solution. In a similar manner, we find the corresponding y-values for x-values of -2, -1, 0, 1, 2, and 3. If we plot the ordered pairs listed in the table in Figure 3-14 and join the points with a smooth curve, we get the graph shown in the figure, which is called a **parabola.**

FIGURE 3-14

$$y = x^2$$

x	y	(x, y)
-3	9	$(-3, 9)$
-2	4	$(-2, 4)$
-1	1	$(-1, 1)$
0	0	$(0, 0)$
1	1	$(1, 1)$
2	4	$(2, 4)$
3	9	$(3, 9)$

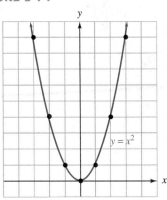

SELF CHECK   Graph $y = x^2 - 2$ and compare the result to the graph of $y = x^2$. What do you notice?

Answer: The graph has the same shape, but is 2 units lower.

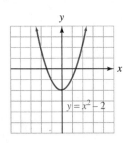

E X A M P L E 4	**Graphing equations.** Graph $y = |x|$.

Solution   To make a table of values, we will choose numbers for x and find the corresponding values of y. If $x = -5$, we have

$y = |x|$	The original equation.
$y = |-5|$	Substitute the input -5 for x.
$y = 5$	The output is 5.

The ordered pair $(-5, 5)$ satisfies the equation. This pair and several others that satisfy the equation are listed in the table of values in Figure 3-15. If we plot the ordered pairs in the table, we see that they lie in a "V" shape. We join the points to complete the graph shown in the figure.

FIGURE 3-15

$y = |x|$

x	y	(x, y)
-5	5	$(-5, 5)$
-4	4	$(-4, 4)$
-3	3	$(-3, 3)$
-2	2	$(-2, 2)$
-1	1	$(-1, 1)$
0	0	$(0, 0)$
1	1	$(1, 1)$
2	2	$(2, 2)$
3	3	$(3, 3)$

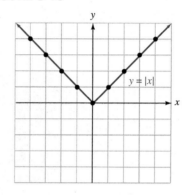

SELF CHECK   Graph $y = |x| + 2$ and compare the result to the graph of $y = |x|$. What do you notice?

Answer: The graph has the same shape, but is 2 units higher.

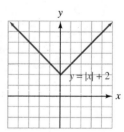

E X A M P L E 5	**Graphing equations.** Graph $y = x^3$.

Solution   If we let $x = -2$, we have

$y = x^3$	The original equation.
$y = (-2)^3$	Substitute the input -2 for x.
$y = -8$	The output is -8.

The ordered pair $(-2, -8)$ satisfies the equation. This ordered pair and several others that satisfy the equation are listed in the table of values in Figure 3-16. Plotting the ordered pairs and joining them with a smooth curve gives us the graph shown in the figure.

FIGURE 3-16

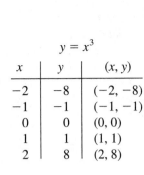

$y = x^3$

x	y	(x, y)
-2	-8	$(-2, -8)$
-1	-1	$(-1, -1)$
0	0	$(0, 0)$
1	1	$(1, 1)$
2	8	$(2, 8)$

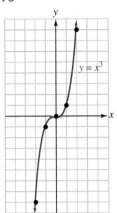

SELF CHECK   Graph $y = (x - 2)^3$ and compare the result to the graph of $y = x^3$. What do you notice?

Answer: The graph has the same shape but is 2 units to the right.

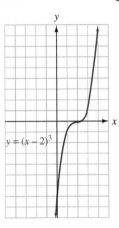

ACCENT ON TECHNOLOGY *Using a Graphing Calculator to Graph an Equation*

So far, we have graphed equations by making tables of values and plotting points. The task of graphing is made much easier when we use a graphing calculator. The instructions in this discussion will be general in nature. For specific details about your calculator, please consult your owner's manual.

The Viewing Window   All graphing calculators have a viewing **window,** used to display graphs. The **standard window** has settings of

$$\text{Xmin} = -10, \qquad \text{Xmax} = 10, \qquad \text{Ymin} = -10, \qquad \text{and} \qquad \text{Ymax} = 10$$

which indicate that the minimum x- and y-coordinates used in the graph will be -10, and that the maximum x- and y-coordinates will be 10.

Graphing an Equation   To graph the equation $y = x - 1$ using a graphing calculator, we press the $\boxed{\text{Y=}}$ key and enter the right-hand side of the equation after the symbol Y_1. The display will show the equation

$$Y_1 = x - 1$$

Then we press the $\boxed{\text{GRAPH}}$ key to produce the graph shown in Figure 3-17.

(continued)

Next, we will graph the equation $y = |x - 4|$. Since absolute values are always nonnegative, the minimum y-value is zero. To obtain a reasonable viewing window, we set the Ymin value slightly lower, at Ymin = −3. We set Ymax to be 10 units greater than Ymin, at Ymax = 7. The minimum value of y occurs when $x = 4$. To center the graph in the viewing window, we set the Xmin and Xmax values 5 units to the left and right of 4. Therefore, Xmin = −1 and Xmax = 9.

After entering the right-hand side of the equation, we obtain the graph shown in Figure 3-18. Consult your owner's manual to learn how to enter an absolute value.

FIGURE 3-17   **FIGURE 3-18**

Changing the Viewing Window

The choice of viewing windows is extremely important when graphing equations. To show this, let's graph $y = x^2 - 25$ with x-values from −1 to 6 and y-values from −5 to 5.

To graph this equation, we set the x and y window values and enter the right-hand side of the equation. The display will show

$$Y_1 = x^2 - 25$$

Then we press the ⎹ GRAPH ⎸ key to produce the graph shown in Figure 3-19(a). Although the graph appears to be a straight line, it is not. Actually, we are only seeing part of a parabola. If we pick a viewing window with x-values of −6 to 6 and y-values of −30 to 2, as in Figure 3-19(b), we can see that the graph is a parabola.

FIGURE 3-19

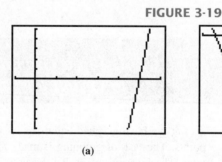
(a)

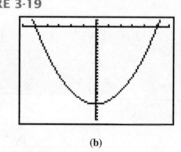
(b)

Using Different Variables

We will often encounter equations with variables other than x and y. When we make tables of values and graph these equations, we must know which is the independent variable (the input values) and which is the dependent variable (the output values). The independent variable is usually associated with the horizontal axis of the coordinate system, and the dependent variable is usually associated with the vertical axis.

EXAMPLE 6

Advertising. The profit a certain company can make depends on the amount of money it spends on advertising. This relationship is described by the equation $P = 6m - m^2$, where P represents the profit (in millions of dollars) and m represents the amount spent on advertising (in tens of thousands of dollars). Graph the equation and interpret the results.

Solution Since P depends on m in the equation $P = 6m - m^2$, m is the independent variable (the input) and P is the dependent variable (the output). Therefore, we will choose values for m and find the corresponding values of P. Since m represents the amount of money spent on advertising, we will not choose negative values for m.

If $m = 0$, we have

$$P = 6m - m^2$$
$$P = 6(0) - (0)^2 \quad \text{Substitute the input 0 for } x.$$
$$= 6(0) - 0 \quad \text{Find the power: } (0)^2 = 0.$$
$$= 0 - 0 \quad \text{Do the multiplication: } 6(0) = 0.$$
$$= 0 \quad \text{The output is 0.}$$

The pair $m = 0$ and $P = 0$, or $(0, 0)$, is a solution. This ordered pair and others that satisfy the equation are listed in the table of values shown in Figure 3-20. If we plot the ordered pairs as in Figure 3-20(a) and join them with a smooth curve, we will obtain the graph shown in Figure 3-20(b). Here the graph is a parabola opening downward.

From the graph, we see that the profit increases to a point and then decreases. Since the graph peaks at the point $(3, 9)$, we know that $30,000 spent on advertising will generate a maximum profit of $9 million.

FIGURE 3-20

Plot the profit on this axis in millions of dollars.

$P = 6m - m^2$

m	P	(m, P)
0	0	$(0, 0)$
1	5	$(1, 5)$
2	8	$(2, 8)$
3	9	$(3, 9)$
4	8	$(4, 8)$
5	5	$(5, 5)$
6	0	$(6, 0)$

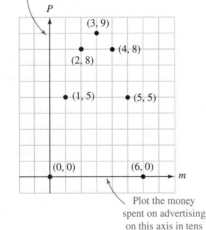

Plot the money spent on advertising on this axis in tens of thousands of dollars

(a)

The profit is a maximum at this "peak" in the graph, which occurs at $(3, 9)$.

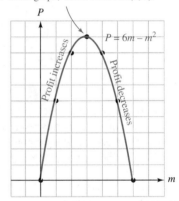

(b)

STUDY SET

Section 3.2

VOCABULARY

In Exercises 1–4, fill in the blanks to make the statements true.

1. The equation $y = x + 1$ is an equation in <u>two</u> variables.

2. An ordered pair is a <u>solution</u> of an equation if the numbers in the ordered pair satisfy the equation.

3. In equations containing the variables x and y, x is called the <u>independent</u> variable and y is called the <u>dependent</u> variable.

4. When constructing a <u>table</u> of values, the values of x are the <u>input</u> values and the values of y are the <u>output</u> values.

CONCEPTS

5. Consider the equation $y = -2x + 6$.
  a. How many variables does the equation contain? 2
  b. Does the ordered pair $(4, -2)$ satisfy the equation?
    yes
  c. Is $x = -3$ and $y = 12$ a solution? yes
  d. How many solutions does this equation have?
    infinitely many

6. Consider the equation $8 = -2x + 6$.
  a. How many variables does the equation contain? 1
  b. Is $x = -1$ a solution of the equation? yes
  c. How many solutions does this equation have?
    one

7. Consider the graph of an equation shown in Illustration 1.

ILLUSTRATION 1

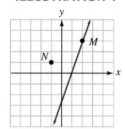

  a. If the coordinates of point M are substituted into
    the equation, is the result a true statement? yes
  b. If the coordinates of point N are substituted into the
    equation, is the result a true statement? no

8. Complete each table of values.
  a. $y = x^3$

x (inputs)	y (outputs)
0	0
-1	-1
-2	-8
1	1
2	8

  b. $y = x^4$

x (inputs)	y (outputs)
0	0
-1	1
-2	16
1	1
2	16

9. To graph $y = x^2 - 4$, a table of values is constructed
  and a graph is drawn, as shown in Illustration 2. Explain the error made here.
    Not enough ordered pairs were found—the correct graph is
    not a line.

ILLUSTRATION 2

$y = x^2 - 4$

x	y	(x, y)
0	-4	$(0, -4)$
2	0	$(2, 0)$

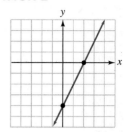

10. Several solutions of an equation are listed in the table
  of values. When graphing them, how should the horizontal and vertical axes of the graph be labeled?
    horizontal: t; vertical: s

t	s	(t, s)
0	4	$(0, 4)$
1	5	$(1, 5)$
2	10	$(2, 10)$

11. How many variables does the equation $x + 2 = 6$
  have? How many solutions does it have? Graph the
  solution(s).
    one, one,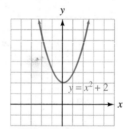

12. How many variables does the equation $x + 3 = 1$
  have? How many solutions does it have? Graph the
  solution(s).
    one, one,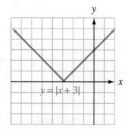

13. How many variables does the equation $y = x^2 + 2$
  have? How many solutions does it have? Graph the
  solution(s). two, infinitely many

$y = x^2 + 2$

14. How many variables does the equation $y = |x + 3|$
  have? How many solutions does it have? Graph the
  solution(s). two, infinitely many

$y = |x + 3|$

NOTATION

In Exercises 15–16, complete each solution.

15. Verify that $(-2, 6)$ satisfies $y = -x + 4$.

$$y = -x + 4$$
$$6 \stackrel{?}{=} -(\boxed{-2}) + 4$$
$$6 \stackrel{?}{=} \boxed{2} + 4$$
$$6 = 6$$

16. For the equation $y = |x - 2|$, if $x = -3$, find y.

$$y = |x - 2|$$
$$y = |\boxed{-3} - 2|$$
$$y = |\boxed{-5}|$$
$$y = 5$$

PRACTICE

In Exercises 17–20, tell whether the ordered pair satisfies the equation.

17. $x - 2y = -4$; $(4, 4)$
   yes

18. $y = 2x - x^2$; $(8, 48)$
   no

19. $y = |5 - 2x|$; $(4, -3)$
   no

20. $y = 3 - x^3$; $(-2, 11)$
   yes

In Exercises 21–24, complete each table of values.

21. $y = x - 3$

x	y
0	-3
1	-2
-2	-5

22. $y = |x - 3|$

x	$|x - 3|$
0	3
-1	4
3	0

23. $y = x^2 - 3$

Input	Output
0	-3
2	1
-2	1

24. $y = x + 1$

Input	Output
0	1
2	3
-1	0

In Exercises 25–28, construct a table of values, then graph each equation.

25. $y = 2x - 3$

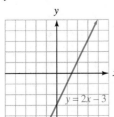

26. $y = 3x + 1$

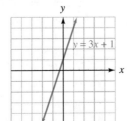

27. $y = -2x + 1$

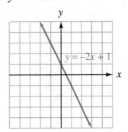

28. $y = -3x + 2$

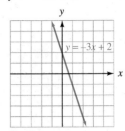

In Exercises 29–32, construct a table of values, then graph each equation. Compare it to the graph of $y = x^2$.

29. $y = x^2 + 1$
   1 unit higher

30. $y = -x^2$
   It is turned upside down.

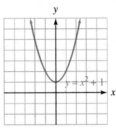

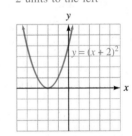

31. $y = (x - 2)^2$
   2 units to the right

32. $y = (x + 2)^2$
   2 units to the left

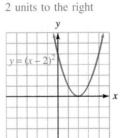

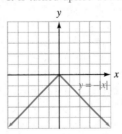

In Exercises 33–36, construct a table of values, then graph each equation. Compare it to the graph of $y = |x|$.

33. $y = -|x|$
   It is turned upside down.

34. $y = |x| - 2$
   2 units lower

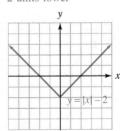

35. $y = |x + 2|$
   2 units to the left

36. $y = |x - 2|$
   2 units to the right

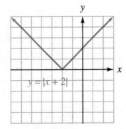

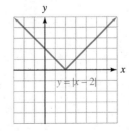

In Exercises 37–40, construct a table of values, then graph each equation. Compare it to the graph of $y = x^3$.

37. $y = -x^3$

It is turned upside down.

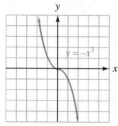

38. $y = x^3 + 2$

2 units higher

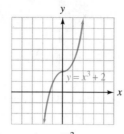

39. $y = x^3 - 2$

2 units lower

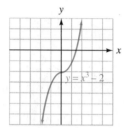

40. $y = (x + 2)^3$

2 units to the left

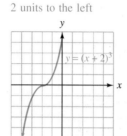

41. 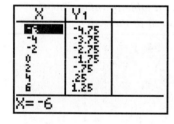 The table of values for $y = 0.5x - 1.75$ shown in Illustration 3 was generated by a graphing calculator.

   a. What increment was used to generate the x-values?   2

   b. Graph the equation by plotting the ordered pairs represented in the table.

ILLUSTRATION 3

42. Use a graphing calculator to create a table of values for $y = -0.25x^2 + 0.75$. In the table, begin with an x-value of -2 and use an increment of 0.5. Then graph the equation by plotting the ordered pairs represented in the table.

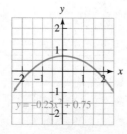

In Exercises 43–46, use a graphing calculator to graph each equation. Use a viewing window of $x = -5$ to 5 and $y = -5$ to 5.

43. $y = 2.1x - 1.1$

44. $y = 1.12x^2 - 1$

45. $y = |x + 0.7|$

46. $y = 0.1x^3 + 1$

In Exercises 47–50, graph each equation in a viewing window of $x = -4$ to 4 and $y = -4$ to 4. This graph is not what it appears to be. Pick a better viewing window and find a better representation of the true graph.

47. $y = -x^3 - 8.2$

48. $y = -|x - 4.01|$

49. $y = x^2 + 5.9$

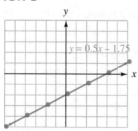

50. $y = -x + 7.95$

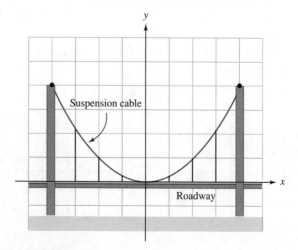

APPLICATIONS

51. SUSPENSION BRIDGES The suspension cables of a bridge hang in the shape of a parabola, as shown in Illustration 4. Use the information in the illustration to complete the table of values.

ILLUSTRATION 4

x	0	2	4	-2	-4
y	0	1	4	1	4

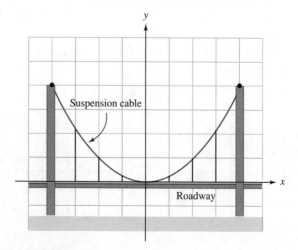

52. FIRE BOAT   A stream of water from a high-pressure hose on a fire boat travels in the shape of a parabola, as shown in Illustration 5. Use the information in the graph to complete the table of values.

ILLUSTRATION 5

x	1	2	3	4
y	3	4	3	0

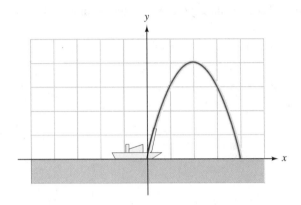

53. MANUFACTURING   The graph in Illustration 6 shows the relationship between the length l (in inches) of a machine bolt and the cost C (in cents) to manufacture it.
   a. What information does the point (2, 8) on the graph give us?   It costs 8¢ to make a 2-in. bolt.
   b. How much does it cost to make a 7-inch bolt?   12¢
   c. What length bolt is the least expensive to make?
      a 4-in. bolt
   d. Describe how the cost changes as the length of the bolt increases.
      It decreases as the length approaches 4 in., then increases as the length increases to 7 in.

ILLUSTRATION 6

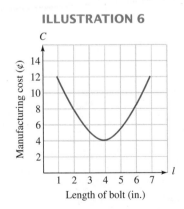

54. SOFTBALL   The graph in Illustration 7 shows the relationship between the distance d (in feet) traveled by a batted softball and the height h (in feet) it attains.
   a. What information does the point (40, 40) on the graph give us?
      After the ball has traveled 40 ft, its height is 40 ft.

   b. At what distance from home plate does the ball reach its maximum height?   100 ft
   c. Where will the ball land?   200 ft from home plate

ILLUSTRATION 7

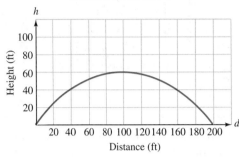

Distance (ft)

55. MARKET VALUE OF A HOUSE   The graph in Illustration 8 shows the relationship between the market value v of a house and the time t since it was purchased.

ILLUSTRATION 8

Time since purchase (years)

   a. What was the purchase price of the house? $90,000
   b. When did the value of the house reach its lowest point?   the 3rd yr after being bought
   c. When did the value of the house begin to surpass the purchase price?   after the 6th yr
   d. Describe how the market value of the house changed over the 8-year period.
      It decreased in value for 3 yr, then increased in value for 5 yr.

56. POLITICAL SURVEY   The graph in Illustration 9 shows the relationship between the percent P of those surveyed who rated their senator's job performance as satisfactory or better and the time t she had been in office.
   a. When did her job performance rating reach a maximum?   the 8th month after being elected
   b. When was her job performance rating at or above the 60% mark?   between the 4th and 12th months
   c. Describe how her job performance rating changed over the 12-month period.
      After the election, it increased for 8 mo to a high of 70%. Then it decreased for 4 mo.

ILLUSTRATION 9

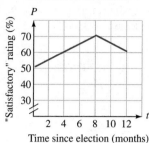

Time since election (months)

60. What does it mean when we say that an equation in two variables has infinitely many solutions?

REVIEW

61. Solve $\dfrac{x}{8} = -12$. -96

62. Combine like terms: $3t - 4T + 5T - 6t$. $-3t + T$

63. Is $\dfrac{x + 5}{6}$ an expression or an equation?   an expression

64. What formula is used to find the perimeter of a rectangle? $P = 2l + 2w$

65. What number is 0.5% of 250?   1.25

66. Solve $-3x + 5 > -7$. $x < 4$

67. Find $-2.5 - (-2.6)$.   0.1

68. Evaluate $(-5)^3$. -125

WRITING

57. What is a table of values? Why is it often called a table of solutions?

58. To graph an equation in two variables, how many solutions of the equation must be found?

59. Give an example of an equation in one variable and an equation in two variables. How do their solutions differ?

▶ 3.3 Graphing Linear Equations

In this section, you will learn about

> Linear equations ■ Solutions of linear equations ■ Graphing linear equations ■ The intercept method ■ Graphing horizontal and vertical lines ■ An application of linear equations ■ Solving equations graphically

Introduction   In Section 3.2, we graphed the equations shown in Figure 3-21. Because the graph of the equation $y = x - 1$ is a line, we call it a *linear equation*. Since the graphs of $y = x^2$, $y = |x|$, and $y = x^3$ are *not* lines, they are *nonlinear equations*.

FIGURE 3-21

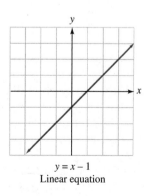

$y = x - 1$
Linear equation

(a)

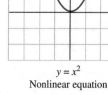

$y = x^2$
Nonlinear equation

(b)

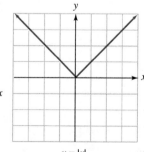

$y = |x|$
Nonlinear equation

(c)

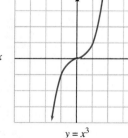

$y = x^3$
Nonlinear equation

(d)

In this section, we will discuss how to graph linear equations and show how to use their graphs to solve problems.

Linear Equations

Any equation, such as $y = x - 1$, whose graph is a line is called a **linear equation in x and y.** Some other examples of linear equations are

$$y = \frac{1}{2}x + 2, \qquad 2x + 3y = 6, \qquad 5y - x + 2 = 0, \qquad \text{and} \qquad y = -2$$

A linear equation in x and y is any equation that can be written in a special form, called **general** (or **standard**) form.

General form of a linear equation

> If A, B, and C are real numbers, the equation
>
> $\quad Ax + By = C \qquad$ (A and B are not both zero)
>
> is called the **general form** (or **standard form**) of the equation of a line.

Whenever possible, we will write the general form $Ax + By = C$ so that A, B, and C are integers and $A \geq 0$. Note that in a linear equation in x and y, the exponents on x and y are 1.

EXAMPLE 1

Identifying linear equations. Which of the following equations are linear equations?   **a.** $3x = 1 - 2y$ **b.** $y = x^3 + 1$ **c.** $y = -\frac{1}{2}x$

Solution   **a.** Since the equation $3x = 1 - 2y$ can be written in $Ax + By = C$ form, it is a linear equation.

$$\begin{aligned} 3x &= 1 - 2y & &\text{The original equation.} \\ 3x + 2y &= 1 - 2y + 2y & &\text{Add } 2y \text{ to both sides.} \\ 3x + 2y &= 1 & &\text{Simplify the right-hand side: } -2y + 2y = 0. \end{aligned}$$

Here $A = 3$, $B = 2$, and $C = 1$.

b. Since the exponent on x in $y = x^3 + 1$ is 3, the equation is a nonlinear equation.

c. Since the equation $y = -\frac{1}{2}x$ can be written in $Ax + By = C$ form, it is a linear equation.

$$\begin{aligned} y &= -\frac{1}{2}x & &\text{The original equation.} \\ -2(y) &= -2\left(-\frac{1}{2}x\right) & &\text{Multiply both sides by } -2 \text{ so that the coefficient of } x \text{ will be 1.} \\ -2y &= x & &\text{Simplify the right-hand side: } -2\left(-\frac{1}{2}\right) = 1. \\ 0 &= x + 2y & &\text{Add } 2y \text{ to both sides.} \\ x + 2y &= 0 & &\text{Write the equation in general form.} \end{aligned}$$

Here $A = 1$, $B = 2$, and $C = 0$.

SELF CHECK   Tell which of the following are linear equations and which are nonlinear: **a.** $y = |x|$,   **b.** $-x = 6 - y$,   and   **c.** $y = x$.

Answers: **a.** nonlinear,   **b.** linear,   **c.** linear

Solutions of Linear Equations

To find solutions of linear equations, we substitute arbitrary values for one variable and solve for the other.

EXAMPLE 2

Finding solutions of linear equations. Complete the table of values for $3x + 2y = 5$.

x	y	(x, y)
7		(7,)
	4	(, 4)

Solution   In the first row, we are given the x-value of 7. To find the corresponding y-value, we substitute 7 for x and solve for y.

$3x + 2y = 5$	The original equation.
$3(7) + 2y = 5$	Substitute 7 for x.
$21 + 2y = 5$	Do the multiplication: $3(7) = 21$.
$2y = -16$	Subtract 21 from both sides: $5 - 21 = -16$.
$y = -8$	Divide both sides by 2.

A solution of $3x + 2y = 5$ is $(7, -8)$.

In the second row, we are given a y-value of 4. To find the corresponding x-value, we substitute 4 for y and solve for x.

$3x + 2y = 5$	The original equation.
$3x + 2(4) = 5$	Substitute 4 for y.
$3x + 8 = 5$	Do the multiplication: $2(4) = 8$.
$3x = -3$	Subtract 8 from both sides: $5 - 8 = -3$.
$x = -1$	Divide both sides by 3.

Another solution is $(-1, 4)$. The completed table is as follows:

x	y	(x, y)
7	-8	$(7, -8)$
-1	4	$(-1, 4)$

SELF CHECK   Complete the table of values for $3x + 2y = 5$:

x	y	(x, y)
	-2	(, -2)
5		(5,)

Answer:

x	y	(x, y)
3	-2	$(3, -2)$
5	-5	$(5, -5)$

■

Graphing Linear Equations

Since two points determine a line, only two points are needed to graph a linear equation. However, we will often plot a third point as a check. If the three points do not lie on a line, then at least one of them is in error.

Graphing linear equations

1. Find three pairs (x, y) that satisfy the equation by picking arbitrary numbers for x and finding the corresponding values of y.

2. Plot each resulting pair (x, y) on a rectangular coordinate system. If the three points do not lie on a line, check your computations.

3. Draw the line passing through the points.

EXAMPLE 3

Graphing linear equations. Graph $y = -3x$.

Solution   To find three ordered pairs that satisfy the equation, we begin by choosing three x-values: -2, 0, and 2.

If x = −2	*If x = 0*	*If x = 2*
$y = -3x$	$y = -3x$	$y = -3x$
$y = -3(-2)$	$y = -3(0)$	$y = -3(2)$
$y = 6$	$y = 0$	$y = -6$

We enter the results in a table of values, plot the points, and draw a line through the points. The graph appears in Figure 3-22. Check this work with a graphing calculator.

FIGURE 3-22

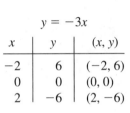

$$y = -3x$$

x	y	(x, y)
-2	6	$(-2, 6)$
0	0	$(0, 0)$
2	-6	$(2, -6)$

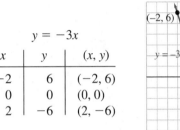

SELF CHECK   Graph $y = -3x + 2$ and compare the result to the graph of $y = -3x$. What do you notice?

Answer: It is a line 2 units above the graph of $y = -3x$.

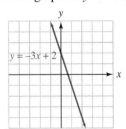

When graphing linear equations, it is often easier to find solutions of the equation if it is first solved for y.

EXAMPLE 4

Solving for y. Graph $2y = 4 - x$.

Solution   We first solve the equation for y.

$$2y = 4 - x$$

$$\frac{2y}{2} = \frac{4}{2} - \frac{x}{2} \qquad \text{To isolate } y, \text{ divide both sides by 2.}$$

$$y = 2 - \frac{x}{2} \qquad \text{Simplify: } \frac{2y}{2} = y \text{ and } \frac{4}{2} = 2.$$

Since each value of x will be divided by 2, we will choose values of x that are divisible by 2: -4, 0, and 4. If $x = -4$, we have

$$y = 2 - \frac{x}{2}$$

$$y = 2 - \frac{-4}{2} \quad \text{Substitute } -4 \text{ for } x.$$

$$y = 2 - (-2) \quad \text{Divide: } \frac{-4}{2} = -2.$$

$$y = 4 \quad \text{Simplify.}$$

A solution is $(-4, 4)$. This pair and two others satisfying the equation are shown in the table of values in Figure 3-23. If we plot the points and draw a line through them, we will obtain the graph shown in the figure. Check this work with a graphing calculator.

FIGURE 3-23

$$y = 2 - \frac{x}{2}$$

x	y	(x, y)
-4	4	$(-4, 4)$
0	2	$(0, 2)$
4	0	$(4, 0)$

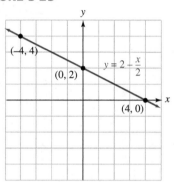

SELF CHECK   Solve $3y = 3 + x$ for y, then graph the equation.   *Answer:* $y = 1 + \frac{x}{3}$

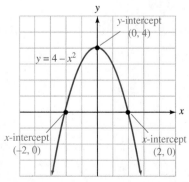

The Intercept Method

In Figure 3-24(a), the graph of $3x + 4y = 12$ intersects the y-axis at the point $(0, 3)$; we call this point the **y-intercept** of the graph. Since the graph intersects the x-axis at

FIGURE 3-24

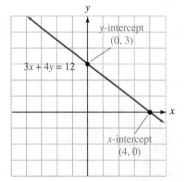

(a)

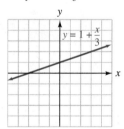

(b)

(4, 0), the point (4, 0) is the **x-intercept.** In Figure 3-24(b), we see that the graph of $y = 4 - x^2$ has two x-intercepts and one y-intercept.

In general, we have the following definitions.

y- and x-intercepts

The **y-intercept** of a line is the point $(0, b)$ where the line intersects the y-axis. To find b, substitute 0 for x in the equation of the line and solve for y.

The **x-intercept** of a line is the point $(a, 0)$ where the line intersects the x-axis. To find a, substitute 0 for y in the equation of the line and solve for x.

Plotting the x- and y-intercepts of a graph and drawing a line through them is called the **intercept method of graphing a line.** This method is useful when graphing equations written in general form.

E X A M P L E   5

The intercept method. Graph $3x - 2y = 8$.

Solution   To find the x-intercept, we let $y = 0$ and solve for x.

$$3x - 2y = 8$$
$$3x - 2(0) = 8 \qquad \text{Substitute 0 for } y.$$
$$3x = 8 \qquad \text{Simplify the left-hand side.}$$
$$x = \frac{8}{3} \qquad \text{Divide both sides by 3.}$$
$$x = 2\frac{2}{3} \qquad \text{Write } \tfrac{8}{3} \text{ as a mixed number.}$$

The x-intercept is $\left(2\frac{2}{3}, 0\right)$. This ordered pair is entered in the table in Figure 3-25. To find the y-intercept, we let $x = 0$ and solve for y.

$$3x - 2y = 8$$
$$3(0) - 2y = 8 \qquad \text{Substitute 0 for } x.$$
$$-2y = 8 \qquad \text{Simplify the left-hand side.}$$
$$y = -4 \qquad \text{Divide both sides by } -2.$$

The y-intercept is $(0, -4)$. It is entered in the table below. As a check, we find one more point on the line. If $x = 4$, then $y = 2$. We plot these three points and draw a line through them. The graph of $3x - 2y = 8$ is shown in Figure 3-25.

FIGURE 3-25

$$3x - 2y = 8$$

x	y	(x, y)
$2\frac{2}{3}$	0	$\left(2\frac{2}{3}, 0\right)$
0	-4	$(0, -4)$
4	2	$(4, 2)$

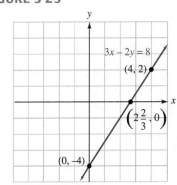

SELF CHECK   Graph $4x + 3y = 6$ using the intercept method.   *Answer:*

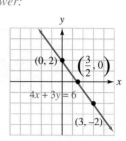

Graphing Horizontal and Vertical Lines

Equations such as $y = 4$ and $x = -3$ are linear equations, because they can be written in the general form $Ax + By = C$.

$y = 4$ is equivalent to $0x + 1y = 4$

$x = -3$ is equivalent to $1x + 0y = -3$

We now discuss how to graph these types of linear equations.

EXAMPLE 6

Graphing horizontal lines. Graph $y = 4$.

Solution   We can write the equation in general form as $0x + y = 4$. Since the coefficient of x is zero, the numbers chosen for x have no effect on y. The value of y is always 4. For example, if $x = 2$, we have

$0x + y = 4$ The original equation written in general form.

$0(2) + y = 4$ Substitute 2 for x.

$y = 4$ Simplify the left-hand side.

The table of values shown in Figure 3-26 contains three ordered pairs that satisfy the equation $y = 4$. If we plot the points and draw a line through them, the result is a horizontal line. The y-intercept is $(0, 4)$, and there is no x-intercept.

FIGURE 3-26

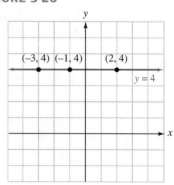

$y = 4$

x	y	(x, y)
2	4	$(2, 4)$
-1	4	$(-1, 4)$
-3	4	$(-3, 4)$

SELF CHECK   Graph $y = -2$.   *Answer:*

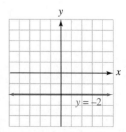

EXAMPLE 7

Graphing vertical lines. Graph $x = -3$.

Solution   We can write the equation in general form as $x + 0y = -3$. Since the coefficient of y is zero, the numbers chosen for y have no effect on x. The value of x is always -3. For example, if $y = -2$, we have

$$x + 0y = -3 \quad \text{The original equation written in general form.}$$
$$x + 0(-2) = -3 \quad \text{Substitute } -2 \text{ for } y.$$
$$x = -3 \quad \text{Simplify the left-hand side.}$$

The table of values shown in Figure 3-27 contains three ordered pairs that satisfy the equation $x = -3$. If we plot the points and draw a line through them, the result is a vertical line. The x-intercept is $(-3, 0)$, and there is no y-intercept.

FIGURE 3-27

$x = -3$

x	y	(x, y)
-3	-2	$(-3, -2)$
-3	0	$(-3, 0)$
-3	3	$(-3, 3)$

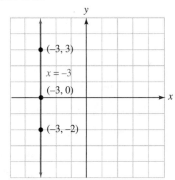

SELF CHECK   Graph $x = 4$.   *Answer:*

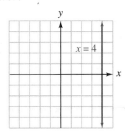

From the results of Examples 6 and 7, we have the following facts.

Equation of horizontal and vertical lines

The equation $y = b$ represents the horizontal line that intersects the y-axis at $(0, b)$. If $b = 0$, the line is the x-axis.

The equation $x = a$ represents the vertical line that intersects the x-axis at $(a, 0)$. If $a = 0$, the line is the y-axis.

An Application of Linear Equations

EXAMPLE 8

Birthday parties. A restaurant offers a party package that includes food, drinks, cake, and party favors for a cost of $25 plus $3 per child. Write a linear equation that will give the cost for a party of any size, and then graph the equation.

Solution　　We can let c represent the cost of the party. The cost c is the sum of the basic charge of \$25 and the cost per child times the number of children attending. If the number of children attending is n, at \$3 per child, the total cost for the children is \$3n.

The cost	is	the basic \$25 charge	plus	\$3	times	the number of children.
c	$=$	25	$+$	3	$\cdot$	n

For the equation $c = 25 + 3n$, the independent variable (input) is n, the number of children. The dependent variable (output) is c, the cost of the party. We will find three points on the graph of the equation by choosing n-values of 0, 5, and 10 and finding the corresponding c-values. The results are recorded in the table.

If $n = 0$	**If $n = 5$**	**If $n = 10$**
$c = 25 + 3(0)$	$c = 25 + 3(5)$	$c = 25 + 3(10)$
$c = 25$	$c = 25 + 15$	$c = 25 + 30$
	$c = 40$	$c = 55$

$c = 25 + 3n$

n	c
0	25
5	40
10	55

Next, we graph the points and draw a line through them (Figure 3-28). Note that the c-axis is scaled in units of \$5 to accommodate costs ranging from \$0 to \$65. We don't draw an arrowhead on the left, because it doesn't make sense to have a negative number of children attend a party. We can use the graph to determine the cost of a party of any size. For example, to find the cost of a party with 8 children, we locate 8 on the horizontal axis and then move up to find a point on the graph directly above the 8. Since the coordinates of that point are $(8, 49)$, the cost for 8 children would be \$49.

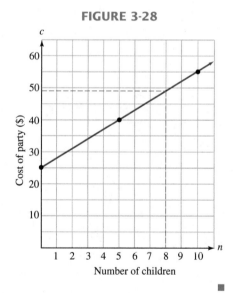

FIGURE 3-28

■

Solving Equations Graphically

Some of the graphing concepts discussed in this chapter can be used to solve equations. For example, the solution of $-2x - 4 = 0$ is the number x that will make y equal to 0 in the equation $y = -2x - 4$. To find this number, we inspect the graph of $y = -2x - 4$ and locate the point on the graph that has a y-coordinate of 0. In Figure 3-29, we see that the point is $(-2, 0)$, which is the x-intercept of the graph. We can conclude that the x-coordinate of the x-intercept, $x = -2$, is the solution of $-2x - 4 = 0$.

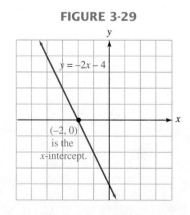

FIGURE 3-29

ACCENT ON TECHNOLOGY *Solving Equations Graphically*

To solve the equation $2(x - 3) - 8 = -2x$ using a graphing calculator, we first add $2x$ to both sides so that the right-hand side is 0.

$$2(x - 3) - 8 = -2x$$
$$2(x - 3) - 8 + 2x = -2x + 2x$$
$$2(x - 3) - 8 + 2x = 0$$

Next, we enter the left-hand side of the equation into the calculator in the form $y = 2(x - 3) - 8 + 2x$. We do not need to simplify the left-hand side to graph it. See Figure 3-30(a).

If we use window settings where x is between -5 and 5 and y is between -5 and 5 and press $\boxed{\text{GRAPH}}$, we will obtain the graph in Figure 3-30(b).

To find the x-intercept, we trace as shown in Figure 3-30(c). After repeated zooms and traces, we will see that the x-coordinate of the x-intercept is 3.5, so $x = 3.5$ is the solution of $2(x - 3) - 8 = -2x$.

FIGURE 3-30

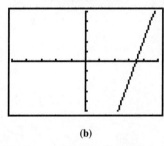

(a)

(b)

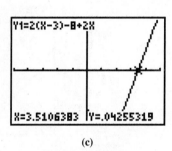

(c)

An easier way to find the solution of $2(x - 3) - 8 + 2x = -2x$ is using the zero feature, found under the CALC menu. With this option, the cursor automatically locates the x-intercept of the graph of $y = 2(x - 3) - 8 + 2x$ and displays its coordinates. See Figure 3-30(d). The x-coordinate of the x-intercept of the graph is called a *zero* of $y = 2(x - 3) - 8 + 2x$, and the zero (in this case, 3.5) is the solution of the given equation. Consult your owner's manual for specific instructions on how to use this feature.

FIGURE 3-30(d)

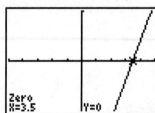

STUDY SET

Section 3.3

VOCABULARY

In Exercises 1–6, fill in the blanks to make the statements true.

1. An equation whose graph is a line and whose variables are to the first power is called a ___linear___ equation.

2. The equation $Ax + By = C$ is the ___standard or general___ form of the equation of a line.

3. The ___y-intercept___ of a line is the point $(0, b)$ where the line intersects the y-axis.

4. The ___x-intercept___ of a line is the point $(a, 0)$ where the line intersects the x-axis.

5. Lines _____parallel_____ to the y-axis are vertical lines.

6. Lines parallel to the x-axis are _____horizontal_____ lines.

CONCEPTS

7. Classify each equation as linear or nonlinear.
  a. $y = x^3$ nonlinear
  b. $2x + 3y = 6$ linear
  c. $y = |x + 2|$ nonlinear
  d. $x = -2$ linear
  e. $y = 2x - x^2$ nonlinear

8. What information can be obtained by finding the x- and y-intercepts of the graph shown in Illustration 1?
  The equipment originally cost $25,000; in 10 yr, it will be worth nothing.

ILLUSTRATION 1

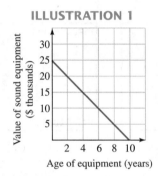

In Exercises 9–12, complete each table of values.

9. $5y = 2x + 10$

x	y
10	6
-5	0
5	4

10. $2x + 4y = 24$

x	y
4	4
-2	7
-4	8

11. $x - 2y = 4$

x	y
0	-2
4	0
1	$-\frac{3}{2}$

12. $5x - y = 3$

x	y
0	-3
$\frac{3}{5}$	0
1	2

In Exercises 13–14, consider the graph of a linear equation shown in Illustration 2.

13. Why will the coordinates of point A, when substituted into the equation, yield a true statement?
  because A is on the line

14. Why will the coordinates of point B, when substituted into the equation, yield a false statement?
  because B is not on the line

ILLUSTRATION 2

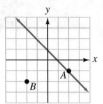

15. To what power is each variable in the equation $y = 2x - 6$?   1st power

16. To what power is each variable in the equation $y = x^2 - 6$? y: 1st power; x: 2nd power

17. Give the x- and y-intercepts of the graph in Illustration 3. x-intercept: $(-3, 0)$; y-intercept: $(0, -1)$

ILLUSTRATION 3

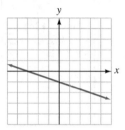

18. Give the x- and y-intercepts of the graph in Illustration 4. x-intercepts: $(4, 0)$, $(-4, 0)$; y-intercept: $(0, 2)$

ILLUSTRATION 4

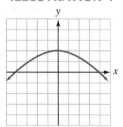

19. What is the general form of an equation of a horizontal line? $y = b$

20. What is the general form of an equation of a vertical line? $x = a$

21. A student found three solutions of a linear equation and plotted them as shown in Illustration 5. What conclusion can be made?
  The student made a mistake; the points should lie on a line.

ILLUSTRATION 5

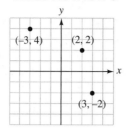

22. On the same coordinate system:
   a. Draw the graph of a line with no x-intercept.
   b. Draw the graph of a line with no y-intercept.
   c. Draw a line with an x-intercept of $(2, 0)$.
   d. Draw a line with a y-intercept of $\left(0, -\frac{5}{2}\right)$.

Answers may vary.

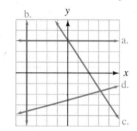

In Exercises 23–24, fill in the blanks to make the statements true.

23. To find the y-intercept of the graph of a linear equation, we let $\boxed{x} = 0$ and solve for $\boxed{y}$.

24. To find the x-intercept of the graph of a linear equation, we let $\boxed{y} = 0$ and solve for $\boxed{x}$.

25. Consider the linear equation $y = 6x$.
   a. Find the x-intercept of the graph. $(0, 0)$
   b. Find the y-intercept of the graph. $(0, 0)$
   c. Explain why your answers to parts a and b are not enough information to graph the line.
      It takes two distinct points to determine a line.

26. a. What is another name for the line $x = 0$?
      the y-axis
   b. What is another name for the line $y = 0$?
      the x-axis

27. How can the solution of the equation $-3x - 3 = 0$ be determined from the graphing calculator display of the graph of $y = -3x - 3$? What is the solution?
   The x-coordinate of the x-intercept is the solution: -1.

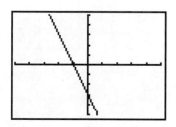

28. What is the solution of the equation $5x - 3(x + 1) = x$ if the calculator display shows the graph of $y = 5x - 3(x + 1) - x$? 3

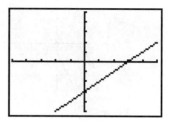

NOTATION

29. Write each equation in general form.
   a. $-4x = -y - 6$ $4x - y = 6$
   b. $y = \dfrac{1}{2}x$ $x - 2y = 0$
   c. $3 = \dfrac{x}{3} + y$ $x + 3y = 9$
   d. $x = 12$ $x + 0y = 12$

30. Solve each equation for y.
   a. $x + y = 8$ $y = 8 - x$
   b. $2x - y = 8$ $y = 2x - 8$
   c. $3x + \dfrac{y}{2} = 4$ $y = -6x + 8$
   d. $y - 2 = 0$ $y = 2$

PRACTICE

In Exercises 31–34, find three solutions of the equation, then graph it.

31. $y = -x + 2$

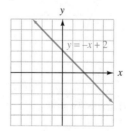

32. $y = -x - 1$

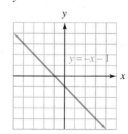

33. $y = 2x + 1$

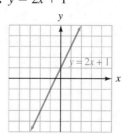

34. $y = 3x - 2$

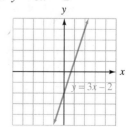

In Exercises 35–38, solve each equation for y, find three solutions of the equation, and then graph it.

35. $2y = 4x - 6$

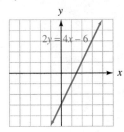

36. $3y = 6x - 3$

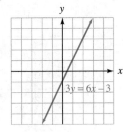

37. $2y = x - 4$

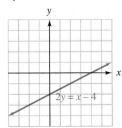

38. $4y = x + 16$

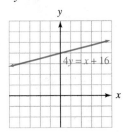

In Exercises 39–46, graph each equation using the intercept method.

39. $2y - 2x = 6$

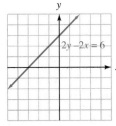

40. $3x - 3y = 9$

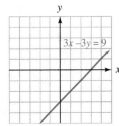

41. $15y + 5x = -15$

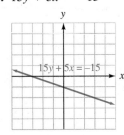

42. $8x + 4y = -24$

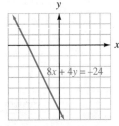

43. $3x + 4y = 8$

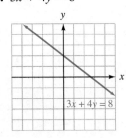

44. $2x + 3y = 9$

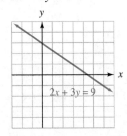

45. $-4y + 9x = -9$

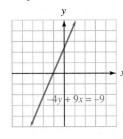

46. $-4y + 5x = -15$

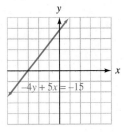

In Exercises 47–58, graph each equation.

47. $y = 4$

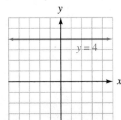

48. $y = -3$

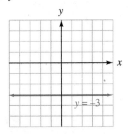

49. $x = -2$

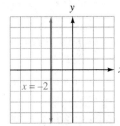

50. $x = 5$

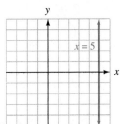

51. $y = -\dfrac{1}{2}$

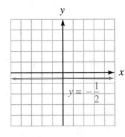

52. $y = \dfrac{5}{2}$

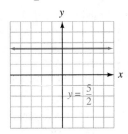

53. $x = \dfrac{4}{3}$

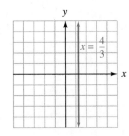

54. $x = -\dfrac{5}{3}$

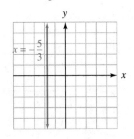

55. $y = 2x$

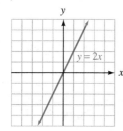

56. $y = 3x$

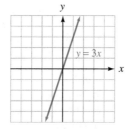

57. $y = -2x$

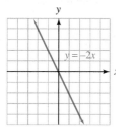

58. $y = -3x$

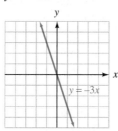

 In Exercises 59–62, solve each equation graphically.

59. $3(x + 2) - x = 12$ 3

60. $-(3x + 1) + x = 2$ −1.5

61. $10 - 5x = 3x + 18$ −1

62. $0.5x + 0.2(12 - x) = 0.3(12)$ 4

APPLICATIONS

63. EDUCATION COSTS Each semester, a college charges a services fee of $50 plus $25 for each unit taken by a student.

   a. Write a linear equation that gives the total enrollment cost c for a student taking u units.
   $c = 50 + 25u$

   b. Complete the table of values and graph the equation. (See Illustration 6.)

   c. What does the y-intercept of the line tell you?
   The service fee is $50.

   d. Use the graph to find the total cost for a student taking 18 units the first semester and 12 units the second semester. $850

ILLUSTRATION 6

u	c
4	150
8	250
14	400

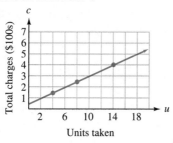

Units taken

64. GROUP RATES To promote the sale of tickets for a cruise to Alaska, a travel agency reduces the regular

ticket price of $3,000 by $5 for each individual traveling in the group.

   a. Write a linear equation that would find the ticket price t for the cruise if a group of p people travel together. $t = 3,000 - 5p$

   b. Complete the table of values and graph the equation. (See Illustration 7.)

   c. As the size of the group increases, what happens to the ticket price?   It decreases.

   d. Use the graph to determine the cost of an individual ticket if a group of 25 will be traveling together. $2,875

ILLUSTRATION 7

p	t
10	2,950
30	2,850
60	2,700

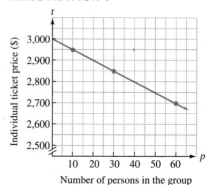

Number of persons in the group

65. PHYSIOLOGY Physiologists have found that a woman's height h in inches can be approximated using the linear equation $h = 3.9r + 28.9$, where r represents the length of her radius bone in inches.

   a. Complete the table of values in Illustration 8. Round to the nearest tenth and then graph the equation.

   b. Complete this sentence: From the graph, we see that the longer the radius bone, the . . . .
   taller the woman is.

   c. From the graph, estimate the height of a woman whose radius bone is 7.5 inches long.   58 in.

ILLUSTRATION 8

r	h
7	56.2
8.5	62.1
9	64.0

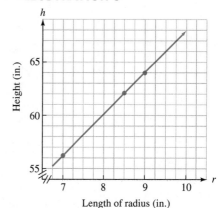

Length of radius (in.)

66. RESEARCH EXPERIMENT A psychology major found that the time t (in seconds) that it took a white rat to complete a maze was related to the number of trials n the rat had been given. The resulting equation was $t = 25 - 0.25n$.

   a. Complete the table of values in Illustration 9 and then graph the equation.

   b. Complete this sentence: From the graph, we see that the more trials the rat had, the . . . .

   less time it took it to complete the maze.

   c. From the graph, estimate the time it will take the rat to complete the maze on its 32nd trial.   17 sec

ILLUSTRATION 9

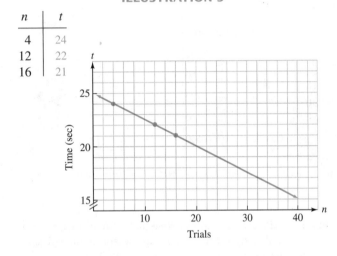

n	t
4	24
12	22
16	21

WRITING

67. A linear equation and a graph are two ways of mathematically describing a relationship between two quantities. Which do you think is more informative and why?

68. From geometry, we know that two points determine a line. Explain why it is a good practice when graphing linear equations to find and plot three points instead of just two.

69. How can we tell by looking at an equation if its graph will be a straight line?

70. Can the x-intercept and the y-intercept of a line be the same point? Explain.

REVIEW

71. Simplify $-(-5 - 4c)$. $5 + 4c$

72. List the integers. $\{ . . . , -3, -2, -1, 0, 1, 2, 3, . . . \}$

73. Solve $\dfrac{x + 6}{2} = 1$. -4

74. Evaluate $-2^2 + 2^2$. 0

75. Write a formula that relates profit, revenue, and costs.
   profit = revenue − costs

76. Find the volume, to the nearest tenth, of a sphere with radius 6 feet.   904.8 ft³

77. Evaluate $1 + 2[-3 - 4(2 - 8^2)]$.   491

78. Evaluate $\dfrac{x + y}{x - y}$ if $x = -2$ and $y = -4$. -3

▶ **3.4**

Rate of Change and the Slope of a Line

In this section, you will learn about

   Rates of change ▪ Slope of a line ▪ The slope formula ▪ Positive and negative slope ▪ Slopes of horizontal and vertical lines ▪ Using slope to graph a line

Introduction   Since our world is one of constant change, we must be able to describe change so that we can plan effectively for the future. In this section, we will show how to describe the amount of change of one quantity in relation to the amount of change of another quantity by finding a *rate of change*.

Rates of Change

The line graph in Figure 3-31(a) shows the number of business permits issued each month by a city over a 12-month period. From the shape of the graph, we can see that the number of permits issued *increased* each month.

For situations such as the one graphed in Figure 3-31(a), it is often useful to calculate a rate of increase (called a **rate of change**). We do so by finding the **ratio** of the change in the number of business permits issued each month to the number of months over which that change took place.

FIGURE 3-31

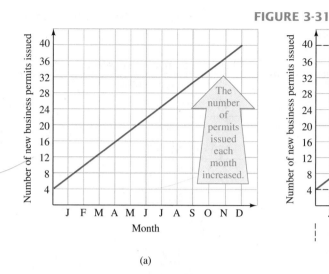

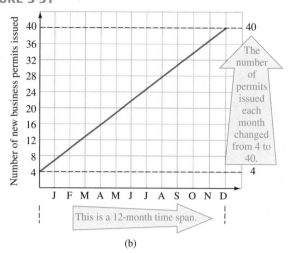

(a)

(b)

Ratios and rates

A **ratio** is a comparison of two numbers by their indicated quotient. In symbols, if a and b are two numbers, the ratio of a to b is $\frac{a}{b}$. Ratios that are used to compare quantities with different units are called **rates.**

In Figure 3-31(b), we see that the number of permits issued prior to the month of January was 4. By the end of the year, the number of permits issued during the month of December was 40. This is a change of $40 - 4$, or 36, over a 12-month period. So we have

$$\text{Rate of change} = \frac{\text{change in number of permits issued each month}}{\text{change in time}} \qquad \text{The rate of change is a ratio.}$$

$$= \frac{36 \text{ permits}}{12 \text{ months}}$$

$$= \frac{\overset{1}{\cancel{12}} \cdot 3 \text{ permits}}{\underset{1}{\cancel{12}} \text{ months}} \qquad \text{Factor 36 as } 12 \cdot 3 \text{ and divide out the common factor of 12.}$$

$$= \frac{3 \text{ permits}}{1 \text{ month}}$$

The number of business permits being issued increased at a rate of 3 per month, denoted as 3 permits/month.

FIGURE 3-32

EXAMPLE 1

Finding rate of change.
The graph in Figure 3-32 shows the number of subscribers to a newspaper. Find the rate of change in the number of subscribers over the first 5-year period. Write the rate in simplest form.

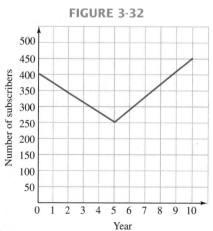

Solution   We need to write the ratio of the change in the number of subscribers over the change in time.

$$\text{Rate of change} = \frac{\text{change in number of subscribers}}{\text{change in time}} \qquad \text{Set up the ratio.}$$

$$= \frac{(250 - 400) \text{ subscribers}}{5 \text{ years}} \qquad \begin{array}{l}\text{Subtract the earlier number of}\\\text{subscribers from the later}\\\text{number of subscribers.}\end{array}$$

$$= \frac{-150 \text{ subscribers}}{5 \text{ years}} \qquad 250 - 400 = -150$$

$$= \frac{-30 \cdot \overset{1}{\cancel{5}} \text{ subscribers}}{\underset{1}{\cancel{5}} \text{ years}} \qquad \begin{array}{l}\text{Factor } -150 \text{ as } -30 \cdot 5 \text{ and}\\\text{divide out the common factor}\\\text{of 5.}\end{array}$$

$$= \frac{-30 \text{ subscribers}}{1 \text{ year}}$$

The number of subscribers for the first 5 years *decreased* by 30 per year, as indicated by the negative sign in the result. We can write this as -30 subscribers/year.

SELF CHECK   Find the rate of change in the number of subscribers over the second 5-year period. Write the rate in simplest form.   *Answer:* 40 subscribers/year   ∎

Slope of a Line

The **slope** of a nonvertical line is a number that measures the line's steepness. We can calculate the slope by picking two points on the line and writing the ratio of the vertical change (called the **rise**) to the corresponding horizontal change (called the **run**) as we move from one point to the other. As an example, we will find the slope of the line that was used to describe the number of building permits issued and show that it gives the rate of change.

In Figure 3-33 (a modified version of Figure 3-31(a)), the line passes through points $P(0, 4)$ and $Q(12, 40)$. Moving along the line from point P to point Q causes the value of y to change from $y = 4$ to $y = 40$, an increase of $40 - 4 = 36$ units. We say that the *rise* is 36.

FIGURE 3-33

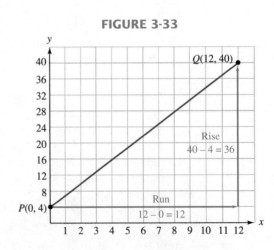

Moving from point P to point Q, the value of x increases from $x = 0$ to $x = 12$, an increase of $12 - 0 = 12$ units. We say that the *run* is 12. The slope of a line, usually denoted with the letter m, is defined to be the ratio of the change in y to the change in x.

$$m = \frac{\text{change in } y\text{-values}}{\text{change in } x\text{-values}} \qquad \text{Slope is a ratio.}$$

$$= \frac{40 - 4}{12 - 0} \qquad \begin{array}{l}\text{To find the change in } y \text{ (the rise), subtract the } y\text{-values.}\\ \text{To find the change in } x \text{ (the run), subtract the } x\text{-values.}\end{array}$$

$$= \frac{36}{12} \qquad \text{Do the subtractions.}$$

$$= 3 \qquad \text{Do the division.}$$

This is the same value we obtained when we found the rate of change of the number of business permits issued over the 12-month period. Therefore, by finding the slope of the line, we found a rate of change.

The Slope Formula

The slope of a line can be described in several ways.

$$\text{Slope} = m = \frac{\text{vertical change}}{\text{horizontal change}} = \frac{\text{rise}}{\text{run}} = \frac{\text{change in } y}{\text{change in } x}$$

To distinguish between the coordinates of two points, say points P and Q (see Figure 3-34), we often use **subscript notation.**

■ Point P is denoted as $P(x_1, y_1)$. Read as "point P with coordinates of x sub 1 and y sub 1."

■ Point Q is denoted as $Q(x_2, y_2)$. Read as "point Q with coordinates of x sub 2 and y sub 2."

FIGURE 3-34

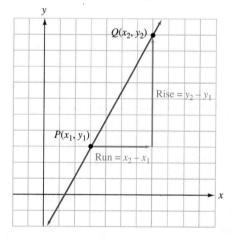

As a point on the line in Figure 3-34 moves from P to Q, its y-coordinate changes by the amount $y_2 - y_1$ (the rise), while its x-coordinate changes by $x_2 - x_1$ (the run). Since the slope is the ratio $\frac{\text{rise}}{\text{run}}$, we have the following formula for calculating slope.

Slope of a nonvertical line

If $P(x_1, y_1)$ and $Q(x_2, y_2)$ are two points on a nonvertical line, the slope m of line PQ is given by the formula

$$m = \frac{y_2 - y_1}{x_2 - x_1}$$

EXAMPLE 2

Using the slope formula. Find the slope of line l_1 shown in Figure 3-35.

FIGURE 3-35

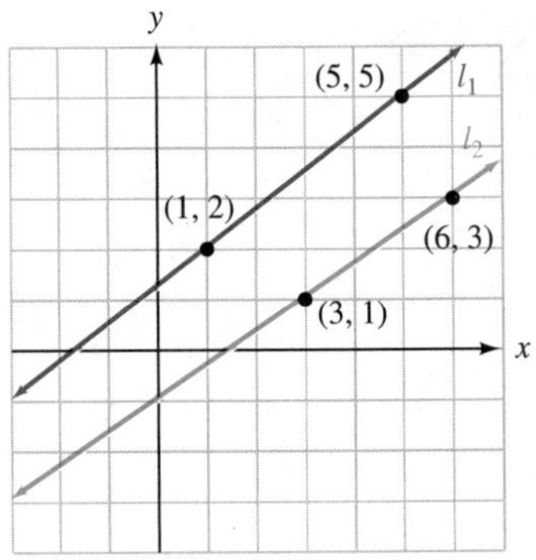

Solution   To find the slope of l_1, we will use two points on the line whose coordinates are given: $(1, 2)$ and $(5, 5)$. If (x_1, y_1) is $(1, 2)$ and (x_2, y_2) is $(5, 5)$, then

$$x_1 = 1 \quad \text{and} \quad x_2 = 5$$
$$y_1 = 2 \qquad\qquad y_2 = 5$$

To find the slope of line l_1, we substitute these values into the formula for slope and simplify.

$$m = \frac{y_2 - y_1}{x_2 - x_1} \quad \text{The slope formula.}$$

$$= \frac{5 - 2}{5 - 1} \quad \text{Substitute 5 for } y_2, \text{ 2 for } y_1, \text{ 5 for } x_2, \text{ and 1 for } x_1.$$

$$= \frac{3}{4} \quad \text{Do the subtractions.}$$

The slope of l_1 is $\frac{3}{4}$. We would have obtained the same result if we had let $(x_1, y_1) = (5, 5)$ and $(x_2, y_2) = (1, 2)$.

SELF CHECK   Find the slope of line l_2 shown in Figure 3-35.   *Answer:* $\frac{2}{3}$ ■

WARNING! When finding the slope of a line, always subtract the y-values and the x-values in the same order. Otherwise your answer will have the wrong sign:

$$m = \frac{y_2 - y_1}{x_2 - x_1} \quad \text{or} \quad m = \frac{y_1 - y_2}{x_1 - x_2}$$

However,

$$m \neq \frac{y_2 - y_1}{x_1 - x_2} \quad \text{and} \quad m \neq \frac{y_1 - y_2}{x_2 - x_1}$$

EXAMPLE 3

Using the slope formula. Find the slope of the line that passes through $(-2, 4)$ and $(5, -6)$ and draw its graph.

Solution   Since we know the coordinates of two points on the line, we can find its slope. If (x_1, y_1) is $(-2, 4)$ and (x_2, y_2) is $(5, -6)$, then

$$x_1 = -2 \quad \text{and} \quad x_2 = 5$$
$$y_1 = 4 \qquad\qquad y_2 = -6$$

FIGURE 3-36

$$m = \frac{y_2 - y_1}{x_2 - x_1} \quad \text{The slope formula.}$$

$$m = \frac{-6 - 4}{5 - (-2)} \quad \text{Substitute } -6 \text{ for } y_2, \text{ 4 for } y_1, \text{ 5 for } x_2, \text{ and } -2 \text{ for } x_1.$$

$$m = -\frac{10}{7} \quad \begin{array}{l}\text{Simplify the numerator: } -6 - 4 = -10.\\ \text{Simplify the denominator: } 5 - (-2) = 7.\end{array}$$

The slope of the line is $-\frac{10}{7}$. Figure 3-36 shows the graph of the line. Note that the line "falls" from left to right—a fact that is indicated by its negative slope.

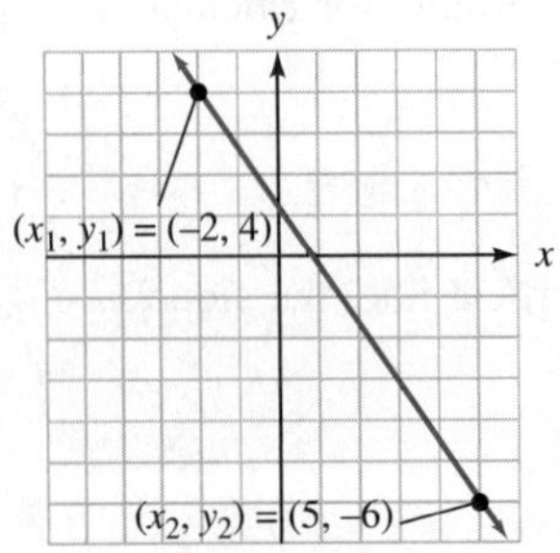

SELF CHECK Find the slope of the line that passes through $(-1, -2)$ and $(1, -7)$. *Answer:* $-\frac{5}{2}$ ∎

Positive and Negative Slope

In Example 2, the slope of line l_1 was positive $\left(\frac{3}{4}\right)$. In Example 3, the slope of the line was negative $\left(-\frac{10}{7}\right)$. In general, lines that rise from left to right have a positive slope, and lines that fall from left to right have a negative slope, as shown in Figure 3-37.

FIGURE 3-37

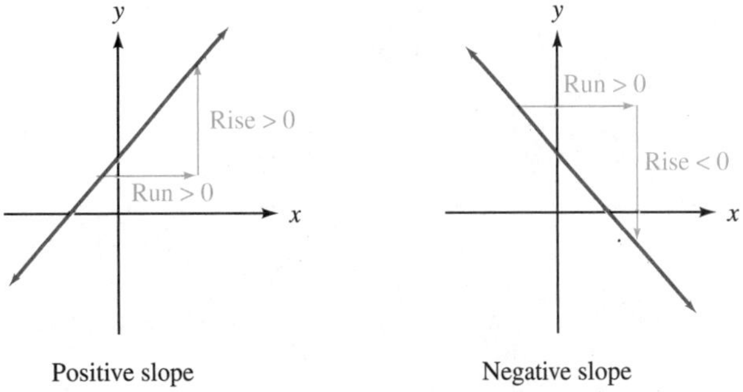

Positive slope

(a)

Negative slope

(b)

Slopes of Horizontal and Vertical Lines

In the next two examples, we will calculate the slope of a horizontal line and show that a vertical line has no defined slope.

EXAMPLE 4

Slope of a horizontal line. Find the slope of the line $y = 3$.

Solution To find the slope of the line $y = 3$, we need to know two points on the line. In Figure 3-38, we graph the horizontal line $y = 3$ and label two points on the line: $(-2, 3)$ and $(3, 3)$.

If (x_1, y_1) is $(-2, 3)$ and (x_2, y_2) is $(3, 3)$, we have

$m = \dfrac{y_2 - y_1}{x_2 - x_1}$ The slope formula.

$m = \dfrac{3 - 3}{3 - (-2)}$ Substitute 3 for y_2, 3 for y_1, 3 for x_2, and -2 for x_1.

$m = \dfrac{0}{5}$ Simplify the numerator and the denominator.

$m = 0$

FIGURE 3-38

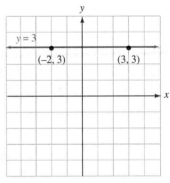

The slope of the line $y = 3$ is 0.

SELF CHECK Find the slope of the line $y = -5$. *Answer:* $m = 0$ ∎

The y-values of any two points on any horizontal line will be the same, and the x-values will be different. Thus, the numerator of

$\dfrac{y_2 - y_1}{x_2 - x_1}$

will always be zero, and the denominator will always be nonzero. Therefore, the slope of a horizontal line is zero.

| EXAMPLE 5 | **Slope of a vertical line.** If possible, find the slope of the line $x = -2$. |

FIGURE 3-39

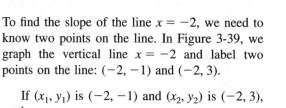

Solution To find the slope of the line $x = -2$, we need to know two points on the line. In Figure 3-39, we graph the vertical line $x = -2$ and label two points on the line: $(-2, -1)$ and $(-2, 3)$.

If (x_1, y_1) is $(-2, -1)$ and (x_2, y_2) is $(-2, 3)$, we have

$$m = \frac{y_2 - y_1}{x_2 - x_1} \qquad \text{The slope formula.}$$

$$m = \frac{3 - (-1)}{-2 - (-2)} \qquad \begin{array}{l}\text{Substitute 3 for } y_2, -1 \text{ for } y_1,\\ -2 \text{ for } x_2, \text{ and } -2 \text{ for } x_1.\end{array}$$

$$m = \frac{4}{0} \qquad \text{Simplify the numerator and the denominator.}$$

Since division by zero is undefined, $\frac{4}{0}$ has no meaning. The slope of the line $x = -2$ is undefined.

SELF CHECK If possible, find the slope of the line $x = 1$. *Answer:* undefined slope ■

The y-values of any two points on a vertical line will be different, and the x-values will be the same. Thus, the numerator of

$$\frac{y_2 - y_1}{x_2 - x_1}$$

will always be nonzero, and the denominator will always be zero. Therefore, the slope of a vertical line is undefined.

We now summarize the results from Examples 4 and 5.

Slopes of horizontal and vertical lines

Horizontal lines (lines with equations of the form $y = b$) have a slope of zero.

Vertical lines (lines with equations of the form $x = a$) have undefined slope.

FIGURE 3-40

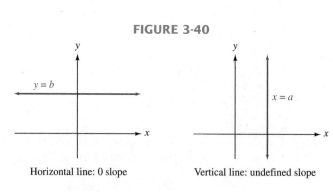

Horizontal line: 0 slope

(a)

Vertical line: undefined slope

(b)

FIGURE 3-41

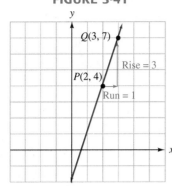

Using Slope to Graph a Line

We can graph a line whenever we know the coordinates of one point on the line and the slope of the line. For example, to graph the line that passes through $P(2, 4)$ and has a slope of 3, we first plot $P(2, 4)$, as in Figure 3-41. We can express the slope of 3 as a fraction: $3 = \frac{3}{1}$. Therefore, the line *rises* 3 units for every 1 unit it *runs* to the right. We can find a second point on the line by starting at $P(2, 4)$ and moving 1 unit to the right (run) and then 3 units up (rise). This brings us to a point that we will call Q with coordinates $(2 + \mathbf{1}, 4 + \mathbf{3})$ or $(3, 7)$. The required line must pass through points P and Q.

EXAMPLE 6

Using slope to graph a line. Graph the line that passes through the point $(-3, 4)$ with slope $-\frac{2}{5}$.

Solution We plot the point $(-3, 4)$ as shown in Figure 3-42. Then, after writing slope $-\frac{2}{5}$ as $\frac{-2}{5}$, we see that the *rise* is -2 and the *run* is 5. From the point $(-3, 4)$, we can find a second point on the line by moving 5 units to the right (run) and then 2 units down (a rise of -2 means to move down 2 units). This brings us to the point with coordinates of $(-3 + 5, 4 - 2) = (2, 2)$. We then draw a line that passes through the two points.

FIGURE 3-42

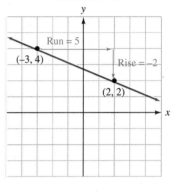

SELF CHECK

Graph the line that passes through the point $(-4, 2)$ with slope -4.

Answer:

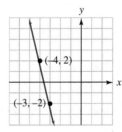

STUDY SET

Section 3.4

VOCABULARY

In Exercises 1–6, fill in the blanks to make the statements true.

1. A ____ratio____ is a comparison of two numbers by their indicated quotient.

2. Ratios used to compare quantities with different units are called ____rates____.

3. The ____slope____ of a line is defined to be the ratio of the change in y to the change in x.

4. $m = \dfrac{\text{vertical change}}{\text{horizontal change}} = \dfrac{\text{rise}}{\text{run}} = \dfrac{\text{change in } y}{\text{change in } x}$

5. The rate of ____change____ of a linear relationship can be found by finding the slope of the graph of the line.

6. ____Horizontal____ lines have a slope of zero. Vertical lines have ____undefined____ slope.

CONCEPTS

7. Which line graphed in Illustration 1 has

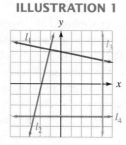

ILLUSTRATION 1

   a. a positive slope? l_2
   b. a negative slope? l_1
   c. zero slope? l_4
   d. undefined slope? l_3

8. For the line graphed in Illustration 2:
   a. Find its slope using points A and B. $\frac{1}{2}$
   b. Find its slope using points B and C. $\frac{1}{2}$
   c. Find its slope using points A and C. $\frac{1}{2}$
   d. What observation is suggested by your answers to parts a, b, and c?

> When finding the slope of a line, any two points on the line give the same result.

ILLUSTRATION 2

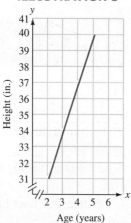

9. Use the information in the table of values for a linear equation to determine what the slope of the line would be if it was graphed. -1

x	y
-4	2
5	-7

10. Fill in the blanks to make the statements true.
   a. A line with positive slope ____rises____ from left to right.
   b. A line with negative slope ____falls____ from left to right.

11. GROWTH RATE Use the graph in Illustration 3 to find the rate of change of a boy's height over the time period shown. 3 in./yr

ILLUSTRATION 3

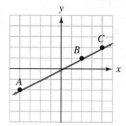

Age (years)

12. IRRIGATION The graph in Illustration 4 shows the number of gallons of water remaining in a reservoir as water is discharged from it to irrigate a field. Find the rate of change in the number of gallons of water for the time the field was being irrigated.

-875 gal/hr

ILLUSTRATION 4

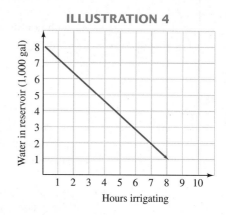

13. RECORD-SETTING FILMS Prior to the release of the movie *Titanic*, in 1997, the three movies that earned $100 million at the box office the fastest are shown in Illustration 5.
   a. Which film reached the $100-million mark fastest? Explain how you can tell.

> "The Lost World"; it reached $100 million in the fewest days, 6.

   b. Find the rate of earnings of each film by finding the slope of each line. Round to the nearest tenth.

> LW: $\frac{100}{6} \approx \$16.7$ million/day; ID: $\frac{100}{7} \approx \$14.3$ million/day; JP: $\frac{100}{9} \approx \$11.1$ million/day

ILLUSTRATION 5

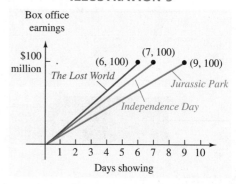

14. WAL-MART On the graph in Illustration 6, draw a straight line through the points (1991, 34) and (1998, 118). This line approximates Wal-Mart's annual net sales for the years 1991–1998. Find the rate of increase in sales over this period by finding the slope of the line. $12 billion/yr

ILLUSTRATION 6

Based on data from Wal-Mart and *USA TODAY* (November 6, 1998)

ILLUSTRATION 8

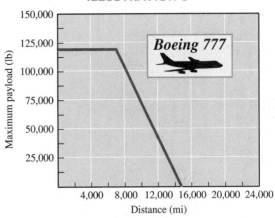

Based on data from Lawrence Livermore National Laboratory and *Los Angeles Times* (October 22, 1998)

15. THE UNCOLA On the graph in Illustration 7, draw a straight line through the points (90, 200) and (97, 200). This line approximates the number of cases of 7-Up sold annually for the years 1990–1997. Find the slope of the line. What important information does the slope give about 7-Up sales?

0; sales of 7-Up are not changing—each year the same number cases have been sold.

ILLUSTRATION 7

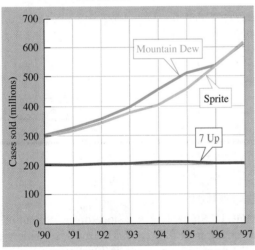

Based on data from *Beverage Digest* and *Los Angeles Times* (November 5, 1998)

16. COMMERCIAL JETS Examine the graph in Illustration 8 and consider trips of more than 7,000 miles by a Boeing 777. Use a rate of change to estimate how the maximum payload decreases as the distance traveled increases. Explain your result in words.

−15 lb/mi; for every mile, over 7,000 miles, that the plane travels, the maximum payload is reduced by about 15 pounds.

NOTATION

17. What is the formula used to find the slope of the line passing through (x_1, y_1) and (x_2, y_2)?

$m = \dfrac{y_2 - y_1}{x_2 - x_1}$ or $m = \dfrac{y_1 - y_2}{x_1 - x_2}$

18. Explain the difference between y^2 and y_2.

y^2 means $y \cdot y$ and y_2 means y sub 2.

PRACTICE

In Exercises 19–36, find the slope of the line passing through the given points, when possible.

19. (2, 4) and (1, 3)   1

20. (1, 3) and (2, 5)   2

21. (3, 4) and (2, 7)   −3

22. (3, 6) and (5, 2)   −2

23. (0, 0) and (4, 5) $\frac{5}{4}$

24. (4, 3) and (7, 8) $\frac{5}{3}$

25. (−3, 5) and (−5, 6) $-\frac{1}{2}$

26. (6, −2) and (−3, 2) $-\frac{4}{9}$

27. (−2, −2) and (−12, −8) $\frac{3}{5}$

28. (−1, −2) and (−10, −5) $\frac{1}{3}$

29. (5, 7) and (−4, 7)   0

30. (−1, −12) and (6, −12)   0

31. (8, −4) and (8, −3)   undefined

32. (−2, 8) and (−2, 15)   undefined

33. (−6, 0) and (0, −4) $-\frac{2}{3}$

34. (0, −9) and (−6, 0) $-\frac{3}{2}$

35. (−2.5, 1.75) and (−0.5, −7.75)   −4.75

36. (6.4, −7.2) and (−8.8, 4.2)   −0.75

In Exercises 37–40, find the slope of each line.

37. $m = \dfrac{2}{3}$

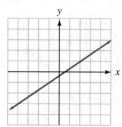

38. $m = \dfrac{4}{3}$

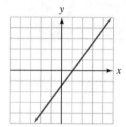

39. $m = -\dfrac{7}{8}$

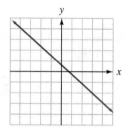

40. $m = -\dfrac{1}{5}$

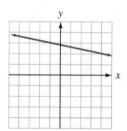

In Exercises 41–52, graph the line that passes through the given point and has the given slope.

41. $(0, 1)$, $m = 2$ **42.** $(-4, 1)$, $m = -3$

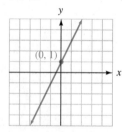   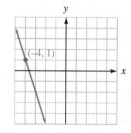

43. $(-3, -3)$, $m = -\dfrac{3}{2}$ **44.** $(-2, -1)$, $m = \dfrac{4}{3}$

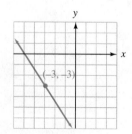   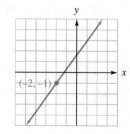

45. $(5, -3)$, $m = \dfrac{3}{4}$ **46.** $(2, -4)$, $m = \dfrac{2}{3}$

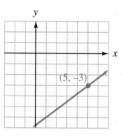   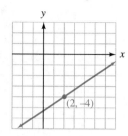

47. $(0, 0)$, $m = -4$ **48.** $(0, 0)$, $m = 5$

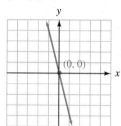   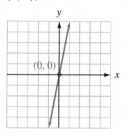

49. $(-5, 1)$, $m = 0$ **50.** $(0, 3)$, undefined slope

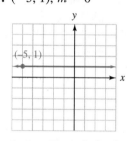   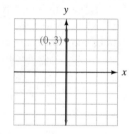

51. $(-1, -4)$, undefined slope   **52.** $(-3, -2)$, $m = 0$

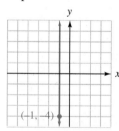   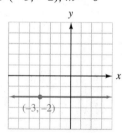

APPLICATIONS

53. POOL DESIGN Find the slope of the bottom of the swimming pool as it drops off from the shallow end to the deep end, as shown in Illustration 9. $\frac{2}{5}$

ILLUSTRATION 9

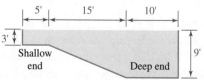

54. DRAINAGE To measure the amount of fall (slope) of a concrete patio slab in Illustration 10, a 10-foot-long 2-by-4, a 1-foot ruler, and a level were used. Find the amount of fall in the slab. Explain what it means. $\frac{1}{40}$; 1-in. fall for every 40 in. of horizontal run

ILLUSTRATION 10

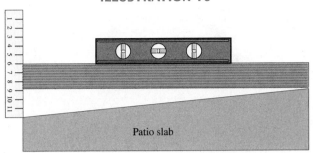

Patio slab

55. GRADE OF A ROAD The vertical fall of the road shown in Illustration 11 is 264 feet for a horizontal run of 1 mile. Find the slope of the decline and use that fact to complete the roadside warning sign for truckers. (*Hint:* 1 mile = 5,280 feet.) $\frac{1}{20}$; 5%

ILLUSTRATION 11

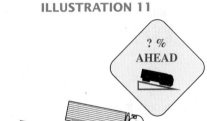

? %
AHEAD

264 ft

1 mi

56. TREADMILL For each height setting listed in the table, find the resulting slope of the jogging surface of the treadmill shown in Illustration 12. Express each incline as a percent.

ILLUSTRATION 12

Height setting	% incline
2 inches	4%
4 inches	8%
6 inches	12%

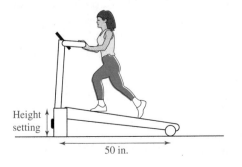

Height setting

50 in.

57. ACCESSIBILITY Illustration 13 shows two designs to make the upper level of a stadium wheelchair-accessible.
  a. Find the slope of the ramp in design 1. $\frac{1}{8}$
  b. Find the slopes of the ramps in design 2. $\frac{1}{12}$
  c. Give one advantage and one drawback of each design.
    1: less expensive, steeper; 2: not as steep, more expensive

ILLUSTRATION 13

Design #1

Upper level

Ground level

2 ft

16 ft

Design #2

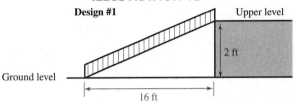

Upper level

Ground level

1 ft

1 ft

4 ft

58. ARCHITECTURE Since the slope of the roof of the house shown in Illustration 14 is to be $\frac{2}{5}$, there will be a 2-foot rise for every 5-foot run. Draw the roof line if it is to pass through the given black points. Find the coordinates of the peak of the roof. (10, 10)

ILLUSTRATION 14

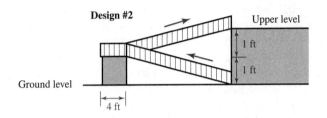

59. ENGINE OUTPUT Use the graph in Illustration 15 to find the rate of change in the horsepower (hp) produced by an automobile engine for engine speeds in the range of 2,400–4,800 revolutions per minute (rpm). 3 hp/40 rpm

ILLUSTRATION 15

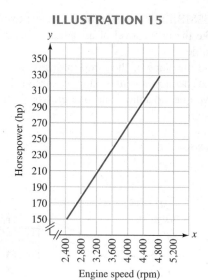

ILLUSTRATION 16

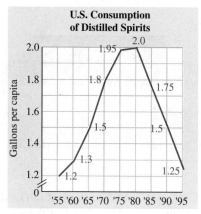

Based on data from *Business Week* (Nov. 25, 1996), p. 46

60. ▦ LIQUOR CONSUMPTION Refer to Illustration 16.

a. During what 5-year period did the per capita consumption of distilled spirits increase the most? Find the rate of change.   1965–1970; 0.06 gal/yr

b. In what year did per capita consumption reach a maximum? What was it?   1980; 2.0 gal

c. Find the rate of change in per capita consumption for 1980–1995. What does it mean?

   −0.05 gal/yr; per capita consumption was decreasing.

WRITING

61. Explain why the slope of a vertical line is undefined.

62. How do we distinguish between a line with positive slope and a line with negative slope?

63. Give an example of a rate of change that government officials might be interested in knowing so they can plan for the future needs of our country.

64. Explain the difference between a rate of change that is positive and one that is negative. Give an example of each.

REVIEW

65. In what quadrant does the point $(-3, 6)$ lie?
   quadrant II

66. What is the name given the point $(0, 0)$?   origin

67. Is $(-1, -2)$ a solution of $y = x^2 + 1$?   no

68. What basic shape does the graph of the equation $y = |x - 2|$ have?   V-shape

69. Is the equation $y = 2x + 2$ linear or nonlinear?   linear

70. Solve $-3x \le 15$. $x \ge -5$

▶ 3.5   Describing Linear Relationships

In this section, you will learn about

Linear relationships ■ Slope–intercept form of the equation of a line ■ Parallel lines ■ Perpendicular lines

---

Introduction   We have seen that numerical relationships are often presented in tables and graphs. In this section, we will begin a discussion of a special type of relationship between two quantities whose graph is a straight line. Our objective is to learn how to write an equation that describes these *linear relationships* using a given table or graph.

Linear Relationships

A company manufactures concrete drain pipe. Various lengths of pipe and their corresponding weights are listed in the table in Figure 3-43. When this information is plotted as ordered pairs of the form (length, weight), we see that the points lie in a line. We say that the relationship between length and weight is *linear*.

FIGURE 3-43

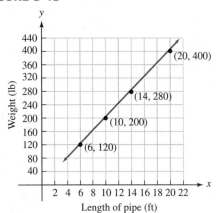

Length (ft)	Weight (lb)
6	120
10	200
14	280
20	400

Figure 3-44 shows a graph of the *time* a cup of coffee has been sitting on a kitchen counter and its *temperature*. Since the graph is not a straight line, the relationship between time and temperature in this example is not linear.

FIGURE 3-44

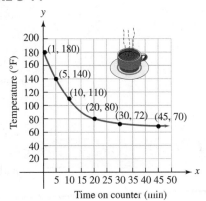

Time on counter (min)	Temperature of coffee (°F)
1	180
5	140
10	110
20	80
30	72
45	70

Slope–Intercept Form of the Equation of a Line

The graph of $2x + 3y = 12$ shown in Figure 3-45 enables us to see that the slope of the line is $-\frac{2}{3}$ and that the y-intercept is $(0, 4)$.

FIGURE 3-45

$$2x + 3y = 12$$

x	y	(x, y)
6	0	$(6, 0)$
0	4	$(0, 4)$

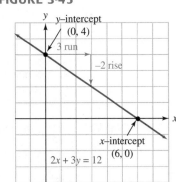

If we solve the equation for y, we will observe some interesting results.

$$2x + 3y = 12$$

$$3y = -2x + 12 \quad \text{Subtract } 2x \text{ from both sides.}$$

$$\frac{3y}{3} = \frac{-2x}{3} + \frac{12}{3} \quad \text{Divide both sides by 3.}$$

$$y = -\frac{2}{3}x + 4 \quad \text{Do the divisions. Rewrite } \tfrac{-2x}{3} \text{ as } -\tfrac{2}{3}x.$$

In the equation $y = -\frac{2}{3}x + 4$, the *slope* of the graph $\left(-\frac{2}{3}\right)$ is the coefficient of x, and the constant (4) is the y-coordinate of the *y-intercept* of the graph.

$$y = -\frac{2}{3}x + 4$$

The slope of the line.   The y-intercept is $(0, 4)$.

These observations suggest the following form of an equation of a line.

Slope–intercept form of the equation of a line

If a linear equation is written in the form

$$y = mx + b$$

where m and b are constants, the graph of the equation is a line with slope m and y-intercept $(0, b)$.

EXAMPLE 1

Slope–intercept form. Find the slope and the y-intercept of the graph of each equation: **a.** $y = 6x - 2$ and **b.** $y = -\frac{5}{4}x$.

Solution **a.** If we write the subtraction as the addition of the opposite, the equation will be in $y = mx + b$ form:

$$y = 6x + (-2)$$

Since $m = 6$ and $b = -2$, the slope of the line is 6 and the y-intercept is $(0, -2)$.

b. Writing $y = -\frac{5}{4}x$ in slope–intercept form, we have

$$y = -\frac{5}{4}x + 0$$

Since $m = -\frac{5}{4}$ and $b = 0$, the slope of the line is $-\frac{5}{4}$ and the y-intercept is $(0, 0)$.

SELF CHECK   Find the slope and the y-intercept:

a. $y = -5x - 1$, **b.** $y = \frac{7}{8}x$, and

c. $y = 5 - 2x$.

Answers: **a.** $m = -5$, $(0, -1)$; **b.** $m = \frac{7}{8}$, $(0, 0)$; **c.** $m = -2$, $(0, 5)$ ■

EXAMPLE 2

Slope–intercept form. Find the slope and the y-intercept of the line determined by $6x - 3y = 9$. Then graph it.

Solution

To find the slope and the y-intercept of the line, we need to write the equation in slope–intercept form. We do this by solving for y.

$$6x - 3y = 9$$
$$-3y = -6x + 9 \qquad \text{Subtract } 6x \text{ from both sides.}$$
$$\frac{-3y}{-3} = \frac{-6x}{-3} + \frac{9}{-3} \qquad \text{Divide both sides by } -3.$$
$$y = 2x - 3 \qquad \begin{array}{l}\text{Do the divisions.}\\ \text{Note: } m = 2 \text{ and}\\ b = -3.\end{array}$$

FIGURE 3-46

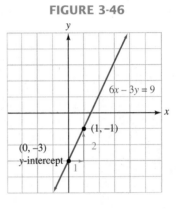

From the equation, we see that the slope is 2 and the y-intercept is $(0, -3)$.

To graph $y = 2x - 3$, we plot the y-intercept $(0, -3)$, as shown in Figure 3-46. Since the slope is $\frac{\text{rise}}{\text{run}} = 2 = \frac{2}{1}$, the line rises 2 units for every unit it moves to the right. If we begin at $(0, -3)$ and move 1 unit to the right (run) and then 2 units up (rise), we locate the point $(1, -1)$, which is a second point on the line. We then draw a line through $(0, -3)$ and $(1, -1)$.

SELF CHECK

Find the slope and the y-intercept of the line determined by $8x - 2y = -2$. Then graph it.

Answer: $m = 4$, $(0, 1)$

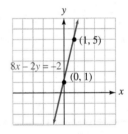

If we are given the slope and y-intercept of a line, we can write its equation, as in the next example.

EXAMPLE 3

Limo service. On weekends, a limousine service charges a fee of $100, plus 50¢ per mile, for the rental of a stretch limo. Write a linear equation that describes the relationship between the rental cost and the number of miles driven. Graph the result.

Solution

To write an equation describing this relationship, we will let x represent the number of miles driven and y represent the cost (in dollars). We can make two observations:

■ The cost increases by $0.50 (50¢) for each mile driven. This is the *rate of change* of the rental cost to miles driven, and it will be the *slope* of the graph of the equation. Thus, $m = 0.50$.

■ The basic fee is $100. Before driving any miles (that is, when $x = 0$), the cost y is 100. The ordered pair $(0, 100)$ will be the y-intercept of the graph of the equation. So we know that $b = 100$.

We substitute 0.50 for m and 100 for b in the slope–intercept form to get

$$y = 0.50x + 100$$

Here the cost y depends on x (the number of miles driven).

$m = 0.50 \qquad b = 100$

To graph $y = 0.50x + 100$, we plot its y-intercept, $(0, 100)$, as shown in Figure 3-47. Since the slope is $0.50 = \frac{50}{100} = \frac{5}{10}$, we can start at $(0, 100)$ and locate a second point on the line by moving 10 units to the right (run) and then 5 units up (rise). This point will have coordinates $(0 + 10, 100 + 5)$ or $(10, 105)$. We draw a line through these two points to get a graph that illustrates the relationship between the rental cost and the number of miles driven.

FIGURE 3-47

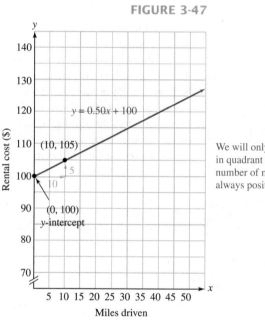

We will only draw the graph in quadrant I, because the number of miles driven is always positive.

EXAMPLE 4

Videotapes. A VHS videocassette contains 800 feet of tape. In the long play (LP) mode, it plays 10 feet of tape every 3 minutes. Write a linear equation that relates the number of feet of tape yet to be played and the number of minutes the tape has been playing. Graph the equation.

Solution

The number of feet yet to be played depends on the time the tape has been playing. To write an equation describing this relationship, we let x represent the number of minutes the tape has been playing and y represent the number of feet of tape yet to be played. We can make two observations:

■ Since the VCR plays 10 feet of tape every 3 minutes, the number of feet remaining is constantly *decreasing*. This rate of change $\left(-\frac{10}{3}\right.$ feet per minute$\left.\right)$ will be the slope of the graph of the equation. Thus, $m = -\frac{10}{3}$.

■ The cassette tape is 800 feet long. Before any of the tape is played (that is, when $x = 0$), the amount of tape yet to be played is $y = 800$. Written as an ordered pair, we have $(0, 800)$. Thus, $b = 800$.

Writing the equation in slope–intercept form, we have $y = -\frac{10}{3}x + 800$. Its graph is shown in Figure 3-48.

FIGURE 3-48

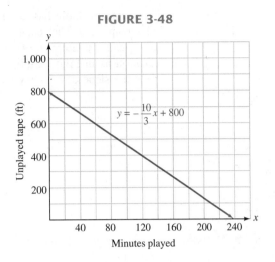

$$y = -\frac{10}{3}x + 800$$

Minutes played

SELF CHECK   Answer the problem in Example 4, where the VCR is in super long play (SLP) mode, which plays 11 feet every 5 minutes. Graph the equation on the graph in Figure 3-48 and make an observation.

Answer: $y = -\frac{11}{5}x + 800$; the graphs have the same y-intercept but different slopes.   ∎

Parallel Lines

Suppose it costs $75, plus 50¢ per mile, to rent the limo discussed in Example 3 on a weekday. If we substitute 0.50 for m and 75 for b in the slope–intercept form of a line, we have

$$y = 0.50x + 75$$

The graph of this equation and the graph of the equation

$$y = 0.50x + 100$$

appear in Figure 3-49.

From the figure, we see that the lines, each with slope 0.50, are parallel (do not intersect). This observation suggests the following fact.

FIGURE 3-49

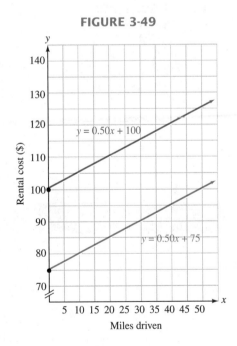

$y = 0.50x + 100$

$y = 0.50x + 75$

Miles driven

Slopes of parallel lines

Two lines with the same slope are parallel.

EXAMPLE 5

Parallel lines. Graph $y = -\frac{2}{3}x$ and $y = -\frac{2}{3}x + 3$ on the same set of axes.

Solution The first equation has a slope of $-\frac{2}{3}$ and a y-intercept of 0. The second equation has a slope of $-\frac{2}{3}$ and a y-intercept of (0, 3). We graph each equation as in Figure 3-50. Since the lines have the same slope of $-\frac{2}{3}$, they are parallel.

FIGURE 3-50

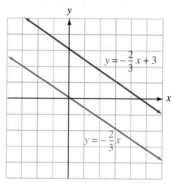

SELF CHECK Graph $y = \frac{5}{2}x - 2$ and $y = \frac{5}{2}x$ on the same set of axes.

Answer:

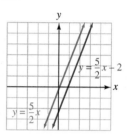

FIGURE 3-51

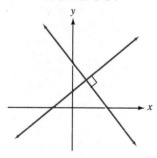

Perpendicular Lines

The two lines shown in Figure 3-51 meet at right angles and are called **perpendicular lines**. In the figure, the symbol ⌐ is used to denote a right angle. Each of the four angles that are formed has a measure of 90°.

The slopes of two (nonvertical) perpendicular lines are related by the following fact.

Slopes of perpendicular lines

The product of the slopes of perpendicular lines is −1.

If the product of two numbers is −1, they are called **negative reciprocals**. For example, 3 and $-\frac{1}{3}$ are negative reciprocals, because their product is −1.

$$3\left(-\frac{1}{3}\right) = -\frac{3}{3} = -1$$

Perpendicular lines have slopes that are negative reciprocals.

ACCENT ON TECHNOLOGY *Exploring Graphs with a Graphing Calculator*

Graphing calculators can be used to explore some important features of the graph of a linear equation, such as its x- and y-intercepts and its slope. To graph a linear equation such as $2x - 3y = 12$ using a graphing calculator, we must enter the equation into the calculator using the $\boxed{Y=}$ key. This requires that the equation be solved for y.

$$2x - 3y = 12$$

$$-3y = 12 - 2x \qquad \text{To eliminate } 2x \text{ from the left-hand side, subtract } 2x \text{ from both sides.}$$

$$\frac{-3y}{-3} = \frac{12}{-3} - \frac{2x}{-3} \qquad \text{To undo the multiplication by } -3, \text{ divide both sides by } -3.$$

$$y = \frac{2}{3}x - 4 \qquad \text{Simplify and write the equation in slope-intercept form: } y = mx + b.$$

Using the standard viewing window, we press the $\boxed{Y=}$ key and enter the equation as $(2/3)x - 4$. The display will show

$$Y_1 = (2/3)\, x - 4$$

Then we press the $\boxed{\text{GRAPH}}$ key to obtain the graph shown in Figure 3-52.

FIGURE 3-52

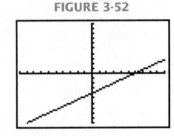

Determining the Intercepts of a Graph

To determine the x-intercept of the graph of $y = \frac{2}{3}x - 4$, we can use the zero feature, found under the CALC menu. After we guess left and right bounds, the cursor automatically moves to the x-intercept of the graph, and the coordinates of that point are displayed at the bottom of the screen. See Figure 3-53(a).

To determine the y-intercept of the graph, we can use the value feature, which is also found under the CALC menu. With this option, we first enter an x-value of 0, as shown in Figure 3-53(b). Then the cursor highlights the y-intercept and the coordinates of the y-intercept are given. See Figure 3-53(c).

FIGURE 3-53

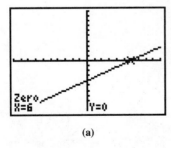 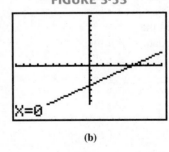 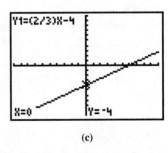

(a)   (b)   (c)

Inputting Several Equations

Graphing calculators can graph several equations on the same set of axes. For example, we can graph the equations $y = -2x$, $y = -2x - 4$, and $y = -2x + 5$ by pressing the $\boxed{Y=}$ key and entering the equations, one at a time, on the first three lines.

$$Y_1 = -2x$$

$$Y_2 = -2x - 4$$

$$Y_3 = -2x + 5$$

(continued)

The resulting three graphs are shown in Figure 3-54. The lines are parallel, which is a fact we can confirm by examining their equations. In each case, $m = -2$, which implies that each line has a slope of -2.

FIGURE 3-54

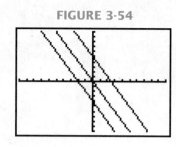

EXAMPLE 6

Parallel and perpendicular lines. Determine whether the graphs of $y = -5x + 6$ and $y = \frac{x}{5} - 2$ are parallel, perpendicular, or neither.

Solution   The slope of the line $y = -5x + 6$ is -5. The slope of the line $y = \frac{x}{5} - 2$ is $\frac{1}{5}$. (Recall that $\frac{x}{5} = \frac{1}{5}x$.) Since the slopes are not equal, the lines are not parallel. If we find the product of their slopes, we have

$$-5\left(\frac{1}{5}\right) = -\frac{5}{5} = -1$$

Since the product of their slopes is -1, the lines are perpendicular.

SELF CHECK   Determine whether the graphs of $y = 4x + 4$ and   *Answer:* neither
$y = \frac{1}{4}x$ are parallel, perpendicular, or neither.

STUDY SET

Section 3.5

VOCABULARY

In Exercises 1–6, fill in the blanks to make the statements true.

1. The equation $y = mx + b$ is called the ___slope–intercept___ form for the equation of a line.

2. The graph of the linear equation $y = mx + b$ has a ___y-intercept___ of $(0, b)$ and a ___slope___ of m.

3. ___Parallel___ lines do not intersect.

4. The slope of a line is a ___rate___ of change.

5. The numbers $\frac{5}{6}$ and $-\frac{6}{5}$ are called negative ___reciprocals___. Their product is -1.

6. The product of the slopes of ___perpendicular___ lines is -1.

CONCEPTS

7. TREE GROWTH   Graph the values shown in Illustration 1 and connect the points with a smooth curve.

Does the graph indicate a linear relationship between the age of the tree and its height? Explain your answer.   no, because the graph is not a straight line.

ILLUSTRATION 1

Age	Height
0	0
5	8
10	15
15	28
20	45
25	62
30	85
35	100
40	112
45	118

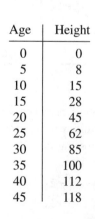
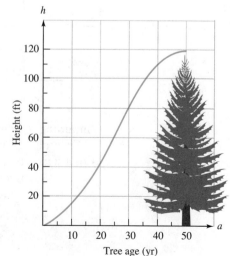

Tree age (yr)

8. See Illustration 2.

ILLUSTRATION 2

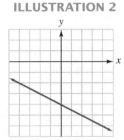

   a. What is the slope of the line? $-\frac{1}{2}$
   b. What is the y-intercept of the line? $(0, -4)$
   c. Write the equation of the line. $y = -\frac{1}{2}x - 4$

9. NAVIGATION The performance graph in Illustration 3 shows the recommended speed at which a ship should proceed into head waves of various heights.

   a. What information does the y-intercept of the graph give?

   When there are no head waves, the ship could travel at 18 knots.

   b. What is the rate of change in the recommended speed of the ship as the wave height increases? $-\frac{1}{2}$ knot/ft

   c. Write the equation of the graph. $y = -\frac{1}{2}x + 18$

ILLUSTRATION 3

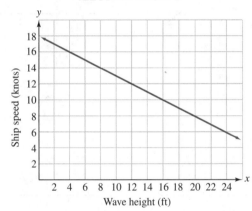

10. In Illustration 4, the slope of line l_1 is 2.

   a. What is the slope of line l_2? $-\frac{1}{2}$
   b. What is the slope of line l_3? 2
   c. What is the slope of line l_4? $-\frac{1}{2}$
   d. Which lines have the same y-intercept? l_1 and l_2

ILLUSTRATION 4

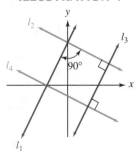

11. a. What is the y-intercept of line l_1 graphed in Illustration 5? $(0, 0)$

   b. What do lines l_1 and l_2 have in common? How are they different? same slope; different y-intercepts

ILLUSTRATION 5

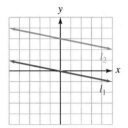

12. Use the graph in Illustration 6 to find m and b; then write the equation of the line in slope–intercept form. $m = \frac{5}{4}$, $b = 0$; $y = \frac{5}{4}x$

ILLUSTRATION 6

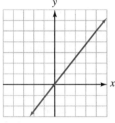

13. What is the slope of the line defined by each equation?

   a. $y = \dfrac{-2x}{3} - 2$ $\;-\frac{2}{3}$ **b.** $y = \frac{x}{4} + 1$ $\;\frac{1}{4}$
   c. $y = 2 - 8x$ $\;-8$ **d.** $y = 3x$ $\;3$
   e. $y = x$ $\;1$ **f.** $y = -x$ $\;-1$

 & y intercept?
graph it.

14. Without graphing, tell whether the graphs of each pair of lines are parallel, perpendicular, or neither.

   a. $y = 0.5x - 3$; $y = \dfrac{1}{2}x + 3$ parallel

   b. $y = 0.75x$; $y = -\dfrac{4}{3}x + 2$ perpendicular

   c. $y = -x$; $y = x$ perpendicular

In Exercises 15–16, a graphing calculator was used to graph $y = -2.5x - 1.25$, and the TRACE key was pressed. What important feature of the graph does the TRACE cursor show?

15. It shows that the y-intercept is $(0, -1.25)$

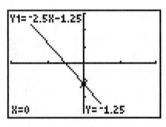

16. It suggests that the x-intercept is $(-0.5, 0)$

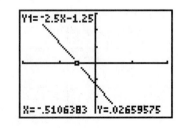

NOTATION

In Exercises 17–18, complete each solution by solving the equation for y. Then find the slope and the y-intercept of its graph.

17. $6x - 2y = 10$

$6x - \boxed{6x} - 2y = -6x + 10$

$-2y = \boxed{-6x} + 10$

$y = \boxed{3x} - 5$

The slope is $\boxed{3}$ and the y-intercept is $\boxed{(0, -5)}$.

18. $2x + 5y = 15$

$2x + 5y - \boxed{2x} = -\boxed{2x} + 15$

$\boxed{5y} = -2x + 15$

$y = -\dfrac{2}{5}x + 3$

The slope is $\boxed{-\frac{2}{5}}$ and the y-intercept is $\boxed{(0, 3)}$.

PRACTICE

In Exercises 19–20, find the slope and the y-intercept of the graph of each equation.

19. a. $y = 4x + 2$
4, (0, 2)

b. $y = -4x - 2$
−4, (0, −2)

c. $4y = x - 2$
$\frac{1}{4}$, $\left(0, -\frac{1}{2}\right)$

d. $4x - 2 = y$ 4, (0, −2)

20. a. $y = \dfrac{1}{2}x + 6$
$\frac{1}{2}$, (0, 6)

b. $y = 6 - x$ −1, (0, 6)

c. $6y = x - 6$
$\frac{1}{6}$, (0, −1)

d. $6x - 1 = y$ 6, (0, −1)

In Exercises 21–26, write the equation of the line with the given slope and y-intercept. Then graph it.

21. $m = 5$, (0, −3)
$y = 5x - 3$

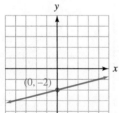

22. $m = -2$, (0, 1)
$y = -2x + 1$

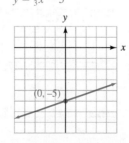

23. $m = \dfrac{1}{4}$, (0, −2)
$y = \frac{1}{4}x - 2$

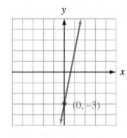

24. $m = \dfrac{1}{3}$, (0, −5)
$y = \frac{1}{3}x - 5$

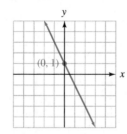

25. $m = -\dfrac{8}{3}$, (0, 5)
$y = -\frac{8}{3}x + 5$

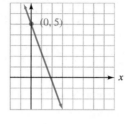

26. $m = -\dfrac{7}{6}$, (0, 2)
$y = -\frac{7}{6}x + 2$

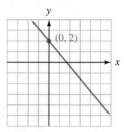

In Exercises 27–30, find the slope and the y-intercept of the graph of each equation. Then graph it.

27. $3x + 4y = 16$
$m = -\frac{3}{4}$, (0, 4)

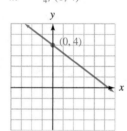

28. $2x + 3y = 9$
$m = -\frac{2}{3}$, (0, 3)

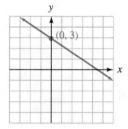

29. $10x - 5y = 5$
$m = 2$, (0, −1)

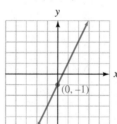

30. $4x - 2y = 6$
$m = 2$, (0, −3)

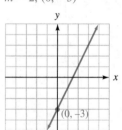

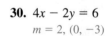 *In Exercises 31–32, solve each equation for y. Then use a graphing calculator to graph it. Use a viewing window of x = −5 to 5 and y = −5 to 5.*

31. $2y + 5x = 4$

32. $1.2x - 3.2y + 4.7 = 0$

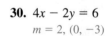 *In Exercises 33–34, estimate the x-intercept of the graph of each equation to the nearest hundredth.*

33. $y = 2.3x + 3.76$
(−1.63, 0)

34. $y = \dfrac{7}{8}x + 8$ (−9.14, 0)

In Exercises 35–36, graph each equation on the same set of axes. In each case, what do the graphs have in common?

35. $y = 4x - 2.1$ same slope

$y = 4x$

$y = 4x + 3.75$

36. $y = -x + 3$ same y-intercept

$y = 5.34x + 3$

$y = \dfrac{7}{2}x + 3$

APPLICATIONS

37. PRODUCTION COSTS A television production company charges a basic fee of $5,000 and then $2,000 an hour when filming a commercial.
   a. Write a linear equation that describes the relationship between the total production costs y and the hours of filming x. $y = 2,000x + 5,000$
   b. Use your answer to part a to find the production costs if a commercial required 8 hours of filming. $21,000

38. COLLEGE FEES Each semester, students enrolling at a community college must pay tuition costs of $20 per unit as well as a $40 student services fee.
   a. Write a linear equation that gives the total fees y to be paid by a student enrolling at the college and taking x units. $y = 20x + 40$
   b. Use your answer to part a to find the enrollment cost for a student taking 12 units. $280

39. CHEMISTRY EXPERIMENT A portion of a student's chemistry lab manual is shown in Illustration 7. Use the information to write a linear equation relating the temperature y (in degrees Fahrenheit) of the compound to the time x (in minutes) elapsed during the lab procedure. $y = 5x - 10$

ILLUSTRATION 7

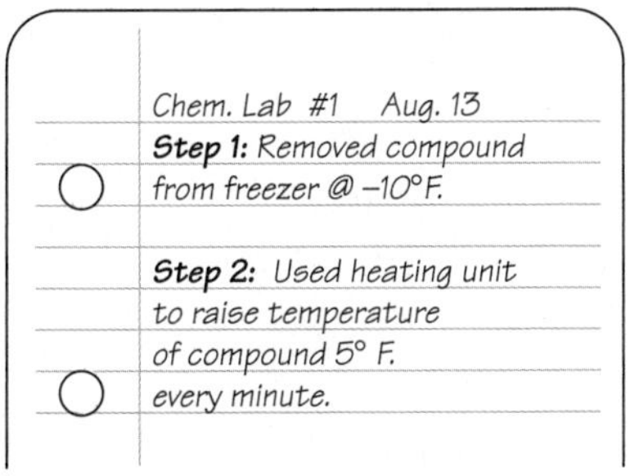

Chem. Lab #1 Aug. 13

Step 1: Removed compound from freezer @ –10° F.

Step 2: Used heating unit to raise temperature of compound 5° F. every minute.

40. INCOME PROPERTY
   Use the information in the newspaper advertisement in Illustration 8 to write a linear equation that gives the amount of income y (in dollars) the apartment owner will receive when the unit is rented for x months. $y = 500x + 250$

ILLUSTRATION 8

APARTMENT FOR RENT

1 bedroom/1 bath, with garage

$500 per month + $250 nonrefundable security fee.

41. SALAD BAR For lunch, a delicatessen offers a "Salad and Soda" special where customers serve themselves at a well-stocked salad bar. The cost is $1.00 for the drink and 20¢ an ounce for the salad.
   a. Write a linear equation that will find the cost y of a "Salad and Soda" lunch when a salad weighing x ounces is purchased. $y = 0.20x + 1.00$
   b. Graph the equation (see Illustration 9).
   c. How would the graph from part b change if the delicatessen began charging $2.00 for the drink? same slope, different y-intercept
   d. How would the graph from part b change if the cost of the salad changed to 30¢ an ounce? same y-intercept, steeper slope

ILLUSTRATION 9

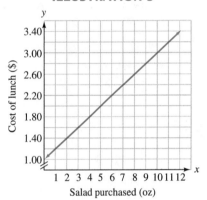

Cost of lunch ($) vs. Salad purchased (oz)

42. SEWING COSTS A tailor charges a basic fee of $20.00 plus $2.50 per letter to sew an athlete's name on the back of a jacket.
   a. Write a linear equation that will find the cost y to have a name containing x letters sewn on the back of a jacket. $y = 2.50x + 20.00$
   b. Graph the equation (see Illustration 10).
   c. Suppose the tailor raises the basic fee to $30. On your graph from part b, draw the new graph showing the increased cost.

ILLUSTRATION 10

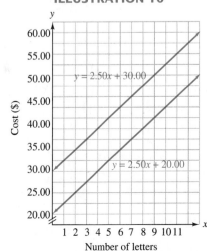

$y = 2.50x + 30.00$

$y = 2.50x + 20.00$

Cost ($) vs. Number of letters

43. EMPLOYMENT SERVICE The policy statement of LIZCO, Inc., is shown in Illustration 11. Suppose a secretary had to pay an employment service $500 to get placed in a new job at LIZCO. Write a linear equation that tells the secretary the actual cost y of the employment service to her x months after being hired. $y = -20x + 500$

ILLUSTRATION 11

> **Policy no. 23452**– A new hire will be reimbursed by LIZCO for any employment service fees paid by the employee at the rate of $20 per month.

44. POPULATION Use the data in Illustration 12 to write a linear equation that approximates South Korea's population y (in millions) for the years 1980–2000. On the horizontal axis, $x = 0$ represents 1980, $x = 5$ represents 1985, and so on. $y = \frac{1}{2}x + 40$

ILLUSTRATION 12

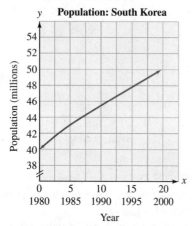

Based on data from *Los Angeles Times* (Apr. 9, 1997), p. A5

45. 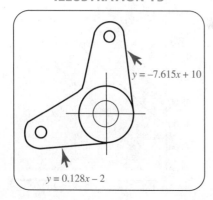 **COMPUTER DRAFTING** Illustration 13 shows a computer-generated drawing of an automobile engine mount. When the designer clicks the mouse on a line of the drawing, the computer finds the equation of the line. Determine whether the two lines selected in the drawing are perpendicular.

not quite: $(0.128)(-7.615) = -0.97472 \neq -1$

ILLUSTRATION 13

46. CALLING CARDS Use the data in Illustration 14 to write a linear equation that approximates the sales y (in billions of dollars) of prepaid calling cards for the years 1995–2000. On the horizontal axis, $x = 0$ represents 1995, $x = 1$ represents 1996, and so on.

$y = 0.36x + 0.7$

ILLUSTRATION 14

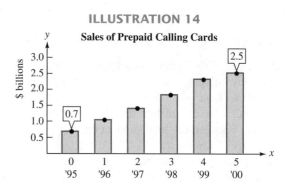

Based on data from *Los Angeles Times* (Apr. 17, 1997), p. D1

WRITING

47. Explain the advantages of writing the equation of a line in slope–intercept form ($y = mx + b$) as opposed to general form ($Ax + By = C$).

48. To describe a linear relationship between two quantities, we can use a graph or an equation. Which method do you think is better? Explain why.

49. What is the minimum number of points needed to draw the graph of a line? Explain why.

50. List some examples of parallel and perpendicular lines that you see in your daily life.

REVIEW

51. Find the slope of the line passing through the points $(6, -2)$ and $(-6, 1)$. $-\frac{1}{4}$

52. Is $(3, -7)$ a solution of $y = 3x - 2$?   no

53. Evaluate $-4 - (-4)$.   0

54. Solve $2(x - 3) = 3x$. -6

55. To evaluate $[-2(4 - 8) + 4^2]$, which operation should be performed first?   subtraction

56. Translate to mathematical symbols: four less than twice the price p. $2p - 4$

57. What percent of 6 is 1.5?   25%

58. Does $x = -6.75$ make $x + 1 > -9$ true?   yes

▶ 3.6   Writing Linear Equations

In this section, you will learn about

Point–slope form of the equation of a line ■ Writing the equation of a line through two points ■ Horizontal and vertical lines

Introduction   If we know the slope of a line and its *y*-intercept, we can use the slope–intercept form to write the equation of the line. The question that now arises is, can *any* point on the line be used in combination with its slope to write its equation? In this section, we will answer this question.

Point–Slope Form of the Equation of a Line

FIGURE 3-55

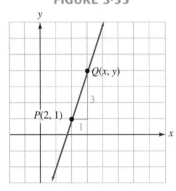

For the line shown in Figure 3-55, suppose we know that it has a slope of 3 and that it passes through the point $P(2, 1)$. If we pick another point on the line and call it $Q(x, y)$, we can find the slope of the line by using the coordinates of points P and Q. Using the slope formula, we have

$$\frac{y_2 - y_1}{x_2 - x_1} = m \quad \text{The slope formula.}$$

$$\frac{y - 1}{x - 2} = m \quad \text{Substitute } y \text{ for } y_2, 1 \text{ for } y_1, x \text{ for } x_2, \text{ and } 2 \text{ for } x_1.$$

Since the slope of the line is given to be 3, we can substitute 3 for m in the previous equation.

$$\frac{y - 1}{x - 2} = m$$

$$\frac{y - 1}{x - 2} = 3$$

We then multiply both sides by $(x - 2)$ to get

$$\frac{y - 1}{x - 2}(x - 2) = 3(x - 2) \quad \text{Clear the equation of the fraction.}$$

$$y - 1 = 3(x - 2) \quad \text{Simplify the left-hand side.}$$

The resulting equation displays the slope of the line and the coordinates of one point on the line:

$$
\begin{array}{c}
\text{Slope} \\
\text{of the line} \\
\downarrow \\
y - 1 = 3(x - 2) \\
\uparrow \qquad\quad \uparrow \\
y\text{-coordinate} \quad x\text{-coordinate} \\
\text{of the point} \quad\ \text{of the point}
\end{array}
$$

In general, suppose we know that the slope of a line is m and that the line passes through the point (x_1, y_1). Then if (x, y) is any other point on the line, we can use the definition of slope to write

$$\frac{y - y_1}{x - x_1} = m$$

If we multiply both sides by $x - x_1$, we have

$$y - y_1 = m(x - x_1)$$

This form of a linear equation is called the **point–slope form.** It can be used to write the equation of a line when the slope and one point on the line are known.

If a line with slope m passes through the point (x_1, y_1), the equation of the line is

$$y - y_1 = m(x - x_1)$$

EXAMPLE 1

Point–slope form. Write the equation of a line that has a slope of -3 and passes through $(-1, 5)$. Express the result in slope–intercept form.

Solution   Since we are given the slope and a point on the line, we will use the point–slope form.

$\qquad y - y_1 = m(x - x_1)$ The point–slope form.

$\qquad y - 5 = -3[x - (-1)]$ Substitute -3 for m, -1 for x_1, and 5 for y_1.

$\qquad y - 5 = -3(x + 1)$ Simplify inside the brackets.

We can write this result in slope–intercept form, as follows:

$\qquad y - 5 = -3x - 3$ Distribute the -3.

$\qquad\qquad y = -3x + 2$ Add 5 to both sides: $-3 + 5 = 2$.

In slope–intercept form, the equation is $y = -3x + 2$.

SELF CHECK

Write the equation of a line that has a slope of -2 and passes through $(4, -3)$. Write the result in slope–intercept form.   *Answer:* $y = -2x + 5$ ∎

EXAMPLE 2

Temperature drop. A refrigeration unit can lower the temperature in a railroad car by 6°F every 5 minutes. One day, the temperature in a car was 76°F after the cooler had run for 10 minutes. Find a linear equation that describes the relationship between the time the cooler has been running and the temperature in the car.

Graph the equation and use it to find the temperature in the car before the cooler was turned on and the temperature in the car after the cooler had run for 25 minutes.

Solution   We will let x represent the time, in minutes, that the cooler was running, and y will represent the air temperature in the car. We can make two observations:

■ With the cooler on, the temperature in the railroad car drops 6° every 5 minutes. The rate of change of $-\frac{6}{5}$ degrees per minute is the slope of the graph of the linear equation that we want to find. Thus, $m = -\frac{6}{5}$.

■ We know that after the cooler had been running for 10 minutes ($x = 10$), the temperature in the car was 76° ($y = 76$). We can express these facts with the ordered pair $(10, 76)$.

To write the linear equation, we substitute $-\frac{6}{5}$ for m, 10 for x_1, and 76 for y_1, into the point–slope form of the equation of a line.

$\qquad y - y_1 = m(x - x_1)$ The point–slope form.

$\qquad y - 76 = -\dfrac{6}{5}(x - 10)$ Substitute: $m = -\frac{6}{5}$, $x_1 = 10$, and $y_1 = 76$.

$\qquad 5(y - 76) = 5\left(-\dfrac{6}{5}\right)(x - 10)$ Multiply both sides by 5 to eliminate the fraction.

$$5y - 380 = -6(x - 10)$$ Distribute the 5 on the left-hand side. On the right-hand side, $5\left(-\frac{6}{5}\right) = -6$.

$$5y - 380 = -6x + 60$$ Distribute the -6.

$$5y = -6x + 440$$ Add 380 to both sides: $60 + 380 = 440$.

$$\frac{5y}{5} = \frac{-6x}{5} + \frac{440}{5}$$ Divide both sides by 5.

$$y = -\frac{6}{5}x + 88$$ Do the divisions. Write $\frac{-6x}{5}$ as $-\frac{6x}{5}$.

The graph of $y = -\frac{6}{5}x + 88$ is shown in Figure 3-56. From the graph, we see that the temperature in the railroad car before the cooler was turned on was 88°F. This is given by the y-intercept of the graph, $(0, 88)$. If we locate 25 on the x-axis and move straight up to intersect the graph, we will see that the temperature in the car was 58°F. This shows that after the cooler ran for 25 minutes, the temperature was 58°F.

FIGURE 3-56

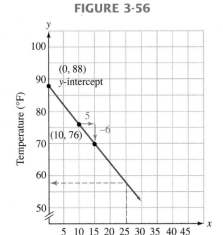

Writing the Equation of a Line Through Two Points

In the next example, we will show that it is possible to write the equation of a line when we know the coordinates of two points on the line.

EXAMPLE 3

Given two points on a line. Write the equation of the line passing through $P(4, 0)$ and $Q(6, -8)$.

Solution   First we find the slope of the line.

$$m = \frac{y_2 - y_1}{x_2 - x_1}$$ The slope formula.

$$= \frac{-8 - 0}{6 - 4}$$ Substitute -8 for y_2, 0 for y_1, 6 for x_2, and 4 for x_1.

$$= \frac{-8}{2}$$ Simplify.

$$= -4$$

Since the line passes through both P and Q, we can choose either point and substitute its coordinates into the point–slope form. If we choose $P(4, 0)$, we substitute 4 for x_1, 0 for y_1, and -4 for m and proceed as follows.

$$y - y_1 = m(x - x_1)$$ Point–slope form.

$$y - 0 = -4(x - 4)$$ Substitute -4 for m, 4 for x_1, and 0 for y_1.

$$y = -4x + 16$$ Remove parentheses: distribute -4.

The equation of the line is $y = -4x + 16$.

SELF CHECK

Write the equation of the line passing through $R(0, -3)$ and $S(2, 1)$.

Answer: $y = 2x - 3$ ∎

EXAMPLE 4

Market research. A company that makes a breakfast cereal has found that the number of discount coupons redeemed for its product is linearly related to the coupon's value. In one advertising campaign, 10,000 "10¢ off" coupons were redeemed. In another campaign, 45,000 "50¢ off" coupons were redeemed. How many coupons can the company expect to be redeemed if it issues a "35¢ off" coupon?

Solution

If we let x represent the value of a coupon and y represent the number of coupons that will be redeemed, ordered pairs will have the form

(coupon value, number redeemed)

FIGURE 3-57

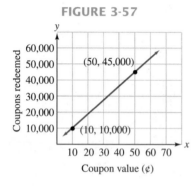

Two points on the graph of the equation are (10, 10,000) and (50, 45,000). These points are plotted on the graph shown in Figure 3-57. To write the equation of the line passing through the points, we first find the slope of the line.

$$m = \frac{y_2 - y_1}{x_2 - x_1} \qquad \text{The slope formula.}$$

$$= \frac{45,000 - 10,000}{50 - 10} \qquad \text{Substitute 45,000 for } y_2, \text{ 10,000 for } y_1, \text{ 50 for } x_2, \text{ and 10 for } x_1.$$

$$= \frac{35,000}{40}$$

$$= 875$$

We then substitute 875 for m and the coordinates of one known point—say, (10, 10,000)—into the point–slope form of the equation of a line and proceed as follows:

$$y - y_1 = m(x - x_1) \qquad \text{The point–slope form.}$$
$$y - 10,000 = 875(x - 10) \qquad \text{Substitute for } m, x_1, \text{ and } y_1.$$
$$y - 10,000 = 875x - 8,750 \qquad \text{Distribute the 875.}$$
$$y = 875x + 1,250 \qquad \text{Add 10,000 to both sides.}$$

To find the expected number of coupons that will be redeemed, we substitute the value of the coupon, 35¢, into the equation $y = 875x + 1,250$ and find y.

$$y = 875x + 1,250$$
$$y = 875(35) + 1,250 \qquad \text{Substitute 35 for } x.$$
$$y = 30,625 + 1,250 \qquad \text{Do the multiplication.}$$
$$y = 31,875$$

The company can expect 31,875 of the 35¢ coupons to be redeemed. ∎

Horizontal and Vertical Lines

We have graphed horizontal and vertical lines. We will now discuss how to write their equations.

EXAMPLE 5

Equations of horizontal and vertical lines. Write the equation of each line and then graph it: **a.** A horizontal line passing through $(-2, -4)$ and **b.** A vertical line passing through $(1, 3)$.

Solution
a. The equation of a horizontal line can be written in the form $y = b$. Since the y-coordinate of $(-2, -4)$ is -4, the equation of the line is $y = -4$. The graph is shown in Figure 3-58.

b. The equation of a vertical line can be written in the form $x = a$. Since the x-coordinate of $(1, 3)$ is 1, the equation of the line is $x = 1$. The graph is shown in Figure 3-58.

FIGURE 3-58

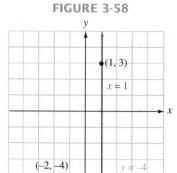

SELF CHECK
Write the equation of each line and then graph it:
a. a horizontal line passing through $(3, 2)$ and
b. a vertical line passing through $(-1, -3)$.

Answers: **a.** $y = 2$, **b.** $x = -1$

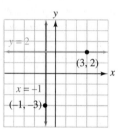

STUDY SET

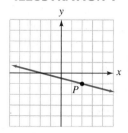

Section 3.6

VOCABULARY

In Exercises 1–4, fill in the blanks to make the statements true.

1. $y - y_1 = m(x - x_1)$ is called the ____point–slope____ form of the equation of a line.

2. The line in Illustration 1 ____passes____ through point P.

3. In Illustration 1, point P has an ____x-coordinate____ of 2 and a ____y-coordinate____ of -1.

4. The ____slope____ of a line gives a rate of change.

ILLUSTRATION 1

CONCEPTS

5. a. The linear equation $y = 2x - 3$ is written in *slope–intercept* form. What are the slope and the y-intercept of the graph of this line?
The slope is 2; the y-intercept is $(0, -3)$.

b. The linear equation $y - 4 = 6(x - 5)$ is written in *point–slope* form. What point does the graph of this equation pass through, and what is the line's slope?
The graph passes through $(5, 4)$; the slope is 6.

6. Is the following statement true or false? The equations

$$y - 1 = 2(x - 2)$$
$$y = 2x - 3$$
$$2x - y = 3$$

all describe the same line.   true

7. See Illustration 2.

a. What information can be obtained from the fact that the line passes through the point $(8, 108)$?
In a group of 8,000 women, 108 would have given birth in 1997.

b. What is the slope of the line and what does it tell us?
$\frac{108}{8,000} = \frac{27}{2,000}$; the birth rate is 27 per 2,000 or 13.5 per 1,000.

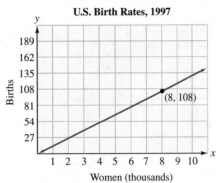

ILLUSTRATION 2

U.S. Birth Rates, 1997

Based on data from the National Center for Health Statistics

8. In each of the following cases, a linear relationship between two quantities is described. If the relationship were graphed, what would be the slope of the line?

a. The sales of new cars increased by 15 every 2 months. $\frac{15}{2}$

b. There were 35 fewer robberies for each dozen police officers added to the force. $-\frac{35}{12}$

c. Withdrawals were occurring at the rate of $700 every 45 minutes. $-\frac{700}{45} = -\frac{140}{9}$

d. One acre of forest is being destroyed every 30 seconds. $-\frac{1}{30}$

9. In each of the following cases, is the given information sufficient to write the equation of the line?

a. It passes through $(2, -7)$.   no

b. Its slope is $-\frac{3}{4}$.   no

c. It has the following table of values: yes

x	y
2	3
-3	-6

10. In each of the following cases, is the given information sufficient to write the equation of the line?

a. It is horizontal.   no

b. It is vertical and passes through $(-1, 1)$.   yes

c. It has the following table of values: no

x	y
4	5

NOTATION

11. Fill in the blank. In $y - y_1 = m(x - x_1)$, we read x_1 as "x _sub_ one."

12. Write the equation of a horizontal line passing through $(0, b)$. $y = b$

In Exercises 13–14, write the equation in slope–intercept form.

13. $y - 2 = -3(x - 4)$

$y - \boxed{2} = \boxed{-3x} + 12$

$y = -3x + 14$

14. $y + 2 = \dfrac{1}{2}(x + 2)$

$\boxed{2}\,(y + 2) = \boxed{2}\left(\dfrac{1}{2}\right)(x + 2)$

$2y + \boxed{4} = \boxed{x} + 2$

$2y = x - \boxed{2}$

$y = \dfrac{1}{2}x - 1$

In Exercises 15–16, complete each solution.

15. Write the equation of the line with slope -2 that passes through the point $(-1, 5)$.

$y - y_1 = m(x - x_1)$

$y - \boxed{5} = -2\big[x - \big(\boxed{-1}\big)\big]$

$y - 5 = \boxed{-2x} - 2$

$y = -2x + 3$

16. Write the equation of the line with slope 4 that passes through the point $(0, 3)$.

$y - y_1 = m(x - x_1)$

$y - \boxed{3} = 4\big(x - \boxed{0}\big)$

$y - 3 = \boxed{4x}$

$y = 4x + 3$

PRACTICE

In Exercises 17–20, use the point–slope form to write the equation of the line with the given slope and point.

17. $m = 3$, passes through $(2, 1)$ $y - 1 = 3(x - 2)$

18. $m = 2$, passes through $(4, 3)$ $y - 3 = 2(x - 4)$

19. $m = -\dfrac{4}{5}$, passes through $(-5, -1)$ $y + 1 = -\frac{4}{5}(x + 5)$

20. $m = -\dfrac{7}{8}$, passes through $(-2, -9)$ $y + 9 = -\frac{7}{8}(x + 2)$

In Exercises 21–32, use the point–slope form to first write the equation of the line with the given slope and point. Then write your result in slope–intercept form.

21. $m = \dfrac{1}{5}$, passes through $(10, 1)$ $y = \frac{1}{5}x - 1$

22. $m = \dfrac{1}{4}$, passes through $(8, 1)$ $y = \frac{1}{4}x - 1$

23. $m = -5$, passes through $(-9, 8)$ $y = -5x - 37$

24. $m = -4$, passes through $(-2, 10)$ $y = -4x + 2$

25. $m = -\dfrac{4}{3}$,

x	y
6	-4

$y = -\frac{4}{3}x + 4$

26. $m = -\dfrac{3}{2}$,

x	y
-2	1

$y = -\frac{3}{2}x - 2$

27. $m = -\dfrac{2}{3}$, passes through $(3, 0)$ $y = -\frac{2}{3}x + 2$

28. $m = -\dfrac{2}{5}$, passes through $(15, 0)$ $y = -\frac{2}{5}x + 6$

29. $m = 8$, passes through $(0, 4)$ $y = 8x + 4$

30. $m = 6$, passes through $(0, -4)$ $y = 6x - 4$

31. $m = -3$, passes through the origin $y = -3x$

32. $m = -1$, passes through the origin $y = -x$

In Exercises 33–38, write the equation of the line that passes through the two given points. Write your result in slope–intercept form.

33. Passes through $(1, 7)$ and $(-2, 1)$ $y = 2x + 5$

34. Passes through $(-2, 2)$ and $(2, -8)$ $y = -\frac{5}{2}x - 3$

35.
x	y
-4	3
2	0
  $y = -\frac{1}{2}x + 1$

36.
x	y
-1	-4
1	-2
  $y = x - 3$

37. Passes through $(5, 5)$ and $(7, 5)$ $y = 5$

38. Passes through $(-2, 1)$ and $(-2, 15)$ $x = -2$

In Exercises 39–42, write the equation of the line with the given characteristics.

39. vertical, passes through $(4, 5)$ $x = 4$

40. vertical, passes through $(-2, -5)$ $x = -2$

41. horizontal, passes through $(4, 5)$ $y = 5$

42. horizontal, passes through $(-2, -5)$ $y = -5$

APPLICATIONS

43. POLE VAULT See Illustration 3.
    a. For each of the four positions of the vault shown, give two points that the pole passes through.
    position 1: $(0, 0)$, $(-5, 2)$; position 2: $(0, 0)$, $(-3, 6)$;
    position 3: $(0, 0)$, $(-1, 7)$; position 4: $(0, 0)$, $(0, 10)$

ILLUSTRATION 3

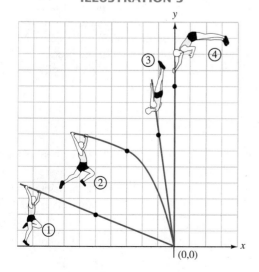

b. Write the equations of the lines that describe the position of the pole for parts 1, 3, and 4 of the jump. $y = -\frac{2}{5}x, \ y = -7x, \ x = 0$

c. Why can't we write a linear equation describing the position of the pole for part 2?
The pole is not in the shape of a straight line.

44. FREEWAY DESIGN The graph in Illustration 4 shows the route of a proposed freeway.
    a. Give the coordinates of the points where the proposed freeway will join Interstate 25 and Highway 40. $(-3, -4), (6, 2)$
    b. Write the equation of the line that mathematically describes the route of the proposed freeway. Answer in slope–intercept form. $y = \frac{2}{3}x - 2$

ILLUSTRATION 4

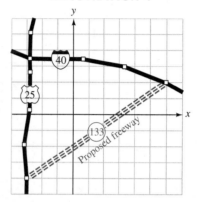

45. TOXIC CLEANUP Three months after cleanup began at a dump site, 800 cubic yards of toxic waste had yet to be removed. Two months later, that number had been lowered to 720 cubic yards.
    a. Write an equation that mathematically describes the linear relationship between the length of time x (in months) the cleanup crew has been working and the number of cubic yards y of toxic waste remaining. $y = -40x + 920$
    b. Use your answer to part a to predict the number of cubic yards of waste that will still be on the site one year after the cleanup project began. 440 yd^3

46. DEPRECIATION To lower its corporate income tax, accountants of a large company depreciated a word processing system over several years using a linear model, as shown in the worksheet in Illustration 5.

ILLUSTRATION 5

Tax Worksheet

Method of depreciation: *Linear*

Property	Value	Years after purchase
Word processing system	$60,000	2
"	$30,000	4

a. Use the information in Illustration 5 to write a linear equation relating the years since the system was purchased x and its value y, in dollars.
   $y = -15,000x + 90,000$

b. Find the purchase price of the system by substituting $x = 0$ into your answer from part a. $90,000

47. COUNSELING In the first year of her practice, a family counselor saw 75 clients. In her second year, the number of clients grew to 105. If a linear trend continues, write an equation that gives the number of clients c the counselor will have t years after beginning her practice. $c = 30t + 45$

48. HEALTH CARE SPENDING The graph in Illustration 6 can be approximated by a straight line.
   a. Use the given data to write a linear equation that approximates the per-person health care expenditures for the years 1992–1997. Let $x = 0$ represent 1992, $x = 2$ represent 1994, and so on.
      $y = 145x + 3,205$
   b. Use the linear model from part a to predict the per person health care expenditure in the year 2020.
      $7,265

ILLUSTRATION 6

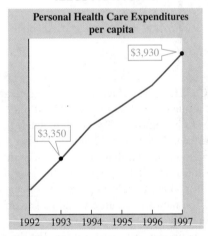

Based on data from the Health Care Financing Administration

49. CONVERTING TEMPERATURES The relationship between Fahrenheit temperature, F, and Celsius temperature, C, is linear.
   a. Use the data in Illustration 7 to write two ordered pairs of the form (C, F). $(0, 32)$; $(100, 212)$

ILLUSTRATION 7

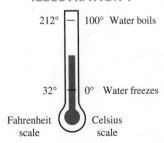

212° — 100° Water boils

32° — 0° Water freezes

Fahrenheit   Celsius
scale   scale

b. Use your answer to part a to write a linear equation relating the Fahrenheit and Celsius scales.
   $F = \frac{9}{5}C + 32$

50. TRAMPOLINE The relationship between the circumference of a circle and its radius is linear. For instance, the length of the protective pad that wraps around a trampoline is related to the radius of the trampoline. Use the data in Illustration 8 to write a linear equation that approximates the length of pad needed for any trampoline radius. $y = 6.25x + 0.25$

ILLUSTRATION 8

Radius (ft)	Approximate length of padding (ft)
3	19
7	44

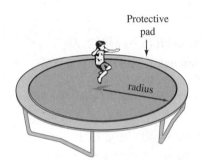

Protective pad
↓

radius

51. AIR CONDITIONING An air-conditioning unit can lower the air temperature in a classroom 4° every 15 minutes. After the air conditioner had been running for half an hour, the air temperature in the room was 75°F. Write a linear equation relating the time in minutes the unit had been on and the temperature of the classroom. (*Hint:* How many minutes are there in half an hour?) $y = -\frac{4}{15}x + 83$

52. AUTOMATION An automated production line uses distilled water at a rate of 300 gallons every 2 hours to make shampoo. After the line had run for 7 hours, planners noted that 2,500 gallons of distilled water remained in the storage tank. Write a linear equation relating the time in hours since the production line began and the number of gallons of distilled water in the storage tank. $y = -150x + 3,550$

WRITING

53. Why is $y - y_1 = m(x - x_1)$ called the point–slope form of the equation of a line?

54. If we know two points that a line passes through, we can write its equation. Explain how this is done.

55. If we know the slope of a line and a point it passes through, we can write its equation. Explain how this is done.

56. Think of several points on the graph of the horizontal line $y = 4$. What do the points have in common? How do they differ?

REVIEW

57. Find the slope of the line passing through the points $(2, 4)$ and $(-6, 8)$. $-\frac{1}{2}$

58. Is the graph of $y = x^2$ a line? no

59. Find the area of a circle with a diameter of 12 feet. Round to the nearest tenth. 113.1 ft²

60. If a 15-foot board is cut into two pieces and we let x represent the length of one piece (in feet), how long is the other piece? $(15 - x)$ ft

61. Evaluate $(-1)^5$. -1

62. Solve $\dfrac{x - 3}{4} = -4$. -13

63. What is the coefficient of the second term of $-4x^2 + 6x - 13$? 6

64. Simplify $(-2p)(-5)(4x)$. $40px$

▶ 3.7 Functions

In this section, you will learn about

Functions ■ Domain and range of a function ■ Function notation ■ Graphs of functions ■ The vertical line test ■ Determining the domain and range ■ Direct variation

---

Introduction In everyday life, we see a wide variety of situations where one quantity depends on another:

- The distance traveled by a car depends on its speed.
- The cost of renting a video depends on the number of days it is rented.
- A state's number of representatives in Congress depends on the state's population.

In this section, we will discuss many situations where one quantity depends on another according to a specific rule, called a *function*. For example, the equation $y = 2x - 3$ sets up a rule where each value of y depends on the choice of some number x. The rule is: *To find y, double the value of x and subtract* 3. In this case, y (the *dependent variable*) depends on x (the *independent variable*).

Functions

We have previously described relationships between two quantities in different ways:

The number of tires to order	is	two	times	the number of bicycles to be manufactured.

Here words are used to state that the number of bicycle tires to order depends on the number of bicycles to be manufactured.

This rectangular coordinate graph shows many ordered pairs (x, y) that satisfy the equation $y = x^2$, where the value of the y-coordinate depends on the value of the x-coordinate.

Acres	Schools
400	4
800	8
1,000	10
2,000	20

This table shows that the number of schools needed depends on the size of the housing development.

$$t = 1,500 - d$$

This equation describes how the amount of take-home pay t depends on the amount of deductions d.

Two observations can be made about these examples:

- Each one establishes a relationship between two sets of values. For example, the number of bicycle tires that must be ordered depends on the number of bicycles to be manufactured.

- In these relationships, each value in one set is assigned a *single* value of a second set. For example, for each number of bicycles to be manufactured, there is exactly one number of tires to order.

Relationships between two quantities that exhibit both of these characteristics are called **functions.**

Functions

A **function** is a rule that assigns to each number x (the input) a single value y (the output). In this case, we say y is a *function* of x.

We can restate the previous definition as follows: For y to be a function of x, each value of x must determine exactly one value of y.

EXAMPLE 1

Identifying functions. a. Does $y = 4x + 1$ define a function? **b.** Is age a function of body weight?

Solution   **a.** For each number x, we apply the rule: *Multiply x by 4 and add 1.* Since this arithmetic gives a single value of y, the equation defines a function.

b. To answer this question, we ask ourselves: *Does each body weight determine exactly one age?* Since a person weighing 130 pounds could be almost any age, the answer is no. This statement does not define a function.

SELF CHECK   **a.** Does $y = 2 - x^2$ define a function? **b.** Is the temperature in a city a function of the time of day?   *Answers:* **a.** yes, **b.** yes ■

Domain and Range of a Function

We have seen that functions can be represented by equations in two variables. Most often, x and y are used, but any letters are acceptable. Some examples of functions are

$$y = 2x - 10, \qquad y = x^2 + 2x - 3, \qquad \text{and} \qquad s = 5 - 16t$$

For a function, the set of all possible values of the independent variable x (the inputs) is called the **domain of the function.** The set of all possible values of the dependent variable y (the outputs) is called the **range of the function.**

EXAMPLE 2

Finding the domain and range of a function. Find the domain and range of $y = |x|$.

Solution   To find the domain of $y = |x|$, we determine which real numbers are allowable inputs for x. Since we can find the absolute value of any real number, the domain is the set of all real numbers. Since the absolute value of any real number x is greater than or equal to zero, the range of $y = |x|$ is the set of all real numbers greater than or equal to zero.

Find the domain and range of the function $y = -x$.   *Answer:* domain: all real numbers; range: all real numbers ∎

Function Notation

There is a special notation that we use to denote functions.

Function notation

The notation $y = f(x)$ denotes that y is a function of x.

The notation $y = f(x)$ is read as "y equals f of x." Note that y and $f(x)$ are two different notations for the same quantity. Thus, the equations $y = 4x + 1$ and $f(x) = 4x + 1$ represent the same relationship.

 WARNING! The symbol $f(x)$ denotes a function. It does not mean "f times x."

The notation $y = f(x)$ provides a way of denoting the value of y that corresponds to some number x. For example, if $f(x) = 4x + 1$, the value of y that is determined when $x = 2$ is denoted by $f(2)$.

$$f(x) = 4x + 1 \qquad \text{The function.}$$
$$f(2) = 4(2) + 1 \quad \text{Replace } x \text{ with 2.}$$
$$= 8 + 1$$
$$= 9$$

Thus, $f(2) = 9$.

The letter f used in the notation $y = f(x)$ represents the word *function*. However, other letters can be used to represent functions. For example, $y = g(x)$ and $y = h(x)$ also denote functions involving the variable x.

EXAMPLE 3

Evaluating functions. For $g(x) = 3 - 2x$ and $h(x) = x^3 - 1$, find   **a.** $g(3)$ and   **b.** $h(-2)$.

Solution   **a.** To find $g(3)$, we use the function rule $g(x) = 3 - 2x$ and replace x with 3.

$$g(x) = 3 - 2x$$
$$g(3) = 3 - 2(3)$$
$$= 3 - 6$$
$$= -3$$

b. To find $h(-2)$, we use the function rule $h(x) = x^3 - 1$ and replace x with -2.

$$h(x) = x^3 - 1$$
$$h(-2) = (-2)^3 - 1$$
$$= -8 - 1$$
$$= -9$$

Find $g(0)$ and $h(4)$ using the functions in Example 3.   *Answers:* **a.** 3, **b.** 63 ∎

We can think of a function as a machine that takes some input x and turns it into some output $f(x)$, as shown in Figure 3-59(a). The machine in Figure 3-59(b) turns the input value of -2 into the output value of -9, and we can write $f(-2) = -9$.

FIGURE 3-59

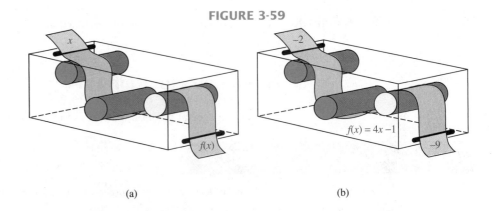

(a)   (b)

ACCENT ON TECHNOLOGY *Predicting Business Profits*

Accountants have found that the function $f(x) = -0.000065x^2 + 12x - 278,000$ estimates the profit a bowling alley will make when x games are bowled per year. Suppose that management predicts that 90,000 games will be bowled in the upcoming year. The expected profit for that year can be found by evaluating $f(90,000)$.

$$f(90,000) = -0.000065(90,000)^2 + 12(90,000) - 278,000$$

On a scientific calculator, we enter these numbers and press these keys:

Keystrokes   .000065 $\boxed{+/-}$ $\boxed{\times}$ 90000 $\boxed{x^2}$ $\boxed{+}$ 12 $\boxed{\times}$ 90000 $\boxed{-}$ 278000 $\boxed{=}$

$$\boxed{275500}$$

The expected profit is \$275,500.

Using the Table mode on a graphing calculator, we can quickly evaluate a function for several values of the independent variable. For example, to predict the profit that the bowling alley would make if 93,000, 94,000 or 95,000 games are bowled, we first press the $\boxed{Y=}$ key and enter the function, as shown in Figure 3-60(a).

Next, we press $\boxed{2nd}$ $\boxed{TBLSET}$ and use the $\boxed{\blacktriangledown}$ key and the $\boxed{\blacktriangleright}$ key, in combination, to highlight the Ask option on the line labeled Indpnt and press ENTER. See Figure 3-60(b).

Finally, we press $\boxed{2nd}$ $\boxed{TABLE}$, enter the first value for x (93,000), and press $\boxed{ENTER}$. The predicted profit will appear under the column headed Y_1. See Figure 3-60(c). The dark cursor will drop down a line, and the calculator will "ask" you for another value of the independent variable. Enter 94,000, and then 95,000. In each case, the corresponding profit (275,660 and 275,375) will be displayed to the right of the input value. This process can be used to evaluate the function for any value of the independent variable.

FIGURE 3-60

      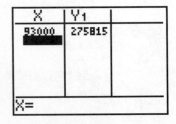

(a)   (b)   (c)

Graphs of Functions

The graph of the function $f(x) = 4x + 1$ is the same as the graph of the equation $y = 4x + 1$. So we can graph the function by making a table of values, plotting the points, and drawing the graph. A table of values and the graph of $f(x) = 4x + 1$ are shown in Figure 3-61.

FIGURE 3-61

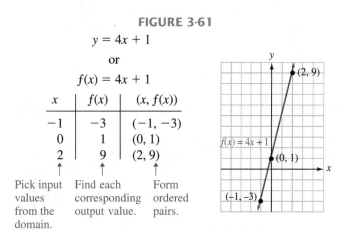

$$y = 4x + 1$$

or

$$f(x) = 4x + 1$$

x	$f(x)$	$(x, f(x))$
-1	-3	$(-1, -3)$
0	1	$(0, 1)$
2	9	$(2, 9)$

Pick input values from the domain.   Find each corresponding output value.   Form ordered pairs.

Any linear equation, except those of the form $x = a$, can be written using function notation by writing it in slope–intercept form ($y = mx + b$) and then replacing y with $f(x)$. We call this type of function a **linear function.**

Figure 3-62 shows the graphs of four basic functions.

FIGURE 3-62

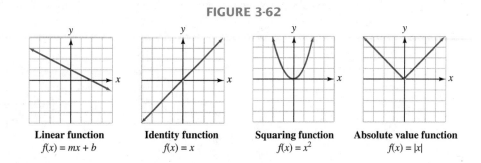

Linear function
$f(x) = mx + b$

Identity function
$f(x) = x$

Squaring function
$f(x) = x^2$

Absolute value function
$f(x) = |x|$

The Vertical Line Test

We can use the **vertical line test** to determine whether a given graph is the graph of a function. If any vertical line intersects a graph more than once, the graph cannot represent a function, because to one value of x, there corresponds more than one value of y. The graph in Figure 3-63(a), shown in red, is not the graph of a function, because the x-value -1 is assigned to three different y-values: 3, -1, and -4.

The graph shown in Figure 3-63(b) does represent a function, because every vertical line intersects the graph exactly once.

FIGURE 3-63

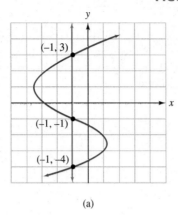

(a)

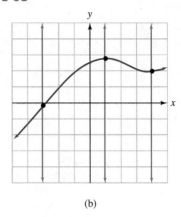

(b)

EXAMPLE 4

The vertical line test. Which of the following graphs are graphs of functions?

a.

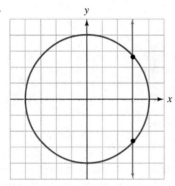

b.

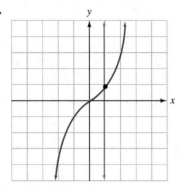

Solution   **a.** This graph is not the graph of a function, because the vertical line intersects the graph at more than one point.

b. This graph is the graph of a function, because no vertical line will intersect the graph at more than one point.

SELF CHECK   Which of the following graphs are graphs of functions?

a.

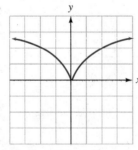

b.

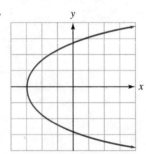

Answers:
a. function,
b. not a function ■

FIGURE 3-64

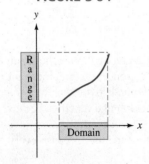

Determining the Domain and Range

In a function of the form $y = f(x)$, the numbers that we can substitute for x make up the domain of the function. The resulting values of y make up the range.

In a table of values, the numbers in the x-column are a partial listing of the numbers in the domain of a function. The numbers in the y-column are a partial listing of the numbers in the range of the function.

We can also determine the domain and range by looking at the graph of a function. For the function graphed in Figure 3-64, the domain is highlighted on the x-axis, and the range is highlighted on the y-axis.

| EXAMPLE 5 | **Finding the domain and range.** Find the domain and range of the function $f(x) = 4 - x^2$. | **FIGURE 3-65** |

Solution   The graph of $f(x) = 4 - x^2$ is shown in Figure 3-65. From the graph, we see that all numbers x on the x-axis are used. Thus, the domain is the set of all real numbers.

Since the values for y are always less than or equal to 4, the range is the set of real numbers less than or equal to 4.

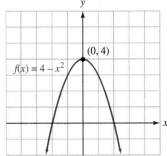

SELF CHECK   Graph $f(x) = x^2 + 2$. Find the domain and range of the function.

Answer: domain: all real numbers; range: all real numbers greater than or equal to 2   ∎

Direct Variation

In many applications, the relationship between two quantities can be described using the language of variation:

- The sales tax on an item *varies* with the price.
- The intensity of light *varies* with the distance from its source.
- The number of hours it takes to paint a house *varies* with the size of the painting crew.

One type of variation, called **direct variation,** is represented by a linear function of the form $y = kx$, where k is a constant.

Direct variation

The words *y varies directly with x* mean that

$$y = kx$$

for some constant k, called the **constant of variation.**

Scientists have found that the distance a spring will stretch *varies directly* with the force applied to it. The more force applied to the spring, the more it will stretch. If d represents distance and f represents force, this relationship can be expressed by the equation

$d = kf$ where k is the constant of variation

Suppose that a 150-pound wooden garage door stretches a spring 18 inches when closed. (See Figure 3-66.) We can find the constant of variation for the spring by substituting 150 for f and 18 for d in the formula $d = kf$ and solving for k:

FIGURE 3-66

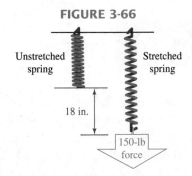

Unstretched spring   Stretched spring

18 in.

150-lb force

$$d = kf$$

$$18 = k(150)$$

$$\frac{18}{150} = k \qquad \text{Divide both sides by 150 to isolate } k.$$

$$\frac{3}{25} = k \qquad \text{Simplify the fraction: } \frac{18}{150} = \frac{\overset{1}{\cancel{6} \cdot 3}}{\cancel{6} \cdot 25} = \frac{3}{25}.$$

Therefore, the *function* describing the relationship between the distance the spring will stretch and the amount of force applied to it is $d = \frac{3}{25}f$. To find the distance that the same spring will stretch when a new 50-pound aluminum garage door is installed, we substitute 50 for f in $d = \frac{3}{25}f$.

$$d = \frac{3}{25}f \qquad \text{The function describing the direct variation.}$$
$$d = \frac{3}{25}(50)$$
$$d = 6 \qquad \text{Simplify the right-hand side.}$$

The spring will stretch 6 inches when the new door is closed.

E X A M P L E   6

Direct variation. The weight of an object on earth varies directly with its weight on the moon. If a rock weighs 5 pounds on the moon and 30 pounds on earth, what would be the weight on earth of a larger rock weighing 26 pounds on the moon?

Solution

Step 1: We let e represent the weight of the object on the earth and m the weight of the object on the moon. Translating the words *weight on earth varies directly with weight on the moon,* we get the equation

$$e = km$$

Step 2: To find the constant of variation, k, we substitute 30 for e and 5 for m.

$$e = km$$
$$30 = k(5)$$
$$6 = k \qquad \text{Divide both sides by 5.}$$

Step 3: The function describing the relationship between the weight of an object on the earth and on the moon is

$$e = 6m$$

Step 4: We can find the weight of the larger rock on the earth by substituting 26 for m in $e = 6m$.

$$e = 6m$$
$$e = 6(26)$$
$$e = 156$$

The rock would weigh 156 pounds on earth.

SELF CHECK

The cost of a bus ticket varies directly with the number of miles traveled. If a ticket for a 180-mile trip cost $45, what would a ticket for a 1,500-mile trip cost?

Answer: $375 ■

S T U D Y   S E T

Section 3.7

VOCABULARY

In Exercises 1–6, fill in the blanks to make the statements true.

1. A _____ function _____ is a rule that assigns to each value of the input set a single value of the output set.

2. The set of all possible input values for a function is called the _____ domain _____, and the set of all possible output values is called the _____ range _____.

3. For $y = 2x + 8$, x is called the _____ independent _____ variable, and y is called the _____ dependent _____ variable.

4. ___Direct___ variation is represented by a linear function of the form $y = kx$.

5. In $y = kx$, k is called the ___constant___ of variation.

6. $f(x) = 6 - 5x$ is an example of ___function___ notation.

CONCEPTS

7. Consider the function $f(x) = x^2$.
  a. If positive real numbers are substituted for x, what type of numbers result? positive numbers
  b. If negative real numbers are substituted for x, what type of numbers result? positive numbers
  c. If zero is substituted for x, what number results? 0
  d. What are the domain and range of the function? D: all reals; R: real numbers greater than or equal to 0

8. Consider the function $g(x) = x^4$.
  a. What type of numbers can be input in this function? What is the special name for this set? all real numbers; domain
  b. What type of numbers will be output by this function? What is the special name for this set? real numbers greater than or equal to 0; range

9. See Illustration 1.
  a. Give the coordinates of the points where the given vertical line intersects the graph. $(-2, 4), (-2, -4)$
  b. Is this the graph of a function? Explain your answer. No; the x-value -2 is assigned to more than one y-value (4 and -4).

ILLUSTRATION 1

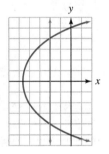

10. A function can be thought of as a machine that converts inputs into outputs. Use the terms *domain, range, input,* and *output* to label the diagram of a function machine in Illustration 2. Then find $f(2)$. $f(2) = 4$

ILLUSTRATION 2

domain   input   output   range

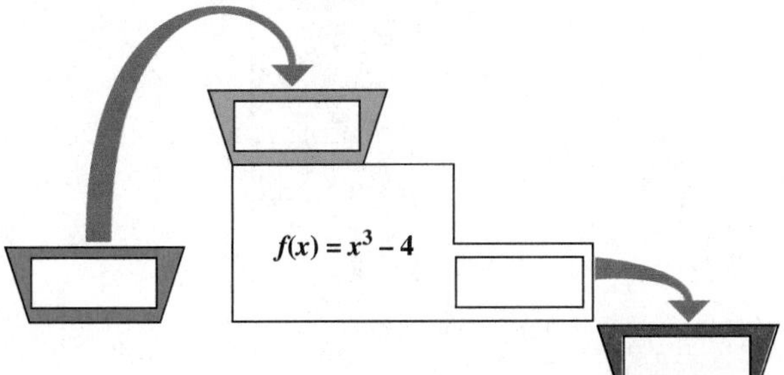

$f(x) = x^3 - 4$

11. Use the graph in Illustration 3 to find:
  a. $f(2)$ 0 **b.** $f(0)$ 1
  c. $f(-4)$ 3

ILLUSTRATION 3

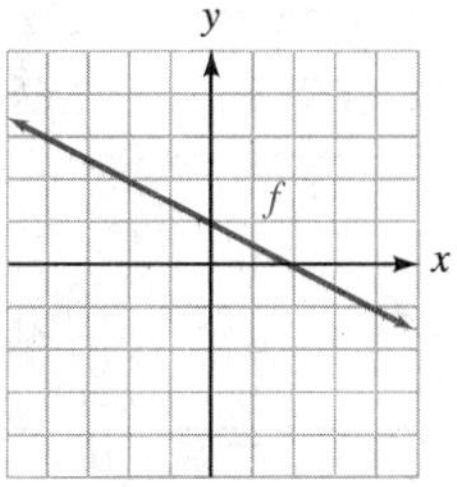

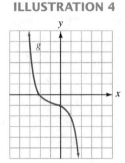

12. Use the graph in Illustration 4 to find:
  a. $g(-2)$ 0 **b.** $g(1)$ -2
  c. $g(-3)$ 5

ILLUSTRATION 4

13. Tell whether the equation defines direct variation.
  a. $y = kx$ yes **b.** $y = kx^2$ no
  c. $y = kx^3$ no **d.** $f(x) = kx$ yes

14. a. Translate to mathematical symbols: $h = ka$

| A farmer's harvest h | varies directly | with the number of acres planted a. |

  b. If the constant of variation for part a is $k = 10{,}000$, what will happen to the size of the harvest as the number of acres planted increases? It will increase.

15. Express this relationship using an equation: The number of gallons g of paint needed to paint a room varies directly with the number of square feet f to be painted. $g = kf$

16. Express this relationship using an equation: The amount of sales tax t varies directly with the purchase price p of a new car. $t = kp$

NOTATION

17. In the function $f(x) = x^2 + 1$, what is the input (independent) variable? x

18. In the function $v(t) = -32t + 1{,}000$, what is the input (independent) variable? t

19. Fill in the blanks to make the statements true. The function notation $f(4) = -5$ states that when 4 is substituted for x in function f, the result is -5. This fact can be illustrated graphically by plotting the point $(\,4\,,\,-5\,)$.

20. Fill in the blank: $f(x) = 6 - 5x$ is read as "f _of_ x is $6 - 5x$."

21. Fill in the blanks: If $f(x) = 6 - 5x$, then $f(0) = 6$ is read as "f <u>of</u> zero <u>is</u> 6."

22. Tell whether this statement is true or false: The equations $y = 3x + 5$ and $f(x) = 3x + 5$ are the same.
true

PRACTICE

In Exercises 23–36, tell whether a function is defined. If it is not, indicate an input for which there is more than one output.

23. $y = 2x + 10$ yes

24. $y = x - 15$ yes

25. $y = x^2$ yes

26. $y = |x|$ yes

27. $y^2 = x$
no; (4, 2), (4, −2)

28. $|y| = x$
no; (1, 1), (1, −1)

29. $y = x^3$ yes

30. $y = -x$ yes

31. yes

32. yes

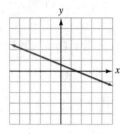

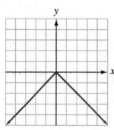

33. no; (3, 4), (3, −1)

34. no; (−1, 3), (−1, −3)

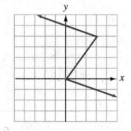

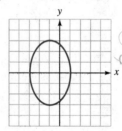

35. Is your age a function of your shoe size?
No; you could have a shoe size of 10 when you are 25 and 26 years old: (10, 25), (10, 26).

36. Is the number of phone calls you receive during the day a function of the time you wake up?
No; on Monday and Tuesday you could wake up at 7 A.M. but receive a different number of calls each day: (7, 3), (7, 5).

In Exercises 37–40, find the domain and range of the function.

37. $y = x + 1$
D: all reals; R: all reals

38. $y = 3x - 2$
D: all reals; R: all reals

39. $y = x^2$
D: all reals; R: real numbers greater than or equal to 0

40. $y = -|x|$
D: all reals; R: real numbers less than or equal to 0

In Exercises 41–48, find each value.

41. $f(x) = 4x - 1$
  a. $f(1)$ 3   **b.** $f(-2)$ −9
  c. $f\left(\dfrac{1}{4}\right)$ 0   **d.** $f(50)$ 199

42. $g(x) = 1 - 5x$
  a. $g(0)$ 1   **b.** $g(-75)$ 376
  c. $g(0.2)$ 0   **d.** $g\left(-\dfrac{4}{5}\right)$ 5

43. $h(t) = 2t^2$
  a. $h(0.4)$ 0.32   **b.** $h(-3)$ 18
  c. $h(1,000)$ 2,000,000   **d.** $h\left(\dfrac{1}{8}\right)$ $\frac{1}{32}$

44. $v(t) = 6 - t^2$
  a. $v(30)$ −894   **b.** $v(6)$ −30
  c. $v(-1)$ 5   **d.** $v(0.5)$ 5.75

45. $s(x) = |x - 7|$
  a. $s(0)$ 7   **b.** $s(-7)$ 14
  c. $s(7)$ 0   **d.** $s(8)$ 1

46. $f(x) = |2 + x|$
  a. $f(0)$ 2   **b.** $f(2)$ 4
  c. $f(-2)$ 0   **d.** $f(-99)$ 97

47. $f(x) = x^3 - x$
  a. $f(1)$ 0   **b.** $f(10)$ 990
  c. $f(-3)$ −24   **d.** $f(6)$ 210

48. $g(x) = x^4 + x$
  a. $g(1)$ 2   **b.** $g(-2)$ 14
  c. $g(0)$ 0   **d.** $g(10)$ 10,010

In Exercises 49–52, complete the table and graph the function. Then give the domain and range of the function.

49. $f(x) = -2 - 3x$

x	$f(x)$
0	−2
1	−5
−1	1
−2	4

D: all reals; R: all reals

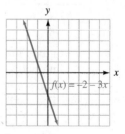

50. $h(x) = |1 - x|$

x	$h(x)$
0	1
1	0
2	1
3	2
−1	2
−2	3

D: all reals; R: real numbers greater than or equal to 0

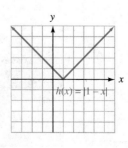

51. $s(x) = 2 - x^2$

x	$s(x)$
0	2
1	1
2	-2
-1	1
-2	-2

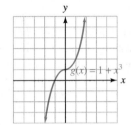

$s(x) = 2 - x^2$

D: all reals; R: real numbers less than or equal to 2

52. $g(x) = 1 + x^3$

x	$g(x)$
0	1
1	2
2	9
-1	0
-2	-7

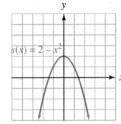

$g(x) = 1 + x^3$

D: all reals; R: all reals

53. Assume that y varies directly with x and $y = kx$.
  a. If $y = 10$ when $x = 2$, find k. 5
  b. Find y when $x = 7$. 35

54. Assume that t varies directly with s and $t = ks$.
  a. If $t = 21$ when $s = 6$, find k. $\frac{7}{2}$
  b. Find t when $s = 12$. 42

  use k that you found

APPLICATIONS

55. REFLECTIONS When a beam of light hits a mirror, it is reflected off the mirror at the same angle that the incoming beam struck the mirror, as shown in Illustration 5. What type of function could serve as a mathematical model for the path of the light beam shown here? $f(x) = |x|$

ILLUSTRATION 5

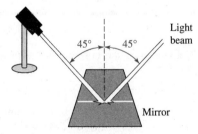

45°   45°   Light beam

Mirror

56. MATHEMATICAL MODELS Illustration 6 shows the path of a basketball shot taken by a player. What type of function could be used to mathematically model the path of the basketball? $f(x) = -x^2$

57. TIDES Illustration 7 shows the graph of a function f, which gives the height of the tide for a 24-hour period in Seattle, Washington. (Note that military time is used on the x-axis: 3 A.M. = 3, noon = 12, 3 P.M. = 15, 9 P.M. = 21, and so on.)
  a. Find the domain of the function. $0 \le x \le 24$
  b. Find $f(3)$. 0.5

ILLUSTRATION 6

  c. Find $f(6)$. 1.5
  d. Find $f(15)$. -1.5
  e. What information does $f(12)$ give?
    The low tide mark was -2.5 m.
  f. Estimate $f(21)$. 1.6

ILLUSTRATION 7

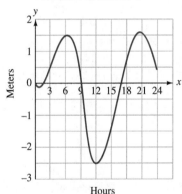

Meters
Hours

58. SOCCER Illustration 8 (on the next page) shows the graphs of three functions on the same coordinate system: $g(x)$ represents the number of girls, $f(x)$ represents the number of boys, and $t(x)$ represents the total number playing high school soccer in year x.
  a. What is the domain of each of these functions?
    77 to 95
  b. Find $g(87)$, $f(86)$, and $t(93)$.
    100,000, 200,000, and 400,000
  c. Estimate $g(95)$, $f(95)$, and $t(95)$.
    190,000, 280,000, 470,000
  d. For what year x was $g(x) = 75,000$? 1984
  e. For what year x was $f(x) = 225,000$? 1990
  f. For what year x was $t(x)$ first greater than 350,000? 1991

59. LAWN SPRINKLERS The function $A(r) = \pi r^2$ can be used to determine the area that will be watered by a rotating sprinkler that sprays out a stream of water r feet. See Illustration 9 on the next page. Find $A(5)$, $A(10)$, and $A(20)$. Round to the nearest tenth. 78.5 ft^2, 314.2 ft^2, 1,256.6 ft^2

ILLUSTRATION 8

U.S. High School Soccer Participation
1977-1995

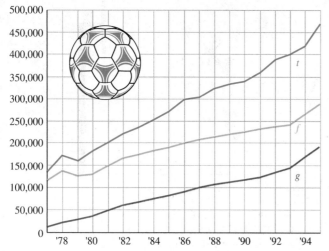

Based on data from *Los Angeles Times* (Dec. 12, 1996), p. A5

ILLUSTRATION 9

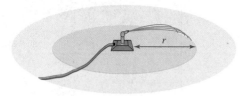

60. ⊞ PARTS LIST The function

$$f(r) = 2.30 + 3.25(r + 0.40)$$

approximates the length (in feet) of the belt that joins the two pulleys shown in Illustration 10. *r* is the radius (in feet) of the smaller pulley. Find the belt length needed for each pulley in the parts list.

ILLUSTRATION 10

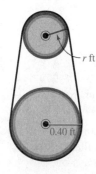

PARTS LIST		
Pulley	**r**	**Belt length**
P-45M	0.32	4.64
P-08D	0.24	4.38
P-00A	0.18	4.185
P-57X	0.38	4.835

61. COMMUTING DISTANCE The distance that a car can travel without refueling varies directly with the number of gallons of gasoline in the tank. If a car can go 360 miles on a full tank of gas (15 gallons), how far can it go on 7 gallons of gas? 168 mi

62. COMPUTING FORCES The force of gravity acting on an object varies directly with the mass of the object. The force on a mass of 5 kilograms is 49 newtons. What is the force acting on a mass of 12 kilograms? 117.6 newtons $g = km, \ 49 = k5$

63. MEDICATION To fight ear infections in children, doctors often prescribe the antibiotic Ceclor. The recommended dose in milligrams varies directly with the child's body weight in pounds. The correct dosage for a 20-pound child is 124 milligrams. What would be the correct dosage for a 28-pound child? 173.6 mg

64. DOSAGE The recommended dose (in milligrams) of Demerol, a preoperative medication given to children, varies directly with the child's weight in pounds. The proper dosage for a child weighing 30 pounds is 18 milligrams. What would be the correct dosage for a child weighing 45 pounds? 27 mg

WRITING

65. What is a function? Give an example.

66. Give an example of two quantities that vary directly and two that do not.

67. In the function $y = -5x + 2$, why do you think x is called the *independent* variable and y the *dependent* variable?

68. Explain what a politician meant when she said, "The speed at which the downtown area will be redeveloped is a function of the number of low-interest loans made available to the property owners."

REVIEW

69. Give the equation of the horizontal line passing through $(-3, 6)$. $y = 6$

70. Is $t = -3$ a solution of $t^2 - t + 1 = 13$? yes

71. Write the formula that relates profit, revenue, and costs. profit = revenue − costs

72. What is the word used to represent the perimeter of a circle? circumference

73. Remove the parentheses in $-3(2x - 4)$. $-6x + 12$

74. Evaluate $r^2 - r$ for $r = -0.5$. 0.75

75. Write an expression for how many eggs there are in d dozen. $12d$

76. On a rectangular coordinate graph, what variable is associated with the horizontal axis? x

Describing Linear Relationships

In Chapter 3, we discussed four ways to mathematically describe linear relationships between two quantities.

Equations in Two Variables

The general form of the equation of a line is $Ax + By = C$. Two very useful forms of the equation of a line are the slope–intercept form and the point–slope form.

1. Write the equation of a line with a slope of -3 and a y-intercept of $(0, -4)$. $y = -3x - 4$

2. Write the equation of the line that passes through $(5, 2)$ and $(-5, 0)$. Answer in slope–intercept form. $y = \frac{1}{5}x + 1$

Rectangular Coordinate Graphs

The graph of an equation is a "picture" of all of its solutions (x, y). Important information can be obtained from a graph.

3. Complete the table of solutions for $2x - 4y = 8$. Then graph the equation.

4. See Illustration 1.
   a. What information does the y-intercept of the graph give us? When new, the press cost $40,000.
   b. What is the slope of the line and what does it tell us? $-5,000$; the value of the press decreased $5,000/yr

$2x - 4y = 8$

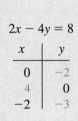

x	y
0	-2
4	0
-2	-3

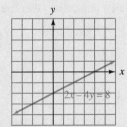

ILLUSTRATION 1

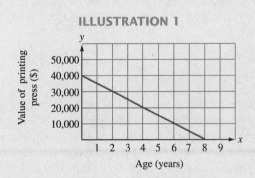

Linear Functions

We can use the notation $f(x) = mx + b$ to describe linear functions.

5. The function $f(x) = 25 + 35x$ gives the cost (in dollars) to rent a cement mixer for x days. Find $f(3)$. What does it represent?
130; the cost to rent the mixer for 3 days

6. The function $T(c) = \frac{1}{4}c + 40$ predicts the outdoor temperature T in degrees Fahrenheit using the number of cricket chirps c per minute. Find $T(160)$. $80°F$

Direct Variation

We can describe direct variation by using an equation of the form $y = kx$.

7. The number of purchases n made at a toy store in a mall varies directly with the number of people p entering the mall. Write an equation describing this relationship if 75 purchases were made on a weekday when 5,000 people visited the mall. $n = \frac{3}{200}p$

8. Use your result from Exercise 7 to find the number of purchases the toy store should make if an attendance of 9,000 is predicted for a special Saturday promotion in the mall. 135

Accent on Teamwork

Section 3.1

Daily high temperature For a 2-week period, plot the daily high temperature for your city on a rectangular coordinate system. You can normally find this information in a local newspaper. Label the *x*-axis "observation day" and the *y*-axis "daily high temperature in degrees Fahrenheit." For example, the ordered pair (3, 72) indicates that on day 3 of the observation period, the high temperature was 72°F. At the end of the 2-week period, see whether any temperature trend is apparent from the graph.

Section 3.2

Translations On a piece of graph paper, sketch the graph of $y = |x|$ with a black marker. Using a different color, sketch the graphs of $y = |x| + 2$ and $y = |x| - 2$ on the same coordinate system. On another piece of graph paper, do the same for $y = |x|$ and $y = |x + 2|$ and $y = |x - 2|$. Make some observations about how the graph of $y = |x|$ is "moved" or "translated" by the addition or subtraction of 2. Use what you have learned to discuss the graphs of $y = x^2$, $y = x^2 + 2$, $y = x^2 - 2$, $y = (x + 2)^2$, and $y = (x - 2)^2$.

Section 3.3

Computer graphing programs If your school has a mathematics computer lab, ask the lab supervisor whether there is a graphing program on the system. If so, familiarize yourself with the operation of the program and then graph each of the equations from Figure 3-21 on page 210 and from Examples 3–6 in Section 3.3. Print out each graph and compare with those in the textbook.

Section 3.4

Measuring slope Use a tape measure (and a level if necessary) to find the slopes of five objects by finding $\frac{\text{rise}}{\text{run}}$. See the applications in Study Set 3.4 for some ideas about what you can measure. Record your results in a chart like the one shown in Illustration 1. List the examples in increasing order of magnitude, starting with the smallest slope.

Section 3.5

Shopping Visit a local grocery store and find the price per pound of bananas. Make a rectangular coordinate graph that

ILLUSTRATION 1

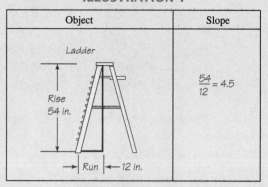

could be posted next to the scale in the produce area so that shoppers could determine from the graph the cost of a banana purchase up to 8 pounds in weight. Label the *x*-axis in quarters of a pound and label the *y*-axis in cents.

Section 3.6

Matching game Have a student in your group write 10 linear equations on 3×5 note cards, one equation per card. Then have him or her graph each equation on a separate set of 10 cards. Shuffle each set of cards. Then put all the equation cards on one side of a table and all the cards with graphs on the other side. Work together to match each equation with its proper graph.

Section 3.7

Direct variation Purchase a small spring at a hardware store and borrow a 5- and a 10-pound weight from someone you know who lifts weights. Attach the spring to a table and measure the number of inches (the distance *d*) the spring is stretched when the 5-pound weight (the force *f*) is attached to it. Use this information to determine *k* in $d = kf$. Then substitute 10 pounds for *f* in the equation and determine the distance a 10-pound weight should stretch the spring. Finally, attach the 10-pound weight to the spring and measure the distance the spring is stretched. Are the two numbers the same? If not, what could be causing the difference in values?

Section 3.1

Graphing Using the Rectangular Coordinate System

CONCEPTS

A *rectangular coordinate system* is composed of a horizontal number line called the *x*-axis and a vertical number line called the *y*-axis.

The coordinates of the *origin* are (0, 0).

To *graph* ordered pairs means to locate their position on a coordinate system.

The two axes divide the co-ordinate plane into four distinct regions called *quadrants*.

REVIEW EXERCISES

1. a. Graph the points with coordinates $(-1, 3)$, $(0, 1.5)$, $(-4, -4)$, $\left(2, \frac{7}{2}\right)$, and $(4, 0)$.

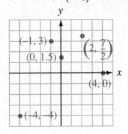

b. Use the graph in Illustration 1 to complete the table.

x	y
3	-1
0	0
-3	1

ILLUSTRATION 1

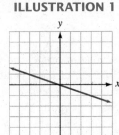

2. In what quadrant does the point $(-3, -4)$ lie?   quadrant III

3. SNOWFALL The amount of snow on the ground at a mountain resort was measured once each day over a 7-day period. (See Illustration 2.)
   a. On the first day, how much snow was on the ground?   2 ft
   b. What was the difference in the amount of snow on the ground when the measurements were taken the second and third day?   2 ft
   c. How much snow was on the ground on the sixth day?   6 ft

ILLUSTRATION 2

ILLUSTRATION 3

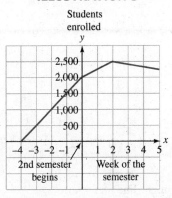

4. COLLEGE ENROLLMENT The graph in Illustration 3 gives the number of students enrolled at a college for the period from 4 weeks before to 5 weeks after the semester began.
   a. What was the maximum enrollment and when did it occur?   2,500; week 2
   b. How many students had enrolled 2 weeks before the semester began?
      1,000
   c. When was enrollment 2,250?   1st week and 5th week

Section 3.2

Equations Containing Two Variables

An ordered pair is a *solution* if, after substituting the values of the ordered pair for the variables in the equation, the result is a true statement.

5. Check to see whether $(-3, 5)$ is a solution of $y = |2 + x|$.   not a solution

Solutions of an equation can be shown in a *table of values*.

In an equation in x and y, x is called the *independent variable,* or *input,* and y is called the *dependent variable,* or *output.*

6. a. Complete the table of values and graph the equation $y = -x^3$.

b. How would the graph of $y = -x^3 + 2$ compare to the graph of the equation given in part a?

It would be 2 units higher.

$$y = -x^3$$

x	y	(x, y)
-2	8	$(-2, 8)$
-1	1	$(-1, 1)$
0	0	$(0, 0)$
1	-1	$(1, -1)$
2	-8	$(2, -8)$

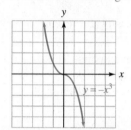

To graph an equation in two variables:
1. Make a table of values that contains several solutions written as ordered pairs.
2. Plot each ordered pair.
3. Draw a line or smooth curve through the points.

In many application problems, we encounter equations that contain variables other than x and y.

7. The graph in Illustration 4 shows the relationship between the number of oranges O an acre of land will yield if t orange trees are planted on it.
a. If $t = 70$, what is O?   9,000
b. What importance does the point $(40, 18)$ on the graph have?

It tells us that 40 trees on an acre give the highest yield, 18,000 oranges.

ILLUSTRATION 4

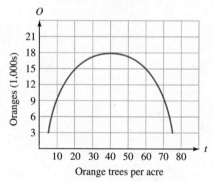

Section 3.3

Graphing Linear Equations

An equation whose graph is a straight line and whose variables are raised to the first power is called a *linear equation.*

8. Classify each equation as either linear or nonlinear.
a. $y = |x + 2|$ nonlinear
b. $3x + 4y = 12$ linear

c. $y = 2x - 3$ linear
d. $y = x^2 - x$ nonlinear

The *general* or *standard form* of a linear equation is $Ax + By = C$ where A, B, and C are real numbers and A and B are not both zero.

9. The equation $5x + 2y = 10$ is in general form; what are A, B, and C?
$A = 5$, $B = 2$, $C = 10$

10. Complete the table of solutions for the
equation $3x + 2y = -18$.

x	y	(x, y)
-2	-6	$(-2, -6)$
-8	3	$(-8, 3)$

To graph a linear equation:
1. Find three (x, y) pairs that
   satisfy the equation by
   picking three arbitrary
   x-values and finding their
   corresponding y-values.
2. Plot each ordered pair.
3. Draw a line through the
   points.

11. Solve the equation $x + 2y = 6$ for y, find three
solutions, and then graph it. $y = -\frac{1}{2}x + 3$

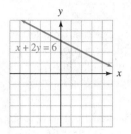

To find the *y-intercept* of a
linear equation, substitute
zero for x in the equation of
the line and solve for y. To
find the *x-intercept* of a linear equation, substitute zero
for y in the equation of the
line and solve for x.

12. Graph $-4x + 2y = 8$ by finding its x- and
y-intercepts.
x-intercept: $(-2, 0)$; y-intercept: $(0, 4)$

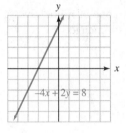

The equation $y = b$ represents the horizontal line that
intersects the y-axis at $(0, b)$.
The equation $x = a$ represents the vertical line that
intersects the x-axis at $(a, 0)$.

13. Graph each equation.
a. $y = 4$

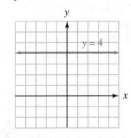

b. $x = -1$

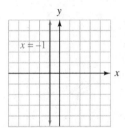

We can solve equations
graphically by determining
the x-coordinate of the
x-intercept of the associated
graph.

14. What is the solution of the equation
$-3(x - 1) - 4x = 17$ if the graphing
calculator display shows the graph of
$y = -3(x - 1) - 4x - 17$? -2

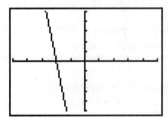

Section 3.4   Rate of Change and the Slope of a Line

The *slope m* of a nonvertical
line is a number that measures "steepness" by finding
the ratio $\frac{\text{rise}}{\text{run}}$.

$$m = \frac{\text{change in the } y\text{-values}}{\text{change in the } x\text{-values}}$$

15. In each case, find the slope of the line.
a. $\frac{1}{4}$

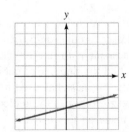

b. The line with the table of values
shown here. -7

x	y	(x, y)
2	-3	$(2, -3)$
4	-17	$(4, -17)$

If $P(x_1, y_1)$ and $Q(x_2, y_2)$ are two points on a nonvertical line, the slope m of line PQ is

$$m = \frac{y_2 - y_1}{x_2 - x_1}$$

Lines that rise from left to right have a *positive slope,* and lines that fall from left to right have a *negative slope.*

Horizontal lines have a slope of zero. Vertical lines have *undefined* slope.

The slope of a line gives a rate of change.

c. The line passing through the points $(2, -5)$ and $(5, -5)$.   0

d. The line passing through the points $(1, -4)$ and $(3, -7)$. $-\frac{3}{2}$

16. Graph the line that passes through $(-2, 4)$ and has slope $m = -\frac{4}{5}$.

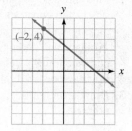

17. SALMON FISHING The graph in Illustration 5 shows the annual British Columbia salmon catch, in millions of fish.

a. When did the largest decline in the number of salmon caught occur? What was the rate of change over this time period?
1985–1987; −8.5 million salmon/yr

b. When did the largest increase in the number of salmon caught occur? What was the rate of change over this time period?
1987–1989; 6 million salmon/yr

ILLUSTRATION 5

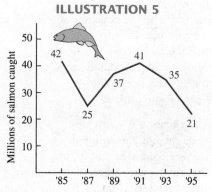

Based on data from *Los Angeles Times* (June 13, 1996)

Section 3.5   Describing Linear Relationships

If a linear equation is written in *slope–intercept* form,

$y = mx + b$

the graph of the equation is a line with slope m and y-intercept $(0, b)$.

18. Find the slope and the y-intercept of each line.

a. $y = \frac{3}{4}x - 2$

$m = \frac{3}{4}$; y-intercept: $(0, -2)$

b. $y = -4x$ $m = -4$; y-intercept: $(0, 0)$

19. Find the slope and the y-intercept of the line determined by $9x - 3y = 15$. Then graph it.
$m = 3$; y-intercept: $(0, -5)$

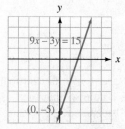

The *rate of change* is the slope of the graph of a linear equation.

20. COPIER USAGE A business buys a used copy machine that, when purchased, has already produced 75,000 copies.

a. If the business plans to run 300 copies a week, write a linear equation that would find the number of copies c the machine has made in its lifetime after the business has used it for w weeks. $c = 300w + 75{,}000$

b. Use your result to part a to predict the total number of copies that will have been made on the machine 1 year, or 52 weeks, after being purchased by the business.   90,600

Two lines with the same slope are *parallel*.

The product of the slopes of *perpendicular* lines is -1.

21. Without graphing, tell whether graphs of the given pair of lines would be parallel, perpendicular, or neither.

a. $y = -\dfrac{2}{3}x + 6$

$y = -\dfrac{2}{3}x - 6$ parallel

b. $x + 5y = -10$

$y = 5x$

perpendicular

Section 3.6

If a line with slope m passes through the point (x_1, y_1), the equation of the line in *point–slope* form is

$$y - y_1 = m(x - x_1)$$

Writing Linear Equations

22. Write the equation of a line with the given slope that passes through the given point. Express the result in slope–intercept form and graph the equation.

a. $m = 3$, $(1, 5)$ $y = 3x + 2$

b. $m = -\dfrac{1}{2}$, $(-4, -1)$ $y = -\frac{1}{2}x - 3$

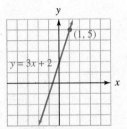

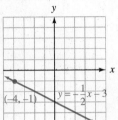

23. Write the equation of the line with the following characteristics. Express the result in slope–intercept form.

a. passing through $(3, 7)$ and $(-6, 1)$ $y = \frac{2}{3}x + 5$

b. horizontal, passing through $(6, -8)$

$y = -8$

24. CAR REGISTRATION   When it was 2 years old, the annual registration fee for a Dodge Caravan was \$380. When it was 4 years old, the registration fee dropped to \$310. If the relationship is linear, write an equation that gives the registration fee f in dollars for the van when it is x years old.

$f = -35x + 450$

Section 3.7

A *function* is a rule that assigns to each input value a single output value.

For a function, the set of all possible values of the independent variable x (the inputs) is called the *domain*, and the set of all possible values of the dependent variable y (the outputs) is called the *range*.

The notation $y = f(x)$ denotes that y is a function of x.

Functions

25. In each case, tell whether a function is defined.

a. $y = 3x - 2$ yes

b. Is your age a function of your height?   no

26. Find the domain and range of each function.

a.

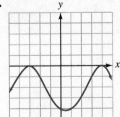

D: all reals; R: real numbers less than or equal to 0

b. $y = x^2$ D: all reals; R: real numbers greater than or equal to 0

27. For the function $g(x) = 1 - 6x$, find each value.

a. $g(1)$ -5

b. $g(-6)$ 37

c. $g(0.5)$ -2

d. $g\left(\dfrac{3}{2}\right)$ -8

Four basic functions are

Linear: $f(x) = mx + b$
Identity: $f(x) = x$
Squaring: $f(x) = x^2$
Absolute value: $f(x) = |x|$

28. Complete the table and graph the function.

$h(x) = 1 - |x|$

x	$h(x)$
0	1
1	0
2	−1
−1	0
−2	−1
−3	−2

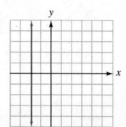

We can use the *vertical line test* to determine whether a graph is the graph of a function.

29. Tell whether each graph is the graph of a function.

a. no

b. yes

30. The function $f(r) = 15.7r^2$ estimates the volume in cubic inches of a can 5 inches tall with a radius of r inches. Find the volume of the can in Illustration 6.   1,004.8 in.^{3}

ILLUSTRATION 6

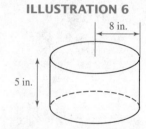

8 in.

5 in.

The words *y varies directly with x* mean that $y = kx$ for some constant k, called the *constant of variation*.

31. PROFIT The profit made by a strawberry farm varies directly with the number of baskets of strawberries sold. If a profit of $500 was made from the sale of 750 baskets, what is the profit when 1,250 baskets are sold? $833.33

CHAPTER 3

Test

The graph in Illustration 1 shows the number of dogs being boarded in a kennel over a 3-day holiday weekend. Use the graph to answer Problems 1–4.

1. How many dogs were in the kennel 2 days before the holiday?   10

2. What is the maximum number of dogs that were boarded on the holiday weekend?   60

3. When were there 30 dogs in the kennel?

1 day before and the 3rd day of the holiday

ILLUSTRATION 1

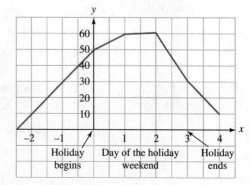

4. What information does the *y*-intercept of the graph give? 50 dogs were in the kennel when the holiday began.

5. Graph $y = x^2 - 4$.

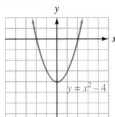

6. Graph $8x + 4y = -24$.

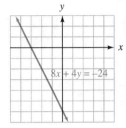

7. Is $(-3, -4)$ a solution of $3x - 4y = 7$? yes

8. Is $y = x^3$ a linear equation? no

9. What are the *x*- and *y*-intercepts of the graph of $2x - 3y = 6$? *x*-intercept: $(3, 0)$; *y*-intercept: $(0, -2)$

10. Find the slope and the *y*-intercept of $x + 2y = 8$. $m = -\frac{1}{2}$; $(0, 4)$

11. Graph $x = -4$.

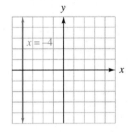

12. Graph the line passing through $(-2, -4)$ having a slope of $\frac{2}{3}$.

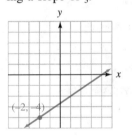

13. What is the slope of the line passing through $(-1, 3)$ and $(3, -1)$? -1

14. What is the slope of a vertical line? undefined

15. What is the slope of a line that is perpendicular to a line with slope $-\frac{7}{8}$? $\frac{8}{7}$

16. When graphed, are the lines $y = 2x + 6$ and $6x - 3y = 0$ parallel, perpendicular, or neither? parallel

In Problems 17–18, refer to the graph in Illustration 2, which shows the elevation changes in a 26-mile marathon course. Give the rate of change of the part of the course that has . . .

17. the steepest incline
the 15–20 mi segment: 30 ft/mi

18. the steepest decline
the 22–25 mi segment: $-\frac{100}{3}$ ft/mi $= -33\frac{1}{3}$ ft/mi

19. DEPRECIATION After it is purchased, a \$15,000 computer loses \$1,500 in resale value every year. Write an equation that gives the resale value *v* of the computer *x* years after being purchased. $v = 15{,}000 - 1{,}500x$

20. Write the equation of the line passing through $(-2, 5)$ and $(-3, -2)$. Answer in slope–intercept form. $y = 7x + 19$

21. Is this the graph of a function? no

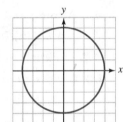

22. Find the domain and range of the function.
D: all reals; R: real numbers less than or equal to 0

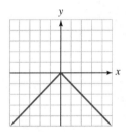

23. Does the equation $y = 2x - 8$ define a function? yes

24. If $f(x) = 2x - 7$, find $f(-3)$. -13

25. If $g(s) = 3.5s^3$, find $g(6)$. 756

26. If *y* varies directly with *x*, and $y = 32$ when $x = 8$, find *y* when $x = 1$. 4

27. Explain what is meant by the statement slope $= \frac{\text{rise}}{\text{run}}$.

28. Give an example of two quantities that vary directly.

ILLUSTRATION 2

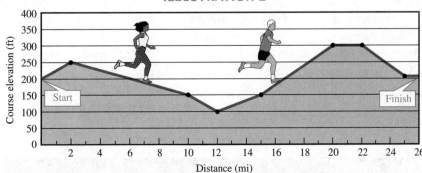

4

Exponents and Polynomials

CAMPUS CONNECTION

The Administration of Justice Department

People studying to become police officers learn about traffic law enforcement. As their course of study progresses, they use algebra in several aspects of traffic control and accident investigation. For example, the number of feet a car travels before stopping can be found using the *polynomial* function $f(v) = 0.04v^2 + 0.9v$, where v is the velocity of the car. For a car traveling at 30 mph, we can find its stopping distance by substituting 30 for v and evaluating $0.04(30)^2 + 0.9(30)$. In this chapter, polynomial functions are introduced. You will see that they have applications in many different areas, including administration of justice.

> *In this chapter, we introduce the rules for exponents*
> *and use them to add, subtract, multiply,*
> *and divide polynomials.*

▶ 4.1   Natural-Number Exponents

In this section, you will learn about

> Powers of expressions ■ The product rule for exponents ■ The power rules for exponents ■ The quotient rule for exponents

---

Introduction   We have used natural-number exponents to indicate repeated multiplication. For example,

$$2^5 = 2 \cdot 2 \cdot 2 \cdot 2 \cdot 2 = 32 \qquad (-7)^3 = (-7)(-7)(-7) = -343$$
$$x^4 = x \cdot x \cdot x \cdot x \qquad\qquad -y^5 = -y \cdot y \cdot y \cdot y \cdot y$$

These examples suggest a definition for x^n, where n is a natural number.

Natural-number exponents

If n is a natural number, then

$$x^n = \overbrace{x \cdot x \cdot x \cdots \cdots x}^{n \text{ factors of } x}$$

In the exponential expression x^n, x is called the **base** and n is called the **exponent.** The entire expression is called a **power of x.**

$$\text{Base} \rightarrow x^n \leftarrow \text{Exponent}$$

Powers of Expressions

If an exponent is a natural number, it tells how many times its base is to be used as a factor. An exponent of 1 indicates that its base is to be used one time as a factor, an exponent of 2 indicates that its base is to be used two times as a factor, and so on.

$$3^1 = 3, \qquad (-y)^1 = -y, \qquad (-4z)^2 = (-4z)(-4z), \qquad \text{and} \qquad (t^2)^3 = t^2 \cdot t^2 \cdot t^2$$

EXAMPLE 1   **Powers of integers.** Show that -2^4 and $(-2)^4$ have different values.

Solution   We find each power and show that the results are different. In the first expression, the base is 2. In the second expression, the base is -2.

$$-2^4 = -(2^4) \qquad\qquad (-2)^4 = (-2)(-2)(-2)(-2)$$
$$= -(2 \cdot 2 \cdot 2 \cdot 2) \qquad\qquad = 16$$
$$= -16$$

Since $-16 \neq 16$, it follows that $-2^4 \neq (-2)^4$.

SELF CHECK   Show that $(-4)^3$ and -4^3 have the same value.   *Answer:* $(-4)^3 = -4^3 = -64$ ■

EXAMPLE 2

Powers of algebraic expressions. Write each expression without using exponents: **a.** r^3, **b.** $(-2s)^4$, and **c.** $\left(\frac{1}{3}ab\right)^5$.

Solution

a. $r^3 = r \cdot r \cdot r$ Write the base r as a factor three times.

b. $(-2s)^4 = (-2s)(-2s)(-2s)(-2s)$ Write the base $-2s$ as a factor four times.

c. $\left(\frac{1}{3}ab\right)^5 = \left(\frac{1}{3}ab\right)\left(\frac{1}{3}ab\right)\left(\frac{1}{3}ab\right)\left(\frac{1}{3}ab\right)\left(\frac{1}{3}ab\right)$ Write the base $\frac{1}{3}ab$ as a factor five times.

SELF CHECK

Write each expression without using exponents: *Answers:*   **a.** $x \cdot x \cdot x \cdot x$,
a. x^4 and   **b.** $\left(-\frac{1}{2}xy\right)^3$.   **b.** $\left(-\frac{1}{2}xy\right)\left(-\frac{1}{2}xy\right)\left(-\frac{1}{2}xy\right)$ ∎

The Product Rule for Exponents

To develop a rule for multiplying exponential expressions with the same base, we consider the product $x^2 \cdot x^3$. Since the expression x^2 means that x is to be used as a factor two times and the expression x^3 means that x is to be used as a factor three times, we have

$$x^2 \cdot x^3 = \overbrace{x \cdot x}^{2 \text{ factors of } x} \cdot \overbrace{x \cdot x \cdot x}^{3 \text{ factors of } x}$$

$$= \overbrace{x \cdot x \cdot x \cdot x \cdot x}^{5 \text{ factors of } x}$$

$$= x^5$$

In general,

$$x^m \cdot x^n = \overbrace{x \cdot x \cdot x \cdots \cdot x}^{m \text{ factors of } x} \overbrace{x \cdot x \cdot x \cdot x \cdots \cdot x}^{n \text{ factors of } x}$$

$$= \overbrace{x \cdot x \cdot x \cdot x \cdot x \cdot x \cdots \cdot x \cdot x \cdot x}^{m + n \text{ factors of } x}$$

$$= x^{m+n}$$

This discussion suggests the following rule: *To multiply two exponential expressions with the same base, keep the common base and add the exponents.*

Product rule for exponents

If m and n are natural numbers, then

$$x^m x^n = x^{m+n}$$

EXAMPLE 3

The product rule for exponents. Simplify by writing each expression using one base and one exponent: **a.** x^3x^4 and   **b.** y^2y^4y.

a. $x^3x^4 = x^{3+4}$ Keep the base and add the exponents.

     $= x^7$ Do the addition: $3 + 4 = 7$.

b. $y^2y^4y = (y^2y^4)y$ Use the associative property to group y^2 and y^4.

     $= (y^{2+4})y$ Keep the base and add the exponents.

     $= y^6y$ Do the addition: $2 + 4 = 6$.

     $= y^{6+1}$ Keep the base and add the exponents. Recall that $y = y^1$.

     $= y^7$ Do the addition: $6 + 1 = 7$.

SELF CHECK Simplify **a.** zz^3 and **b.** $x^2x^3x^6$. *Answers:* **a.** z^4, **b.** x^{11} ∎

WARNING! We cannot simplify $x^3 + x^4$ or $x^3 - x^4$ because x^3 and x^4 are not like terms—they don't have exactly the same variables *and* exponents. However, we can simplify $x^3 \cdot x^4$ using the product rule, because the exponential expressions have the same base.

EXAMPLE 4

The product rule for exponents. Simplify $(2y^3)(3y^2)$.

Solution

$(2y^3)(3y^2) = 2(3)y^3y^2$ Use the commutative and associative properties to group the coefficients together and the variables together.

$= 6y^{3+2}$ Multiply 2 and 3. Keep the base y and add the exponents.

$= 6y^5$ Do the addition: $3 + 2 = 5$.

SELF CHECK Simplify $(4x)(-3x^2)$. *Answer:* $-12x^3$ ∎

WARNING! The product rule for exponents applies to exponential expressions with the same base. An expression such as x^2y^3 cannot be simplified, because x^2 and y^3 have different bases.

The Power Rules for Exponents

To find another rule for exponents, we consider the expression $(x^3)^4$, which can be written as $x^3 \cdot x^3 \cdot x^3 \cdot x^3$. Because each of the four factors of x^3 contains three factors of x, there are $4 \cdot 3$ (or 12) factors of x. This product can be written as x^{12}.

$$(x^3)^4 = x^3 \cdot x^3 \cdot x^3 \cdot x^3$$

$$= \underbrace{\underbrace{x \cdot x \cdot x}_{x^3} \cdot \underbrace{x \cdot x \cdot x}_{x^3} \cdot \underbrace{x \cdot x \cdot x}_{x^3} \cdot \underbrace{x \cdot x \cdot x}_{x^3}}_{\text{12 factors of } x}$$

$$= x^{12}$$

In general,

$$(x^m)^n = \overbrace{x^m \cdot x^m \cdot x^m \cdot \cdots \cdot x^m}^{n \text{ factors of } x^m}$$

$$= \overbrace{x \cdot x \cdot x \cdot x \cdot x \cdot x \cdot x \cdots \cdots x}^{m \cdot n \text{ factors of } x}$$

$$= x^{m \cdot n}$$

This discussion suggests the following rule: *To raise an exponential expression to a power, keep the base and multiply the exponents.*

The first power rule for exponents

If m and n are natural numbers, then

$$(x^m)^n = x^{m \cdot n}$$

EXAMPLE 5

The first power rule for exponents. Simplify by writing each expression using one base and one exponent.

a. $(2^3)^7 = 2^{3 \cdot 7}$ Keep the base and multiply the exponents.

$= 2^{21}$ Do the multiplication: $3 \cdot 7 = 21$.

b. $(z^7)^7 = z^{7 \cdot 7}$ Keep the base and multiply the exponents.

$= z^{49}$ Do the multiplication: $7 \cdot 7 = 49$.

SELF CHECK

Write each expression using one exponent:
a. $(y^5)^2$ and   **b.** $(u^x)^y$.

Answers: **a.** y^{10}, **b.** u^{xy} ■

EXAMPLE 6

Applying two rules for exponents. Use the product and power rules of exponents to write each expression using one base and one exponent.

a. $(x^2 x^5)^2 = (x^7)^2$

$= x^{14}$

b. $(y^6 y^2)^3 = (y^8)^3$

$= y^{24}$

c. $(z^2)^4 (z^3)^3 = z^8 z^9$

$= z^{17}$

d. $(x^3)^2 (x^5 x^2)^3 = x^6 (x^7)^3$

$= x^6 x^{21}$

$= x^{27}$

SELF CHECK

Write each expression using one exponent:
a. $(a^4 a^3)^3$ and   **b.** $(a^3)^3 (a^4)^2$.

Answers: **a.** a^{21}, **b.** a^{17} ■

To find two more rules for exponents, we consider the expressions $(2x)^3$ and $\left(\frac{2}{x}\right)^3$.

$(2x)^3 = (2x)(2x)(2x)$

$\left(\frac{2}{x}\right)^3 = \left(\frac{2}{x}\right)\left(\frac{2}{x}\right)\left(\frac{2}{x}\right)$ $(x \neq 0)$

$= (2 \cdot 2 \cdot 2)(x \cdot x \cdot x)$

$= \dfrac{2 \cdot 2 \cdot 2}{x \cdot x \cdot x}$ Multiply the numerators.
Multiply the denominators.

$= 2^3 x^3$

$= \dfrac{2^3}{x^3}$

$= 8x^3$

$= \dfrac{8}{x^3}$

These examples suggest the following rules: *To raise a product to a power, we raise each factor of the product to that power,* and *to raise a fraction to a power, we raise both the numerator and the denominator to that power.*

More power rules for exponents

If n is a natural number, then

$(xy)^n = x^n y^n$ and if $y \neq 0$, then $\left(\dfrac{x}{y}\right)^n = \dfrac{x^n}{y^n}$

EXAMPLE 7

Power rules for exponents. Simplify by writing each expression without using parentheses.

a. $(ab)^4 = a^4 b^4$

b. $(3c)^3 = 3^3 c^3$

$= 27c^3$

c. $(x^2 y^3)^5 = (x^2)^5 (y^3)^5$

$= x^{10} y^{15}$

d. $(-2x^3 y)^2 = (-2)^2 (x^3)^2 y^2$

$= 4x^6 y^2$

e. $\left(\dfrac{4}{k}\right)^3 = \dfrac{4^3}{k^3}$

$= \dfrac{64}{k^3}$

f. $\left(\dfrac{3x^2}{2y^3}\right)^5 = \dfrac{3^5(x^2)^5}{2^5(y^3)^5}$

$= \dfrac{243x^{10}}{32y^{15}}$

SELF CHECK   Write each expression without using parentheses:

a. $(3x^2y)^2$ and   **b.** $\left(\dfrac{2x^3}{3y^2}\right)^4$ *Answers:* **a.** $9x^4y^2$, **b.** $\dfrac{16x^{12}}{81y^8}$

■

The Quotient Rule for Exponents

We now consider the fraction

$$\dfrac{4^5}{4^2}$$

where the exponent in the numerator is greater than the exponent in the denominator. We can simplify this fraction as follows:

$$\dfrac{4^5}{4^2} = \dfrac{4 \cdot 4 \cdot 4 \cdot 4 \cdot 4}{4 \cdot 4}$$

$$= \dfrac{\overset{1}{\cancel{4}} \cdot \overset{1}{\cancel{4}} \cdot 4 \cdot 4 \cdot 4}{\underset{1}{\cancel{4}} \cdot \underset{1}{\cancel{4}}}$$ Divide out the common factors of 4.

$$= 4^3$$

The result of 4^3 has a base of 4 and an exponent of $5 - 2$ (or 3). This suggests that *to divide exponential expressions with the same base, we keep the common base and subtract the exponents.*

Quotient rule for exponents

If m and n are natural numbers, $m > n$ and $x \neq 0$, then

$$\dfrac{x^m}{x^n} = x^{m-n}$$

EXAMPLE 8   **Quotient rule for exponents.** Simplify each expression. Assume that there are no divisions by zero.

a. $\dfrac{x^4}{x^3} = x^{4-3}$

$= x^1$

$= x$

b. $\dfrac{8y^2y^6}{4y^3} = \dfrac{8y^8}{4y^3}$

$= \dfrac{8}{4}y^{8-3}$

$= 2y^5$

c. $\dfrac{a^3a^5a^7}{a^4a} = \dfrac{a^{15}}{a^5}$

$= a^{15-5}$

$= a^{10}$

d. $\dfrac{(a^3b^4)^2}{ab^5} = \dfrac{a^6b^8}{ab^5}$

$= a^{6-1}b^{8-5}$

$= a^5b^3$

SELF CHECK Simplify each expression: **a.** $\dfrac{a^5}{a^3}$,

b. $\dfrac{6b^2b^3}{2b^4}$, and **c.** $\dfrac{(x^2y^3)^2}{x^3y^4}$.

Answers: **a.** a^2, **b.** $3b$, **c.** xy^2

The rules for natural-number exponents are summarized as follows.

Rules for exponents

If n is a natural number, then

$$x^n = \overbrace{x \cdot x \cdot x \cdot \cdots \cdot x}^{n \text{ factors of } x}$$

If m and n are natural numbers and there are no divisions by zero, then

$$x^m x^n = x^{m+n} \qquad (x^m)^n = x^{m \cdot n} \qquad (xy)^n = x^n y^n$$

$$\left(\frac{x}{y}\right)^n = \frac{x^n}{y^n} \qquad \frac{x^m}{x^n} = x^{m-n} \quad (\text{provided } m > n)$$

STUDY SET

Section 4.1

VOCABULARY

In Exercises 1–4, fill in the blanks to make the statements true.

1. The _____base_____ of the exponential expression $(-5)^3$ is -5. The _____exponent_____ is 3.

2. The _____exponential_____ expression x^4 represents a repeated multiplication where x is to be written as a _____factor_____ four times.

3. x^n is called a _____power_____ of x.

4. $\{1, 2, 3, 4, 5, \ldots\}$ is the set of _____natural_____ numbers.

CONCEPTS

In Exercises 5–14, fill in the blanks to make the statements true.

5. $(3x)^4$ means $\boxed{3x} \cdot \boxed{3x} \cdot \boxed{3x} \cdot \boxed{3x}$

6. Using an exponent, $(-5y)(-5y)(-5y)$ can be written as $\boxed{(-5y)^3}$.

7. $x^m x^n = \boxed{x^{m+n}}$

8. $(xy)^n = \boxed{x^n y^n}$

9. $\left(\dfrac{a}{b}\right)^n = \boxed{\dfrac{a^n}{b^n}}$

10. $(a^b)^c = \boxed{a^{bc}}$

11. $\dfrac{x^m}{x^n} = \boxed{x^{m-n}}$

12. $x = x^{\boxed{1}}$

13. $(xy) = (xy)^{\boxed{1}}$

14. $(t^3)^2 = \boxed{t^3} \cdot \boxed{t^3}$

In Exercises 15–18, simplify each expression, if possible.

15. a. $x^2 + x^2$ $2x^2$ **b.** $x^2 - x^2$ 0

     c. $x^2 \cdot x^2$ x^4 **d.** $\dfrac{x^2}{x^2}$ 1

16. a. $x^2 + x$ **b.** $x^2 - x$ doesn't simplify
     doesn't simplify

     c. $x^2 \cdot x$ x^3 **d.** $\dfrac{x^2}{x}$ x

17. a. $6x^3 + 2x^2$ **b.** $6x^3 - 2x^2$
     doesn't simplify   doesn't simplify

     c. $6x^3 \cdot 2x^2$ $12x^5$ **d.** $\dfrac{6x^3}{2x^2}$ $3x$

18. a. $-8x^4 + 4x^4$ $-4x^4$ **b.** $-8x^4 - 4x^4$ $-12x^4$

     c. $-8x^4(4x^4)$ $-32x^8$ **d.** $\dfrac{-8x^4}{4x^4}$ -2

In Exercises 19–22, find the area or volume of each figure, whichever is appropriate. You may leave π in your answer.

19. a^{10} mi^2

a^5 mi

a^5 mi

20. $16y^6\pi$ yd^2

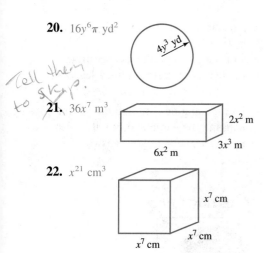
$4y^3$ yd

Tell them to skip.

21. $36x^7$ m^3

$2x^2$ m
$3x^3$ m
$6x^2$ m

22. x^{21} cm^3

x^7 cm
x^7 cm
x^7 cm

In Exercises 23–26, complete each table.

23.

x	3^x
1	3
2	9
3	27
4	81

24.

x	6^x
1	6
2	36
3	216
4	1,296

N/A

25.

x	$(-4)^x$
1	-4
2	16
3	-64
4	256

26.

x	$(-5)^x$
1	-5
2	25
3	-125
4	625

In Exercises 27–28, use the graph to determine the missing y-coordinates in the table.

27.

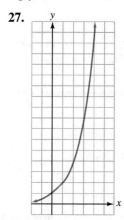

x	y
1	2
2	4
3	8
4	16

28.

x	y
-1	16
0	8
1	4
2	2

NOTATION

In Exercises 29–30, complete each solution.

29. $(x^4x^2)^3 = \left(x^6 \right)^3$
$= x^{18}$

30. $\dfrac{12a^3a^4}{2a^2} = \dfrac{12\,a^7}{2a^2}$
$= 6a^{\,7\,-2}$
$= 6a^5$

PRACTICE

In Exercises 31–38, identify the base and the exponent in each expression.

31. 4^3 base 4, exponent 3

32. $(-8)^2$ base -8, exponent 2

33. x^5 base x, exponent 5

34. $(2y)^3$ base $2y$, exponent 3

35. $(-3x)^2$ base $-3x$, exponent 2

36. $-x^4$ base x, exponent 4

37. $-\dfrac{1}{3}y^6$ base y, exponent 6

38. $3.14r^4$ base r, exponent 4

why?
Answ. ( )

In Exercises 39–46, write each expression without using exponents.

39. 5^3 $5 \cdot 5 \cdot 5$

40. -4^5 $-4 \cdot 4 \cdot 4 \cdot 4 \cdot 4$

41. x^6 $x \cdot x \cdot x \cdot x \cdot x \cdot x$

42. $3x^3$ $3 \cdot x \cdot x \cdot x$

43. $-\dfrac{3}{4}x^5$
$-\frac{3}{4} \cdot x \cdot x \cdot x \cdot x \cdot x$

44. $(2.1y)^4$
$(2.1y)(2.1y)(2.1y)(2.1y)$

45. $\left(\dfrac{1}{3}t\right)^3$ $\left(\frac{1}{3}t\right)\left(\frac{1}{3}t\right)\left(\frac{1}{3}t\right)$

46. a^3b^2 $a \cdot a \cdot a \cdot b \cdot b$

In Exercises 47–50, write each expression using exponents.

47. $4t(4t)(4t)(4t)$ $(4t)^4$

48. $-5u(-5u)$ $(-5u)^2$

49. $-4 \cdot t \cdot t \cdot t$ $-4t^3$

50. $-5 \cdot u \cdot u$ $-5u^2$

In Exercises 51–54, evaluate each expression using the rules for the order of operations.

51. $2(5^4 - 4^3)$ 1,122

52. $2(4^3 + 3^2)$ 146

53. $-5^2(3^4 + 4^3)$ $-3,625$

54. $-5^2(4^3 - 2^6)$ 0

In Exercises 55–74, write each expression as an expression involving only one base and one exponent.

55. x^4x^3 x^7

56. y^5y^2 y^7

57. a^3aa^5 a^9

58. b^2b^3b b^6

59. $y^3(y^2y^4)$ y^9

60. $(y^4y)y^6$ y^{11}

61. $4x^2(3x^5)$ $12x^7$

62. $-2y(y^3)$ $-2y^4$

63. $(-y^2)(4y^3)$ $-4y^5$

64. $(-4x^3)(-5x)$ $20x^4$

65. $(3^2)^4$ 3^8

66. $(4^3)^3$ 4^9

67. $(y^5)^3$ y^{15}

68. $(b^3)^6$ b^{18}

69. $(x^2x^3)^5$ x^{25}

70. $(y^3y^4)^4$ y^{28}

71. $(3zz^2z^3)^5$ $243x^{30}$

72. $(4t^3t^6t^2)^2$ $16t^{22}$

73. $(x^5)^2(x^7)^3$ x^{31}

74. $(y^3y)^2(y^2)^2$ y^{12}

In Exercises 75–90, simplify by writing each expression without using parentheses.

75. $(xy)^3$ x^3y^3

76. $(uv)^4$ u^4v^4

77. $(r^3s^2)^2$ r^6s^4

78. $(a^3b^2)^3$ a^9b^6

79. $(4ab^2)^2$ $16a^2b^4$

80. $(3x^2y)^3$ $27x^6y^3$

81. $(-2r^2s^3)^3$ $-8r^6s^9$

82. $(-3x^2y^4)^2$ $9x^4y^8$

83. $\left(\dfrac{a}{b}\right)^3$ $\dfrac{a^3}{b^3}$

84. $\left(\dfrac{r}{s}\right)^4$ $\dfrac{r^4}{s^4}$

85. $\left(\dfrac{x^2}{y^3}\right)^5$ $\dfrac{x^{10}}{y^{15}}$

86. $\left(\dfrac{u^4}{v^2}\right)^6$ $\dfrac{u^{24}}{v^{12}}$

87. $\left(\dfrac{-2a}{b}\right)^5$ $\dfrac{-32a^5}{b^5}$

88. $\left(\dfrac{-2t}{3}\right)^4$ $\dfrac{16t^4}{81}$

89. $\left(\dfrac{b^2}{3a}\right)^3$ $\dfrac{b^6}{27a^3}$

90. $\left(\dfrac{a^3b}{c^4}\right)^5$ $\dfrac{a^{15}b^5}{c^{20}}$

In Exercises 91–102, simplify each expression.

91. $\dfrac{x^5}{x^3}$ x^2

92. $\dfrac{a^6}{a^3}$ a^3

93. $\dfrac{y^3y^4}{yy^2}$ y^4

94. $\dfrac{b^4b^5}{b^2b^3}$ b^4

95. $\dfrac{12a^2a^3a^4}{4(a^4)^2}$ $3a$

96. $\dfrac{16(aa^2)^3}{2a^2a^3}$ $8a^4$

97. $\dfrac{(ab^2)^3}{(ab)^2}$ ab^4

98. $\dfrac{(m^3n^4)^3}{(mn^2)^3}$ m^6n^6

99. $\dfrac{20(r^4s^3)^4}{6(rs^3)^3}$ $\dfrac{10r^{13}s^3}{3}$

100. $\dfrac{15(x^2y^5)^5}{21(x^3y)^2}$ $\dfrac{5x^4y^{23}}{7}$

101. $\left(\dfrac{y^3y}{2yy^2}\right)^3$ $\dfrac{y^3}{8}$

102. $\left(\dfrac{3t^3t^4t^5}{4t^2t^6}\right)^3$ $\dfrac{27t^{12}}{64}$

APPLICATIONS

103. ART HISTORY Leonardo da Vinci's drawing relating a human figure to a square and a circle is shown in Illustration 1.

ILLUSTRATION 1

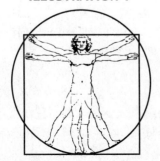

a. Find the area of the square if the man's height is $5x$ feet. $25x^2$ ft^2

b. Find the area of the circle if the distance from his waist to his feet is $3x$ feet. You may leave π in your answer. $9\pi x^2$ ft^2

104. PACKAGING Use Illustration 2 to find the volume of the bowling ball and the cardboard box it is packaged in. You may leave π in your answer. $36\pi x^3$ in.^{3}, $216x^3$ in.^{3}

ILLUSTRATION 2

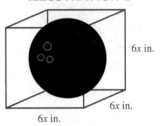

$6x$ in.

$6x$ in.

$6x$ in.

105. BOUNCING BALL A ball is dropped from a height of 32 feet. Each rebound is one-half of its previous height.

a. Draw a diagram of the path of the ball, showing four bounces.

b. Explain why the expressions $32\left(\tfrac{1}{2}\right)$, $32\left(\tfrac{1}{2}\right)^2$, $32\left(\tfrac{1}{2}\right)^3$, and $32\left(\tfrac{1}{2}\right)^4$ represent the height of the ball on the first, second, third, and fourth bounces, respectively. Find the heights of the first four bounces. 16 ft, 8 ft, 4 ft, 2 ft

106. HAVING BABIES The probability that a couple will have n baby boys in a row is given by the formula $\left(\tfrac{1}{2}\right)^n$. Find the probability that a couple will have four baby boys in a row. $\tfrac{1}{16}$

107. PACKAGING Use a power of 12 to express the number of pencils in a case if

■ each package contains a dozen pencils,

■ each box contains a dozen packages,

■ each carton contains a dozen boxes, and

■ each case contains a dozen cartons. 12^4

108. COMPUTERS Text is stored by computers using a sequence of eight 0's and 1's. Such a sequence is called a **byte.** An example of a byte is 10101110.

a. Write four other bytes, all ending in 1.
11000001, 11010001, 11001101, 11000011 (answers may vary)

b. Each of the eight digits of a byte can be chosen in *two* ways (either 0 or 1). The total number of different bytes can be represented by an exponential expression with base 2. What is it? 2^8

109. INVESTMENT On average, a certain investment doubles every 7 years. Use this information to find the value of a $1,000 investment after 28 years. $16,000

110. INVESTING Guess the answer to the following problem. Then use a calculator to find the correct answer. Were you close?

If the value of 1¢ is to double every day, what will the penny be worth after 31 days? $21,474,836.48

a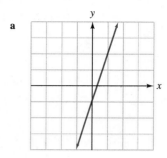

WRITING

111. Describe how you would multiply two exponential expressions with like bases.

112. Are the expressions $2x^3$ and $(2x)^3$ equivalent? Explain.

113. Is the operation of raising to a power commutative? That is, is $a^b = b^a$? Explain.

114. When a number is raised to a power, is the result always larger than the original number? Support your answer with some examples.

b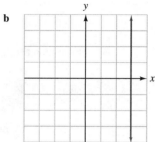

REVIEW

In Exercises 115–118, match each equation with its graph in the right column.

115. $y = 2x - 1$ c

116. $y = 3x - 1$ a

117. $y = 3$ d

118. $x = 3$ b

c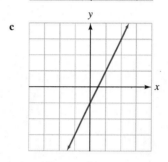

d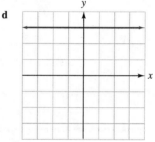

▶ **4.2**

Zero and Negative Integer Exponents

In this section, you will learn about

Zero exponents ■ Negative integer exponents ■ Variable exponents

---

Introduction In the previous section, we discussed natural-number exponents. We now extend the discussion to include exponents that are zero and exponents that are negative integers.

Zero Exponents

When we discussed the quotient rule for exponents in the previous section, the exponent in the numerator was always greater than the exponent in the denominator. We now consider what happens when the exponents are equal.

If we apply the quotient rule to the fraction

$$\frac{5^3}{5^3}$$

where the exponents in the numerator and denominator are equal, we obtain 5^0. However, because any nonzero number divided by itself equals 1, we also obtain 1.

$$\frac{5^3}{5^3} = 5^{3-3} = 5^0 \qquad \frac{5^3}{5^3} = \frac{\overset{1}{\cancel{5}} \cdot \overset{1}{\cancel{5}} \cdot \overset{1}{\cancel{5}}}{\underset{1}{\cancel{5}} \cdot \underset{1}{\cancel{5}} \cdot \underset{1}{\cancel{5}}} = 1$$

These are equal.

For this reason, we will define 5^0 to be equal to 1. This example suggests the following.

Zero exponents

If x is any nonzero real number, then

$$x^0 = 1$$

EXAMPLE 1

Zero exponents. Write each expression without using exponents.

a. $\left(\dfrac{1}{13}\right)^0 = 1$ **b.** $\dfrac{x^5}{x^5} = x^{5-5}$ $(x \neq 0)$

$$= x^0$$
$$= 1$$

c. $3x^0 = 3(1)$ **d.** $(3x)^0 = 1$
$$= 3$$

Parts c and d point out that $3x^0 \neq (3x)^0$.

SELF CHECK

Write each expression without using exponents:
a. $(-0.115)^0$ and **b.** $-5a^0b$. *Answers:* **a.** 1, **b.** $-5b$ ∎

Negative Integer Exponents

If we apply the quotient rule to

$$\frac{6^2}{6^5}$$

where the exponent in the numerator is less than the exponent in the denominator, we obtain 6^{-3}. However, by dividing out two factors of 6, we also obtain $\frac{1}{6^3}$.

$$\frac{6^2}{6^5} = 6^{2-5} = 6^{-3} \qquad \frac{6^2}{6^5} = \frac{\overset{1}{\cancel{6}} \cdot \overset{1}{\cancel{6}}}{\underset{1}{\cancel{6}} \cdot \underset{1}{\cancel{6}} \cdot 6 \cdot 6 \cdot 6} = \frac{1}{6^3}$$

These are equal.

For these reasons, we define 6^{-3} to be equal to $\dfrac{1}{6^3}$.

$$6^{-3} = \frac{1}{6^3}$$

Negative exponents

If x is any nonzero number and n is a natural number, then

$$x^{-n} = \frac{1}{x^n}$$

EXAMPLE 2

Negative exponents. Simplify by using the definition of negative exponents.

a. $3^{-5} = \dfrac{1}{3^5}$

$= \dfrac{1}{243}$

b. $(-2)^{-3} = \dfrac{1}{(-2)^3}$

$= -\dfrac{1}{8}$

c. $\dfrac{1}{5^{-2}} = \dfrac{1}{\dfrac{1}{5^2}}$

$= 1 \div \dfrac{1}{5^2}$

$= 1 \cdot \dfrac{5^2}{1}$

$= 5^2$

$= 25$

d. $\dfrac{2^{-3}}{3^{-4}} = \dfrac{\dfrac{1}{2^3}}{\dfrac{1}{3^4}}$

$= \dfrac{1}{2^3} \cdot \dfrac{3^4}{1}$

$= \dfrac{3^4}{2^3}$

$= \dfrac{81}{8}$

SELF CHECK

Simplify by using the definition of negative exponents: **a.** 4^{-4}, **b.** $(-5)^{-3}$, and **c.** $\dfrac{8^{-2}}{7^{-1}}$.

Answers: **a.** $\frac{1}{256}$, **b.** $-\frac{1}{125}$, **c.** $\frac{7}{64}$

The results from parts c and d of Example 2 suggest that in a fraction, we can move factors that have negative exponents between the numerator and denominator if we change the sign of their exponents. For example,

$$\frac{3^{-3}}{b^{-1}} = \frac{b^1}{3^3} = \frac{b}{27}$$ This property applies only if there is one term in the numerator and one term in the denominator.

EXAMPLE 3

Negative exponents. Simplify by using the definition of negative exponents. Assume that no denominators are zero.

a. $x^{-4} = \dfrac{1}{x^4}$

b. $\dfrac{x^{-3}}{y^{-7}} = \dfrac{y^7}{x^3}$

c. $(2x)^{-2} = \dfrac{1}{(2x)^2}$

$= \dfrac{1}{4x^2}$

d. $2x^{-2} = 2\left(\dfrac{1}{x^2}\right)$

$= \dfrac{2}{x^2}$

e. $(-3a)^{-4} = \dfrac{1}{(-3a)^4}$

$= \dfrac{1}{81a^4}$

f. $(x^3x^2)^{-3} = (x^5)^{-3}$

$= \dfrac{1}{(x^5)^3}$

$= \dfrac{1}{x^{15}}$

Simplify by using the definition of negative exponents: **a.** a^{-5}, **b.** $3y^{-3}$, and **c.** $(a^4a^3)^{-2}$.   *Answers:*   **a.** $\dfrac{1}{a^5}$,   **b.** $\dfrac{3}{y^3}$,   **c.** $\dfrac{1}{a^{14}}$ ■

The rules for exponents discussed in Section 4.1 (the product, power, and quotient rules) are also true for zero and negative exponents.

Rules for exponents

If m and n are integers and there are no divisions by zero, then

$$x^m x^n = x^{m+n} \qquad (x^m)^n = x^{m \cdot n} \qquad (xy)^n = x^n y^n \qquad \left(\frac{x}{y}\right)^n = \frac{x^n}{y^n}$$

$$x^0 = 1 \quad (x \neq 0) \qquad x^{-n} = \frac{1}{x^n} \qquad \frac{x^m}{x^n} = x^{m-n}$$

EXAMPLE 4

Applying rules for exponents. Write $\left(\frac{5}{16}\right)^{-1}$ without using exponents.

Solution

$$\left(\frac{5}{16}\right)^{-1} = \frac{5^{-1}}{16^{-1}}$$

$$= \frac{16^1}{5^1}$$

$$= \frac{16}{5}$$

Write $\left(\frac{3}{7}\right)^{-2}$ without using exponents.   *Answer:* $\frac{49}{9}$ ■

EXAMPLE 5

Applying rules for exponents. Simplify and write the result without using negative exponents. Assume that no denominators are zero.

a. $(x^{-3})^2 = x^{-6}$

$\qquad\qquad = \dfrac{1}{x^6}$

b. $\dfrac{x^3}{x^7} = x^{3-7}$

$\qquad = x^{-4}$

$\qquad = \dfrac{1}{x^4}$

c. $\dfrac{y^{-4}y^{-3}}{y^{-20}} = \dfrac{y^{-7}}{y^{-20}}$

$\qquad\qquad = y^{-7-(-20)}$

$\qquad\qquad = y^{-7+20}$

$\qquad\qquad = y^{13}$

d. $\dfrac{12a^3b^4}{4a^5b^2} = 3a^{3-5}b^{4-2}$

$\qquad\qquad = 3a^{-2}b^2$

$\qquad\qquad = \dfrac{3b^2}{a^2}$

e. $\left(-\dfrac{x^3y^2}{xy^{-3}}\right)^{-2} = (-x^{3-1}y^{2-(-3)})^{-2}$

$\qquad\qquad\qquad = (-x^2y^5)^{-2}$

$\qquad\qquad\qquad = \dfrac{1}{(-x^2y^5)^2}$

$\qquad\qquad\qquad = \dfrac{1}{x^4y^{10}}$

SELF CHECK

Simplify and write the result without using negative exponents: **a.** $(x^4)^{-3}$, **b.** $\dfrac{a^4}{a^8}$, **c.** $\dfrac{a^{-4}a^{-5}}{a^{-3}}$, and **d.** $\dfrac{20x^5y^3}{5x^3y^6}$.

Answers: **a.** $\dfrac{1}{x^{12}}$, **b.** $\dfrac{1}{a^4}$, **c.** $\dfrac{1}{a^6}$, **d.** $\dfrac{4x^2}{y^3}$

$\blacksquare$

Variable Exponents

We can apply the rules for exponents to simplify expressions involving variable exponents.

EXAMPLE 6

Variable exponents. Simplify each expression.

a. $\dfrac{6^n}{6^n} = 6^{n-n}$

$\qquad = 6^0$

$\qquad = 1$

b. $\dfrac{y^m}{y^m} = y^{m-m} \quad (y \neq 0)$

$\qquad = y^0$

$\qquad = 1$

c. $x^{2m}x^{3m} = x^{2m+3m}$

$\qquad = x^{5m}$

d. $\dfrac{y^{2m}}{y^{4m}} = y^{2m-4m} \quad (y \neq 0)$

$\qquad = y^{-2m}$

$\qquad = \dfrac{1}{y^{2m}}$

e. $a^{2m-1}a^{2m} = a^{2m-1+2m}$

$\qquad = a^{4m-1}$

f. $(b^{m+1})^{2m} = b^{(m+1)2m}$

$\qquad = b^{2m^2+2m}$

SELF CHECK

Simplify each expression: **a.** $\dfrac{x^m}{x^m}$, **b.** $z^{3n}z^{2n}$, **c.** $\dfrac{z^{3n}}{z^{5n}}$, and **d.** $(x^{m+2})^{3m}$

Answers: **a.** 1, **b.** z^{5n}, **c.** $\dfrac{1}{z^{2n}}$, **d.** x^{3m^2+6m}

$\blacksquare$

ACCENT ON TECHNOLOGY *Finding Present Value*

As a gift for their newborn grandson, the grandparents want to deposit enough money in the bank now so that when he turns 18, the young man will have a college fund of $20,000 waiting for him. How much should they deposit now if the money will earn 6% annually?

To find how much money P must be invested at an annual rate i (expressed as a decimal) to have A in n years, we use the formula $P = A(1 + i)^{-n}$. If we substitute 20,000 for A, 0.06 (6%) for i, and 18 for n, we have

$$P = A(1 + i)^{-n} \qquad \text{P is called the \textit{present value}.}$$
$$P = 20{,}000(1 + 0.06)^{-18}$$

To find P with a scientific calculator, we enter these numbers and press these keys:

Keystrokes $\boxed{(}$ 1 $\boxed{+}$ .06 $\boxed{)}$ $\boxed{y^x}$ 18 $\boxed{+/-}$ $\boxed{\times}$ 20000 $\boxed{=}$ $\boxed{\text{7006.875823}}$

(continued)

To evaluate the expression with a graphing calculator, we use the following keystrokes:

Keystrokes   20000 $\boxed{\times}$ $\boxed{(}$ 1 $\boxed{+}$ .06 $\boxed{)}$ $\boxed{\wedge}$ $\boxed{(-)}$ 18 $\boxed{\text{ENTER}}$

```
20000*(1+.06)^-1
8
        7006.875823
```

They must invest approximately $7,006.88 to have $20,000 in 18 years.

STUDY SET

Section 4.2

VOCABULARY

In Exercises 1–4, fill in the blanks to make the statements true.

1. In the exponential expression 8^{-3}, 8 is the _____base_____ and -3 is the _____exponent_____.

2. The set $\{. . . , -3, -2, -1, 0, 1, 2, 3, . . .\}$ is the set of _____integers_____.

3. In the exponential expression 5^{-1}, the exponent is a _____negative_____ integer.

4. In the exponential expression z^m, the exponent is a _____variable_____.

CONCEPTS

5. In parts a and b, fill in the blanks as you simplify the fraction in two different ways. Then complete the sentence in part c.

a. $\dfrac{6^4}{6^4} = 6^{\boxed{4-4}}$

$= 6^{\boxed{0}}$

b. $\dfrac{6^4}{6^4} = \dfrac{\boxed{6} \cdot \boxed{6} \cdot \boxed{6} \cdot \boxed{6}}{6 \cdot 6 \cdot 6 \cdot 6}$

$= \boxed{1}$

c. So we define 6^0 to be $\boxed{1}$, and in general, if x is any nonzero real number, then $x^0 = \boxed{1}$.

6. In parts a and b, fill in the blanks as you simplify the fraction in two different ways. Then complete the sentence in part c.

a. $\dfrac{8^3}{8^5} = 8^{\boxed{3-5}}$

$= 8^{\boxed{-2}}$

b. $\dfrac{8^3}{8^5} = \dfrac{\boxed{8} \cdot \boxed{8} \cdot \boxed{8}}{8 \cdot 8 \cdot 8 \cdot 8 \cdot 8}$

$= \dfrac{1}{8^{\boxed{2}}}$

c. So we define 8^{-2} to be $\boxed{\dfrac{1}{8^2}}$, and in general, if x is any nonzero real number, then $x^{-n} = \boxed{\dfrac{1}{x^n}}$.

In Exercises 7–10, complete each table.

7.

x	3^x
2	9
1	3
0	1
-1	$\frac{1}{3}$
-2	$\frac{1}{9}$

8.

x	4^x
2	16
1	4
0	1
-1	$\frac{1}{4}$
-2	$\frac{1}{16}$

9.

x	$(-9)^x$
2	81
1	-9
0	1
-1	$-\frac{1}{9}$
-2	$\frac{1}{81}$

10.

x	$(-5)^x$
2	25
1	-5
0	1
-1	$-\frac{1}{5}$
-2	$\frac{1}{25}$

In Exercises 11–12, first use the graph to determine the missing y-coordinates in the table. Then express each y-coordinate as a power of 2.

11.

x	y	Power
2	4	2^2
1	2	2^1
0	1	2^0
-1	$\frac{1}{2}$	2^{-1}
-2	$\frac{1}{4}$	2^{-2}

12.

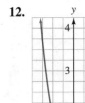

x	y	Power
1	$\frac{1}{4}$	2^{-2}
0	$\frac{1}{2}$	2^{-1}
-1	1	2^{0}
-2	2	2^{1}
-3	4	2^{2}

NOTATION

In Exercises 13–14, complete each solution.

13. $(y^5 y^3)^{-5} = \left(y^{\boxed{8}} \right)^{-5}$

$= y^{\boxed{-40}}$

$= \dfrac{1}{y^{40}}$

14. $\left(\dfrac{a^2 b^3}{a^{-3} b} \right)^{-3} = \left(a^{2-(-3)} b^{\boxed{3}\,-1} \right)^{-3}$

$= \left(a^{\boxed{5}} b^{\boxed{2}} \right)^{-3}$

$= \dfrac{1}{(a^5 b^2)^{\boxed{3}}}$

$= \dfrac{1}{a^{15} b^6}$

15. In the expression $3x^{-2}$, what is the base and what is the exponent?   base x, exponent -2

16. In the expression $-3x^{-2}$, what is the base and what is the exponent?   base x, exponent -2

17. First tell the base and the exponent, then evaluate each expression.
  a. -4^2 $4; 2; -16$
  b. 4^{-2} $4; -2; \frac{1}{16}$
  c. -4^{-2} $4; -2; -\frac{1}{16}$

18. First tell the base and the exponent, then evaluate each expression.
  a. $(-7)^2$ $-7; 2; 49$
  b. -7^{-2} $7; -2; -\frac{1}{49}$
  c. $(-7)^{-2}$ $-7; -2; \frac{1}{49}$

PRACTICE

In Exercises 19–84, simplify each expression. Write each answer without using parentheses or negative exponents.

19. 7^0 1

20. 9^0 1

21. $\left(\dfrac{1}{4} \right)^0$ 1

22. $\left(\dfrac{3}{8} \right)^0$ 1

23. $2x^0$ 2

24. $(2x)^0$ 1

25. $(-x)^0$ 1

26. $-x^0$ -1

27. $\left(\dfrac{a^2 b^3}{a b^4} \right)^0$ 1

28. $\dfrac{2}{3} \left(\dfrac{xyz}{x^2 y} \right)^0$ $\frac{2}{3}$

29. $\dfrac{5}{2x^0}$ $\frac{5}{2}$

30. $\dfrac{4}{3a^0}$ $\frac{4}{3}$

31. 12^{-2} $\frac{1}{144}$

32. 11^{-2} $\frac{1}{121}$

33. $(-4)^{-1}$ $-\frac{1}{4}$

34. $(-8)^{-1}$ $-\frac{1}{8}$

35. $\dfrac{1}{5^{-3}}$ 125

36. $\dfrac{1}{3^{-3}}$ 27

37. $\dfrac{2^{-4}}{3^{-1}}$ $\frac{3}{16}$

38. $\dfrac{7^{-2}}{2^{-3}}$ $\frac{8}{49}$

39. -4^{-3} $-\frac{1}{64}$

40. -6^{-3} $-\frac{1}{216}$

41. $-(-4)^{-3}$ $\frac{1}{64}$

42. $-(-4)^{-2}$ $-\frac{1}{16}$

43. x^{-2} $\frac{1}{x^2}$

44. y^{-3} $\frac{1}{y^3}$

45. $-b^{-5}$ $-\frac{1}{b^5}$

46. $-c^{-4}$ $-\frac{1}{c^4}$

47. $(2y)^{-4}$ $\frac{1}{16y^4}$

48. $(-3x)^{-1}$ $-\frac{1}{3x}$

49. $(ab^2)^{-3}$ $\frac{1}{a^3 b^6}$

50. $(m^2 n^3)^{-2}$ $\frac{1}{m^4 n^6}$

51. $2^5 \cdot 2^{-2}$ 8

52. $10^2 \cdot 10^{-4}$ $\frac{1}{100}$

53. $4^{-3} \cdot 4^{-2} \cdot 4^5$ 1

54. $3^{-4} \cdot 3^5 \cdot 3^{-3}$ $\frac{1}{9}$

55. $\left(\dfrac{7}{8} \right)^{-1}$ $\frac{8}{7}$

56. $\left(\dfrac{16}{5} \right)^{-1}$ $\frac{5}{16}$

57. $\dfrac{3^5 \cdot 3^{-2}}{3^3}$ 1

58. $\dfrac{6^2 \cdot 6^{-3}}{6^{-2}}$ 6

59. $\dfrac{y^4}{y^5}$ $\frac{1}{y}$

60. $\dfrac{t^7}{t^{10}}$ $\frac{1}{t^3}$

61. $\dfrac{(r^2)^3}{(r^3)^4}$ $\frac{1}{r^6}$

62. $\dfrac{(b^3)^4}{(b^5)^4}$ $\frac{1}{b^8}$

63. $\dfrac{y^4 y^3}{y^4 y^{-2}}$ y^5

64. $\dfrac{x^{12} x^{-7}}{x^3 x^4}$ $\frac{1}{x^2}$

65. $\dfrac{10a^4 a^{-2}}{5a^2 a^0}$ 2

66. $\dfrac{9b^0 b^3}{3b^{-3} b^4}$ $3b^2$

67. $(ab^2)^{-2}$ $\frac{1}{a^2 b^4}$

68. $(c^2 d^3)^{-2}$ $\frac{1}{c^4 d^6}$

69. $(x^2 y)^{-3}$ $\frac{1}{x^6 y^3}$

70. $(-xy^2)^{-4}$ $\frac{1}{x^4 y^8}$

71. $(x^{-4} x^3)^3$ $\frac{1}{x^3}$

72. $(y^{-2} y)^3$ $\frac{1}{y^3}$

73. $(a^{-2}b^3)^{-4}$ $\dfrac{a^8}{b^{12}}$

74. $(y^{-3}z^5)^{-6}$ $\dfrac{y^{18}}{z^{30}}$

75. $(-2x^3y^{-2})^{-5}$ $-\dfrac{y^{10}}{32x^{15}}$

76. $(-3u^{-2}v^3)^{-3}$ $-\dfrac{u^6}{27v^9}$

77. $\left(\dfrac{a^3}{a^{-4}}\right)^2$ a^{14}

78. $\left(\dfrac{a^4}{a^{-3}}\right)^3$ a^{21}

79. $\left(\dfrac{b^5}{b^{-2}}\right)^{-2}$ $\dfrac{1}{b^{14}}$

80. $\left(\dfrac{b^{-2}}{b^3}\right)^3$ $\dfrac{1}{b^{15}}$

81. $\left(\dfrac{4x^2}{3x^{-5}}\right)^4$ $\dfrac{256x^{28}}{81}$

82. $\left(\dfrac{-3r^4r^{-3}}{r^{-3}r^7}\right)^3$ $-\dfrac{27}{r^9}$

83. $\left(\dfrac{12y^3z^{-2}}{3y^{-4}z^3}\right)^2$ $\dfrac{16y^{14}}{z^{10}}$

84. $\left(\dfrac{6xy^3}{3x^{-1}y}\right)^3$ $8x^6y^6$

In Exercises 85–96, write each expression with a single exponent.

85. $x^{2m}x^m$ x^{3m}

86. $y^{3m}y^{2m}$ y^{5m}

87. $u^{2m}u^{-3m}$ $\dfrac{1}{u^m}$

88. $r^{5m}r^{-6m}$ $\dfrac{1}{r^m}$

89. $y^{3m+2}y^{-m}$ y^{2m+2}

90. $x^{m+1}x^m$ x^{2m+1}

91. $\dfrac{y^{3m}}{y^{2m}}$ y^m

92. $\dfrac{z^{4m}}{z^{2m}}$ z^{2m}

93. $\dfrac{x^{3n}}{x^{6n}}$ $\dfrac{1}{x^{3n}}$

94. $\dfrac{x^m}{x^{5m}}$ $\dfrac{1}{x^{4m}}$

95. $(x^{m+1})^2$ x^{2m+2}

96. $(y^2)^{m+1}$ y^{2m+2}

APPLICATIONS

97. THE DECIMAL NUMERATION SYSTEM Decimal numbers are written by putting digits into place-value columns that are separated by a decimal point. Express the value of each of the columns shown in Illustration 1 using a power of 10. $10^2, 10^1, 10^0, 10^{-1}, 10^{-2}, 10^{-3}, 10^{-4}$

ILLUSTRATION 1

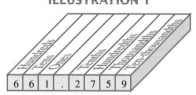

98. UNIT COMPARISON Consider the relative sizes of the items listed in the table in Illustration 2. In the column titled "measurement," write the most appropriate number from the following list. Each number is used only once.

10^0 meter

10^{-1} meter

10^{-2} meter

10^{-3} meter

10^{-4} meter

10^{-5} meter

ILLUSTRATION 2

Item	Measurement (m)
Thickness of a dime	10^{-3}
Height of a bathroom sink	10^0
Length of a pencil eraser	10^{-2}
Thickness of soap bubble film	10^{-5}
Width of a video cassette	10^{-1}
Thickness of a piece of paper	10^{-4}

99. RETIREMENT YEARS How much money should a young married couple invest now at an 8% annual rate if they want to have $100,000 in the bank when they reach retirement age in 40 years? (See the Accent on Technology in this section for the formula.) approximately $4,603.09

100. BIOLOGY During bacterial reproduction, the time required for a population to double is called the **generation time.** If b bacteria are introduced into a medium, then after the generation time has elapsed, there will be $2b$ bacteria. After n generations, there will be $b \cdot 2^n$ bacteria. Explain what this expression represents when $n = 0$. It gives the initial number of bacteria b.

WRITING

101. Explain how you would help a friend understand that 2^{-3} is not equal to -8.

102. Describe how you would verify on a calculator that
$$2^{-3} = \frac{1}{2^3}$$

REVIEW

103. IQ TEST An IQ (intelligence quotient) is a score derived from the formula
$$IQ = \frac{\text{mental age}}{\text{chronological age}} \cdot 100$$

Find the mental age of a 10-year-old girl if she has an IQ of 135. 13.5 yr

104. DIVING When under water, the pressure you feel in your ears is given by the formula
$$\text{Pressure} = \text{depth} \cdot \text{density of water}$$

Find the density of water (in lb/ft^3) if, at a depth of 9 feet, the pressure on your eardrum is 561.6 lb/ft^2. 62.4 lb/ft^3

105. Write the equation of the line having slope $\frac{3}{4}$ and y-intercept -5. $y = \frac{3}{4}x - 5$

106. Find $f(-6)$ if $f(x) = x^2 - 3x + 1$. 55

▶ 4.3 Scientific Notation

In this section, you will learn about

Scientific notation ■ Writing numbers in scientific notation ■ Changing from scientific notation to standard notation ■ Using scientific notation to simplify computations

Introduction Scientists often deal with extremely large and extremely small numbers. Two examples are shown in Figure 4-1.

FIGURE 4-1

The distance from the earth to the sun is approximately 150,000,000 kilometers.

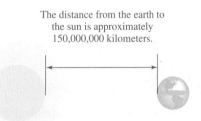

The influenza virus, which causes "flu" symptoms of cough, sore throat, headache, and congestion, has a diameter of 0.00000256 inch.

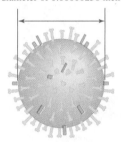

The large number of zeros in 150,000,000 and 0.00000256 makes them difficult to read and hard to remember. In this section, we will discuss a notation that will make these numbers easier to work with.

Scientific Notation

Scientific notation provides a compact way of writing large and small numbers.

Scientific notation

> A number is written in **scientific notation** if it is written as the product of a number between 1 (including 1) and 10 and an integer power of 10.

These numbers are written in scientific notation:

$$3.67 \times 10^6, \qquad 2.24 \times 10^{-4}, \qquad \text{and} \qquad 9.875 \times 10^{22}$$

Every number written in scientific notation has the following form:

An integer exponent

$$\boxed{} \cdot \boxed{} \times 10^{}$$

A decimal between 1 and 10

Writing Numbers in Scientific Notation

EXAMPLE 1

Writing numbers in scientific notation. Change 150,000,000 to scientific notation.

Solution We note that 1.5 lies between 1 and 10. To obtain 150,000,000, the decimal point in 1.5 must be moved eight places to the right.

$$1\underset{\text{8 places to the right}}{.50000000}$$

Because multiplying a number by 10 moves the decimal point one place to the right, we can accomplish this by multiplying 1.5 by 10 eight times. We can show the multiplication of 1.5 by 10 eight times using the notation 10^8. Thus, 150,000,000 written in scientific notation is 1.5×10^8.

SELF CHECK The distance from the earth to the sun is approximately 93,000,000 miles. Write this number in scientific notation. *Answer:* 9.3×10^7 ∎

EXAMPLE 2 **Writing numbers in scientific notation.** Change 0.00000256 to scientific notation.

Solution We note that 2.56 is between 1 and 10. To obtain 0.00000256, the decimal point in 2.56 must be moved six places to the left.

$$\underset{\text{6 places to the left}}{000002}.56$$

We can accomplish this by dividing 2.56 by 10^6, which is equivalent to multiplying 2.56 by $\frac{1}{10^6}$ (or by 10^{-6}). Thus, 0.00000256 written in scientific notation is 2.56×10^{-6}.

SELF CHECK The *Salmonella* bacterium, which causes food poisoning, is 0.00009055 inch long. Write this number in scientific notation. *Answer:* 9.055×10^{-5} ∎

EXAMPLE 3 **Writing numbers in scientific notation.** Write **a.** 235,000 and **b.** 0.0000073 in scientific notation.

Solution **a.** $235,000 = 2.35 \times 10^5$ Because $2.35 \times 10^5 = 235,000$ and 2.35 is between 1 and 10.

b. $0.0000073 = 7.3 \times 10^{-6}$ Because $7.3 \times 10^{-6} = 0.0000073$ and 7.3 is between 1 and 10.

SELF CHECK Write **a.** 17,500 and **b.** 0.657 in scientific notation. *Answers:* **a.** 1.75×10^4, **b.** 6.57×10^{-1} ∎

From Examples 1, 2, and 3, we see that in scientific notation, a positive exponent is used when writing a number that is greater than 1. A negative exponent is used when writing a number that is between 0 and 1.

EXAMPLE 4 **Writing numbers in scientific notation.** Write 432.0×10^5 in scientific notation.

Solution The number 432.0×10^5 is not written in scientific notation, because 432.0 is not a number between 1 and 10. To write this number in scientific notation, we proceed as follows:

$$432.0 \times 10^5 = 4.32 \times 10^2 \times 10^5 \quad \text{Write 432.0 in scientific notation.}$$
$$= 4.32 \times 10^7 \quad\quad 10^2 \times 10^5 = 10^{2+5} = 10^7.$$

SELF CHECK   Write 85×10^{-3} in scientific notation.   *Answer:* 8.5×10^{-2} ∎

ACCENT ON TECHNOLOGY *Calculators and Scientific Notation*

When displaying a very large or a very small number as an answer, most scientific calculators express it in scientific notation. To show this, we will find the values of $(453.46)^5$ and $(0.0005)^{12}$. To find these powers of decimals, we enter these numbers and press these keys:

Keystrokes $453.46 \boxed{y^x} 5 \boxed{=}$ $\boxed{\text{1.917321395 }^{13}}$

.0005 $\boxed{y^x} 12 \boxed{=}$ $\boxed{\text{2.44140625 }^{-40}}$

Since these numbers in standard notation require more space than the calculator display has, the calculator gives each result in scientific notation. The first display represents $1.917321395 \times 10^{13}$, and the second display represents $2.44140625 \times 10^{-40}$.

If we evaluate the same two expressions using a graphing calculator, we see that the letter E is used when displaying a number in scientific notation.

Keystrokes $453.46 \boxed{\wedge} 5 \boxed{\text{ENTER}}$ $\boxed{\begin{array}{l}\text{453.46^5}\\ \quad\text{1.917321395e13}\end{array}}$

.0005 $\boxed{\wedge} 12 \boxed{\text{ENTER}}$ $\boxed{\begin{array}{l}\text{.0005^12}\\ \quad\text{2.44140625e-40}\end{array}}$

Changing from Scientific Notation to Standard Notation

We can change a number written in scientific notation to **standard notation.** For example, to write 9.3×10^7 in standard notation, we multiply 9.3 by 10^7.

$$9.3 \times 10^7 = 9.3 \times 10,000,000 \quad \text{10^7 is equal to 1 followed by 7 zeros.}$$
$$= 93,000,000$$

EXAMPLE 5   **Writing numbers in standard notation.** Write   **a.** 3.4×10^5 and **b.** 2.1×10^{-4} in standard notation.

Solution   **a.** $3.4 \times 10^5 = 3.4 \times 100,000$ **b.** $2.1 \times 10^{-4} = 2.1 \times \dfrac{1}{10^4}$
$$= 340,000$$
$$= 2.1 \times \dfrac{1}{10,000}$$
$$= 2.1 \times 0.0001$$
$$= 0.00021$$

SELF CHECK   Write   **a.** 4.76×10^5 and   **b.** 9.8×10^{-3} in standard notation.   *Answers:*   **a.** 476,000,   **b.** 0.0098   ∎

The following numbers are written in both scientific and standard notation. In each case, the exponent gives the number of places that the decimal point moves, and the sign of the exponent indicates the direction that it moves.

$5.32 \times 10^5 = 5\,3\,2\,0\,0\,0.$ 5 places to the right.

$2.37 \times 10^6 = 2\,3\,7\,0\,0\,0\,0.$ 6 places to the right.

$8.95 \times 10^{-4} = 0.0\,0\,0\,8\,9\,5$ 4 places to the left.

$8.375 \times 10^{-3} = 0.0\,0\,8\,3\,7\,5$ 3 places to the left.

$9.77 \times 10^0 = 9.77$ No movement of the decimal point.

Using Scientific Notation to Simplify Computations

Another advantage of scientific notation becomes apparent when we evaluate products or quotients that contain very large or very small numbers.

EXAMPLE 6

Stars. Except for the sun, the nearest star visible to the naked eye from most parts of the United States is Sirius. Light from Sirius reaches the earth in about 70,000 hours. If light travels at approximately 670,000,000 miles per hour, how far from the earth is Sirius?

Solution

We are given the *rate* at which light travels (670,000,000 mi/hr) and the *time* it takes the light to travel from Sirius to earth (70,000 hr). We can find the *distance* the light travels using the formula $d = rt$.

$d = rt$

$d = \mathbf{670{,}000{,}000 \cdot 70{,}000}$ Substitute 670,000,000 for r and 70,000 for t.

$\quad = 6.7 \times 10^8 \cdot 7.0 \times 10^4$ Write each number in scientific notation.

$\quad = (6.7 \cdot 7.0) \times (10^8 \cdot 10^4)$ Apply the commutative and associative properties of multiplication to group the numbers together and the powers of 10 together.

$\quad = (6.7 \cdot 7.0) \times 10^{8+4}$ For the powers of 10, keep the base and add the exponents.

$\quad = 46.9 \times 10^{12}$ Do the multiplication. Do the addition.

We note that 46.9 is not between 0 and 1, so 46.9×10^{12} is not written in scientific notation. To answer in scientific notation, we proceed as follows.

$\quad = 4.69 \times 10^1 \times 10^{12}$ Write 46.9 in scientific notation as 4.69×10^1.

$\quad = 4.69 \times 10^{13}$ Keep the base of 10 and add the exponents.

Sirius is approximately 4.69×10^{13} or 46,900,000,000,000 miles from the earth   ∎

EXAMPLE 7

Atoms Scientific notation is used in chemistry. As an example, the approximate weight (in grams) of one atom of the heaviest naturally occurring element, uranium, is given by

$$\frac{2.4 \times 10^2}{6.0 \times 10^{23}}$$

Evaluate the expression.

Solution

$$\frac{2.4 \times 10^2}{6.0 \times 10^{23}} = \frac{2.4}{6.0} \times \frac{10^2}{10^{23}}$$

Divide the numbers and the powers of 10 separately.

$$= \frac{2.4}{6.0} \times 10^{2-23}$$

For the powers of 10, keep the base and subtract the exponents.

$$= 0.4 \times 10^{-21}$$

Do the division. Subtract the exponents: $2 - 23 = -21$.

$$= 4.0 \times 10^{-1} \times 10^{-21}$$

Write 0.4 in scientific notation as 4.0×10^{-1}.

$$= 4.0 \times 10^{-22}$$

For the powers of 10, keep the base and add the exponents: $-1 + (-21) = -22$.

One atom of uranium weighs 4.0×10^{-22} gram. Written in standard notation, this is 0.00000000000000000000004 g.

SELF CHECK Find the approximate weight (in grams) of one atom of gold by evaluating $\dfrac{1.98 \times 10^2}{6.0 \times 10^{23}}$.

Answer: 3.3×10^{-22} g ∎

ACCENT ON TECHNOLOGY *Electric Charges*

The electric charges on each of the 0.20-gram balls that are suspended from 50-cm-long strings in Figure 4-2 are the same. For this reason, the balls repel each other. The charge q (in coulombs) on each ball is given by

$$q = \frac{2.97 \times 10^{-2}}{9.49 \times 10^4}$$

FIGURE 4-2

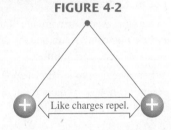

Like charges repel.

We can evaluate the expression by entering the numbers written in scientific notation with the ⎡EE⎤ key on a scientific calculator.

Keystrokes   2.97 ⎡EE⎤ ⎡+/−⎤ 2 ÷ 9.49 ⎡EE⎤ 4 ⎡=⎤   ⎡0.000000313⎤

In standard notation, the charge on each ball is 0.000000313 coulomb. This could be written as 3.13×10^{-7} coulomb using scientific notation.

If we use a graphing calculator, the keystrokes are similar. The result is displayed in scientific notation.

Keystrokes   2.97 ⎡2nd⎤ ⎡EE⎤ ⎡(−)⎤ 2 ÷ 9.49 ⎡2nd⎤ ⎡EE⎤ 4 ⎡ENTER⎤

```
2.97E−2/9.49E4
        3.129610116E−7
```

STUDY SET

Section 4.3

VOCABULARY

In Exercises 1–2, fill in the blanks to make the statements true.

1. A number is written in ___scientific___ notation when it is written as the product of a number between 1 (including 1) and 10.

2. The number 125,000 is written in ___standard___ notation.

CONCEPTS

In Exercises 3–16, fill in the blanks to make the statements true.

3. $2.5 \times 10^2 =$ 250

4. $2.5 \times 10^{-2} =$ 0.025

5. $2.5 \times 10^{-5} =$ 0.000025

6. $2.5 \times 10^5 =$ 250,000

7. $387,000 = 3.87 \times$ 10^5

8. $38.7 = 3.87 \times$ 10^1

9. $0.00387 = 3.87 \times$ 10^{-3}

10. $0.000387 = 3.87 \times$ 10^{-4}

11. When we multiply a decimal by 10^5, the decimal point moves 5 places to the _____right_____.

12. When we multiply a decimal by 10^{-7}, the decimal point moves 7 places to the _____left_____.

13. Dividing a decimal by 10^4 is equivalent to multiplying it by 10^{-4} .

14. Multiplying a decimal by 10^0 does not move the decimal point, because $10^0 =$ 1 .

15. When a real number greater than 1 is written in scientific notation, the exponent on 10 is a _____positive_____ number.

16. When a real number between 0 and 1 is written in scientific notation, the exponent on 10 is a _____negative_____ number.

NOTATION

In Exercises 17–18, complete each solution.

17. Write 63.7×10^5 in scientific notation.

$$63.7 \times 10^5 = 6.37 \times 10^1 \times 10^5$$
$$= 6.37 \times 10^{1 + 5}$$
$$= 6.37 \times 10^6$$

18. Simplify $\dfrac{64,000}{0.00004}$.

$$\frac{64,000}{0.00004} = \frac{6.4 \times 10^4}{4 \times 10^{-5}}$$
$$= \frac{6.4}{4} \times \frac{10^4}{10^{-5}}$$
$$= 1.6 \times 10^{4 - (-5)}$$
$$= 1.6 \times 10^9$$

PRACTICE

In Exercises 19–30, write each number in scientific notation.

19. 23,000 2.3×10^4

20. 4,750 4.75×10^3

21. 1,700,000 1.7×10^6

22. 290,000 2.9×10^5

23. 0.062 6.2×10^{-2}

24. 0.00073 7.3×10^{-4}

25. 0.0000051 5.1×10^{-6}

26. 0.04 4×10^{-2}

27. 42.5×10^2 4.25×10^3

28. 0.3×10^3 3×10^2

29. 0.25×10^{-2}
   2.5×10^{-3}

30. 25.2×10^{-3}
   2.52×10^{-2}

In Exercises 31–42, write each number in standard notation.

31. 2.3×10^2 230

32. 3.75×10^4 37,500

33. 8.12×10^5 812,000

34. 1.2×10^3 1,200

35. 1.15×10^{-3} 0.00115

36. 4.9×10^{-2} 0.049

37. 9.76×10^{-4} 0.000976

38. 7.63×10^{-5} 0.0000763

39. 25×10^6 25,000,000

40. 0.07×10^3 70

41. 0.51×10^{-3} 0.00051

42. 617×10^{-2} 6.17

43. ASTRONOMY The distance from earth to Alpha Centauri (the nearest star outside our solar system) is about 25,700,000,000,000 miles. Express this number in scientific notation. 2.57×10^{13} mi

44. SPEED OF SOUND The speed of sound in air is 33,100 centimeters per second. Express this number in scientific notation. 3.31×10^4 cm/sec

45. GEOGRAPHY The largest ocean in the world is the Pacific Ocean, which covers 6.38×10^7 square miles. Express this number in standard notation.
63,800,000 mi^2

46. ATOMS The number of atoms in 1 gram of iron is approximately 1.08×10^{22}. Express this number in standard notation.   10,800,000,000,000,000,000,000

47. LENGTH OF A METER One meter is approximately 0.00622 mile. Use scientific notation to express this number. 6.22×10^{-3} mi

48. ANGSTROM One angstrom is 1.0×10^{-7} millimeter. Express this number in standard notation.
0.0000001 mm

In Exercises 49–54, use scientific notation and the rules for exponents to simplify each expression. Give all answers in standard notation. Use a calculator to check each result.

49. $(3.4 \times 10^2)(2.1 \times 10^3)$ 714,000

50. $(4.1 \times 10^{-3})(3.4 \times 10^4)$ 139.4

51. $\dfrac{9.3 \times 10^2}{3.1 \times 10^{-2}}$ 30,000

52. $\dfrac{7.2 \times 10^6}{1.2 \times 10^8}$ 0.06

53. $\dfrac{96,000}{(12,000)(0.00004)}$ 200,000

54. $\dfrac{(0.48)(14,400,000)}{96,000,000}$ 0.072

In Exercises 55–60, use a calculator to evaluate each expression.

55. $(456.4)^6$ $9.038030748 \times 10^{15}$

56. $(0.053)^8$ $6.225969041 \times 10^{-11}$

57. $(0.009)^{-6}$ $1.881676423 \times 10^{12}$

58. 225^{-5} $1.734152992 \times 10^{-12}$

59. $\dfrac{(3.12 \times 10^{16})(4.50 \times 10^{-6})}{2.40 \times 10^{-5}}$ 5.85×10^{15}

60. $(7.35 \times 10^5)(3.84 \times 10^{-7}) \cdot \dfrac{1}{2.10 \times 10^{12}}$

1.344×10^{-13}

APPLICATIONS

61. WAVELENGTH Transmitters, vacuum tubes, and lights emit energy that can be modeled as a wave, as shown in Illustration 1. Examples of the most common types of electromagnetic waves are given in the table. List the wavelengths in order from shortest to longest.   g, x, u, v, i, m, r

ILLUSTRATION 1

This distance between the two crests of the wave is called the wavelength.

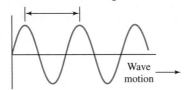

Wave motion →

Type	Use	Wavelength (m)
visible light	lighting	9.3×10^{-6}
infrared	photography	3.7×10^{-5}
x-ray	medical	2.3×10^{-11}
radio wave	communication	3.0×10^2
gamma ray	treating cancer	8.9×10^{-14}
microwave	cooking	1.1×10^{-2}
ultraviolet	sun lamp	6.1×10^{-8}

62. EXPLORATION On July 4, 1997, the Pathfinder, carrying the rover vehicle called Sojourner, landed on Mars to perform a scientific investigation of the planet. The distance from Mars to the earth is approximately 3.5×10^7 miles. Use scientific notation to express this distance in feet. (*Hint:* 5,280 feet = 1 mile.) 1.848×10^{11} ft

63. PROTON The mass of one proton is approximately 1.7×10^{-24} gram. Use scientific notation to express the mass of 1 million protons. 1.7×10^{-18} g

64. SPEED OF SOUND The speed of sound in air is approximately 3.3×10^4 centimeters per second. Use scientific notation to express this speed in kilometers per second. (*Hint:* 100 centimeters = 1 meter and 1,000 meters = 1 kilometer.) 3.3×10^{-1} km/sec

65. LIGHT YEAR One light year is about 5.87×10^{12} miles. Use scientific notation to express this distance in feet. (*Hint:* 5,280 feet = 1 mile.)
3.099363×10^{16} ft

66. OIL RESERVES Saudi Arabia is believed to have crude oil reserves of about 2.61×10^{11} barrels. A barrel contains 42 gallons of oil. Use scientific notation to express its oil reserves in gallons.
1.0962×10^{13} gal

67. INTEREST EARNED As of December 31, 1997, the Federal Deposit Insurance Corporation (FDIC) reported that the total insured deposits in U.S. banks and savings and loans was approximately 3.42×10^{12} dollars. If this money was invested at a rate of 4% simple annual interest, how much would it earn in one year? (Use scientific notation to express the answer.) 1.368×10^{11} dollars

68. CURRENCY As of March 31, 1998, the U.S. Treasury reported that the number of $20 bills in circulation was approximately 4.136×10^9. What was the total value of the currency? (Use scientific notation to express the answer.) 8.272×10^{10} dollars

69. SIZE OF THE MILITARY The graph in Illustration 2 shows the number of U.S. troops for 1979–1997. Estimate each of the following and express your answers in scientific and standard notation.
  a. The number of troops in 1993
   1.7×10^6; 1,700,000
  b. The smallest and largest numbers of troops during these years
   1.5×10^6, 1,500,000; 2.05×10^6; 2,050,000

ILLUSTRATION 2

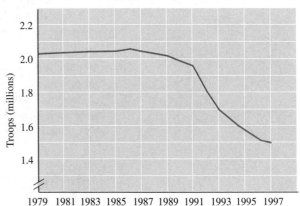

Based on data from the U.S. Department of Defense

70. THE NATIONAL DEBT The graph in Illustration 3 shows the growth of the national debt for the fiscal years 1992–1997.

ILLUSTRATION 3

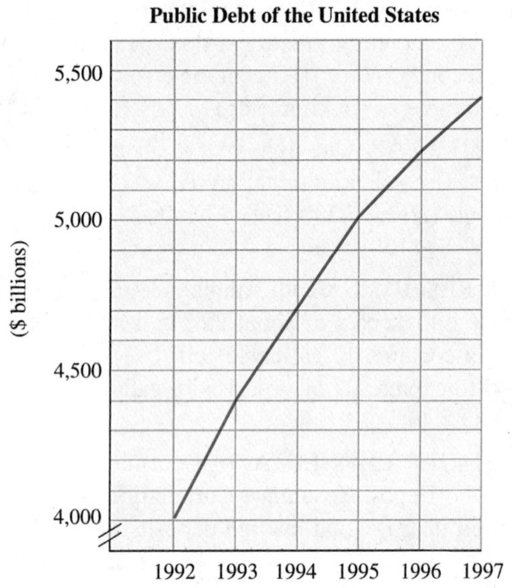

Public Debt of the United States

Based on data from the U.S. Department of the Treasury

a. Use scientific notation to express the debt as of 1994, 1995, and 1997.
  4.7×10^{12} dollars, 5.0×10^{12} dollars, 5.4×10^{12} dollars

b. In 1997, the population of the United States was about 2.68×10^8. To the nearest dollar, what was the share of the debt for each man, woman, and child in the United States? Answer in standard notation. $20,149

WRITING

71. In what situations would scientific notation be more convenient than standard notation?

72. To multiply a number by a power of 10, we move the decimal point. Which way, and how far? Explain.

73. 2.3×10^{-3} contains a negative sign but represents a positive number. Explain.

74. Is this a true statment? $2.0 \times 10^3 = 2 \times 10^3$. Explain.

REVIEW

75. If $y = -1$, find the value of $-5y^{55}$.   5

76. What is the y-intercept of the graph of $y = -3x - 5$?
  $(0, -5)$

In Exercises 77–78, tell which property of real numbers justifies each statement.

77. $5 + z = z + 5$ commutative property of addition

78. $7(u + 3) = 7u + 7 \cdot 3$ distributive property

In Exercises 79–80, solve each equation.

79. $3(x - 4) - 6 = 0$ 6

80. $8(3x - 5) - 4(2x + 3) = 12$ 4

▶ **4.4**

Polynomials and Polynomial Functions

In this section, you will learn about

> Polynomials ■ Monomials, binomials, and trinomials ■ Degree of a polynomial ■ Evaluating polynomial functions ■ Making tables ■ Graphing polynomial functions ■ Graphs of polynomial functions

Introduction   In arithmetic, we learned how to add, subtract, multiply, divide, and find powers of numbers. In algebra, we will learn how to perform these operations on *polynomials*. In this section, we will introduce polynomials, classify them into groups, define their degrees, and show how to evaluate them at specific values of their variables. Finally, we will show how to graph polynomial functions.

Polynomials

Recall that a **term** is a number or a product of a number and one or more variables, which may be raised to powers. Examples of terms are

$$3x, \qquad -4y^2, \qquad \frac{1}{2}a^2b^3, \qquad t, \qquad \text{and} \qquad 25$$

The **numerical coefficients,** or simply **coefficients,** of the first four of these terms are $3, -4, \frac{1}{2},$ and 1, respectively. Because $25 = 25x^0$, 25 is considered to be the numerical coefficient of the term 25.

Polynomials

A **polynomial** is a term or a sum of terms in which all variables have whole-number exponents.

Here are some examples of polynomials:

$$3x + 2, \qquad -4y^2 - 2y - 3, \qquad 8xy^2, \qquad \text{and} \qquad 3a - 4b - 4c + 8d$$

The polynomial $3x + 2$ has two terms, and we say that it is a **polynomial in** x. Since $-4y^2 - 2y - 3$ can be written as $-4y^2 + (-2y) + (-3)$, it is the sum of three terms, $-4y^2, -2y, -3$. It is written in **decreasing powers** of y, because the powers on y decrease from left to right. $8xy^2$ is a single term, and it is in two variables, x and y.

 WARNING! The expression $2x^3 - 3x^{-2} + 5$ is not a polynomial, because the second term contains a variable with an exponent that is not a whole number. Similarly, $y^2 - \frac{7}{y}$ is not a polynomial, because $\frac{7}{y}$ can be written $7y^{-1}$.

EXAMPLE 1

Identifying polynomials. Tell whether each expression is a polynomial.

a. $x^2 + 2x + 1$ Yes.

b. $3a^{-1} - 2a - 3$ No. In the first term, the exponent on the variable is not a whole number.

c. $\dfrac{1}{2}x^3 - 2.3x$ Yes, since it can be written as the sum $\frac{1}{2}x^3 + (-2.3x)$.

d. $\dfrac{p + 3}{p - 1}$ No. Variables cannot be in the denominator of a fraction.

SELF CHECK   Tell whether each expression is a polynomial:
a. $3x^{-4} + 2x^2 - 3$ and   **b.** $7.5p^3 - 4p^2 - 3p + 4$.   *Answers:*   **a.** no,   **b.** yes ■

Monomials, Binomials, and Trinomials

We classify some polynomials by the number of terms they contain. A polynomial with one term is called a **monomial.** A polynomial with two terms is called a **binomial.** A polynomial with three terms is called a **trinomial.** Here are some examples. There is no special name for a polynomial with four or more terms.

Monomials	Binomials	Trinomials
$5x^2y$	$3u^3 - 4u^2$	$-5t^2 + 4t + 3$
$-6x$	$18a^2b + 4ab$	$27x^3 - 6x - 2$
29	$-29z^{17} - 1$	$-32r^6 + 7y^3 - z$

E X A M P L E   2

Classifying polynomials. Classify each polynomial as a monomial, binomial, or trinomial.

a. $5.2x^4 + 3.1x$ Since the polynomial has two terms, $5.2x^4$ and $3.1x$, it is a binomial.

b. $7g^4 - 5g^3 - 2$ Since the polynomial has three terms, $7g^4$, $-5g^3$, and -2, it is a trinomial.

c. $-5x^2y^3$ Since the polynomial has one term, it is a monomial.

SELF CHECK

Classify each polynomial as a monomial, binomial, or trinomial: **a.** $5x$,   **b.** $-5x^2 + 2x - 0.5$, and   **c.** $16x^2 - 9y^2$.

Answers: **a.** monomial,   **b.** trinomial,   **c.** binomial

Degree of a Polynomial

The monomial $7x^6$ is called a **monomial of sixth degree** or a **monomial of degree 6,** because the variable x occurs as a factor six times. The monomial $3x^3y^4$ is a monomial of seventh degree, because the variables x and y occur as factors a total of seven times. Here are some more examples:

    $2.7a$ is a monomial of degree 1.

    $-2x^3$ is a monomial of degree 3.

    $47x^2y^3$ is a monomial of degree 5.

    8 is a monomial of degree 0, because $8 = 8x^0$.

These examples illustrate the following definition.

Degree of a monomial

If a is a nonzero constant, the **degree of the monomial** ax^n is n.

The **degree of a monomial** in several variables is the sum of the exponents on those variables.

 WARNING! Note that the degree of ax^n is not defined when $a = 0$. Since $ax^n = 0$ when $a = 0$, the constant 0 has no defined degree.

Because each term of a polynomial is a monomial, we define the degree of a polynomial by considering the degrees of each of its terms.

Degree of a polynomial

The **degree of a polynomial** is determined by the term with the largest degree.

Here are some examples:

$x^2 + 2x$ is a binomial of degree 2, because the degree of its first term is 2 and the degree of its second term is less than 2.

$d^3 - 3d^2 + 1$ is a trinomial of degree 3, because the degree of its first term is 3 and the degree of each of its other terms is less than 3.

$25y^{13} - 15y^8z^{10} - 32y^{10}z^8 + 4$ is a polynomial of degree 18, because its second and third terms are of degree 18. Its other terms have degree less than 18.

EXAMPLE 3

Degree of a polynomial. Find the degree of each polynomial:
a. $-4x^3 - 5x^2 + 3x$, **b.** $1.6w - 1.6$, and **c.** $-17a^2b^3 + 12ab^6$.

Solution

a. The trinomial $-4x^3 - 5x^2 + 3x$ has terms of degree 3, 2, and 1. Therefore, its degree is 3.

b. The first term of $1.6w - 1.6$ has degree 1 and the second term has degree 0, so the binomial has degree 1.

c. The degree of the first term of $-17a^2b^3 + 12ab^6$ is 5 and the degree of the second term is 7, so the binomial has degree 7.

SELF CHECK

Find the degree of each polynomial:
a. $15p^3 - 25p^2 - 3p + 4$ and **b.** $-14st^4 + 12s^3t$. *Answers:* **a.** 3, **b.** 5 ∎

If written in descending powers of the variable, the **leading term** of a polynomial is the term of highest degree. For example, the leading term of $-4x^3 - 5x^2 + 3x$ is $-4x^3$. The coefficient of the leading term (in this case, -4) is called the **leading coefficient.**

Evaluating Polynomial Functions

Each of the equations below defines a function, because each input x-value gives exactly one output value. Since the right-hand side of each equation is a polynomial, these functions are called **polynomial functions.**

$$f(x) = \underbrace{6x + 4} \qquad g(x) = \underbrace{3x^2 + 4x - 5} \qquad h(x) = \underbrace{-x^3 + x^2 - 2x + 3}$$

| This polynomial has two terms. Its degree is 1. | This polynomial has three terms. Its degree is 2. | This polynomial has four terms. Its degree is 3. |

To evaluate a polynomial function for a specific value, we replace the variable in the defining equation with the value, called the **input**. Then we simplify the resulting expression to find the **output**. For example, suppose we wish to evaluate the polynomial function $f(x) = 6x + 4$ for $x = 1$. Then $f(1)$ (read as "f of 1") represents the value of $f(x) = 6x + 4$ when $x = 1$. We find $f(1)$ as follows.

$f(x) = 6x + 4$	The given function.
$f(1) = 6(1) + 4$	Substitute 1 for x. The number 1 is the input.
$= 6 + 4$	Do the multiplication.
$= 10$	Do the addition. 10 is the output.

Thus, $f(1) = 10$.

EXAMPLE 4

Evaluating polynomial functions. Consider the function $g(x) = 3x^2 + 4x - 5$. Find **a.** $g(0)$ and **b.** $g(-2)$.

Solution **a.** $g(x) = 3x^2 + 4x - 5$ The given function.

$g(0) = 3(0)^2 + 4(0) - 5$ To find $g(0)$, substitute 0 for x.

$= 3(0) + 4(0) - 5$ Evaluate the power.

$= 0 + 0 - 5$ Do the multiplications.

$g(0) = -5$

b. $g(x) = 3x^2 + 4x - 5$ The given function.

$g(-2) = 3(-2)^2 + 4(-2) - 5$ To find $g(-2)$, substitute -2 for x.

$= 3(4) + 4(-2) - 5$ Evaluate the power.

$= 12 + (-8) - 5$ Do the multiplications.

$g(-2) = -1$

SELF CHECK

Consider the function $h(x) = -x^3 + x - 2x + 3$.
Find **a.** $h(0)$ and **b.** $h(-3)$.   *Answers:* **a.** 3, **b.** 33   ∎

EXAMPLE 5

Supermarket display. The polynomial function

$$f(c) = \frac{1}{3}c^3 + \frac{1}{2}c^2 + \frac{1}{6}c$$

gives the number of cans used in a display shaped like a square pyramid, having a square base formed by c cans per side. Find the number of cans of soup used in the display shown in Figure 4-3.

FIGURE 4-3

Solution Since each side of the square base of the display is formed by 4 cans, $c = 4$. We can find the number of cans used in the display by finding $f(4)$.

$f(c) = \frac{1}{3}c^3 + \frac{1}{2}c^2 + \frac{1}{6}c$ The given function.

$f(4) = \frac{1}{3}(4)^3 + \frac{1}{2}(4)^2 + \frac{1}{6}(4)$ Substitute 4 for c.

$= \frac{1}{3}(64) + \frac{1}{2}(16) + \frac{1}{6}(4)$ Find the powers.

$= \frac{64}{3} + 8 + \frac{2}{3}$ Do the multiplication, then simplify: $\frac{4}{6} = \frac{2}{3}$.

$= \frac{66}{3} + 8$ Add the fractions.

$= 22 + 8$

$= 30$

30 cans of soup were used in the display.   ∎

ACCENT ON TECHNOLOGY *100-Meter Sprint*

In the 1996 Olympics, Donovan Bailey of Canada set the world record in the 100 meters with a time of 9.84 seconds. Suppose the polynomial function $f(t) = -0.2t^2 + 11.12t$ gives the distance in meters run in t seconds by another sprinter in the 100 meters. To predict how far behind Bailey this sprinter would have finished in the Olympic competition, we can substitute Bailey's time of 9.84 seconds for t and evaluate $f(9.84)$.

$$f(t) = -0.2t^2 + 11.12t$$
$$f(9.84) = -0.2(9.84)^2 + 11.12 (9.84)$$

To evaluate $f(9.84)$ using a scientific calculator, we enter these numbers and press these keys:

Keystrokes .2 $\boxed{+/-}$ $\boxed{\times}$ 9.84 $\boxed{x^2}$ $\boxed{+}$ 11.12 $\boxed{\times}$ 9.84 $\boxed{=}$ $\boxed{\text{90.05568}}$

In 9.84 seconds, the sprinter can run about 90 meters and would finish approximately 10 meters behind Bailey.

Making Tables

When we evaluate a polynomial function for several values of its variable, we can write the results in a table.

EXAMPLE 6

Constructing a table. Find $f(-2), f(-1)$, and $f(0)$, where $f(x) = x^3 - 3x^2 + 4$. Write the results in a table.

Solution

In the first column of the table, we write each of the input values for x. (These are the numbers inside the parentheses of the function notation: $-2, -1, 0$.) We then find each corresponding output value by substituting an input value for x in $x^3 - 3x^2 + 4$ and evaluating the expression. To review how to generate a table using a graphing calculator, see page 199.

x	$f(x)$
-2	-16
-1	0
0	4

$f(-2) = (-2)^3 - 3(-2)^2 + 4$
$\quad = -8 - 3(4) + 4$
$\quad = -8 - 12 + 4$
$\quad = -16$

$f(-1) = (-1)^3 - 3(-1)^2 + 4$
$\quad = -1 - 3(1) + 4$
$\quad = -1 - 3 + 4$
$\quad = 0$

$f(0) = (0)^3 - 3(0)^2 + 4$
$\quad = 0 - 3(0) + 4$
$\quad = 0 - 0 + 4$
$\quad = 4$

SELF CHECK

Find $f(1), f(2)$, and $f(3)$, where $f(x) = x^3 - 3x^2 + 4$. Write the results in a table. *Answers:*

x	$f(x)$
1	2
2	0
3	4

Graphing Polynomial Functions

We can graph polynomial functions by making a table of values, plotting points, and drawing a smooth curve that passes through those points.

E X A M P L E   7

Graphing a polynomial function. Graph $f(x) = x^3 - 3x^2 + 4$.

Solution
We substitute numbers for x, compute the corresponding values of $f(x)$, and list the results in a table as ordered pairs. (Note that this work was done in Example 6 and the Self Check.) We then plot the pairs (x, y) and draw a smooth curve through the points, as shown in Figure 4-4.

FIGURE 4-4

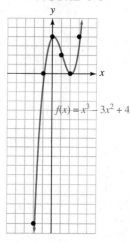

$$f(x) = x^3 - 3x^2 + 4$$

x	$f(x)$	(x, y)
-2	-16	$(-2, -16)$
-1	0	$(-1, 0)$
0	4	$(0, 4)$
1	2	$(1, 2)$
2	0	$(2, 0)$
3	4	$(3, 4)$

This column can also be labeled $(x, f(x))$.

The value of $f(x)$ is the y-coordinate of the point.

ACCENT ON TECHNOLOGY *Graphing Polynomial Functions*

It is possible to use graphing calculators to generate tables and graphs for polynomial functions. For example, Figure 4-5 (a) shows how to enter the function from Example 7, $f(x) = x^3 - 3x^2 + 4$. Figures 4-5(b) and (c) show the calculator display of a table and the graph.

FIGURE 4-5

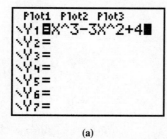

(a)

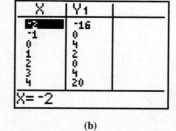

(b)

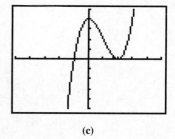

(c)

Graphs of Polynomial Functions

The graphs of two polynomial functions are shown in Figure 4-6. These graphs have characteristic "peaks" and "valleys." When such graphs describe real-life situations, locating the highest and lowest points on the graph can give valuable information.

FIGURE 4-6

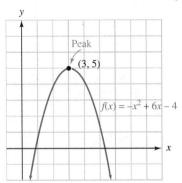

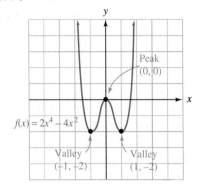

EXAMPLE 8

Immigration. The polynomial function graphed in Figure 4-7 approximates the percent of the U.S. population that was foreign born for each of the years 1900–1997. What are the coordinates of the highest point on the graph? Explain its significance.

Solution The highest point on the graph has coordinates (10, 14.5). This means that in 1910, 14.5% of the population of the United States was foreign born. This percent is the largest for the 97-year span shown on the graph.

FIGURE 4-7

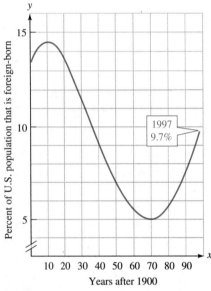

Based on data from Bureau of the Census, U.S. Department of Commerce

SELF CHECK What are the coordinates of the lowest point on the graph in Figure 4-7? Explain its significance.

Answer: (70, 5). In 1970, 5% of the population was foreign born. This percent is the smallest for the 97 years shown. ∎

STUDY SET

Section 4.4

VOCABULARY

In Exercises 1–10, fill in the blanks to make the statements true.

1. A ___polynomial___ is an algebraic expression that is the sum of one or more terms containing whole-number exponents.

2. The numerical ___coefficient___ of the term $-25x^2y^3$ is -25.

3. The degree of a polynomial is the same as the degree of its ___term___ with the largest degree.

4. A ___monomial___ is a polynomial with one term. A ___binomial___ is a polynomial with two terms.

5. The _____degree_____ of the monomial $3x^7$ is 7.

6. For the polynomial $6x^2 + 3x - 1$, the
_____leading_____ term is $6x^2$, and the
_____leading_____ coefficient is 6.

7. $-x^3 - 6x^2 + 9x - 2$ is a polynomial __in__ x and is
written in _____decreasing_____ powers of x.

8. A _____trinomial_____ is a polynomial with three terms.

9. The notation $f(x)$ is read as f _of_ x.

10. $f(2)$ represents the _____value_____ of a function
when $x = 2$.

CONCEPTS

In Exercises 11–14, tell whether each expression is a
polynomial.

11. $x^3 - 5x^2 - 2$ yes

12. $x^{-4} - 5x$ no

13. $\dfrac{1}{2x} + 3$ no

14. $x^3 - 1$ yes

In Exercises 15–26, classify each polynomial as a mono-
mial, binomial, or trinomial, if possible.

15. $3x + 7$ binomial

16. $3y - 5$ binomial

17. $y^2 + 4y + 3$ trinomial

18. $3xy$ monomial

19. $3z^2$ monomial

20. $3x^4 - 2x^3 + 3x - 1$
none of these

21. $5t - 32$ binomial

22. $9x^2y^3z^4$ monomial

23. $s^2 - 23s + 31$
trinomial

24. $2x^3 - 5x^2 + 6x - 3$
none of these

25. $3x^5 - x^4 - 3x^3 + 7$
none of these

26. x^3 monomial

In Exercises 27–38, find the degree of each polynomial.

27. $3x^4$ 4th

28. $3x^5$ 5th

29. $-2x^2 + 3x + 1$ 2nd

30. $-5x^4 + 3x^2 - 3x$ 4th

31. $3x - 5$ 1st

32. $y^3 + 4y^2$ 3rd

33. $-5r^2s^2 - r^3s + 3$ 4th

34. $4r^2s^3 - 5r^2s^8$ 10th

35. $x^{12} + 3x^2y^3$ 12th

36. $17ab^5 - 12a^3b$ 6th

37. 38   0th

38. -25 0th

39. Give the coordinates of the "valley" of the graph of
the polynomial function shown in Illustration 1.
$(-2, -5)$

ILLUSTRATION 1

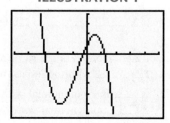

40. Give the coordinates of the "peaks" of the graph of
the polynomial function shown in Illustration 2.
$(-1, 3), (1, 3)$

ILLUSTRATION 2

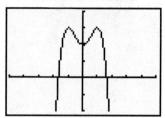

NOTATION

In Exercises 41–42, complete each solution.

41. If $f(x) = -2x^2 + 3x - 1$, find $f(2)$.
$$f(2) = -2(\;2\;)^2 + 3(\;2\;) - 1$$
$$= -2(\;4\;) + \;6\; - 1$$
$$= -8 + 6 - \;1\;$$
$$= \;-2\; - 1$$
$$= -3$$

42. If $f(x) = -2x^2 + 3x - 1$, find $f(-2)$.
$$f(-2) = -2(\;-2\;)^2 + 3(\;-2\;) - 1$$
$$= -2(\;4\;) + (\;-6\;) - 1$$
$$= \;-8\; + (-6) - 1$$
$$= \;-14\; - 1$$
$$= -15$$

43. Explain why $f(x) = x^3 + 2x^2 - 3$ is called a polynomial function.   Because $x^3 + 2x^2 - 3$ is a polynomial.

44. Give another way to label the last column of the
table shown below. (x, y)

x	$f(x)$	$(x, f(x))$

PRACTICE

In Exercises 45–52, let $f(x) = 5x - 3$. Find each value.

45. $f(2)$ 7

46. $f(0)$ -3

47. $f(-1)$ -8

48. $f(-2)$ -13

49. $f\left(\dfrac{1}{5}\right)$ -2

50. $f\left(\dfrac{4}{5}\right)$ 1

51. $f(-0.9)$ -7.5

52. $f(-1.2)$ -9

In Exercises 53–60, let $g(x) = -x^2 - 4$. Find each
value.

53. $g(0)$ -4

54. $g(1)$ -5

55. $g(-1)$ -5

56. $g(-2)$ -8

57. $g(1.3)$ -5.69 **58.** $g(2.4)$ -9.76

59. $g(-13.6)$ -188.96 **60.** $g(-25.3)$ -644.09

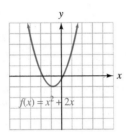 *In Exercises 61–68, let $h(x) = x^3 - 2x + 3$. Find each value.*

61. $h(0)$ 3   **62.** $h(3)$ 24

63. $h(-2)$ -1 **64.** $h(-1)$ 4

65. $h(0.9)$ 1.929   **66.** $h(0.4)$ 2.264

67. $h(-8.1)$ -512.241 **68.** $h(-7.7)$ -438.133

In Exercises 69–72, complete each table and then graph the polynomial function.

69. $f(x) = x^2 + 2x$

x	$f(x)$
-3	3
-2	0
-1	-1
0	0
1	3

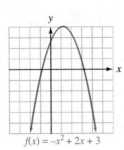

$f(x) = x^2 + 2x$

70. $f(x) = -x^2 + 2x + 3$

x	$f(x)$
-2	-5
-1	0
0	3
1	4
2	3
3	0
4	-5

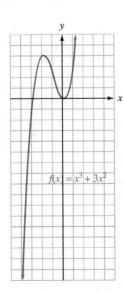

$f(x) = -x^2 + 2x + 3$

71. $f(x) = x^3 + 3x^2$

x	$f(x)$
-4	-16
-3	0
-2	4
-1	2
0	0
1	4

$f(x) = x^3 + 3x^2$

72. $f(x) = -x^3 + 3x^2 - 4$

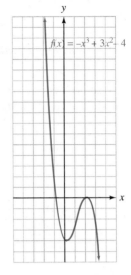

$f(x) = -x^3 + 3x^2 - 4$

x	$f(x)$
-2	16
-1	0
0	-4
1	-2
2	0
3	-4

 In Exercises 73–74, use a graphing calculator to graph each polynomial function. Use window settings of $x = -5$ to 5 and $y = -10$ to 10.

73. $f(x) = -0.5x^2 + 1.2x - 2.5$

74. $f(x) = x^3 + x^2 - 6x$

APPLICATIONS

In Exercises 75–82, use a calculator to help solve each problem.

75. TRACK RECORDS By how many meters would the slower sprinter mentioned in the Accent on Technology feature in this section finish behind Florence Griffith Joyner if she ran her world record time of 10.49 seconds in the 100 meters?   about 5.4 m

76. MAXIMIZING REVENUE The revenue (in dollars) that a manufacturer of office desks receives is given by the polynomial function

$$f(d) = -0.08d^2 + 100d$$

where d is the number of desks manufactured.
a. Find the total revenue if 625 desks are manufactured. $31,250
b. Does increasing the number of desks being manufactured to 650 increase the revenue?   no

77. WATER BALLOONS Some college students launched water balloons from the balcony of their dormitory on unsuspecting sunbathers on the college quad. The height in feet of the balloons at a time t seconds after being launched is given by the polynomial function

$$f(t) = -16t^2 + 12t + 20$$

What was the height of the balloons 0.5 second and 1.5 seconds after being launched?   22 ft, 2 ft

78. STOPPING DISTANCE The number of feet that a car travels before stopping depends on the driver's reaction time and the braking distance, as shown in Illustration 3. For one driver, the stopping distance is given by the polynomial function

$$f(v) = 0.04v^2 + 0.9v$$

where v is the velocity of the car. Find the stopping distance when the driver is traveling at 30 mph. 63 ft

ILLUSTRATION 3

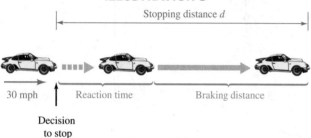

Stopping distance d

30 mph   Reaction time   Braking distance

Decision
to stop

79. SUSPENSION BRIDGE See Illustration 4. The function

$$f(s) = 400 + 0.0066667s^2 - 0.0000001s^4$$

approximates the length of the cable between the two vertical towers of a suspension bridge, where s is the sag in the cable. Estimate the length of the cable if the sag is 24.6 feet. about 404 ft

ILLUSTRATION 4

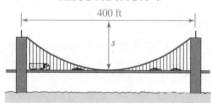

400 ft

s

80. PRODUCE DEPARTMENT Suppose a grocer is going to set up a pyramid-shaped display of cantaloupes like that shown in Figure 4–3 in Example 5. If each side of the square base of the display is made of 6 cantaloupes, how many will be used in the display?
91

81. DOLPHINS At a marine park, three trained dolphins jump in unison over an arching stream of water whose path can be described by the polynomial function $f(x) = -0.05x^2 + 2x$. See Illustration 5. Given the takeoff points for each dolphin, how high must each dolphin jump to clear the stream of water?
18.75 ft, 20 ft, 15 ft

82. TUNNEL The arch at the entrance to a tunnel is described by the polynomial function $f(x) = -0.25x^2 + 23$. See Illustration 6. What is the height of the arch at the edge of the pavement? 14 ft

ILLUSTRATION 5

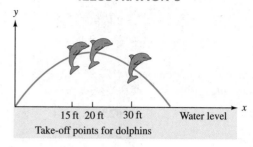

15 ft 20 ft   30 ft   Water level
Take-off points for dolphins

ILLUSTRATION 6

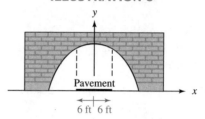

Pavement

6 ft 6 ft

83. IMPORTED STEEL Plot the data for each quarter on the graph in Illustration 7. Connect the points with a smooth curve. When were the imports the lowest? When were they the highest?
the 4th quarter of 1995; the 3rd quarter of 1998

U.S. Steel Imports (100,000 tons)			
1995	**1996**	**1997**	**1998**
Q1 5.0	Q1 2.0	Q1 8.0	Q1 10.0
Q2 3.5	Q2 2.5	Q2 9.0	Q2 18.0
Q3 2.0	Q3 4.0	Q3 8.0	Q3 23.5
Q4 1.5	Q4 7.0	Q4 7.5	

ILLUSTRATION 7

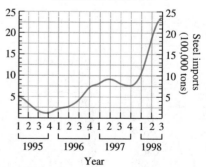

1 2 3 4 1 2 3 4 1 2 3 4 1 2 3
1995   1996   1997   1998
Year
Steel imports (100,000 tons)

Based on data from the U.S. Department of Commerce

84. TIDES The graph in Illustration 8 shows how the water level in a certain bay changed over a 24-hour period. Estimate the coordinates of each of the following: the lower high water mark and the higher low water mark. (16, −0.1), (11, −0.7)

ILLUSTRATION 8

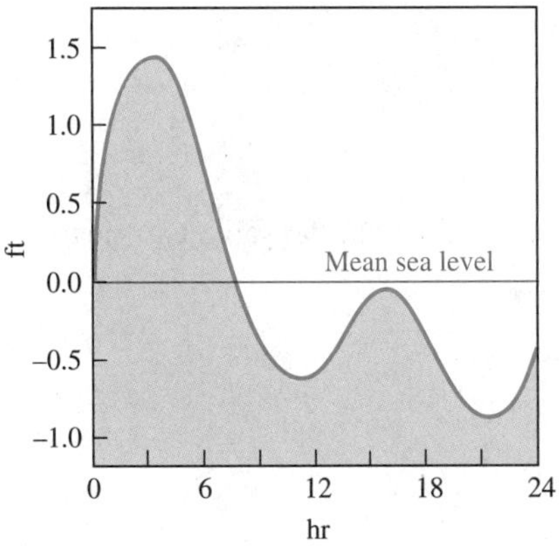

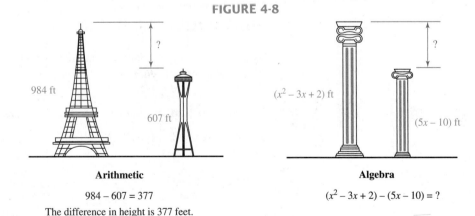

REVIEW

In Exercises 87–88, solve each inequality and graph the solution set.

87. $-4(3y + 2) \le 28$
$y \ge -3$

88. $-5 < 3t + 4 \le 13$
$-3 < t \le 3$

In Exercises 89–92, write each expression without using parentheses or negative exponents.

89. $(x^2 x^4)^3$ x^{18}

90. $(a^2)^3(a^3)^2$ a^{12}

91. $\left(\dfrac{y^2 y^5}{y^4}\right)^3$ y^9

92. $\left(\dfrac{2t^3}{t}\right)^{-4}$ $\dfrac{1}{16t^8}$

WRITING

85. Describe how to determine the degree of a polynomial.

86. List some words that contain the prefixes *mono*, *bi*, or *tri*.

▶ 4.5 Adding and Subtracting Polynomials

In this section, you will learn about

Adding monomials ■ Subtracting monomials ■ Adding polynomials ■ Subtracting polynomials ■ Adding and subtracting multiples of polynomials ■ An application of adding polynomials

Introduction   In Figure 4-8(a), the heights of the Seattle Space Needle and the Eiffel Tower in Paris are given. Using rules from arithmetic, we can find the difference in the heights of the towers by subtracting two numbers.

FIGURE 4-8

984 ft

607 ft

$(x^2 - 3x + 2)$ ft

$(5x - 10)$ ft

Arithmetic

$984 - 607 = 377$

The difference in height is 377 feet.

(a)

Algebra

$(x^2 - 3x + 2) - (5x - 10) = ?$

(b)

In Figure 4-8(b), the heights of two types of classical Greek columns are expressed using *polynomials*. To find the difference in their heights, we must subtract the polynomials. In this section, we will discuss the algebraic rules that are used to do this. Since any subtraction can be written in terms of addition, we will consider the procedures used to add polynomials first. We begin with monomials, which are polynomials with just one term.

Adding Monomials

Recall that like terms have the same variables with the same exponents:

Like terms	*Unlike terms*
$-7x$ and $15x$	$-7x$ and $15a$
$4y^3$ and $16y^3$	$4y^3$ and $16y^2$
$\frac{1}{2}xy^2$ and $-\frac{1}{3}xy^2$	$\frac{1}{2}xy^2$ and $-\frac{1}{3}x^2y$

Also recall that to combine like terms, we combine their coefficients and keep the same variables with the same exponents. For example,

$$4y + 5y = (4 + 5)y \qquad \text{and} \qquad 8x^2 - x^2 = (8 - 1)x^2$$
$$= 9y \qquad\qquad\qquad\qquad = 7x^2$$

Likewise,

$$3a + 4b - 6a + 3b = -3a + 7b \qquad \text{and} \qquad -4cd^3 + 9cd^3 = 5cd^3$$

These examples suggest that to add like monomials, we simply combine like terms.

EXAMPLE 1

Adding monomials. Do the following additions.

a. $4x^4 + 81x^4 = 85x^4$

b. $-8x^2y^2 + 6x^2y^2 + x^2y^2 = -2x^2y^2 + x^2y^2$ Work from left to right.

$$= -x^2y^2 \qquad\qquad \text{Combine like terms.}$$

c. $32c^2 + 10c + 4c^2 = 32c^2 + 4c^2 + 10c$ Group like terms together.

$$= 36c^2 + 10c \qquad\qquad \text{Combine like terms. The terms of the}$$
answer are written in descending order.

SELF CHECK   Do the following additions: **a.** $27x^6 + 8x^6$,   *Answers:*   **a.** $35x^6$,
b. $-12pq^2 + 5pq^2 + 8pq^2$,   and   **c.** $6a^3 + 15a + a^3$.   **b.** pq^2,   **c.** $7a^3 + 15a$ ∎

Subtracting Monomials

To subtract one monomial from another, we add the opposite of the monomial that is to be subtracted. In symbols, $x - y = x + (-y)$.

EXAMPLE 2

Subtracting monomials. Find each difference.

a. $8x^2 - 3x^2 = 8x^2 + (-3x^2)$ Add the opposite of $3x^2$, which is $-3x^2$.

$$= 5x^2 \qquad\qquad\qquad \text{Combine like terms.}$$

b. $6xy - 9xy = 6xy + (-9xy)$

$$= -3xy$$

c. $-3r - 5 - 4r = -3r + (-5) + (-4r)$ Add the opposite of 5 and $4r$.

$$= -3r + (-4r) + (-5) \qquad \text{Group like terms together.}$$

$$= -7r - 5 \qquad\qquad\qquad \text{Combine like terms. Write the addition of}$$
-5 as a subtraction of 5.

SELF CHECK   Find each difference: **a.** $12m^3 - 7m^3$ and   *Answers:*   **a.** $5m^3$,
b. $-4pq - 27p - 8pq$.   **b.** $-12pq - 27p$ ∎

Adding Polynomials

Because of the distributive property, we can remove parentheses enclosing several terms when the sign preceding the parentheses is a + sign. We simply drop the parentheses.

$$
\begin{aligned}
+(3x^2 + 3x - 2) &= +1(3x^2 + 3x - 2) \\
&= 1(3x^2) + 1(3x) + 1(-2) \quad \text{Use the distributive property.} \\
&= 3x^2 + 3x + (-2) \\
&= 3x^2 + 3x - 2
\end{aligned}
$$

We can add polynomials by removing parentheses, if necessary, and then combining any like terms that are contained within the polynomials.

EXAMPLE 3

Adding polynomials. Add $(3x^2 - 3x + 2) + (2x^2 + 7x - 4)$.

Solution

$$
\begin{aligned}
(3x^2 - 3x + 2) &+ (2x^2 + 7x - 4) \\
&= 3x^2 - 3x + 2 + 2x^2 + 7x - 4 \quad \text{Drop the parentheses.} \\
&= 3x^2 + 2x^2 - 3x + 7x + 2 - 4 \quad \text{Write the like terms together.} \\
&= 5x^2 + 4x - 2 \quad \text{Combine like terms.}
\end{aligned}
$$

SELF CHECK Add $(2a^2 - a + 4) + (5a^2 + 6a - 5)$.   *Answer:* $7a^2 + 5a - 1$ ∎

Problems such as Example 3 are often written with like terms aligned vertically. We can then add column by column.

$$
\begin{array}{r}
3x^2 - 3x + 2 \\
+\ 2x^2 + 7x - 4 \\
\hline
5x^2 + 4x - 2
\end{array}
$$

EXAMPLE 4

Adding polynomials vertically. Add $(4x^2 - 3)$ and $(3x^2 - 8x + 8)$.

Solution

Since the first polynomial does not have an x-term, we leave a space so that the constant terms can be aligned.

$$
\begin{array}{r}
4x^2 \qquad - 3 \\
+\ 3x^2 - 8x + 8 \\
\hline
7x^2 - 8x + 5
\end{array}
$$

SELF CHECK Add $(4q^2 - 7)$ and $(2q^2 - 8q + 9)$ vertically.   *Answer:* $6q^2 - 8q + 2$ ∎

Subtracting Polynomials

Because of the distributive property, we can remove parentheses enclosing several terms when the sign preceding the parentheses is a − sign. We simply drop the minus sign and the parentheses, and *change the sign of every term within the parentheses.*

$$
\begin{aligned}
-(3x^2 + 3x - 2) &= -1(3x^2 + 3x - 2) \\
&= -1(3x^2) + (-1)(3x) + (-1)(-2) \\
&= -3x^2 + (-3x) + 2 \\
-(3x^2 + 3x - 2) &= -3x^2 - 3x + 2
\end{aligned}
$$

This suggests that the way to subtract polynomials is to remove parentheses and combine like terms.

EXAMPLE 5

Subtracting polynomials. Find each difference.

a. $(3x - 4) - (5x + 7) = 3x - 4 - 5x - 7$ Change the sign of each term inside $(5x + 7)$.

$$= -2x - 11 \qquad \text{Combine like terms.}$$

b. $(3x^2 - 4x - 6) - (2x^2 - 6x) = 3x^2 - 4x - 6 - 2x^2 + 6x$

$$= x^2 + 2x - 6$$

c. $(-t^3 - 2t^2 - 1) - (-t^3 - 2t^2) = -t^3 - 2t^2 - 1 + t^3 + 2t^2$

$$= -1$$

SELF CHECK

Find the difference: $(-2a^2 + 5) - (-5a^2 - 7)$.   *Answer:* $3a^2 + 12$ ∎

To subtract polynomials in vertical form, we add the opposite of the **subtrahend** (the bottom polynomial) to the **minuend** (the top polynomial).

EXAMPLE 6

Subtracting polynomials vertically. Subtract $3x^2 - 2x$ from $2x^2 + 4x$.

Solution

Since $3x^2 - 2x$ is to be subtracted from $2x^2 + 4x$, we write $3x^2 - 2x$ below $2x^2 + 4x$ in vertical form. Then we change the signs of the terms of $3x^2 - 2x$ and add:

$$
\begin{array}{r}
2x^2 + 4x \\
-\ \underline{3x^2 - 2x}
\end{array}
\quad\longrightarrow\quad
\begin{array}{r}
2x^2 + 4x \\
+\ \underline{-3x^2 + 2x} \\
-x^2 + 6x
\end{array}
$$

SELF CHECK

Subtract $2p^2 + 2p - 8$ from $5p^2 - 6p + 7$.   *Answer:* $3p^2 - 8p + 15$ ∎

EXAMPLE 7

Combining polynomials. Subtract $(12a - 7)$ from the sum of $(6a + 5)$ and $(4a - 10)$.

Solution

We will use brackets to show that $(12a - 7)$ is to be subtracted from the *sum* of $(6a + 5)$ and $(4a - 10)$.

$$[(6a + 5) + (4a - 10)] - (12a - 7)$$

Next, we remove the grouping symbols to obtain

$$= 6a + 5 + 4a - 10 - 12a + 7$$
$$= -2a + 2 \qquad \text{Combine like terms.}$$

SELF CHECK

Subtract $(-2q^2 - 2q)$ from the sum of $(q^2 - 6q)$ and $(3q^2 + q)$.   *Answer:* $6q^2 - 3q$ ∎

Adding and Subtracting Multiples of Polynomials

Because of the distributive property, we can remove parentheses enclosing several terms when a monomial precedes the parentheses. We simply multiply every term within the parentheses by that monomial. For example, to add $3(2x + 5)$ and $2(4x - 3)$, we proceed as follows:

$$3(2x + 5) + 2(4x - 3) = 6x + 15 + 8x - 6 \qquad \text{Use the distributive property to remove parentheses.}$$

$$= 6x + 8x + 15 - 6 \qquad 15 + 8x = 8x + 15.$$

$$= 14x + 9 \qquad \text{Combine like terms.}$$

EXAMPLE 8

Adding and subtracting multiples of polynomials. Remove parentheses and simplify.

a. $3(x^2 + 4x) + 2(x^2 - 4) = 3x^2 + 12x + 2x^2 - 8$
$$= 5x^2 + 12x - 8$$

b. $-8(y^2 - 2y + 3) - 4(2y^2 + y - 6) = -8y^2 + 16y - 24 - 8y^2 - 4y + 24$
$$= -16y^2 + 12y$$

c. $-4x(x^2 - x + 3) - x(x^2 - 2) + 3(x^2 + 2x)$
$$= -4x^3 + 4x^2 - 12x - x^3 + 2x + 3x^2 + 6x$$
$$= -5x^3 + 7x^2 - 4x$$

SELF CHECK

Remove parentheses and simplify:
a. $2(a^2 - 3a) + 5(a^2 + 2a)$ and
b. $5x(x^2 + 2x + 1) - x(x - 3)$.

Answers: **a.** $7a^2 + 4a$,
b. $5x^3 + 9x^2 + 8x$ ■

An Application of Adding Polynomials

EXAMPLE 9

Property values. A house purchased for $95,000 is expected to appreciate according to the polynomial function $y = 2{,}500x + 95{,}000$, where y is the value of the house after x years. A second house purchased for $125,000 is expected to appreciate according to the equation $y = 4{,}500x + 125{,}000$. Find one polynomial function that will give the total value of both properties after x years.

Solution

The value of the first house after x years is given by the polynomial $2{,}500x + 95{,}000$. The value of the second house after x years is given by the polynomial $4{,}500x + 125{,}000$. The value of both houses will be the sum of these two polynomials.

$$2{,}500x + 95{,}000 + 4{,}500x + 125{,}000 = 7{,}000x + 220{,}000$$

The total value y of the properties is given by the polynomial function $y = 7{,}000x + 220{,}000$. ■

STUDY SET

Section 4.5

VOCABULARY

In Exercises 1–4, fill in the blanks to make the statements true.

1. The expression $(x^2 - 3x + 2) + (x^2 - 4x)$ is the sum of two ___polynomials___.

2. ___Like___ terms have the same variables and the same exponents.

3. "To add or subtract like terms" means to combine their ___coefficients___ and keep the same variables with the same exponents.

4. If two polynomials are subtracted in vertical form, the bottom polynomial is called the ___subtrahend___, and the top polynomial is called the ___minuend___.

CONCEPTS

In Exercises 5–12, fill in the blanks to make the statements true.

5. To add like monomials, combine like ___terms___.

6. $a - b = a + $ ___(−b)___

7. To add two polynomials, combine any ___like___ terms contained in the polynomials.

8. To subtract two polynomials, remove parentheses and combine ___like___ terms.

9. When the sign preceding parentheses is a − sign, we can remove the parentheses by dropping the sign and the parentheses, and ___changing___ the sign of every term within the parentheses.

10. When a monomial precedes parentheses, we can remove the parentheses by __multiplying__ every term within the parentheses by that monomial.

11. $-(-2x^2 - 3x + 4) =$ $2x^2 + 3x - 4$

12. $-3(-2x^2 - 3x + 4) =$ $6x^2 + 9x - 12$

13. JETS Find the polynomial representing the length of the passenger jet in Illustration 1. $(11x - 12)$ ft

ILLUSTRATION 1

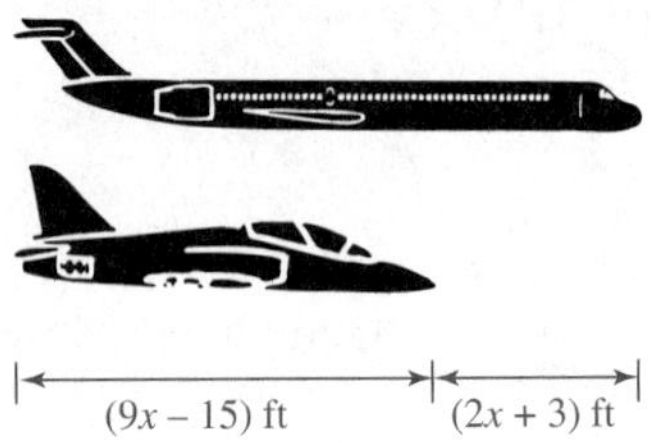

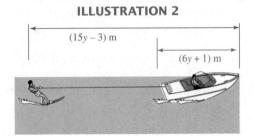

$(9x - 15)$ ft $(2x + 3)$ ft

14. WATER SKIING Find the polynomial representing the distance of the water skier from the boat in Illustration 2. $(9y - 4)$ m

ILLUSTRATION 2

$(15y - 3)$ m

$(6y + 1)$ m

NOTATION

In Exercises 15–16, complete each solution.

15. $(5x^2 + 3x) - (7x^2 - 2x)$
$= 5x^2 +$ $3x$ $- 7x^2 +$ $2x$
$= 5x^2 -$ $7x^2$ $+ 3x + 2x$
$= -2x^2 + 5x$

16. $4(3x^2 - 2x) - x(2x + 4)$
$= 12x^2 -$ $8x$ $-$ $2x^2$ $- 4x$
$= 12x^2 - 2x^2 -$ $8x$ $- 4x$
$= 10x^2 - 12x$

PRACTICE

In Exercises 17–32, simplify each expression, if possible.

17. $4y + 5y$ $9y$

18. $-2x + 3x$ x

19. $8t^2 + 4t^2$ $12t^2$

20. $15x^2 + 10x^2$ $25x^2$

21. $-32u^3 - 16u^3$ $-48u^3$

22. $-25x^3 - 7x^3$ $-32x^3$

23. $1.8x - 1.9x$ $-0.1x$

24. $1.7y - 2.2y$ $-0.5y$

25. $\frac{1}{2}s + \frac{3}{2}s$ $2s$

26. $\frac{2}{5}a + \frac{1}{5}a$ $\frac{3}{5}a$

27. $3r - 4r + 7r$ $6r$

28. $-2b + 7b - 3b$ $2b$

29. $-4ab + 4ab - ab$ $-ab$

30. $xy - 4xy - 2xy$ $-5xy$

31. $(3x)^2 - 4x^2 + 10x^2$ $15x^2$

32. $(2x)^4 - (3x^2)^2$ $7x^4$

In Exercises 33–46, do the operations.

33. $(3x + 7) + (4x - 3)$ $7x + 4$

34. $(2y - 3) + (4y + 7)$ $6y + 4$

35. $(4a + 3) - (2a - 4)$ $2a + 7$

36. $(5b - 7) - (3b - 5)$ $2b - 2$

37. $(2x + 3y) + (5x - 10y)$ $7x - 7y$

38. $(5x - 8y) - (-2x + 5y)$ $7x - 13y$

39. $(-8x - 3y) - (-11x + y)$ $3x - 4y$

40. $(-4a + b) + (5a - b)$ a

41. $(3x^2 - 3x - 2) + (3x^2 + 4x - 3)$ $6x^2 + x - 5$

42. $(3a^2 - 2a + 4) - (a^2 - 3a + 7)$ $2a^2 + a - 3$

43. $(2b^2 + 3b - 5) - (2b^2 - 4b - 9)$ $7b + 4$

44. $(4c^2 + 3c - 2) + (3c^2 + 4c + 2)$ $7c^2 + 7c$

45. $(2x^2 - 3x + 1) - (4x^2 - 3x + 2) + (2x^2 + 3x + 2)$
$3x + 1$

46. $(-3z^2 - 4z + 7) + (2z^2 + 2z - 1) - (2z^2 - 3z + 7)$
$-3z^2 + z - 1$

In Exercises 47–52, add the polynomials.

47. $\begin{array}{r} 3x^2 + 4x + 5 \\ + \underline{2x^2 - 3x + 6} \\ 5x^2 + x + 11 \end{array}$

48. $\begin{array}{r} 2x^3 + 2x^2 - 3x + 5 \\ + \underline{3x^3 - 4x^2 - x - 7} \\ 5x^3 - 2x^2 - 4x - 2 \end{array}$

49. $\begin{array}{r} 2x^3 - 3x^2 + 4x - 7 \\ + \underline{-9x^3 - 4x^2 - 5x + 6} \\ -7x^3 - 7x^2 - x - 1 \end{array}$

50. $\begin{array}{r} -3x^3 + 4x^2 - 4x + 9 \\ + \underline{2x^3 \qquad + 9x - 3} \\ -x^3 + 4x^2 + 5x + 6 \end{array}$

51. $\begin{array}{r} -3x^2 + 4x + 25 \\ + \underline{5x^2 \qquad - 12} \\ 2x^2 + 4x + 13 \end{array}$

52. $\begin{array}{r} -6x^3 - 4x^2 + 7 \\ + \underline{-7x^3 + 9x^2} \\ -13x^3 + 5x^2 + 7 \end{array}$

In Exercises 53–58, find each difference.

53. $\begin{array}{r} 3x^2 + 4x - 5 \\ - \underline{-2x^2 - 2x + 3} \\ 5x^2 + 6x - 8 \end{array}$

54. $\begin{array}{r} 3y^2 - 4y + 7 \\ - \underline{6y^2 - 6y - 13} \\ -3y^2 + 2y + 20 \end{array}$

55. $\begin{array}{r} 4x^3 + 4x^2 - 3x + 10 \\ -\quad 5x^3 - 2x^2 - 4x - \ 4 \\ \hline -x^3 + 6x^2 + x + 14 \end{array}$

56. $\begin{array}{r} 3x^3 + 4x^2 + 7x + 12 \\ -\quad -4x^3 + 6x^2 + 9x - \ 3 \\ \hline 7x^3 - 2x^2 - 2x + 15 \end{array}$

57. $\begin{array}{r} -2x^2y^2 \qquad\ + 12y^2 \\ -\quad 10x^2y^2 + 9xy - 24y^2 \\ \hline -12x^2y^2 - 9xy + 36y^2 \end{array}$

58. $\begin{array}{r} 25x^3 \qquad\ + 31xz^2 \\ -\quad 12x^3 + 27x^2z - 17xz^2 \\ \hline 13x^3 - 27x^2z + 48xz^2 \end{array}$

59. Find the difference when $t^3 - 2t^2 + 2$ is subtracted from the sum of $3t^3 + t^2$ and $-t^3 + 6t - 3$. $t^3 + 3t^2 + 6t - 5$

60. Find the difference when $-3z^3 - 4z + 7$ is subtracted from the sum of $2z^2 + 3z - 7$ and $-4z^3 - 2z - 3$. $-z^3 + 2z^2 + 5z - 17$

61. Find the sum when $3x^2 + 4x - 7$ is added to the sum of $-2x^2 - 7x + 1$ and $-4x^2 + 8x - 1$. $-3x^2 + 5x - 7$

62. Find the difference when $32x^2 - 17x + 45$ is subtracted from the sum of $23x^2 - 12x - 7$ and $-11x^2 + 12x + 7$. $-20x^2 + 17x - 45$

In Exercises 63–70, simplify each expression.

63. $2(x + 3) + 4(x - 2)$ $6x - 2$

64. $3(y - 4) - 5(y + 3)$ $-2y - 27$

65. $-2(x^2 + 7x - 1) - 3(x^2 - 2x + 2)$ $-5x^2 - 8x - 4$

66. $-5(y^2 - 2y - 6) + 6(2y^2 + 2y - 5)$ $7y^2 + 22y$

67. $2(2y^2 - 2y + 2) - 4(3y^2 - 4y - 1) + 4y(y^2 - y - 1)$ $4y^3 - 12y^2 + 8y + 8$

68. $-4(z^2 - 5z) - 5(4z^2 - 1) + 6(2z - 3)$ $-24z^2 + 32z - 13$

69. $2a(ab^2 - b) - 3b(a + 2ab) + b(b - a + a^2b)$ $3a^2b^2 - 6ab + b^2 - 6ab^2$

70. $3y(xy + y) - 2y^2(x - 4 + y) + 2(y^3 + y^2)$ $xy^2 + 13y^2$

In Exercises 71–72, find the polynomial that represents the perimeter of the figure.

71. $(3x^2 + 6x - 2)$ yd

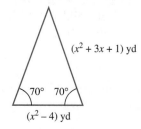

72.

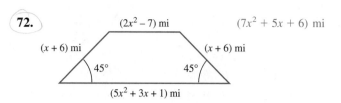

APPLICATIONS

73. GREEK ARCHITECTURE Find the difference in the heights of the columns shown in Figure 4-8 at the beginning of this section. $(x^2 - 8x + 12)$ ft

74. CLASSICAL GREEK COLUMNS If the columns shown in Figure 4-8 at the beginning of this section were stacked one atop the other, to what height would they reach? $(x^2 + 2x - 8)$ ft

75. AUTO MECHANICS Find the polynomial representing the length of the fan belt shown in Illustration 3. The dimensions are in inches. Your answer will involve π. $(3x^2 + 11x + 4.5\pi)$ in.

ILLUSTRATION 3

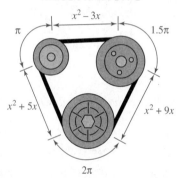

76. READING BLUEPRINTS
  a. What is the difference in the length and width of the one-bedroom apartment shown in Illustration 4? $(6x + 5)$ ft
  b. Find the perimeter of the apartment. $(4x^2 + 26)$ ft

ILLUSTRATION 4

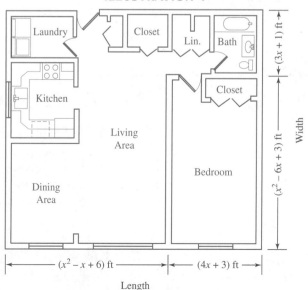

In Exercises 77–80, consider the following information: If a house is purchased for $105,000 and is expected to appreciate $900 per year, its value y after x years is given by the polynomial function y = 900x + 105,000.

77. VALUE OF A HOUSE Find the expected value of the house in 10 years. $114,000

78. VALUE OF A HOUSE A second house is purchased for $120,000 and is expected to appreciate $1,000 per year.
   a. Find a polynomial function that will give the value y of the house in x years.
      $y = 1,000x + 120,000$
   b. Find the value of this second house after 12 years.
      $132,000

79. VALUE OF TWO HOUSES Find one polynomial function that will give the combined value y of both houses after x years. $y = 1,900x + 225,000$

80. VALUE OF TWO HOUSES Find the value of the two houses after 20 years by
   a. substituting 20 into the polynomial functions $y = 900x + 105,000$ and $y = 1,000x + 120,000$ and adding. $263,000
   b. substituting into the result of Exercise 79.
      $263,000

In Exercises 81–84, consider the following information: A business purchases two computers, one for $6,600 and the other for $9,200. The first computer is expected to depreciate $1,100 per year and the second $1,700 per year.

81. VALUE OF A COMPUTER Write a polynomial function that gives the value of the first computer after x years. $y = -1,100x + 6,600$

82. VALUE OF A COMPUTER Write a polynomial function that gives the value of the second computer after x years. $y = -1,700x + 9,200$

83. VALUE OF TWO COMPUTERS Find one polynomial function that gives the combined value of both computers after x years. $y = -2,800x + 15,800$

84. VALUE OF TWO COMPUTERS In two ways, find the combined value of the two computers after 3 years. $7,400

WRITING

85. How do you recognize like terms?

86. How do you add like terms?

87. Explain the concept that is illustrated by the statement
$$-(x^2 + 3x - 1) = -1(x^2 + 3x - 1)$$

88. Explain the mistake made in the solution. Simplify: $(12x - 4) - (3x - 1)$.

REVIEW

89. What is the sum of the measures of the angles of a triangle?   180°

90. What is the sum of the measures of two complementary angles?   90°

91. Solve the inequality $-4(3x - 3) \geq -12$ and graph the solution. $x \leq 2$

   ←———|———→
        2

92. CURLING IRON A curling iron is plugged into a 110-volt electrical outlet and used for $\frac{1}{4}$ hour. If its resistance is 10 ohms, find the electrical power (in kilowatt hours, kwh) used by the curling iron by applying the formula
$$kwh = \frac{(volts)^2}{1,000 \cdot ohms} \cdot hours$$
   0.3025 kwh

▶ **4.6**   # Multiplying Polynomials

In this section, you will learn about

> Multiplying monomials ■ Multiplying a polynomial by a monomial
> ■ Multiplying a binomial by a binomial ■ The FOIL method
> ■ Multiplying a polynomial by a binomial ■ Multiplying binomials
> to solve equations ■ An application of multiplying polynomials

---

Introduction   In Figure 4-9(a), the length and width of a dollar bill are given. We can find the area of the bill by multiplying its length and width.

FIGURE 4-9

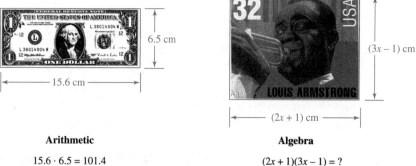

Arithmetic	Algebra
$15.6 \cdot 6.5 = 101.4$	$(2x + 1)(3x - 1) = ?$
The area is 101.4 cm^2.	
(a)	(b)

In Figure 4-9(b), the length and the width of a postage stamp are represented by two-term polynomials called *binomials*. To find the area of the stamp, we must multiply the binomials. In this section, we will discuss the algebraic rules that are used to do this. We begin the discussion of multiplication of polynomials with the simplest case, the product of two monomials.

Multiplying Monomials

In Section 4.1, we multiplied monomials by other monomials. For example, to multiply $8x^2$ by $-3x^4$, we use the commutative and associative properties of multiplication to group the numerical factors and the variable factors. Then we multiply the numerical factors and multiply the variable factors.

$$8x^2(-3x^4) = 8(-3)x^2x^4$$
$$= -24x^6$$

This example suggests the following rule.

Multiplying monomials

To multiply two monomials, multiply the numerical factors and then multiply the variable factors.

EXAMPLE 1

Multiplying monomials. Multiply **a.** $3x^4(2x^5)$, **b.** $-2a^2b^3(5ab^2)$, and **c.** $-4y^5z^2(2y^3z^3)(3yz)$.

Solution **a.** $3x^4(2x^5) = 3(2)x^4x^5$ Multiply the numerical factors, 3 and 2. Multiply the variable factors, x^4 and x^5. Use the product rule for exponents: $x^4x^5 = x^{4+5}$.

$$= 6x^9$$

b. $-2a^2b^3(5ab^2) = -2(5)a^2ab^3b^2$
$$= -10a^3b^5$$

c. $-4y^5z^2(2y^3z^3)(3yz) = -4(2)(3)y^5y^3yz^2z^3z$
$$= -24y^9z^6$$

SELF CHECK   Multiply: **a.** $(5a^2b^3)(6a^3b^4)$ and   *Answers:* **a.** $30a^5b^7$, **b.** $(-15p^3q^2)(5p^3q^2)$.   **b.** $-75p^6q^4$ ∎

Multiplying a Polynomial by a Monomial

To find the product of a polynomial (with more than one term) and a monomial, we use the distributive property. To multiply $2x + 4$ by $5x$, for example, we proceed as follows:

$$5x\,(2x + 4) = 5x \cdot 2x + 5x \cdot 4 \quad \text{Use the distributive property.}$$
$$= 10x^2 + 20x \qquad \text{Multiply the monomials:}$$
$$5x \cdot 2x = 10x^2 \text{ and}$$
$$5x \cdot 4 = 20x.$$

This example suggests the following rule.

Multiplying polynomials by monomials

To multiply a polynomial with more than one term by a monomial, use the distributive property to remove parentheses and simplify.

E X A M P L E 2

Multiplying a polynomial by a monomial. Multiply **a.** $3a^2(3a^2 - 5a)$ and **b.** $-2xz^2(2x - 3z + 2z^2)$.

Solution

a. $3a^2(3a^2 - 5a) = 3a^2 \cdot 3a^2 - 3a^2 \cdot 5a$ Use the distributive property.
$$= 9a^4 - 15a^3 \qquad \text{Multiply: } 3a^2 \cdot 3a^2 = 9a^4 \text{ and } 3a^2 \cdot 5a = 15a^3.$$

b. $-2xz^2(2x - 3z + 2z^2)$
$$= -2xz^2 \cdot 2x - (-2xz^2) \cdot 3z + (-2xz^2) \cdot 2z^2 \quad \text{Use the distributive property.}$$
$$= -4x^2z^2 - (-6xz^3) + (-4xz^4) \qquad \text{Multiply: } -2xz^2 \cdot 2x = -4x^2z^2,$$
$$-2xz^2 \cdot 3z = -6xz^3, \text{ and}$$
$$-2xz^2 \cdot 2z^2 = -4xz^4.$$

$$= -4x^2z^2 + 6xz^3 - 4xz^4$$

SELF CHECK

Multiply: **a.** $2p^3(3p^2 - 5p)$ and
b. $-5a^2b(3a + 2b - 4ab)$.

Answers: **a.** $6p^5 - 10p^4$,
b. $-15a^3b - 10a^2b^2 + 20a^3b^2$

■

Multiplying a Binomial by a Binomial

To multiply two binomials, we must use the distributive property more than once. For example, to multiply $2a - 4$ by $3a + 5$, we proceed as follows.

$$(2a - 4)\,(3a + 5) = (2a - 4) \cdot 3a + (2a - 4) \cdot 5 \qquad \begin{array}{l}\text{Distribute the binomial}\\(2a - 4).\end{array}$$

$$= 3a(2a - 4) + 5(2a - 4) \qquad \begin{array}{l}\text{Use the commutative}\\\text{property of multiplication.}\end{array}$$

$$= 3a \cdot 2a - 3a \cdot 4 + 5 \cdot 2a - 5 \cdot 4 \quad \begin{array}{l}\text{Use the distributive property}\\\text{twice.}\end{array}$$

$$= 6a^2 - 12a + 10a - 20 \qquad \text{Do the multiplications.}$$

$$= 6a^2 - 2a - 20 \qquad \begin{array}{l}\text{Combine like terms:}\\-12a + 10a = -2a\end{array}$$

This example suggests the following rule.

Multiplying two binomials

To multiply two binomials, multiply each term of one binomial by each term of the other binomial and combine like terms.

The FOIL Method

We can use a shortcut method, called the **FOIL** method, to multiply binomials. FOIL is an acronym for **F**irst terms, **O**uter terms, **I**nner terms, and **L**ast terms. To use the FOIL method to multiply $2a - 4$ by $3a + 5$, we

1. multiply the **F**irst terms $2a$ and $3a$ to obtain $6a^2$,
2. multiply the **O**uter terms $2a$ and 5 to obtain $10a$,
3. multiply the **I**nner terms -4 and $3a$ to obtain $-12a$, and
4. multiply the **L**ast terms -4 and 5 to obtain -20.

Then we simplify the resulting polynomial, if possible.

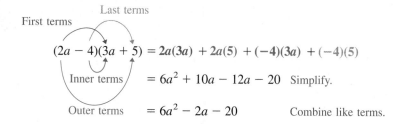

$$(2a - 4)(3a + 5) = 2a(3a) + 2a(5) + (-4)(3a) + (-4)(5)$$

$$= 6a^2 + 10a - 12a - 20 \quad \text{Simplify.}$$

$$= 6a^2 - 2a - 20 \qquad \text{Combine like terms.}$$

EXAMPLE 3

Using the FOIL method. Find each product.

a. $(3x + 4)(2x - 3) = 3x(2x) + 3x(-3) + 4(2x) + 4(-3)$

$$= 6x^2 - 9x + 8x - 12$$

$$= 6x^2 - x - 12$$

b. $(2y - 7)(5y - 4) = 2y(5y) + 2y(-4) + (-7)(5y) + (-7)(-4)$

$$= 10y^2 - 8y - 35y + 28$$

$$= 10y^2 - 43y + 28$$

c. $(2r - 3s)(2r + t) = 2r(2r) + 2r(t) - 3s(2r) - 3s(t)$

$$= 4r^2 + 2rt - 6rs - 3st$$

SELF CHECK

Find each product: **a.** $(2a - 1)(3a + 2)$ and **b.** $(5y - 2z)(2y + 3z)$.

Answers: **a.** $6a^2 + a - 2$, **b.** $10y^2 + 11yz - 6z^2$ ∎

EXAMPLE 4

Simplifying expressions. Simplify each expression.

   a. $3(2x - 3)(x + 1) = 3(2x^2 + 2x - 3x - 3)$ Use FOIL to multiply the binomials.

                    $= 3(2x^2 - x - 3)$ Combine like terms.

                    $= 6x^2 - 3x - 9$ Use the distributive property to remove parentheses.

   b. $(x + 1)(x - 2) - 3x(x + 3) = x^2 - 2x + x - 2 - 3x^2 - 9x$ Use FOIL to find $(x + 1)(x + 2)$.

                              $= -2x^2 - 10x - 2$ Combine like terms.

SELF CHECK   Simplify $(x + 3)(2x - 1) + 2x(x - 1)$.   *Answer:* $4x^2 + 3x - 3$ ■

The products discussed in Example 5 are called **special products.**

EXAMPLE 5

Special products. Find each product.

   a. The square of the sum of two quantities has three terms:

$$(x + y)^2 = (x + y)(x + y)$$
$$= x^2 + xy + xy + y^2$$
$$= \quad x^2 \quad + \quad 2xy \quad + \quad y^2$$

             The square of   Twice the product   The square of the
             the first quantity   of the quantities   second quantity

   b. The square of the difference of two quantities has three terms:

$$(x - y)^2 = (x - y)(x - y)$$
$$= x^2 - xy - xy + y^2$$
$$= \quad x^2 \quad - \quad 2xy \quad + \quad y^2$$

             The square of   Twice the product   The square of the
             the first quantity   of the quantities   second quantity

   c. The product of a sum and a difference of two quantities is a binomial.

$$(x + y)(x - y) = x^2 - xy + xy - y^2$$
$$= \quad x^2 \quad - \quad y^2$$

              The product of the   The product of the
              first quantities   second quantities

SELF CHECK   Find each product: **a.** $(p + 2)^2$, **b.** $(p - 2)^2$,
         and **c.** $(p + 2q)(p - 2q)$.   *Answers:* **a.** $p^2 + 4p + 4$,
                                         b. $p^2 - 4p + 4$, **c.** $p^2 - 4q^2$ ■

Because the products discussed in Example 5 occur so often, it is wise to learn their forms.

Special products

$$(x + y)^2 = x^2 + 2xy + y^2$$
$$(x - y)^2 = x^2 - 2xy + y^2$$
$$(x + y)(x - y) = x^2 - y^2$$

EXAMPLE 6

Road signs.   The area of a regular octagon is approximately $3.314r^2$, where r is as shown in Figure 4-10(a). Find the polynomial that approximates the area of the stop sign shown in Figure 4-10(b).

FIGURE 4-10

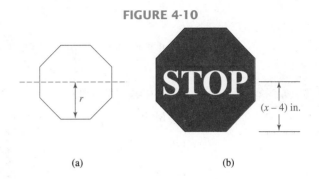

(a)   (b)

Solution   To find the approximate area of the stop sign, we substitute $(x - 4)$ for r in the formula.

$$A \approx 3.314r^2$$
$$A \approx 3.314(x - 4)^2$$

We note that $(x - 4)^2$ is a special product. The result is a trinomial consisting of the first term (x) squared, minus twice the product of x and 4, plus the square of 4.

$$A \approx 3.314[x^2 - 2(x)(4) + (4)^2]$$
$$\approx 3.314[x^2 - 8x + 16] \qquad \text{Simplify inside the brackets.}$$
$$\approx 3.314x^2 - 26.512x + 53.024 \quad \text{Distribute 3.314.}$$

The approximate area of the stop sign is $(3.314x^2 - 26.512x + 53.024)$ in.^{2}

SELF CHECK   See Example 6. Find the approximate area of a stop sign where $r = (y + 6)$ inches.

Answer:
$(3.314y^2 + 39.768y + 119.304)$ in.^{2} ■

WARNING!   A common error when squaring a binomial is to forget the middle term of the product. For example, $(x + 2)^2 \neq x^2 + 4$ and $(x - 2)^2 \neq x^2 + 4$. Applying the special product formulas, we have $(x + 2)^2 = x^2 + 4x + 4$ and $(x - 2)^2 = x^2 - 4x + 4$.

Multiplying a Polynomial by a Binomial

We must use the distributive property more than once to multiply a polynomial by a binomial. For example, to multiply $3x^2 + 3x - 5$ by $2x + 3$, we proceed as follows:

$$(2x + 3)(3x^2 + 3x - 5) = (2x + 3)3x^2 + (2x + 3)3x - (2x + 3)5$$
$$= 3x^2(2x + 3) + 3x(2x + 3) - 5(2x + 3)$$
$$= 6x^3 + 9x^2 + 6x^2 + 9x - 10x - 15$$
$$= 6x^3 + 15x^2 - x - 15$$

This example suggests the following rule.

Multiplying polynomials

To multiply one polynomial by another, multiply each term of one polynomial by each term of the other polynomial and combine like terms.

It is often convenient to organize the work vertically.

EXAMPLE 7

Multiplying polynomials using vertical form.

a. Multiply:

$$
\begin{array}{r}
3a^2 - 4a + 7 \\
2a + 5 \\
\hline
\end{array}
$$

$$
\begin{array}{ll}
2a(3a^2 - 4a + 7) \rightarrow & 6a^3 - 8a^2 + 14a \\
5(3a^2 - 4a + 7) \rightarrow & \underline{+\ 15a^2 - 20a + 35} \\
& 6a^3 + 7a^2 - 6a + 35
\end{array}
$$

b. Multiply:

$$
\begin{array}{r}
3y^2 - 5y + 4 \\
-4y^2 - 3 \\
\hline
\end{array}
$$

$$
\begin{array}{ll}
-4y^2(3y^2 - 5y + 4) \rightarrow & -12y^4 + 20y^3 - 16y^2 \\
-3(3y^2 - 5y + 4) \rightarrow & \underline{\qquad\qquad - 9y^2 + 15y - 12} \\
& -12y^4 + 20y^3 - 25y^2 + 15y - 12
\end{array}
$$

SELF CHECK

Multiply: **a.** $(3x + 2)(2x^2 - 4x + 5)$ and
b. $(-2x^2 + 3)(2x^2 - 4x - 1)$.

Answers:
a. $6x^3 - 8x^2 + 7x + 10$,
b. $-4x^4 + 8x^3 + 8x^2 - 12x - 3$

■

Multiplying Binomials to Solve Equations

To solve an equation such as $(x + 2)(x + 3) = x(x + 7)$, we can first use the FOIL method to remove the parentheses on the left-hand side, then use the distributive property to remove parentheses on the right-hand side, and proceed as follows:

$$(x + 2)(x + 3) = x(x + 7)$$

$$x^2 + 3x + 2x + 6 = x^2 + 7x$$

$$x^2 + 3x + 2x + 6 - x^2 = x^2 + 7x - x^2 \qquad \text{Subtract } x^2 \text{ from both sides.}$$

$$5x + 6 = 7x \qquad \text{Combine like terms: } x^2 - x^2 = 0 \text{ and } 3x + 2x = 5x.$$

$$6 = 2x \qquad \text{Subtract } 5x \text{ from both sides.}$$

$$3 = x \qquad \text{Divide both sides by 2.}$$

Check: $(x + 2)(x + 3) = x(x + 7)$

$$(3 + 2)(3 + 3) \stackrel{?}{=} 3(3 + 7) \qquad \text{Replace } x \text{ with 3.}$$

$$5(6) \stackrel{?}{=} 3(10) \qquad \text{Do the additions within parentheses.}$$

$$30 = 30$$

An Application of Multiplying Polynomials

EXAMPLE 8

A square painting is surrounded by a border 2 inches wide. If the area of the border is 96 square inches, find the dimensions of the painting.

ANALYZE THE PROBLEM | Refer to Figure 4-11, which shows a square painting surrounded by a border 2 inches wide. We know that the area of this border is 96 square inches, and we are to find the dimensions of the painting.

FORM AN EQUATION | Let x represent the length of each side of the square painting. Since the border is 2 inches wide, the length and the width of the outer rectangle are both $(x + 2 + 2)$ inches. Then the outer rectangle is also a square, and its dimensions are $(x + 4)$ by $(x + 4)$ inches. Since the area of a square is the product of its length and width, the area of the larger square is $(x + 4)(x + 4)$, and the area of the painting is $x \cdot x$. If we subtract the area of the painting from the area of the larger square, the difference is 96.

FIGURE 4-11

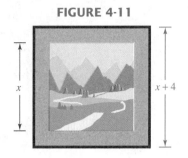

The area of the large square	minus	the area of the square painting	is	the area of the border.
$(x + 4)(x + 4)$	$-$	$x \cdot x$	$=$	96

SOLVE THE EQUATION

$(x + 4)(x + 4) - x^2 = 96$ $x \cdot x = x^2$.

$x^2 + 8x + 16 - x^2 = 96$ $(x + 4)(x + 4) = x^2 + 8x + 16$.

$8x + 16 = 96$ Combine like terms: $x^2 - x^2 = 0$.

$8x = 80$ Subtract 16 from both sides.

$x = 10$ Divide both sides by 8.

STATE THE CONCLUSION | The dimensions of the painting are 10 inches by 10 inches.

CHECK THE RESULT | Verify that the 2-inch-wide border of a 10-inch-square painting would have an area of 96 square inches. ∎

STUDY SET

Section 4.6

VOCABULARY

In Exercises 1–4, fill in the blanks to make the statements true.

1. The expression $(2a - 4)(3a + 5)$ is the product of two ___binomials___.

2. The expression $(2a - 4)(3a^2 + 5a - 1)$ is the product of a ___binomial___ and a ___trinomial___.

3. When multiplying a monomial and a polynomial, the ___distributive___ property is used to remove parentheses.

4. In the acronym FOIL, F stands for ___first___ terms, O for ___outer___ terms, I for ___inner___ terms, and L for ___last___ terms.

CONCEPTS

In Exercises 5–8, consider the product $(2x + 5)(3x - 4)$.

5. The product of the first terms is $6x^2$.

6. The product of the outer terms is $-8x$.

7. The product of the inner terms is $15x$.

8. The product of the last terms is -20 .

9. STAMPS Find the area of the stamp shown in Figure 4-9 at the beginning of this section. $(6x^2 + x - 1)$ cm^2

10. LUGGAGE Find the volume of the garment bag shown in Illustration 1. $(2x^3 - 4x^2 - 6x)$ in.^{3}

ILLUSTRATION 1

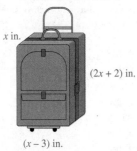

x in.

$(2x + 2)$ in.

$(x - 3)$ in.

NOTATION

In Exercises 11–12, complete each solution.

11. $7x(3x^2 - 2x + 5) = \boxed{7x} \cdot 3x^2 - \boxed{7x} \cdot 2x + \boxed{7x} \cdot 5$

$\quad = 21x^3 - 14x^2 + 35x$

12. $(2x + 5)(3x - 2) = 2x \cdot 3x - \boxed{2x} \cdot 2 + \boxed{5} \cdot 3x - \boxed{5} \cdot 2$

$\quad = 6x^2 - \boxed{4x} + \boxed{15x} - 10$

$\quad = 6x^2 + 11x - 10$

PRACTICE

In Exercises 13–20, find each product.

13. $(3x^2)(4x^3)$ $12x^5$

14. $(-2a^3)(3a^2)$ $-6a^5$

15. $(3b^2)(-2b)(4b^3)$ $-24b^6$

16. $(3y)(2y^2)(-y^4)$ $-6y^7$

17. $(2x^2y^3)(3x^3y^2)$ $6x^5y^5$

18. $(-5x^3y^6)(x^2y^2)$ $-5x^5y^8$

19. $(x^2y^5)(x^2z^5)(-3z^3)$ $-3x^4y^5z^8$

20. $(-r^4st^2)(2r^2st)(rst)$ $-2r^7s^3t^4$

In Exercises 21–34, find each product.

21. $3(x + 4)$ $3x + 12$

22. $-3(a - 2)$ $-3a + 6$

23. $-4(t + 7)$ $-4t - 28$

24. $6(s^2 - 3)$ $6s^2 - 18$

25. $3x(x - 2)$ $3x^2 - 6x$

26. $4y(y + 5)$ $4y^2 + 20y$

27. $-2x^2(3x^2 - x)$ $-6x^4 + 2x^3$

28. $4b^3(2b^2 - 2b)$ $8b^5 - 8b^4$

29. $3xy(x + y)$ $3x^2y + 3xy^2$

30. $-4x^2z(3x^2 - z)$ $-12x^4z + 4x^2z^2$

31. $2x^2(3x^2 + 4x - 7)$ $6x^4 + 8x^3 - 14x^2$

32. $3y^3(2y^2 - 7y - 8)$ $6y^5 - 21y^4 - 24y^3$

33. $(3x)(-2x^2)(x + 4)$ $-6x^4 - 24x^3$

34. $(-2a^2)(-3a^3)(3a - 2)$ $18a^6 - 12a^5$

In Exercises 35–50, find each product.

35. $(a + 4)(a + 5)$ $a^2 + 9a + 20$

36. $(y - 3)(y + 5)$ $y^2 + 2y - 15$

37. $(3x - 2)(x + 4)$ $3x^2 + 10x - 8$

38. $(t + 4)(2t - 3)$ $2t^2 + 5t - 12$

39. $(2a + 4)(3a - 5)$ $6a^2 + 2a - 20$

40. $(2b - 1)(3b + 4)$ $6b^2 + 5b - 4$

41. $(3x - 5)(2x + 1)$ $6x^2 - 7x - 5$

42. $(2y - 5)(3y + 7)$ $6y^2 - y - 35$

43. $(x + 3)(2x - 3)$ $2x^2 + 3x - 9$

44. $(2x + 3)(2x - 5)$ $4x^2 - 4x - 15$

45. $(2t + 3s)(3t - s)$ $6t^2 + 7st - 3s^2$

46. $(3a - 2b)(4a + b)$ $12a^2 - 5ab - 2b^2$

47. $(x + y)(x + z)$ $x^2 + xz + xy + yz$

48. $(a - b)(x + y)$ $ax + ay - bx - by$

49. $(4t - u)(-3t + u)$ $-12t^2 + 7tu - u^2$

50. $(-3t + 2s)(2t - 3s)$ $-6t^2 + 13st - 6s^2$

In Exercises 51–58, simplify each expression.

51. $4(2x + 1)(x - 2)$ $8x^2 - 12x - 8$

52. $-5(3a - 2)(2a + 3)$ $-30a^2 - 25a + 30$

53. $3a(a + b)(a - b)$ $3a^3 - 3ab^2$

54. $-2r(r + s)(r + s)$ $-2r^3 - 4r^2s - 2rs^2$

55. $2t(t + 2) + 3t(t - 5)$ $5t^2 - 11t$

56. $3y(y + 2) + (y + 1)(y - 1)$ $4y^2 + 6y - 1$

57. $(x + y)(x - y) + x(x + y)$ $2x^2 + xy - y^2$

58. $(3x + 4)(2x - 2) - (2x + 1)(x + 3)$ $4x^2 - 5x - 11$

In Exercises 59–76, find each special product.

59. $(x + 4)(x + 4)$ $x^2 + 8x + 16$

60. $(a + 3)(a + 3)$ $a^2 + 6a + 9$

61. $(t - 3)(t - 3)$ $t^2 - 6t + 9$

62. $(z - 5)(z - 5)$ $z^2 - 10z + 25$

63. $(r + 4)(r - 4)$ $r^2 - 16$

64. $(b + 2)(b - 2)$ $b^2 - 4$

65. $(4x + 5)(4x - 5)$ $16x^2 - 25$

66. $(5z + 1)(5z - 1)$ $25z^2 - 1$

67. $(2s + 1)(2s + 1)$ $4s^2 + 4s + 1$

68. $(3t - 2)(3t - 2)$ $9t^2 - 12t + 4$

69. $(x + 5)^2$ $x^2 + 10x + 25$

70. $(y - 6)^2$ $y^2 - 12y + 36$

71. $(x - 2y)^2$ $x^2 - 4xy + 4y^2$

72. $(3a + 2b)^2$ $9a^2 + 12ab + 4b^2$

73. $(2a - 3b)^2$ $4a^2 - 12ab + 9b^2$

74. $(2x + 5y)^2$ $4x^2 + 20xy + 25y^2$

75. $(4x + 5y)^2$ $16x^2 + 40xy + 25y^2$

76. $(6p - 5q)^2$ $36p^2 - 60pq + 25q^2$

In Exercises 77–80, find the area of each figure. You may leave π in your answer.

77.

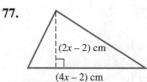

$(2x - 2)$ cm

$(4x - 2)$ cm

$(4x^2 - 6x + 2)$ cm^2

78.

$(2x + 1)$ cm

$(3x - 4)$ cm

$(6x^2 - 5x - 4)$ m^2

79.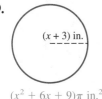

$(x + 3)$ in.

$(x^2 + 6x + 9)\pi$ in.^{2}

80.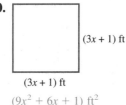

$(3x + 1)$ ft

$(3x + 1)$ ft

$(9x^2 + 6x + 1)$ ft^2

In Exercises 81–86, find each product.

81. $(x + 2)(x^2 - 2x + 3)$ $x^3 - x + 6$

82. $(x - 5)(x^2 + 2x - 3)$ $x^3 - 3x^2 - 13x + 15$

83. $(4t + 3)(t^2 + 2t + 3)$ $4t^3 + 11t^2 + 18t + 9$

84. $(3x + 1)(2x^2 - 3x + 1)$ $6x^3 - 7x^2 + 1$

85. $(-3x + y)(x^2 - 8xy + 16y^2)$
$-3x^3 + 25x^2y - 56xy^2 + 16y^3$

86. $(3x - y)(x^2 + 3xy - y^2)$ $3x^3 + 8x^2y - 6xy^2 + y^3$

In Exercises 87–90, find each product.

87. $x^2 - 2x + 1$
$\underline{\quad x + 2}$
$x^3 - 3x + 2$

88. $5r^2 + r + 6$
$\underline{\quad 2r - 1}$
$10r^3 - 3r^2 + 11r - 6$

89. $4x^2 + 3x - 4$
$\underline{\quad 3x + 2}$
$12x^3 + 17x^2 - 6x - 8$

90. $x^2 - x + 1$
$\underline{\quad x + 1}$
$x^3 + 1$

In Exercises 91–98, solve each equation.

91. $(s - 4)(s + 1) = s^2 + 5$ -3

92. $(y - 5)(y - 2) = y^2 - 4$ 2

93. $z(z + 2) = (z + 4)(z - 4)$ -8

94. $(z + 3)(z - 3) = z(z - 3)$ 3

95. $(x + 4)(x - 4) = (x - 2)(x + 6)$ -1

96. $(y - 1)(y + 6) = (y - 3)(y - 2) + 8$ 2

97. $(a - 3)^2 = (a + 3)^2$ 0

98. $(b + 2)^2 = (b - 1)^2$ $-\frac{1}{2}$

APPLICATIONS

99. TOYS Find the perimeter and the area of the screen of the Etch A Sketch® shown in Illustration 2.
$(24x + 14)$ cm, $(35x^2 + 43x + 12)$ cm^2

ILLUSTRATION 2

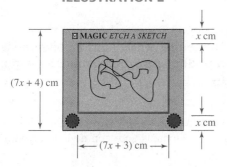

$(7x + 4)$ cm

x cm

x cm

$(7x + 3)$ cm

100. SUNGLASSES An ellipse is an oval-shaped closed curve. The area of an ellipse is approximately $3.14ab$, where a is its length and b is its width. Find the polynomial that approximates the total area of the elliptical-shaped lenses of the sunglasses shown in Illustration 3. $(6.28x^2 - 6.28)$ in.^{2}

ILLUSTRATION 3

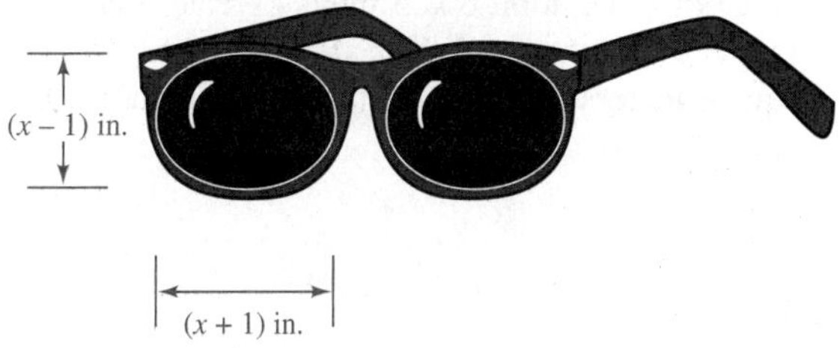

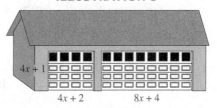

$(x - 1)$ in.

$(x + 1)$ in.

101. GARDENING See Illustration 4.
   a. What is the area of the region planted with corn? tomatoes? beans? carrots? Use your answers to find the total area of the garden.
   x^2 ft^2, $6x$ ft^2, $5x$ ft^2, 30 ft^2; $(x^2 + 11x + 30)$ ft^2
   b. What is the length of the garden? What is its width? Use your answers to find its area.
   $(x + 6)$ ft, $(x + 5)$ ft; $(x^2 + 11x + 30)$ ft^2
   c. How do the answers from parts a and b for the area of the garden compare? They are the same.

ILLUSTRATION 4

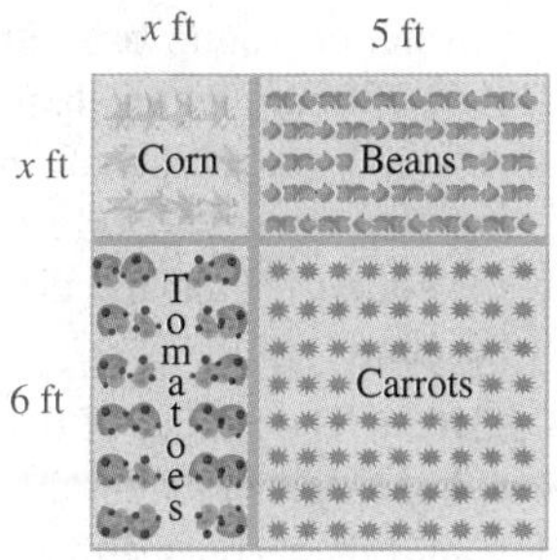

x ft   5 ft

x ft   Corn   Beans

6 ft   Tomatoes   Carrots

102. PAINTING See Illustration 5. To purchase the correct amount of enamel to paint these two garage doors, a painter must find their areas. Find a polynomial that gives the number of square feet to be painted. All dimensions are in feet, and the windows are squares with sides of x feet. $(36x^2 + 36x + 6)$ ft^2

ILLUSTRATION 5

$4x + 1$

$4x + 2$ $8x + 4$

103. INTEGER PROBLEM The difference between the squares of two consecutive integers is 11. Find the integers. 5 and 6

104. INTEGER PROBLEM If 3 less than a certain integer is multiplied by 4 more than the integer, the product is 6 less than the square of the integer. Find the integer. 6

105. STONE-GROUND FLOUR The radius of one millstone in Illustration 6 is 3 meters greater than the radius of another, and their areas differ by 15π square meters. Find the radius of the larger millstone. 4 m

ILLUSTRATION 6

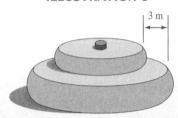

106. BOOKBINDING Two square sheets of cardboard used for making book covers differ in area by 44 square inches. An edge of the larger square is 2 inches greater than an edge of the smaller square. Find the length of an edge of the smaller square. 10 in.

107. BASEBALL In major league baseball, the distance between bases is 30 feet greater than it is in softball. The bases in major league baseball mark the corners of a square that has an area 4,500 square feet greater than for softball. Find the distance between the bases in baseball. 90 ft

108. PULLEY DESIGN The radius of one pulley in Illustration 7 is 1 inch greater than the radius of the second pulley, and their areas differ by 4π square inches. Find the radius of the smaller pulley. $\frac{3}{2}$ in.

ILLUSTRATION 7

WRITING

109. Describe the steps involved in finding the product of $(x + 2)$ and $(x - 2)$.

110. Writing $(x + y)^2$ as $x^2 + y^2$ illustrates a common error. Explain.

REVIEW

In Exercises 111–116, refer to Illustration 8.

111. What is the slope of line AB?   1

112. What is the slope of line BC?   undefined

113. What is the slope of line CD? $-\frac{2}{3}$

114. What is the slope of the x-axis?   0

115. What is the y-intercept of line AB? $(0, 2)$

116. What is the x-intercept of line AB? $(-2, 0)$

ILLUSTRATION 8

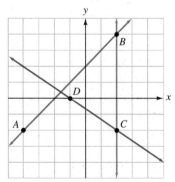

▶ 4.7　Dividing Polynomials by Monomials

In this section, you will learn about

Dividing a monomial by a monomial ■ Dividing a polynomial by a monomial ■ An application of dividing a polynomial by a monomial

Introduction In this section, we will discuss how to divide polynomials by monomials. We will first divide monomials by monomials and then divide polynomials with more than one term by monomials.

Dividing a Monomial by a Monomial

Recall that to simplify a fraction, we write both its numerator and denominator as the product of several factors and then divide out all common factors:

$$\frac{4}{6} = \frac{2 \cdot 2}{2 \cdot 3} \quad \text{Factor: } 4 = 2 \cdot 2 \text{ and } 6 = 2 \cdot 3.$$

$$= \frac{\overset{1}{\cancel{2}} \cdot 2}{\underset{1}{\cancel{2}} \cdot 3} \quad \text{Divide out the common factor of 2.}$$

$$= \frac{2}{3} \quad \tfrac{2}{2} = 1.$$

$$\frac{20}{25} = \frac{4 \cdot 5}{5 \cdot 5} \quad \text{Factor: } 20 = 4 \cdot 5 \text{ and } 25 = 5 \cdot 5.$$

$$= \frac{4 \cdot \overset{1}{\cancel{5}}}{\underset{1}{\cancel{5}} \cdot 5} \quad \text{Divide out the common factor of 5.}$$

$$= \frac{4}{5} \quad \tfrac{5}{5} = 1.$$

We can use the same method to simplify algebraic fractions that contain variables.

$$\frac{3p^2}{6p} = \frac{3 \cdot p \cdot p}{2 \cdot 3 \cdot p} \quad \text{Factor: } p^2 = p \cdot p \text{ and } 6 = 2 \cdot 3.$$

$$= \frac{\overset{1}{\cancel{3}} \cdot \overset{1}{\cancel{p}} \cdot p}{2 \cdot \underset{1}{\cancel{3}} \cdot \underset{1}{\cancel{p}}} \quad \text{Divide out the common factors of 3 and } p.$$

$$= \frac{p}{2} \quad \tfrac{3}{3} = 1 \text{ and } \tfrac{p}{p} = 1.$$

To divide monomials, we can use either the preceding method for simplifying arithmetic fractions or the rules for exponents.

EXAMPLE 1

Dividing monomials. Simplify: **a.** $\dfrac{x^2y}{xy^2}$ and **b.** $\dfrac{-8a^3b^2}{4ab^3}$.

Solution

By simplifying fractions

a. $\dfrac{x^2y}{xy^2} = \dfrac{x \cdot x \cdot y}{x \cdot y \cdot y}$

$$= \frac{\overset{1}{\cancel{x}} \cdot x \cdot \overset{1}{\cancel{y}}}{\underset{1}{\cancel{x}} \cdot y \cdot \underset{1}{\cancel{y}}}$$

$$= \frac{x}{y}$$

b. $\dfrac{-8a^3b^2}{4ab^3} = \dfrac{-2 \cdot 4 \cdot a \cdot a \cdot a \cdot b \cdot b}{4 \cdot a \cdot b \cdot b \cdot b}$

$$= \frac{-2 \cdot \overset{1}{\cancel{4}} \cdot \overset{1}{\cancel{a}} \cdot a \cdot a \cdot \overset{1}{\cancel{b}} \cdot \overset{1}{\cancel{b}}}{\underset{1}{\cancel{4}} \cdot \underset{1}{\cancel{a}} \cdot \underset{1}{\cancel{b}} \cdot \underset{1}{\cancel{b}} \cdot b}$$

$$= -\frac{2a^2}{b}$$

Using the rules of exponents

a. $\dfrac{x^2y}{xy^2} = x^{2-1}y^{1-2}$

$$= x^1y^{-1}$$

$$= \frac{x}{y}$$

b. $\dfrac{-8a^3b^2}{4ab^3} = \dfrac{-2^3a^3b^2}{2^2ab^3}$

$$= -2^{3-2}a^{3-1}b^{2-3}$$

$$= -2^1a^2b^{-1}$$

$$= -\frac{2a^2}{b}$$

SELF CHECK Simplify $\dfrac{-5p^2q^3}{10pq^4}$.

Answer: $-\dfrac{p}{2q}$

EXAMPLE 2

Dividing monomials. Simplify $\dfrac{25(s^2t^3)^2}{15(st^3)^3}$. Write the result using positive exponents only.

Solution To divide these monomials, we will use the method for simplifying fractions and several rules for exponents.

$$\dfrac{25(s^2t^3)^2}{15(st^3)^3} = \dfrac{25s^4t^6}{15s^3t^9} \qquad \text{Use the power rules for exponents:} \\ (xy)^n = x^ny^n \text{ and } (x^m)^n = x^{m\cdot n}.$$

$$= \dfrac{5\cdot 5\cdot s^{4-3}t^{6-9}}{5\cdot 3} \qquad \text{Factor 25 and 15. Use the quotient rule for exponents:} \\ \dfrac{x^m}{x^n} = x^{m-n}.$$

$$= \dfrac{5\cdot \overset{1}{\cancel{5}}\cdot s^1t^{-3}}{\underset{1}{\cancel{5}}\cdot 3} \qquad \text{Divide out the common factors of 5. Do the subtractions.}$$

$$= \dfrac{5s}{3t^3} \qquad \text{Use the negative integer exponent rule: } x^{-n} = \dfrac{1}{x^n}.$$

SELF CHECK Simplify $\dfrac{-24(h^3p)^5}{20(h^2p^2)^3}$.   *Answer:* $-\dfrac{6h^9}{5p}$ ∎

Dividing a Polynomial by a Monomial

Dividing by a number is equivalent to multiplying by its reciprocal. For example, dividing the number 8 by 2 gives the same answer as multiplying 8 by $\frac{1}{2}$.

$$\dfrac{8}{2} = 4 \qquad \text{and} \qquad \dfrac{1}{2}\cdot 8 = 4$$

In general, the following is true.

Division

$$\dfrac{a}{b} = \dfrac{1}{b}\cdot a \qquad (b \neq 0)$$

To divide a polynomial with more than one term by a monomial, we write the division as a product of the numerator times the reciprocal of the denominator, use the distributive property to remove parentheses, and then simplify each resulting fraction.

EXAMPLE 3

Dividing a binomial by a monomial. Simplify $\dfrac{9x + 6}{3}$.

Solution

$$\dfrac{9x + 6}{3} = \dfrac{1}{3}(9x + 6) \qquad \text{Division by 3 is equivalent to multiplication by } \tfrac{1}{3}.$$

$$= \dfrac{9x}{3} + \dfrac{6}{3} \qquad \text{Remove parentheses.}$$

$$= 3x + 2 \qquad \text{Simplify each fraction.}$$

SELF CHECK Simplify $\dfrac{4 - 8b}{4}$.   *Answer:* $1 - 2b$ ∎

| E X A M P L E 4 | **Dividing a trinomial by a monomial.** Simplify $\dfrac{6x^2y^2 + 4x^2y - 2xy}{2xy}$. |

Solution

$$\frac{6x^2y^2 + 4x^2y - 2xy}{2xy}$$

$$= \frac{1}{2xy}(6x^2y^2 + 4x^2y - 2xy) \quad \text{Multiply by the reciprocal of } 2xy, \text{ which is } \frac{1}{2xy}.$$

$$= \frac{6x^2y^2}{2xy} + \frac{4x^2y}{2xy} - \frac{2xy}{2xy} \quad \text{Remove parentheses.}$$

$$= 3xy + 2x - 1 \quad \text{Simplify each fraction.}$$

SELF CHECK Simplify $\dfrac{9a^2b - 6ab^2 + 3ab}{3ab}$.   *Answer:* $3a - 2b + 1$ ∎

| E X A M P L E 5 | **Dividing a trinomial by a monomial.** Simplify $\dfrac{12a^3b^2 - 4a^2b + a}{6a^2b^2}$. |

Solution

$$\frac{12a^3b^2 - 4a^2b + a}{6a^2b^2}$$

$$= \frac{1}{6a^2b^2}(12a^3b^2 - 4a^2b + a) \quad \text{Multiply by the reciprocal of } 6a^2b^2.$$

$$= \frac{12a^3b^2}{6a^2b^2} - \frac{4a^2b}{6a^2b^2} + \frac{a}{6a^2b^2} \quad \text{Remove parentheses.}$$

$$= 2a - \frac{2}{3b} + \frac{1}{6ab^2} \quad \text{Simplify each fraction.}$$

SELF CHECK Simplify $\dfrac{14p^3q + pq^2 - p}{7p^2q}$.   *Answer:* $2p + \dfrac{q}{7p} - \dfrac{1}{7pq}$ ∎

| E X A M P L E 6 | **Dividing by a monomial.** Simplify $\dfrac{(x - y)^2 - (x + y)^2}{xy}$. |

Solution

$$\frac{(x - y)^2 - (x + y)^2}{xy}$$

$$= \frac{x^2 - 2xy + y^2 - (x^2 + 2xy + y^2)}{xy} \quad \text{Square the binomials in the numerator.}$$

$$= \frac{x^2 - 2xy + y^2 - x^2 - 2xy - y^2}{xy} \quad \text{Remove parentheses.}$$

$$= \frac{-4xy}{xy} \quad \text{Combine like terms.}$$

$$= -4 \quad \text{Divide out } xy.$$

SELF CHECK Simplify $\dfrac{(x + y)^2 - (x - y)^2}{xy}$.   *Answer:* 4   ∎

An Application of Dividing a Polynomial by a Monomial

FIGURE 4-12

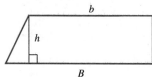

The area of the trapezoid shown in Figure 4-12 is given by the formula $A = \frac{1}{2}h(B + b)$, where B and b are its bases and h is its height. To solve the formula for b, we proceed as follows.

$$A = \frac{1}{2}h(B + b)$$

$$2 \cdot A = 2 \cdot \frac{1}{2}h(B + b) \qquad \text{Multiply both sides by 2 to clear the equation of the fraction.}$$

$$2A = h(B + b) \qquad \text{Simplify: } 2 \cdot \frac{1}{2} = \frac{2}{2} = 1.$$

$$2A = hB + hb \qquad \text{Use the distributive property to remove parentheses.}$$

$$2A - hB = hB + hb - hB \qquad \text{Subtract } hB \text{ from both sides.}$$

$$2A - hB = hb \qquad \text{Combine like terms: } hB - hB = 0.$$

$$\frac{2A - hB}{h} = \frac{hb}{h} \qquad \text{Divide both sides by } h.$$

$$\frac{2A - hB}{h} = b \qquad \frac{h}{h} = 1.$$

EXAMPLE 7

Confirming answers. Another student worked the previous problem in a different way and got a result of $b = \frac{2A}{h} - B$. Is this result correct?

Solution   To determine whether this result is correct, we must show that

$$\frac{2A - hB}{h} = \frac{2A}{h} - B$$

We can do this by dividing $2A - hB$ by h.

$$\frac{2A - hB}{h} = \frac{1}{h}(2A - hB)$$

$$= \frac{2A}{h} - \frac{hB}{h} \qquad \text{Use the distributive property to remove parentheses.}$$

$$= \frac{2A}{h} - B \qquad \text{Simplify: } \frac{\overset{1}{\cancel{h}}B}{\underset{1}{\cancel{h}}} = B.$$

The results are the same.

SELF CHECK   Suppose another student got $b = 2A - \frac{hB}{h}$. Is this result correct?   *Answer:* no

---

STUDY SET

Section 4.7

VOCABULARY

In Exercises 1–6, fill in the blanks to make the statements true.

1. A <u>polynomial</u> is an algebraic expression that is the sum of one or more terms containing whole-number exponents.

2. A <u>monomial</u> is a polynomial with one term.

3. A binomial is a polynomial with <u>two</u> terms.

4. A trinomial is a polynomial with <u>three</u> terms.

5. Dividing by a number is equivalent to multiplying by its ___reciprocal___.

6. To ___simplify___ a fraction, we divide out common factors of the numerator and denominator.

CONCEPTS

In Exercises 7–8, fill in the blanks to make the statements true.

7. $\dfrac{1}{b} \cdot a = \dfrac{a}{b}$

8. $\dfrac{15x - 6y}{6xy} = \dfrac{1}{6xy}(15x - 6y)$

9. What do the slashes and the small 1's mean?

$$\dfrac{4}{6} = \dfrac{\overset{1}{\cancel{2}} \cdot 2}{\underset{1}{\cancel{2}} \cdot 3}$$ In the numerator and denominator, a common factor of 2 was divided out.

10. Complete each rule of exponents.

a. $\dfrac{x^m}{x^n} = x^{m-n}$

b. $x^{-n} = \dfrac{1}{x^n}$

11. **a.** Solve the formula $d = rt$ for t. $t = \dfrac{d}{r}$

b. Use your answer from part a to complete the table.

	r	$\cdot$ t	$= d$
Motorcycle	$2x$	$3x^2$	$6x^3$

12. **a.** Solve the formula $I = Prt$ for r. $r = \dfrac{I}{Pt}$

b. Use your answer from part a to complete the table.

	P $\cdot$	r $\cdot$	t $=$	I
Savings account	$8x^3$	$\frac{3}{2}x^2$	$2x$	$24x^6$

13. How many nickels would have a value of $(10x + 35)$ cents? $2x + 7$

14. How many twenty-dollar bills would have a value of $\$(60x - 100)$? $3x - 5$

NOTATION

In Exercises 15–16, complete each solution.

15. $\dfrac{a^2b^3}{a^3b^2} = \dfrac{a \cdot a \cdot b \cdot b \cdot b}{a \cdot a \cdot a \cdot b \cdot b}$

$$= \dfrac{\overset{1}{\cancel{a}} \cdot \overset{1}{\cancel{a}} \cdot \overset{1}{\cancel{b}} \cdot \overset{1}{\cancel{b}} \cdot b}{\underset{1}{\cancel{a}} \cdot \underset{1}{\cancel{a}} \cdot a \cdot \underset{1}{\cancel{b}} \cdot \underset{1}{\cancel{b}}}$$

$$= \dfrac{b}{a}$$

16. $\dfrac{6pq^2 - 9p^2q^2 + pq}{3p^2q} = \dfrac{1}{3p^2q}(6pq^2 - 9p^2q^2 + pq)$

$$= \dfrac{6pq^2}{3p^2q} - \dfrac{9p^2q^2}{3p^2q} + \dfrac{pq}{3p^2q}$$

$$= \dfrac{6 \cdot p \cdot q \cdot q}{3 \cdot p \cdot p \cdot q} - \dfrac{9 \cdot p \cdot p \cdot q \cdot q}{3 \cdot p \cdot p \cdot q} + \dfrac{p \cdot q}{3 \cdot p \cdot p \cdot q}$$

$$= \dfrac{2q}{p} - 3q + \dfrac{1}{3p}$$

PRACTICE

In Exercises 17–24, simplify each fraction.

17. $\dfrac{5}{15}$ $\frac{1}{3}$

18. $\dfrac{64}{128}$ $\frac{1}{2}$

19. $\dfrac{-125}{75}$ $-\frac{5}{3}$

20. $\dfrac{-98}{21}$ $-\frac{14}{3}$

21. $\dfrac{120}{160}$ $\frac{3}{4}$

22. $\dfrac{70}{420}$ $\frac{1}{6}$

23. $\dfrac{-3,612}{-3,612}$ 1

24. $\dfrac{-288}{-112}$ $\frac{18}{7}$

In Exercises 25–48, do each division by simplifying the fraction.

25. $\dfrac{x^5}{x^2}$ x^3

26. $\dfrac{a^{12}}{a^8}$ a^4

27. $\dfrac{r^3s^2}{rs^3}$ $\frac{r^2}{s}$

28. $\dfrac{y^4z^3}{y^2z^2}$ y^2z

29. $\dfrac{8x^3y^2}{4xy^3}$ $\frac{2x^2}{y}$

30. $\dfrac{-3y^3z}{6yz^2}$ $-\frac{y^2}{2z}$

31. $\dfrac{12u^5v}{-4u^2v^3}$ $-\frac{3u^3}{v^2}$

32. $\dfrac{16rst^2}{-8rst^3}$ $-\frac{2}{t}$

33. $\dfrac{-16r^3y^2}{-4r^2y^4}$ $\frac{4r}{y^2}$

34. $\dfrac{35xyz^2}{-7x^2yz}$ $-\frac{5z}{x}$

35. $\dfrac{-65rs^2t}{15r^2s^3t}$ $-\frac{13}{3rs}$

36. $\dfrac{112u^3z^6}{-42u^3z^6}$ $-\frac{8}{3}$

37. $\dfrac{x^2x^3}{xy^6}$ $\frac{x^4}{y^6}$

38. $\dfrac{x^2y^2}{x^2y^3}$ $\frac{1}{y}$

39. $\dfrac{(a^3b^4)^3}{ab^4}$ a^8b^8

40. $\dfrac{(a^2b^3)^3}{a^6b^6}$ b^3

41. $\dfrac{15(r^2s^3)^2}{-5(rs^5)^3}$ $-\frac{3r}{s^9}$

42. $\dfrac{-5(a^2b)^3}{10(ab^2)^3}$ $-\frac{a^3}{2b^3}$

43. $\dfrac{-32(x^3y)^3}{128(x^2y^2)^3}$ $\frac{x^3}{4y^3}$

44. $\dfrac{68(a^6b^7)^2}{-96(abc^2)^3}$ $-\dfrac{17a^9b^{11}}{24c^6}$

45. $\dfrac{-(4x^3y^3)^2}{(x^2y^4)^3}$ $-\frac{16}{y^6}$

46. $\dfrac{(2r^3s^2)^2}{-(4r^2s^2)^2}$ $-\frac{r^2}{4}$

47. $\dfrac{(a^2a^3)^4}{(a^4)^3}$ a^8

48. $\dfrac{(b^3b^4)^5}{(bb^2)^2}$ b^{29}

In Exercises 49–62, do each division.

49. $\dfrac{6x + 9}{3}$ $2x + 3$

50. $\dfrac{8x + 12y}{4}$ $2x + 3y$

51. $\dfrac{5x - 10y}{25xy}$ $\dfrac{1}{5y} - \dfrac{2}{5x}$

52. $\dfrac{2x - 32}{16x}$ $\dfrac{1}{8} - \dfrac{2}{x}$

53. $\dfrac{3x^2 + 6y^3}{3x^2y^2}$ $\dfrac{1}{y^2} + \dfrac{2y}{x^2}$

54. $\dfrac{4a^2 - 9b^2}{12ab}$ $\dfrac{a}{3b} - \dfrac{3b}{4a}$

55. $\dfrac{15a^3b^2 - 10a^2b^3}{5a^2b^2}$

$3a - 2b$

56. $\dfrac{9a^4b^3 - 16a^3b^4}{12a^2b}$

$\dfrac{3a^2b^2}{4} - \dfrac{4ab^3}{3}$

57. $\dfrac{4x - 2y + 8z}{4xy}$

$\dfrac{1}{y} - \dfrac{1}{2x} + \dfrac{2z}{xy}$

58. $\dfrac{5a^2 + 10b^2 - 15ab}{5ab}$

$\dfrac{a}{b} + \dfrac{2b}{a} - 3$

59. $\dfrac{12x^3y^2 - 8x^2y - 4x}{4xy}$

$3x^2y - 2x - \dfrac{1}{y}$

60. $\dfrac{12a^2b^2 - 8a^2b - 4ab}{4ab}$

$3ab - 2a - 1$

61. $\dfrac{-25x^2y + 30xy^2 - 5xy}{-5xy}$ $5x - 6y + 1$

62. $\dfrac{-30a^2b^2 - 15a^2b - 10ab^2}{-10ab}$ $3ab + \dfrac{3a}{2} + b$

In Exercises 63–72, simplify each numerator and do the division.

63. $\dfrac{5x(4x - 2y)}{2y}$ $\dfrac{10x^2}{y} - 5x$

64. $\dfrac{9y^2(x^2 - 3xy)}{3x^2}$ $3y^2 - \dfrac{9y^3}{x}$

65. $\dfrac{(-2x)^3 + (3x^2)^2}{6x^2}$ $-\dfrac{4x}{3} + \dfrac{3x^2}{2}$

66. $\dfrac{(-3x^2y)^3 + (3xy^2)^3}{27x^3y^4}$ $-\dfrac{x^3}{y} + y^2$

67. $\dfrac{4x^2y^2 - 2(x^2y^2 + xy)}{2xy}$ $xy - 1$

68. $\dfrac{-5a^3b - 5a(ab^2 - a^2b)}{10a^2b^2}$ $-\dfrac{1}{2}$

69. $\dfrac{(3x - y)(2x - 3y)}{6xy}$ $\dfrac{x}{y} - \dfrac{11}{6} + \dfrac{y}{2x}$

70. $\dfrac{(2m - n)(3m - 2n)}{-3m^2n^2}$ $-\dfrac{2}{n^2} + \dfrac{7}{3mn} - \dfrac{2}{3m^2}$

71. $\dfrac{(a + b)^2 - (a - b)^2}{2ab}$ 2

72. $\dfrac{(x - y)^2 + (x + y)^2}{2x^2y^2}$ $\dfrac{1}{y^2} + \dfrac{1}{x^2}$

APPLICATIONS

73. POOL The rack shown in Illustration 1 is used to set up the balls when beginning a game of pool. If the perimeter of the rack, in inches, is given by the polynomial $6x^2 - 3x + 9$, what is the length of one side? $(2x^2 - x + 3)$ in.

ILLUSTRATION 1

74. CHECKERBOARD If the perimeter (in inches) of the checkerboard, shown in Illustration 2 is $12x^2 - 8x + 32$, what is the length of one side? $(3x^2 - 2x + 8)$ in.

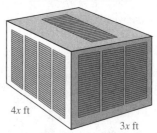
ILLUSTRATION 2

75. AIR CONDITIONING If the volume occupied by the air conditioning unit shown in Illustration 3 is $(36x^3 - 24x^2)$ cubic feet, find its height. $(3x - 2)$ ft

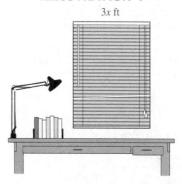
ILLUSTRATION 3

$4x$ ft

$3x$ ft

76. MINI BLINDS The area covered by the mini blinds shown in Illustration 4 is $(3x^3 - 6x)$ square feet. How long are the blinds? $(x^2 - 2)$ ft

ILLUSTRATION 4

$3x$ ft

77. CONFIRMING FORMULAS Are these formulas the same?

$$l = \dfrac{P - 2w}{2} \quad \text{and} \quad l = \dfrac{P}{2} - w \quad \text{yes}$$

78. CONFIRMING FORMULAS Are these formulas the same?

$$r = \frac{G + 2b}{2b} \quad \text{and} \quad r = \frac{G}{2b} + b \quad \text{no}$$

79. ELECTRIC BILLS On an electric bill, the following two formulas are used to compute the average cost of x kwh of electricity. Are the formulas equivalent? no

$$\frac{0.08x + 5}{x} \quad \text{and} \quad 0.08x + \frac{5}{x}$$

80. PHONE BILLS On a phone bill, the following two formulas are used to compute the average cost per minute of x minutes of phone usage. Are the formulas equivalent? yes

$$\frac{0.15x + 12}{x} \quad \text{and} \quad 0.15 + \frac{12}{x}$$

WRITING

81. What would you say to a student to explain the error in the following solution?

Simplify $\dfrac{3x + 5}{5}$.

$$\frac{3x + 5}{5} = \frac{3x + \overset{1}{\cancel{5}}}{\underset{1}{\cancel{5}}}$$
$$= 3x$$

82. How do you simplify this fraction?

$$\frac{4x^2y + 8xy^2}{4xy}$$

REVIEW

In Exercises 83–86, identify each polynomial as a monomial, binomial, trinomial, or none of the above.

83. $5a^2b + 2ab^2$ binomial

84. $-3x^3y$ monomial

85. $-2x^3 + 3x^2 - 4x + 12$ none of the above

86. $17t^2 - 15t + 27$ trinomial

87. What is the degree of the trinomial $3x^2 - 2x + 4$?   2

88. What is the numerical coefficient of the the second term of the trinomial $-7t^2 + 5t + 17$?   5

▶ 4.8   Dividing Polynomials by Polynomials

In this section, you will learn about

> Dividing polynomials by polynomials ■ Writing powers in descending order ■ Missing terms

Introduction In this section, we will conclude our discussion on operations with polynomials by discussing how to divide one polynomial by another.

Dividing Polynomials by Polynomials

To divide one polynomial by another, we use a method similar to long division in arithmetic. We illustrate the method with several examples.

EXAMPLE 1

Dividing polynomials. Divide $x^2 + 5x + 6$ by $x + 2$.

Solution Here the divisor is $x + 2$, and the dividend is $x^2 + 5x + 6$.

Step 1:
$$\begin{array}{r} x \\ x + 2 \overline{)x^2 + 5x + 6} \end{array}$$

How many times does x divide x^2? $x^2 \div x = x$. Place the x above the division symbol.

Step 2:
$$\begin{array}{r} x \\ x + 2 \overline{)x^2 + 5x + 6} \\ x^2 + 2x \end{array}$$

Multiply each term in the divisor by x. Place the product under $x^2 + 5x$ and draw a line.

$$\begin{array}{r} x \\ x + 2{\overline{\smash{\big)}\,x^2 + 5x + 6}} \\ \underline{x^2 + 2x} \\ 3x + 6 \end{array}$$

Step 3: Subtract $x^2 + 2x$ from $x^2 + 5x$. Work vertically, column by column: $x^2 - x^2 = 0$ and $5x - 2x = 3x$.

Bring down the 6.

$$\begin{array}{r} x + 3 \\ x + 2{\overline{\smash{\big)}\,x^2 + 5x + 6}} \\ \underline{x^2 + 2x} \\ 3x + 6 \end{array}$$

Step 4: How many times does x divide $3x$? $3x \div x = +3$. Place the $+3$ above the division symbol.

$$\begin{array}{r} x + 3 \\ x + 2{\overline{\smash{\big)}\,x^2 + 5x + 6}} \\ \underline{x^2 + 2x} \\ 3x + 6 \\ \underline{3x + 6} \end{array}$$

Step 5: Multiply each term in the divisor by 3. Place the product under $3x + 6$ and draw a line.

$$\begin{array}{r} x + 3 \\ x + 2{\overline{\smash{\big)}\,x^2 + 5x + 6}} \\ \underline{x^2 + 2x} \\ 3x + 6 \\ \underline{3x + 6} \\ 0 \end{array}$$

Step 6: Subtract $3x + 6$ from $3x + 6$. Work vertically: $3x - 3x = 0$ and $6 - 6 = 0$.

The quotient is $x + 3$ and the remainder is 0.

Step 7: Check the work by verifying that $x + 2$ times $x + 3$ is $x^2 + 5x + 6$.

$$(x + 2)(x + 3) = x^2 + 3x + 2x + 6$$
$$= x^2 + 5x + 6$$

The answer checks.

SELF CHECK Divide $x^2 + 7x + 12$ by $x + 3$.   *Answer: $x + 4$* ∎

E X A M P L E   2 **Dividing polynomials.** Divide $\dfrac{6x^2 - 7x - 2}{2x - 1}$.

Solution Here the divisor is $2x - 1$ and the dividend is $6x^2 - 7x - 2$.

$$\begin{array}{r} 3x \\ 2x - 1{\overline{\smash{\big)}\,6x^2 - 7x - 2}} \end{array}$$

Step 1: How many times does $2x$ divide $6x^2$? $6x^2 \div 2x = 3x$. Place the $3x$ above the division symbol.

$$\begin{array}{r} 3x \\ 2x - 1{\overline{\smash{\big)}\,6x^2 - 7x - 2}} \\ \underline{6x^2 - 3x} \end{array}$$

Step 2: Multiply each term in the divisor by $3x$. Place the product under $6x^2 - 7x$ and draw a line.

$$\begin{array}{r} 3x \\ 2x - 1{\overline{\smash{\big)}\,6x^2 - 7x - 2}} \\ \underline{6x^2 - 3x} \\ -4x - 2 \end{array}$$

Step 3: Subtract $6x^2 - 3x$ from $6x^2 - 7x$. Work vertically: $6x^2 - 6x^2 = 0$ and $-7x - (-3x) = -7x + 3x = -4x$.

Bring down the -2.

$$\begin{array}{r} 3x - 2 \\ 2x - 1{\overline{\smash{\big)}\,6x^2 - 7x - 2}} \\ \underline{6x^2 - 3x} \\ -4x - 2 \end{array}$$

Step 4: How many times does $2x$ divide $-4x$? $-4x \div 2x = -2$. Place the -2 above the division symbol.

Step 5: $\overset{\displaystyle 3x\ \ -2}{2x-1\overline{)6x^2-7x-2}}$ Multiply each term in the divisor by -2. Place the product under $-4x-2$ and draw a line.

$$\underline{6x^2-3x}$$
$$-4x-2$$
$$\underline{-4x+2}$$

Step 6: $\overset{\displaystyle 3x\ \ -2}{2x-1\overline{)6x^2-7x-2}}$ Subtract $-4x+2$ from $-4x-2$. Work vertically: $-4x-(-4x)=-4x+4x=0$ and $-2-2=-4$.

$$\underline{6x^2-3x}$$
$$-4x-2$$
$$\underline{-4x+2}$$
$$-4$$

Here the quotient is $3x-2$ and the remainder is -4. It is common to write the answer in this form:

$$3x-2+\frac{-4}{2x-1} \quad \text{Quotient} + \frac{\text{remainder}}{\text{divisor}}.$$

Step 7: To check the answer, we multiply

$$3x-2+\frac{-4}{2x-1} \quad \text{by} \quad 2x-1$$

The product should be the dividend.

$$(2x-1)\left(3x-2+\frac{-4}{2x-1}\right) = (2x-1)(3x-2)+(2x-1)\left(\frac{-4}{2x-1}\right)$$
$$= (2x-1)(3x-2)-4$$
$$= 6x^2-4x-3x+2-4$$
$$= 6x^2-7x-2$$

Because the result is the dividend, the answer checks.

SELF CHECK   Divide $\dfrac{8x^2+6x-3}{2x+3}$.   *Answer:* $4x-3+\dfrac{6}{2x+3}$ ∎

Writing Powers in Descending Order

The division method works best when the exponents of the terms in the divisor and the dividend are written in descending order. This means that the term involving the highest power of x appears first, the term involving the second-highest power of x appears second, and so on. For example, the terms in

$$3x^3+2x^2-7x+5$$

have their exponents written in descending order.

If the powers in the dividend or divisor are not in descending order, we use the commutative property of addition to write them that way.

E X A M P L E   3   **Dividing polynomials.** Divide $4x^2+2x^3+12-2x$ by $x+3$.

Solution   We write the dividend so that the exponents are in descending order.

$$\overset{\displaystyle 2x^2-2x\ \ +4}{x+3\overline{)2x^3+4x^2-2x+12}}$$
$$\underline{2x^3+6x^2}$$
$$-2x^2-2x$$
$$\underline{-2x^2-6x}$$
$$4x+12$$
$$\underline{4x+12}$$

Check: $(x + 3)(2x^2 - 2x + 4) = 2x^3 - 2x^2 + 4x + 6x^2 - 6x + 12$
$$= 2x^3 + 4x^2 - 2x + 12$$

SELF CHECK   Divide $x^2 - 10x + 6x^3 + 4$ by $2x - 1$.   *Answer:* $3x^2 + 2x - 4$ ∎

Missing Terms

When we write the terms of a dividend in descending powers of x, we must determine whether some powers of x are missing. For example, in the dividend of

$$x + 1 \overline{)3x^4 - 7x^2 - 3x + 15}$$

the term involving x^3 is missing. When this happens, we should either write the term with a coefficient of 0 or leave a blank space for it. In this case, we would write the dividend as

$$3x^4 + 0x^3 - 7x^2 - 3x + 15 \qquad \text{or} \qquad 3x^4 \qquad - 7x^2 - 3x + 15$$

EXAMPLE 4   **Dividing polynomials.** Divide $\dfrac{x^2 - 4}{x + 2}$.

Solution   Since $x^2 - 4$ does not have a term involving x, we must either include the term $0x$ or leave a space for it.

$$
\begin{array}{r}
x - 2 \\
x + 2 \overline{)x^2 + 0x - 4} \\
\underline{x^2 + 2x} \\
-2x - 4 \\
\underline{-2x - 4}
\end{array}
$$

Check: $(x + 2)(x - 2) = x^2 - 2x + 2x - 4$
$$= x^2 - 4$$

SELF CHECK   Divide $\dfrac{x^2 - 9}{x - 3}$ *Answer:* $x + 3$ ∎

STUDY SET

Section 4.8

1–12, 13–17 odd, 19–39 evod, 51–55 odd

VOCABULARY

In Exercises 1–4, fill in the blanks to make the statements true.

1. In the division $x + 1 \overline{)x^2 + 2x + 1}$, $x + 1$ is called the _____divisor_____ and $x^2 + 2x + 1$ is called the _____dividend_____.

2. The answer to a division problem is called the _____quotient_____.

3. If a division does not come out even, the leftover part is called a _____remainder_____.

4. The exponents in $2x^4 + 3x^3 + 4x^2 - 7x - 2$ are said to be written in _____descending_____ order.

CONCEPTS

In Exercises 5–8, write each polynomial with the powers in descending order.

5. $4x^3 + 7x - 2x^2 + 6$ $4x^3 - 2x^2 + 7x + 6$

6. $5x^2 + 7x^3 - 3x - 9$ $7x^3 + 5x^2 - 3x - 9$

7. $9x + 2x^2 - x^3 + 6x^4$ $6x^4 - x^3 + 2x^2 + 9x$

8. $7x^5 + x^3 - x^2 + 2x^4$ $7x^5 + 2x^4 + x^3 - x^2$

In Exercises 9–10, identify the missing terms in each polynomial.

9. $5x^4 + 2x^2 - 1$ $0x^3$ and $0x$

10. $-3x^5 - 2x^3 + 4x - 6$ $0x^4$ and $0x^2$

In Exercises 11–12, without doing the division, determine which of the three possible quotients seems reasonable.

11. $\dfrac{x^4 - 81}{x - 3}$ $x^2 + 3x + 9$
$x^3 + 3x^2 + 9x + 27$ ← quotient
$x^4 + 3x^3 + 9x^2 + 27x + 1$

12. $\dfrac{8x^3 - 27}{2x - 3}$ $4x^2 + 6x + 9$ ← quotient
$4x^3 - 6x^2 - 9$
$4x^4 - 6x^3 - 9x^2 + 1$

13. a. Solve $d = rt$ for r. $r = \frac{d}{t}$
   b. Use your answer to part a and the long division method to complete the chart.

	r	$\cdot$ t	$=$ d
Subway	$x - 3$	$x + 4$	$x^2 + x - 12$

14. a. Solve $I = Prt$ for P. $P = \frac{I}{rt}$
   b. Use your answer to part a and the long division method to complete the chart.

	P $\cdot$	r $\cdot$	$t =$	I
Bonds	$x + 3$	$x + 4$	1	$x^2 + 7x + 12$

15. Using long division, a student found that

$$\frac{3x^2 + 8x + 4}{3x + 2} = x + 2$$

Use multiplication to see whether the result is correct.
It is correct.

16. Using long division, a student found that

$$\frac{x^2 + 4x - 21}{x - 3} = x - 7$$

Use multiplication to see whether the result is correct.
It is incorrect.

NOTATION

In Exercises 17–18, complete each division.

17.
$$
\begin{array}{r}
\boxed{x} \;+ 2 \\
x + 2 \overline{)\, x^2 + 4x + 4} \\
\underline{x^2 + \boxed{2x}} \\
2x + 4 \\
\underline{2x + 4} \\
0
\end{array}
$$

18.
$$
\begin{array}{r}
\boxed{x^2} \;+ x \;- 2 + \frac{7}{2x+1} \\
2x + 1 \overline{)\, 2x^3 + 3x^2 - 3x + 5} \\
\underline{2x^3 + x^2} \\
2x^2 - 3x \\
\underline{2x^2 + \boxed{x}} \\
-4x + 5 \\
\underline{-4x - \boxed{2}} \\
7
\end{array}
$$

PRACTICE

In Exercises 19–26, do each division.

19. Divide $x^2 + 4x - 12$ by $x - 2$. $x + 6$

20. Divide $x^2 - 5x + 6$ by $x - 2$. $x - 3$

21. Divide $y^2 + 13y + 12$ by $y + 1$. $y + 12$

22. Divide $z^2 - 7z + 12$ by $z - 3$. $z - 4$

23. $\dfrac{6a^2 + 5a - 6}{2a + 3}$ $3a - 2$

24. $\dfrac{8a^2 + 2a - 3}{2a - 1}$ $4a + 3$

25. $\dfrac{3b^2 + 11b + 6}{3b + 2}$ $b + 3$

26. $\dfrac{3b^2 - 5b + 2}{3b - 2}$ $b - 1$

In Exercises 27–34, write the terms so that the powers of x are in descending order. Then do each division.

27. $5x + 3 \overline{)\, 11x + 10x^2 + 3}$ $2x + 1$

28. $2x - 7 \overline{)\, -x - 21 + 2x^2}$ $x + 3$

29. $4 + 2x \overline{)\, -10x - 28 + 2x^2}$ $x - 7$

30. $1 + 3x \overline{)\, 9x^2 + 1 + 6x}$ $3x + 1$

31. $2x - 1 \overline{)\, x - 2 + 6x^2}$ $3x + 2$

32. $2 + x \overline{)\, 3x + 2x^2 - 2}$ $2x - 1$

33. $3 + x \overline{)\, 2x^2 - 3 + 5x}$ $2x - 1$

34. $x - 3 \overline{)\, 2x^2 - 3 - 5x}$ $2x + 1$

In Exercises 35–40, do each division.

35. $2x + 3 \overline{)\, 2x^3 + 7x^2 + 4x - 3}$ $x^2 + 2x - 1$

36. $2x - 1 \overline{)\, 2x^3 - 3x^2 + 5x - 2}$ $x^2 - x + 2$

37. $3x + 2 \overline{)\, 6x^3 + 10x^2 + 7x + 2}$ $2x^2 + 2x + 1$

38. $4x + 3 \overline{)\, 4x^3 - 5x^2 - 2x + 3}$ $x^2 - 2x + 1$

39. $2x + 1 \overline{)\, 2x^3 + 3x^2 + 3x + 1}$ $x^2 + x + 1$

40. $3x - 2 \overline{)\, 6x^3 - x^2 + 4x - 4}$ $2x^2 + x + 2$

In Exercises 41–50, do each division. If there is a remainder, write the answer in quotient $+ \dfrac{remainder}{divisor}$ form.

41. $\dfrac{2x^2 + 5x + 2}{2x + 3}$

$x + 1 + \dfrac{-1}{2x + 3}$

42. $\dfrac{3x^2 - 8x + 3}{3x - 2}$

$x - 2 + \dfrac{-1}{3x - 2}$

43. $\dfrac{4x^2 + 6x - 1}{2x + 1}$

$2x + 2 + \dfrac{-3}{2x + 1}$

44. $\dfrac{6x^2 - 11x + 2}{3x - 1}$

$2x - 3 + \dfrac{-1}{3x - 1}$

45. $\dfrac{x^3 + 3x^2 + 3x + 1}{x + 1}$

$x^2 + 2x + 1$

46. $\dfrac{x^3 + 6x^2 + 12x + 8}{x + 2}$

$x^2 + 4x + 4$

47. $\dfrac{2x^3 + 7x^2 + 4x + 3}{2x + 3}$

$x^2 + 2x - 1 + \dfrac{6}{2x + 3}$

48. $\dfrac{6x^3 + x^2 + 2x + 1}{3x - 1}$

$2x^2 + x + 1 + \dfrac{2}{3x - 1}$

49. $\dfrac{2x^3 + 4x^2 - 2x + 3}{x - 2}$

$2x^2 + 8x + 14 + \dfrac{31}{x - 2}$

50. $\dfrac{3y^3 - 4y^2 + 2y + 3}{y + 3}$

$3y^2 - 13y + 41 + \dfrac{-120}{y + 3}$

In Exercises 51–60, do each division.

51. $\dfrac{x^2 - 1}{x - 1}$ $x + 1$

52. $\dfrac{x^2 - 9}{x + 3}$ $x - 3$

53. $\dfrac{4x^2 - 9}{2x + 3}$ $2x - 3$

54. $\dfrac{25x^2 - 16}{5x - 4}$ $5x + 4$

55. $\dfrac{x^3 + 1}{x + 1}$ $x^2 - x + 1$

56. $\dfrac{x^3 - 8}{x - 2}$ $x^2 + 2x + 4$

57. $\dfrac{a^3 + a}{a + 3}$

$a^2 - 3a + 10 + \dfrac{-30}{a + 3}$

58. $\dfrac{y^3 - 50}{y - 5}$

$y^2 + 5y + 25 + \dfrac{75}{y - 5}$

59. $3x - 4 \overline{)15x^3 - 23x^2 + 16x}$ $5x^2 - x + 4 + \dfrac{16}{3x - 4}$

60. $2y + 3 \overline{)21y^2 + 6y^3 - 20}$ $3y^2 + 6y - 9 + \dfrac{7}{2y + 3}$

APPLICATIONS

61. FURNACE FILTER The area of the furnace filter shown in Illustration 1 is $(x^2 - 2x - 24)$ square inches.
   a. Find its length. $(x - 6)$ in.
   b. Find its perimeter.
      $(4x - 4)$ in.

ILLUSTRATION 1

$(x + 4)$ in.

62. SHELF SPACE The formula $V = Bh$ gives the volume of a cylinder where B is the area of the base and h is the height. Find the amount of shelf space that the container of potato chips shown in Illustration 2 occupies if its volume is $(2x^3 - 4x - 2)$ cubic inches.
$(x^2 - x - 1)$ in.

ILLUSTRATION 2

$(2x + 2)$ in.

$2.09

63. COMMUNICATION See Illustration 3. Telephone poles were installed every $(2x - 3)$ feet along a stretch of railroad track $(8x^3 - 6x^2 + 5x - 21)$ feet long. How many poles were used? $4x^2 + 3x + 7$

ILLUSTRATION 3

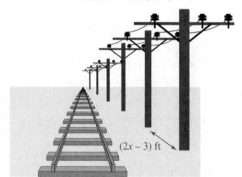

$(2x - 3)$ ft

64. CONSTRUCTION COSTS Find the price per square foot to remodel each of the three rooms listed in the chart.

Room	Remodeling cost	Area (ft²)	Cost (per ft²)
Kitchen	$(3x^3 - 9x - 6)$	$3x + 3$	$(x^2 - x - 2)$
Bathroom	$(2x^2 + x - 6)$	$2x - 3$	$(x + 2)$
Bedroom	$(x^2 + 9x + 20)$	$x + 4$	$(x + 5)$

WRITING

65. Distinguish between *dividend, divisor, quotient,* and *remainder.*

66. How would you check the results of a division?

REVIEW

67. Simplify $(x^5 x^6)^2$. x^{22}

68. Simplify $(a^2)^3 (a^3)^4$. a^{18}

In Exercises 69–70, simplify each expression.

69. $3(2x^2 - 4x + 5) + 2(x^2 + 3x - 7)$ $8x^2 - 6x + 1$

70. $-2(y^3 + 2y^2 - y) - 3(3y^3 + y)$ $-11y^3 - 4y^2 - y$

71. What can be said about the slopes of two parallel lines?   They are the same.

72. What is the slope of a line perpendicular to a line with a slope of $\frac{3}{4}$? $-\frac{4}{3}$

Polynomials

A **polynomial** is an algebraic expression that is the sum of one or more terms containing whole-number exponents. Some examples are

$$-16a^2b, \qquad y + 2, \qquad x^2 - 3x + 9, \qquad \text{and} \qquad 3st - 6r + 5st - 8r$$

Operations with Polynomials

Polynomials are the numbers of algebra. Just like numbers in arithmetic, they can be added, subtracted, multiplied, divided, and raised to powers. In Chapter 4, we have discussed some rules to be used when performing operations with polynomials.

In Exercises 1–4, fill in the blanks to make the statements true. They are rules for working with polynomials containing more than one term.

1. To add polynomials, remove the parentheses and then _____combine_____ any like terms.

2. To subtract polynomials, drop the minus sign and the parentheses, and _____change_____ the sign of every term within the parentheses. Then combine like terms.

3. To multiply polynomials, multiply _____each_____ term of one polynomial by _____each_____ term of the other polynomial and then combine like terms.

4. To divide polynomials, use the _____long_____ division method.

In Exercises 5–12, do the operations.

5. $(2x + 3) + (x - 8)$ $3x - 5$

6. $(2x + 3) - (x - 8)$ $x + 11$

7. $(2x + 3)(x - 8)$ $2x^2 - 13x - 24$

8. $(2x^2 + 3)^2$ $4x^4 + 12x^2 + 9$

9. $(y^2 + y - 6) + (y + 3)$ $y^2 + 2y - 3$

10. $(y^2 + y - 6) - (y + 3)$ $y^2 - 9$

11. $(y^2 + y - 6)(y + 3)$ $y^3 + 4y^2 - 3y - 18$

12. $(y^2 + y - 6) \div (y + 3)$ $y - 2$

Polynomial Functions

Polynomial functions can be used to mathematically describe such situations as the stopping distance of a car, the appreciation of a house, and the area of a geometric figure.

13. Complete the table and then graph the polynomial function $f(x) = -x^3 - 1$.

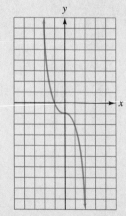

x	$f(x)$	(x, y)
-2	7	$(-2, 7)$
-1	0	$(-1, 0)$
0	-1	$(0, -1)$
1	-2	$(1, -2)$
2	-9	$(2, -9)$

14. YOUTH DRUG USE Find the coordinates of the lowest point on the graph in Illustration 1 and explain its significance.

$(2, 5.5)$; for 1990–1997, youth drug use was the lowest in 1992, at 5.5%

ILLUSTRATION 1

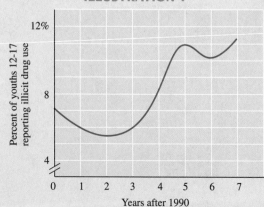

Years after 1990

Accent on Teamwork

Section 4.1

Rules for exponents Have a student in your group write each of the five rules for exponents listed on page 284 on separate 3×5 cards. On a second set of cards, write an explanation of each rule using words. On a third set of cards, write a separate example of the use of each rule for exponents. Shuffle the cards and work together to match the symbolic description, the word description, and the example for each of the five rules for exponents.

Section 4.2

Graphing Complete Table 1, plot the ordered pairs on a rectangular coordinate system, and then draw a smooth curve through the points. Do the same for Table 2, using the same coordinate system. Compare the graphs. How are they similar and how do they differ?

TABLE 1		TABLE 2	
x	2^x	x	2^{-x}
-2		-2	
-1		-1	
0		0	
1		1	
2		2	
3		3	

Section 4.3

Scientific notation Go to the library and find five examples of extremely large and five examples of extremely small numbers. Encyclopedias, government statistics books, and science books are good places to look. Write each number in scientific notation on a separate piece of paper. Include a brief explanation of what the number represents. Present the ten examples in numerical order, beginning with the smallest number first.

Section 4.4

Polynomial functions The height (in feet) of a rock from the floor of the Grand Canyon t seconds after being thrown downward from the rim with an initial velocity of 6 feet per second is given by the polynomial

$$f(t) = -16t^2 - 6t + 5{,}292$$

a. Find $f(0)$ and $f(18)$ and explain their significance.

b. Find $f(3)$, $f(6)$, $f(9)$, $f(12)$, and $f(15)$. Use this information to show the position of the rock for these times on the scale shown in Illustration 1.

c. Are the distances the rock fell during each 3-second time interval the same?

ILLUSTRATION 1

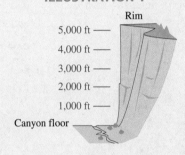

Section 4.5

Adding polynomials An old adage is that "You can't add apples and oranges." Give an example of how this concept applies when adding two polynomials.

Section 4.6

Multiplying binomials Use colored construction paper to make a model that can be used in a presentation explaining what it means to multiply the binomials $x + 3$ and $x + 4$. See Exercise 101 in Study Set 4.6 for an example.

Section 4.7

Working with monomials For the monomials $15a^3$ and $5a^2$, show, if possible, how they are added, subtracted, multiplied, and divided. If an operation cannot be done, explain why this is so.

Section 4.8

Working with polynomials Pick a certain binomial that divides a certain trinomial evenly. Then show how they are added, subtracted, multiplied, and divided.

Section 4.1

CONCEPTS

If n is a natural number, then

$$x^n = \overbrace{x \cdot x \cdot x \cdot \cdots \cdot x}^{n \text{ factors of } x}$$

where x is called the *base* and n is called the *exponent*.

Rules for exponents:
If m and n are integers, then

$$x^m x^n = x^{m+n}$$

$$(x^m)^n = x^{m \cdot n}$$

$$(xy)^n = x^n y^n$$

$$\left(\frac{x}{y}\right)^n = \frac{x^n}{y^n} \quad (y \neq 0)$$

$$\frac{x^m}{x^n} = x^{m-n} \quad (x \neq 0)$$

Natural-Number Exponents

REVIEW EXERCISES

1. Write each expression without using exponents.

  a. $-3x^4$ $-3 \cdot x \cdot x \cdot x \cdot x$ **b.** $\left(\frac{1}{2}pq\right)^3$ $(\frac{1}{2}pq)(\frac{1}{2}pq)(\frac{1}{2}pq)$

2. Evaluate each expression.

  a. 5^3 125 **b.** $(-8)^2$ 64

  c. -8^2 -64 **d.** $(5 - 3)^2$ 4

3. Simplify each expression.

  a. $x^3 x^2$ x^5 **b.** $-3y(y^5)$ $-3y^6$

  c. $(y^7)^3$ y^{21} **d.** $(3x)^4$ $81x^4$

  e. $b^3 b^4 b^5$ b^{12} **f.** $-z^2(z^3 y^2)$ $-y^2 z^5$

  g. $(-16s)^2 s$ $256s^3$ **h.** $(2x^2 y)^2$ $4x^4 y^2$

  i. $(x^2 x^3)^3$ x^{15} **j.** $\left(\frac{x^2 y}{xy^2}\right)^2$ $\frac{x^2}{y^2}$

  k. $\frac{x^7}{x^3}$ x^4 **l.** $\frac{(5y^2 z^3)^3}{25(yz)^5}$ $5yz^4$

4. Find the area or the volume of each figure, whichever is appropriate.

  a. $64x^{12}$ in.^{3} **b.** y^4 m^2

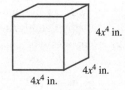

$4x^4$ in. $4x^4$ in. $4x^4$ in.

y^2 m y^2 m

Section 4.2

Zero exponents:
$$x^0 = 1 \quad (x \neq 0)$$

Negative integer exponents:
$$x^{-n} = \frac{1}{x^n} \quad (x \neq 0)$$

Zero and Negative Integer Exponents

5. Write each expression without using negative exponents or parentheses.

  a. x^0 1 **b.** $(3x^2 y^2)^0$ 1

  c. $(3x^0)^2$ 9 **d.** 10^{-3} $\frac{1}{1,000}$

  e. $\left(\frac{3}{4}\right)^{-1}$ $\frac{4}{3}$ **f.** -5^{-2} $-\frac{1}{25}$

  g. x^{-5} $\frac{1}{x^5}$ **h.** $-6y^4 y^{-5}$ $-\frac{6}{y}$

  i. $\frac{x^{-3}}{x^7}$ $\frac{1}{x^{10}}$ **j.** $(x^{-3} x^{-4})^{-2}$ x^{14}

  k. $\left(\frac{x^2}{x}\right)^{-5}$ $\frac{1}{x^5}$ **l.** $\left(\frac{15z^4}{5z^3}\right)^{-2}$ $\frac{1}{9z^2}$

6. Write each expression with a single exponent.

  a. $y^{3n} y^{4n}$ y^{7n} **b.** $\frac{z^{8c}}{z^{10c}}$ $\frac{1}{z^{2c}}$

Section 4.3

A number is written in *scientific notation* if it is written as the product of a number between 1 (including 1) and 10 and an integer power of 10.

Scientific notation provides an easier way to do some computations.

Scientific Notation

7. Write each number in scientific notation.
   a. 728 7.28×10^2
   b. 9,370,000 9.37×10^6
   c. 0.0136 1.36×10^{-2}
   d. 0.00942 9.42×10^{-3}
   e. 0.018×10^{-2} 1.8×10^{-4}
   f. 753×10^3 7.53×10^5

8. Write each number in standard notation.
   a. 7.26×10^5 726,000
   b. 3.91×10^{-4} 0.000391
   c. 2.68×10^0 2.68
   d. 5.76×10^1 57.6

9. Simplify each fraction by first writing each number in scientific notation, then do the arithmetic. Express the result in standard notation.
   a. $\dfrac{(0.00012)(0.00004)}{0.00000016}$ 0.03
   b. $\dfrac{(4,800)(20,000)}{600,000}$ 160

10. WORLD POPULATION As of 1998, the world's population was estimated to be 5.927 billion. Write this number in standard notation and in scientific notation. $5,927,000,000;\ 5.927 \times 10^9$

11. ATOMS Illustration 1 shows a cross section of an atom. How many nuclei, placed end-to-end, would it take to stretch across the atom? $1.0 \times 10^5 = 100,000$

ILLUSTRATION 1

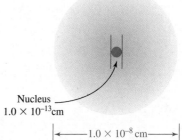

Nucleus
1.0×10^{-13}cm

$\longleftarrow 1.0 \times 10^{-8}$ cm $\longrightarrow$

Section 4.4

A *polynomial* is a term or a sum of terms in which all variables have whole-number exponents.

The *degree of a monomial* ax^n is n. The *degree of a monomial* in several variables is the sum of the exponents on those variables. The *degree of a polynomial* is the same as the degree of its term with the largest degree.

If $f(x)$ is a polynomial function in x, then $f(3)$ is the value of the function when $x = 3$.

Polynomials and Polynomial Functions

12. Tell whether each expression is a polynomial.
    a. $x^3 - x^2 - x - 1$ yes
    b. $x^{-2} - x^{-1} - 1$ no
    c. $\dfrac{11}{y} + 4y$ no
    d. $-16x^2y + 5xy^2$ yes

13. Find the degree of each polynomial and classify it as a monomial, binomial, trinomial, or none of these.
    a. $13x^7$ 7th, monomial
    b. $-16a^2b$ 3rd, monomial
    c. $5^3x + x^2$ 2nd, binomial
    d. $-3x^5 + x - 1$ 5th, trinomial
    e. $9xy^2 + 21x^3y^3$ 6th, binomial
    f. $4s^4 - 3s^2 + 5s + 4$
       4th, none of these

14. Let $f(x) = 3x^2 + 2x + 1$. Find each value.
    a. $f(3)$ 34
    b. $f(0)$ 1
    c. $f(-2)$ 9
    d. $f(-0.2)$ 0.72

15. Complete the table and then graph the polynomial function $f(x) = x^3 - 3x + 2$.

x	$f(x)$	(x, y)
-3	-16	$(-3, -16)$
-2	0	$(-2, 0)$
-1	4	$(-1, 4)$
0	2	$(0, 2)$
1	0	$(1, 0)$
2	4	$(2, 4)$
3	20	$(3, 20)$

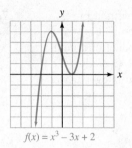

$f(x) = x^3 - 3x + 2$

16. DIVING See Illustration 2. The number of inches that the woman deflects the diving board is given by the function

$$f(x) = 0.1875x^2 - 0.0078125x^3$$

where x is the number of feet that she stands from the front anchor point of the board. Find the amount of deflection if she stands on the end of the diving board, 8 feet from the anchor point.   8 in.

ILLUSTRATION 2

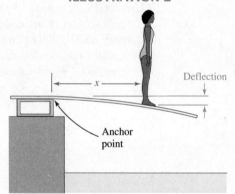

Deflection

x

Anchor
point

Section 4.5 — Adding and Subtracting Polynomials

When *adding* or *subtracting polynomials,* add or subtract like terms by combining the numerical coefficients and using the same variables and the same exponents.

17. Simplify each expression.

a. $3x^6 + 5x^5 - x^6$ $2x^6 + 5x^5$ **b.** $x^2y^2 - 3x^2y^2$ $-2x^2y^2$

c. $(3x^2 + 2x) + (5x^2 - 8x)$ $8x^2 - 6x$

d. $3(9x^2 + 3x + 7) - 2(11x^2 - 5x + 9)$ $5x^2 + 19x + 3$

Polynomials can be added or subtracted *vertically.*

18. Do the operations.

a.
$$\begin{array}{r} 3x^2 + 5x + 2 \\ + \quad x^2 - 3x + 6 \\ \hline 4x^2 + 2x + 8 \end{array}$$

b.
$$\begin{array}{r} 20x^3 \qquad\quad + 12x \\ - \quad 12x^3 + 7x^2 - \ 7x \\ \hline 8x^3 - 7x^2 + 19x \end{array}$$

Section 4.6 — Multiplying Polynomials

To multiply two monomials, first multiply the numerical factors and then multiply the variable factors.

19. Find each product.

a. $(2x^2)(5x)$ $10x^3$ **b.** $(-6x^4z^3)(x^6z^2)$ $-6x^{10}z^5$

c. $(2rst)(-3r^2s^3t^4)$ $-6r^3s^4t^5$ **d.** $5b^3 \cdot 6b^2 \cdot 4b^6$ $120b^{11}$

To multiply a polynomial with more than one term by a monomial, multiply each term of the polynomial by the monomial and simplify.

20. Find each product.
  a. $5(x + 3)$ $5x + 15$
  b. $x^2(3x^2 - 5)$ $3x^4 - 5x^2$
  c. $x^2y(y^2 - xy)$ $x^2y^3 - x^3y^2$
  d. $-2y^2(y^2 - 5y)$ $-2y^4 + 10y^3$
  e. $2x(3x^4)(x + 2)$ $6x^6 + 12x^5$
  f. $-3x(x^2 - x + 2)$ $-3x^3 + 3x^2 - 6x$

To multiply two binomials, use the *FOIL method*:
  F: First
  O: Outer
  I: Inner
  L: Last

21. Find each product.
  a. $(x + 3)(x + 2)$ $x^2 + 5x + 6$
  b. $(2x + 1)(x - 1)$ $2x^2 - x - 1$
  c. $(3a - 3)(2a + 2)$ $6a^2 - 6$
  d. $6(a - 1)(a + 1)$ $6a^2 - 6$
  e. $(a - b)(2a + b)$ $2a^2 - ab - b^2$
  f. $(-3x - y)(2x + y)$ $-6x^2 - 5xy - y^2$

Special products:
  $(x + y)^2 = x^2 + 2xy + y^2$
  $(x - y)^2 = x^2 - 2xy + y^2$
  $(x + y)(x - y) = x^2 - y^2$

22. Find each product.
  a. $(x + 3)(x + 3)$ $x^2 + 6x + 9$
  b. $(x + 5)(x - 5)$ $x^2 - 25$
  c. $(a - 3)^2$ $a^2 - 6a + 9$
  d. $(x + 4)^2$ $x^2 + 8x + 16$
  e. $(-2y + 1)^2$ $4y^2 - 4y + 1$
  f. $(y^2 + 1)(y^2 - 1)$ $y^4 - 1$

To multiply one polynomial by another, multiply each term of one polynomial by each term of the other polynomial, and simplify.

23. Find each product.
  a. $(3x + 1)(x^2 + 2x + 1)$
    $3x^3 + 7x^2 + 5x + 1$
  b. $(2a - 3)(4a^2 + 6a + 9)$ $8a^3 - 27$

24. Solve each equation.
  a. $x^2 + 3 = x(x + 3)$ 1
  b. $x^2 + x = (x + 1)(x + 2)$ -1
  c. $(x + 2)(x - 5) = (x - 4)(x - 1)$ 7
  d. $(x + 5)(3x + 1) = x^2 + (2x - 1)(x - 5)$ 0

25. APPLIANCES Find the perimeter of the base, the area of the base, and the volume occupied by the dishwasher shown in Illustration 3.
  $(6x + 10)$ in.; $(2x^2 + 11x - 6)$ in.^{2};
  $(6x^3 + 33x^2 - 18x)$ in.^{3}

ILLUSTRATION 3

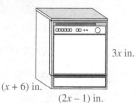

$3x$ in.

$(x + 6)$ in.

$(2x - 1)$ in.

Section 4.7

Dividing Polynomials by Monomials

To divide monomials, use the method for simplifying fractions or use the rules for exponents.

26. Simplify each expression ($x > 0$, $y > 0$).
  a. $\dfrac{-14x^2y}{21xy^3}$ $-\dfrac{2x}{3y^2}$
  b. $\dfrac{(x^2)^2}{xx^4}$ $\dfrac{1}{x}$

To divide a polynomial by a monomial, write the division as a product, use the distributive property to remove parentheses, and simplify each resulting fraction.

27. Do each division. All the variables represent positive numbers.
  a. $\dfrac{8x + 6}{2}$ $4x + 3$
  b. $\dfrac{14xy - 21x}{7xy}$ $2 - \dfrac{3}{y}$
  c. $\dfrac{15a^2b + 20ab^2 - 25ab}{5ab}$
    $3a + 4b - 5$
  d. $\dfrac{(x + y)^2 + (x - y)^2}{-2xy}$ $-\dfrac{x}{y} - \dfrac{y}{x}$

28. SAVINGS BONDS How many \$50 savings bonds would have a total value of \$$(50x + 250)$? $x + 5$

Section 4.8   Dividing Polynomials by Polynomials

Long division is used to divide one polynomial by another.

When a division has a remainder, write the answer in the form

$$\text{Quotient} + \frac{\text{remainder}}{\text{divisor}}$$

The division method works best when the exponents of the terms of the divisor and the dividend are written in descending order.

When the dividend is missing a term, write it with a coefficient of zero or leave a blank space.

29. Do each division.

  a. $x + 2\overline{)x^2 + 3x + 5}$

       $x + 1 + \dfrac{3}{x + 2}$

  b. $x - 1\overline{)x^2 - 6x + 5}$ $x - 5$

  c. $\dfrac{2x^2 + 3 + 7x}{x + 3}$ $2x + 1$

  d. $\dfrac{3x^2 + 14x - 2}{3x - 1}$ $x + 5 + \dfrac{3}{3x - 1}$

  e. $2x - 1\overline{)6x^3 + x^2 + 1}$

       $3x^2 + 2x + 1 + \dfrac{2}{2x - 1}$

  f. $3x + 1\overline{)-13x - 4 + 9x^3}$

       $3x^2 - x - 4$

30. Use multiplication to show that the answer when dividing $3y^2 + 11y + 6$ by $y + 3$ is $3y + 2$.

31. ZOOLOGY The distance in inches traveled by a certain type of snail in $(2x - 1)$ minutes is given by the polynomial $8x^2 + 2x - 3$. At what rate did the snail travel? $(4x + 3)$ in./min

---

CHAPTER 4

Test

1. Use exponents to rewrite $2xxxyyy$. $2x^3y^4$

2. Evaluate $(3 + 5)^2$. 64

In Problems 3–4, write each expression as an expression containing only one exponent.

3. $y^2(yy^3)$ y^6

4. $(2x^3)^5(x^2)^3$ $32x^{21}$

In Problems 5–8, simplify each expression. Write answers without using parentheses or negative exponents.

5. $3x^0$ 3

6. $2y^{-5}y^2$ $\dfrac{2}{y^3}$

7. $\dfrac{y^2}{yy^{-2}}$ y^3

8. $\left(\dfrac{a^2b^{-1}}{4a^3b^{-2}}\right)^{-3}$ $\dfrac{64a^3}{b^3}$

9. What is the volume of a cube that has sides of length $10y^4$ inches? $1{,}000y^{12}$ in.^{3}

10. Rewrite 4^{-2} using a positive exponent and then evaluate the result. $\dfrac{1}{4^2}, \dfrac{1}{16}$

11. ELECTRICITY One ampere (amp) corresponds to the flow of 6,250,000,000,000,000,000 electrons per second past any point in a direct current (DC) circuit. Write this number in scientific notation. 6.25×10^{18}

12. Write 9.3×10^{-5} in standard notation. 0.000093

13. Identify $3x^2 + 2$ as a monomial, binomial, or trinomial. binomial

14. Find the degree of the polynomial $3x^2y^3 + 2x^3y - 5x^2y$. 5th degree

15. If $f(x) = x^2 + x - 2$, find $f(-2)$. 0

16. Simplify $(xy)^2 + 5x^2y^2 - (3x)^2y^2$. $-3x^2y^2$

17. Simplify $-6(x - y) + 2(x + y) - 3(x + 2y)$. $-7x + 2y$

18. Subtract: $2x^2 - 7x + 3$
           $3x^2 - 2x - 1$
         $\overline{-x^2 - 5x + 4}$

In Problems 19–24, find each product.

19. $(-2x^3)(2x^2y)$ $-4x^5y$

20. $3y^2(y^2 - 2y + 3)$ $3y^4 - 6y^3 + 9y^2$

21. $(x - 9)(x + 9)$ $x^2 - 81$

22. $(3y - 4)^2$ $9y^2 - 24y + 16$

23. $(2x - 5)(3x + 4)$ $6x^2 - 7x - 20$

24. $(2x - 3)(x^2 - 2x + 4)$ $2x^3 - 7x^2 + 14x - 12$

25. Solve the equation $(a + 2)^2 = (a - 3)^2$. $\frac{1}{2}$

26. Simplify $\dfrac{8x^2y^3z^4}{16x^3y^2z^4}$ $\frac{y}{2x}$

27. Simplify $\dfrac{6a^2 - 12b^2}{24ab}$. $\frac{a}{4b} - \frac{b}{2a}$

28. Divide $2x + 3\overline{)2x^2 - x - 6}$. $x - 2$

29. In your own words, explain this rule for exponents:

$$x^{-n} = \frac{1}{x^n}$$

30. A rectangle has an area of $(x^2 - 6x + 5)$ ft^2 and a length of $(x - 1)$ feet. Show how division can be used to find the width of the rectangle. Explain your steps. $(x - 5)$ ft

31. Complete the table and then graph the polynomial function $f(x) = -x^3 + 3x^2 - 5$.

x	$f(x)$	(x, y)
-2	15	$(-2, 15)$
-1	-1	$(-1, -1)$
0	-5	$(0, -5)$
1	-3	$(1, -3)$
2	-1	$(2, -1)$
3	-5	$(3, -5)$

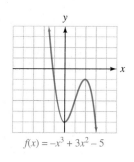

$f(x) = -x^3 + 3x^2 - 5$

32. ROCK SALT Illustration 1 shows U.S. sales of rock salt, which is used for road de-icing. Find the coordinates of the lowest and highest points on the graph and explain their significance.

$(1, 9), (6, 21)$; the smallest number of tons, about 9 million, was sold in 1991. The largest number of tons, about 21 million, was sold in 1996.

ILLUSTRATION 1

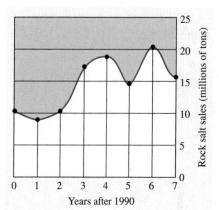

CHAPTERS 1–4

Cumulative Review Exercises

1. PERSONAL SAVINGS RATE The graph in Illustration 1 shows a situation occurring in September of 1998 that hadn't occurred since the Great Depression. Explain what was unusual.

The negative savings rate means that Americans spent more than they earned that month.

ILLUSTRATION 1

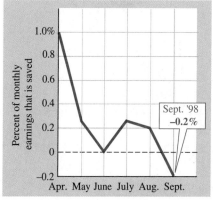

Based on data from the U.S. Department of Commerce

2. CLINICAL TRIALS In a clinical test of Aricept, a drug to treat Alzheimer's disease, one group of patients took a placebo (a sugar pill) while another group took the actual medication. See Illustration 2. Find the number of patients in each group who experienced nausea. Round to the nearest whole number.   19, 16

ILLUSTRATION 2
Comparison of rates of adverse events in patients

Adverse event	Group 1—Placebo (number = 315)	Group 2—Aricept (number = 311)
Nausea	6%	5%

In Exercises 3–4, consider the algebraic expression $3x^3 + 5x^2y + 37y$.

3. Find the coefficient of the second term.   5

4. What is the third term? $37y$

In Exercises 5–8, simplify each expression.

5. $3x - 5x + 2y$ $-2x + 2y$

6. $3(x - 7) + 2(8 - x)$ $x - 5$

7. $2x^2y^3 - xy(xy^2)$ x^2y^3

8. $x^2(3 - y) + x(xy + x)$ $4x^2$

In Exercises 9–10, solve each equation.

9. $3(x - 5) + 2 = 2x$ 13   **10.** $\dfrac{x - 5}{3} - 5 = 7$ 41

In Exercises 11–12, solve each formula for the variable indicated.

11. $A = \dfrac{1}{2}h(b + B)$; for h $h = \frac{2A}{b + B}$

12. $y = mx + b$; for x $x = \frac{y - b}{m}$

In Exercises 13–16, evaluate each expression.

13. $4^2 - 5^2$ -9 **14.** $(4 - 5)^2$ 1

15. $\dfrac{-3 - (-7)}{2^2 - 3}$ 4   **16.** $12 - 2[1 - (-8 + 2)]$ -2

In Exercises 17–18, solve each inequality and graph the solution set.

17. $8(4 + x) > 10(6 + x)$

$x < -14$

18. $-9 < 3(x + 2) \le 3$

$-5 < x \le -1$

In Exercises 19–22, graph each equation.

19. $y = x^2$

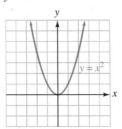

20. $y = |x|$

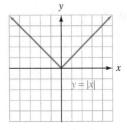

21. $4x - 3y = 12$

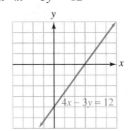

22. $3x = 12$

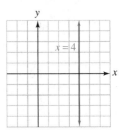

In Exercises 23–26, find the slope of the line with the given properties.

23. Passing through $(-2, 4)$ and $(6, 8)$ $\frac{1}{2}$

24. A line that is horizontal   0

25. An equation of $y = -4x + 3$ -4

26. An equation of $2x - 3y = 12$ $\frac{2}{3}$

In Exercises 27–30, write the equation of the line with the following properties.

27. Slope $= \dfrac{2}{3}$, y-intercept $= (0, 5)$ $y = \frac{2}{3}x + 5$

28. Passing through $(-2, 4)$ and $(6, 10)$ $3x - 4y = -22$

29. A horizontal line passing through $(2, 4)$ $y = 4$

30. A vertical line passing through $(2, 4)$ $x = 2$

In Exercises 31–32, are the graphs of the lines parallel or perpendicular?

31. $\begin{cases} y = -\dfrac{3}{4}x + \dfrac{15}{4} \\ 4x - 3y = 25 \end{cases}$ perpendicular

32. $\begin{cases} y = -\dfrac{3}{4}x + \dfrac{15}{4} \\ 6x = 15 - 8y \end{cases}$ parallel

In Exercises 33–34, tell whether each equation defines a function.

33. $y = x^3 - 4$ yes   **34.** $x = |y|$ no

In Exercises 35–38, $f(x) = 2x^2 - 3$. Find each value.

35. $f(0)$ -3 **36.** $f(3)$ 15

37. $f(-2)$ 5   **38.** $f(0.5)$ -2.5

39. Find the domain and range of the function graphed in Illustration 3.

D: all real numbers, R: real numbers greater than or equal to 0

ILLUSTRATION 3

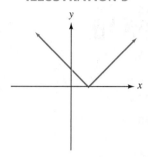

40. Tell whether the graph is the graph of a function. no

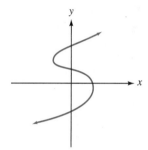

41. Give an example of a positive rate of change and a negative rate of change.

The temperature is rising at a rate of 3°/hr; the temperature is falling at a rate of −3°/hr.

42. Assume that y varies directly with x. If $y = 4$ when $x = 10$, find y when $x = 30$. 12

In Exercises 43–50, write each expression using one positive exponent.

43. $(y^3y^5)y^6$ y^{14}

44. $(x^3x^4)^2$ x^{14}

45. $\dfrac{x^3x^4}{x^2x^3}$ x^2

46. $\dfrac{a^4b^0}{a^{-3}}$ a^7

47. x^{-5} $\dfrac{1}{x^5}$

48. $(-2y)^{-4}$ $\dfrac{1}{16y^4}$

49. $(x^{-4})^2$ $\dfrac{1}{x^8}$

50. $\left(-\dfrac{x^3}{x^{-2}}\right)^3$ $-x^{15}$

In Exercises 51–52, write each number in scientific notation.

51. 615,000 6.15×10^5

52. 0.0000013 1.3×10^{-6}

In Exercises 53–54, write each number in standard notation.

53. 5.25×10^{-4} 0.000525

54. 2.77×10^3 2,770

In Exercises 55–56, give the degree of each polynomial.

55. $3x^2 + 2x - 5$ 2

56. $-3x^3y^2 + 3x^2y^2 - xy$ 5

57. MUSICAL INSTRUMENTS The gong shown in Illustration 4 is a percussion instrument used throughout Southeast Asia. The amount of deflection of the horizontal support (in inches) is given by the polynomial function

$$f(x) = 0.01875x^4 - 0.15x^3 + 1.2x$$

where x is the distance (in feet) that the gong is hung from one end of the support. Find the deflection if the gong is hung in the middle of the support. 1.5 in.

ILLUSTRATION 4

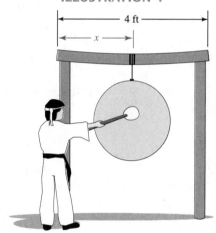

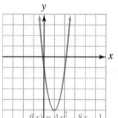

58. Complete the table and then graph the function $f(x) = 4x^2 - 8x - 1$.

x	$f(x)$	(x, y)
-1	11	$(-1, 11)$
0	-1	$(0, -1)$
1	-5	$(1, -5)$
2	-1	$(2, -1)$
3	11	$(3, 11)$

$f(x) = 4x^2 - 8x - 1$

In Exercises 59–68, do the operations.

59. $(3x^2 + 2x - 7) - (2x^2 - 2x + 7)$ $x^2 + 4x - 14$

60. $(2x^2 - 3x + 4) + (2x^2 + 2x - 5)$ $4x^2 - x - 1$

61. $-5x^2(7x^3 - 2x^2 - 2)$ $-35x^5 + 10x^4 + 10x^2$

62. $(3x^3y^2)(-4x^2y^3)$ $-12x^5y^5$

63. $(3x - 7)(2x + 8)$ $6x^2 + 10x - 56$

64. $(5x - 4y)(3x + 2y)$ $15x^2 - 2xy - 8y^2$

65. $(3x + 1)^2$ $9x^2 + 6x + 1$

66. $(x - 2)(x^2 + 2x + 4)$ $x^3 - 8$

67. $\dfrac{6x^2 - 8x}{2x}$ $3x - 4$

68. $x - 3\overline{)2x^2 - 5x - 3}$ $2x + 1$

5

Roots and Radicals

CAMPUS CONNECTION

The Electronics Department

To determine the power used by a household iron, electronics students use the *radical* equation

$$I = \sqrt{\dfrac{P}{R}}$$

where I is the current (in amperes), R is the resistance (in ohms), and P is the power (in watts). Electronics students need a solid understanding of algebraic concepts, because they encounter a variety of formulas and must solve several different types of equations. In this chapter, we introduce radicals and radical equations and show how they have applications in fields such as science, engineering, and electronics.

To solve many applied problems, we must determine what number x must be squared to obtain another number n. We call x the square root of n.

▶ 5.1 Square Roots and the Pythagorean Theorem

In this section, you will learn about

Square roots ■ Approximating square roots ■ Rational, irrational, and imaginary numbers ■ The square root function ■ Right triangles ■ The Pythagorean theorem

FIGURE 5-1

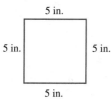

5 in.

5 in.   5 in.

5 in.

Introduction To find the area A of the square shown in Figure 5-1, we multiply its length by its width.

$$A = l \cdot w$$
$$A = 5 \cdot 5$$
$$= 25$$

The area is 25 square inches.

We have seen that the product $5 \cdot 5$ can be denoted by the exponential expression 5^2, where 5 is raised to the second power. Whenever we raise a number to the second power, we are squaring it, or finding its **square.** This example illustrates that the formula for the area of a square with sides of length s is $A = s^2$.

Here are some more squares of numbers:

■ The square of 3 is 9, because $3^2 = 9$.

■ The square of -3 is 9, because $(-3)^2 = 9$.

■ The square of 12 is 144, because $12^2 = 144$.

■ The square of -12 is 144, because $(-12)^2 = 144$.

■ The square of $\frac{1}{8}$ is $\frac{1}{64}$, because $\left(\frac{1}{8}\right)^2 = \frac{1}{8} \cdot \frac{1}{8} = \frac{1}{64}$.

■ The square of $-\frac{1}{8}$ is $\frac{1}{64}$, because $\left(-\frac{1}{8}\right)^2 = \left(-\frac{1}{8}\right)\left(-\frac{1}{8}\right) = \frac{1}{64}$.

■ The square of 0 is 0, because $0^2 = 0$.

In this section, we will reverse the squaring process and find **square roots** of numbers. We will consider the square root function and introduce the Pythagorean theorem. Finally, we will solve several application problems.

Square Roots

FIGURE 5-2

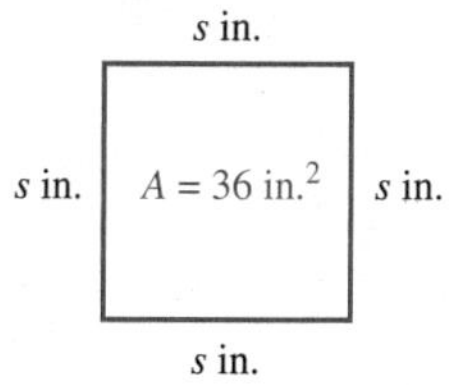

s in.

s in. $A = 36$ in.^{2} s in.

s in.

Suppose we know that the area of the square shown in Figure 5-2 is 36 square inches. To find the length of each side, we substitute 36 for A in the formula $A = s^2$ and solve for s.

$$A = s^2$$
$$36 = s^2$$

To solve for s, we must find a positive number whose square is 36. Since 6 is such a number, the sides of the square are 6 inches long. The number 6 is called a *square root* of 36, because 6 is the positive number that we square to get 36.

- 3 is a square root of 9, because $3^2 = 9$.
- -3 is a square root of 9, because $(-3)^2 = 9$.
- 12 is a square root of 144, because $12^2 = 144$.
- -12 is a square root of 144, because $(-12)^2 = 144$.
- $\frac{1}{8}$ is a square root of $\frac{1}{64}$, because $\left(\frac{1}{8}\right)^2 = \left(\frac{1}{8}\right)\left(\frac{1}{8}\right) = \frac{1}{64}$.
- $-\frac{1}{8}$ is a square root of $\frac{1}{64}$, because $\left(-\frac{1}{8}\right)^2 = \left(-\frac{1}{8}\right)\left(-\frac{1}{8}\right) = \frac{1}{64}$.
- 0 is a square root of 0, because $0^2 = 0$.

In general, we have the following definition.

Square root

The number b is a **square root** of a if $b^2 = a$.

All positive numbers have two square roots, one that is positive and one that is negative. The two square roots of 9 are 3 and -3, and the two square roots of 144 are 12 and -12. The number 0 is the only number that has one square root, which is 0.

The **principal square root** of a positive number is its positive square root. Although 3 and -3 are both square roots of 9, only 3 is the principal square root. The symbol $\sqrt{}$, called a **radical sign,** is used to represent the principal square root of a number, and $-\sqrt{}$ is used to represent the negative square root of a number. For example, $\sqrt{9} = 3$ and $-\sqrt{9} = -3$. Likewise, $\sqrt{144} = 12$ and $-\sqrt{144} = -12$.

Principal square root

If $a > 0$, the expression $\sqrt{a}$ represents the principal (or positive) square root of a.

The principal square root of 0 is 0: $\sqrt{0} = 0$.

The number (or expression) under a radical sign is called the **radicand.** In $\sqrt{9}$, 9 is the radicand, and the entire symbol $\sqrt{9}$ is called a **radical.** We read $\sqrt{9}$ as either "the square root of 9" or as "radical 9."

An algebraic expression containing a radical is called a **radical expression.** In this chapter, we will consider radical expressions such as

$$\sqrt{49}, \qquad \frac{5}{\sqrt{3}}, \qquad -2\sqrt{x + 1}, \qquad \text{and} \qquad \sqrt{28y^2} - 2y\sqrt{63}$$

EXAMPLE 1   **Finding square roots.** Find each square root.

a. $\sqrt{0} = 0$ **b.** $\sqrt{1} = 1$ **c.** $\sqrt{225} = 15$ **d.** $\sqrt{1.44} = 1.2$

e. $\sqrt{576} = 24$ **f.** $\sqrt{1,600} = 40$ **g.** $-\sqrt{4} = -2$ **h.** $-\sqrt{900} = -30$

i. $\sqrt{\dfrac{1}{4}} = \dfrac{1}{2}$ **j.** $\sqrt{\dfrac{4}{9}} = \dfrac{2}{3}$

SELF CHECK   Find each square root: **a.** $\sqrt{121}$, **b.** $-\sqrt{49}$, **c.** $\sqrt{0.64}$, **d.** $\sqrt{256}$, **e.** $\sqrt{\frac{1}{25}}$, and **f.** $\sqrt{\frac{9}{49}}$.   *Answers:* **a.** 11, **b.** -7, **c.** 0.8, **d.** 16, **e.** $\frac{1}{5}$, **f.** $\frac{3}{7}$ ∎

Square roots of certain numbers, such as 7, are hard to compute by hand. However, we can find $\sqrt{7}$ with a calculator or with a table of square roots.

Approximating Square Roots

TABLE 5-1

n	n^2	$\sqrt{n}$	n^3	$\sqrt[3]{n}$
5	25	2.236	125	1.710
6	36	2.449	216	1.817
7	49	2.646	343	1.913
8	64	2.828	512	2.000

To find the principal square root of 7, we can enter 7 into a scientific calculator and press the $\boxed{\sqrt{x}}$ key. The approximate value of $\sqrt{7}$ will appear on the display.

$\sqrt{7} \approx 2.6457513$ Read $\approx$ as "is approximately equal to."

To find the principal square root of 7, we can also look in a table of square roots, a portion of which is shown in Table 5-1. In the left column, headed by n, we locate the number 7. The column headed $\sqrt{n}$ contains the approximate value of $\sqrt{7}$.

$\sqrt{7} \approx 2.646$

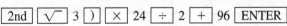

ACCENT ON TECHNOLOGY *Freeway Road Signs*

To find the total height of the sign and post shown in Figure 5-3, we observe that

The total height	is	the height of the sign	plus	the height of the post.

The sign is in the shape of an equilateral triangle, and we can find its height h using the formula

$$h = \frac{\sqrt{3}s}{2}$$

where s is the length of a side of the triangle. In this case, $s = 24$ inches, so we have

$$\text{Total height} = \frac{\sqrt{3}(24)}{2} + 96$$

To evaluate this expression with a scientific calcula-tor, we enter these numbers and press these keys:

FIGURE 5-3

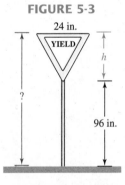

24 in.

YIELD

h

?

96 in.

Keystrokes $\boxed{(}\ 3\ \boxed{\sqrt{x}}\ \boxed{\times}\ 24\ \boxed{)}\ \boxed{\div}\ 2\ \boxed{+}\ 96\ \boxed{=}$ $\boxed{\text{116.7846097}}$

To evaluate this expression using a graphing calculator, we press these keys:

Keystrokes $\boxed{\text{2nd}}\ \boxed{\sqrt{}}\ 3\ \boxed{)}\ \boxed{\times}\ 24\ \boxed{\div}\ 2\ \boxed{+}\ 96\ \boxed{\text{ENTER}}$

$\boxed{\begin{array}{l}\sqrt{}(3)*24/2+96 \\ \quad\quad 116.7846097\end{array}}$

The total height of the sign and the post is approximately 116.8 inches.

Rational, Irrational, and Imaginary Numbers

Whole numbers such as 4, 9, 16, and 49 are called **integer squares,** because each one is the square of an integer. The square root of any integer square is an integer and there-fore a rational number:

$\sqrt{4} = 2$, $\sqrt{9} = 3$, $\sqrt{16} = 4$,   and $\sqrt{49} = 7$

The square root of any whole number that is not an integer square is an **irrational number.** For example, $\sqrt{7}$ is an irrational number. Recall that the set of rational num-bers and the set of irrational numbers together make up the set of real numbers.

WARNING! Square roots of negative numbers are not real numbers. For example, $\sqrt{-4}$ is nonreal, because the square of no real number is -4. The number $\sqrt{-4}$ is an example from a set of numbers called **imaginary numbers.** Remember: *The square root of a negative number is not a real number.*

If we attempt to evaluate $\sqrt{-4}$ using a calculator, an error message like the ones shown below will be displayed.

Error	ERR:NONREAL ANS **1:** Quit 2: Goto
Scientific calculator	**Graphing calculator**

In this chapter, we will assume that *all radicands under the square root symbols are either positive or zero.* Thus, all square roots will be real numbers.

The Square Root Function

Since there is one principal square root for every nonnegative real number x, the equation $f(x) = \sqrt{x}$ determines a square root function. For example, the value that is determined by $f(x) = \sqrt{x}$ when $x = 4$ is denoted by $f(4)$, and we have $f(4) = \sqrt{4} = 2$.

To graph this function, we make a table of values and plot each ordered pair. In the table, we chose five values for x that are perfect squares. This made computing $f(x)$ quite simple. The graph appears in Figure 5-4.

FIGURE 5-4

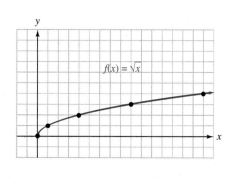

$$f(x) = \sqrt{x}$$

x	$f(x)$	$(x, f(x))$
0	0	$(0, 0)$
1	1	$(1, 1)$
4	2	$(4, 2)$
9	3	$(9, 3)$
16	4	$(16, 4)$

Values to be input into $\sqrt{x}$ Output values   Ordered pairs to plot

ACCENT ON TECHNOLOGY *Graphing Square Root Functions*

We can use a graphing calculator to generate tables and graphs for square root functions. For example, Figure 5-5(a) shows how to enter the function $f(x) = \sqrt{x}$. In Figure 5-5(b), the Indpnt Ask option in TABLSET mode has been used to select specific x-values of -1, 0, 1, 4, 9, and 16. We entered $x = -1$ to illustrate that the calculator displays an error message when the output $\left(\text{in this case, } \sqrt{-1}\right)$ is not a real number. Figure 5-5(c) shows the graph, using window settings of $x = -1$ to 16 and $y = -5$ to 5.

FIGURE 5-5

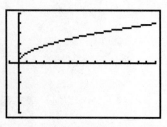

(a)   (b)   (c)

<table>
<tr><td>E X A M P L E   2</td></tr>
</table>

Period of a pendulum. The *period* of a pendulum is the time required for the pendulum to swing back and forth to complete one cycle. (See Figure 5-6.) The period (in seconds) of a pendulum having length L (in feet) is approximated by the function

FIGURE 5-6

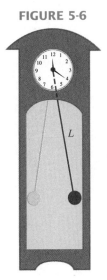

$$f(L) = 1.11\sqrt{L}$$

Find the period of a pendulum that is 5 feet long.

Solution   We substitute 5 for L in the formula and multiply using a calculator.

$$f(L) = 1.11\sqrt{L}$$
$$f(5) = 1.11\sqrt{5} \qquad 1.11\sqrt{5} \text{ means } 1.11 \cdot \sqrt{5}.$$
$$\approx 2.482035455$$

The period is approximately 2.5 seconds.

SELF CHECK   Find the period of a pendulum that is 3 feet long.   *Answer:* about 1.9 sec   ■

FIGURE 5-7

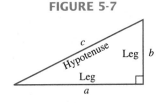

Right Triangles

A triangle that contains a 90° angle is called a **right triangle.** The longest side of a right triangle is the **hypotenuse,** which is the side opposite the right angle. The remaining two sides are the **legs** of the triangle. In the right triangle shown in Figure 5-7, side c is the hypotenuse and sides a and b are legs.

The Pythagorean Theorem

The **Pythagorean theorem** provides a formula relating the lengths of the three sides of a right triangle.

The Pythagorean theorem

If the length of the hypotenuse of a right triangle is c and the lengths of the two legs are a and b, then

$$c^2 = a^2 + b^2$$

Since the lengths of the sides of a triangle are positive numbers, we can use the **square root property of equality** and the Pythagorean theorem to find the length of the third side of any right triangle when the measures of two sides are given.

Square root property of equality

If a and b are positive numbers, then

if $a = b$, then $\sqrt{a} = \sqrt{b}$.

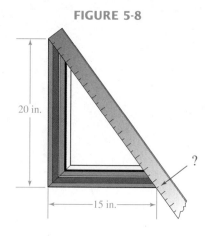

FIGURE 5-8

EXAMPLE 3

Picture frame. After gluing and nailing two pieces of picture frame molding together, a frame maker checks her work by making a diagonal measurement. (See Figure 5-8.) If the sides of the frame form a right angle, what measurement should the frame maker read on the yardstick?

Solution   If the sides of the frame form a right angle, the sides and the diagonal form a right triangle. The lengths of the legs of the right triangle are 15 inches and 20 inches. We can find c, the length of the hypotenuse, using the Pythagorean theorem.

$c^2 = a^2 + b^2$ The Pythagorean theorem.

$c^2 = 15^2 + 20^2$ Substitute 15 for a and 20 for b.

$c^2 = 225 + 400$ $15^2 = 225$ and $20^2 = 400$.

$c^2 = 625$ Do the addition: $225 + 400 = 625$.

$\sqrt{c^2} = \sqrt{625}$ To find c, we isolate it by "undoing" the operation performed on it. Since c is squared, we take the positive square root of both sides.

$c = 25$ $\sqrt{625} = 25$ and $\sqrt{c^2} = c$, because $(c)^2 = c^2$.

The diagonal distance should measure 25 inches. If it does not, the sides of the frame do not form a right angle.   ∎

EXAMPLE 4

Building a high ropes adventure course. The builder of a high ropes course wants to use a 25-foot cable to stabilize the pole shown in Figure 5-9. To be safe, the ground anchor must be farther than 18 feet from the base of the pole. Is the cable long enough to use?

Solution   We can use the Pythagorean theorem, with $b = 16$ and $c = 25$, to find a.

FIGURE 5-9

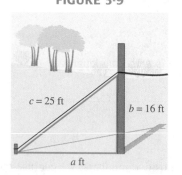

$c^2 = a^2 + b^2$

$25^2 = a^2 + 16^2$ Substitute 25 for c and 16 for b.

$625 = a^2 + 256$ $25^2 = 625$ and $16^2 = 256$.

$369 = a^2$ To isolate a^2, subtract 256 from both sides.

$\sqrt{369} = \sqrt{a^2}$ To find a, we undo the operation that is performed on it (squaring) by taking the positive square root of both sides.

$19.209372 \approx a$ Use a calculator to approximate $\sqrt{369}$.

Since the anchor will be more than 18 feet from the base, the cable is long enough.   ∎

EXAMPLE 5

Reach of a ladder. A 26-foot ladder rests against the side of a building. If the base of the ladder is 10 feet from the wall, how far up the building will the ladder reach?

ANALYZE THE PROBLEM   The wall, the ground, and the ladder form a right triangle, as shown in Figure 5-10. In this triangle, the hypotenuse is 26 feet, and one of the legs is the base-to-wall distance of 10 feet. We can let x represent the length of the other leg, which is the distance that the ladder will reach up the wall.

FORM AN EQUATION   We can use the Pythagorean theorem to form the equation.

The hypotenuse squared	is	one leg squared	plus	the other leg squared.
26^2	$=$	10^2	$+$	x^2

SOLVE THE EQUATION

FIGURE 5-10

$$26^2 = 10^2 + x^2$$

$676 = 100 + x^2$ $26^2 = 676$ and $10^2 = 100$.

$676 - 100 = x^2$ To isolate x^2, subtract 100 from both sides.

$576 = x^2$ $676 - 100 = 576$.

$\sqrt{576} = \sqrt{x^2}$ Take the positive square root of both sides.

$24 = x$ $\sqrt{576} = 24$ and $\sqrt{x^2} = x$, because $x \cdot x = x^2$.

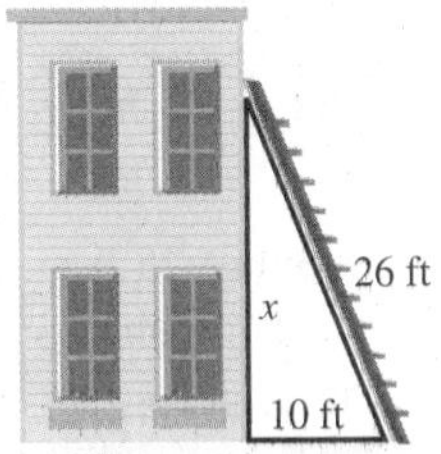

STATE THE CONCLUSION   The ladder will reach 24 feet up the side of the building.

EXAMPLE 6

Roof design. The gable end of the roof shown in Figure 5-11 is an isosceles right triangle with a span of 48 feet. Find the distance from the eaves to the peak.

ANALYZE THE PROBLEM   The two equal sides of the isosceles triangle are the two legs of the right triangle, and the span of 48 feet is the length of the hypotenuse. We can let x represent the length of each leg, which is the distance from eaves to peak.

FORM AN EQUATION   We can use the Pythagorean theorem to form the equation.

The hypotenuse squared	is	one leg squared	plus	the other leg squared.
48^2	$=$	x^2	$+$	x^2

SOLVE THE EQUATION

FIGURE 5-11

$$48^2 = x^2 + x^2$$

$2{,}304 = 2x^2$ $48^2 = 2{,}304$ and $x^2 + x^2 = 2x^2$.

$1{,}152 = x^2$ To isolate x^2, divide both sides by 2.

$\sqrt{1{,}152} = \sqrt{x^2}$ Take the positive square root of both sides.

$33.9411255 \approx x$ Use a calculator to find the approximate value of $\sqrt{1{,}152}$.

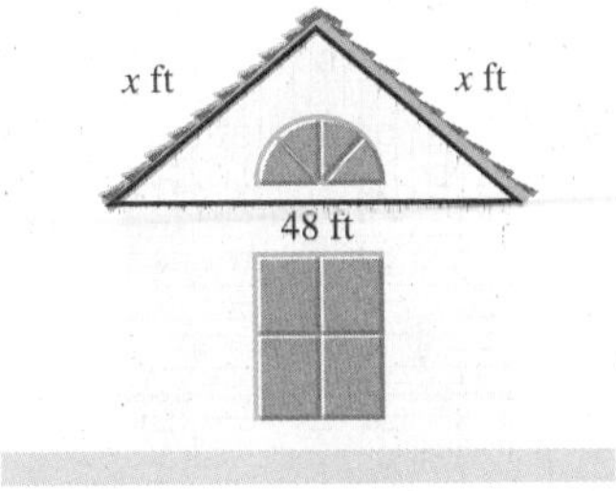

STATE THE CONCLUSION   The eaves-to-peak distance of the roof is approximately 34 feet.

STUDY SET

Section 5.1

VOCABULARY

In Exercises 1–6, fill in the blanks to make the statements true.

1. b is a ___square root___ of a if $b^2 = a$.

2. The symbol $\sqrt{}$ is called a ___radical___ sign.

3. The principal square root of a positive number is a ___positive___ number.

4. The number under the radical sign is called the ___radicand___.

5. If a triangle has a right angle, it is called a
_____right_____ triangle.

6. The longest side of a right triangle is called the
_____hypotenuse_____, and the other two sides are called
_____legs_____.

CONCEPTS

In Exercises 7–12, fill in the blanks to make the statements true.

7. The number 25 has _____two_____ square roots.
They are ⟨5⟩ and ⟨−5⟩.

8. $\sqrt{-11}$ is not a _____real_____ number.

9. If the length of the hypotenuse of a right triangle is c
and the legs are a and b, then $c^2 = $ ⟨$a^2 + b^2$⟩.

10. ⟨The hypotenuse squared⟩ is ⟨one leg squared⟩ plus ⟨the other leg squared⟩.

11. If a and b are positive numbers and $a = b$, then
$\sqrt{a} = $ ⟨$\sqrt{b}$⟩.

12. The _____square_____ of 2 is 4, because $2^2 = 4$, and
2 is a _____square root_____ of 4, because $2^2 = 4$.

13. To isolate x, what step should be used to "undo" the
operation performed on it? (Assume that x is a positive number.)
a. $2x = 16$ Divide both sides by 2.
b. $x^2 = 16$ Take the positive square root of both sides.

14. Graph each number on the number line.
$$\left\{ \sqrt{16}, -\sqrt{\tfrac{9}{4}}, \sqrt{1.8}, \sqrt{6}, -\sqrt{23} \right\}$$

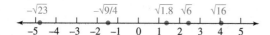

15. Complete the table of values.
Do not use a calculator.

x	$\sqrt{x}$
0	0
$\frac{1}{81}$	$\frac{1}{9}$
0.16	0.4
36	6
400	20

16. If $f(x) = \sqrt{x}$, find each value. Do not use a calculator.
a. $f\left(\frac{1}{121}\right)$ $\frac{1}{11}$ **b.** $f(1)$ 1
c. $f(0.25)$ 0.5   **d.** $f(81)$ 9
e. $f(900)$ 30

17. a. What do the dashed lines in the graph in Illustration 1 help to approximate? $\sqrt{5} \approx 2.2$
b. Use the graph to approximate $\sqrt{3}$ and $\sqrt{8}$.
$\sqrt{3} \approx 1.7$; $\sqrt{8} \approx 2.8$

ILLUSTRATION 1

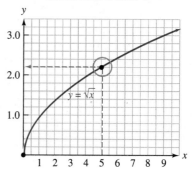

18. A table for the square root function $f(x) = \sqrt{x}$ is
shown in Illustration 2. Explain why the word
"ERROR" appears in the output column.
$\sqrt{-1}$ is not a real number.

ILLUSTRATION 2

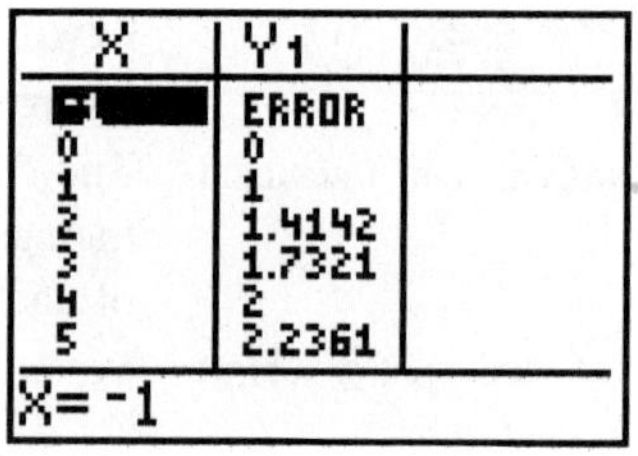

X	Y₁	
−1	ERROR	
0	0	
1	1	
2	1.4142	
3	1.7321	
4	2	
5	2.2361	

X = −1

NOTATION

In Exercises 19–20, complete each solution.

19. If the legs of a right triangle measure 5 and 12 centimeters, find the length of the hypotenuse.
$$c^2 = a^2 + b^2$$
$$c^2 = \boxed{5}^2 + \boxed{12}^2$$
$$c^2 = 25 + \boxed{144}$$
$$c^2 = \boxed{169}$$
$$\boxed{\sqrt{c^2}} = \sqrt{169}$$
$$c = 13$$

20. If the hypotenuse of a right triangle measures 25 centimeters and one leg measures 24 centimeters, find
the length of the other leg.
$$c^2 = a^2 + b^2$$
$$\boxed{25}^2 = \boxed{24}^2 + b^2$$
$$625 = \boxed{576} + b^2$$
$$\boxed{49} = b^2$$
$$\sqrt{49} = \boxed{\sqrt{b^2}}$$
$$7 = b$$

21. Is the statement $-\sqrt{9} = \sqrt{-9}$ true or false? Explain
your answer.
False; $-\sqrt{9} = -3$, $\sqrt{-9}$ is not a real number.

22. Consider the statement $\sqrt{26} \approx 5.1$. Explain why an $\approx$ symbol is used instead of an $=$ sign.

Since $(5.1)^2$ is 26.01, rather than exactly 26, we write $\sqrt{26} \approx 5.1$.

PRACTICE

In Exercises 23–38, find each square root without using a calculator.

23. $\sqrt{25}$ 5

24. $\sqrt{49}$ 7

25. $-\sqrt{81}$ -9

26. $-\sqrt{36}$ -6

27. $\sqrt{1.21}$ 1.1

28. $\sqrt{1.69}$ 1.3

29. $\sqrt{196}$ 14

30. $\sqrt{169}$ 13

31. $\sqrt{\dfrac{9}{256}}$ $\frac{3}{16}$

32. $\sqrt{\dfrac{49}{225}}$ $\frac{7}{15}$

33. $-\sqrt{289}$ -17

34. $-\sqrt{324}$ -18

35. $-\sqrt{2,500}$ -50

36. $-\sqrt{625}$ -25

37. $\sqrt{3,600}$ 60

38. $\sqrt{1,600}$ 40

In Exercises 39–58, use a calculator to evaluate each expression to three decimal places.

39. $\sqrt{2}$ 1.414

40. $\sqrt{3}$ 1.732

41. $\sqrt{11}$ 3.317

42. $\sqrt{53}$ 7.280

43. $\sqrt{95}$ 9.747

44. $\sqrt{99}$ 9.950

45. $\sqrt{428}$ 20.688

46. $\sqrt{844}$ 29.052

47. $-\sqrt{9,876}$ -99.378

48. $-\sqrt{3,619}$ -60.158

49. $\sqrt{21.35}$ 4.621

50. $\sqrt{13.78}$ 3.712

51. $\sqrt{0.3588}$ 0.599

52. $\sqrt{0.9999}$ 1.000

53. $-\sqrt{0.8372}$ -0.915

54. $-\sqrt{0.4279}$ -0.654

55. $2\sqrt{3}$ 3.464

56. $3\sqrt{2}$ 4.243

57. $\dfrac{2 + \sqrt{3}}{2}$ 1.866

58. $\dfrac{2 - \sqrt{3}}{2}$ 0.134

In Exercises 59–60, tell whether each number in each set is rational, irrational, or imaginary.

59. $\left\{\sqrt{9}, \sqrt{17}, \sqrt{49}, \sqrt{-49}\right\}$

rational; irrational; rational; imaginary

60. $\left\{-\sqrt{5}, \sqrt{0}, \sqrt{-100}, -\sqrt{225}\right\}$

irrational; rational; imaginary; rational

In Exercises 61–64, complete the table and then graph the function. Check your work with a graphing calculator.

61. $f(x) = 1 + \sqrt{x}$

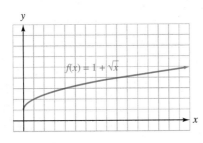

x	$f(x)$
0	1
1	2
4	3
9	4
16	5

62. $f(x) = -1 + \sqrt{x}$

x	$f(x)$
0	-1
1	0
4	1
9	2
16	3

63. $f(x) = -\sqrt{x}$

x	$f(x)$
0	0
1	-1
4	-2
9	-3
16	-4

64. $f(x) = 1 - \sqrt{x}$

x	$f(x)$
0	1
1	0
4	-1
9	-2
16	-3

 In Exercises 65–66, use a graphing calculator to graph the square root function. Use window settings of $x = -5$ to 16 and $y = -5$ to 5.

65. $f(x) = \sqrt{x + 3}$

66. $f(x) = -\sqrt{x - 2}$

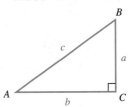

In Exercises 67–72, refer to the right triangle in Illustration 3. Find the length of the unknown side.

ILLUSTRATION 3

67. Find c where $a = 4$ and $b = 3$.   5

68. Find c where $a = 5$ and $b = 12$.   13

69. Find b where $a = 15$ and $c = 17$.   8

70. Find b where $a = 21$ and $c = 29$.   20

71. Find a where $b = 16$ and $c = 34$.   30
72. Find a where $b = 45$ and $c = 53$.   28

APPLICATIONS

In Exercises 73–88, use a calculator to help solve each problem. If an answer is not exact, give it to the nearest tenth.

73. ADJUSTING A LADDER A 20-foot ladder reaches a window 16 feet above the ground. How far from the wall is the base of the ladder?   12 ft

74. LINE OF SIGHT A movie viewer in a car parked at a drive-in theater sits 600 feet from the base of the vertical screen. What is the line-of-sight distance for the viewer to the middle of the screen, which is 35 feet above the base?   601.0 ft

75. QUALITY CONTROL How can a tool manufacturer use the Pythagorean theorem to verify that the two sides of the carpenter's square shown in Illustration 4 meet to form a 90° angle?

The diagonal measurement should be $\sqrt{16^2 + 30^2} = 34$ in.

ILLUSTRATION 4

76. GARDENING A rectangular garden has sides of 28 and 45 feet. Find the length of a path that extends from one corner to the opposite corner.   53 ft

77. BASEBALL A baseball diamond is a square, with each side 90 feet long, as shown in Illustration 5. How far is it from home plate to second base?
127.3 ft

ILLUSTRATION 5

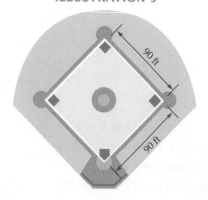

78. TELEVISION The *size* of a television screen is the diagonal distance from the upper left to the lower right corner. What is the size of the screen shown in Illustration 6?   27.0 in.

ILLUSTRATION 6

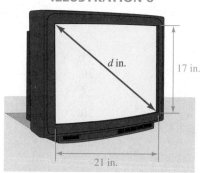

79. FINDING LOCATION A team of archaeologists travels 4.2 miles east and then 4.0 miles north of their base camp to explore some ancient ruins. "As the crow flies," how far from their base camp are they?   5.8 mi

80. TAKING A SHORTCUT Instead of walking on the sidewalk, students take a diagonal shortcut across the vacant lot shown in Illustration 7. How much distance do they save?   44 ft

ILLUSTRATION 7

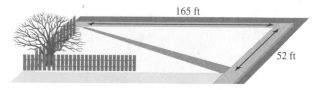

81. FOOTBALL On first down and ten, a quarterback tells his tight end to go out 6 yards, cut 45° to the right, and run 5 yards, as shown in Illustration 8. The tight end follows instructions, catches a pass, and is tackled immediately. Does he gain the necessary 10 yards for a first down?   no

ILLUSTRATION 8

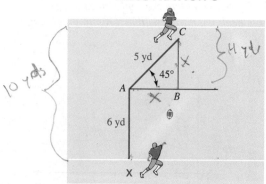

82. GEOMETRY The legs of a right triangle are equal, and the hypotenuse is 2.82843 units long. Find the length of each leg. 2.0 units

83. PROFESSIONAL WRESTLING The sides of a square wrestling ring are 18 feet long. Find the distance from one corner to the opposite corner. 25.5 ft

84. PERIMETER OF A SQUARE The diagonal of a square is 3 feet long. Find its perimeter. 8.5 ft

85. ALTITUDE OF A TRIANGLE Find the area of the isosceles triangle shown in Illustration 9. 240 in.²

ILLUSTRATION 9

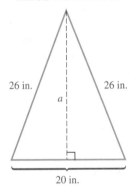

26 in.   26 in.

a

20 in.

86. INTERIOR DECORATING The square table in Illustration 10 is covered by a circular tablecloth. If the sides of the table are 2 feet long, find the area of the tablecloth. 6.3 ft²

ILLUSTRATION 10

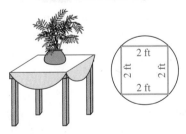

2 ft
2 ft   2 ft
2 ft

87. DRAFTING Among the tools used in drafting are the 30–60–90 and the 45–45–90 triangles shown in Illustration 11.

  a. Find the length of the hypotenuse of the 45–45–90 triangle if it is $\sqrt{2}$ times as long as a leg. 8.5 in.

  b. Find the length of the side opposite the 60° angle of the other triangle if it is $\frac{\sqrt{3}}{2}$ times as long as the hypotenuse. 7.8 in.

ILLUSTRATION 11

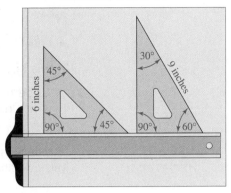

88. ORGAN PIPES The design for a set of brass pipes for a church organ is shown in Illustration 12. Find the length of each pipe (to the nearest tenth of a foot), and then find the total length of pipe needed to construct this set. 2, 2.8, 3.5, 4, 4.5, 4.9, 5.3, 5.7, and 6 ft; 38.7 ft

ILLUSTRATION 12

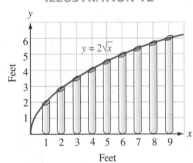

$y = 2\sqrt{x}$

Feet

Feet

WRITING

89. Explain why the square root of a negative number cannot be a real number.

90. Explain the Pythagorean theorem.

91. Suppose you are told that $\sqrt{10} \approx 3.16$. Explain how another key on your calculator, besides the square root key $\sqrt{}$, could be used to see whether this is a reasonable approximation.

92. Explain the difference between the *square* of a number and the *square root* of a number.

REVIEW

93. Add: $(3s^2 - 3s - 2) + (3s^2 + 4s - 3)$. $6s^2 + s - 5$

94. Subtract: $(3c^2 - 2c + 4) - (c^2 - 3c + 7)$. $2c^2 + c - 3$

95. Multiply: $(3x - 2)(x + 4)$. $3x^2 + 10x - 8$

96. Divide: $x^2 + 13x + 12$ by $x + 1$. $x + 12$

*n*th Roots and Radicands That Contain Variables

In this section, you will learn about

> Cube roots ■ Approximating cube roots ■ The cube root function ■ *n*th roots ■ Radicands that contain variables

FIGURE 5-12

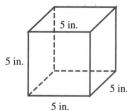

Introduction To find the volume *V* of the cube shown in Figure 5-12, we multiply its length, width, and height.

$$V = l \cdot w \cdot h$$
$$V = 5 \cdot 5 \cdot 5$$
$$= 125$$

The volume is 125 cubic inches.

We have seen that $5 \cdot 5 \cdot 5$ can be denoted by the exponential expression 5^3, where 5 is raised to the third power. Whenever we raise a number to the third power, we are cubing it, or finding its **cube.** This example illustrates that the formula for the volume of a cube with each side of length *s* is $V = s^3$.

Here are some more cubes of numbers:

- The cube of 3 is 27, because $3^3 = 27$.
- The cube of -3 is -27, because $(-3)^3 = -27$.
- The cube of 12 is 1,728, because $12^3 = 1,728$.
- The cube of -12 is $-1,728$, because $(-12)^3 = -1,728$.
- The cube of $\frac{1}{4}$ is $\frac{1}{64}$, because $\left(\frac{1}{4}\right)^3 = \frac{1}{4} \cdot \frac{1}{4} \cdot \frac{1}{4} = \frac{1}{64}$.
- The cube of $-\frac{1}{4}$ is $-\frac{1}{64}$, because $\left(-\frac{1}{4}\right)^3 = \left(-\frac{1}{4}\right)\left(-\frac{1}{4}\right)\left(-\frac{1}{4}\right) = -\frac{1}{64}$.
- The cube of 0 is 0, because $0^3 = 0$.

In this section, we will reverse the cubing process and find **cube roots** of numbers. We will also consider fourth roots, fifth roots, and so on. After graphing the cube root function, we will work with radical expressions having radicands containing variables.

Cube Roots

FIGURE 5-13

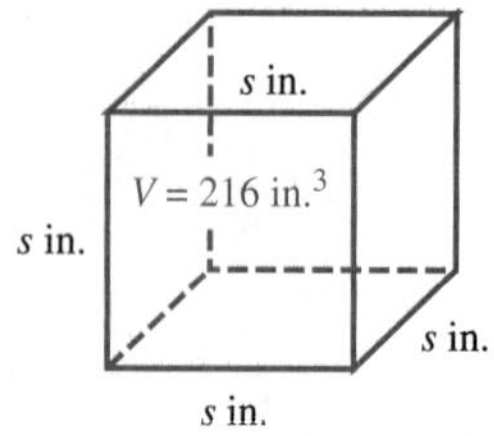

Suppose we know that the volume of the cube shown in Figure 5-13 is 216 cubic inches. To find the length of each side, we substitute 216 for *V* in the formula $V = s^3$ and solve for *s*.

$$V = s^3$$
$$216 = s^3$$

To solve for *s*, we must find a number whose cube is 216. Since 6 is such a number, the sides of the cube are 6 inches long. The number 6 is called a *cube root* of 216, because $6^3 = 216$.

Here are more examples of cube roots:

- 3 is a cube root of 27, because $3^3 = 27$.
- -3 is a cube root of -27, because $(-3)^3 = -27$.
- 12 is a cube root of 1,728, because $12^3 = 1,728$.
- -12 is a cube root of $-1,728$, because $(-12)^3 = -1,728$.
- $\frac{1}{4}$ is a cube root of $\frac{1}{64}$, because $\left(\frac{1}{4}\right)^3 = \left(\frac{1}{4}\right)\left(\frac{1}{4}\right)\left(\frac{1}{4}\right) = \frac{1}{64}$.
- $-\frac{1}{4}$ is a cube root of $-\frac{1}{64}$, because $\left(-\frac{1}{4}\right)^3 = \left(-\frac{1}{4}\right)\left(-\frac{1}{4}\right)\left(-\frac{1}{4}\right) = -\frac{1}{64}$.
- 0 is a cube root of 0, because $0^3 = 0$.

In general, we have the following definition.

Cube root

The number b is a **cube root** of a if $b^3 = a$.

All real numbers have one real cube root. As the preceding examples show, a positive number has a positive cube root, a negative number has a negative cube root, and the cube root of 0 is 0.

Cube root notation

The **cube root of a** is denoted by $\sqrt[3]{a}$. By definition,

$$\sqrt[3]{a} = b \qquad \text{if} \qquad b^3 = a$$

EXAMPLE 1

Finding cube roots. Find each cube root.

a. $\sqrt[3]{8} = 2$, because $2^3 = 8$

b. $\sqrt[3]{343} = 7$, because $7^3 = 343$

c. $\sqrt[3]{-8} = -2$, because $(-2)^3 = -8$

d. $\sqrt[3]{-125} = -5$ because $(-5)^3 = -125$

SELF CHECK Find each cube root: **a.** $\sqrt[3]{64}$, **b.** $\sqrt[3]{-64}$, and **c.** $\sqrt[3]{216}$.   *Answers:* **a.** 4, **b.** -4, **c.** 6   ∎

EXAMPLE 2

Finding cube roots. Find each cube root.

a. $\sqrt[3]{\dfrac{1}{8}} = \dfrac{1}{2}$, because $\left(\dfrac{1}{2}\right)^3 = \dfrac{1}{2} \cdot \dfrac{1}{2} \cdot \dfrac{1}{2} = \dfrac{1}{8}$

b. $\sqrt[3]{-\dfrac{125}{27}} = -\dfrac{5}{3}$, because $\left(-\dfrac{5}{3}\right)^3 = \left(-\dfrac{5}{3}\right)\left(-\dfrac{5}{3}\right)\left(-\dfrac{5}{3}\right) = -\dfrac{125}{27}$

SELF CHECK Find each cube root: **a.** $\sqrt[3]{\frac{1}{27}}$ and **b.** $\sqrt[3]{-\frac{8}{125}}$.   *Answers:* **a.** $\frac{1}{3}$, **b.** $-\frac{2}{5}$ ∎

Cube roots of numbers such as 7 are hard to compute by hand. However, we can find $\sqrt[3]{7}$ with a calculator or with a table of cube roots.

Approximating Cube Roots

To find $\sqrt[3]{7}$, we can enter 7 into a scientific calculator, press the $\boxed{\sqrt[x]{y}}$ key, enter 3, and press the $\boxed{=}$ key. The approximate value of $\sqrt[3]{7}$ will appear on the calculator's display.

$$\sqrt[3]{7} \approx 1.912931183$$

If your scientific calculator doesn't have a $\boxed{\sqrt[x]{y}}$ key, you can use the $\boxed{y^x}$ key. We will see later that $\sqrt[3]{7} = 7^{1/3}$. To find the value of $7^{1/3}$, we enter 7 into the calculator and press these keys:

$$7 \;\; \boxed{y^x} \;\; \boxed{(} \;\; 1 \;\; \boxed{\div} \;\; 3 \;\; \boxed{)} \;\; \boxed{=}$$

The display will read 1.912931183.

To use a graphing calculator to find $\sqrt[3]{7}$, we press $\boxed{\text{MATH}}$, use the arrow down key $\boxed{\blacktriangledown}$ to highlight $\sqrt[3]{\ }($, and press $\boxed{\text{ENTER}}$. We then enter 7, press $\boxed{)}$, and finally press $\boxed{\text{ENTER}}$. As before, we obtain $\sqrt[3]{7} \approx 1.912931183$.

TABLE 5-2

n	n²	√n	n³	∛n
5	25	2.236	125	1.710
6	36	2.449	216	1.817
7	49	2.646	343	1.913
8	64	2.828	512	2.000

To find the cube root of 7, we can also look in a table of cube roots, a portion of which is shown in Table 5-2. In the left column, headed by n, we locate the number 7. The column headed $\sqrt[3]{n}$ contains the approximate value of $\sqrt[3]{7}$.

$$\sqrt[3]{7} \approx 1.913$$

Numbers such as 8, 27, 64, and 125 are called **integer cubes,** because each one is the cube of an integer. The cube root of any integer cube is an integer and therefore a rational number:

$$\sqrt[3]{8} = 2, \qquad \sqrt[3]{27} = 3, \qquad \sqrt[3]{64} = 4, \qquad \text{and} \qquad \sqrt[3]{125} = 5$$

Cube roots of integers such as 7 and -10, that are not integer cubes, are irrational numbers. For example, $\sqrt[3]{7}$ and $\sqrt[3]{10}$ are irrational numbers.

ACCENT ON TECHNOLOGY *Radius of a Water Tank*

Engineers want to design a spherical tank that will hold 33,500 cubic feet of water, as shown in Figure 5-14. They know that the formula for the radius r of a sphere with volume V is given by the formula

$$r = \sqrt[3]{\frac{3V}{4\pi}} \qquad \text{Where } \pi = 3.14159. \ldots$$

FIGURE 5-14

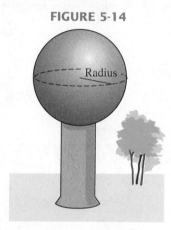

Radius

To use a scientific calculator to find the radius r, we substitute 33,500 for V and enter these numbers and press these keys:

Keystrokes $3 \boxed{\times} 33500 \boxed{\div} \boxed{(} 4 \boxed{\times} \boxed{\pi} \boxed{)} \boxed{=} \boxed{\sqrt[n]{x}} 3 \boxed{=}$

$$\boxed{19.99794636}$$

To evaluate this expression using a graphing calculator, we press the $\boxed{\text{MATH}}$ key. In this mode, arrow down $\boxed{\blacktriangledown}$ to highlight the option $\sqrt[3]{\ }($ and $\boxed{\text{ENTER}}$. Then press the following keys:

Keystrokes $3 \boxed{\times} 33500 \boxed{\div} \boxed{(} 4 \boxed{\times} \boxed{\text{2nd}} \boxed{\pi} \boxed{)} \boxed{)} \boxed{\text{ENTER}}$

```
3√¯(3*33500/(4*π)
)
        19.99794636
```

The result is 19.99794636, so the engineers should design a tank with a radius of 20 feet.

The Cube Root Function

Since every real number has one real-number cube root, there is a cube root function $f(x) = \sqrt[3]{x}$. For example, the value that is determined by $f(x) = \sqrt[3]{x}$ when $x = 8$ is denoted as $f(8)$, and we have $f(8) = \sqrt[3]{8} = 2$.

To graph this function, we substitute numbers for x, compute $f(x)$, plot the resulting ordered pairs, and connect them with a smooth curve, as shown in Figure 5-15.

FIGURE 5-15

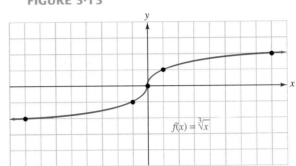

$$f(x) = \sqrt[3]{x}$$

x	$f(x)$	$(x, f(x))$
-8	-2	$(-8, -2)$
-1	-1	$(-1, -1)$
0	0	$(0, 0)$
1	1	$(1, 1)$
8	2	$(8, 2)$

ACCENT ON TECHNOLOGY *Graphing Cube Root Functions*

We can use a graphing calculator to generate tables and graphs for cube root functions. For example, Figure 5-16(a) shows how to enter the function $f(x) = \sqrt[3]{x}$. In Figure 5-16(b), the Indpnt Ask option in the TABLSET mode has been used to select specific x-values of -8, -1, 0, 1, and 8. Figure 5-16(c) shows the graph, using window settings of $x = -10$ to 10 and $y = -5$ to 5.

FIGURE 5-16

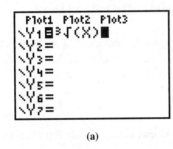

(a)

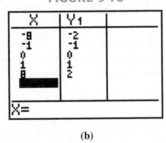

(b)

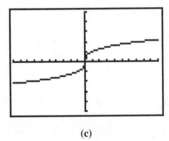

(c)

*n*th Roots

Just as there are square roots and cube roots, there are also fourth roots, fifth roots, sixth roots, and so on. In general, we have this definition.

nth roots of *a*

The **nth root of *a*** is denoted by $\sqrt[n]{a}$, and

$$\sqrt[n]{a} = b \qquad \text{if} \qquad b^n = a$$

The number n is called the **index** of the radical. If n is an even natural number, a must be positive or zero, and b must be positive.

In the square root symbol $\sqrt{}$, the unwritten index is understood to be 2.

$$\sqrt{a} = \sqrt[2]{a}$$

| E X A M P L E   3 | **Finding fourth and fifth roots.** Find each root. |

a. $\sqrt[4]{81} = 3$, because $3^4 = 81$. **b.** $\sqrt[5]{32} = 2$, because $2^5 = 32$.

c. $\sqrt[5]{-32} = -2$, because $(-2)^5 = -32$. **d.** $\sqrt[4]{-81}$ is not a real number, because no number raised to the fourth power is -81.

SELF CHECK Find each root: **a.** $\sqrt[4]{16}$, **b.** $\sqrt[5]{243}$, and **c.** $\sqrt[5]{-1,024}$.

Answers: **a.** 2, **b.** 3, **c.** -4 ∎

| E X A M P L E   4 | **Finding fourth and fifth roots.** Find each root. |

a. $\sqrt[4]{\dfrac{1}{81}} = \dfrac{1}{3}$, because $\left(\dfrac{1}{3}\right)^4 = \dfrac{1}{81}$.

b. $\sqrt[5]{-\dfrac{32}{243}} = -\dfrac{2}{3}$, because $\left(-\dfrac{2}{3}\right)^5 = -\dfrac{32}{243}$.

SELF CHECK Find each root: **a.** $\sqrt[4]{\frac{1}{16}}$ and **b.** $\sqrt[5]{-\frac{243}{32}}$.

Answers: **a.** $\frac{1}{2}$, **b.** $-\frac{3}{2}$ ∎

Radicands That Contain Variables

When n is even and $x \geq 0$, we say that the radical $\sqrt[n]{x}$ represents an **even root.** We can find even roots of many quantities that contain variables, provided that these variables represent positive numbers or zero.

| E X A M P L E   5 | **Finding even roots.** Assume that each variable represents a positive number and find each root. |

a. $\sqrt{x^2} = x$, because $(x)^2 = x^2$. **b.** $\sqrt{x^4} = x^2$, because $(x^2)^2 = x^4$.

c. $\sqrt{x^4y^2} = x^2y$, because $(x^2y)^2 = x^4y^2$. **d.** $\sqrt[4]{81x^{12}} = 3x^3$, because $(3x^3)^4 = 81x^{12}$.

SELF CHECK Find each root: **a.** $\sqrt{a^4}$, **b.** $\sqrt{m^6n^8}$, and **c.** $\sqrt[4]{16y^8}$.

Answers: **a.** a^2, **b.** m^3n^4, **c.** $2y^2$ ∎

When n is odd, we say that the radical expression $\sqrt[n]{x}$ represents an **odd root.**

| E X A M P L E   6 | **Finding odd roots.** Find each root. |

a. $\sqrt[3]{y^3} = y$, because $(y)^3 = y^3$. **b.** $\sqrt[3]{64x^6} = 4x^2$, because $(4x^2)^3 = 64x^6$.

c. $\sqrt[5]{x^{10}} = x^2$, because $(x^2)^5 = x^{10}$.

SELF CHECK Find each root: **a.** $\sqrt[3]{p^6}$, **b.** $\sqrt[3]{-27p^9}$, and **c.** $\sqrt[5]{\frac{1}{32}n^{15}}$.

Answers: **a.** p^2, **b.** $-3p^3$, **c.** $\frac{1}{2}n^3$ ∎

STUDY SET

STUDY SET

Section 5.2

VOCABULARY

In Exercises 1–4, fill in the blanks to make the statements true.

1. If $p^3 = q$, p is called a _____cube_____ root of q.

2. If $p^4 = q$, p is called a _____fourth_____ root of q.

3. We denote the cube root _____function_____ with the notation $f(x) = \sqrt[3]{x}$.

4. If the index of a radical is an even number, the root is called an _____even_____ root.

ILLUSTRATION 1

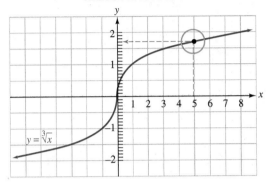

CONCEPTS

In Exercises 5–8, fill in the blanks to make the statements true.

5. The _____cube_____ of -4 is -64, because $(-4)^3 = -64$. -3 is a cube _____root_____ of -27, because $(-3)^3 = -27$.

6. $\sqrt[n]{a} = b$ if $b^n = a$.

7. $\sqrt[3]{-216} = -6$, because $(-6)^3 = -216$.

8. $\sqrt[5]{32x^5} = 2x$, because $(2x)^5 = 32x^5$.

9. To isolate x, what step should be used to "undo" the operation performed on it?
   a. $3x = 27$ Divide both sides by 3.
   b. $x^3 = 27$ Take the cube root of both sides.

10. Graph each number on the number line.
$$\left\{ \sqrt[3]{16}, -\sqrt[4]{100}, \sqrt[3]{-1.8}, \sqrt[4]{0.6} \right\}$$

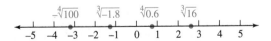

11. If $f(x) = \sqrt[3]{x}$, find each value. Do not use a calculator.
   a. $f(1)$ 1
   b. $f\left(-\frac{1}{27}\right)$ $-\frac{1}{3}$
   c. $f(125)$ 5
   d. $f(0.008)$ 0.2
   e. $f(1,000)$ 10

12. a. What do the dashed lines in the graph in Illustration 1 help to approximate? $\sqrt[3]{5} \approx 1.7$
   b. Use the graph to approximate $\sqrt[3]{4}$ and $\sqrt[3]{-6}$.
   $\sqrt[3]{4} \approx 1.6$; $\sqrt[3]{-6} \approx -1.8$

NOTATION

In Exercises 13–16, fill in the blanks to make the statements true.

13. In the notation $\sqrt[3]{x^6}$, 3 is called the _____index_____ and x^6 is called the _____radicand_____.

14. $\sqrt{}$ is called a _____radical_____ symbol.

15. The "understood" index of the radical expression $\sqrt{55}$ is _2_ .

16. In reading $f(x) = \sqrt[3]{x}$, we say "f _of_ x equals the cube root _of_ x."

PRACTICE

In Exercises 17–32, find each value without using a calculator.

17. $\sqrt[3]{8}$ 2

18. $\sqrt[3]{27}$ 3

19. $\sqrt[3]{0}$ 0

20. $\sqrt[3]{1}$ 1

21. $\sqrt[3]{-8}$ -2

22. $\sqrt[3]{-1}$ -1

23. $\sqrt[3]{-64}$ -4

24. $\sqrt[3]{-27}$ -3

25. $\sqrt[3]{\dfrac{1}{125}}$ $\frac{1}{5}$

26. $\sqrt[3]{\dfrac{1}{1,000}}$ $\frac{1}{10}$

27. $-\sqrt[3]{-1}$ 1

28. $-\sqrt[3]{-27}$ 3

29. $-\sqrt[3]{64}$ -4

30. $-\sqrt[3]{343}$ -7

31. $\sqrt[3]{729}$ 9

32. $\sqrt[3]{512}$ 8

In Exercises 33–36, use a calculator to find each cube root to the nearest hundredth.

33. $\sqrt[3]{32,100}$ 31.78

34. $\sqrt[3]{-25,713}$ -29.52

35. $\sqrt[3]{-0.11324}$ -0.48

36. $\sqrt[3]{0.875}$ 0.96

In Exercises 37–40, complete the table and then graph the function. Check your work with a graphing calculator.

37. $f(x) = \sqrt[3]{x} + 1$

x	$f(x)$
-8	-1
-1	0
0	1
1	2
8	3

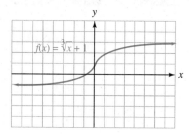

38. $f(x) = \sqrt[4]{x}$

x	$f(x)$
0	0
1	1
16	2

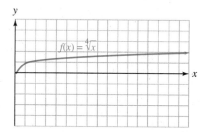

39. $f(x) = -\sqrt[3]{x}$

x	$f(x)$
-8	2
-1	1
0	0
1	-1
8	-2

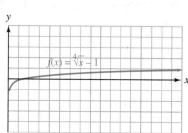

40. $f(x) = \sqrt[4]{x} - 1$

x	$f(x)$
0	-1
1	0
16	1

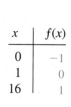

In Exercises 41–42, use a graphing calculator to graph the cube root function. Use window settings of $x = -10$ to 10 and $y = -5$ to 5.

41. $f(x) = \sqrt[3]{x - 2}$ **42.** $f(x) = -\sqrt[3]{x + 1}$

In Exercises 43–50, find each value without using a calculator.

43. $\sqrt[4]{16}$ 2   **44.** $\sqrt[4]{81}$ 3
45. $-\sqrt[5]{32}$ -2 **46.** $-\sqrt[5]{243}$ -3
47. $\sqrt[6]{1}$ 1   **48.** $\sqrt[6]{0}$ 0
49. $\sqrt[5]{-32}$ -2 **50.** $\sqrt[7]{-1}$ -1

In Exercises 51–54, use a calculator to find each root to the nearest hundredth.

51. $\sqrt[4]{125}$ 3.34   **52.** $\sqrt[5]{12,450}$ 6.59
53. $\sqrt[5]{-6,000}$ -5.70 **54.** $\sqrt[6]{0.5}$ 0.89

In Exercises 55–82, write each expression without a radical sign. All variables represent positive numbers.

55. $\sqrt{x^2}$ x **56.** $\sqrt{y^4}$ y^2
57. $\sqrt{x^6}$ x^3 **58.** $\sqrt{b^8}$ b^4
59. $\sqrt{x^{10}}$ x^5 **60.** $\sqrt{y^{12}}$ y^6
61. $\sqrt[4]{x^4}$ x **62.** $\sqrt[4]{x^8}$ x^2
63. $\sqrt{4z^2}$ $2z$ **64.** $\sqrt{9t^6}$ $3t^3$
65. $-\sqrt{x^4y^2}$ $-x^2y$ **66.** $-\sqrt{x^2y^4}$ $-xy^2$
67. $-\sqrt{0.04y^2}$ $-0.2y$ **68.** $-\sqrt{0.81b^6}$ $-0.9b^3$
69. $-\sqrt{25x^4z^{12}}$ $-5x^2z^6$ **70.** $-\sqrt{100a^6b^4}$ $-10a^3b^2$
71. $\sqrt{36z^{36}}$ $6z^{18}$ **72.** $\sqrt{64y^{64}}$ $8y^{32}$
73. $-\sqrt{625z^2}$ $-25z$ **74.** $-\sqrt{729x^8}$ $-27x^4$
75. $\sqrt[3]{y^6}$ y^2 **76.** $\sqrt[3]{c^3}$ c
77. $\sqrt[5]{f^5}$ f **78.** $\sqrt[5]{y^{20}}$ y^4
79. $\sqrt[3]{27y^3}$ $3y$ **80.** $\sqrt[3]{64y^6}$ $4y^2$
81. $\sqrt[3]{-p^6q^3}$ $-p^2q$ **82.** $\sqrt[3]{-r^{12}t^6}$ $-r^4t^2$

APPLICATIONS

In Exercises 83–88, use a calculator to help solve each problem. Give your answers to the nearest hundredth.

83. PACKAGING A cubical box has a volume of 2 cubic feet. Substitute 2 for V in the formula $V = s^3$ and solve for s to find the length of each side of the box. 1.26 ft

84. HOT-AIR BALLOONS If a hot-air balloon is in the shape of a sphere and has a volume of 15,000 cubic feet, what is its radius? (*Hint:* See the first Accent on Technology feature in this section.) 15.30 ft

85. WINDMILLS The power generated by a windmill is related to the speed of the wind by the formula

$$S = \sqrt[3]{\frac{P}{0.02}}$$

where S is the speed of the wind (in mph) and P is the power (in watts). Find the speed of the wind when the windmill is producing 400 watts of power. 27.14 mph

86. ASTRONOMY In the early 17th century, Johannes Kepler, a German astronomer, discovered that a planet's mean distance R from the sun (in millions of miles) is related to the time T (in years) it takes the planet to orbit the sun by the formula

$$R = 93\sqrt[3]{\frac{T^2}{1.002}}$$

Use the information in Illustration 2 to find R for each planet shown.

Mercury, 35.89; Venus, 67.58; Earth, 92.94; Mars, 141.57; Jupiter, 483.34

ILLUSTRATION 2

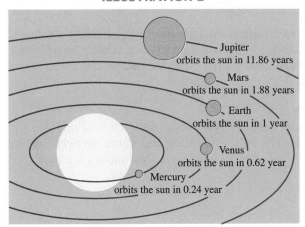

Jupiter
orbits the sun in 11.86 years

Mars
orbits the sun in 1.88 years

Earth
orbits the sun in 1 year

Venus
orbits the sun in 0.62 year

Mercury
orbits the sun in 0.24 year

87. DEPRECIATION The formula

$$r = 1 - \sqrt[n]{\frac{S}{C}}$$

gives the annual depreciation rate r of an item that had an original cost of C dollars and has a useful life of n years and a salvage value of S dollars. Use the information in Illustration 3 to find the annual depreciation rate for the new piece of sound equipment. 26.22%

ILLUSTRATION 3

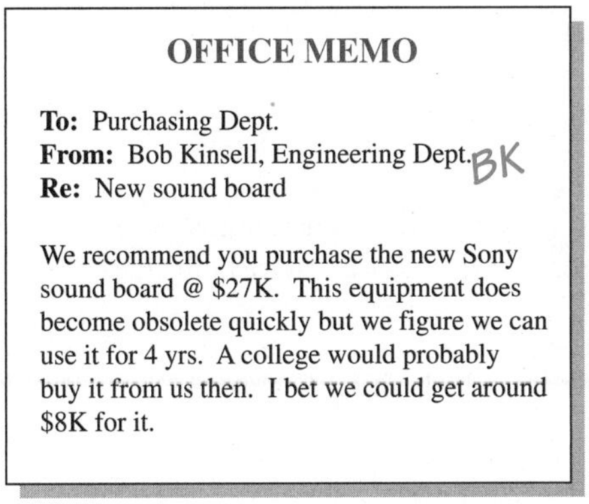

OFFICE MEMO

To: Purchasing Dept.
From: Bob Kinsell, Engineering Dept. BK
Re: New sound board

We recommend you purchase the new Sony sound board @ $27K. This equipment does become obsolete quickly but we figure we can use it for 4 yrs. A college would probably buy it from us then. I bet we could get around $8K for it.

---

88. SAVINGS ACCOUNT The interest rate r earned by a savings account after n compoundings is given by the formula

$$\sqrt[n]{\frac{V}{P}} - 1 = r$$

where V is the current value and P is the original principal. What interest rate r was paid on an account in which a deposit of $1,000 grew to $1,338.23 after 5 compoundings?   6.00%

WRITING

89. Explain why a negative number can have a real number for its cube root yet cannot have a real number for its fourth root.

90. To find $\sqrt[3]{15}$, we can use the $\boxed{\sqrt[x]{x}}$ key on a calculator to obtain 2.466212074. Explain how a key other than $\boxed{\sqrt[x]{x}}$ can be used to check the validity of this result.

REVIEW

In Exercises 91–98, write each expression as an expression involving only one exponent.

91. $m^5 m^2$ m^7

92. $(-5x^3)(-5x)$ $25x^4$

93. $(3^2)^4$ 3^8 or 9^4

94. $r^3 r r^5$ r^9

95. $(x^2 x^3)^5$ x^{25}

96. $(3aa^2a^3)^5$ $243a^{30}$

97. $4x^3(6x^5)$ $24x^8$

98. $-2x(5x^3)$ $-10x^4$

▶ **5.3**

Solving Equations Containing Radicals; the Distance Formula

In this section, you will learn about

The squaring property of equality ■ Checking solutions ■ Solving equations containing one square root ■ Solving equations containing two square roots ■ Solving equations containing cube roots ■ Solving radical equations graphically ■ The distance formula

Introduction   Many situations can be modeled mathematically by equations that contain radicals. In this section, we will develop techniques to solve such equations. Then we will consider a special formula called the *distance formula*.

The Squaring Property of Equality

The equation $\sqrt{x} = 6$ is called a **radical equation,** because it contains a radical expression with a variable radicand. To solve this equation, we isolate x by undoing the operation performed on it. Recall that $\sqrt{x}$ represents the number that, when squared, gives x. Therefore, if we *square* $\sqrt{x}$, we will obtain x.

$$\left(\sqrt{x}\right)^2 = x$$

Using this observation, we can eliminate the radical on the left-hand side of $\sqrt{x} = 6$ by squaring that side. Intuition tells us that we should also square the right-hand side. This is a valid step, because if two numbers are equal, their squares are equal.

Squaring property of equality

If $a = b$, then $a^2 = b^2$.

We can now solve $\sqrt{x} = 6$ by applying the squaring property of equality.

$\sqrt{x} = 6$ The original equation to solve.

$\left(\sqrt{x}\right)^2 = (6)^2$ Square both sides of the equation to eliminate the radical.

$x = 36$ Simplify each side: $\left(\sqrt{x}\right)^2 = x$ and $(6)^2 = 36$.

Checking this result, we have

$\sqrt{x} = 6$

$\sqrt{36} \overset{?}{=} 6$ Substitute 36 for x.

$6 = 6$ Simplify the left-hand side: $\sqrt{36} = 6$. We obtain a true statement, so $x = 36$ is a solution.

Checking Solutions

If we square both sides of an equation, the resulting equation may or may not have the same solutions as the original one. For example, if we square both sides of the equation

1. $x = 2$

with the solution 2, we obtain $(x)^2 = 2^2$, which simplifies to

2. $x^2 = 4$

with solutions 2 and -2, since $2^2 = 4$ and $(-2)^2 = 4$.

Equations 1 and 2 are not equivalent, because they have a different set of solutions. The solution -2 of Equation 2 does not satisfy Equation 1. Because squaring both sides of an equation can produce an equation with solutions that don't satisfy the original one, we must always check each potential solution in the original equation.

Solving Equations Containing One Square Root

To solve an equation containing square root radicals, we follow these steps.

Solving radical equations

1. Whenever possible, isolate a single radical expression on one side of the equation.
2. Square both sides of the equation and solve the resulting equation.
3. Check the solution in the original equation. This step is required.

| EXAMPLE 1 | **Solving radical equations.** Solve $\sqrt{x+2} = 3$. |

Solution To solve the equation $\sqrt{x+2} = 3$, we note that the radical is already isolated on one side. We proceed to step 2 and square both sides to eliminate the radical. Since this might produce an equation with more solutions than the original one, we must check each solution.

$$\sqrt{x+2} = 3$$
$$\left(\sqrt{x+2}\right)^2 = (3)^2 \quad \text{Square both sides.}$$
$$x + 2 = 9 \quad \text{Simplify each side: } \left(\sqrt{x+2}\right)^2 = x + 2 \text{ and } 3^2 = 9.$$
$$x = 7 \quad \text{Subtract 2 from both sides.}$$

We check by substituting 7 for x in the original equation.

$$\sqrt{x+2} = 3$$
$$\sqrt{7+2} \stackrel{?}{=} 3 \quad \text{Substitute 7 for } x.$$
$$\sqrt{9} \stackrel{?}{=} 3 \quad \text{Do the addition within the radical symbol.}$$
$$3 = 3$$

The solution $x = 7$ checks.

SELF CHECK Solve $\sqrt{x-4} = 9$. *Answer:* 85 ■

| EXAMPLE 2 | **A radical equation with no solution.** Solve $\sqrt{x+1} + 5 = 3$. |

Solution We isolate the radical on one side and proceed as follows:

$$\sqrt{x+1} + 5 = 3$$
$$\sqrt{x+1} = -2 \quad \text{Subtract 5 from both sides.}$$
$$\left(\sqrt{x+1}\right)^2 = (-2)^2 \quad \text{Square both sides to eliminate the radical.}$$
$$x + 1 = 4 \quad \text{Simplify: } \left(\sqrt{x+1}\right)^2 = x + 1 \text{ and } (-2)^2 = 4.$$
$$x = 3 \quad \text{Subtract 1 from both sides.}$$

We check by substituting 3 for x in the original equation.

$$\sqrt{x+1} + 5 = 3$$
$$\sqrt{3+1} + 5 \stackrel{?}{=} 3 \quad \text{Substitute 3 for } x.$$
$$\sqrt{4} + 5 \stackrel{?}{=} 3 \quad \text{Do the addition within the radical symbol.}$$
$$2 + 5 \stackrel{?}{=} 3$$
$$7 \neq 3$$

Since $7 \neq 3$, 3 is not a solution. In fact, the equation has no solution. This result was obvious in step 2 of the solution. There is no number x that could make the nonnegative number $\sqrt{x+1}$ equal to -2.

SELF CHECK Solve $\sqrt{x-2} + 5 = 2$. *Answer:* no solution ■

Example 2 shows that squaring both sides of an equation can lead to false solutions, called **extraneous solutions.** These solutions do not satisfy the original equation and must be discarded.

EXAMPLE 3

Height of a bridge. The distance d (in feet) that an object will fall in t seconds is given by the formula

$$t = \sqrt{\frac{d}{16}}$$

To find the height of the bridge shown in Figure 5-17, a man drops a stone into the water. If it takes the stone 3 seconds to hit the water, how high is the bridge?

FIGURE 5-17

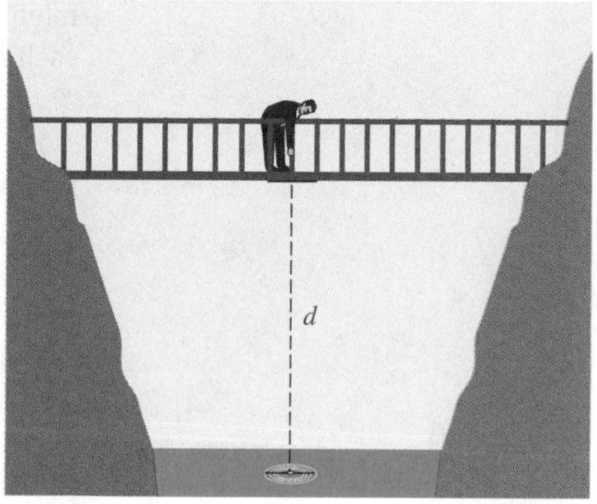

Solution   We substitute 3 for t in the formula and solve for d.

$$t = \sqrt{\frac{d}{16}}$$

$$3 = \sqrt{\frac{d}{16}} \qquad \text{Substitute } 3 \text{ for } t.$$

$$(3)^2 = \left(\sqrt{\frac{d}{16}}\right)^2 \qquad \text{Square both sides to eliminate the radical.}$$

$$9 = \frac{d}{16} \qquad \text{Simplify: } 3^2 = 9 \text{ and } \left(\sqrt{\frac{d}{16}}\right)^2 = \frac{d}{16}.$$

$$144 = d \qquad \text{Multiply both sides by 16.}$$

The bridge is 144 feet above the water. Check this result in the original equation.

SELF CHECK

If it takes 4 seconds for the stone in Example 3 to hit the water, how high is the bridge?   *Answer:* 256 feet   ■

EXAMPLE 4

Solving radical equations. Solve $a + 2 = \sqrt{a^2 + 3a + 3}$.

Solution   The radical is isolated on the right-hand side, so we proceed by squaring both sides to eliminate it.

$$a + 2 = \sqrt{a^2 + 3a + 3}$$

$$(a + 2)^2 = \left(\sqrt{a^2 + 3a + 3}\right)^2 \quad \text{Square both sides.}$$

$$a^2 + 4a + 4 = a^2 + 3a + 3 \qquad \begin{array}{l}\text{On the left-hand side, use the FOIL} \\ \text{method: } (a + 2)^2 = a^2 + 4a + 4. \\ \text{On the right-hand side, simplify:} \\ \left(\sqrt{a^2 + 3a + 3}\right)^2 = a^2 + 3a + 3.\end{array}$$

$$a^2 + 4a + 4 - a^2 = a^2 + 3a + 3 - a^2 \qquad \begin{array}{l}\text{To eliminate } a^2, \text{ subtract } a^2 \text{ from} \\ \text{both sides.}\end{array}$$

$$4a + 4 = 3a + 3 \qquad \begin{array}{l}\text{On each side, combine like terms:} \\ a^2 - a^2 = 0.\end{array}$$

$$a + 4 = 3 \qquad \text{Subtract } 3a \text{ from both sides.}$$

$$a = -1 \qquad \text{Subtract 4 from both sides.}$$

We check by substituting -1 for x in the original equation.

$$a + 2 = \sqrt{a^2 + 3a + 3}$$

$$-1 + 2 \stackrel{?}{=} \sqrt{(-1)^2 + 3(-1) + 3} \quad \text{Substitute } -1 \text{ for } a.$$

$$1 \stackrel{?}{=} \sqrt{1 - 3 + 3} \qquad \begin{array}{l}\text{Within the radical symbol, first find the} \\ \text{power, then do the multiplication.}\end{array}$$

$$1 \stackrel{?}{=} \sqrt{1} \qquad \text{Simplify within the radical symbol.}$$

$$1 = 1$$

The solution checks.

SELF CHECK   Solve $b + 4 = \sqrt{b^2 + 6b + 12}$.   *Answer:* -2 ■

Solving Equations Containing Two Square Roots

In the next example, the equation contains two square roots.

EXAMPLE 5   **Solving an equation containing two square roots.**   Solve $\sqrt{x + 12} = 3\sqrt{x + 4}$.

Solution   Note that each radical is isolated on one side of the equation. We begin by squaring both sides to eliminate them.

$$\sqrt{x + 12} = 3\sqrt{x + 4}$$

$$\left(\sqrt{x + 12}\right)^2 = \left(3\sqrt{x + 4}\right)^2 \quad \text{Square both sides.}$$

$$x + 12 = 9(x + 4) \quad \begin{array}{l}\text{On the left-hand side: } \left(\sqrt{x + 12}\right)^2 = x + 12.\\ \text{On the right-hand side: } \left(3\sqrt{x + 4}\right)^2 = 3^2\left(\sqrt{x + 4}\right)^2 = \\ 9(x + 4).\end{array}$$

$$x + 12 = 9x + 36 \quad \text{Remove parentheses.}$$

$$-8x = 24 \quad \text{Subtract } 9x \text{ and } 12 \text{ from both sides.}$$

$$x = -3 \quad \text{Divide both sides by } -8.$$

We check the solution by substituting -3 for x in the original equation.

$$\sqrt{x + 12} = 3\sqrt{x + 4}$$

$$\sqrt{-3 + 12} \stackrel{?}{=} 3\sqrt{-3 + 4} \quad \text{Substitute } -3 \text{ for } x.$$

$$\sqrt{9} \stackrel{?}{=} 3\sqrt{1} \quad \text{Simplify within the radical symbols.}$$

$$3 = 3$$

The solution checks.

SELF CHECK   Solve $\sqrt{x - 4} = 2\sqrt{x - 16}$.   *Answer:* 20   ■

Solving Equations Containing Cube Roots

In the next example, we cube both sides of an equation to eliminate a cube root.

EXAMPLE 6   **Solving an equation containing a cube root.**   Solve $\sqrt[3]{2x + 10} = 2$.

Solution   To undo the operation performed on $2x + 10$, we cube both sides and proceed as follows:

$$\sqrt[3]{2x + 10} = 2$$

$$\left(\sqrt[3]{2x + 10}\right)^3 = (2)^3 \quad \text{Cube both sides.}$$

$$2x + 10 = 8 \quad \text{Simplify: } \left(\sqrt[3]{2x + 10}\right)^3 = 2x + 10 \text{ and } (2)^3 = 8.$$

$$2x = -2 \quad \text{Subtract 10 from both sides.}$$

$$x = -1 \quad \text{Divide both sides by 2.}$$

Check the result.

SELF CHECK   Solve $\sqrt[3]{3x - 3} = 3$.   *Answer:* 10   ■

Solving Radical Equations Graphically

Radical equations can be solved graphically. For example, the solution of the radical equation $\sqrt{x-1} = 0$ is the number x that will make y equal to 0 in the equation $y = \sqrt{x-1}$. To find this number, we inspect the graph of $y = \sqrt{x-1}$ and locate the point on the graph that has a y-coordinate of 0. In Figure 5-18, we see that the point is $(1, 0)$, and we note that it is the x-intercept of the graph. We can then conclude that the x-coordinate of the x-intercept, $x = 1$, is the solution of the radical equation $\sqrt{x-1} = 0$.

FIGURE 5-18

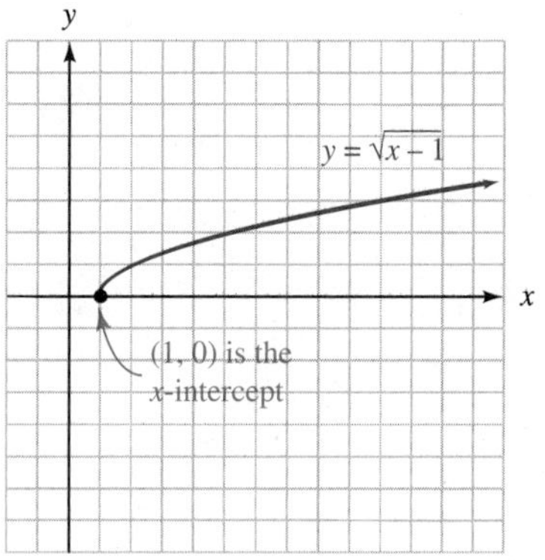

$y = \sqrt{x-1}$

$(1, 0)$ is the x-intercept

ACCENT ON TECHNOLOGY *Solving Radical Equations Graphically*

To solve the radical equation $\sqrt{x+1} = 2$ using a graphing calculator, we subtract 2 from both sides so that the right-hand side is 0.

$$\sqrt{x+1} = 2$$
$$\sqrt{x+1} - 2 = 2 - 2$$
$$\sqrt{x+1} - 2 = 0$$

Next, we enter the left-hand side of the equation into the calculator in the form $y = \sqrt{x+1} - 2$. See Figure 5-19(a). We then use window settings of $x = -5$ to 10 and $y = -4$ to 4 and press GRAPH to get Figure 5-19(b).

To find the x-intercept, we trace as shown in Figure 5-19(c). We zoom and trace again to get a y-coordinate close to 0. See Figure 5-19(d). From the display, we conclude that the x-coordinate of the x-intercept is 3. So $x = 3$ is the solution of $\sqrt{x+1} = 2$. Verify that 3 is a solution of the equation.

FIGURE 5-19

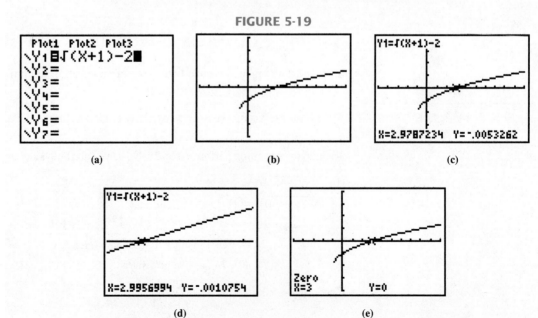

(a)

(b)

(c)

(d)

(e)

Another way to find the solution of $\sqrt{x + 1} = 2$ is using the zero feature of the calculator. With this feature, the x-coordinate of the x-intercept of the graph of $y = \sqrt{x + 1} - 2$ can be found quickly. See Figure 5-19(e). The *zero*, in this case 3, is the solution of the $\sqrt{x + 1} = 2$. Consult your owner's manual for the specific instructions to use the zero feature.

The Distance Formula

We can use the Pythagorean theorem to derive a formula for finding the distance between two points $P(x_1, y_1)$ and $Q(x_2, y_2)$ on a rectangular coordinate system. The distance d between points P and Q is the length of the hypotenuse of the triangle in Figure 5-20. The two legs have lengths $x_2 - x_1$ and $y_2 - y_1$.

By the Pythagorean theorem, we have

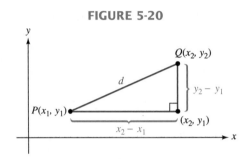

FIGURE 5-20

$$d^2 = (x_2 - x_1)^2 + (y_2 - y_1)^2$$

We can take the positive square root of both sides of this equation to get the **distance formula.**

$$d = \sqrt{(x_2 - x_1)^2 + (y_2 - y_1)^2}$$

The distance formula

The distance d between points $P(x_1, y_1)$ and $Q(x_2, y_2)$ is given by

$$d = \sqrt{(x_2 - x_1)^2 + (y_2 - y_1)^2}$$

EXAMPLE 7

Using the distance formula. Find the distance between points $P(1, 5)$ and $Q(4, 9)$. (See Figure 5-21.)

FIGURE 5-21

Solution We use the distance formula and substitute 1 for x_1, 5 for y_1, 4 for x_2, and 9 for y_2. Then we evaluate the expression under the radical symbol.

$$d = \sqrt{(x_2 - x_1)^2 + (y_2 - y_1)^2}$$
$$= \sqrt{(4 - 1)^2 + (9 - 5)^2} \qquad \text{Substitute.}$$
$$= \sqrt{3^2 + 4^2} \qquad \text{Do the subtractions within the parentheses first.}$$
$$= \sqrt{9 + 16} \qquad \text{Evaluate the powers.}$$
$$= \sqrt{25} \qquad \text{Do the addition.}$$
$$= 5 \qquad \text{Find the square root.}$$

The distance between points P and Q is 5 units.

SELF CHECK   Find the distance between $P(-2, 1)$ and $Q(4, 9)$.   *Answer:* 10

EXAMPLE 8

Marching band. Figure 5-22(a) shows the configuration that a marching band forms when performing "The Star Spangled Banner" before a college football game. How far from the drum major is the leader of the tuba section?

FIGURE 5-22

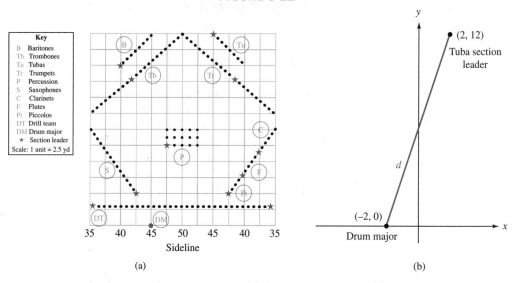

(a)

(b)

Solution   In Figure 5-22(b), we model the situation using a coordinate system where the *x*-axis represents the sideline, the *y*-axis represents the 50-yard line, and the scale is 1 unit = 2.5 yards. On the grid, the drum major is at $(-2, 0)$ and the tuba section leader is at $(2, 12)$. To find the distance between them, we use the distance formula, where $(x_1, y_1) = (-2, 0)$ and $(x_2, y_2) = (2, 12)$.

$$d = \sqrt{(x_2 - x_1)^2 + (y_2 - y_1)^2}$$

$$= \sqrt{[2 - (-2)]^2 + (12 - 0)^2} \quad \text{Substitute for } x_2, x_1, y_2, \text{ and } y_1.$$

$$= \sqrt{4^2 + 12^2} \quad \text{Do the subtractions: } 2 - (-2) = 2 + 2 = 4.$$

$$= \sqrt{16 + 144} \quad \text{Evaluate the exponential expressions.}$$

$$= \sqrt{160} \quad \text{Do the addition.}$$

$$\approx 12.64911064 \quad \text{Use a calculator to approximate the square root.}$$

To the nearest tenth, the tuba section leader is 12.6 units from the drum major. Because each unit on the grid represents 2.5 yards, the distance between these two band members is about 12.6(2.5) or 31.5 yards.   ∎

STUDY SET

Section 5.3

VOCABULARY

In Exercises 1–4, fill in the blanks to make the statements true.

1. A _____radical_____ equation contains one or more radical expressions with a variable radicand.

2. "To _____isolate_____ the radical expression" in $\sqrt{x} + 1 = 10$ means to get $\sqrt{x}$ all by itself on one side of the equation.

3. A false solution that occurs when you square both sides of an equation is called an _____extraneous_____ solution.

4. The squaring property of equality states that if two numbers are equal, their _____squares_____ are equal.

CONCEPTS

In Exercises 5–6, fill in the blanks to make the statements true.

5. The squaring property of equality states that

If $a = b$, then $a^2 = b^2$.

6. The distance formula states that

$$d = \sqrt{(x_2 - x_1)^2 + (y_2 - y_1)^2}$$

7. To isolate x, what step should be used to undo the operation performed on it? (Assume that x is a positive number.)

a. $x^2 = 4$

Take the positive square root of both sides.

b. $\sqrt{x} = 4$

Square both sides.

8. Simplify each expression.

a. $\left(\sqrt{x}\right)^2$ x

b. $\left(\sqrt{x-1}\right)^2$ $x - 1$

c. $\left(2\sqrt{x}\right)^2$ $4x$

d. $\left(2\sqrt{x-1}\right)^2$ $4x - 4$

e. $\left(\sqrt{2x}\right)^2$ $2x$

f. $\left(\sqrt[3]{x}\right)^3$ x

In Exercises 9–10, tell what is wrong with each solution.

9. Solve $\sqrt{x - 2} = 3$.

$\sqrt{x - 2} = 3$

$x - 2 = 3$

$x = 5$

Both sides of the equation were not squared—only the left-hand side.

10. Solve $2 = \sqrt{x - 9}$.

$2 = \sqrt{x - 9}$

$4 = x - 9$

$-5 = x$

$x = -5$

On the third line, 9 should have been added to *both* sides.

11. a. What type of geometric figure is figure *ABCD* shown in Illustration 1?   trapezoid

b. Give the coordinates of points *A*, *B*, *C*, and *D*.

$(-2, 5)$, $(2, 5)$, $(8, -3)$, $(-8, -3)$

c. Find the length of each side of the figure.

AB: 4; *BC*: 10; *CD*: 16; *DA*: 10

d. Find the area of the figure.   80 square units

ILLUSTRATION 1

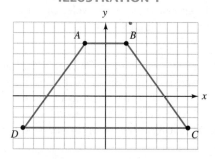

12. a. On the graph in Illustration 2, plot the points $A(-4, 6)$, $B(4, 0)$, $C(1, -4)$, and $D(-7, 2)$.

b. Draw figure *ABCD*. What type of geometric figure is it?   rectangle

c. Find the length of each side of the figure.

AB: 10; *BC*: 5; *CD*: 10; *DA*: 5

d. Find the perimeter of the figure.   30 units

ILLUSTRATION 2

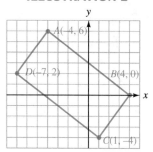

13. Use the graph of $y = -2\sqrt{x + 5} + 2$, shown in Illustration 3, to determine the solution of the equation $-2\sqrt{x + 5} = -2$. Check your answer. -4

ILLUSTRATION 3

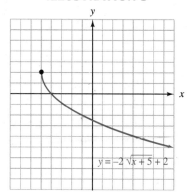

14. Illustration 4 shows the graph of $y = 3\sqrt{4 - x} - 2$. Use the graph to estimate the solution of the equation $3\sqrt{4 - x} = 2$. Check your answer to see if it is reasonable.   about 3.5

ILLUSTRATION 4

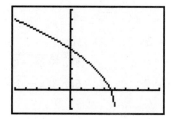

NOTATION

In Exercises 15–16, complete each solution.

15. Solve $\sqrt{x-3} = 5$.

$$\sqrt{x-3} = 5$$
$$\left(\boxed{\sqrt{x-3}}\right)^2 = \boxed{5}^2$$
$$x - 3 = \boxed{25}$$
$$x = 28$$

16. Solve $\sqrt{2x-18} = \sqrt{x-1}$.

$$\sqrt{2x-18} = \sqrt{x-1}$$
$$\left(\boxed{\sqrt{2x-18}}\right)^2 = \left(\sqrt{x-1}\right)^2$$
$$\boxed{2x-18} = x - 1$$
$$\boxed{x} - 18 = -1$$
$$x = 17$$

PRACTICE

In Exercises 17–56, solve each equation. Check all solutions. If an equation has no solutions, write "none."

17. $\sqrt{x} = 3$ 9
18. $\sqrt{x} = 5$ 25
19. $\sqrt{2a} = 4$ 8
20. $\sqrt{3a} = 9$ 27
21. $\sqrt{r} + 4 = 0$ none
22. $\sqrt{r} + 1 = 0$ none
23. $-\sqrt{x} = -5$ 25
24. $-\sqrt{x} = -12$ 144
25. $10 - \sqrt{s} = 7$ 9
26. $-4 = 6 - \sqrt{s}$ 100
27. $\sqrt{x+3} = 2$ 1
28. $\sqrt{x-2} = 3$ 11
29. $\sqrt{3-T} = -2$ none
30. $\sqrt{5-T} = 10$ −95
31. $\sqrt{6+2x} = 4$ 5
32. $\sqrt{7+2x} = -4$ none
33. $\sqrt{5x-5} - 5 = 0$ 6
34. $\sqrt{6x+19} - 7 = 0$ 5
35. $\sqrt{x+3} + 5 = 12$ 46
36. $\sqrt{x-5} - 3 = 4$ 54
37. $x - 3 = \sqrt{x^2 - 15}$ 4
38. $v - 2 = \sqrt{v^2 - 16}$ 5
39. $\sqrt{3t-9} = \sqrt{t+1}$ 5
40. $\sqrt{a-3} = \sqrt{2a-8}$ 5
41. $\sqrt{10-3x} = \sqrt{2x+20}$ −2
42. $\sqrt{1-2x} = \sqrt{x+10}$ −3
43. $\sqrt{3c-8} - \sqrt{c} = 0$ 4
44. $\sqrt{2x} - \sqrt{x+8} = 0$ 8
45. $x - 1 = \sqrt{x^2 - 4x + 9}$ 4
46. $3d = \sqrt{9d^2 - 2d + 8}$ 4
47. $\sqrt{4m^2 + 6m + 6} = -2m$ −1
48. $\sqrt{9t^2 + 4t + 20} = -3t$ −5
49. $\sqrt{3x+3} = 3\sqrt{x-1}$ 2
50. $2\sqrt{4x+5} = 5\sqrt{x+4}$ none
51. $2\sqrt{3x+4} = \sqrt{5x+9}$ −1
52. $\sqrt{3x+6} = 2\sqrt{2x-11}$ 10
53. $\sqrt[3]{x-1} = 4$ 65
54. $\sqrt[3]{2x+5} = 3$ 11
55. $\sqrt[3]{\frac{1}{2}x - 3} = 2$ 22
56. $\sqrt[3]{x+4} = 1$ −3

In Exercises 57–60, use a graphing calculator to solve each equation graphically.

57. $\sqrt{x+0.01} = 0.9$ 0.8
58. $\sqrt{2x+1} = \sqrt{x+0.75}$ −0.25
59. $\sqrt{4x+23} = 5$ 0.5
60. $x + 2 = \sqrt{x^2 + 6x - 2}$ 3

In Exercises 61–68, find the distance between points P and Q. If an answer is not exact, round to the nearest hundredth.

61. $P(3, -4)$ and $Q(0, 0)$ 5
62. $P(0, 0)$ and $Q(-6, 8)$ 10
63. $P(2, 4)$ and $Q(5, 9)$ 5.83
64. $P(5, 9)$ and $Q(9, 13)$ 5.66
65. $P(-2, -8)$ and $Q(3, 4)$ 13
66. $P(-5, -2)$ and $Q(7, 3)$ 13
67. $P(6, 8)$ and $Q(12, 16)$ 10
68. $P(10, 4)$ and $Q(2, -2)$ 10

▦ APPLICATIONS

In Exercises 69–80, use a calculator to help solve each problem.

69. NIAGARA FALLS The distance s (in feet) that an object will fall in t seconds is given by the formula

$$t = \frac{\sqrt{s}}{4}$$

The time it took a stuntman to go over the Niagara Falls in a barrel was 3.25 seconds. Substitute 3.25 for t and solve the equation for s to find the height of the waterfall.   169 ft

70. WASHINGTON MONUMENT Gabby Street, a professional baseball player of the 1920s, was known for once catching a ball dropped from the top of the Washington Monument in Washington, D.C. If the ball fell for slightly less than 6 seconds before it was caught, find the approximate height of the monument. (*Hint:* See Exercise 69.)   about 576 ft

71. FOUCAULT PENDULUM The time t (in seconds) required for a pendulum of length L feet to swing through one back-and-forth cycle, called its period, is given by the formula

$$t = 1.11\sqrt{L}$$

The Foucault pendulum in Chicago's Museum of Science and Industry, shown in Illustration 5, is used to demonstrate the rotation of the earth. It completes one cycle in 8.91 seconds. To the nearest tenth of a foot, how long is the pendulum?   64.4 ft

ILLUSTRATION 5

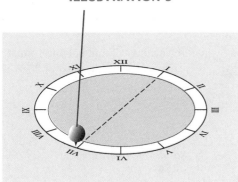

72. POWER USAGE The current I (in amperes), the resistance R (in ohms), and the power P (in watts) are related by the formula

$$I = \sqrt{\dfrac{P}{R}}$$

Find the power (to the nearest watt) used by a space heater that draws 7 amps when the resistance is 10.2 ohms. 500 watts

73. ROAD SAFETY The formula $s = k\sqrt{d}$ relates the speed s (in mph) of a car and the distance d of the skid when a driver hits the brakes. On wet pavement, $k = 3.24$. How far will a car skid if it is going 55 mph? about 288 ft

74. ROAD SAFETY How far will the car in Exercise 73 skid if it is traveling on dry pavement? On dry pavement, $k = 5.34$. about 106 ft

75. SATELLITE ORBITS The orbital speed s of an earth satellite is related to its distance r from the earth's center by the formula

$$\sqrt{r} = \dfrac{2.029 \times 10^7}{s}$$

If the satellite's orbital speed is 7×10^3 meters per second, find its altitude a (in meters) above the earth's surface, as shown in Illustration 6. about 2×10^6 m

ILLUSTRATION 6

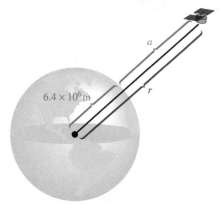

76. HIGHWAY DESIGN A highway curve banked at 8° will accommodate traffic traveling at speed s (in mph) if the radius of the curve is r (feet), according to the equation $s = 1.45\sqrt{r}$. If highway engineers expect traffic to travel at 65 mph, to the nearest foot, what radius should they specify? (See Illustration 7.) 2,010 ft

ILLUSTRATION 7

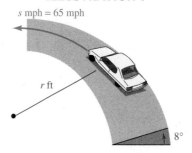

77. GEOMETRY The radius of a cone with volume V and height h is given by the formula

$$r = \sqrt{\dfrac{3V}{\pi h}}$$

Solve the equation for V. $\quad V = \dfrac{\pi r^2 h}{3}$

78. WINDMILLS The power produced by a certain windmill is related to the speed of the wind by the formula

$$s = \sqrt[3]{\dfrac{P}{0.02}}$$

where P is the power (in watts) and s is the speed of the wind (in mph). How much power will the windmill produce if the wind is blowing at 30 mph? 540 watts

79. NAVIGATION An oil tanker is to travel from Tunisia to Italy, as shown in Illustration 8. The captain wants to travel a course that is *always* the same distance from a point on the coast of Sardinia as it is from a point on the coast of Sicily (both denoted in red). How far will the tanker be from these points when it reaches
a. position 1? $\sqrt{2} \approx 1.4$ units
b. position 2? $\sqrt{10} \approx 3.2$ units

ILLUSTRATION 8

80. DECK DESIGN The plans for a patio deck shown in Illustration 9 call for three 6×6 redwood support braces directly under the hot tub. Find the length of each support. Round to the nearest tenth of a foot.

brace 1: 4.2 units; brace 2: 6.7 units; brace 3: 2.2 units

ILLUSTRATION 9

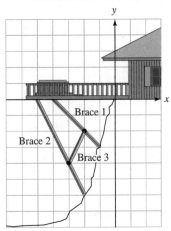

81. Explain why a check is necessary when solving radical equations.

82. How would you know, without solving it, that the equation $\sqrt{x + 2} = -4$ has no solutions?

In Exercises 83–86, do the operations.

83. $(3x^2 + 2x) + (5x^2 - 8x)$ $8x^2 - 6x$

84. $(7a^2 + 2a - 5) - (3a^2 - 2a + 1)$ $4a^2 + 4a - 6$

85. $(x + 3)(x + 3)$ $x^2 + 6x + 9$

86. $x - 1\overline{)x^2 - 6x + 5}$ $x - 5$

▶ **5.4**

Simplifying Radical Expressions

In this section, you will learn about

The multiplication property of radicals ▪ Simplifying square root radicals ▪ The division property of radicals ▪ Simplifying cube roots

---

Introduction   Square dancing is a traditional American folk dance in which four couples, arranged in a square, perform various moves. Figure 5-23 shows a group as they promenade around a square.

If the square shown in the figure has an area of 12 square yards, the length of a side is $\sqrt{12}$ yards. We can use the formula for the area of a square and the concept of square root to show that this is so.

FIGURE 5-23

$A = s^2$ *s* is the length of a side of the square.

$12 = s^2$ Substitute 12 for *A*, the area of the square.

$\sqrt{12} = \sqrt{s^2}$ Take the positive square root of both sides.

$\sqrt{12} = s$ The length of a side of the square is $\sqrt{12}$ yards.

The form in which we express the length of a side of the square depends on the situation. If an approximation is acceptable, we can use a calculator to find that

$\sqrt{12} \approx 3.464101615$, and we can then round to a specified degree of accuracy. For example, to the nearest tenth, each side is 3.5 yards long.

If the situation calls for the *exact* length, we must use a radical expression. As you will see in this section, it is common practice to write a radical expression such as $\sqrt{12}$ in *simplified form*. To simplify radicals, we will use the multiplication and division properties of radicals.

The Multiplication Property of Radicals

We introduce the first of two properties of radicals with the following examples:

$$\sqrt{4 \cdot 25} = \sqrt{100} \qquad \text{and} \qquad \sqrt{4}\sqrt{25} = 2 \cdot 5$$
$$= 10 \qquad\qquad\qquad\qquad = 10$$

Read as "the square root of 4 times the square root of 25."

In each case, the answer is 10. Thus, $\sqrt{4 \cdot 25} = \sqrt{4}\sqrt{25}$. Likewise,

$$\sqrt{9 \cdot 16} = \sqrt{144} \qquad \text{and} \qquad \sqrt{9}\sqrt{16} = 3 \cdot 4$$
$$= 12 \qquad\qquad\qquad\qquad = 12$$

In each case, the answer is 12. Thus, $\sqrt{9 \cdot 16} = \sqrt{9}\sqrt{16}$. These results suggest the **multiplication property of radicals.**

The multiplication property of radicals

If a and b are positive or zero, then

$$\sqrt{ab} = \sqrt{a}\sqrt{b}$$

In words, *the square root of the product of two nonnegative numbers is equal to the product of their square roots.*

Simplifying Square Root Radicals

A square root radical is in **simplified form** when each of the following statements is true.

Simplified form of a radical

1. Except for 1, the radicand has no perfect square factors.
2. No fraction appears in a radicand.
3. No radical appears in the denominator of a fraction.

We can use the multiplication property of radicals to simplify square roots whose radicands have perfect square factors. For example, we can simplify $\sqrt{12}$ as follows:

$$\sqrt{12} = \sqrt{4 \cdot 3} \qquad \text{Factor 12 as } 4 \cdot 3.$$
$$= \sqrt{4}\sqrt{3} \qquad \text{Use the multiplication property of radicals.}$$
$$= 2\sqrt{3} \qquad \text{Write } \sqrt{4} \text{ as 2. Read as "2 times the square root of 3."}$$

The square in Figure 5-23, which we considered in the introduction to this section, has a side length of $\sqrt{12}$ yards. We now see that the *exact* length of a side can be expressed in simplified form as $2\sqrt{3}$ yards.

To simplify more difficult square roots, we need to know the integers that are perfect squares. For example, 81 is a perfect square, because $9^2 = 81$. The first 20 integer squares are

1, 4, 9, 16, 25, 36, 49, 64, 81, 100, 121, 144, 169, 196, 225, 256, 289, 324, 361, 400

EXAMPLE 1	**Simplifying square roots.** Simplify $\sqrt{27}$.

Solution   Since the greatest perfect square that divides 27 exactly is 9, we will factor 27 as $9 \cdot 3$ and apply the multiplication property of radicals.

$$\sqrt{27} = \sqrt{9 \cdot 3}$$
$$= \sqrt{9}\sqrt{3} \quad \text{The square root of a product (that is, } \sqrt{9 \cdot 3} \text{) is equal}$$
$$\text{to the product of the square roots, } \sqrt{9} \; \sqrt{3}.$$
$$= 3\sqrt{3} \quad \sqrt{9} = 3.$$

SELF CHECK   Simplify $\sqrt{500}$.   *Answer:* $10\sqrt{5}$ ∎

EXAMPLE 2	**Teaching drawing.** The grid in Figure 5-24 is used as an instructional aid in teaching students how to draw horses. Artists have found that if the axes are scaled using a unit based on one-half the length of the horse's head, the correct proportions are obtained. What is the number of units from the hind quarter of the horse to the nose?

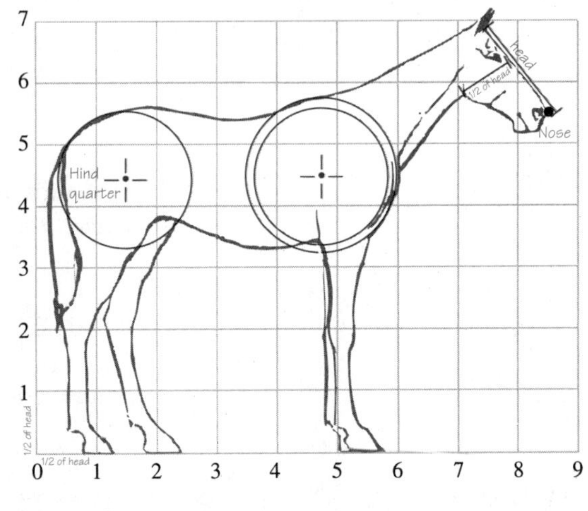

Solution   The hind quarter is located at (1.5, 4.5), and the nose is at (8.5, 5.5). To find the distance between the two points, we substitute 1.5 for x_1, 4.5 for y_1, 8.5 for x_2, and 5.5 for y_2 in the distance formula.

FIGURE 5-24

$$d = \sqrt{(x_2 - x_1)^2 + (y_2 - y_1)^2} \quad \text{The distance formula.}$$
$$= \sqrt{(8.5 - 1.5)^2 + (5.5 - 4.5)^2} \quad \text{Substitute.}$$
$$= \sqrt{7^2 + 1^2} \quad \text{Do the subtractions.}$$
$$= \sqrt{49 + 1} \quad \text{Evaluate the exponential expressions.}$$
$$= \sqrt{50} \quad \text{Do the addition.}$$

The expression $\sqrt{50}$ is not in simplified form, because the radicand, 50, has a perfect square factor of 25. To simplify the radical, we write 50 in factored form as $25 \cdot 2$.

$$\sqrt{50} = \sqrt{25 \cdot 2}$$
$$= \sqrt{25}\sqrt{2} \quad \text{The square root of a product is equal to the product of the square}$$
$$\text{roots.}$$
$$= 5\sqrt{2} \quad \text{Simplify: } \sqrt{25} = 5.$$

The distance from the hind quarter to the nose is $5\sqrt{2}$ units, which is approximately 7.1 units when rounded to the nearest tenth.   ∎

Expressions containing variables can also be perfect squares. For example, $36x^2$ is a perfect square, because

$$36x^2 = (6x)^2 \quad \text{Think, "What was squared to obtain } 36x^2\text{?"}$$

We can use this observation to help simplify radicals involving variable radicands. We will assume that all of the variables in the following examples represent positive numbers.

EXAMPLE 3

Simplifying radicals involving variable radicands. Simplify $\sqrt{b^3}$.

Solution   To write the $\sqrt{b^3}$ in simplified form, we factor b^3 into two factors, one of which is the greatest perfect square that divides b^3. The greatest perfect square that divides b^3 is b^2, so such a factorization is $b^3 = b^2 \cdot b$. We then proceed as follows:

$$\sqrt{b^3} = \sqrt{b^2 \cdot b} \quad \text{Write } \sqrt{b^3} \text{ as } \sqrt{b^2 \cdot b}.$$
$$= \sqrt{b^2}\sqrt{b} \quad \text{The square root of a product is equal to the product of the square roots.}$$
$$= b\sqrt{b} \quad \sqrt{b^2} = b.$$

SELF CHECK   Simplify $\sqrt{y^5}$.   *Answer:* $y^2\sqrt{y}$ ■

EXAMPLE 4

Simplifying radicals involving variable radicands. Simplify $-7\sqrt{8m}$.

Solution   We first write $\sqrt{8m}$ in simplified form, and then we multiply the result by -7. By inspection, we see that the radicand, $8m$, has a perfect square factor of 4. We can write $8m$ in factored form as $4 \cdot 2m$.

$$-7\sqrt{8m} = -7\sqrt{4 \cdot 2m} \quad \text{Write } \sqrt{8m} \text{ as } \sqrt{4 \cdot 2m}.$$
$$= -7\sqrt{4}\sqrt{2m} \quad \text{The square root of a product is equal to the product of the square roots.}$$
$$= -7(2\sqrt{2m}) \quad \sqrt{4} = 2.$$
$$= -14\sqrt{2m} \quad -7(2) = -14.$$

SELF CHECK   Simplify $2\sqrt{18c}$.   *Answer:* $6\sqrt{2c}$ ■

WARNING! When writing radical expressions such as $-14\sqrt{2m}$, be sure to extend the radical symbol completely over $2m$, because the expressions $-14\sqrt{2m}$ and $-14\sqrt{2}m$ are not the same. Similar care should be taken when writing expressions such as $\sqrt{3}x$. To avoid any misinterpretation, $\sqrt{3}x$ can be written as $x\sqrt{3}$.

EXAMPLE 5

Simplifying radicals involving variable radicands. Simplify $\sqrt{72x^3}$.

Solution   We factor $72x^3$ into two factors, one of which is the greatest perfect square that divides $72x^3$. Since the greatest perfect square that divides $72x^3$ is $36x^2$, such a factorization is $72x^3 = 36x^2 \cdot 2x$. We now use the multiplication property of radicals to get

$$\sqrt{72x^3} = \sqrt{36x^2 \cdot 2x}$$
$$= \sqrt{36x^2}\sqrt{2x} \quad \text{The square root of a product is equal to the product of the square roots.}$$
$$= 6x\sqrt{2x} \quad \sqrt{36x^2} = 6x.$$

SELF CHECK   Simplify $\sqrt{48y^3}$.   *Answer:* $4y\sqrt{3y}$ ■

| EXAMPLE 6 | **Simplifying radicals involving variable radicands.** Simplify $3a\sqrt{288a^4b^7}$. |

Solution   We first simplify $\sqrt{288a^4b^7}$. Then we multiply the result by $3a$. To simplify $\sqrt{288a^4b^7}$, we look for the greatest perfect square that divides $288a^4b^7$. Because

- 144 is the greatest perfect square that divides 288,
- a^4 is the greatest perfect square that divides a^4, and
- b^6 is the greatest perfect square that divides b^7,

the factor $144a^4b^6$ is the greatest perfect square that divides $288a^4b^7$.

We can now use the multiplication property of radicals to simplify the radical.

$$
\begin{aligned}
3a\sqrt{288a^4b^7} &= 3a\sqrt{144a^4b^6 \cdot 2b} & \text{Factor } 288a^4b^7. \\
&= 3a\sqrt{144a^4b^6}\sqrt{2b} & \text{The square root of a product is equal} \\
& & \text{to the product of the square roots.} \\
&= 3a\left(12a^2b^3\sqrt{2b}\right) & \sqrt{144a^4b^6} = 12a^2b^3. \\
&= 36a^3b^3\sqrt{2b} & \text{Multiply: } 3a(12a^2b^3) = 36a^3b^3.
\end{aligned}
$$

SELF CHECK   Simplify $5q\sqrt{63p^5q^4}$.   *Answer:* $15p^2q^3\sqrt{7p}$ ■

The Division Property of Radicals

To introduce the second property of radicals, we consider these examples.

$$\sqrt{\frac{100}{25}} = \sqrt{4} \qquad \text{and} \qquad \frac{\sqrt{100}}{\sqrt{25}} = \frac{10}{5} \quad \begin{array}{l}\text{Read as "the square root of 100 divided}\\\text{by the square root of 25."}\end{array}$$

$$= 2 \qquad\qquad\qquad\qquad = 2$$

Since the answer is 2 in each case,

$$\sqrt{\frac{100}{25}} = \frac{\sqrt{100}}{\sqrt{25}}$$

Likewise,

$$\sqrt{\frac{36}{4}} = \sqrt{9} \qquad \text{and} \qquad \frac{\sqrt{36}}{\sqrt{4}} = \frac{6}{2}$$

$$= 3 \qquad\qquad\qquad = 3$$

Since the answer is 3 in each case,

$$\sqrt{\frac{36}{4}} = \frac{\sqrt{36}}{\sqrt{4}}$$

These results suggest the **division property of radicals.**

The division property of radicals

If $a \geq 0$ and $b > 0$, then

$$\sqrt{\frac{a}{b}} = \frac{\sqrt{a}}{\sqrt{b}}$$

In words, *the square root of the quotient of two numbers is the quotient of their square roots.*

We can use the division property of radicals to simplify radicals that have fractions in their radicands. For example,

$$\sqrt{\frac{59}{49}} = \frac{\sqrt{59}}{\sqrt{49}}$$

$$= \frac{\sqrt{59}}{7} \quad \text{Simplify the denominator: } \sqrt{49} = 7.$$

EXAMPLE 7

Simplifying radicals involving fractions. Simplify $\sqrt{\dfrac{108}{25}}$.

Solution

$$\sqrt{\frac{108}{25}} = \frac{\sqrt{108}}{\sqrt{25}} \quad \text{The square root of a quotient is equal to the quotient of the square roots.}$$

$$= \frac{\sqrt{36 \cdot 3}}{5} \quad \text{Factor 108 using the largest perfect square factor of 108, which is 36. Write } \sqrt{25} \text{ as 5.}$$

$$= \frac{\sqrt{36}\sqrt{3}}{5} \quad \text{The square root of a product is equal to the product of the square roots.}$$

$$= \frac{6\sqrt{3}}{5} \quad \sqrt{36} = 6. \text{ This result can also be written as } \frac{6}{5}\sqrt{3}.$$

SELF CHECK   Simplify $\sqrt{\dfrac{20}{81}}$.   *Answer:* $\dfrac{2\sqrt{5}}{9}$ ∎

EXAMPLE 8

Simplifying radicals involving fractions. Simplify $\sqrt{\dfrac{44x^3}{9xy^2}}$.

Solution

$$\sqrt{\frac{44x^3}{9xy^2}} = \sqrt{\frac{44x^2}{9y^2}} \quad \text{Simplify the fraction by dividing out the common factor of } x: \frac{44x^3}{9xy^2} = \frac{44x^2\overset{1}{x}}{9\underset{1}{x}y^2} = \frac{44x^2}{9y^2}.$$

$$= \frac{\sqrt{44x^2}}{\sqrt{9y^2}} \quad \text{The square root of a quotient is equal to the quotient of the square roots.}$$

$$= \frac{\sqrt{4x^2}\sqrt{11}}{\sqrt{9y^2}} \quad \text{Factor } 44x^2 \text{ as } 4x^2 \cdot 11. \text{ The square root of a product is equal to the product of the square roots.}$$

$$= \frac{2x\sqrt{11}}{3y} \quad \sqrt{4x^2} = 2x \text{ and } \sqrt{9y^2} = 3y.$$

SELF CHECK   Simplify $\sqrt{\dfrac{99b^3}{16a^2b}}$.   *Answer:* $\dfrac{3b\sqrt{11}}{4a}$ ∎

Simplifying Cube Roots

The multiplication and division properties of radicals are also true for cube roots and higher. To simplify cube roots, we must know the following perfect cube integers:

8, 27, 64, 125, 216, 343, 512, 729, 1,000

| EXAMPLE 9 | **Simplifying cube roots.** Simplify $\sqrt[3]{54}$. |

Solution   The greatest perfect cube that divides 54 is 27.

$$\sqrt[3]{54} = \sqrt[3]{27 \cdot 2} \qquad \text{Factor 54: } 54 = 27 \cdot 2.$$
$$= \sqrt[3]{27}\sqrt[3]{2} \qquad \text{The square root of a product is equal}$$
$$\text{to the product of the square roots.}$$
$$= 3\sqrt[3]{2} \qquad \sqrt[3]{27} = 3.$$

SELF CHECK   Simplify $\sqrt[3]{250}$. *Answer:* $5\sqrt[3]{2}$ ∎

Expressions containing variables can also be perfect cubes. For example, $8x^3y^3$ is a perfect cube, because

$$8x^3y^3 = (2xy)^3 \qquad \text{Think, "What was cubed to obtain } 8x^3y^3\text{?"}$$

| EXAMPLE 10 | **Simplifying cube roots involving variables.** Simplify   **a.** $\sqrt[3]{16x^3y^4}$ and   **b.** $\sqrt[3]{\dfrac{64n^4}{27m^3}}$. |

Solution   **a.** We factor $16x^3y^4$ into two factors, one of which is the greatest perfect cube that divides $16x^3y^4$. Since $8x^3y^3$ is the greatest perfect cube that divides $16x^3y^4$, the factorization is $16x^3y^4 = 8x^3y^3 \cdot 2y$.

$$\sqrt[3]{16x^3y^4} = \sqrt[3]{8x^3y^3 \cdot 2y} \qquad \text{Factor } 16x^3y^4.$$
$$= \sqrt[3]{8x^3y^3}\sqrt[3]{2y} \qquad \text{The cube root of a product is equal to the product}$$
$$\text{of the cube roots.}$$
$$= 2xy\sqrt[3]{2y} \qquad \sqrt[3]{8x^3y^3} = 2xy.$$

b. $\sqrt[3]{\dfrac{64n^4}{27m^3}} = \dfrac{\sqrt[3]{64n^4}}{\sqrt[3]{27m^3}}$ The cube root of a quotient is equal to the quotient of the cube roots.

$$= \dfrac{\sqrt[3]{64n^3}\sqrt[3]{n}}{3m} \qquad \text{In the numerator, use the multiplication property}$$
$$\text{of radicals. In the denominator, } \sqrt[3]{27m^3} = 3m.$$

$$= \dfrac{4n\sqrt[3]{n}}{3m} \qquad \sqrt[3]{64n^3} = 4n.$$

SELF CHECK   Simplify: **a.** $\sqrt[3]{54a^3b^5}$ and   **b.** $\sqrt[3]{\dfrac{27q^5}{64p^3}}$.   *Answers:*   **a.** $3ab\sqrt[3]{2b^2}$,   **b.** $\dfrac{3q\sqrt[3]{q^2}}{4p}$ ∎

   WARNING! Note that $\sqrt{a+b} \neq \sqrt{a} + \sqrt{b}$ and $\sqrt{a-b} \neq \sqrt{a} - \sqrt{b}$. To see that this is true, we consider these correct simplifications:

$$\sqrt{9+16} = \sqrt{25} = 5 \qquad \text{and} \qquad \sqrt{25-16} = \sqrt{9} = 3$$

It is incorrect to write

$$\sqrt{9+16} = \sqrt{9} + \sqrt{16} \qquad \text{or} \qquad \sqrt{25-16} = \sqrt{25} - \sqrt{16}$$
$$= 3 + 4 \qquad\qquad\qquad\qquad\qquad = 5 - 4$$
$$= 7 \qquad\qquad\qquad\qquad\qquad\qquad = 1$$

Section 5.4

VOCABULARY

In Exercises 1–4, fill in the blanks to make the statements true.

1. Squares of integers such as 4, 9, and 16 are called _____perfect_____ squares.

2. Cubes of integers such as 8, 27, and 64 are called perfect _____cubes_____.

3. "To _____simplify_____ $\sqrt{8}$" means to write it as $2\sqrt{2}$.

4. The word *product* is associated with the operation of _____multiplication_____ and the word *quotient* with _____division_____.

CONCEPTS

In Exercises 5–6, fill in the blanks to make the statements true.

5. The square root of the product of two positive numbers is equal to the _____product_____ of their square roots. In symbols,
$$\sqrt{ab} = \sqrt{a}\sqrt{b}$$

6. The square root of the quotient of two positive numbers is equal to the _____quotient_____ of their square roots. In symbols,
$$\sqrt{\frac{a}{b}} = \frac{\sqrt{a}}{\sqrt{b}}$$

*In Exercises 7–8, tell what is **wrong** with each solution.*

7. Simplify $\sqrt{20}$.

$$\sqrt{20} = \sqrt{16+4}$$
$$= \sqrt{16} + \sqrt{4}$$
$$= 4 + 2$$
$$= 6$$

Line 2 is not true. There is no addition property of radicals.

8. Simplify $\sqrt{27}$.

$$\sqrt{27} = \sqrt{36-9}$$
$$= \sqrt{36} - \sqrt{9}$$
$$= 6 - 3$$
$$= 3$$

Line 2 is not true. There is no subtraction property of radicals.

9. A crossword puzzle in a newspaper occupies an area of 28 square inches. See Illustration 1.
   a. Express the exact length of a side of the square-shaped puzzle in simplified radical form? $2\sqrt{7}$ in.

ILLUSTRATION 1

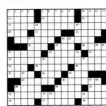

   b. What is the length of a side to the nearest tenth of an inch? 5.3 in.

10. See Illustration 2.
    a. What is the exact length of a side of the cube written in simplified radical form? $2\sqrt[3]{5}$ ft
    b. What is the length of a side to the nearest tenth of a foot? 3.4 ft

ILLUSTRATION 2

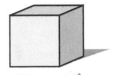

Volume = 40 ft^3

In Exercises 11–14, evaluate the expression $\sqrt{b^2 - 4ac}$ for the given values. Do the operations within the radical first, and then simplify the radical.

11. $a = 5, b = 10, c = 3$ $2\sqrt{10}$

12. $a = 2, b = 6, c = 1$ $2\sqrt{7}$

13. $a = -1, b = 6, c = 9$ $6\sqrt{2}$

14. $a = 1, b = -2, c = -11$ $4\sqrt{3}$

NOTATION

In Exercises 15–16, a radical is simplified. Complete each solution.

15.
$$\sqrt{80a^3b^2} = \sqrt{16 \cdot \boxed{5} \cdot a^2 \cdot a \cdot b^2}$$
$$= \sqrt{16a^2b^2 \cdot \boxed{5a}}$$
$$= \sqrt{\boxed{16a^2b^2}}\,\sqrt{5a}$$
$$= 4ab\sqrt{5a}$$

16.
$$\sqrt[3]{\frac{27a^4b^2}{64}} = \frac{\sqrt[3]{27a^4b^2}}{\boxed{\sqrt[3]{64}}}$$
$$= \frac{\sqrt[3]{27a^3 \cdot \boxed{ab^2}}}{\sqrt[3]{64}}$$
$$= \frac{\sqrt[3]{\boxed{27a^3}}\,\sqrt[3]{ab^2}}{\sqrt[3]{64}}$$
$$= \frac{3a\sqrt[3]{ab^2}}{4}$$

17. What operation is indicated between the two radicals in the expression $\sqrt{4}\sqrt{3}$? multiplication

18. Fill in each blank to make a true statement.
    a. $16x^2 = \left(\boxed{4x}\right)^2$ b. $27a^3b^6 = \left(\boxed{3ab^2}\right)^3$

19. Write each expression in a better form.
    a. $\sqrt{5} \cdot 2$ $2\sqrt{5}$ b. $\sqrt{7}a$ $a\sqrt{7}$
    c. $9\sqrt{x^2}\sqrt{6}$ $9x\sqrt{6}$ d. $\sqrt{y}\sqrt{25z^4}$ $5z^2\sqrt{y}$

20. a. Explain the difference between $\sqrt{5x}$ and $\sqrt{5}x$.
$\sqrt{5x} = \sqrt{5 \cdot x}$; $\sqrt{5}x = \sqrt{5} \cdot x$

b. Why do you think it is better to write $\sqrt{5}x$
as $x\sqrt{5}$? $\sqrt{5}x$ could be mistaken for $\sqrt{5x}$.

PRACTICE

In Exercises 21–60, simplify each radical. Assume that all variables represent positive numbers.

21. $\sqrt{20}$ $\quad 2\sqrt{5}$

22. $\sqrt{18}$ $\quad 3\sqrt{2}$

23. $\sqrt{50}$ $\quad 5\sqrt{2}$

24. $\sqrt{75}$ $\quad 5\sqrt{3}$

25. $\sqrt{45}$ $\quad 3\sqrt{5}$

26. $\sqrt{54}$ $\quad 3\sqrt{6}$

27. $\sqrt{98}$ $\quad 7\sqrt{2}$

28. $\sqrt{147}$ $\quad 7\sqrt{3}$

29. $\sqrt{48}$ $\quad 4\sqrt{3}$

30. $\sqrt{128}$ $\quad 8\sqrt{2}$

31. $-\sqrt{200}$ $\quad -10\sqrt{2}$

32. $-\sqrt{300}$ $\quad -10\sqrt{3}$

33. $\sqrt{192}$ $\quad 8\sqrt{3}$

34. $\sqrt{88}$ $\quad 2\sqrt{22}$

35. $\sqrt{250}$ $\quad 5\sqrt{10}$

36. $\sqrt{1,000}$ $\quad 10\sqrt{10}$

37. $2\sqrt{24}$ $\quad 4\sqrt{6}$

38. $3\sqrt{32}$ $\quad 12\sqrt{2}$

39. $-2\sqrt{28}$ $\quad -4\sqrt{7}$

40. $-3\sqrt{72}$ $\quad -18\sqrt{2}$

41. $\sqrt{n^3}$ $\quad n\sqrt{n}$

42. $\sqrt{x^5}$ $\quad x^2\sqrt{x}$

43. $\sqrt{4k}$ $\quad 2\sqrt{k}$

44. $\sqrt{9p}$ $\quad 3\sqrt{p}$

45. $\sqrt{12x}$ $\quad 2\sqrt{3x}$

46. $\sqrt{20y}$ $\quad 2\sqrt{5y}$

47. $6\sqrt{75t}$ $\quad 30\sqrt{3t}$

48. $2\sqrt{24s}$ $\quad 4\sqrt{6s}$

49. $\sqrt{25x^3}$ $\quad 5x\sqrt{x}$

50. $\sqrt{36y^3}$ $\quad 6y\sqrt{y}$

51. $\sqrt{a^2b}$ $\quad a\sqrt{b}$

52. $\sqrt{rs^4}$ $\quad s^2\sqrt{r}$

53. $\sqrt{9x^4y}$ $\quad 3x^2\sqrt{y}$

54. $\sqrt{16xy^2}$ $\quad 4y\sqrt{x}$

55. $\dfrac{1}{5}x^2y\sqrt{50x^2y^2}$ $\quad x^3y^2\sqrt{2}$

56. $\dfrac{1}{5}x^5y\sqrt{75x^3y^2}$ $\quad x^6y^2\sqrt{3x}$

57. $-12x\sqrt{16x^2y^3}$ $\quad -48x^2y\sqrt{y}$

58. $-4x^5y^3\sqrt{36x^3y^3}$ $\quad -24x^6y^4\sqrt{xy}$

59. $-\dfrac{2}{5}\sqrt{80mn^4}$ $\quad -\dfrac{8n^2\sqrt{5m}}{5}$

60. $\dfrac{5}{6}\sqrt{180ab^6}$ $\quad 5b^3\sqrt{5a}$

In Exercises 61–72, write each quotient as the quotient of two radicals and simplify.

61. $\sqrt{\dfrac{25}{9}}$ $\quad \dfrac{5}{3}$

62. $\sqrt{\dfrac{36}{49}}$ $\quad \dfrac{6}{7}$

63. $\sqrt{\dfrac{81}{64}}$ $\quad \dfrac{9}{8}$

64. $\sqrt{\dfrac{121}{144}}$ $\quad \dfrac{11}{12}$

65. $\sqrt{\dfrac{26}{25}}$ $\quad \dfrac{\sqrt{26}}{5}$

66. $\sqrt{\dfrac{17}{169}}$ $\quad \dfrac{\sqrt{17}}{13}$

67. $-\sqrt{\dfrac{20}{49}}$ $\quad -\dfrac{2\sqrt{5}}{7}$

68. $-\sqrt{\dfrac{50}{9}}$ $\quad -\dfrac{5\sqrt{2}}{3}$

69. $\sqrt{\dfrac{48}{81}}$ $\quad \dfrac{4\sqrt{3}}{9}$

70. $\sqrt{\dfrac{27}{64}}$ $\quad \dfrac{3\sqrt{3}}{8}$

71. $\sqrt{\dfrac{32}{25}}$ $\quad \dfrac{4\sqrt{2}}{5}$

72. $\sqrt{\dfrac{75}{16}}$ $\quad \dfrac{5\sqrt{3}}{4}$

In Exercises 73–80, simplify each expression. All variables represent positive numbers.

73. $\sqrt{\dfrac{72x^3}{y^2}}$ $\quad \dfrac{6x\sqrt{2x}}{y}$

74. $\sqrt{\dfrac{108b^2}{d^4}}$ $\quad \dfrac{6b\sqrt{3}}{d^2}$

75. $\sqrt{\dfrac{125n^5}{64n}}$ $\quad \dfrac{5n^2\sqrt{5}}{8}$

76. $\sqrt{\dfrac{72q^7}{25q^3}}$ $\quad \dfrac{6q^2\sqrt{2}}{5}$

77. $\sqrt{\dfrac{128m^3n^5}{81mn^7}}$ $\quad \dfrac{8m\sqrt{2}}{9n}$

78. $\sqrt{\dfrac{75p^3q^2}{p^5q^4}}$ $\quad \dfrac{5\sqrt{3}}{pq}$

79. $\sqrt{\dfrac{12r^7s^7}{r^5s^2}}$ $\quad 2rs^2\sqrt{3s}$

80. $\sqrt{\dfrac{m^2n^9}{100mn^3}}$ $\quad \dfrac{n^3\sqrt{m}}{10}$

In Exercises 81–96, simplify each cube root.

81. $\sqrt[3]{24}$ $\quad 2\sqrt[3]{3}$

82. $\sqrt[3]{32}$ $\quad 2\sqrt[3]{4}$

83. $\sqrt[3]{-128}$ $\quad -4\sqrt[3]{2}$

84. $\sqrt[3]{-250}$ $\quad -5\sqrt[3]{2}$

85. $\sqrt[3]{8x^3}$ $\quad 2x$

86. $\sqrt[3]{27x^3}$ $\quad 3x$

87. $\sqrt[3]{-64x^5}$ $\quad -4x\sqrt[3]{x^2}$

88. $\sqrt[3]{-16x^4}$ $\quad -2x\sqrt[3]{2x}$

89. $\sqrt[3]{54x^3z^6}$ $\quad 3xz^2\sqrt[3]{2}$

90. $\sqrt[3]{-24x^3y^5}$ $\quad -2xy\sqrt[3]{3y^2}$

91. $\sqrt[3]{-81x^2y^3}$ $\quad -3y\sqrt[3]{3x^2}$

92. $\sqrt[3]{81y^2z^3}$ $\quad 3z\sqrt[3]{3y^2}$

93. $\sqrt[3]{\dfrac{27m^3}{8n^6}}$ $\quad \dfrac{3m}{2n^2}$

94. $\sqrt[3]{\dfrac{125t^9}{27s^6}}$ $\quad \dfrac{5t^3}{3s^2}$

95. $\sqrt[3]{\dfrac{r^4s^5}{1,000t^3}}$ $\quad \dfrac{rs\sqrt[3]{rs^2}}{10t}$

96. $\sqrt[3]{\dfrac{54m^4n^3}{r^3s^6}}$ $\quad \dfrac{3mn\sqrt[3]{2m}}{rs^2}$

🖩 APPLICATIONS

In Exercises 97–100, use a calculator to help solve each problem.

97. AMUSEMENT PARK RIDE Illustration 3 shows the "Swashbuckler" pirate ship ride. The time (in seconds) it takes to swing from one extreme to the other is given by

$$t = \pi\sqrt{\dfrac{L}{32}}$$

a. Find t and express it in simplified radical form. Leave π in your answer. $\dfrac{3\pi\sqrt{3}}{4}$ sec

b. Express your answer to part a as a decimal. Round to the nearest tenth of a second. 4.1 sec

ILLUSTRATION 3

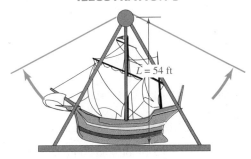

$L = 54$ ft

98. HERB GARDEN The perimeter of the herb garden shown in Illustration 4 is given by

$$p = 2\pi \sqrt{\frac{a^2 + b^2}{2}}$$

a. Find the length of fencing (in meters) needed to enclose the garden. Express the result in simplified radical form. Leave π in your answer. $10\pi\sqrt{2}$ m
b. Express the result from part a as a decimal. Round to the nearest tenth of a meter. 44.4 m

ILLUSTRATION 4

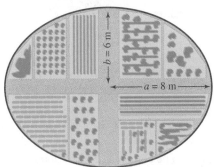

$b = 6$ m

$a = 8$ m

99. ARCHAEOLOGY Framed grids, made up of 20 cm × 20 cm squares, are often used to record the location of artifacts found during an excavation. (See Illustration 5.)

a. Use the distance formula to determine the *exact* distance between a piece of pottery found at point A and a cooking utensil found at point B. $60\sqrt{2}$ cm
b. Approximate the distance to the nearest tenth of a centimeter. 84.9 cm

ILLUSTRATION 5

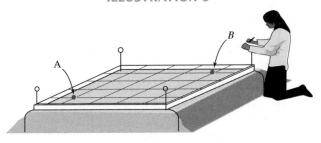

A

B

100. ENVIRONMENTAL PROTECTION A new campground is to be constructed 2 miles from a major highway, as shown in Illustration 6. The proposed entrance, although longer than the direct route, bypasses a grove of old-growth redwood trees.

a. Use the Pythagorean theorem to find the length of the entrance road. Express the result as a radical in simplified form. $2\sqrt{5}$ mi
b. Express the result from part a as a decimal. Round to the nearest hundredth of a mile.
   4.47 mi
c. How much longer is the proposed entrance as compared to the direct route into the campground? about 2.47 mi

ILLUSTRATION 6

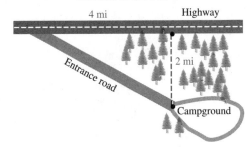

4 mi   Highway

Entrance road

2 mi

Campground

WRITING

101. State the multiplication property of radicals.
102. When comparing $\sqrt{8}$ and $2\sqrt{2}$, why is $2\sqrt{2}$ called simplified radical form?

REVIEW

103. Multiply $(-2a^3)(3a^2)$. $-6a^5$
104. Find the slope of the line passing through $(-6, 0)$ and $(0, -4)$. $-\frac{2}{3}$
105. Write the equation of the line passing through $(0, 3)$ with slope -2. $y = -2x + 3$
106. Solve $-x = -5$. 5
107. Solve $-x > -5$. $x < 5$
108. What is the slope of a line perpendicular to a line with a slope of 2? $-\frac{1}{2}$

▶ 5.5   Adding and Subtracting Radical Expressions

In this section, you will learn about

Combining like radicals ▪ Combining expressions containing higher-order radicals

Introduction   We have previously discussed how to add and subtract like terms. In this section, we will introduce a similar topic: how to add and subtract expressions that contain like radicals.

Combining Like Radicals

When adding monomials, we can often combine **like terms.** For example,

$$3x + 5x = (3 + 5)x \quad \text{Use the distributive property.}$$
$$= 8x \qquad\qquad \text{Do the addition.}$$

 WARNING! The expression $3x + 5y$ cannot be simplified, because $3x$ and $5y$ are not like terms.

It is often possible to combine terms that contain *like radicals.*

> **Like radicals**

> Radicals are called **like radicals** when they have the same index and the same radicand.

Expressions that contain like radicals can be combined by addition and subtraction. For example, we have

$$3\sqrt{2} + 5\sqrt{2} = (3+5)\sqrt{2} \quad \text{Use the distributive property.}$$
$$= 8\sqrt{2} \qquad\qquad \text{Do the addition.}$$

Likewise, we can simplify the expression $5x\sqrt{3y} - 2x\sqrt{3y}$.

$$5x\sqrt{3y} - 2x\sqrt{3y} = (5x - 2x)\sqrt{3y} \quad \text{Use the distributive property.}$$
$$= 3x\sqrt{3y} \qquad\qquad \text{Do the subtraction.}$$

 WARNING! The expression $3\sqrt{2} + 5\sqrt{7}$ cannot be simplified, because the radicals are unlike.

EXAMPLE 1   **Combining like radicals.** Simplify   **a.** $\sqrt{6} + 6 + 5\sqrt{6}$ and **b.** $-2\sqrt{m} - 3\sqrt{m}$.

Solution   **a.** The expression contains three terms: $\sqrt{6}$, 6, and $5\sqrt{6}$. The first and third terms have like radicals, and they can be combined.

$$\sqrt{6} + 6 + 5\sqrt{6} = 6 + \left(\sqrt{6} + 5\sqrt{6}\right) \quad \text{Group the expressions with like radicals.}$$
$$= 6 + (1 + 5)\sqrt{6} \qquad \text{Write } \sqrt{6} \text{ as } 1\sqrt{6}. \text{ Use the distributive property.}$$
$$= 6 + 6\sqrt{6} \qquad\qquad \text{Do the addition.}$$

Note that 6 and $6\sqrt{6}$ do not contain like radicals and cannot be combined.

Fractions - need common denom. to add/subtract

b. Since the expressions $-2\sqrt{m}$ and $-3\sqrt{m}$ contain like radicals, we can combine them.

$$-2\sqrt{m} - 3\sqrt{m} = (-2 - 3)\sqrt{m} \quad \text{Use the distributive property.}$$
$$= -5\sqrt{m} \qquad\quad \text{Do the subtraction: } -2 - 3 = -5.$$

SELF CHECK   Simplify **a.** $\sqrt{7} + 7 + 7\sqrt{7}$ and   *Answer:* **a.** $8\sqrt{7} + 7$,
b. $24\sqrt{m} - 25\sqrt{m}$.   **b.** $-\sqrt{m}$ ∎

Radical expressions such as $3\sqrt{18}$ and $5\sqrt{8}$ can be simplified so that they contain like radicals. They can then be combined.

EXAMPLE 2

Adding radicals. Simplify $3\sqrt{18} + 5\sqrt{8}$.

Solution   The radical $\sqrt{18}$ is not in simplified form, because 18 has a perfect square factor of 9. The radical $\sqrt{8}$ is not in simplified form either, because 8 has a perfect square factor of 4. To simplify the radicals and add the expressions, we proceed as follows.

$$3\sqrt{18} + 5\sqrt{8}$$
$$= 3\sqrt{9 \cdot 2} + 5\sqrt{4 \cdot 2} \quad \text{Factor 18 and 8 using perfect square factors.}$$
$$= 3\sqrt{9}\sqrt{2} + 5\sqrt{4}\sqrt{2} \quad \text{The square root of a product is equal to the product of the square roots.}$$
$$= 3(3)\sqrt{2} + 5(2)\sqrt{2} \quad \sqrt{9} = 3 \text{ and } \sqrt{4} = 2.$$
$$= 9\sqrt{2} + 10\sqrt{2} \quad 3(3) = 9 \text{ and } 5(2) = 10.$$
$$= 19\sqrt{2} \quad \text{Combine the expressions: } 9 + 10 = 19.$$

SELF CHECK   Simplify $2\sqrt{50} + \sqrt{32}$.   *Answer:* $14\sqrt{2}$ ∎

EXAMPLE 3

Orthopedics. Doctors sometimes use traction to help align a broken bone so that a fracture can heal properly. Figure 5-25 shows how traction is applied by fixing a weight, two pulleys, and some stainless steel cable to a broken leg. How many feet of cable are used in the setup shown in the figure?

FIGURE 5-25

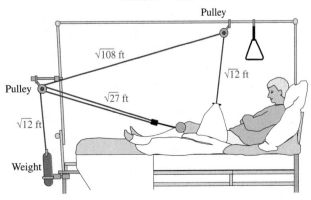

Solution   Two segments of the cable are $\sqrt{12}$ feet long, another is $\sqrt{108}$ feet long, and two others are $\sqrt{27}$ feet long. The total number of feet of cable can be found by adding

$$2\sqrt{12} + \sqrt{108} + 2\sqrt{27}$$

Since $\sqrt{12}$, $\sqrt{108}$, $\sqrt{27}$ are not like radicals, we cannot do the addition at this time. First, we need to write each radical in simplified form. Then we can add any expressions that contain like radicals.

$$2\sqrt{12} + \sqrt{108} + 2\sqrt{27}$$

$= 2\sqrt{4 \cdot 3} + \sqrt{36 \cdot 3} + 2\sqrt{9 \cdot 3}$	Factor 12, 108, and 27 using perfect squares.
$= 2\sqrt{4}\sqrt{3} + \sqrt{36}\sqrt{3} + 2\sqrt{9}\sqrt{3}$	The square root of a product is equal to the product of the square roots.
$= 2(2\sqrt{3}) + 6\sqrt{3} + 2(3\sqrt{3})$	Simplify: $\sqrt{4} = 2$, $\sqrt{36} = 6$, and $\sqrt{9} = 3$.
$= 4\sqrt{3} + 6\sqrt{3} + 6\sqrt{3}$	Do the multiplications.
$= 16\sqrt{3}$	Add the expressions that contain like radicals: $4 + 6 + 6 = 16$.

The traction setup uses $16\sqrt{3}$ feet of cable.   ∎

EXAMPLE 4   **Adding radicals.** Simplify $\sqrt{44x^2y} + x\sqrt{99y}$.

Solution   We simplify each radical and then add the expressions containing like radicals.

$$\sqrt{44x^2y} + x\sqrt{99y}$$

$= \sqrt{4x^2 \cdot 11y} + x\sqrt{9 \cdot 11y}$	Factor $44x^2y$ and $99y$.
$= \sqrt{4x^2}\sqrt{11y} + x\sqrt{9}\sqrt{11y}$	The square root of a product is equal to the product of the square roots.
$= 2x\sqrt{11y} + 3x\sqrt{11y}$	Simplify: $\sqrt{4x^2} = 2x$ and $\sqrt{9} = 3$.
$= 5x\sqrt{11y}$	Combine the expressions: $2x + 3x = 5x$.

SELF CHECK   Simplify $\sqrt{12xy^2} + \sqrt{27xy^2}$.   *Answer:* $5y\sqrt{3x}$ ∎

EXAMPLE 5   **Subtracting radicals.** Simplify $\sqrt{28x^2y} - 2\sqrt{63y^3}$.

Solution   We begin by simplifying each radical.

$$\sqrt{28x^2y} - 2\sqrt{63y^3}$$

$= \sqrt{4x^2 \cdot 7y} - 2\sqrt{9y^2 \cdot 7y}$	Factor $28x^2y$ and $63y^3$.
$= \sqrt{4x^2}\sqrt{7y} - 2\sqrt{9y^2}\sqrt{7y}$	The square root of a product is equal to the product of the square roots.
$= 2x\sqrt{7y} - 2(3y)\sqrt{7y}$	$\sqrt{4x^2} = 2x$ and $\sqrt{9y^2} = 3y$.
$= 2x\sqrt{7y} - 6y\sqrt{7y}$	

Since $2x$ and $6y$ are not like terms, the expression does not simplify further.

SELF CHECK   Simplify $\sqrt{20mn^2} - \sqrt{80m^3}$.   *Answer:* $2n\sqrt{5m} - 4m\sqrt{5m}$ ∎

EXAMPLE 6   **Adding radicals.** Simplify $\sqrt{27xy} + \sqrt{20xy}$.

Solution

$\sqrt{27xy} + \sqrt{20xy} = \sqrt{9 \cdot 3xy} + \sqrt{4 \cdot 5xy}$	Factor $27xy$ and $20xy$.
$= \sqrt{9}\sqrt{3xy} + \sqrt{4}\sqrt{5xy}$	The square root of a product is equal to the product of the square roots.
$= 3\sqrt{3xy} + 2\sqrt{5xy}$	$\sqrt{9} = 3$ and $\sqrt{4} = 2$.

Since the terms have unlike radicals, the expression does not simplify further.

SELF CHECK Simplify $\sqrt{75ab} + \sqrt{72ab}$.   *Answer:* $5\sqrt{3ab} + 6\sqrt{2ab}$ ∎

EXAMPLE 7

Adding and subtracting radicals. Simplify
$\sqrt{8x} + \sqrt{3y} - \sqrt{50x} + \sqrt{27y}$.

Solution   We simplify the radicals and then combine like radicals, where possible.

$$\sqrt{8x} + \sqrt{3y} - \sqrt{50x} + \sqrt{27y}$$
$$= \sqrt{4 \cdot 2x} + \sqrt{3y} - \sqrt{25 \cdot 2x} + \sqrt{9 \cdot 3y} \quad \text{Factor } 8x, 50x, \text{ and } 27y.$$
$$= \sqrt{4}\sqrt{2x} + \sqrt{3y} - \sqrt{25}\sqrt{2x} + \sqrt{9}\sqrt{3y}$$
$$= 2\sqrt{2x} + \sqrt{3y} - 5\sqrt{2x} + 3\sqrt{3y}$$
$$= -3\sqrt{2x} + 4\sqrt{3y} \quad \begin{array}{l}\text{Combine the expressions that}\\ \text{contain like radicals:}\\ 2 - 5 = -3 \text{ and } 1 + 3 = 4.\end{array}$$

SELF CHECK Simplify $\sqrt{32x} - \sqrt{5y} - \sqrt{200x} + \sqrt{125y}$.   *Answer:* $-6\sqrt{2x} + 4\sqrt{5y}$ ∎

Combining Expressions Containing Higher-Order Radicals

We can extend the concepts used to combine square roots to radicals with higher order.

EXAMPLE 8

Subtracting cube roots. Simplify $\sqrt[3]{81x^4} - x\sqrt[3]{24x}$.

Solution   We simplify each radical and then combine like radicals.

$$\sqrt[3]{81x^4} - x\sqrt[3]{24x} = \sqrt[3]{27x^3 \cdot 3x} - x\sqrt[3]{8 \cdot 3x} \quad \text{Factor } 81x^4 \text{ and } 24x.$$
$$= \sqrt[3]{27x^3}\sqrt[3]{3x} - x\sqrt[3]{8}\sqrt[3]{3x}$$
$$= 3x\sqrt[3]{3x} - 2x\sqrt[3]{3x}$$
$$= x\sqrt[3]{3x}$$

SELF CHECK Simplify $\sqrt[3]{24a^4} + a\sqrt[3]{81a}$.   *Answer:* $5a\sqrt[3]{3a}$ ∎

STUDY SET

Section 5.5

VOCABULARY

In Exercises 1–4, fill in the blanks to make the statements true.

1. Like ____radicals____ have the same index and the same radicand.

2. Like ____terms____ have the same variables with the same exponents.

3. Radical expressions such as $\sqrt{8}$ and $\sqrt{18}$ can be ____simplified____ so that they contain like radicals.

4. The expression $3\sqrt{2} + \sqrt{8} - 2$ contains three ____terms____.

CONCEPTS

In Exercises 5–8, tell whether the expressions contain like radicals.

5. $5\sqrt{2}$ and $2\sqrt{3}$ no

6. $7\sqrt{3x}$ and $3\sqrt{3x}$ yes

7. $125\sqrt[3]{13a}$ and $-\sqrt[3]{13a}$ yes

8. $-17\sqrt[4]{5x}$ and $25\sqrt[3]{5x}$ no

In Exercises 9–12, tell what is wrong with the work shown.

9. $7\sqrt{5} - 3\sqrt{2} = 4\sqrt{3}$

The radicals don't have the same radicand, so they can't be combined.

10. $12\sqrt{7} + 20\sqrt{11} = 32\sqrt{18}$

The radicals don't have the same radicand, so they can't be combined.

11. $7 - 3\sqrt{2} = 4\sqrt{2}$

The two terms are not like terms—they cannot be combined.

12. $12 + 20\sqrt{11} = 32\sqrt{11}$

The two terms are not like terms—they cannot be combined.

In Exercises 13–14, complete the table of values.

13.

x	$\sqrt{x} + \sqrt{3}$
3	$2\sqrt{3}$
12	$3\sqrt{3}$
27	$4\sqrt{3}$
48	$5\sqrt{3}$

14.

x	$3\sqrt{x} - \sqrt{2}$
2	$2\sqrt{2}$
8	$5\sqrt{2}$
18	$8\sqrt{2}$
32	$11\sqrt{2}$

NOTATION

In Exercises 15–16, complete each solution.

15. Add: $3\sqrt{80} + 4\sqrt{125}$.

$$3\sqrt{80} + 4\sqrt{125} = 3\sqrt{\boxed{16} \cdot 5} + 4\sqrt{\boxed{25} \cdot 5}$$
$$= 3\sqrt{16}\,\boxed{\sqrt{5}} + 4\sqrt{25}\,\boxed{\sqrt{5}}$$
$$= 3\left(\boxed{4}\right)\sqrt{5} + 4(5)\sqrt{5}$$
$$= 12\sqrt{5} + \boxed{20}\,\sqrt{5}$$
$$= 32\sqrt{5}$$

16. Subtract: $3\sqrt{125} - 2\sqrt{80}$.

$$3\sqrt{125} - 2\sqrt{80} = 3\sqrt{25 \cdot \boxed{5}} - 2\sqrt{\boxed{16} \cdot 5}$$
$$= 3\,\boxed{\sqrt{25}}\,\sqrt{5} - 2\sqrt{16}\sqrt{5}$$
$$= 3(5)\sqrt{5} - 2\left(\boxed{4}\right)\sqrt{5}$$
$$= \boxed{15}\,\sqrt{5} - 8\sqrt{5}$$
$$= 7\sqrt{5}$$

PRACTICE

In Exercises 17–50, simplify each expression.

17. $5\sqrt{7} + 4\sqrt{7}$ $9\sqrt{7}$

18. $3\sqrt{10} + 4\sqrt{10}$ $7\sqrt{10}$

19. $\sqrt{x} - 4\sqrt{x}$ $-3\sqrt{x}$

20. $\sqrt{t} - 9\sqrt{t}$ $-8\sqrt{t}$

21. $5 + 3\sqrt{3} + 3\sqrt{3}$ $5 + 6\sqrt{3}$

22. $\sqrt{5} + 2 + 3\sqrt{5}$ $2 + 4\sqrt{5}$

23. $-1 + 2\sqrt{r} - 3\sqrt{r}$ $-1 - \sqrt{r}$

24. $-8 - 5\sqrt{c} + 4\sqrt{c}$ $-8 - \sqrt{c}$

25. $\sqrt{12} + \sqrt{27}$ $5\sqrt{3}$

26. $\sqrt{20} + \sqrt{45}$ $5\sqrt{5}$

27. $\sqrt{18} - \sqrt{8}$ $\sqrt{2}$

28. $\sqrt{32} - \sqrt{18}$ $\sqrt{2}$

29. $2\sqrt{45} + 2\sqrt{80}$ $14\sqrt{5}$

30. $3\sqrt{80} + 3\sqrt{125}$ $27\sqrt{5}$

31. $2\sqrt{80} - 3\sqrt{125}$ $-7\sqrt{5}$

32. $3\sqrt{245} - 2\sqrt{180}$ $9\sqrt{5}$

33. $\sqrt{20} + \sqrt{180}$ $8\sqrt{5}$ **34.** $2\sqrt{28} + 7\sqrt{63}$ $25\sqrt{7}$

35. $\sqrt{12} - \sqrt{48}$ $-2\sqrt{3}$ **36.** $\sqrt{48} - \sqrt{75}$ $-\sqrt{3}$

37. $\sqrt{288} - 3\sqrt{200}$ $-18\sqrt{2}$

38. $\sqrt{80} - \sqrt{245}$ $-3\sqrt{5}$

39. $2\sqrt{28} + 2\sqrt{112}$ $12\sqrt{7}$

40. $4\sqrt{63} + 6\sqrt{112}$ $36\sqrt{7}$

41. $\sqrt{20} + \sqrt{45} + \sqrt{80}$ $9\sqrt{5}$

42. $\sqrt{48} + \sqrt{27} + \sqrt{75}$ $12\sqrt{3}$

43. $\sqrt{200} - \sqrt{75} + \sqrt{48}$ $10\sqrt{2} - \sqrt{3}$

44. $\sqrt{20} + \sqrt{80} - \sqrt{125}$ $\sqrt{5}$

45. $8\sqrt{6} - 5\sqrt{2} - 3\sqrt{6}$ $5\sqrt{6} - 5\sqrt{2}$

46. $3\sqrt{2} - 3\sqrt{15} - 4\sqrt{15}$ $3\sqrt{2} - 7\sqrt{15}$

47. $\sqrt{24} + \sqrt{150} + \sqrt{240}$ $7\sqrt{6} + 4\sqrt{15}$

48. $\sqrt{28} + \sqrt{63} + \sqrt{18}$ $5\sqrt{7} + 3\sqrt{2}$

49. $\sqrt{48} - \sqrt{8} + \sqrt{27} - \sqrt{32}$ $7\sqrt{3} - 6\sqrt{2}$

50. $\sqrt{162} + \sqrt{50} - \sqrt{75} - \sqrt{108}$ $14\sqrt{2} - 11\sqrt{3}$

In Exercises 51–64, simplify each expression. All variables represent positive numbers.

51. $\sqrt{2x^2} + \sqrt{8x^2}$ $3x\sqrt{2}$

52. $\sqrt{3y^2} - \sqrt{12y^2}$ $-y\sqrt{3}$

53. $\sqrt{2d^3} + \sqrt{8d^3}$ $3d\sqrt{2d}$

54. $\sqrt{3a^3} - \sqrt{12a^3}$ $-a\sqrt{3a}$

55. $\sqrt{18x^2y} - \sqrt{27x^2y}$ $3x\sqrt{2y} - 3x\sqrt{3y}$

56. $\sqrt{49xy} + \sqrt{xy}$ $8\sqrt{xy}$

57. $\sqrt{32x^5} - \sqrt{18x^5}$ $x^2\sqrt{2x}$

58. $\sqrt{27xy^3} - \sqrt{48xy^3}$ $-y\sqrt{3xy}$

59. $3\sqrt{54b^2} + 5\sqrt{24b^2}$ $19b\sqrt{6}$

60. $3\sqrt{24x^4y^3} + 2\sqrt{54x^4y^3}$ $12x^2y\sqrt{6y}$

61. $y\sqrt{490y} - 2\sqrt{360y^3}$ $-5y\sqrt{10y}$

62. $3\sqrt{20x} + 2\sqrt{63y}$ $6\sqrt{5x} + 6\sqrt{7y}$

63. $\sqrt{20x^3y} + \sqrt{45x^5y^3} - \sqrt{80x^7y^5}$

$2x\sqrt{5xy} + 3x^2y\sqrt{5xy} - 4x^3y^2\sqrt{5xy}$

64. $x\sqrt{48xy^2} - y\sqrt{27x^3} + \sqrt{75x^3y^2}$ $6xy\sqrt{3x}$

In Exercises 65–80, simplify each expression.

65. $\sqrt[3]{3} + \sqrt[3]{3}$ $2\sqrt[3]{3}$ **66.** $\sqrt[3]{2} + 5\sqrt[3]{2}$ $6\sqrt[3]{2}$

67. $2\sqrt[3]{x} - 3\sqrt[3]{x}$ $-\sqrt[3]{x}$ **68.** $4\sqrt[3]{s} - 5\sqrt[3]{s}$ $-\sqrt[3]{s}$

69. $\sqrt[3]{16} + \sqrt[3]{54}$ $5\sqrt[3]{2}$ **70.** $\sqrt[3]{24} - \sqrt[3]{81}$ $-\sqrt[3]{3}$

71. $\sqrt[3]{81} - \sqrt[3]{24}$ $\sqrt[3]{3}$ **72.** $\sqrt[3]{32} + \sqrt[3]{108}$ $5\sqrt[3]{4}$

73. $\sqrt[3]{40} + \sqrt[3]{125}$

$2\sqrt[3]{5} + 5$

74. $\sqrt[3]{3,000} - \sqrt[3]{192}$

$6\sqrt[3]{3}$

75. $\sqrt[3]{x^4} - \sqrt[3]{x^7}$ $x\sqrt[3]{x} - x^2\sqrt[3]{x}$

76. $\sqrt[3]{8x^5} + \sqrt[3]{27x^8}$ $2x\sqrt[3]{x^2} + 3x^2\sqrt[3]{x^2}$

77. $\sqrt[3]{192x^4y^5} - \sqrt[3]{24x^4y^5}$ $2xy\sqrt[3]{3xy^2}$

78. $\sqrt[3]{24a^5b^4} + \sqrt[3]{81a^5b^4}$ $5ab\sqrt[3]{3a^2b}$

79. $\sqrt[3]{135x^7y^4} - \sqrt[3]{40x^7y^4}$ $x^2y\sqrt[3]{5xy}$

80. $\sqrt[3]{56a^4b^5} + \sqrt[3]{7a^4b^5}$ $3ab\sqrt[3]{7ab^2}$

APPLICATIONS

81. **ANATOMY** Use the measurements in Illustration 1 to determine the length of the patient's arm if he lets it fall to his side. $18\sqrt{3}$ in.

ILLUSTRATION 1

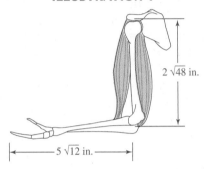

$2\sqrt{48}$ in.

$5\sqrt{12}$ in.

82. **PLAYGROUND EQUIPMENT** Find the total length of pipe necessary to construct the frame of the swing set shown in Illustration 2. $\left(16 + 24\sqrt{5}\right)$ ft

ILLUSTRATION 2

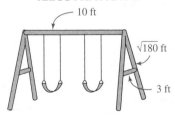

10 ft

$\sqrt{180}$ ft

3 ft

83. **READING BLUEPRINTS** What is the length of the motor on the machine shown in Illustration 3? $27\sqrt{2}$ cm

ILLUSTRATION 3

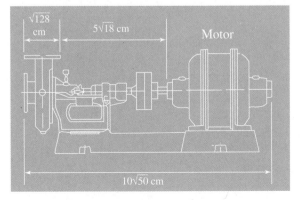

$\sqrt{128}$ cm

$5\sqrt{18}$ cm

Motor

$10\sqrt{50}$ cm

84. **TENTS** The length of a center support pole for the tents shown in Illustration 4 is given by the formula

$$l = 0.5s\sqrt{3}$$

where s is the length of the side of the tent. Find the total length of the four poles needed for the parents' and children's tents. $10\sqrt{3}$ ft

ILLUSTRATION 4

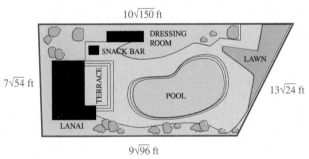

Parents' tent   Children's tent

$s = 6$ ft $s = 4$ ft

85. **FENCING** Find the number of feet of fencing needed to enclose the swimming pool complex shown in Illustration 5. $133\sqrt{6}$ ft

ILLUSTRATION 5

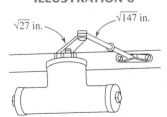

$10\sqrt{150}$ ft

DRESSING ROOM

SNACK BAR

LAWN

$7\sqrt{54}$ ft

TERRACE

POOL

$13\sqrt{24}$ ft

LANAI

$9\sqrt{96}$ ft

86. **HARDWARE** Find the difference in the lengths of the "arms" of the door-closing device shown in Illustration 6. $4\sqrt{3}$ in.

ILLUSTRATION 6

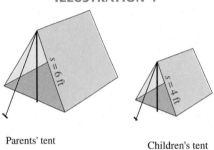

$\sqrt{27}$ in. $\sqrt{147}$ in.

WRITING

87. Explain why $\sqrt{3} + \sqrt{2}$ cannot be combined.

88. Explain why $\sqrt{4x}$ and $\sqrt[3]{4x}$ cannot be combined.

REVIEW

In Exercises 89–96, simplify each expression. Write each answer without using negative exponents.

89. 3^{-2} $\frac{1}{9}$

90. $\dfrac{1}{3^{-2}}$ 9

93. x^{-3} $\frac{1}{x^3}$

94. $\dfrac{1}{x^{-3}}$ x^3

91. -3^2 -9

92. -3^{-2} $-\frac{1}{9}$

95. 3^0 1

96. x^0 1

▶ 5.6   Multiplying and Dividing Radical Expressions

In this section, you will learn about

Multiplying radical expressions ■ Multiplying monomials containing radicals ■ Multiplying polynomials containing radicals by monomials containing radicals ■ Multiplying binomials containing radicals ■ Dividing radical expressions ■ Rationalizing monomial denominators ■ Rationalizing binomial denominators

---

Introduction   In this section, we will discuss how to multiply and divide radical expressions. We will also learn the technique of rationalizing a denominator.

Multiplying Radical Expressions

Recall that the *product of the square roots of two nonnegative numbers is equal to the square root of the product of those numbers.* For example,

$$\sqrt{2}\sqrt{8} = \sqrt{2 \cdot 8} \qquad \sqrt{3}\sqrt{27} = \sqrt{3 \cdot 27} \qquad \sqrt{x}\sqrt{x^3} = \sqrt{x \cdot x^3}$$
$$= \sqrt{16} \qquad\qquad\quad = \sqrt{81} \qquad\qquad\quad = \sqrt{x^4}$$
$$= 4 \qquad\qquad\qquad = 9 \qquad\qquad\qquad = x^2$$

Likewise, the *product of the cube roots of two numbers is equal to the cube root of the product of those numbers.* For example,

$$\sqrt[3]{2}\sqrt[3]{4} = \sqrt[3]{2 \cdot 4} \qquad \sqrt[3]{4}\sqrt[3]{16} = \sqrt[3]{4 \cdot 16} \qquad \sqrt[3]{3x^2}\sqrt[3]{9x} = \sqrt[3]{3x^2 \cdot 9x}$$
$$= \sqrt[3]{8} \qquad\qquad\quad = \sqrt[3]{64} \qquad\qquad\quad = \sqrt[3]{27x^3}$$
$$= 2 \qquad\qquad\qquad = 4 \qquad\qquad\qquad = 3x$$

EXAMPLE 1

Multiplying radicals. Multiply   **a.** $\sqrt{3}\sqrt{2}$,   **b.** $\sqrt{6}\sqrt{8}$, and   **c.** $\sqrt[3]{4}\sqrt[3]{10}$.

Solution   **a.** $\sqrt{3}\sqrt{2} = \sqrt{3 \cdot 2}$ The product of the square roots of two numbers is equal to the square root of the product of those numbers.

$\qquad\qquad\quad = \sqrt{6}$ Do the multiplication within the radical.

b. $\sqrt{6}\sqrt{8} = \sqrt{6 \cdot 8}$ The product of two square roots is equal to the square root of the product.

$\qquad\qquad\quad = \sqrt{48}$ Do the multiplication within the radical. Note that this radical can be simplified.

$\qquad\qquad\quad = \sqrt{16}\sqrt{3}$ Write $\sqrt{48}$ as $\sqrt{16}\sqrt{3}$.

$\qquad\qquad\quad = 4\sqrt{3}$ Simplify: $\sqrt{16} = 4$.

c. $\sqrt[3]{4}\sqrt[3]{10} = \sqrt[3]{4 \cdot 10}$ The product of two cube roots is equal to the cube root of the product.

$\phantom{\textbf{c.} \sqrt[3]{4}\sqrt[3]{10}} = \sqrt[3]{40}$ Do the multiplication within the radical.

$\phantom{\textbf{c.} \sqrt[3]{4}\sqrt[3]{10}} = \sqrt[3]{8}\sqrt[3]{5}$ Write $\sqrt[3]{40}$ as $\sqrt[3]{8}\sqrt[3]{5}$.

$\phantom{\textbf{c.} \sqrt[3]{4}\sqrt[3]{10}} = 2\sqrt[3]{5}$ Simplify: $\sqrt[3]{8} = 2$.

SELF CHECK   Multiply   **a.** $\sqrt{5}\sqrt{3}$,   **b.** $\sqrt{8}\sqrt{9}$,   and   *Answers:*   **a.** $\sqrt{15}$,   **b.** $6\sqrt{2}$,
c. $\sqrt[3]{6}\sqrt[3]{9}$.   **c.** $3\sqrt[3]{2}$.   ■

Multiplying Monomials Containing Radicals

To multiply monomials containing radical expressions, we multiply the coefficients and multiply the radicals separately and then simplify the result, when possible.

EXAMPLE 2

Multiplying radical expressions. Multiply   **a.** $3\sqrt{6}$ by $4\sqrt{3}$ and
b. $-2\sqrt[3]{7x}$ by $6\sqrt[3]{49x^2}$.

Solution   The commutative and associative properties enable us to multiply the coefficients and the radicals separately.

a. $3\sqrt{6} \cdot 4\sqrt{3} = 3(4)\sqrt{6}\sqrt{3}$ Write the coefficients together and the radicals together.

$\phantom{\textbf{a.} 3\sqrt{6} \cdot 4\sqrt{3}} = 12\sqrt{18}$ Multiply the coefficients and multiply the radicals.

$\phantom{\textbf{a.} 3\sqrt{6} \cdot 4\sqrt{3}} = 12\sqrt{9}\sqrt{2}$ Write $\sqrt{18}$ as $\sqrt{9}\sqrt{2}$.

$\phantom{\textbf{a.} 3\sqrt{6} \cdot 4\sqrt{3}} = 12(3)\sqrt{2}$ Simplify: $\sqrt{9} = 3$.

$\phantom{\textbf{a.} 3\sqrt{6} \cdot 4\sqrt{3}} = 36\sqrt{2}$ Do the multiplication: $12(3) = 36$.

b. $-2\sqrt[3]{7x} \cdot 6\sqrt[3]{49x^2} = -2(6)\sqrt[3]{7x}\sqrt[3]{49x^2}$ Write the coefficients together and the radicals together.

$\phantom{\textbf{b.} -2\sqrt[3]{7x} \cdot 6\sqrt[3]{49x^2}} = -12\sqrt[3]{7x \cdot 49x^2}$ Multiply the coefficients and multiply the radicals.

$\phantom{\textbf{b.} -2\sqrt[3]{7x} \cdot 6\sqrt[3]{49x^2}} = -12\sqrt[3]{343x^3}$ Do the multiplication within the radical.

$\phantom{\textbf{b.} -2\sqrt[3]{7x} \cdot 6\sqrt[3]{49x^2}} = -12(7x)$ Simplify: $\sqrt[3]{343x^3} = 7x$.

$\phantom{\textbf{b.} -2\sqrt[3]{7x} \cdot 6\sqrt[3]{49x^2}} = -84x$ Multiply.

SELF CHECK   Multiply   **a.** $\left(2\sqrt{2x}\right)\left(-3\sqrt{3x}\right)$ and
b. $\left(5\sqrt[3]{2}\right)\left(2\sqrt[3]{4}\right)$.   *Answers:*   **a.** $-6x\sqrt{6}$,   **b.** 20   ■

EXAMPLE 3

Powers of radical expressions. Find $\left(2\sqrt{5}\right)^2$.

Solution   Recall that a power is used to indicate repeated multiplication.

$\left(2\sqrt{5}\right)^2 = 2\sqrt{5} \cdot 2\sqrt{5}$ Write $2\sqrt{5}$ as a factor two times.

$\phantom{\left(2\sqrt{5}\right)^2} = 2(2)\sqrt{5}\sqrt{5}$ Multiply the coefficients and the radicals separately.

$\phantom{\left(2\sqrt{5}\right)^2} = 4\sqrt{5 \cdot 5}$ The product of two square roots is equal to the square root of the product.

$\phantom{\left(2\sqrt{5}\right)^2} = 4\sqrt{25}$ Do the multiplication within the radical.

$\phantom{\left(2\sqrt{5}\right)^2} = 4 \cdot 5$ $\sqrt{25} = 5$.

$\phantom{\left(2\sqrt{5}\right)^2} = 20$

SELF CHECK   Find $\left(3\sqrt[3]{-2}\right)^3$.   *Answer:* -54 ■

Multiplying Polynomials Containing Radicals by Monomials Containing Radicals

Recall that to multiply a polynomial by a monomial, we use the distributive property to remove parentheses, and then we combine like terms.

EXAMPLE 4

Using the distributive property. Multiply **a.** $\sqrt{2x}(\sqrt{6x} + \sqrt{8x})$ and
b. $\sqrt[3]{3}(\sqrt[3]{9} - 2)$.

Solution **a.** $\sqrt{2x}(\sqrt{6x} + \sqrt{8x}) = \sqrt{2x}\sqrt{6x} + \sqrt{2x}\sqrt{8x}$ Use the distributive property to remove parentheses.

$$= \sqrt{12x^2} + \sqrt{16x^2}$$ The product of two square roots is equal to the square root of the product.

$$= \sqrt{4x^2 \cdot 3} + \sqrt{16x^2}$$ Factor 12 as $4 \cdot 3$.

$$= \sqrt{4x^2}\sqrt{3} + \sqrt{16x^2}$$ The square root of a product is equal to the product of the square roots.

$$= 2x\sqrt{3} + 4x$$ Simplify: $\sqrt{4x^2} = 2x$ and $\sqrt{16x^2} = 4x$.

b. $\sqrt[3]{3}(\sqrt[3]{9} - 2) = \sqrt[3]{3}\sqrt[3]{9} - 2\sqrt[3]{3}$ Use the distributive property to remove parentheses.

$$= \sqrt[3]{27} - 2\sqrt[3]{3}$$ The product of two cube roots is equal to the cube root of the product.

$$= 3 - 2\sqrt[3]{3}$$ Simplify: $\sqrt[3]{27} = 3$.

SELF CHECK   Multiply **a.** $\sqrt{3}(3\sqrt{6} - \sqrt{3})$ and
b. $\sqrt[3]{2x}(3 - \sqrt[3]{4x^2})$.   *Answers:* **a.** $9\sqrt{2} - 3$,
b. $3\sqrt[3]{2x} - 2x$ ■

Multiplying Binomials Containing Radicals

To multiply two binomials, we multiply each term of one binomial by each term of the other binomial and simplify.

EXAMPLE 5

Using the FOIL method. Multiply $(\sqrt{3} + \sqrt{2})(\sqrt{3} - \sqrt{2})$.

Solution We can find the product of the binomials with the FOIL method.

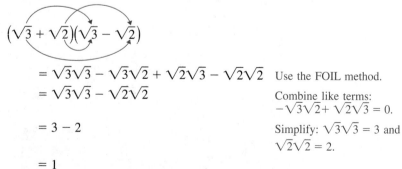

$$= \sqrt{3}\sqrt{3} - \sqrt{3}\sqrt{2} + \sqrt{2}\sqrt{3} - \sqrt{2}\sqrt{2}$$ Use the FOIL method.
$$= \sqrt{3}\sqrt{3} - \sqrt{2}\sqrt{2}$$ Combine like terms: $-\sqrt{3}\sqrt{2} + \sqrt{2}\sqrt{3} = 0$.

$$= 3 - 2$$ Simplify: $\sqrt{3}\sqrt{3} = 3$ and $\sqrt{2}\sqrt{2} = 2$.

$$= 1$$

SELF CHECK   Multiply $(\sqrt{5} + \sqrt{3})(\sqrt{5} - \sqrt{3})$.   *Answer:* 2   ■

EXAMPLE 6

Using the FOIL method. Multiply $(\sqrt{3x} + 1)(\sqrt{3x} + 2)$.

Solution

$$(\sqrt{3x} + 1)(\sqrt{3x} + 2)$$

$$= \sqrt{3x}\sqrt{3x} + 2\sqrt{3x} + \sqrt{3x} + 2 \quad \text{Use the FOIL method.}$$

$$= \sqrt{3x}\sqrt{3x} + 3\sqrt{3x} + 2 \quad \begin{array}{l}\text{Combine like terms:}\\ 2\sqrt{3x} + \sqrt{3x} = 3\sqrt{3x}.\end{array}$$

$$= 3x + 3\sqrt{3x} + 2 \quad \text{Simplify: } \sqrt{3x}\sqrt{3x} = 3x.$$

SELF CHECK   Multiply $(\sqrt{5a} - 2)(\sqrt{5a} + 3)$.   *Answer:* $5a + \sqrt{5a} - 6$ ∎

EXAMPLE 7

Multiplying binomials. Multiply $(\sqrt[3]{4x} - 3)(\sqrt[3]{2x^2} + 1)$.

Solution

$$(\sqrt[3]{4x} - 3)(\sqrt[3]{2x^2} + 1)$$

$$= \sqrt[3]{4x}\sqrt[3]{2x^2} + \sqrt[3]{4x} - 3\sqrt[3]{2x^2} - 3 \quad \text{Use the FOIL method.}$$

$$= \sqrt[3]{8x^3} + \sqrt[3]{4x} - 3\sqrt[3]{2x^2} - 3 \quad \begin{array}{l}\text{The product of two cube roots is}\\ \text{equal to the cube root of the product.}\end{array}$$

$$= 2x + \sqrt[3]{4x} - 3\sqrt[3]{2x^2} - 3 \quad \text{Simplify: } \sqrt[3]{8x^3} = 2x.$$

SELF CHECK   Multiply $(\sqrt[3]{3x} + 1)(\sqrt[3]{9x^2} - 2)$.   *Answer:* $3x - 2\sqrt[3]{3x} + \sqrt[3]{9x^2} - 2$ ∎

Dividing Radical Expressions

To divide radical expressions, we use the division property of radicals. For example, to divide $\sqrt{108}$ by $\sqrt{36}$, we proceed as follows:

$$\frac{\sqrt{108}}{\sqrt{36}} = \sqrt{\frac{108}{36}} \quad \text{The quotient of two square roots is the square root of the quotient.}$$

$$= \sqrt{3} \quad \text{Do the division within the radical: } 108 \div 36 = 3.$$

EXAMPLE 8

Dividing radical expressions. Divide $\dfrac{\sqrt{22a^2}}{\sqrt{99a^4}}$ $(a > 0)$.

Solution

$$\frac{\sqrt{22a^2}}{\sqrt{99a^4}} = \sqrt{\frac{22a^2}{99a^4}}$$

$$= \sqrt{\frac{2}{9a^2}} \quad \text{Simplify the radicand: } \frac{22a^2}{99a^4} = \frac{\overset{1}{\cancel{11}} \cdot 2 \cdot \overset{1}{\cancel{a^2}}}{\underset{1}{\cancel{11}} \cdot 9 \cdot \underset{1}{\cancel{a^2}} \cdot a^2} = \frac{2}{9a^2}.$$

$$= \frac{\sqrt{2}}{\sqrt{9a^2}} \quad \begin{array}{l}\text{The square root of a quotient is equal to the}\\ \text{quotient of the square roots.}\end{array}$$

$$= \frac{\sqrt{2}}{3a} \quad \text{Simplify: } \sqrt{9a^2} = 3a.$$

SELF CHECK   Divide $\dfrac{\sqrt{30y^9}}{\sqrt{160y^5}}$ $(y > 0)$.   *Answer:* $\dfrac{y^2\sqrt{3}}{4}$ ∎

Rationalizing Monomial Denominators

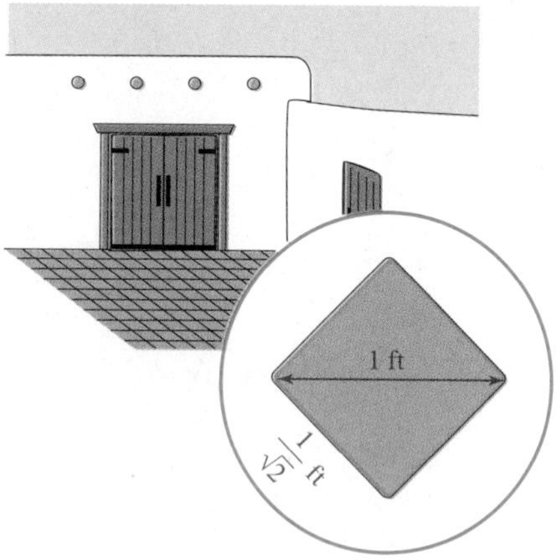

The length of a diagonal of one of the square adobe titles shown in Figure 5-26 is 1 foot. Using the Pythagorean theorem, it can be shown that the length of a side of a tile is $\frac{1}{\sqrt{2}}$ feet. Because the expression $\frac{1}{\sqrt{2}}$ contains a radical in its denominator, it is not in simplified radical form. Since it is often easier to work with a radical expression if the denominator does not contain a radical, we now consider a process in which we change the denominator from a radical that represents an irrational number to a rational number. The process is called **rationalizing the denominator.**

FIGURE 5-26

To rationalize the denominator of $\frac{1}{\sqrt{2}}$, we multiply both the numerator and the denominator by $\sqrt{2}$. Because the expression is multiplied by $\frac{\sqrt{2}}{\sqrt{2}}$, which is 1, the value of $\frac{1}{\sqrt{2}}$ is not changed—only its form.

$$\frac{1}{\sqrt{2}} = \frac{1\sqrt{2}}{\sqrt{2}\sqrt{2}}$$ Multiply both numerator and denominator by $\sqrt{2}$.

$$= \frac{\sqrt{2}}{2}$$ In the numerator, $1\sqrt{2} = \sqrt{2}$. In the denominator, $\sqrt{2}\sqrt{2}$ is the rational number 2.

The length of a side of a patio tile is $\frac{1}{\sqrt{2}} = \frac{\sqrt{2}}{2}$ feet.

This example suggests the following procedure for rationalizing monomial square root denominators.

Rationalizing denominators

Multiply the numerator and the denominator by the smallest factor that gives a perfect integer square radicand in the denominator.

EXAMPLE 9

Rationalizing the denominator. Rationalize each denominator:

a. $\sqrt{\dfrac{5}{3}}$ and   **b.** $\dfrac{2}{\sqrt[3]{3}}$.

Solution   **a.** The expression $\sqrt{\dfrac{5}{3}}$ is not in simplified form, because the radicand is a fraction. To write it in simplified form, we apply the division property of radicals. Then we rationalize the denominator by multiplying the numerator and the denominator by $\sqrt{3}$.

$$\sqrt{\frac{5}{3}} = \frac{\sqrt{5}}{\sqrt{3}}$$ The square root of a quotient is the quotient of the square roots. Note that the denominator is the irrational number $\sqrt{3}$.

$$= \frac{\sqrt{5}\sqrt{3}}{\sqrt{3}\sqrt{3}}$$ Multiply the numerator and the denominator by $\sqrt{3}$.

$$= \frac{\sqrt{15}}{3}$$ In the numerator, do the multiplication. Simplify in the denominator: $\sqrt{3}\sqrt{3} = 3$.

b. The denominator contains a cube root. We multiply by the smallest factor that gives a **perfect integer cube** radicand. Since $\sqrt[3]{3}\sqrt[3]{9} = \sqrt[3]{27}$ and 27 is a perfect integer cube, we multiply the numerator and denominator by $\sqrt[3]{9}$ and simplify.

$$\frac{2}{\sqrt[3]{3}} = \frac{2\sqrt[3]{9}}{\sqrt[3]{3}\sqrt[3]{9}}$$

$$= \frac{2\sqrt[3]{9}}{\sqrt[3]{27}} \qquad \text{Multiply: } \sqrt[3]{3}\sqrt[3]{9} = \sqrt[3]{27}.$$

$$= \frac{2\sqrt[3]{9}}{3} \qquad \text{Simplify: } \sqrt[3]{27} = 3. \text{ The denominator is now a rational number.}$$

SELF CHECK Rationalize each denominator: **a.** $\sqrt{\dfrac{2}{7}}$ and
b. $\dfrac{5}{\sqrt[3]{5}}$.   *Answers:* **a.** $\dfrac{\sqrt{14}}{7}$, **b.** $\sqrt[3]{25}$ ∎

EXAMPLE 10   **Rationalizing denominators.** Divide $\dfrac{5\sqrt{y}}{\sqrt{20x}}$ $(x > 0)$.

Solution   To rationalize the denominator, we don't need to multiply numerator and denominator by $\sqrt{20x}$. To keep the numbers small, we can multiply by $\sqrt{5x}$, because $5x \cdot 20x = 100x^2$, which is a perfect square.

$$\frac{5\sqrt{y}}{\sqrt{20x}} = \frac{5\sqrt{y}\sqrt{5x}}{\sqrt{20x}\sqrt{5x}} \qquad \text{Multiply the numerator and denominator by } \sqrt{5x}.$$

$$= \frac{5\sqrt{5xy}}{\sqrt{100x^2}} \qquad \text{Multiply: } \sqrt{y}\sqrt{5x} = \sqrt{5xy} \text{ and } \sqrt{20x}\sqrt{5x} = \sqrt{100x^2}.$$

$$= \frac{5\sqrt{5xy}}{10x} \qquad \text{Simplify: } \sqrt{100x^2} = 10x.$$

$$= \frac{\sqrt{5xy}}{2x} \qquad \text{Simplify: } \tfrac{5}{10} = \tfrac{1}{2}.$$

SELF CHECK   Divide $\dfrac{6\sqrt{z}}{\sqrt{50y}}$ $(y > 0)$.   *Answer:* $\dfrac{3\sqrt{2yz}}{5y}$ ∎

Rationalizing Binomial Denominators

At times, we will encounter fractions such as

$$\frac{2}{\sqrt{3} - 1}$$

whose denominator is a *binomial* that contains a radical. Note that the denominator, $\sqrt{3} - 1$, is an irrational number. Because the denominator is a binomial, multiplying the denominator by $\sqrt{3}$ will not make it a rational number. The key to rationalizing this denominator is to multiply the numerator and denominator by $\sqrt{3} + 1$, because the product $(\sqrt{3} + 1)(\sqrt{3} - 1)$ has no radicals. Radical expressions such as $\sqrt{3} + 1$ and $\sqrt{3} - 1$ are called **conjugates** of each other.

EXAMPLE 11

Multiplying by the conjugate. Divide $\dfrac{2}{\sqrt{3} - 1}$.

Solution We rationalize the denominator by multiplying the numerator and denominator by the conjugate of the denominator.

$$\dfrac{2}{\sqrt{3} - 1} = \dfrac{2(\sqrt{3} + 1)}{(\sqrt{3} - 1)(\sqrt{3} + 1)}$$ Multiply the numerator and denominator by the conjugate of the denominator, which is $\sqrt{3} + 1$.

$$= \dfrac{2(\sqrt{3} + 1)}{3 - 1}$$ Multiply the binomials in the denominator: $(\sqrt{3} - 1)(\sqrt{3} + 1) = 3 + \sqrt{3} - \sqrt{3} - 1 = 3 - 1$.

$$= \dfrac{2(\sqrt{3} + 1)}{2}$$ Subtract. The denominator is a rational number.

$$= \sqrt{3} + 1$$ Divide out the common factor of 2.

SELF CHECK Divide $\dfrac{3}{\sqrt{2} + 1}$. *Answer:* $3(\sqrt{2} - 1)$

EXAMPLE 12

Binomial denominators. Divide $\dfrac{\sqrt{x} + 1}{\sqrt{x} - 1}$ $(x \neq 1)$.

Solution We multiply the numerator and denominator by the conjugate of the denominator, which is $\sqrt{x} + 1$.

$$\dfrac{\sqrt{x} + 1}{\sqrt{x} - 1} = \dfrac{(\sqrt{x} + 1)(\sqrt{x} + 1)}{(\sqrt{x} - 1)(\sqrt{x} + 1)}$$ Multiply numerator and denominator by $\sqrt{x} + 1$.

$$= \dfrac{\sqrt{x}\sqrt{x} + \sqrt{x}(1) + 1(\sqrt{x}) + 1}{\sqrt{x}\sqrt{x} + \sqrt{x}(1) - 1(\sqrt{x}) - 1}$$ Multiply the binomials using the FOIL method.

$$= \dfrac{x + 2\sqrt{x} + 1}{x - 1}$$ Simplify: $\sqrt{x}\sqrt{x} = x$. Combine like terms.

SELF CHECK Divide $\dfrac{\sqrt{x} - 1}{\sqrt{x} + 1}$. *Answer:* $\dfrac{x - 2\sqrt{x} + 1}{x - 1}$

STUDY SET

Section 5.6

VOCABULARY

In Exercises 1–6, fill in the blanks to make the statements true.

1. The method of changing a radical denominator of a fraction into a rational number is called ___rationalizing___ the denominator.

2. The ___numerator___ of the fraction $\dfrac{4}{\sqrt{3}}$ is 4 and the ___denominator___ is $\sqrt{3}$.

3. $3 + \sqrt{2}$ is the ___conjugate___ of $3 - \sqrt{2}$.

4. Algebraic expressions having exactly two terms, such as $1 + \sqrt{3}$, are called ___binomials___.

5. In the radical expression $3\sqrt{7}$, 3 is the ___coefficient___ of the radical.

6. Nonterminating, nonrepeating decimals such as $\sqrt{2} = 1.414213562. . .$ and $\sqrt{3} = 1.732050808. . .$ are ___irrational___ numbers.

CONCEPTS

In Exercises 7–12, fill in the blanks to make the statements true.

7. To change $\sqrt{11}$ into a perfect integer square, we multiply it by $\boxed{\sqrt{11}}$.

8. To change $\sqrt[3]{11}$ into a perfect integer cube, we multiply it by $\boxed{\sqrt[3]{121}}$.

9. To rationalize the denominator of

$$\frac{x}{\sqrt{7}}$$

we multiply the numerator and denominator by $\boxed{\sqrt{7}}$.

10. To rationalize the denominator of

$$\frac{x}{\sqrt{x}+1}$$

we multiply the numerator and denominator by $\boxed{\sqrt{x}-1}$.

11. Explain why each expression is not in simplified radical form.

  a. $\sqrt{\dfrac{3}{4}}$ The radicand is a fraction.   **b.** $\dfrac{1}{\sqrt{10}}$ There is a radical in the denominator.

12. To multiply $2\sqrt{x}$ and $6\sqrt{x}$, we first multiply the $\underline{\text{coefficients}}$, then multiply the $\underline{\text{radicals}}$, and simplify the result.

13. Which fractions have a rational denominator and which have an irrational denominator?

$$\frac{\sqrt{5}}{3}, \quad \frac{2}{\sqrt{6}}, \quad -\frac{\sqrt{2}}{8}, \quad \frac{1+\sqrt{3}}{4}, \quad \frac{9}{7-\sqrt{10}}$$

rational: $\frac{\sqrt{5}}{3}, -\frac{\sqrt{2}}{8}, \frac{1+\sqrt{3}}{4}$; irrational: $\frac{2}{\sqrt{6}}, \frac{9}{7-\sqrt{10}}$

14. To multiply $\left(\sqrt{3}+\sqrt{2}\right)\left(\sqrt{7}+\sqrt{5}\right)$, we use the FOIL method. What are the

  a. First terms? $\sqrt{3}, \sqrt{7}$ **b.** Outer terms? $\sqrt{3}, \sqrt{5}$

  c. Inner terms? $\sqrt{2}, \sqrt{7}$ **d.** Last terms? $\sqrt{2}, \sqrt{5}$

In Exercises 15–16, do each operation, if possible.

15. a. $\sqrt{2}+\sqrt{3}$ **b.** $\sqrt{2}\cdot\sqrt{3}$ $\sqrt{6}$
    not possible

  c. $\sqrt{2}-\sqrt{3}$ **d.** $\dfrac{\sqrt{2}}{\sqrt{3}}$ $\frac{\sqrt{6}}{3}$
    not possible

16. a. $\sqrt{2}+3\sqrt{2}$ $4\sqrt{2}$ **b.** $\sqrt{2}\cdot 3\sqrt{2}$ 6

  c. $\sqrt{2}-3\sqrt{2}$ **d.** $\dfrac{\sqrt{2}}{3\sqrt{2}}$ $\frac{1}{3}$
    $-2\sqrt{2}$

NOTATION

In Exercises 17–18, complete each solution.

17. Multiply $\left(\sqrt{x}+\sqrt{2}\right)\left(\sqrt{x}-3\sqrt{2}\right)$.

$$\left(\sqrt{x}+\sqrt{2}\right)\left(\sqrt{x}-3\sqrt{2}\right)$$
$$= \sqrt{x}\,\boxed{\sqrt{x}} - \sqrt{x}(3\sqrt{2}) + \sqrt{2}\,\boxed{\sqrt{x}} - \sqrt{2}(3\sqrt{2})$$
$$= x - 3\,\boxed{\sqrt{2x}} + \sqrt{2x} - 3\sqrt{2}\sqrt{2}$$
$$= \boxed{x} - 2\sqrt{2x} - 3(2)$$
$$= x - 2\sqrt{2x} - 6$$

18. Divide $\dfrac{x}{\sqrt{x}-2}$.

$$\frac{x}{\sqrt{x}-2} = \frac{x\left(\sqrt{x}+2\right)}{\left(\sqrt{x}-2\right)\boxed{\left(\sqrt{x}+2\right)}}$$
$$= \frac{x\left(\sqrt{x}+2\right)}{\sqrt{x}\sqrt{x}+\boxed{\sqrt{x}}(2)-2\boxed{\sqrt{x}}-(2)(2)}$$
$$= \frac{x\left(\sqrt{x}+2\right)}{x-4}$$

PRACTICE

In Exercises 19–42, do each multiplication. All variables represent positive numbers.

19. $\left(\sqrt{5}\right)^2$ 5 **20.** $\left(\sqrt{11}\right)^2$ 11

21. $\left(3\sqrt{6}\right)^2$ 54 **22.** $\left(-7\sqrt{2}\right)^2$ 98

23. $\sqrt{2}\sqrt{8}$ 4 **24.** $\sqrt{27}\sqrt{3}$ 9

25. $\sqrt{7}\sqrt{3}$ $\sqrt{21}$ **26.** $\sqrt{2}\sqrt{11}$ $\sqrt{22}$

27. $\sqrt{8}\sqrt{7}$ $2\sqrt{14}$ **28.** $\sqrt{6}\sqrt{8}$ $4\sqrt{3}$

29. $3\sqrt{2}\sqrt{x}$ $3\sqrt{2x}$ **30.** $4\sqrt{3x}\sqrt{5y}$ $4\sqrt{15xy}$

31. $\left(-\sqrt[3]{9}\right)^3$ -9 **32.** $\left(\sqrt[3]{3}\right)^3$ 3

33. $\sqrt{x^3}\sqrt{x^5}$ x^4 **34.** $\sqrt{a^7}\sqrt{a^3}$ a^5

35. $\left(-5\sqrt{6}\right)\left(4\sqrt{3}\right)$ **36.** $\left(6\sqrt{3}\right)\left(-7\sqrt{2}\right)$
   $-60\sqrt{2}$ $-42\sqrt{6}$

37. $\left(2\sqrt[3]{4}\right)\left(3\sqrt[3]{3}\right)$ $6\sqrt[3]{12}$ **38.** $\left(-3\sqrt[3]{3}\right)\left(\sqrt[3]{5}\right)$
                                       $-3\sqrt[3]{15}$

39. $\left(4\sqrt{x}\right)\left(-2\sqrt{x}\right)$ $-8x$ **40.** $\left(3\sqrt{y}\right)\left(15\sqrt{y}\right)$ $45y$

41. $\sqrt{8x}\sqrt{2x^3}$ $4x^2$ **42.** $\sqrt{27y}\sqrt{3y^3}$ $9y^2$

In Exercises 43–54, do each multiplication. All variables represent positive numbers.

43. $\sqrt{2}\left(\sqrt{2}+1\right)$ $2+\sqrt{2}$

44. $\sqrt{5}\left(\sqrt{5}+2\right)$ $5+2\sqrt{5}$

45. $3\sqrt{3}\left(\sqrt{27}-1\right)$ $27-3\sqrt{3}$

46. $2\sqrt{2}\left(\sqrt{8}-1\right)$ $8-2\sqrt{2}$

47. $\sqrt{3}\left(\sqrt{6}+1\right)$ $3\sqrt{2}+\sqrt{3}$

48. $\sqrt{2}\left(\sqrt{6}-2\right)$ $2\sqrt{3}-2\sqrt{2}$

49. $\sqrt{x}\left(\sqrt{3x}-2\right)$ $x\sqrt{3}-2\sqrt{x}$

50. $\sqrt{y}\left(\sqrt{y}+5\right)$ $y+5\sqrt{y}$

51. $2\sqrt{x}(\sqrt{9x} + 3)$ $6x + 6\sqrt{x}$

52. $3\sqrt{z}(\sqrt{4z} - \sqrt{z})$ $3z$

53. $\sqrt[3]{7}(\sqrt[3]{49} - 2)$ $7 - 2\sqrt[3]{7}$

54. $\sqrt[3]{5}(\sqrt[3]{25} + 3)$ $5 + 3\sqrt[3]{5}$

In Exercises 55–64, do each multiplication. All variables represent positive numbers.

55. $(\sqrt{2} + 1)(\sqrt{2} - 1)$ 1

56. $(\sqrt{3} - 1)(\sqrt{3} + 1)$ 2 *Conjugates*

57. $(2\sqrt{7} - x)(3\sqrt{2} + x)$ $6\sqrt{14} + 2x\sqrt{7} - 3x\sqrt{2} - x^2$

58. $(4\sqrt{2} - \sqrt{x})(\sqrt{x} + 2\sqrt{3})$
$4\sqrt{2x} + 8\sqrt{6} - x - 2\sqrt{3x}$

59. $(\sqrt{6} + 1)^2$ $7 + 2\sqrt{6}$

60. $(3 - \sqrt{3})^2$ $12 - 6\sqrt{3}$

61. $(\sqrt{2x} + 3)(\sqrt{8x} - 6)$ $4x - 18$

62. $(\sqrt{5y} - 3)(\sqrt{20y} + 6)$ $10y - 18$

63. $(\sqrt[3]{2} + 1)(\sqrt[3]{2} + 3)$ $\sqrt[3]{4} + 4\sqrt[3]{2} + 3$

64. $(\sqrt[3]{5} - 2)(\sqrt[3]{5} - 1)$ $\sqrt[3]{25} - 3\sqrt[3]{5} + 2$

In Exercises 65–72, simplify each expression. Assume that all variables represent positive numbers.

65. $\dfrac{\sqrt{12x^3}}{\sqrt{27x}}$ $\dfrac{2x}{3}$

66. $\dfrac{\sqrt{32}}{\sqrt{98x^2}}$ $\dfrac{4}{7x}$

67. $\dfrac{\sqrt{18x}}{\sqrt{25x}}$ $\dfrac{3\sqrt{2}}{5}$

68. $\dfrac{\sqrt{27y}}{\sqrt{75y}}$ $\dfrac{3}{5}$

69. $\dfrac{\sqrt{196x}}{\sqrt{49x^3}}$ $\dfrac{2}{x}$

70. $\dfrac{\sqrt{50}}{\sqrt{98z^2}}$ $\dfrac{5}{7z}$

71. $\dfrac{\sqrt[3]{16x^6}}{\sqrt[3]{54x^3}}$ $\dfrac{2x}{3}$

72. $\dfrac{\sqrt[3]{128a^6}}{\sqrt[3]{16a^3}}$ $2a$

In Exercises 73–92, do each division by rationalizing the denominator and simplifying. All variables represent positive numbers.

73. $\dfrac{1}{\sqrt{3}}$ $\dfrac{\sqrt{3}}{3}$

74. $\dfrac{1}{\sqrt{5}}$ $\dfrac{\sqrt{5}}{5}$

75. $\sqrt{\dfrac{13}{7}}$ $\dfrac{\sqrt{91}}{7}$

76. $\sqrt{\dfrac{3}{11}}$ $\dfrac{\sqrt{33}}{11}$

77. $\dfrac{9}{\sqrt{27}}$ $\sqrt{3}$

78. $\dfrac{4}{\sqrt{20}}$ $\dfrac{2\sqrt{5}}{5}$

79. $\dfrac{5}{\sqrt[3]{5}}$ $\sqrt[3]{25}$

80. $\dfrac{7}{\sqrt[3]{7}}$ $\sqrt[3]{49}$

81. $\dfrac{3}{\sqrt{32}}$ $\dfrac{3\sqrt{2}}{8}$

82. $\dfrac{5}{\sqrt{18}}$ $\dfrac{5\sqrt{2}}{6}$

83. $\dfrac{4}{\sqrt[3]{4}}$ $2\sqrt[3]{2}$

84. $\dfrac{7}{\sqrt[3]{10}}$ $\dfrac{7\sqrt[3]{100}}{10}$

85. $\sqrt{\dfrac{12}{5}}$ $\dfrac{2\sqrt{15}}{5}$

86. $\sqrt{\dfrac{24}{7}}$ $\dfrac{2\sqrt{42}}{7}$

87. $\dfrac{10}{\sqrt{x}}$ $\dfrac{10\sqrt{x}}{x}$

88. $\dfrac{12}{\sqrt{y}}$ $\dfrac{12\sqrt{y}}{y}$

89. $\dfrac{\sqrt{9y}}{\sqrt{2x}}$ $\dfrac{3\sqrt{2xy}}{2x}$

90. $\dfrac{\sqrt{4t}}{\sqrt{3z}}$ $\dfrac{2\sqrt{3tz}}{3z}$

91. $\dfrac{\sqrt[3]{5}}{\sqrt[3]{2}}$ $\dfrac{\sqrt[3]{20}}{2}$

92. $\dfrac{\sqrt[3]{2}}{\sqrt[3]{5}}$ $\dfrac{\sqrt[3]{50}}{5}$

In Exercises 93–104, do each division by rationalizing the denominator and simplifying. All variables represent positive numbers.

93. $\dfrac{3}{\sqrt{3} - 1}$ $\dfrac{3\sqrt{3} + 3}{2}$

94. $\dfrac{3}{\sqrt{5} - 2}$ $3\sqrt{5} + 6$

95. $\dfrac{3}{\sqrt{7} + 2}$ $\sqrt{7} - 2$

96. $\dfrac{5}{\sqrt{8} + 3}$ $15 - 10\sqrt{2}$

97. $\dfrac{12}{3 - \sqrt{3}}$ $6 + 2\sqrt{3}$

98. $\dfrac{10}{5 - \sqrt{5}}$ $\dfrac{5 + \sqrt{5}}{2}$

99. $\dfrac{-\sqrt{3}}{\sqrt{3} + 1}$ $\dfrac{\sqrt{3} - 3}{2}$

100. $\dfrac{-\sqrt{2}}{\sqrt{2} - 1}$ $-2 - \sqrt{2}$

101. $\dfrac{5}{\sqrt{3} + \sqrt{2}}$ $5\sqrt{3} - 5\sqrt{2}$

102. $\dfrac{3}{\sqrt{3} - \sqrt{2}}$ $3\sqrt{3} + 3\sqrt{2}$

103. $\dfrac{\sqrt{x} + 2}{\sqrt{x} - 2}$ $\dfrac{x + 4\sqrt{x} + 4}{x - 4}$

104. $\dfrac{\sqrt{x} - 3}{\sqrt{x} + 3}$ $\dfrac{x - 6\sqrt{x} + 9}{x - 9}$

APPLICATIONS

105. ROTARY LAWNMOWER See Illustration 1, which shows the blade of a rotary lawnmower. Use the formula for the area of a circle, $A = \pi r^2$, to find the area of lawn covered by one rotation of the blade. Leave π in your answer. 108π in.^{2}

ILLUSTRATION 1

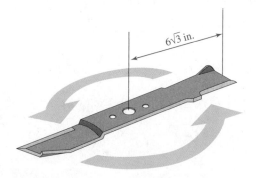

$6\sqrt{3}$ in.

106. AWARDS PLATFORMS Find the total number of cubic feet of concrete needed to construct the Olympic Games awards platforms shown in Illustration 2. $24\sqrt{2}$ ft³

ILLUSTRATION 2

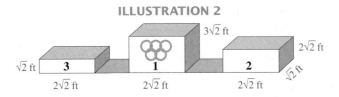

107. AIR HOCKEY GAME Find the area of the playing surface of the air hockey game in Illustration 3. $1{,}800\sqrt{2}$ in.²

ILLUSTRATION 3

108. PROJECTOR SCREEN To find the length l of a rectangle, we can use the formula

$$l = \frac{A}{w}$$

where A is the area of the rectangle and w is its width. Find the length of the screen shown in Illustration 4 if its area is 54 square feet. $\dfrac{9\sqrt{3}}{2}$ ft

ILLUSTRATION 4

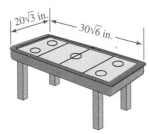

109. COSTUME DESIGN The pattern for one panel of an 1870s English dress is printed on the 1 in. × 1 in. grid shown in Illustration 5. Find the number of square inches of fabric in the trapezoidal-shaped panel. (*Hint:* Use the Pythagorean theorem to determine the lengths of the sides.)   90 in.²

110. SET DESIGN The director of a stage play requested bright downlighting over the portion of the

ILLUSTRATION 5

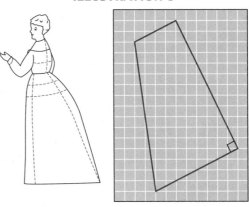

set shown in Illustration 6. Find the area of the rectangle. (*Hint:* Use the Pythagorean theorem to determine the lengths of the sides.)   34 ft²

ILLUSTRATION 6

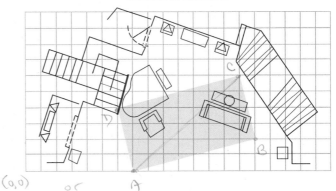

WRITING

Write a paragraph using your own words.

111. When rationalizing the denominator of $\dfrac{5}{\sqrt{6}}$, why must we multiply both the numerator *and* denominator by $\sqrt{6}$?

112. A calculator is used to find decimal approximations for the expressions $\dfrac{2}{\sqrt{6}}$ and $\dfrac{\sqrt{6}}{3}$. In each case, the calculator display reads 0.816496581. Explain why the results are the same.

REVIEW

113. Is $x = -2$ a solution of $3x - 7 = 5x + 1$?   no

114. The graph of a line passes through the point $(2, 0)$. Is this the x- or y-intercept? x-intercept

115. To evaluate the expression $2 - (-3 + 4)^2$, which operation should be performed first?   addition

116. The graph of a straight line rises from left to right. Is the slope of the line positive or negative?   positive

117. Find $(x - 4)(x + 4)$. $x^2 - 16$

118. How far will a car traveling 55 mph go in 3.5 hours?   192.5 mi

▶ **5.7** **Rational Exponents**

In this section, you will learn about

> Fractional exponents with numerators of 1 ■ Fractional exponents with numerators other than 1 ■ Rules for exponents

Introduction   We have seen that a positive integer exponent indicates the number of times that a base is to be used as a factor in a product. For example, x^4 means that x is to be used as a factor four times.

$$x^4 = \overbrace{x \cdot x \cdot x \cdot x}^{4 \text{ factors of } x}$$

Also, recall the following rules for exponents.

Rules for exponents

If m and n are natural numbers and there are no divisions by zero, then

$$x^m x^n = x^{m+n} \qquad (x^m)^n = x^{m \cdot n} \qquad (xy)^n = x^n y^n \qquad \left(\frac{x}{y}\right)^n = \frac{x^n}{y^n}$$

$$x^0 = 1 \qquad\qquad x^{-n} = \frac{1}{x^n} \qquad \frac{x^m}{x^n} = x^{m-n}$$

In this section, we will extend the definition and rules for exponents to cover fractional exponents.

Fractional Exponents with Numerators of 1

It is possible to raise numbers to fractional powers. To give meaning to rational (fractional) exponents, we consider $\sqrt{7}$. Because $\sqrt{7}$ is the positive number whose square is 7, we have

$$\left(\sqrt{7}\right)^2 = 7$$

We now consider the symbol $7^{1/2}$. If fractional exponents are to follow the same rules as integer exponents, the square of $7^{1/2}$ must be 7, because

$$(7^{1/2})^2 = 7^{(1/2)2} \quad \text{Keep the base and multiply the exponents.}$$
$$= 7^1 \qquad \tfrac{1}{2} \cdot 2 = 1.$$
$$= 7$$

Since $(7^{1/2})^2$ and $\left(\sqrt{7}\right)^2$ are both equal to 7, we define $7^{1/2}$ to be $\sqrt{7}$. Similarly, we make these definitions.

$$7^{1/3} = \sqrt[3]{7}$$
$$7^{1/7} = \sqrt[7]{7}$$

and so on.

Rational exponents

If n is a positive integer greater than 1 and $\sqrt[n]{x}$ is a real number, then

$$x^{1/n} = \sqrt[n]{x}$$

EXAMPLE 1

Rational exponents with numerators of 1. Simplify **a.** $64^{1/2}$, **b.** $64^{1/3}$, and **c.** $(-64)^{1/3}$.

Solution   **a.** $64^{1/2} = \sqrt{64} = 8$ The denominator of the fractional exponent is 2. Therefore, we find the square root of the base, which is 64.

b. $64^{1/3} = \sqrt[3]{64} = 4$ The denominator of the fractional exponent is 3. Therefore, we find the cube root of the base, which is 64.

c. $(-64)^{1/3} = \sqrt[3]{-64} = -4$ The denominator of the fractional exponent is 3. Therefore, we find the cube root of the base, which is -64.

SELF CHECK   Simplify **a.** $81^{1/2}$, **b.** $125^{1/3}$, and   *Answers:* **a.** 9, **b.** 5,
c. $(-27)^{1/3}$.   **c.** -3 ∎

Fractional Exponents with Numerators Other Than 1

We can extend the definition of $x^{1/n}$ to cover fractional exponents for which the numerator is not 1. For example, because $4^{3/2}$ can be written as $(4^{1/2})^3$, we have

$$4^{3/2} = (4^{1/2})^3 = \left(\sqrt{4}\right)^3 = 2^3 = 8$$

Because $4^{3/2}$ can also be written as $(4^3)^{1/2}$, we have

$$4^{3/2} = (4^3)^{1/2} = 64^{1/2} = \sqrt{64} = 8$$

In general, $x^{m/n}$ can be written as $(x^{1/n})^m$ or as $(x^m)^{1/n}$. Since $(x^{1/n})^m = \left(\sqrt[n]{x}\right)^m$ and $(x^m)^{1/n} = \sqrt[n]{x^m}$, we make the following definition.

Changing from rational exponents to radicals

If m and n are positive integers ($n \neq 1$) and $\sqrt[n]{x}$ is a real number, then
$$x^{m/n} = \sqrt[n]{x^m} = \left(\sqrt[n]{x}\right)^m$$

EXAMPLE 2

Rational exponents with numerators other than 1. Simplify **a.** $8^{2/3}$ and **b.** $(-27)^{4/3}$.

Solution   These expressions can be simplified in two ways. In the first, we take the root of the base and then we find the power. The second method is to find the power first and then take the root.

a. $8^{2/3} = \left(\sqrt[3]{8}\right)^2$ or $8^{2/3} = \sqrt[3]{8^2}$
       $= 2^2$ $= \sqrt[3]{64}$
       $= 4$ $= 4$

b. $(-27)^{4/3} = \left(\sqrt[3]{-27}\right)^4$ or $(-27)^{4/3} = \sqrt[3]{(-27)^4}$
         $= (-3)^4$ $= \sqrt[3]{531,441}$
         $= 81$ $= 81$

SELF CHECK   Simplify **a.** $16^{3/2}$ and **b.** $(-8)^{4/3}$.   *Answers:* **a.** 64, **b.** 16   ∎

The work in Example 2 suggests that in order to avoid large numbers, it is usually easier to take the root of the base first.

EXAMPLE 3

Rational exponents with numerators other than 1. Simplify
a. $125^{4/3}$, **b.** $9^{5/2}$, **c.** $-25^{3/2}$, and **d.** $(-27)^{2/3}$.

Solution

a. $125^{4/3} = \left(\sqrt[3]{125}\right)^4$
   $= (5)^4$
   $= 625$

b. $9^{5/2} = \left(\sqrt{9}\right)^5$
   $= (3)^5$
   $= 243$

c. $-25^{3/2} = -\left(\sqrt{25}\right)^3$
   $= -(5)^3$
   $= -125$

d. $(-27)^{2/3} = \left(\sqrt[3]{-27}\right)^2$
   $= (-3)^2$
   $= 9$

SELF CHECK   Simplify: **a.** $100^{3/2}$ and **b.** $(-8)^{2/3}$.   *Answers:* **a.** 1,000, **b.** 4 ∎

ACCENT ON TECHNOLOGY *Fractional Exponents*

To use a scientific calculator to approximate an exponential expression containing a fractional exponent, we can use the $\boxed{y^x}$ key. For example, to evaluate $6^{-2/3}$, we enter these numbers and press these keys:

Keystrokes 6 $\boxed{y^x}$ $\boxed{(}$ 2 $\boxed{+/-}$ $\boxed{\div}$ 3 $\boxed{)}$ $\boxed{=}$ $\boxed{0.302853432}$

So $6^{-2/3} \approx 0.302853432$.

To use a graphing calculator to evaluate $6^{-2/3}$, we press the following keys:

Keystrokes 6 $\boxed{\wedge}$ $\boxed{(}$ $\boxed{(-)}$ 2 $\boxed{/}$ 3 $\boxed{)}$ $\boxed{\text{ENTER}}$ $\boxed{\begin{array}{l} 6^{\wedge}(^-2/3) \\ \qquad .3028534321 \end{array}}$

Rules for Exponents

Because of the way in which $x^{1/n}$ and $x^{m/n}$ are defined, the familiar rules for exponents are valid for rational exponents. The following example illustrates the use of each rule.

EXAMPLE 4

Using the rules for exponents. Simplify:

a. $4^{2/5}4^{1/5} = 4^{2/5+1/5} = 4^{3/5}$ $x^m x^n = x^{m+n}$.

b. $(5^{2/3})^{1/2} = 5^{(2/3)(1/2)} = 5^{1/3}$ $(x^m)^n = x^{m\cdot n}$.

c. $(3x)^{2/3} = 3^{2/3}x^{2/3}$ $(xy)^m = x^m y^m$.

d. $\dfrac{4^{3/5}}{4^{2/5}} = 4^{3/5-2/5} = 4^{1/5}$ $\dfrac{x^m}{x^n} = x^{m-n}$.

e. $\left(\dfrac{3}{2}\right)^{2/5} = \dfrac{3^{2/5}}{2^{2/5}}$ $\left(\dfrac{x}{y}\right)^n = \dfrac{x^n}{y^n}$.

f. $4^{-2/3} = \dfrac{1}{4^{2/3}}$ $x^{-n} = \dfrac{1}{x^n}$.

g. $(5^{1/3})^0 = 1$ $x^0 = 1$.

SELF CHECK   Simplify **a.** $5^{1/3}5^{1/3}$, **b.** $(5^{1/3})^4$, **c.** $(3x)^{1/5}$,   *Answers:* **a.** $5^{2/3}$, **b.** $5^{4/3}$,
d. $\dfrac{5^{3/7}}{5^{2/7}}$, **e.** $\left(\dfrac{2}{3}\right)^{2/3}$, **f.** $5^{-2/7}$, and   **c.** $3^{1/5}x^{1/5}$, **d.** $5^{1/7}$, **e.** $\dfrac{2^{2/3}}{3^{2/3}}$,
g. $(12^{1/2})^0$.   **f.** $\dfrac{1}{5^{2/7}}$, **g.** 1 ∎

We can often use the rules for exponents to simplify expressions containing rational exponents.

EXAMPLE 5

Rational exponents. Simplify   **a.** $64^{-2/3}$,   **b.** $(x^2)^{1/2}$,   **c.** $(x^6 y^4)^{1/2}$,   and **d.** $(27x^{12})^{-1/3}$ $(x > 0$ and $y > 0)$.

Solution   **a.** $64^{-2/3} = \dfrac{1}{64^{2/3}}$

$= \dfrac{1}{(64^{1/3})^2}$

$= \dfrac{1}{4^2}$

$= \dfrac{1}{16}$

b. $(x^2)^{1/2} = x^{2(1/2)}$

$= x^1$

$= x$

c. $(x^6 y^4)^{1/2} = x^{6(1/2)} y^{4(1/2)}$

$= x^3 y^2$

d. $(27x^{12})^{-1/3} = \dfrac{1}{(27x^{12})^{1/3}}$

$= \dfrac{1}{27^{1/3} x^{12(1/3)}}$

$= \dfrac{1}{3x^4}$

SELF CHECK   Simplify   **a.** $25^{-3/2}$,   **b.** $(x^3)^{1/3}$,   and **c.** $(x^6 y^9)^{-2/3}$.   *Answers:*   **a.** $\dfrac{1}{125}$,   **b.** x,   **c.** $\dfrac{1}{x^4 y^6}$ ■

EXAMPLE 6

Simplifying expressions containing rational exponents. Simplify **a.** $x^{1/3} x^{1/2}$,   **b.** $\dfrac{3x^{2/3}}{6x^{1/5}}$,   and   **c.** $\dfrac{2x^{-1/2}}{x^{3/4}}$ $(x > 0)$.

Solution   **a.** $x^{1/3} x^{1/2} = x^{2/6} x^{3/6}$ Get a common denominator in the fractional exponents.

$= x^{5/6}$ Keep the base and add the exponents.

b. $\dfrac{3x^{2/3}}{6x^{1/5}} = \dfrac{3x^{10/15}}{6x^{3/15}}$ Get a common denominator in the fractional exponents.

$= \dfrac{1}{2} x^{10/15 - 3/15}$ Simplify $\frac{3}{6}$. Keep the base and subtract the exponents.

$= \dfrac{1}{2} x^{7/15}$

c. $\dfrac{2x^{-1/2}}{x^{3/4}} = \dfrac{2x^{-2/4}}{x^{3/4}}$ Get a common denominator in the fractional exponents.

$= 2x^{-2/4 - 3/4}$ Keep the base and subtract the exponents.

$= 2x^{-5/4}$ Simplify.

$= \dfrac{2}{x^{5/4}}$ $x^{-5/4} = \dfrac{1}{x^{5/4}}$.

SELF CHECK   Simplify   **a.** $x^{2/3} x^{1/2}$ and   **b.** $\dfrac{x^{2/3}}{2x^{1/4}}$.   *Answers:*   **a.** $x^{7/6}$,   **b.** $\frac{1}{2} x^{5/12}$ ■

VOCABULARY

In Exercises 1–2, fill in the blanks to make the statements true.

1. A fractional exponent is also called a
_____rational_____ exponent.

2. In the expression $27^{1/3}$, 27 is called the
_____base_____ and the exponent is $\frac{1}{3}$.

CONCEPTS

In Exercises 3–10, complete each rule for exponents.

3. $x^m x^n = $ x^{m+n}

4. $(x^m)^n = $ $x^{m\cdot n}$

5. $\left(\dfrac{x}{y}\right)^n = $ $\dfrac{x^n}{y^n}$

6. $x^0 = $ 1

7. $x^{-n} = $ $\dfrac{1}{x^n}$

8. $\dfrac{x^m}{x^n} = $ x^{m-n},

9. $x^{1/n} = $ $\sqrt[n]{x}$

10. $x^{m/n} = $ $\sqrt[n]{x^m}$ or $\left(\sqrt[n]{x}\right)^m$

11. Write $\sqrt{5}$ using a fractional exponent. $5^{1/2}$

12. Write $5^{1/3}$ using a radical. $\sqrt[3]{5}$

13. Write $8^{4/3}$ using a radical. $\left(\sqrt[3]{8}\right)^4$

14. Write $\left(\sqrt{8}\right)^3$ using a fractional exponent. $8^{3/2}$

15. Complete the table of values.

x	$x^{1/2}$
0	0
1	1
4	2
9	3

16. Complete the table of values.

x	$x^{1/3}$
0	0
−1	−1
−8	−2
8	2

17. Graph each number on the number line.
$\{8^{1/3}, 17^{1/2}, 2^{3/2}, -5^{2/3}\}$

18. Graph each number on the number line.
$\{4^{-1/2}, 64^{-2/3}, (-8)^{-1/3}\}$

NOTATION

In Exercises 19–20, complete each solution.

19. Simplify $(-216)^{4/3}$.
$$(-216)^{4/3} = \left(\sqrt[3]{(-216)}\right)^4$$
$$= \left(-6\right)^4$$
$$= 1,296$$

20. Simplify $\dfrac{3x^{-2/3}}{x^{3/4}}$.
$$\frac{3x^{-2/3}}{x^{3/4}} = \frac{3x^{-8/12}}{x^{9/12}}$$
$$= 3x^{-8/12-9/12}$$
$$= 3x^{-17/12}$$

PRACTICE

In Exercises 21–36, simplify each expression.

21. $81^{1/2}$ 9   **22.** $100^{1/2}$ 10

23. $-144^{1/2}$ -12 **24.** $-400^{1/2}$ -20

25. $\left(\dfrac{1}{4}\right)^{1/2}$ $\frac{1}{2}$ **26.** $\left(\dfrac{1}{25}\right)^{1/2}$ $\frac{1}{5}$

27. $\left(\dfrac{4}{49}\right)^{1/2}$ $\frac{2}{7}$ **28.** $\left(\dfrac{9}{64}\right)^{1/2}$ $\frac{3}{8}$

29. $27^{1/3}$ 3   **30.** $8^{1/3}$ 2

31. $-125^{1/3}$ -5 **32.** $-1,000^{1/3}$ -10

33. $(-8)^{1/3}$ -2 **34.** $(-125)^{1/3}$ -5

35. $\left(\dfrac{27}{64}\right)^{1/3}$ $\frac{3}{4}$ **36.** $\left(\dfrac{64}{125}\right)^{1/3}$ $\frac{4}{5}$

In Exercises 37–48, simplify each expression.

✓ **37.** $81^{3/2}$ 729   **38.** $16^{3/2}$ 64

39. $25^{3/2}$ 125   **40.** $4^{5/2}$ 32

✓ **41.** $125^{2/3}$ 25   **42.** $8^{4/3}$ 16

43. $1,000^{2/3}$ 100   **44.** $27^{2/3}$ 9

✓ **45.** $(-8)^{2/3}$ 4   **46.** $(-125)^{2/3}$ 25

47. $\left(\dfrac{8}{27}\right)^{2/3}$ $\frac{4}{9}$ **48.** $\left(\dfrac{49}{64}\right)^{3/2}$ $\frac{343}{512}$

In Exercises 49–68, simplify each expression. Write your answers without using negative exponents.

49. $6^{3/5}6^{2/5}$ 6   **50.** $3^{4/7}3^{3/7}$ 3

51. $5^{2/3}5^{4/3}$ 25   **52.** $2^{7/8}2^{9/8}$ 4

✓ **53.** $(7^{2/5})^{5/2}$ 7   **54.** $(8^{1/3})^3$ 8

55. $(5^{2/7})^7$ 25   **56.** $(3^{3/8})^8$ 27

✓ **57.** $\dfrac{8^{3/2}}{8^{1/2}}$ 8   **58.** $\dfrac{11^{9/7}}{11^{2/7}}$ 11

59. $\dfrac{5^{11/3}}{5^{2/3}}$ 125   **60.** $\dfrac{27^{13/15}}{27^{8/15}}$ 3

✓ **61.** $4^{-1/2}$ $\frac{1}{2}$ **62.** $8^{-1/3}$ $\frac{1}{2}$

63. $27^{-2/3}$ $\frac{1}{9}$ **64.** $36^{-3/2}$ $\frac{1}{216}$

✓ **65.** $16^{-3/2}$ $\frac{1}{64}$ **66.** $100^{-5/2}$ $\frac{1}{100,000}$

67. $(-27)^{-4/3}$ $\frac{1}{81}$ **68.** $(-8)^{-4/3}$ $\frac{1}{16}$

In Exercises 69–80, simplify each expression. Assume that all variables represent positive numbers.

✓ **69.** $(x^{1/2})^2$ x **70.** $(x^9)^{1/3}$ x^3

71. $(x^{12})^{1/6}$ x^2 **72.** $(x^{18})^{1/9}$ x^2

✓ **73.** $x^{5/6}x^{7/6}$ x^2 **74.** $x^{2/3}x^{7/3}$ x^3

75. $y^{4/7}y^{10/7}$ y^2 **76.** $y^{5/11}y^{6/11}$ y

✓ **77.** $\dfrac{x^{3/5}}{x^{1/5}}$ $x^{2/5}$ **78.** $\dfrac{x^{4/3}}{x^{2/3}}$ $x^{2/3}$

79. $\dfrac{x^{1/7}x^{3/7}}{x^{2/7}}$ $x^{2/7}$ **80.** $\dfrac{x^{5/6}x^{5/6}}{x^{7/6}}$ $x^{1/2}$

In Exercises 81–88, simplify each expression. Assume that all variables represent positive numbers.

✓ **81.** $x^{2/3}x^{3/4}$ $x^{17/12}$ **82.** $a^{3/5}a^{1/2}$ $a^{11/10}$

83. $(b^{1/2})^{3/5}$ $b^{3/10}$ **84.** $(x^{2/5})^{4/7}$ $x^{8/35}$

✓ **85.** $\dfrac{t^{2/3}}{t^{2/5}}$ $t^{4/15}$ **86.** $\dfrac{p^{3/4}}{p^{1/3}}$ $p^{5/12}$

87. $\left(\dfrac{x^{4/5}}{x^{2/15}}\right)^3$ x^2 **88.** $\left(\dfrac{y^{2/3}}{y^{1/5}}\right)^{15}$ y^7

🖩 **APPLICATIONS**

In Exercises 89–94, use a calculator to help solve each problem. If an answer is not exact, give it to the nearest tenth.

✓ **89.** SPEAKERS The formula $A = V^{2/3}$ can be used to find the area A of one face of a cube if its volume V is known. Find the amount of floor space on the dance floor taken up by the speakers shown in Illustration 1 if each speaker is a cube with a volume of 2,744 cubic inches.   392 in.^{2}

ILLUSTRATION 1

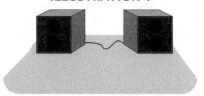

90. MEDICAL TESTS Before a series of X-rays are taken, a patient is injected with a special contrast mixture that highlights obstructions in his blood vessels. The amount of the original dose of contrast material remaining in the patient's bloodstream h hours after it is injected is given by $h^{-3/2}$. How much of the contrast material remains in the patient's bloodstream 4 hours after the injection? $\frac{1}{8}$ of the dose

91. HOLIDAY DECORATING Find the length s of each string of colored lights used to decorate an evergreen tree in the manner shown in Illustration 2 if $s = (r^2 + h^2)^{1/2}$.   26 ft

ILLUSTRATION 2

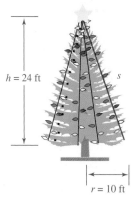

$h = 24$ ft s

$r = 10$ ft

92. VISIBILITY The distance d in miles a person in an airplane can see to the horizon on a clear day is given by the formula $d = 1.22a^{1/2}$, where a is the altitude of the plane in feet. Find d in Illustration 3.   231.5 mi

ILLUSTRATION 3

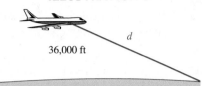

36,000 ft

d

93. TOY DESIGN Knowing the volume V of a sphere, we can find its radius r using the formula

$$r = \left(\frac{3V}{4\pi}\right)^{1/3}$$

If the volume occupied by a ball is 2π cubic inches, find its radius. 1.1 in.

94. EXERCISE EQUIPMENT Find the length l of the incline bench in Illustration 4, using the formula $l = (a^2 + b^2)^{1/2}$. 78.5 in.

ILLUSTRATION 4

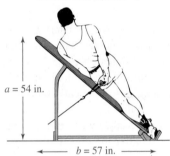

$a = 54$ in.

$b = 57$ in.

WRITING

95. What is a rational exponent? Give several examples.

96. Explain this statement: *In the expression $16^{3/2}$, the number 3/2 requires that two operations be performed on 16.*

REVIEW

In Exercises 97–100, graph each equation.

97. $x = 3$

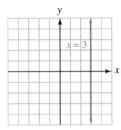

98. $y = -3$

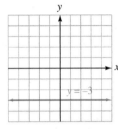

99. $-2x + y = 4$

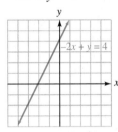

100. $4x - y = 4$

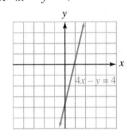

chg to slope-intercept form
$y = 2x + 4$

Inverse Operations

We have performed six operations with real numbers: addition, subtraction, multiplication, division, raising to a power, and finding a root. We have seen that there is a special relationship between *pairs* of operations. That is, subtraction does the opposite of addition, division does the opposite of multiplication, and finding a root does the opposite of raising to a power. Because of this, we call each pair **inverse operations.** Subtraction is the inverse operation of addition, division is the inverse operation of multiplication, and finding a root is the inverse operation of raising to a power.

Solving Equations

When solving equations, we use inverse operations to isolate the variable on one side of the equation.

In Exercises 1–8, tell what operation is performed on the variable and what inverse operation should be used to isolate the variable; then solve the equation.

1. $x + 2 = -4$ addition, subtraction, -6

2. $x - 5 = 10$ subtraction, addition, 15

3. $-6x = 24$ multiplication, division, -4

4. $\dfrac{x}{2} = 40$ division, multiplication, 80

5. $\sqrt{x} = 7$ square root, square, 49

6. $x^2 = 169$ $(x > 0)$ squared, square root, 13

7. $\sqrt[3]{x} = -2$ cube root, cube, -8

8. $x^3 = 64$ cubed, cube root, 4

When solving equations, we must often undo several operations to isolate the variable. Recall that these operations are undone in the *reverse* order of operations.

In Exercises 9–12, solve each equation.

9. $-2x - 4 = 6$ -5

10. $\dfrac{3x}{5} + 3 = 9$ 10

11. $\sqrt{x + 1} = 4$ 15

12. $x^2 + 1 = 10$ $(x > 0)$ 3

Applications

We can use the concept of inverse operation to find the length of a side of the cube in Illustration 1 if we know the area of a face or the volume of the cube.

13. To find the area of a face of the cube, we square the length of a side. How could we find the length of a side, knowing the area of a face?
Find the square root of the area.

14. To find the volume of the cube, we cube the length of a side. How could we find the length of a side if we know the volume of the cube?
Find the cube root of the volume.

ILLUSTRATION 1

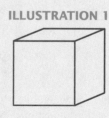

417

Accent on Teamwork

Section 5.1

Approximating square roots Write each of the following numbers on a small piece of masking tape: $\sqrt{2}$, $\sqrt{5}$, $\sqrt{10}$, $\sqrt{20}$, $\sqrt{30}$, $\sqrt{40}$, $\sqrt{50}$, $\sqrt{60}$, $\sqrt{70}$, $\sqrt{80}$, $\sqrt{90}$, $\sqrt{100}$, $\sqrt{110}$, $\sqrt{120}$, $\sqrt{130}$, and $\sqrt{140}$. Stick them on a 12-inch ruler in the approximate positions that show where each of these measurements in inches would fall.

The Pythagorean theorem Put 12 knots in a rope, each 1 foot apart, and connect the ends as shown in Illustration 1. Hammer three tent stakes in the ground so that the rope forms a triangle with sides of length 3, 4, and 5 spaces. Make some observations about the triangle. Use the Pythagorean theorem to prove one of your observations.

ILLUSTRATION 1

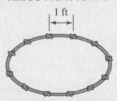

1 ft

Section 5.2

nth roots Use the $\boxed{\sqrt[x]{y}}$ key on a scientific calculator to approximate $\sqrt{2}$, $\sqrt[3]{2}$, $\sqrt[4]{2}$, $\sqrt[5]{2}$, and $\sqrt[6]{2}$. Do you see any pattern? Explain it in words.

Section 5.3

Solving radical equations graphically Complete the table of values for $y = \sqrt{2x + 2}$. Round the y-values to the nearest tenth. Then plot each ordered pair. Draw a smooth line through the points.

x	-1	0	2	4	6	8	10	12
y								

On the graph, draw the horizontal line $y = 2$ and $y = 4$. Use the graph to solve the equations $\sqrt{2x + 2} = 2$ and $\sqrt{2x + 2} = 4$.

Section 5.4

Simplifying radical expressions Suppose you are the algebra instructor of a student whose work is shown here. Write a note to the student explaining how she could save some steps in simplifying $\sqrt{72}$.

$$\sqrt{72} = \sqrt{4 \cdot 18}$$
$$= \sqrt{4}\sqrt{18}$$
$$= 2\sqrt{18}$$
$$= 2\sqrt{9 \cdot 2}$$
$$= 2\sqrt{9}\sqrt{2}$$
$$= 2(3)\sqrt{2}$$
$$= 6\sqrt{2}$$

Section 5.5

Common errors In each addition and subtraction problem below, tell what mistake was made. Compare each problem to a similar one involving variables to clarify your explanation. For example, compare Problem **a** to $2x + 3x$ to help you explain the correct procedure that should be used to simplify the expression.

a. $2\sqrt{5} + 3\sqrt{5} = 5\sqrt{10}$

b. $30 + 2\sqrt{2} = 32\sqrt{2}$

c. $7\sqrt{3} - 5\sqrt{3} = 2$

d. $6\sqrt{7} - 3\sqrt{2} = 3\sqrt{5}$

Section 5.6

Rationalizing numerators Some problems in advanced mathematics require that the numerator of a fraction be rationalized. Extend the concepts studied in this section to develop a method to rationalize the numerators of

$$\frac{\sqrt{5}}{3}, \quad \frac{\sqrt{7}}{\sqrt{5}}, \quad \frac{\sqrt{y}}{6y}, \quad \text{and} \quad \frac{\sqrt{3} - \sqrt{2}}{12}$$

Section 5.7

Graphing Approximate the x- and y-coordinates of the following ordered pairs to the nearest tenth, then graph them on a rectangular coordinate system. (*Hint:* Each quadrant should contain only one point.)

$$A\left(\sqrt{2}, 3^{1/2}\right) \qquad B\left(-\sqrt{6}, 5^{3/2}\right)$$
$$C\left(-16^{2/3}, -\sqrt[3]{25}\right) \qquad D\left(9^{-1/2}, \sqrt[3]{-10}\right)$$

Section 5.1

Square Roots and the Pythagorean Theorem

CONCEPTS

The number b is a *square root* of a if $b^2 = a$.

The *principal square root* of a positive number a, denoted by $\sqrt{a}$, is the positive square root of a.

The expression within a *radical sign* $\sqrt{}$ is called the *radicand*.

Numbers that are not square roots of *integer squares* are *irrational numbers*. Square roots of negative numbers are called *imaginary numbers*.

REVIEW EXERCISES

1. Fill in the blanks to make the statement true: The _____square_____ of 4 is 16, because $4^2 = 16$; 4 is the _____square_____ root of 16, because $4^2 = 16$.

2. Find each square root. Do not use a calculator.
   a. $\sqrt{25}$ 5
   b. $\sqrt{49}$ 7
   c. $-\sqrt{144}$
      -12
   d. $-\sqrt{\dfrac{16}{81}}$ $-\frac{4}{9}$

   e. $\sqrt{900}$ 30
   f. $-\sqrt{0.64}$
      -0.8
   g. $\sqrt{1}$ 1
   h. $\sqrt{0}$ 0

3. ▦ Use a calculator to approximate each expression to three decimal places.
   a. $\sqrt{21}$ b. $-\sqrt{15}$ c. $2\sqrt{7}$ 5.292   d. $\sqrt{751.9}$
   4.583 -3.873 27.421

4. Tell whether each number is rational, irrational, or imaginary. Which is not a real number? $\left\{\sqrt{-2}, \sqrt{68}, \sqrt{81}, \sqrt{3}\right\}$ imag, irr, rat, irr; $\sqrt{-2}$

5. Complete the table of values for each function and then graph it.
   a. $f(x) = \sqrt{x}$
   b. $f(x) = 2 - \sqrt{x}$

x	$f(x)$
0	0
1	1
4	2
9	3

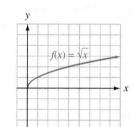

x	$f(x)$
0	2
1	1
4	0
9	-1

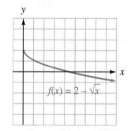

The Pythagorean theorem: If the length of the hypotenuse of a right triangle is c and the lengths of the two legs are a and b, then $c^2 = a^2 + b^2$.

If a and b are positive numbers, and $a = b$, then $\sqrt{a} = \sqrt{b}$.

6. Refer to the right triangle shown in Illustration 1.
   a. Find c where $a = 21$ and $b = 28$.   35
   b. Find b where $a = 1$ and $c = \sqrt{2}$.   1
   c. Find a where $b = 5$ and $c = 7$. $2\sqrt{6}$

ILLUSTRATION 1

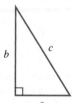

7. ▦ THEATER SEATING For the theater seats shown in Illustration 2, how much higher is the seat at the top of the incline compared to the one at the bottom?   3.5 ft

ILLUSTRATION 2

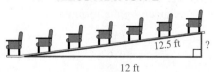

12.5 ft   ?

12 ft

ILLUSTRATION 3

? mph

8. ROAD SIGNS To find the maximum velocity a car can safely travel around a curve without skidding, we can use the formula $v = \sqrt{2.5r}$, where v is the velocity in miles per hour and r is the radius of the curve in feet. How should the road sign in Illustration 3 be labeled if it is to be posted in front of a curve with a radius of 360 feet? 30 mph

Section 5.2

The number b is a *cube root* of a if $b^3 = a$.

The cube root of a is denoted by $\sqrt[3]{a}$. By definition, $\sqrt[3]{a} = b$ if $b^3 = a$.

The number b is an *nth root* of a if $b^n = a$.

In $\sqrt[n]{a}$, the number n is called the *index* of the radical.

When n is even, we say that the radical $\sqrt[n]{x}$ is an *even root*. When n is odd, $\sqrt[n]{x}$ is an *odd root*.

$$\sqrt{a} = \sqrt[2]{a}$$

*n*th Roots and Radicands That Contain Variables

9. Fill in the blanks to make the statement true: $\sqrt[3]{125} = 5$, because $\boxed{5^3} = 125$; 5 is called the ____cube____ root of 125.

10. Find each root. Do not use a calculator.
    a. $\sqrt[3]{-27}$ -3
    b. $-\sqrt[3]{125}$ -5
    c. $\sqrt[4]{81}$ 3
    d. $\sqrt[5]{32}$ 2
    e. $\sqrt[3]{0}$ 0
    f. $\sqrt[3]{-1}$ -1
    g. $\sqrt[3]{\dfrac{1}{64}}$ $\frac{1}{4}$
    h. $\sqrt[3]{1}$ 1

11. [calculator] Use a calculator to find each root to three decimal places.
    a. $\sqrt[3]{16}$ 2.520
    b. $\sqrt[3]{-102.35}$ -4.678
    c. $\sqrt[4]{6}$ 1.565
    d. $\sqrt[5]{34,500}$ 8.083

12. Find each root. Each variable represents a positive number.
    a. $\sqrt{x^2}$ x
    b. $\sqrt{4b^4}$ $2b^2$
    c. $\sqrt{x^4y^4}$ x^2y^2
    d. $-\sqrt{y^{12}}$ $-y^6$
    e. $\sqrt[3]{x^3}$ x
    f. $\sqrt[3]{y^6}$ y^2
    g. $\sqrt[3]{27x^3}$ $3x$
    h. $\sqrt[3]{-r^{12}}$ $-r^4$

13. DICE Find the length of an edge of one of the dice shown in Illustration 4 if each one has a volume of 1,728 cubic millimeters. 12 mm

ILLUSTRATION 4

Section 5.3

Solving Equations Containing Radicals; the Distance Formula

14. Simplify each expression. All variables represent positive numbers.
    a. $\left(\sqrt{x}\right)^2$ x
    b. $\left(\sqrt[3]{x}\right)^3$ x
    c. $\left(2\sqrt{t}\right)^2$ $4t$
    d. $\left(\sqrt{e-1}\right)^2$ $e-1$

To solve an equation containing square root radicals:
1. Isolate the radicals.
2. Square both sides and solve the resulting equation.
3. Check the solution. Discard any *extraneous* solutions.

Squaring property of equality:
If $a = b$, then $a^2 = b^2$.

15. Solve each equation and check all solutions.
    a. $\sqrt{x} = 9$ 81
    b. $\sqrt{2x + 10} = 2$ -3
    c. $\sqrt{3x + 4} + 5 = 3$ none
    d. $\sqrt{2(r + 4)} = 2\sqrt{r}$ 4
    e. $\sqrt{p^2 - 3} = p + 3$ -2
    f. $\sqrt[3]{x - 1} = 3$ 28

16. FERRIS WHEEL The distance d in feet that an object will fall in t seconds is given by the formula

$$t = \sqrt{\dfrac{d}{16}}$$

If a person drops a coin from the top of a Ferris wheel and it takes 2 seconds to hit the ground, how tall is the Ferris wheel? 64 ft

17. Illustration 5 shows the graph of $y = 2\sqrt{x - 2} - 4$. Use it to solve the radical equation $2\sqrt{x - 2} = 4$. 6

ILLUSTRATION 5

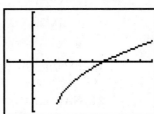

The distance formula:

$$d = \sqrt{(x_2 - x_1)^2 + (y_2 - y_1)^2}$$

18. Find the distance between the points. If an answer is not exact, round to the nearest hundredth.

a. $(-7, 12)$, $(-4, 8)$ 5

b. $(-15, -3)$, $(-10, -16)$ 13.93

Section 5.4

Simplifying Radical Expressions

The *multiplication property* of radicals: If a and b are positive or zero, then

$$\sqrt{ab} = \sqrt{a}\sqrt{b}$$

Simplified form of a radical:

1. Except for 1, the radicand has no perfect square factors.

2. No fraction appears in the radicand.

3. No radical appears in the denominator.

The *division property* of radicals:

$$\sqrt{\frac{a}{b}} = \frac{\sqrt{a}}{\sqrt{b}} \quad (b \neq 0)$$

19. Simplify each expression. All variables represent positive numbers.

a. $\sqrt{32}$ $4\sqrt{2}$

b. $\sqrt{500}$ $10\sqrt{5}$

c. $\sqrt{80x^2}$ $4x\sqrt{5}$

d. $-2\sqrt{63}$ $-6\sqrt{7}$

e. $-\sqrt{250t^3}$ $-5t\sqrt{10t}$

f. $-\sqrt{700z^5}$ $-10z^2\sqrt{7z}$

g. $\sqrt{200x^2y}$ $10x\sqrt{2y}$

h. $\frac{1}{5}\sqrt{75y^4}$ $y^2\sqrt{3}$

i. $\sqrt[3]{8x^2y^3}$ $2y\sqrt[3]{x^2}$

j. $\sqrt[3]{250x^4y^3}$ $5xy\sqrt[3]{2x}$

20. Simplify each expression. All variables represent positive numbers.

a. $\sqrt{\dfrac{16}{25}}$ $\frac{4}{5}$

b. $\sqrt{\dfrac{60}{49}}$ $\frac{2\sqrt{15}}{7}$

c. $\sqrt[3]{\dfrac{1,000}{27}}$ $\frac{10}{3}$

d. $\sqrt{\dfrac{242x^4}{169x^2}}$ $\frac{11x\sqrt{2}}{13}$

21. FITNESS EQUIPMENT The length of the sit-up board in Illustration 6 can be found using the Pythagorean theorem.

a. Find its length. Express the answer in simplified radical form.
$2\sqrt{10}$ ft

b. Express your result to part a as a decimal approximation rounded to the nearest tenth. 6.3 ft

ILLUSTRATION 6

2 ft

6 ft

Section 5.5

Adding and Subtracting Radical Expressions

Radical expressions can be added or subtracted if they contain like radicals.

Radicals are called *like* radicals when they have the same index and the same radicand.

22. Do the operations. All variables represent positive numbers.

a. $\sqrt{2} + \sqrt{8} - \sqrt{18}$ 0

b. $\sqrt{3} + 4 + \sqrt{27} - 7$ $-3 + 4\sqrt{3}$

c. $5\sqrt{28} - 3\sqrt{63}$ $\sqrt{7}$

d. $3y\sqrt{5xy^3} - y^2\sqrt{20xy}$ $y^2\sqrt{5xy}$

e. $\sqrt[3]{16} + \sqrt[3]{54}$ $5\sqrt[3]{2}$

f. $\sqrt[3]{2,000x^3} - \sqrt[3]{128x^3}$ $6x\sqrt[3]{2}$

23. Explain why we cannot add $3\sqrt{5}$ and $5\sqrt{3}$.

They do not contain like radicals—the radicands are different.

24. GARDENING Find the difference in the lengths of the two wires used to secure the tree shown in Illustration 7. $13\sqrt{5}$ in.

ILLUSTRATION 7

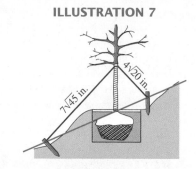

Section 5.6

Multiplying and Dividing Radical Expressions

The product of the square roots of two nonnegative numbers is equal to the square root of the product of those numbers.

To multiply monomials containing radicals, first multiply the coefficients, then multiply the radicals separately, and simplify the result.

Use the FOIL method to multiply two binomials containing radicals.

25. Do the operations.

a. $\sqrt{2}\sqrt{3}$ $\sqrt{6}$

b. $\left(-5\sqrt{5}\right)\left(-2\sqrt{2}\right)$ $10\sqrt{10}$

c. $\left(3\sqrt{3x}\right)\left(4\sqrt{6x}\right)$ $36x\sqrt{2}$

d. $\left(\sqrt[3]{4}\right)\left(2\sqrt[3]{4}\right)$ $4\sqrt[3]{2}$

e. $\sqrt{2}\left(\sqrt{8} - \sqrt{18}\right)$ -2

f. $\left(\sqrt{3} + \sqrt{5}\right)\left(\sqrt{3} - \sqrt{5}\right)$ -2

g. $\left(\sqrt{15} + 3x\right)^2$
$15 + 6x\sqrt{15} + 9x^2$

h. $\left(\sqrt[3]{3} + 2\right)\left(\sqrt[3]{3} - 1\right)$
$\sqrt[3]{9} + \sqrt[3]{3} - 2$

26. VACUUM CLEANER NOZZLE Illustration 8 shows the amount of surface area of a rug suctioned by a vacuum nozzle attachment.

a. Find the perimeter and area of this section of rug. Express the answers in simplified radical form. $\left(4\sqrt{6} + 10\sqrt{3}\right)$ in.; $30\sqrt{2}$ in.^{2}

b. Express your results to part a as decimal approximations to the nearest tenth.
27.1 in.; 42.4 in.^{2}

ILLUSTRATION 8

5 $\sqrt{3}$ in.

2 $\sqrt{6}$ in.

If a square root appears as a monomial in the denominator of a fraction, *rationalize* the denominator by multiplying the numerator and denominator by some appropriate square root.

If the denominator of a fraction contains radicals within a binomial, multiply the numerator and denominator by the *conjugate* of the denominator.

27. Rationalize each denominator.

a. $\dfrac{1}{\sqrt{7}}$ $\dfrac{\sqrt{7}}{7}$

b. $\sqrt{\dfrac{3}{7}}$ $\dfrac{\sqrt{21}}{7}$

c. $\dfrac{\sqrt{9}}{\sqrt{18}}$ $\dfrac{\sqrt{2}}{2}$

d. $\dfrac{8}{\sqrt[3]{16}}$ $2\sqrt[3]{4}$

e. $\dfrac{7}{\sqrt{2} + 1}$ $7\sqrt{2} - 7$

f. $\dfrac{\sqrt{c} - 4}{\sqrt{c} + 4}$ $\dfrac{c - 8\sqrt{c} + 16}{c - 16}$

Section 5.7

Real numbers can be raised to fractional powers.

Rational exponents:

$$x^{1/n} = \sqrt[n]{x}$$
$$x^{m/n} = \sqrt[n]{x^m} = \left(\sqrt[n]{x}\right)^m$$

The rules for exponents can be used to simplify expressions involving rational exponents.

Rational Exponents

28. Simplify each expression. Write answers without using negative exponents.

a. $49^{1/2}$ 7
b. $(-1,000)^{1/3}$ -10
c. $36^{3/2}$ 216
d. $\left(\dfrac{8}{27}\right)^{2/3}$ $\frac{4}{9}$

e. $4^{-3/2}$ $\frac{1}{8}$
f. $8^{2/3}8^{4/3}$ 64
g. $(3^{2/3})^3$ 9
h. $(a^4b^8)^{-1/2}$ $\frac{1}{a^2b^4}$

i. $x^{1/3}x^{2/5}$ $x^{11/15}$
j. $\dfrac{t^{3/4}}{t^{2/3}}$ $t^{1/12}$
k. $\dfrac{x^{2/5}x^{1/5}}{x^{-2/5}}$ x
l. $\dfrac{x^{17/7}}{x^{3/7}}$ x^2

29. Graph each number on the number line: $\{4^{-1/2}, 12^{1/2}, 9^{1/3}, -2^{2/3}\}$.

30. DENTISTRY The fractional amount of painkiller remaining in the system of a patient h hours after the original dose was injected into her gums is given by $h^{-3/2}$. How much of the original dose is in the patient's system 16 hours after the injection? $\frac{1}{64}$ of the original dose

CHAPTER 5

Test

In Problems 1–4, simplify each radical.

1. $\sqrt{100}$ 10

2. $-\sqrt{\dfrac{400}{9}}$ $-\frac{20}{3}$

3. $\sqrt[3]{-27}$ -3

4. $\sqrt{\dfrac{50}{49}}$ $\frac{5\sqrt{2}}{7}$

5. Evaluate $\sqrt{b^2 - 4ac}$ for $a = 2$, $b = 10$, and $c = 6$. Round to the nearest tenth.   7.2

6. A 26-foot ladder reaches a point on a wall 24 feet above the ground. How far from the wall is the ladder's base?   10 ft

In Problems 7–10, simplify each expression. Assume that x and y represent positive numbers.

7. $\sqrt{4x^2}$ $2x$

8. $\sqrt{54x^3}$ $3x\sqrt{6x}$

9. $\sqrt{\dfrac{18x^2y^3}{2xy}}$ $3y\sqrt{x}$

10. $\sqrt[3]{x^6y^3}$ x^2y

11. A square has an area of 24 square yards.
   a. Express the length of a side of the square in simplified radical form. $2\sqrt{6}$ yd

b. Round the length of a side of the square to the nearest tenth.   4.9 yd

12. Find the distance between points $(-2, -3)$ and $(-8, 5)$.   10

In Problems 13–16, solve each equation.

13. $\sqrt{x} = 15$ 225

14. $\sqrt{2 - x} - 2 = 6$ -62

15. $\sqrt{3x + 9} = 2\sqrt{x + 1}$ 5

16. $\sqrt[3]{x - 2} = 3$ 29

In Problems 17–22, do each operation and simplify.

17. $\sqrt{12} + \sqrt{27}$ $5\sqrt{3}$

18. $\sqrt{8x^3} - x\sqrt{18x}$ $-x\sqrt{2x}$

19. $\left(-2\sqrt{8x}\right)\left(3\sqrt{12x}\right)$ $-24x\sqrt{6}$

20. $\sqrt{3}\left(\sqrt{8} + \sqrt{6}\right)$ $2\sqrt{6} + 3\sqrt{2}$

21. $\left(\sqrt{2} + \sqrt{3}\right)\left(\sqrt{2} - \sqrt{3}\right)$ -1

22. $\left(2\sqrt{x} + 2\right)\left(\sqrt{x} - 3\right)$ $2x - 4\sqrt{x} - 6$

In Problems 23 and 24, rationalize each denominator.

23. $\dfrac{2}{\sqrt{2}}$ $\sqrt{2}$

24. $\dfrac{\sqrt{3x}}{\sqrt{x}+2}$ $\dfrac{x\sqrt{3}-2\sqrt{3x}}{x-4}$

In Problems 25–28, simplify each expression and write all answers without using negative exponents. All variables represent positive numbers.

25. $121^{1/2}$ 11

26. $27^{-4/3}$ $\frac{1}{81}$

27. $p^{2/3}p^{4/3}$ p^2

28. $\dfrac{x^{1/15}}{x^{1/3}}$ $\dfrac{1}{x^{4/15}}$

29. Is $x = 0$ a solution of the radical equation $\sqrt{3x+1} = x - 1$? Explain your answer.

No; when 0 is substituted for x, the result is not a true statement: $1 \neq -1$.

30. Explain why we cannot do the subtraction $4\sqrt{3} - 7\sqrt{2}$.

They do not contain like radicals—the radicands are different.

31. Complete the table and then graph the function. Round to the nearest tenth when necessary.

$f(x) = \sqrt{x}$

x	$f(x)$
0	0
1	1
2	1.4
3	1.7
4	2
5	2.2
6	2.4
7	2.6
8	2.8
9	3

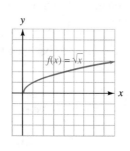

32. CARPENTRY In Illustration 1, a carpenter is using a tape measure to see if the wall he just put up is perfectly "square" with the floor. Explain what mathematical concept he is applying. If the wall is positioned correctly, what should the measurement on the tape read?

the Pythagorean theorem; 5 ft

ILLUSTRATION 1

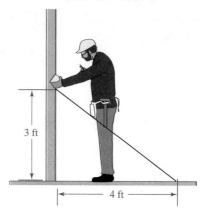

6

Factoring and Quadratic Equations

CAMPUS CONNECTION

The Business Department

The relationship between production and revenue is one that business students study in great detail. For example, a company that manufactures radios has found that as the number of radios produced increases, the amount of revenue increases to reach a maximum, and then it steadily decreases. These observations come from a graph of a *quadratic function,* which mathematically describes the relationship between the number of radios manufactured and the revenue the company receives. In this chapter, we introduce quadratic functions, and we discuss a procedure used to determine maximum values for such situations.

In this chapter, we will discuss three methods used to solve quadratic equations—equations of the form $ax^2 + bx + c = 0$, where a, b, and c are real numbers and $a \neq 0$.

▶6.1 Factoring Out the Greatest Common Factor and Factoring by Grouping

In this section, you will learn about

Prime factorization ▪ The greatest common factor (GCF) ▪ Finding the GCF of several monomials ▪ Factoring out the greatest common factor from a polynomial ▪ Factoring out a negative factor ▪ Factoring by grouping

Introduction   Recall that the distributive property provides a way to multiply the monomial $4y$ and the binomial $3y + 5$.

$$4y(3y + 5) = 4y \cdot 3y + 4y \cdot 5$$
$$= 12y^2 + 20y$$

In this section, we will reverse the operation of multiplication. Given a polynomial such as $12y^2 + 20y$, we will ask ourselves, "What factors were multiplied to obtain $12y^2 + 20y$?" The process of finding the individual factors of a known product is called **factoring**.

The multiplication process	*The factoring process*
Given the factors . . . find the product	Given the product . . . find the factors
$4y(3y + 5) = \;?$	$12y^2 + 20y = \;?(\quad ?\quad)$

Factoring can be used to solve certain types of equations and to simplify certain kinds of algebraic expressions. To begin the discussion of factoring, we consider two procedures that are used to factor natural numbers.

Prime Factorization

Because 4 divides 12 exactly, 4 is called a **factor** of 12. The numbers 1, 2, 3, 4, 6, and 12 are the natural-number factors of 12, because each divides 12 exactly.

Prime numbers

A **prime number** is a natural number greater than 1 whose only factors are 1 and itself.

For example, 17 is a prime number, because

1. 17 is a natural number greater than 1, and
2. the only two natural-number factors of 17 are 1 and 17.

The prime numbers less than 50 are

2, 3, 5, 7, 11, 13, 17, 19, 23, 29, 31, 37, 41, 43, and 47

A natural number is said to be in **prime-factored form** if it is written as the product of factors that are prime numbers.

To find the prime-factored form of a natural number, we can use a **factoring tree.** The following examples show two ways to proceed to find the prime-factored form of 90 using factoring trees. The factoring process stops when a row of the tree contains only prime-number factors.

1. Start with 90.

2. Factor 90 as 9 · 10.

3. Factor 9 and 10.

1. Start with 90.

2. Factor 90 as 6 · 15.

3. Factor 6 and 15.

Since the prime factors in either case are $2 \cdot 3 \cdot 3 \cdot 5$, the prime-factored form, or the **prime factorization**, of 90 is $2 \cdot 3^2 \cdot 5$. This example illustrates the **fundamental theorem of arithmetic,** which states that there is only one prime factorization for every natural number greater than 1.

We can also find the prime factorization of a natural number using the **division method.** For example, to find the prime factorization of 42, we begin by choosing the *smallest* prime number that will divide the given number exactly. We continue this process until the result of the division is a prime number.

Step 1: 2 divides 42 exactly. The result is 21, which is not prime. We continue the process.

$$2\overline{)42} \\ \quad 21$$

Step 2: We choose the smallest prime number that divides 21. The prime number 2 does not divide 21 exactly, but 3 does. The result is 7, which is prime. We are done.

$$2\overline{)42} \\ 3\overline{)21} \\ \quad 7$$

The prime factorization of 42 is $2 \cdot 3 \cdot 7$.

E X A M P L E 1

Prime factorizations. Find the prime factorization of 150.

Solution   We can use a factoring tree or the division method to find the prime factorization of 150.

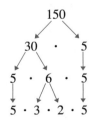

$$2\overline{)150} \\ 3\overline{)\,75} \\ 5\overline{)\,25} \\ \quad\ 5$$

The prime factorization of 150 is $2 \cdot 3 \cdot 5^2$.

SELF CHECK   Find the prime factorization of 225.   *Answer:* $3^2 \cdot 5^2$ ■

The Greatest Common Factor (GCF)

The right-hand sides of the equations

$90 = 2 \cdot 3 \cdot 3 \cdot 5$

$42 = 2 \cdot 3 \cdot 7$

show the prime-factored forms of 90 and 42. The color highlighting indicates that 90 and 42 have one prime factor of 2 and one prime factor of 3 in common. We can conclude that $2 \cdot 3 = 6$ is the largest natural number that divides 90 and 42 exactly, and we say that 6 is their **greatest common factor (GCF).**

$$\frac{90}{6} = 15 \qquad \text{and} \qquad \frac{42}{6} = 7$$

EXAMPLE 2

Finding the GCF of three numbers. Find the greatest common factor of 24, 60, and 96.

Solution

We write each prime factorization and highlight the prime factors the three numbers have in common.

$$24 = 2 \cdot 2 \cdot 2 \cdot 3$$
$$60 = 2 \cdot 2 \cdot 3 \cdot 5$$
$$96 = 2 \cdot 2 \cdot 2 \cdot 2 \cdot 2 \cdot 3$$

Since 24, 60, and 96 each have two factors of 2 and one factor of 3, their greatest common factor is $2 \cdot 2 \cdot 3 = 12$.

SELF CHECK

Find the GCF for 45, 60, 75. *Answer:* $3 \cdot 5 = 15$ ∎

Finding the GCF of Several Monomials

The right-hand sides of the equations

$$12y^2 = 2 \cdot 2 \cdot 3 \cdot y \cdot y$$
$$20y = 2 \cdot 2 \cdot 5 \cdot y$$

show the prime factorizations of $12y^2$ and $20y$. Since the monomials have two factors of 2 and one factor of y in common, their GCF is

$$2 \cdot 2 \cdot y \qquad \text{or} \qquad 4y$$

To find the GCF of several monomials, we follow these steps.

Strategy for finding the greatest common factor (GCF)

1. Find the prime factorization of each monomial.
2. List each common factor the least number of times it appears in any one monomial.
3. Find the product of the factors in the list to obtain the GCF.

EXAMPLE 3

Finding the GCF of three monomials. Find the GCF of $10x^3y^2$, $60x^2y$, and $30xy^2$.

Solution

Step 1: Find the prime factorization of each monomial.

$$10x^3y^2 = 2 \cdot 5 \cdot x \cdot x \cdot x \cdot y \cdot y$$
$$60x^2y = 2 \cdot 2 \cdot 3 \cdot 5 \cdot x \cdot x \cdot y$$
$$30xy^2 = 2 \cdot 3 \cdot 5 \cdot x \cdot y \cdot y$$

Step 2: List each common factor the least number of times it appears in any one monomial: 2, 5, x, and y.

Step 3: Find the product of the factors in the list:

$$2 \cdot 5 \cdot x \cdot y = 10xy \qquad \text{The GCF is } 10xy.$$

SELF CHECK   Find the GCF of $20a^2b^3$, $12ab^4$, and $8a^3b^2$.   *Answer:* $4ab^2$ ∎

Factoring Out the Greatest Common Factor from a Polynomial

To factor $12y^2 + 20y$, we find the GCF of $12y^2$ and $20y$ (which we earlier determined to be $4y$) and use the distributive property.

$$12y^2 + 20y = \mathbf{4y} \cdot 3y + \mathbf{4y} \cdot 5 \qquad \text{Write each term of the polynomial as the product of the GCF, } 4y, \text{ and one other factor.}$$

$$= 4y(3y + 5) \qquad 4y \text{ is a common factor of both terms.}$$

This process is called **factoring out the greatest common factor.**

EXAMPLE 4

Factoring out the greatest common factor. Factor $25 - 5m$.

Solution   To find the GCF of 25 and $5m$, we find their prime factorizations.

$$\left. \begin{array}{l} 25 = 5 \cdot 5 \\ 5m = 5 \cdot m \end{array} \right\} \quad \text{GCF} = 5$$

We can use the distributive property to factor out the GCF.

$$25 - 5m = \mathbf{5} \cdot 5 - \mathbf{5} \cdot m \qquad \text{Factor each monomial using 5 and one other factor.}$$

$$= 5(5 - m) \qquad \text{Factor out the common factor of 5.}$$

We check by verifying that $5(5 - m) = 25 - 5m$.

SELF CHECK   Factor $18x - 24$.   *Answer:* $6(3x - 4)$ ∎

EXAMPLE 5

Factoring out the GCF. Factor $35a^3b^2 - 14a^2b^3$.

Solution   To find the GCF, we find the prime factorizations of $35a^3b^2$ and $14a^2b^3$.

$$\left. \begin{array}{l} 35a^3b^2 = 5 \cdot 7 \cdot a \cdot a \cdot a \cdot b \cdot b \\ 14a^2b^3 = 2 \cdot 7 \cdot a \cdot a \cdot b \cdot b \cdot b \end{array} \right\} \quad \text{GCF} = 7 \cdot a \cdot a \cdot b \cdot b = 7a^2b^2$$

We factor out the GCF of $7a^2b^2$.

$$35a^3b^2 - 14a^2b^3 = 7a^2b^2 \cdot 5a - 7a^2b^2 \cdot 2b$$

$$= 7a^2b^2(5a - 2b)$$

We check by verifying that $7a^2b^2(5a - 2b) = 35a^3b^2 - 14a^2b^3$.

SELF CHECK   Factor $40x^2y^3 + 15x^3y^2$.   *Answer:* $5x^2y^2(8y + 3x)$ ∎

EXAMPLE 6

An implied coefficient of 1. Factor $4x^3y^2z - 2x^2yz + xz$.

Solution   The expression has three terms. We factor out the GCF, which is xz.

$$4x^3y^2z - 2x^2yz + xz = xz \cdot 4x^2y^2 - xz \cdot 2xy + xz \cdot 1$$

$$= xz(4x^2y^2 - 2xy + 1)$$

The last term of $4x^3y^2z - 2x^2yz + xz$ has an implied coefficient of 1. That is, $xz = 1xz = xz \cdot 1$. When xz is factored out, we must write this coefficient of 1. We check by verifying that $xz(4x^2y^2 - 2xy + 1) = 4x^3y^2z - 2x^2yz + xz$.

SELF CHECK   Factor $2ab^2c + 4a^2bc - ab$.   *Answer:* $ab(2bc + 4ac - 1)$ ∎

E X A M P L E   7

Lava lamp. Figure 6-1 shows a lava lamp. The formula that gives the volume of the glass dome of the lamp is given below. Factor the expression on the right-hand side of the equation. (Read $r_1{}^2$ as "r sub 1, squared.")

$$V = \frac{1}{3}\pi r_1{}^2h + \frac{1}{3}\pi r_2{}^2h + \frac{1}{3}\pi r_1 r_2 h$$

FIGURE 6-1

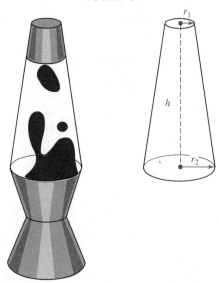

Solution   The expression on the right-hand side of the formula contains three terms. By inspection, we see that each term contains a factor of $\frac{1}{3}$, π, and h. Note that neither r_1 nor r_2 is common to all three terms.

$$V = \frac{1}{3}\pi r_1{}^2h + \frac{1}{3}\pi r_2{}^2h + \frac{1}{3}\pi r_1 r_2 h \quad \text{The given formula.}$$

$$= \frac{1}{3}\pi h(r_1{}^2 + r_2{}^2 + r_1 r_2) \qquad \begin{array}{l}\text{Factor out the}\\ \text{GCF, } \frac{1}{3}\pi h.\end{array}$$ ∎

E X A M P L E   8

Factoring out a common binomial factor. Factor $x(x + 4) + 3(x + 4)$.

Solution   The given polynomial has two terms:

$$\underbrace{x(x + 4)}_{\substack{\text{The first}\\\text{term}}} + \underbrace{3(x + 4)}_{\substack{\text{The second}\\\text{term}}}$$

The GCF of the terms is the binomial $(x + 4)$. We factor it out.

$$x(x + 4) + 3(x + 4) = (x + 4)(x + 3)$$

SELF CHECK   Factor $2y(y - 1) - 7(y - 1)$.   *Answer:* $(y - 1)(2y - 7)$ ■

Factoring Out a Negative Factor

It is often useful to factor out a common factor having a negative coefficient.

EXAMPLE 9   **Factoring out -1.** Factor -1 out of $-a^3 + 2a^2 - 4$.

Solution   First, we write each term of the polynomial as the product of -1 and another factor: $-a^3 = (-1)a^3$, $2a^2 = (-1)(-2a^2)$, and $-4 = (-1)4$. Then we factor out the common factor of -1.

$$-a^3 + 2a^2 - 4 = (-1)a^3 + (-1)(-2a^2) + (-1)4$$
$$= -1(a^3 - 2a^2 + 4) \quad \text{Factor out } -1.$$
$$= -(a^3 - 2a^2 + 4) \quad \text{The coefficient of 1 need not be written.}$$

We check by verifying that $-(a^3 - 2a^2 + 4) = -a^3 + 2a^2 - 4$.

SELF CHECK   Factor -1 out of $-b^4 - 3b^2 + 2$.   *Answer:* $-(b^4 + 3b^2 - 2)$ ■

EXAMPLE 10   **Factoring out the negative of the GCF.** Factor out the negative (opposite) of the GCF in $-18a^2b + 6ab^2 - 12a^2b^2$.

Solution   The GCF is $6ab$. To factor out its negative, we write each term of the polynomial as the product of $-6ab$ and another factor. Then we factor out $-6ab$.

$$-18a^2b + 6ab^2 - 12a^2b^2 = (-6ab)3a - (-6ab)b + (-6ab)2ab$$
$$= -6ab(3a - b + 2ab)$$

We check by verifying that $-6ab(3a - b + 2ab) = -18a^2b + 6ab^2 - 12a^2b^2$.

SELF CHECK   Factor out the negative (opposite) of the GCF in $-27xy^2 - 18x^2y + 36x^2y^2$.   *Answer:* $-9xy(3y + 2x - 4xy)$

■

Factoring by Grouping

Suppose we wish to factor

$$ax + ay + cx + cy$$

Although no factor is common to all four terms, there is a common factor of a in $ax + ay$ and a common factor of c in $cx + cy$. We can factor out the a and c to obtain

$$ax + ay + cx + cy = a(x + y) + c(x + y)$$
$$= (x + y)(a + c) \quad \text{Factor out } x + y.$$

We can check the result by multiplication.

$$(x + y)(a + c) = ax + cx + ay + cy$$
$$= ax + ay + cx + cy \quad \text{Rearrange the terms.}$$

Thus, $ax + ay + cx + cy$ factors as $(x + y)(a + c)$. This type of factoring is called **factoring by grouping.** Polynomials having four terms can be factored by grouping if the polynomials can be split into two groups of terms and both groups share a common factor.

Factoring by grouping

1. Group the terms of the polynomial so that the first two terms have a common factor and the last two terms have a common factor.
2. Factor out the common factor from each group.
3. Factor out the resulting common binomial factor. If there is no common binomial factor, regroup the terms of the polynomial and repeat steps 2 and 3.

EXAMPLE 11

Factoring by grouping. Factor $2c - 2d + cd - d^2$.

Solution No factor is common to all four term, but 2 is common to the first two terms and d is common to the last two terms.

$$2c - 2d + cd - d^2 = 2(c - d) + d(c - d) \qquad \text{Factor out 2 from } 2c - 2d \text{ and } d \text{ from } cd - d^2.$$

$$= (c - d)(2 + d) \qquad \text{Factor out } c - d.$$

We check by verifying that

$$(c - d)(2 + d) = 2c + cd - 2d - d^2$$
$$= 2c - 2d + cd - d^2 \qquad \text{Rearrange the terms.}$$

SELF CHECK Factor $7x - 7y + xy - y^2$. *Answer:* $(x - y)(7 + y)$ ■

EXAMPLE 12

Factoring by grouping. Factor $x^2y - ax - xy + a$.

Solution No factor is common to all four terms, but x is common to $x^2y - ax$.

$$x^2y - ax - xy + a = x(xy - a) - xy + a \qquad \text{Factor out } x \text{ from } x^2y - ax.$$

If we factor -1 from $-xy + a$, a common binomial factor $(xy - a)$ appears.

$$x^2y - ax - xy + a = x(xy - a) - 1(xy - a)$$
$$= (xy - a)(x - 1) \qquad \text{Factor out } xy - a.$$

Check by multiplication.

SELF CHECK Factor $7b + 3c - 7bt - 3ct$. *Answer:* $(7b + 3c)(1 - t)$ ■

 WARNING! When factoring expressions such as those in the previous two examples, don't think that $2(c - d) + d(c - d)$ or $x(xy - a) - 1(xy - a)$ are in factored form. To be in factored form, the result must be a product.

The next example illustrates that when factoring a polynomial, we should always look for a common factor first.

EXAMPLE 13

Factoring out the GCF first. Factor $10k + 10m - 2km - 2m^2$.

Solution The four terms have a common factor of 2. We factor it out first. Then we use factoring by grouping to factor the polynomial in the parentheses. The first two terms have a common factor of 5. The last two terms have a common factor of $-m$.

$$10k + 10m - 2km - 2m^2 = 2(5k + 5m - km - m^2) \qquad \text{Factor out the GCF, 2.}$$

$$= 2[5(k + m) - m(k + m)]$$
$$= 2[(k + m)(5 - m)] \qquad \text{Factor out } k + m.$$
$$= 2(k + m)(5 - m)$$

Use multiplication to check the result.

SELF CHECK   Factor $-4t - 4s - 4tz - 4sz$.   *Answer:* $-4(t + s)(1 + z)$ ∎

STUDY SET

Section 6.1

VOCABULARY

In Exercises 1–6, fill in the blanks to make the statements true.

1. A natural number greater than 1 whose only factors are 1 and itself is called a _____prime_____ number.

2. When we write 24 as $2^3 \cdot 3$, we say that 24 has been written in ___prime-factored___ form.

3. The GCF of several natural numbers is the _____largest_____ number that divides each of the numbers exactly.

4. When we write $15x^2 - 25x$ as $5x(3x - 5)$, we say that we have ___factored out___ the greatest common factor.

5. The process of finding the individual factors of a known product is called ___factoring___.

6. The numbers 1, 2, 3, 4, 6, and 12 are the natural-number ___factors___ of 12.

CONCEPTS

*In Exercises 7–10, explain what is **wrong** with each solution.*

7. Factor $6a + 9b + 3$.
$$6a + 9b + 3 = 3(2a + 3b + 0)$$
$$= 3(2a + 3b)$$
The 0 in the first line should be a 1.

8. Prime factor 100.

$$10\underline{)100}$$
$$5\underline{)10}$$
$$2$$

$$100 = 2 \cdot 5 \cdot 10$$

10 can't be used in the division method; it is not a prime number.

9. Factor out the GCF: $30a^3 - 12a^2$.
$$30a^3 - 12a^2 = 6a(5a^2 - 2a)$$
The GCF is $6a^2$, not $6a$.

10. Factor $ab + b + a + 1$.
$$ab + b + a + 1 = b(a + 1) + (a + 1)$$
$$= (a + 1)b$$
The answer should be $(a + 1)(b + 1)$.

11. a. What property is illustrated here?
$$2x(x - 3) = 2x^2 - 6x$$
the distributive property

   b. Explain how we use the distributive property in reverse to factor $2x^2 - 6x$.

12. a. Complete each tree diagram to prime-factor 30.

   $2 \cdot 3 \cdot 5$ $3 \cdot 2 \cdot 5$ $5 \cdot 3 \cdot 2$

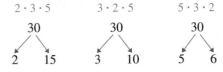

   b. Complete the statement: The fundamental theorem of arithmetic states that every natural number greater than 1 has ___exactly one___ prime factorization.

13. The prime factorizations of three monomials are shown here. Find their GCF. $3x$
$$3 \cdot 3 \cdot 5 \cdot x \cdot x$$
$$2 \cdot 3 \cdot 5 \cdot x \cdot y$$
$$2 \cdot 2 \cdot 3 \cdot x \cdot y \cdot y$$

14. Consider the polynomial $2k - 8 + hk - 4h$.
   a. How many terms does the polynomial have?   4
   b. Is there a common factor of all the terms?   no
   c. What is the common factor of the first two terms?   2
   d. What is the common factor of the last two terms? h

15. How can we check the answer of the factoring problem shown below?

   Factor $3j^3 + 6j^2 + 2j + 4$.
$$3j^3 + 6j^2 + 2j + 4 = 3j^2(j + 2) + 2(j + 2)$$
$$= (j + 2)(3j^2 + 2)$$
Use the FOIL method to multiply $(j + 2)(3j^2 + 2)$. The result should be $3j^3 + 2j + 6j^2 + 4$.

16. List the first 12 prime numbers.
2, 3, 5, 7, 11, 13, 17, 19, 23, 29, 31, 37

NOTATION

In Exercises 17–18, complete each factorization.

17. Factor $b^3 - 6b^2 + 2b - 12$.
$$b^3 - 6b^2 + 2b - 12 = b^2 (b - 6) + 2 (b - 6)$$
$$= (b - 6) (b^2 + 2)$$

18. Factor $12b^3 - 6b^2 + 2b - 12$.

$$12b^3 - 6b^2 + 2b - 12 = \boxed{2}\,(6b^3 - 3b^2 + b - 6)$$

19. In the expression $4x^2y + xy$, what is the implied coefficient of the last term?   1

20. True or false?   true

$$-(x^2 - 3x + 1) = -1(x^2 - 3x + 1)$$

PRACTICE

In Exercises 21–32, find the prime factorization of each number.

✓ **21.** 12 $2^2 \cdot 3$	**22.** 24 $2^3 \cdot 3$
23. 15 $3 \cdot 5$	**24.** 20 $2^2 \cdot 5$
✓ **25.** 40 $2^3 \cdot 5$	**26.** 62 $2 \cdot 31$
27. 98 $2 \cdot 7^2$	**28.** 112 $2^4 \cdot 7$
✓ **29.** 225 $3^2 \cdot 5^2$	**30.** 144 $2^4 \cdot 3^2$
31. 288 $2^5 \cdot 3^2$	**32.** 968 $2^3 \cdot 11^2$

In Exercises 33–36, complete each factorization.

✓ **33.** $4a + 12 = \boxed{4}\,(a + 3)$

34. $r^4 + r^2 = r^2\big(\boxed{r^2} + 1\big)$

✓ **35.** $4y^2 + 8y - 2xy = 2y\big(2y + \boxed{4} - \boxed{x}\big)$

36. $3x^2 - 6xy + 9xy^2 = \boxed{3x}\,\big(\boxed{x} - 2y + 3y^2\big)$

In Exercises 37–60, factor out the GCF.

✓ **37.** $3x + 6$ $3(x + 2)$ **38.** $2y - 10$ $2(y - 5)$

39. $2\pi R - 2\pi r$ $2\pi(R - r)$

40. $\frac{1}{3}\pi R^2 h - \frac{4}{3}\pi R^3$ $\frac{1}{3}\pi R^2(h - 4R)$

✓ **41.** $t^3 + 2t^2$ $t^2(t + 2)$ **42.** $b^3 - 3b^2$ $b^2(b - 3)$

43. $a^3 - a^2$ $a^2(a - 1)$ **44.** $r^3 + r^2$ $r^2(r + 1)$

✓ **45.** $24x^2y^3 + 8xy^2$ $8xy^2(3xy + 1)$

46. $3x^2y^3 - 9x^4y^3$ $3x^2y^3(1 - 3x^2)$

47. $12uvw^3 - 18uv^2w^2$ $6uvw^2(2w - 3v)$

48. $14xyz - 16x^2y^2z$ $2xyz(7 - 8xy)$

✓ **49.** $3x + 3y - 6z$ $3(x + y - 2z)$

50. $2x - 4y + 8z$ $2(x - 2y + 4z)$

51. $ab + ac - ad$ $a(b + c - d)$

52. $rs - rt + ru$ $r(s - t + u)$

✓ **53.** $12r^2 - 3rs + 9r^2s^2$ $3r(4r - s + 3rs^2)$

54. $6a^2 - 12a^3b + 36ab$ $6a(a - 2a^2b + 6b)$

55. $\pi R^2 - \pi ab$ $\pi(R^2 - ab)$

56. $\frac{1}{3}\pi R^2 h - \frac{1}{3}\pi rh$ $\frac{1}{3}\pi h(R^2 - r)$

✓ **57.** $3(x + 2) - x(x + 2)$ $(x + 2)(3 - x)$

58. $t(5 - s) + 4(5 - s)$ $(5 - s)(t + 4)$

59. $h^2(14 + r) + 14 + r$ $(14 + r)(h^2 + 1)$

60. $k^2(14 + v) - 7(14 + v)$ $(14 + v)(k^2 - 7)$

In Exercises 61–68, factor out -1 from each polynomial.

✓ **61.** $-a - b$ $-(a + b)$ **62.** $-x - 2y$ $-(x + 2y)$

63. $-2x + 5y$ $-(2x - 5y)$ **64.** $-3x + 8z$ $-(3x - 8z)$

✓ **65.** $-3m - 4n + 1$ $-(3m + 4n - 1)$

66. $-3r + 2s - 3$ $-(3r - 2s + 3)$

67. $-3ab - 5ac + 9bc$ $-(3ab + 5ac - 9bc)$

68. $-6yz + 12xz - 5xy$ $-(6yz - 12xz + 5xy)$

In Exercises 69–74, factor each polynomial by factoring out the negative of the GCF.

✓ **69.** $-3x^2 - 6x$ $-3x(x + 2)$

70. $-4a^2 + 6a$ $-2a(2a - 3)$

71. $-4a^2b^3 + 12a^3b^2$ $-4a^2b^2(b - 3a)$

72. $-25x^4y^3 + 30x^2y^3$ $-5x^2y^3(5x^2 - 6)$

✓ **73.** $-4a^2b^2c^2 + 14a^2b^2c - 10ab^2c^2$
   $-2ab^2c(2ac - 7a + 5c)$

74. $-10x^4y^3z^2 + 8x^3y^2z - 20x^2y$
   $-2x^2y(5x^2y^2z^2 - 4xyz + 10)$

In Exercises 75–92, factor by grouping.

75. $2x + 2y + ax + ay$ $(x + y)(2 + a)$

76. $bx + bz + 5x + 5z$ $(x + z)(b + 5)$

✓ **77.** $7r + 7s - kr - ks$ $(r + s)(7 - k)$

78. $9p - 9q + mp - mq$ $(p - q)(9 + m)$

79. $xr + xs + yr + ys$ $(r + s)(x + y)$

80. $pm - pn + qm - qn$ $(m - n)(p + q)$

✓ **81.** $2ax + 2bx + 3a + 3b$ $(2x + 3)(a + b)$

82. $3xy + 3xz - 5y - 5z$ $(y + z)(3x - 5)$

83. $2ab + 2ac + 3b + 3c$ $(b + c)(2a + 3)$

84. $3ac + a + 3bc + b$ $(3c + 1)(a + b)$

✓ **85.** $6x^2 - 2x - 15x + 5$ $(3x - 1)(2x - 5)$

86. $6x^2 + 2x + 9x + 3$ $(3x + 1)(2x + 3)$

87. $9mp + 3mq - 3np - nq$ $(3p + q)(3m - n)$

88. $ax + bx - a - b$ $(a + b)(x - 1)$

✓ **89.** $2xy + y^2 - 2x - y$ $(2x + y)(y - 1)$

90. $2xy - 3y^2 + 2x - 3y$ $(2x - 3y)(y + 1)$

91. $8z^5 + 12z^2 - 10z^3 - 15$ $(2z^3 + 3)(4z^2 - 5)$

92. $2a^4 + 2a^3 - 4a - 4$ $(a + 1)(2a^3 - 4)$

In Exercises 93–98, factor by grouping. Factor out the GCF first.

✓ **93.** $ax^3 + bx^3 + 2ax^2y + 2bx^2y$ $x^2(a + b)(x + 2y)$

94. $x^3y^2 - 2x^2y^2 + 3xy^2 - 6y^2$ $y^2(x - 2)(x^2 + 3)$

95. $4a^2b + 12a^2 - 8ab - 24a$ $4a(b + 3)(a - 2)$

96. $-4abc - 4ac^2 + 2bc + 2c^2$ $-2c(b + c)(2a - 1)$

✓ **97.** $x^3y - x^2y - xy^2 + y^2$ $y(x^2 - y)(x - 1)$

98. $2x^3z - 4x^2z + 32xz - 64z$ $2z(x - 2)(x^2 + 16)$

APPLICATIONS

99. PICTURE FRAMING The dimensions of a family portrait and the frame in which it is mounted are given in Illustration 1. Write an algebraic expression that describes
  a. the area of the picture frame. $12x^3$ in.^{2}
  b. the area of the portrait. $20x^2$ in.^{2}
  c. the area of the mat used in the framing. Express the result in factored form. $4x^2(3x - 5)$ in.^{2}

ILLUSTRATION 1

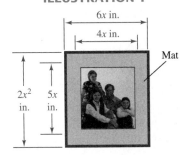

6x in.
4x in.
Mat
$2x^2$ in.
$5x$ in.

100. REARVIEW MIRRORS The dimensions of the three rearview mirrors on an automobile are given in Illustration 2. Write an algebraic expression that gives
  a. the area of the rearview mirror mounted on the windshield. $6x^3$ cm^2
  b. the total area of the two side mirrors. $24x^2$ cm^2
  c. the total area of all three mirrors. Express the result in factored form. $6x^2(x + 4)$ cm^2

ILLUSTRATION 2

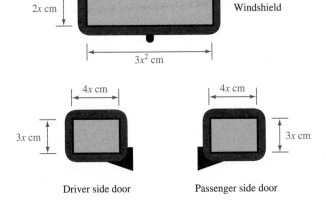

2x cm
Windshield
$3x^2$ cm
4x cm
4x cm
3x cm
3x cm
Driver side door
Passenger side door

101. COOKING See Illustration 3.
  a. What is the length of a side of the square griddle, in terms of r? What is the area of the cooking surface of the griddle, in terms of r? $4r$ in.; $16r^2$ in.^{2}

b. How many square inches of the cooking surface do the pancakes cover, in terms of r? $4\pi r^2$ in.^{2}
c. Find the amount of cooking surface that is not covered by the pancakes. Express the result in factored form. $16r^2 - 4\pi r^2 = 4r^2(4 - \pi)$ in.^{2}

ILLUSTRATION 3

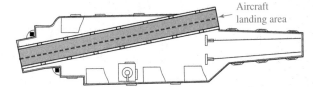

r in.   r in.
r in.   r in.

102. U.S. NAVY Illustration 4 shows the deck of the aircraft carrier *Enterprise*. The rectangular-shaped landing area of $(x^3 + 4x^2 + 5x + 20)$ ft^2 is shaded. What is the width of the landing area? (*Hint:* Factor the expression that represents the area.)
  $(x + 4)$ ft

ILLUSTRATION 4

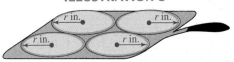

Aircraft landing area

WRITING

103. To add $5x$ and $7x$, we combine like terms: $5x + 7x = 12x$. Explain how this is related to factoring out a common factor.

104. One student commented, "Factoring undoes the distributive property." What do you think she meant? Give an example.

105. If asked to write $ax + ay - bx - by$ in factored form, explain why $a(x + y) - b(x + y)$ is not an acceptable answer.

106. When asked to factor $rx - sy + ry - sx$, a student wrote the expression as $rx + ry - sx - sy$. Then she factored it by grouping. Can the terms be rearranged in this manner? Explain your answer.

REVIEW

107. Find the distance between the points $(3, 5)$ and $(-2, -7)$.   13

108. Find the slope of the line passing through the points $(3, 5)$ and $(-2, -7)$. $\frac{12}{5}$

109. Does the point $(3, 5)$ lie on the graph of the line $4x - y = 7$?   yes

110. Simplify $-\sqrt{100a^8b^4}$. $-10a^4b^2$

▶ **6.2**

Factoring Trinomials of the Form $x^2 + bx + c$

In this section, you will learn about

Factoring trinomials that have a leading coefficient of 1 ■ Factoring out −1 ■ Prime polynomials ■ Factoring completely

Introduction Recall that two binomials can be multiplied by using the FOIL method. For example, to multiply $x + 2$ and $x + 3$, we proceed as follows:

$$(x + 2)(x + 3) = x^2 + 3x + 2x + 6$$
$$= x^2 + 5x + 6$$

In this section, we will reverse the process. Given a trinomial, such as $x^2 + 5x + 6$, we will ask ourselves, "What factors were multiplied to obtain $x^2 + 5x + 6$?" The process of finding the individual factors of a given trinomial is called *factoring the trinomial*. Since the product of two binomials is often a trinomial, we should not be surprised that many trinomials factor into the product of two binomials.

The multiplication process		*The factoring process*	
Given two binomial factors . . .	find the product	Given the product . . .	find the two binomial factors
↓ ↓	↓	↓	↓ ↓
$(x + 2)(x + 3)$	$= ?$	$x^2 + 5x + 6 =$	$(\ ? \ )(\ ? \ )$

We will now consider how to factor trinomials of the form $ax^2 + bx + c$, where a (called the **leading coefficient**) is 1.

Factoring Trinomials That Have a Leading Coefficient of 1

To develop a method for factoring trinomials, we multiply $(x + a)$ and $(x + b)$.

$$(x + a)(x + b) = x \cdot x + bx + ax + ab \quad \text{Use the FOIL method.}$$
$$= x^2 + ax + bx + ab \quad \text{Write } x \cdot x \text{ as } x^2. \text{ Write } bx + ax \text{ as } ax + bx.$$
$$= x^2 + (a + b)x + ab \quad \text{Factor } x \text{ out of } ax + bx. \text{ The result has three terms.}$$
$$\underset{\text{First term}}{\big|} \quad \underset{\text{Middle term}}{\big|} \quad \underset{\text{Last term}}{\big|}$$

From the result, we can see that

■ the first term is the product of x and x,

■ the last term is the product of a and b, and

■ the coefficient of the middle term is the sum of a and b.

We can use these facts to factor trinomials with leading coefficients of 1.

EXAMPLE 1

A positive last term. Factor $x^2 + 5x + 6$.

Solution To factor $x^2 + 5x + 6$, we will write it as the product of two binomials. Since the first term of the trinomial is x^2, the first term of each of its binomial factors must be x.

$$x^2 + 5x + 6 = (x + \underline{})(x + \underline{})$$

To fill in the blanks, we must find two integers such that

■   their *product* is 6 (because the last term of $x^2 + 5x + 6$ is 6), and
■   their *sum* is 5 (because the coefficient of the middle term of $x^2 + 5x + 6$ is 5).

To determine the integers, we list the two-integer factorizations of 6 in a table. Since the integers must have a positive sum, we need not list $(-1)(-6)$ and $(-2)(-3)$, because their sums are -7 and -5, respectively.

Product of the factors of 6	Sum of the factors of 6
1(6)	$1 + 6 = 7$
2(3)	$2 + 3 = 5$

The last row contains the integers 2 and 3, whose product is 6 and whose sum is 5. So we can fill in the blanks with 2 and 3.

$$x^2 + 5x + 6 = \left(x + \boxed{2}\right)\left(x + \boxed{3}\right)$$

To check the result, we find the product of $x + 2$ and $x + 3$ and verify that it is $x^2 + 5x + 6$.

$$\begin{aligned}(x + 2)(x + 3) &= x^2 + 3x + 2x + 6 \\ &= x^2 + 5x + 6\end{aligned}$$

SELF CHECK   Factor $y^2 + 5x + 4$.   *Answer:* $(y + 1)(y + 4)$ ■

In Example 1, the factors can be written in either order. An equivalent factorization is

$$x^2 + 5x + 6 = (x + 3)(x + 2)$$

EXAMPLE 2   **A positive last term.** Factor $y^2 - 7y + 12$.

Solution   Since the first term is y^2, the first term of each factor must be y. The last term of the trinomial is 12 and the coefficient of the middle term is -7. To fill in the blanks, we must find two integers whose product is 12 and whose sum is -7.

$$y^2 - 7y + 12 = \left(y + \boxed{}\right)\left(y + \boxed{}\right)$$

The two-integer factorizations of 12 and the sums of the factors are shown in the following table. Since the integers must have a negative sum, we need not list 1(12), 2(6), and 3(4), because their sums are positive 13, positive 8, and positive 7, respectively.

Product of the factors of 12	Sum of the factors of 12
$-1(-12)$	$-1 + (-12) = -13$
$-2(-6)$	$-2 + (-6) = -8$
$-3(-4)$	$-3 + (-4) = -7$

The last row contains the integers -3 and -4, whose product is 12 and whose sum is -7. So we can fill in the blanks with -3 and -4 and simplify.

$$\begin{aligned}y^2 - 7y + 12 &= \left(y + \boxed{(-3)}\right)\left(y + \boxed{(-4)}\right) \\ &= (y - 3)(y - 4)\end{aligned}$$

To check the result, we multiply $y - 3$ by $y - 4$ and verify that the product is $y^2 - 7y + 12$.

$$(y - 3)(y - 4) = y^2 - 4y - 3y + 12$$
$$= y^2 - 7y + 12$$

SELF CHECK   Factor $p^2 - 5p + 6$.   *Answer:* $(p - 3)(p - 2)$ ∎

The last term of each of the trinomials factored in Examples 1 and 2 was positive. In Example 1, the two integers we chose for the factored form were positive, because the middle term was positive. In Example 2, the two integers we chose for the factored form were negative, because the middle term was negative.

$$x^2 + 5x + 6 = (x + 2)(x + 3) \qquad y^2 - 7y + 12 = (y - 3)(y - 4)$$

Both positive   Both positive   Negative   Positive   Both negative

These observations suggest the following rule.

Factoring $x^2 + bx + c$ ($c > 0$)

To factor $x^2 + bx + c$, where c is positive, find two integers whose product is c and whose sum is b.

1. If b is positive, both the integers are positive.
2. If b is negative, both the integers are negative.

EXAMPLE 3   **A negative last term.** Factor $a^2 + 2a - 15$.

Solution   Since the first term is a^2, the first term of each factor must be a. To fill in the blanks, we must find two integers whose product is -15 and whose sum is 2.

$$a^2 + 2a - 15 = \left(a + \right)\left(a + \right)$$

The possible factorizations of -15 and the sum of the factors are shown in the following table.

Product of the factors of −15	Sum of the factors of −15
1(−15)	$1 + (-15) = -14$
3(−5)	$3 + (-5) = -2$
5(−3)	$5 + (-3) = 2$
15(−1)	$15 + (-1) = 14$

The third row contains the integers 5 and -3, whose product is -15 and whose sum is 2. So we can fill in the blanks with 5 and -3 and simplify.

$$a^2 + 2a - 15 = \left(a + \boxed{5}\right)\left(a + \boxed{(-3)}\right)$$
$$= (a + 5)(a - 3)$$

We can check by multiplying $(a + 5)$ and $(a - 3)$.

$$(a + 5)(a - 3) = a^2 - 3a + 5a - 15$$
$$= a^2 + 2a - 15$$

SELF CHECK | Factor $p^2 + 3p - 18$. | *Answer:* $(p + 6)(p - 3)$ ∎

EXAMPLE 4

A negative last term. Factor $z^2 - 4z - 21$.

Solution   Since the first term is z^2, the first term of each factor must be z. To fill in the blanks, we must find two integers whose product is -21 and whose sum is -4.

$$z^2 - 4y - 21 = (z + \boxed{})(z + \boxed{})$$

The factorizations of -21 and the sums of the factors are shown in the following table.

Product of the factors of -21	Sum of the factors of -21
$1(-21)$	$1 + (-21) = -20$
$3(-7)$	$3 + (-7) = -4$
$7(-3)$	$7 + (-3) = 4$
$21(-1)$	$21 + (-1) = 20$

The second row contains the integers 3 and -7, whose product is -21 and whose sum is -4. So we can fill in the blanks with 3 and -7 and simplify.

$$z^2 - 4z - 21 = (z + \boxed{3})(z + \boxed{(-7)})$$
$$= (z + 3)(z - 7)$$

To check, we multiply $z + 3$ and $z - 7$.

$$(z + 3)(z - 7) = z^2 - 7z + 3z - 21$$
$$= z^2 - 4z - 21$$

SELF CHECK | Factor $q^2 - 2q - 24$. | *Answer:* $(q + 4)(q - 6)$ ∎

In Examples 3 and 4, the last term of the given trinomials was negative. For each factorization, the integers that were chosen had different signs.

$$a^2 + 2a - 15 = (a + 5)(a - 3) \qquad z^2 - 4z - 21 = (z + 3)(z - 7)$$

↑   ↑   ↑   ↑   ↑   ↑
Negative   Different signs   Negative   Different signs

These observations suggest the following rule.

Factoring $x^2 + bx + c$ ($c < 0$)

To factor $x^2 + bx + c$, where c is negative, find two integers (one positive and one negative) whose product is c and whose sum is b.

The trinomials in the next two examples are of a form similar to $x^2 + bx + c$, and we can use the methods of this section to factor them.

EXAMPLE 5

Trinomials containing two variables. Factor $x^2 - 4xy - 5y^2$.

Solution   The trinomial has two variables, x and y. Since the first term is x^2, the first term of each factor must be x.

$$x^2 - 4xy - 5y^2 = (x + \boxed{})(x + \boxed{})$$

To fill in the blanks, we must find two *expressions* whose product is the last term, $-5y^2$, that will give a middle term of $-4xy$. Two such expressions are $-5y$ and y.

$$x^2 - 4xy - 5y^2 = \left(x + \boxed{(-5y)}\right)\left(x + \boxed{y}\right)$$
$$= (x - 5y)(x + y)$$

We check by multiplying.

$$(x - 5y)(x + y) = x^2 + xy - 5xy - 5y^2$$
$$= x^2 - 4xy - 5y^2$$

SELF CHECK   Factor $s^2 + 6st - 7t^2$.   *Answer:* $(s + 7t)(s - t)$ ■

EXAMPLE 6   **A leading term of x^4.** Factor $x^4 - 5x^2 - 150$.

Solution   Since the first term is x^4, the first term of each factor must be x^2.

$$x^4 - 5x^2 - 150 = \left(x^2 + \boxed{}\right)\left(x^2 + \boxed{}\right)$$

To fill in the blanks, we must find two integers whose product is -150 and whose sum is -5. Those two integers are 10 and -15.

$$x^4 - 5x^2 - 150 = \left(x^2 + \boxed{10}\right)\left(x^2 + \boxed{(-15)}\right)$$
$$= (x^2 + 10)(x^2 - 15)$$

We can check by multiplying.

$$(x^2 + 10)(x^2 - 15) = x^2 \cdot x^2 - 15x^2 + 10x^2 - 150$$
$$= x^4 - 5x^2 - 150$$

SELF CHECK   Factor $B^4 - 17B^2 + 70$.   *Answer:* $(B^2 - 7)(B^2 - 10)$ ■

Factoring Out -1

When factoring out trinomials of the form $ax^2 + bx + c$, where $a = -1$, we begin by factoring out -1.

EXAMPLE 7   **Factoring out -1 first.** Factor $-h^2 + 2h + 15$.

Solution   We factor out -1 and then factor $h^2 - 2h - 15$.

$$-h^2 + 2h + 15 = -1(h^2 - 2h - 15) \quad \text{Factor out } -1.$$
$$= -(h^2 - 2h - 15)$$
$$= -(h - 5)(h + 3) \quad \text{Use the integers } -5 \text{ and } 3, \text{ because their product is } -15 \text{ and their sum is } -2.$$

We check by multiplying.

$$-(h - 5)(h + 3) = -(h^2 + 3h - 5h - 15) \quad \text{Use the FOIL method first.}$$
$$= -(h^2 - 2h - 15)$$
$$= -h^2 + 2h + 15$$

SELF CHECK   Factor $-x^2 + 11x - 18$.   *Answer:* $-(x - 9)(x - 2)$ ■

Prime Polynomials

If a trinomial cannot be factored using only integers, it is called a **prime polynomial,** or more specifically, a **prime trinomial.**

EXAMPLE 8

Trinomials that do not factor. Factor $x^2 + 2x + 3$, if possible.

Solution   To factor the trinomial, we must find two integers whose product is 3 and whose sum is 2. The possible factorizations of 3 and the sums of the factors are shown in the following table.

Product of the factors of 3	Sum of the factors of 3
1(3)	$1 + 3 = 4$
$-1(-3)$	$-1 + (-3) = -4$

Since two integers whose product is 3 and whose sum is 2 do not exist, $x^2 + 2x + 3$ cannot be factored. It is a prime trinomial.

SELF CHECK   Factor $x^2 - 4x + 6$, if possible.   *Answer:* not possible; prime trinomial ∎

Factoring Completely

If the terms of a trinomial have a common factor, the GCF should always be factored out before any of the factoring techniques of this section are used. A trinomial is **factored completely** when it is expressed as a product of prime polynomials.

EXAMPLE 9

Factoring out the GCF first. Factor $2x^4 + 26x^3 + 80x^2$.

Solution   We begin by factoring out the GCF of $2x^2$.

$$2x^4 + 26x^3 + 80x^2 = 2x^2(x^2 + 13x + 40)$$

Next, we factor $x^2 + 13x + 40$. The integers 8 and 5 have a product of 40 and a sum of 13, so the completely factored form of the given trinomial is

$$2x^4 + 26x^3 + 80x^2 = 2x^2(x + 8)(x + 5)$$

Check by multiplying $2x^2$, $x + 8$, and $x + 5$.

SELF CHECK   Factor $4m^5 + 8m^4 - 32m^3$.   *Answer:* $4m^3(m + 4)(m - 2)$ ∎

EXAMPLE 10

Writing terms in descending order. Factor $-13g^2 + 36g + g^3$.

Solution   Before factoring the trinomial, we write its terms in descending powers of g.

$$\begin{aligned} -13g^2 + 36g + g^3 &= g^3 - 13g^2 + 36g && \text{Rearrange the terms.} \\ &= g(g^2 - 13g + 36) && \text{Factor out } g, \text{ which is the GCF.} \\ &= g(g - 9)(g - 4) && \text{Factor the trinomial.} \end{aligned}$$

Check by multiplying g, $g - 9$, and $g - 4$.

SELF CHECK   Factor $-12t + t^3 + 4t^2$.   *Answer:* $t(t - 2)(t + 6)$ ∎

Section 6.2

VOCABULARY

In Exercises 1–8, fill in the blanks to make the statements true.

1. A polynomial, such as $x^2 - x - 6$, that has exactly three terms is called a ____trinomial____.

2. A polynomial, such as $x - 3$, that has exactly two terms is called a ____binomial____.

3. The statement $x^2 - x - 12 = (x - 4)(x + 3)$ shows that $x^2 - x - 12$ ____factors____ into the product of two binomials.

4. Since $10 = (-5)(-2)$, we say -5 and -2 are ____factors____ of 10.

5. A ____prime____ polynomial cannot be factored by using only integers.

6. The ____leading____ coefficient of the trinomial $x^2 - 3x + 2$ is 1, the ____coefficient____ of the middle term is -3, and the last ____term____ is 2.

7. To factor $x^2 + x - 56$, we must find two integers whose ____product____ is -56 and whose ____sum____ is 1.

8. A trinomial is factored ____completely____ when it is expressed as a product of prime polynomials.

CONCEPTS

In Exercises 9–12, fill in the blanks to make the statements true.

9. Two factorizations of 4 that involve only positive numbers are $4 \cdot 1$ and $2 \cdot 2$.

10. Two factorizations of 4 that involve only negative numbers are $-4(-1)$ and $-2(-2)$.

11. Before attempting to factor a trinomial, be sure that the exponents are written in ____descending____ order.

12. Before attempting to factor a trinomial into two binomials, always factor out any common ____factors____ first.

13. Complete the table.

Product of the factors of 8	Sum of the factors of 8
1(8)	9
2(4)	6
−1(−8)	−9
−2(−4)	−6

14. If we use the FOIL method to do the multiplication $(x + 5)(x + 4)$, we obtain $x^2 + 9x + 20$.
  a. What step of the FOIL process produced 20?
  Last: $5 \cdot 4$
  b. What steps of the FOIL process produced $9x$?
  Outer: $x \cdot 4$ and Inner: $5 \cdot x$

15. Given $x^2 - 2x - 15$:
  a. What is the coefficient of the x^2-term?   1
  b. What is the last term? The last term is the product of what two integers?
  -15; -5 and 3
  c. What is the coefficient of the middle term? It is the sum of what two integers? -2; -5 and 3

16. Given $x^2 + 8x + 15$:
  a. What is the coefficient of the x^2-term?   1
  b. What is the last term? The last term is the product of what two integers?
  15; 5 and 3
  c. What is the coefficient of the middle term? It is the sum of what two integers?   8; 5 and 3

17. To determine which two integers to use in the factorization of $x^2 + 7x + 10$, a student constructed the following table. Explain why she didn't need to write the last two rows.
  The sum of two negative factors of 10 could not be 7.

Product of the factors of 10	Sum of the factors of 10
1(10)	
2(5)	
−1(−10)	
−2(−5)	

18. Complete the factorization table.
  The order of the entries may vary.

Product of the factors of −9	Sum of the factors of −9
1(−9)	$1 + (-9) = -8$
3(−3)	$3 + (-3) = 0$
−1(9)	$-1 + 9 = 8$

19. Consider factoring a trinomial of the form
$x^2 + bx + c$.
   a. If c is positive, what can be said about the two
   integers that should be chosen for the factorization?
   They are both positive or they are both negative.
   b. If c is negative, what can be said about the two
   integers that should be chosen for the factorization?   One will be positive, the other negative.

20. What polynomial has the factorization of
$(x + 8)(x - 2)$? $x^2 + 6x - 16$

NOTATION

In Exercises 21–22, complete each factorization.

21. $3x^2 + 15x + 18 = \boxed{3}\,(x^2 + 5x + 6)$
$= 3(x + \boxed{3})(x + \boxed{2})$

22. $-a^2 - ab + 20b^2 = \boxed{-}\,(a^2 + ab - 20b^2)$
$= -(a + \boxed{5b})(a - 4b)$

PRACTICE

In Exercises 23–28, complete each factorization.

23. $x^2 + 3x + 2 = (x + \boxed{2})(x + \boxed{1})$

24. $y^2 + 4y + 3 = (y \boxed{+ 3})(y + \boxed{1})$

25. $t^2 - 9t + 14 = (\boxed{t} - 7)(t - \boxed{2})$

26. $c^2 - 9c + 8 = (\boxed{c} - 8)(c - \boxed{1})$

27. $a^2 + 6a - 16 = (a \boxed{+ 8})(a \boxed{-} 2)$

28. $x^2 - 3x - 40 = (x \boxed{-} 8)(x \boxed{+} 5)$

*In Exercises 29–46, factor each trinomial. If a polyno-
mial can't be factored, write "prime."*

No primes on test

29. $z^2 + 12z + 11$ $(z + 11)(z + 1)$

30. $x^2 + 7x + 10$ $(x + 5)(x + 2)$

31. $m^2 - 5m + 6$ $(m - 3)(m - 2)$

32. $n^2 - 7n + 10$ $(n - 5)(n - 2)$

33. $a^2 - 4a - 5$ $(a - 5)(a + 1)$

34. $b^2 + 6b - 7$ $(b + 7)(b - 1)$

35. $x^2 + 5x - 24$ $(x + 8)(x - 3)$

36. $t^2 - 5t - 50$ $(t - 10)(t + 5)$

37. $a^2 - 10a - 39$ $(a - 13)(a + 3)$

38. $r^2 - 9r - 12$ prime

39. $u^2 + 10u + 15$ prime

40. $v^2 + 9v + 15$ prime

41. $s^2 + 11s - 26$ $(s + 13)(s - 2)$

42. $y^2 + 8y + 12$ $(y + 6)(y + 2)$

43. $r^2 - 2r + 4$ prime

44. $m^2 + 3m - 10$ $(m + 5)(m - 2)$

45. $m^2 - m - 12$ $(m - 4)(m + 3)$

46. $u^2 + u - 42$ $(u + 7)(u - 6)$

N/A

*In Exercises 47–54, factor each trinomial in two vari-
ables, if possible.*

47. $x^2 + 4xy + 4y^2$ $(x + 2y)(x + 2y)$

48. $a^2 + 10ab + 9b^2$ $(a + 9b)(a + b)$

49. $m^2 + 3mn - 10n^2$ $(m + 5n)(m - 2n)$

50. $m^2 - mn - 12n^2$ $(m - 4n)(m + 3n)$

51. $a^2 - 4ab - 12b^2$ $(a - 6b)(a + 2b)$

52. $p^2 + pq - 6q^2$ $(p + 3q)(p - 2q)$

53. $r^2 - 2rs + 4s^2$ prime

54. $m^2 + 3mn - 20n^2$ prime

*In Exercises 55–64, factor each trinomial. Factor out -1
first.*

55. $-x^2 - 7x - 10$ $-(x + 5)(x + 2)$

56. $-x^2 + 9x - 20$ $-(x - 5)(x - 4)$

57. $-t^2 - 15t + 34$ $-(t + 17)(t - 2)$

58. $-t^2 - t + 30$ $-(t + 6)(t - 5)$

59. $-r^2 + 14r - 40$ $-(r - 10)(r - 4)$

60. $-r^2 + 14r - 45$ $-(r - 9)(r - 5)$

61. $-a^2 - 4ab - 3b^2$ $-(a + 3b)(a + b)$

62. $-a^2 - 6ab - 5b^2$ $-(a + b)(a + 5b)$

63. $-x^2 + 6xy + 7y^2$ $-(x - 7y)(x + y)$

64. $-x^2 - 10xy + 11y^2$ $-(x + 11y)(x - y)$

*In Exercises 65–74, write each trinomial in descending
powers of one variable and factor.*

65. $4 - 5x + x^2$ $(x - 4)(x - 1)$

66. $y^2 + 5 + 6y$ $(y + 5)(y + 1)$

67. $10y + 9 + y^2$ $(y + 9)(y + 1)$

68. $x^2 - 13 - 12x$ $(x - 13)(x + 1)$

69. $-r^2 + 2 + r$ $-(r - 2)(r + 1)$

70. $u^2 - 3 + 2u$ $(u + 3)(u - 1)$

71. $4rx + r^2 + 3x^2$ $(r + 3x)(r + x)$

72. $a^2 + 5b^2 + 6ab$ $(a + b)(a + 5b)$

73. $-3ab + a^2 + 2b^2$ $(a - 2b)(a - b)$

74. $-13yz + y^2 - 14z^2$ $(y - 14z)(y + z)$

*In Exercises 75–84, completely factor each trinomial.
Factor out any common monomials first (including -1 if
necessary).*

75. $2x^2 + 10x + 12$ $2(x + 3)(x + 2)$

76. $3y^2 - 21y + 18$ $3(y - 6)(y - 1)$

77. $-5a^2 + 25a - 30$ $-5(a - 3)(a - 2)$

78. $-2b^2 + 20b - 18$ $-2(b - 9)(b - 1)$

79. $3z^2 - 15z + 12$ $3(z - 4)(z - 1)$

80. $5m^2 + 45m - 50$ $5(m + 10)(m - 1)$

81. $12xy + 4x^2y - 72y$ $4y(x + 6)(x - 3)$

82. $48xy + 6xy^2 + 96x$ $6x(y + 4)(y + 4)$

83. $-4x^2y - 4x^3 + 24xy^2$ $-4x(x + 3y)(x - 2y)$

84. $3x^2y^3 + 3x^3y^2 - 6xy^4$ $3xy^2(x + 2y)(x - y)$

APPLICATIONS

85. PETS The cage shown in Illustration 1 is used for transporting dogs. Its volume is $(x^3 + 12x^2 + 27x)$ in.^{3}. The dimensions of the cage can be found by factoring this expression. If the cage is longer than it is tall, and taller than it is wide, determine its length, width, and height. $(x + 9)$ in., $(x + 3)$ in., x in.

ILLUSTRATION 1

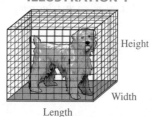

Height

Width

Length

86. GRAPHING In Illustration 2, the graph on the left is that of the polynomial function $f(x) = x^3 + x^2 - 2x$. The graph on the right is that of the function $g(x) = x(x + 2)(x - 1)$.

  a. Do the graphs appear to be the same?   yes

  b. Factor the right-hand side of $f(x) = x^3 + x^2 - 2x$. What can now be said with certainty about the graphs?

    They are the same graphs, because the functions are the same.

ILLUSTRATION 2

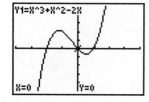 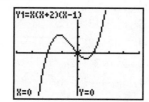

87. GRAPHING CALCULATORS Find the perimeter of the graphing calculator window shown in Illustration 3. Then express the result in factored form.

  $(2x^2 + 10x + 12)$ cm; $2(x + 3)(x + 2)$ cm

ILLUSTRATION 3

$(x^2 + 2x)$ cm

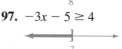

$(3x + 6)$ cm

88. CARPOOLING The average rate at which a carpool van travels and the distance it covers are given in the table in terms of t. Factor the expression representing the distance traveled and then complete the table.

Rate (mi/hr)	Time (hr)	Distance traveled (mi)
$t + 11$	$t + 5$	$t^2 + 16t + 55$

WRITING

89. Explain what it means when we say that a trinomial is the product of two binomials. Give an example.

90. Are $2x^2 - 12x + 16$ and $x^2 - 6x + 8$ factored in the same way? Explain why or why not.

91. When factoring $x^2 - 2x - 3$, one student got $(x - 3)(x + 1)$, and another got $(x + 1)(x - 3)$. Are both answers acceptable? Explain.

92. Explain how to use the FOIL method to check the factorization of a trinomial.

93. In the partial solution shown below, a student began to factor the trinomial and then gave up. Write a brief note to the student explaining his initial mistake.

    Factor $x^2 - 2x - 63$.

$$(x - \ )(x - \ )$$
$$? ? ?$$

94. Explain why the given trinomial is not factored completely.

$$3x^2 - 3x - 60 = 3(x^2 - x - 20)$$

REVIEW

In Exercises 95–98, graph the solution of each inequality on a number line.

95. $x - 3 > 5$

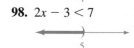

8

96. $x + 4 \le 3$

−1

97. $-3x - 5 \ge 4$

−3

98. $2x - 3 < 7$

5

▶6.3   Factoring Trinomials of the Form $ax^2 + bx + c$

In this section, you will learn about

Observations about multiplying binomials ■ The trial-and-check factoring method ■ The grouping method

Introduction   In the previous section, we factored trinomials having a leading coefficient of 1. In this section, we will consider how to factor trinomials with leading coefficients that are not 1, such as $2x^2 + 5x + 3$, $6a^2 - 17a + 5$, and $3y^2 - 4y - 4$.

Two methods are commonly used to factor these trinomials. With the first method, educated guesses are made. These guesses are checked by multiplication. The correct factorization is determined by a process of elimination. The second method is an extension of factoring by grouping.

Observations about Multiplying Binomials

In the work below, we find the product of $2x + 1$ and $x + 3$. If we interchange the last terms of these binomials and find the product of $2x + 3$ and $x + 1$, several observations can be made when we compare the results.

$$(2x + 1)(x + 3) = 2x^2 + 6x + x + 3 \qquad (2x + 3)(x + 1) = 2x^2 + 2x + 3x + 3$$
$$= 2x^2 + 7x + 3 \qquad\qquad\qquad = 2x^2 + 5x + 3$$

In each case, the result is a trinomial, and

■ the first terms are the same ($2x^2$),
■ the last terms are the same (3), and
■ the middle terms are different ($7x$ and $5x$).

These observations indicate that when the last terms in $(2x + 1)(x + 3)$ are interchanged to form $(2x + 3)(x + 1)$, only the middle terms of the products are different. This fact is helpful when factoring trinomials by using the *trial-and check method*.

The Trial-and-Check Factoring Method

To factor a trinomial with a leading coefficient of 1 (say, $x^2 + 4x + 3$), we begin with a factorization of the form

$$x^2 + 4x + 3 = \left(x + \boxed{}\,\right)\!\left(x + \boxed{}\,\right)$$

and determine which integers to write in the blanks.

To factor a trinomial with a leading coefficient that is not 1 (say, $2x^2 + 5x + 3$), we begin with a factorization of the form

$$2x^2 + 5x + 3 = \left(\,\boxed{}\,x + \boxed{}\,\right)\!\left(\,\boxed{}\,x + \boxed{}\,\right)$$

We must determine what numbers to write in the four blanks. Because of the additional blanks, there are more combinations of factors to consider.

EXAMPLE 1

Factoring trinomials. Factor $2x^2 + 5x + 3$.

Solution Since the first term is $2x^2$, the first terms of the factors must be $2x$ and $1x$. We write the coefficients, 2 and 1, in the proper blanks.

$$2x^2 + 5x + 3 = (\boxed{2}\,x + \boxed{})(\boxed{1}\,x + \boxed{})$$
$$\underbrace{\uparrow \qquad \uparrow}_{\substack{\text{Factors} \\ \text{of } 2}}$$

To fill in the other blanks, we must find two factors of the last term of the trinomial (3) that will give a middle term of $5x$ when the binomials are multiplied.

Because each term of the trinomial is positive, we need only consider positive factors of 3. Since the positive factors of 3 are 1 and 3, we can write 1 and 3 in the blanks—or we can interchange the entries, and write 3 and 1 in the blanks.

$$\big(2x + \boxed{1}\big)\big(x + \boxed{3}\big) \qquad \text{or} \qquad \big(2x + \boxed{3}\big)\big(x + \boxed{1}\big)$$
$$\underbrace{\uparrow \qquad \uparrow}_{\substack{\text{Factors} \\ \text{of } 3}} \qquad\qquad\qquad \underbrace{\uparrow \qquad \uparrow}_{\substack{\text{Factors} \\ \text{of } 3}}$$

The first possibility is incorrect: When we find the outer and inner products and combine like terms, we obtain an incorrect middle term of $7x$.

$$\overset{\displaystyle\text{Outer: } 6x}{(2x + 1)(x + 3)} \quad \text{Apply the FOIL method to find the middle term: } 6x + x = 7x.$$
$$\underset{\displaystyle\text{Inner: } x}{}$$

The second possibility is correct, because the FOIL method gives a middle term of $5x$.

$$\overset{\displaystyle\text{Outer: } 2x}{(2x + 3)(x + 1)} \quad \text{Combine like terms: } 2x + 3x = 5x.$$
$$\underset{\displaystyle\text{Inner: } 3x}{}$$

Thus,

$$2x^2 + 5x + 3 = (2x + 3)(x + 1)$$

SELF CHECK

Factor $3x^2 + 7x + 2$. *Answer:* $(3x + 1)(x + 2)$ ∎

EXAMPLE 2

Factoring trinomials. Factor $6a^2 - 17a + 5$.

Solution Since the first term is $6a^2$, the first terms of the factors must be $6a$ and $1a$ or $3a$ and $2a$. We enter the coefficients in the proper blanks.

$$\big(\boxed{6}\,a + \boxed{}\big)\big(\boxed{1}\,a + \boxed{}\big) \qquad \text{or} \qquad \big(\boxed{3}\,a + \boxed{}\big)\big(\boxed{2}\,a + \boxed{}\big)$$

To fill in the remaining blanks, we must find two integers such that

- their product is 5 and
- the sum of the products of the outer and inner terms will be $-17a$.

Because the last term of the trinomial (5) is positive and the middle term is negative, we need only consider negative factors of the last term. Since the negative factors of 5 are -1 and -5, and their positions in the factored forms can be interchanged, there are two possibilities for each of the factored forms listed earlier.

$$\big(6a + \boxed{(-1)}\big)\big(a + \boxed{(-5)}\big) \qquad \big(6a + \boxed{(-5)}\big)\big(a + \boxed{(-1)}\big)$$
$$\big(3a + \boxed{(-1)}\big)\big(2a + \boxed{(-5)}\big) \qquad \big(3a + \boxed{(-5)}\big)\big(2a + \boxed{(-1)}\big)$$

Next, we rewrite each factorization in a simpler form and determine each middle term.

$$\overset{\overset{-30a}{\frown}}{(6a - 1)(a - 5)} \underset{-a}{} \quad -30a - a = -31a. \qquad \overset{\overset{-6a}{\frown}}{(6a - 5)(a - 1)} \underset{-5a}{} \quad -6a - 5a = -11a.$$

$$\overset{\overset{-15a}{\frown}}{(3a - 1)(2a - 5)} \underset{-2a}{} \quad -15a - 2a = -17a. \qquad \overset{\overset{-3a}{\frown}}{(3a - 5)(2a - 1)} \underset{-10a}{} \quad -3a - 10a = -13a.$$

Only the possibility shown in blue gives the correct middle term of $-17a$. Thus,

$$6a^2 - 17a + 5 = (3a - 1)(2a - 5)$$

SELF CHECK Factor $6h^2 - 7h + 2$. *Answer:* $(3h - 2)(2h - 1)$ ∎

EXAMPLE 3

Factoring trinomials. Factor $3y^2 - 4y - 4$.

Solution Since the first term is $3y^2$, the first terms of the factors must be $3y$ and $1y$. We fill in the first blanks of each binomial with the coefficients, 3 and 1, and simplify the second binomial factor by writing $1y$ as y.

$$3y^2 - 4y - 4 = \left(\; 3 \, y + \; \fbox{} \; \right)\left(\; 1 \, y + \; \fbox{} \; \right)$$
$$= \left(3y + \fbox{} \right)\left(y + \fbox{} \right)$$

To fill in the remaining blanks, we must find two integers that will (in combination with out choice of first terms) give a last term of -4 and a sum of the products of the outer terms and inner terms of $-4y$.

Since the sign of the last term of the trinomial (-4) is negative, the signs inside the binomial factors will be different. For each pair of factors of -4, we must check two possibilities, because we can interchange the position of the factors.

For the factors 1 and −4:

$$\overset{\overset{-12y}{\frown}}{(3y + 1)(y - 4)} \underset{y}{} \quad -12y + y = -11y. \qquad \overset{\overset{3y}{\frown}}{(3y - 4)(y + 1)} \underset{-4y}{} \quad 3y - 4y = -y.$$

For the factors −1 and 4:

$$\overset{\overset{12y}{\frown}}{(3y - 1)(y + 4)} \underset{-y}{} \quad 12y - y = 11y. \qquad \overset{\overset{-3y}{\frown}}{(3y + 4)(y - 1)} \underset{4y}{} \quad -3y + 4y = y.$$

For the factors −2 and 2:

$$\overset{\overset{6y}{\frown}}{(3y - 2)(y + 2)} \underset{-2y}{} \quad 6y - 2y = 4y. \qquad \overset{\overset{-6y}{\frown}}{(3y + 2)(y - 2)} \underset{2y}{} \quad -6y + 2y = -4y.$$

Only the possibility shown in blue gives the correct middle term of $-4y$. Thus,

$$3y^2 - 4y - 4 = (3y + 2)(y - 2)$$

Check the factorization by multiplication.

SELF CHECK Factor $5a^2 - 7a - 6$. *Answer:* $(5a + 3)(a - 2)$ ∎

| EXAMPLE 4 | **Factoring a trinomial in two variables.** Factor $4b^2 + 8bc - 45c^2$. |

Solution Since the first term is $4b^2$, the first terms of the factors must be $4b$ and $1b$ or $2b$ and $2b$.

$$\left(4b + \right)\left(1b + \right) \quad \text{or} \quad \left(2b + \right)\left(2b + \right)$$

To fill in the remaining blanks, we must find two factors of $-45c^2$ that will give a middle term of $8bc$.

Since $-45c^2$ has many factors, there are many possible combinations for the last terms of the binomial factors. The signs of the factors that we try must be different, because the last term of the trinomial is negative.

If we pick factors of $4b$ and b for the first terms, and $-c$ and $45c$ for the last terms, we have

$$\overset{\overset{\displaystyle 180bc}{\frown}}{(4b - c)(b + 45c)} \quad 180bc - bc = 179bc.$$
$$\underset{-bc}{\smile}$$

which gives an incorrect middle term of $179bc$. So the factorization is incorrect. Since the middle term of the trinomial, $8bc$, is small, we see that it was not wise to pick such a large factor as $45c$.

If we pick factors of $4b$ and b for the first terms and $15c$ and $-3c$ for the last terms, we have

$$\overset{\overset{\displaystyle -12bc}{\frown}}{(4b + 15c)(b - 3c)} \quad -12bc + 15bc = 3bc.$$
$$\underset{15bc}{\smile}$$

which gives an incorrect middle term of $3bc$. So it is incorrect.

If we pick factors of $2b$ and $2b$ for the first terms and $-5c$ and $9c$ for the last terms, we have

$$\overset{\overset{\displaystyle 18bc}{\frown}}{(2b - 5c)(2b + 9c)} \quad 18bc - 10bc = 8bc.$$
$$\underset{-10bc}{\smile}$$

which gives the correct middle term of $8bc$. The correct factorization is

$$4b^2 + 8bc - 45c^2 = (2b - 5c)(2b + 9c)$$

Check by multiplication.

SELF CHECK Factor $4x^2 + 4xy - 3y^2$. *Answer:* $(2x + 3y)(2x - y)$ ∎

Because some guesswork is often necessary, it is difficult to give specific rules for factoring trinomials with a leading coefficient that is not 1. However, the following hints are helpful.

Factoring $ax^2 + bx + c$ $(a \neq 1)$

1. Write the trinomial in descending powers of the variable and factor out any GCF (including -1 if that is necessary to make the coefficient of the first term positive).

2. Attempt to write the trinomial as *the product of two binomials.* The coefficients of the first terms of each binomial factor must be factors of a, and the last terms must be factors of c.

$$
\underbrace{(\boxed{\ } x + \boxed{\ })(\boxed{\ } x + \boxed{\ })}
$$

Factors of a

Factors of c

3. If the sign of the third term of the trinomial is positive, the signs between the terms of the binomial factors are the same as the sign of the middle term. If the sign of the third term is negative, the signs between the terms of the binomial factors are opposite.

4. Using multiplication to check your work, try combinations of coefficients of the first terms and last terms until you find one that gives the middle term of the trinomial. If no combination works, the trinomial is prime.

5. Check the factorization by multiplication.

EXAMPLE 5

Writing terms in descending order. Factor $2x^2 - 8x^3 + 3x$.

Solution   We write the trinomial in descending powers of x.

$$-8x^3 + 2x^2 + 3x$$

Then we factor out the negative of the GCF, which is $-x$.

$$-8x^3 + 2x^2 + 3x = -x(8x^2 - 2x - 3)$$

We must now factor $8x^2 - 2x - 3$. Its factorization has the form

$$(\boxed{1} x + \boxed{\ })(\boxed{8} x + \boxed{\ }) \quad \text{or} \quad (\boxed{2} x + \boxed{\ })(\boxed{4} x + \boxed{\ })$$

To fill in the remaining blanks, we find two factors of the last term of the trinomial (-3) that will give a middle term of $-2x$. Because the sign of the last term is negative, the signs within its binomial factors will be different. If we pick factors of $2x$ and $4x$ for the first terms and 1 and -3 for the last terms, we have

$$
\overset{-6x}{\overbrace{(2x + 1)(4x - 3)}} \quad -6x + 4x = -2x.
$$
$$\underset{4x}{}$$

which gives the correct middle term of $-2x$, so it is correct.

$$8x^2 - 2x - 3 = (2x + 1)(4x - 3)$$

We can now give the complete factorization.

$$
\begin{aligned}
-8x^3 + 2x^2 + 3x &= -x(8x^2 - 2x - 3) \\
&= -x(2x + 1)(4x - 3)
\end{aligned}
$$

Check by multiplication.

SELF CHECK   Factor $12y - 2y^3 - 2y^2$.   *Answer:* $-2y(y + 3)(y - 2)$ ∎

The Grouping Method

In the work below, we find the product of $3a - 1$ and $2a - 5$.

$$6(5) = 30$$

$$(3a - 1)(2a - 5) = 6a^2 - 15a - 2a + 5$$

$$-15(-2) = 30$$

We can see that the product of the coefficients of the first and last terms is the same as the product of the coefficients of the two middle terms. This fact is true for any product of two binomials. We use this observation when factoring trinomials with lead coefficients that are not 1 by using the *grouping method*.

To factor a trinomial of the form $ax^2 + bx + c$ ($a \neq 1$) by grouping, we write the trinomial as an equivalent polynomial with four terms. For example, to factor $6x^2 + 11x + 3$, we write it in the form

$$6x^2 + 11x + 3 = 6x^2 + \boxed{}\, x + \boxed{}\, x + 3$$

| First | Middle | Last |
| term | terms | term |

Since the product of the coefficients of the first and last terms is the same as the product of the coefficients of the middle terms, we know that the product of the integers that go in the blanks is $6 \cdot 3$ (or 18). We also know that the sum of the middle terms must be $11x$, so the sum of the integers in the blanks must be 11. Since a positive product and a positive sum are required, the integers must be positive.

Product of the factors of 18	Sum of the factors of 18
1(18)	$1 + 18 = 19$
2(9)	$2 + 9 = 11$
3(6)	$3 + 6 = 9$

Since 2 and 9 have a product of 18 and a sum of 11, we write them in the blanks and factor the four-term polynomial by grouping.

$$6x^2 + 11x + 3 = 6x^2 + \mathbf{2}\, x + \mathbf{9}\, x + 3 \qquad \text{We have written } 2x + 9x \text{ in place of } 11x.$$
$$= 2x(3x + 1) + 3(3x + 1) \qquad \text{Factor out } 2x \text{ from } 6x^2 + 2x. \text{ Factor out } 3 \text{ from } 9x + 3.$$
$$= (3x + 1)(2x + 3) \qquad \text{Factor out } (3x + 1).$$

A check shows that the factorization is correct.

$$(3x + 1)(2x + 3) = 6x^2 + 9x + 2x + 3$$
$$= 6x^2 + 11x + 3$$

EXAMPLE 6   **Factoring a trinomial by grouping.** Factor $6a^2 - 17a + 5$.

Solution   This is the trinomial that was factored in Example 2. To factor this trinomial by grouping, we begin with a four-term polynomial.

$$6a^2 - 17a + 5 = 6a^2 + \boxed{}\, a + \boxed{}\, a + 5$$

To fill in the blanks, we must find two integers whose product is $6 \cdot 5 = 30$. The sum of the integers should be the coefficient of middle term of the trinomial (-17). Both integers must be negative, because their product is positive and their sum is negative.

Product of the factors of 30	Sum of the factors of 30
$-1(-30)$	$-1 + (-30) = -31$
$-2(-15)$	$-2 + (-15) = -17$
$-3(-10)$	$-3 + (-10) = -13$
$-5(-6)$	$-5 + (-6) = -11$

Since -2 and -15 satisfy the two requirements, we write them in the blanks, write the middle terms in simpler form, and factor the polynomial.

$$6x^2 - 17x + 5 = 6x^2 + \boxed{(-2)} x + \boxed{(-15)} x + 5$$
$$= 6x^2 - 2x - 15x + 5$$
$$= 2x(3x - 1) - 5(3x - 1) \qquad \text{Factor out } 2x \text{ from } 6x^2 - 2x.$$
$$\text{Factor out } -5 \text{ from } -15x + 5.$$
$$= (3x - 1)(2x - 5) \qquad \text{Factor out } (3x - 1).$$

Check the result.

SELF CHECK   Factor $21a^2 - 13a + 2$.   *Answer: $(7a - 2)(3a - 1)$*   ■

Factoring $ax^2 + bx + c$ by grouping

1. Write the trinomial in descending powers of the variable and factor out any GCF (including -1 if that is necessary to make the coefficient of the first term positive).
2. Calculate the product of the coefficients of the leading term and the last term of the trinomial.
3. Find two numbers whose product is the number found in step 2 and whose sum is the coefficient of the middle term of the trinomial.
4. Write the numbers in the blanks of the factored form shown below, and then factor the polynomial by grouping.

$$ax^2 + \boxed{} x + \boxed{} x + c$$

5. Check the factorization using multiplication.

EXAMPLE 7   **Factoring by grouping.** Factor $12x^5 - x^4 - 6x^3$.

Solution   First, we factor out the GCF, which is x^3.

$$12x^5 - x^4 - 6x^3 = x^3(12x^2 - x - 6)$$

To factor $12x^2 - x - 6$ by grouping, we must find two integers whose product is $12(-6) = -72$ and whose sum is the coefficient of the middle term of the trinomial, -1. Since their product is negative, the numbers will have opposite signs. The two integers that meet these requirements are -9 and 8.

Product of the factors of -72	Sum of the factors of -72
$-9(8)$	$-9 + 8 = -1$

We enter -9 and 8 in the blanks and factor.

$$12x^2 - x - 6 = 12x^2 + \boxed{(-9)}\ x + \boxed{8}\ x - 6$$
$$= 12x^2 - 9x + 8x - 6$$
$$= 3x(4x - 3) + 2(4x - 3)$$
$$= (4x - 3)(3x + 2)$$

The complete factorization is

$$12x^5 - x^4 - 6x^3 = x^3(4x - 3)(3x + 2)$$

SELF CHECK   Factor $12a^2 - 22a - 20$.   *Answer:* $2(2a - 5)(3a + 2)$ ■

STUDY SET

Section 6.3

VOCABULARY

In Exercises 1–8, fill in the blanks to make the statements true.

1. The trinomial $3x^2 - x - 12$ has a __leading__ coefficient of 3.

2. The numbers 3 and 2 are __factors__ of the first term of the trinomial $6x^2 + x - 12$.

3. Consider $(x - 2)(5x - 1)$. The product of the __outer__ terms is $-x$ and the product of the __inner__ terms is $-10x$.

4. When we write $2x^2 + 7x + 3$ as $(2x + 1)(x + 3)$, we say that we have __factored__ the trinomial—it has been expressed as the product of two __binomials__.

5. The __middle__ term of $4x^2 - 7x + 13$ is $-7x$.

6. The polynomial $6x^2 + 2x + 9x + 3$ has four __terms__.

7. The __sum__ of the middle terms of the polynomial $4a^2 - 12a - a + 3$ is $-13a$.

8. The __GCF__ of the terms of the trinomial $6b^3 - 3b^2 - 12b$ is $3b$.

CONCEPTS

In Exercises 9–10, complete each statement in red.

9.

These coefficients must be factors of $\underset{5}{}$.

$$5x^2 + 6x - 8 = (\ \boxed{}\ x + \boxed{}\ )(\ \boxed{}\ x + \boxed{}\ )$$

These numbers must be factors of $\underset{-8}{}$.

10.

The product of these coefficients must be $\underset{-15}{}$.

$$3x^2 - 14x - 5 = 3x^2 + \boxed{}\ x + \boxed{}\ x - 5$$

The sum of these coefficients must be $\underset{-14}{}$.

In Exercises 11–14, a trinomial has been partially factored. Complete each statement that describes the type of integers we should consider for the blanks.

11. $5y^2 - 13y + 6 = \left(5x + \boxed{}\ \right)\left(x + \boxed{}\ \right)$

Since the last term of the trinomial is __positive__ and the middle term is __negative__, the integers must be __negative__ factors of 6.

12. $5y^2 + 13y + 6 = \left(5x + \boxed{}\ \right)\left(x + \boxed{}\ \right)$

Since the last term of the trinomial is __positive__ and the middle term is __positive__, the integers must be __positive__ factors of 6.

13. $5y^2 + 7y - 6 = \left(5x + \boxed{}\ \right)\left(x + \boxed{}\ \right)$

Since the last term of the trinomial is __negative__, the signs of the integers will be __different__.

14. $5y^2 - 7y - 6 = \left(5x + \boxed{}\ \right)\left(x + \boxed{}\ \right)$

Since the last term of the trinomial is __negative__, the signs of the integers will be __different__.

In Exercises 15–16, a trinomial is to be factored by the grouping method. Complete each statement that describes the type of integers we should consider for the blanks.

15. $8c^2 - 11c + 3 = 8c^2 + \boxed{}c + \boxed{}c + 3$

We need to find two integers whose product is $\underline{24}$ and whose sum is $\underline{-11}$.

16. $15c^2 + 4c - 4 = 15c^2 + \boxed{}c + \boxed{}c - 4$

We need to find two integers whose product is $\underline{-60}$ and whose sum is $\underline{4}$.

NOTATION

17. Write a trinomial of the form $ax^2 + bx + c$
   a. where $a = 1$ $x^2 + 2x + 3$ (answers may vary)
   b. where $a \neq 1$ $2x^2 + 2x + 3$ (answers may vary)

18. Write the terms of the trinomial $40 - t - 4t^2$ in descending powers of the variable. $-4t^2 - t + 40$

PRACTICE

In Exercises 19–24, complete each factorization.

19. $3a^2 + 13a + 4 = \left(3a + \boxed{1}\right)(a + 4)$
20. $2b^2 + 7b + 6 = (2b + 3)\left(b + \boxed{2}\right)$
21. $4z^2 - 13z + 3 = \left(z - \boxed{3}\right)\left(\boxed{4z} - 1\right)$
22. $4t^2 - 4t + 1 = \left(\boxed{2t} - 1\right)\left(\boxed{2t} - 1\right)$
23. $2m^2 + 5m - 12 = \left(2m \boxed{-} 3\right)\left(m \boxed{+} 4\right)$
24. $10u^2 - 13u - 3 = \left(2u \boxed{-} 3\right)\left(5u \boxed{+} 1\right)$

In Exercises 25–26, complete each step of the factorization of the trinomial by grouping.

25. $12t^2 + 17t + 6 = 12t^2 + \boxed{9}\,t + \boxed{8}\,t + 6$
$$= \boxed{3t}\,(4t + 3) + \boxed{2}\,(4t + 3)$$
$$= \left(\boxed{4t + 3}\right)(3t + 2)$$

26. $35t^2 - 11t - 6 = 35t^2 + \boxed{10}\,t - 21t - 6$
$$= 5t(7t + 2) \boxed{-} 3\left(7t \boxed{+} 2\right)$$
$$= \left(\boxed{7t + 2}\right)(5t - 3)$$

In Exercises 27–58, factor each trinomial, if possible.

27. $2x^2 - 3x + 1$ $(2x - 1)(x - 1)$
28. $2y^2 - 7y + 3$ $(2y - 1)(y - 3)$
29. $3a^2 + 13a + 4$ $(3a + 1)(a + 4)$
30. $2b^2 + 7b + 6$ $(2b + 3)(b + 2)$
31. $4z^2 + 13z + 3$ $(z + 3)(4z + 1)$
32. $4t^2 - 4t + 1$ $(2t - 1)(2t - 1)$
33. $6y^2 + 7y + 2$ $(3y + 2)(2y + 1)$
34. $4x^2 + 8x + 3$ $(2x + 3)(2x + 1)$
35. $6x^2 - 7x + 2$ $(3x - 2)(2x - 1)$
36. $4z^2 - 9z + 2$ $(4z - 1)(z - 2)$
37. $3a^2 - 4a - 4$ $(3a + 2)(a - 2)$
38. $8u^2 - 2u - 15$ $(2u - 3)(4u + 5)$
39. $2x^2 - 3x - 2$ $(2x + 1)(x - 2)$
40. $12y^2 - y - 1$ $(4y + 1)(3y - 1)$
41. $2m^2 + 5m - 10$ prime
42. $10u^2 - 13u - 6$ prime
43. $10y^2 - 3y - 1$ $(5y + 1)(2y - 1)$
44. $6m^2 + 19m + 3$ $(6m + 1)(m + 3)$
45. $12y^2 - 5y - 2$ $(3y - 2)(4y + 1)$
46. $10x^2 + 21x - 10$ $(2x + 5)(5x - 2)$
47. $-5t^2 - 13t - 6$ $-(5t + 3)(t + 2)$
48. $-16y^2 - 10y - 1$ $-(8y + 1)(2y + 1)$
49. $-16m^2 + 14m - 3$ $-(8m - 3)(2m - 1)$
50. $-16x^2 - 16x - 3$ $-(4x + 1)(4x + 3)$
51. $4a^2 - 4ab + b^2$ $(2a - b)(2a - b)$
52. $2b^2 - 5bc + 2c^2$ $(2b - c)(b - 2c)$
53. $6r^2 + rs - 2s^2$ $(3r + 2s)(2r - s)$
54. $3m^2 + 5mn + 2n^2$ $(3m + 2n)(m + n)$
55. $4x^2 + 8xy + 3y^2$ $(2x + 3y)(2x + y)$
56. $4b^2 + 15bc - 4c^2$ $(4b - c)(b + 4c)$
57. $4a^2 - 15ab + 9b^2$ $(4a - 3b)(a - 3b)$
58. $12x^2 + 5xy - 3y^2$ $(4x + 3y)(3x - y)$

In Exercises 59–70, write the terms of each trinomial in descending powers of one variable. Then factor the trinomial, if possible.

59. $-13x + 3x^2 - 10$ $(3x + 2)(x - 5)$
60. $-14 + 3a^2 - a$ $(3a - 7)(a + 2)$
61. $15 + 8a^2 - 26a$ $(2a - 5)(4a - 3)$
62. $16 - 40a + 25a^2$ $(5a - 4)(5a - 4)$
63. $12y^2 + 12 - 25y$ $(4y - 3)(3y - 4)$
64. $12t^2 - 1 - 4t$ $(6t + 1)(2t - 1)$
65. $3x^2 + 6 + x$ prime
66. $25 + 2u^2 + 3u$ prime
67. $2a^2 + 3b^2 + 5ab$ $(2a + 3b)(a + b)$
68. $11uv + 3u^2 + 6v^2$ $(3u + 2v)(u + 3v)$
69. $pq + 6p^2 - q^2$ $(3p - q)(2p + q)$
70. $-11mn + 12m^2 + 2n^2$ $(3m - 2n)(4m - n)$

In Exercises 71–82, factor each polynomial. Factor out any common monomial first.

71. $4x^2 + 10x - 6$ $2(2x - 1)(x + 3)$
72. $9x^2 + 21x - 18$ $3(3x - 2)(x + 3)$
73. $-y^3 - 13y^2 - 12y$ $-y(y + 12)(y + 1)$
74. $-2xy^2 - 8xy + 24x$ $-2x(y + 6)(y - 2)$
75. $6x^3 - 15x^2 - 9x$ $3x(2x + 1)(x - 3)$
76. $9y^3 + 3y^2 - 6y$ $3y(3y - 2)(y + 1)$
77. $30r^5 + 63r^4 - 30r^3$ $3r^3(5r - 2)(2r + 5)$

78. $6s^5 - 26s^4 - 20s^3$ $2s^3(3s + 2)(s - 5)$

79. $-16m^3n - 20m^2n^2 - 6mn^3$ $-2mn(4m + 3n)(2m + n)$

80. $-84x^4 - 100x^3y - 24x^2y^2$ $-4x^2(3x + y)(7x + 6y)$

81. $-28u^3v^3 + 26u^2v^4 - 6uv^5$ $-2uv^3(7u - 3v)(2u - v)$

82. $-16x^4y^3 + 30x^3y^4 + 4x^2y^5$ $-2x^2y^3(8x + y)(x - 2y)$

In Exercises 83–86, factor each trinomial.

83. $10x^4 + 3x^2 - 4$ $(2x^2 - 1)(5x^2 + 4)$

84. $7x^4 + 31x^2 + 12$ $(7x^2 + 3)(x^2 + 4)$

85. $5x^4 + 33x^2 + 18$ $(5x^2 + 3)(x^2 + 6)$

86. $20x^4 + 31x^2 - 9$ $(5x^2 + 9)(4x^2 - 1)$

APPLICATIONS

87. GRAPHING In Illustration 1, the graph on the left is that of the polynomial function $f(x) = 8x^2 - 6x - 5$. The graph on the right is that of the function $g(x) = (2x + 1)(4x - 5)$.

a. Do the graphs appear to be the same?   yes

b. Factor the right-hand side of $f(x) = 8x^2 - 6x - 5$. What can now be said with certainty about the graphs?

They are the same graphs, because the functions are the same.

ILLUSTRATION 1

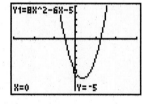

88. OFFICE FURNITURE The area of the desktop shown in Illustration 2 is given by the expression $(4x^2 + 20x - 11)$ in.^{2}. Factor this expression to find the expressions that represent its length and width. Then determine the difference in the length and width of the desktop. $(2x + 11), (2x - 1); 12$ in.

ILLUSTRATION 2

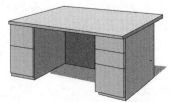

89. STORAGE The volume of the 8-foot-wide portable storage container shown in Illustration 3 is given by the expression $(72x^2 + 120x - 400)$ ft^3. If its dimen-

sions can be determined by factoring the expression, find the height and the length of the container. $(3x - 5)$ ft, $(3x + 10)$ ft

ILLUSTRATION 3

SUPER STORAGE

90. GRAPHIC DESIGN A replica of the Declaration of Independence is to be mounted on cardboard, as shown in Illustration 4. If the border around the document has a uniform width of x inches, the total area of the display is given by $(99 + 40x + 4x^2)$ in.^{2}. Factor this expression to determine the expressions representing its length and width. Then determine the length and the width of the document. $(2x + 11)(2x + 9); 11$ in., 9 in.

ILLUSTRATION 4

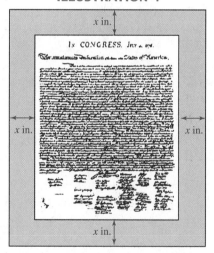

WRITING

91. In the work below, a student began to factor the trinomial and then gave up. Explain his initial mistake.

Factor $3x^2 - 5x - 2$.

$(3x - \quad)(x - 1)$
     ???

92. Two students factor $2x^2 + 20x + 42$ and get two different answers:

$(2x + 6)(x + 7)$ and $(x + 3)(2x + 14)$

Do both answers check? Why don't they agree? Is either answer completely correct? Explain.

93. Why is the process of factoring $6x^2 - 5x - 6$ more complicated than the process of factoring $x^2 - 5x - 6$?

94. How can the factorization shown below be checked?
$$6x^2 - 5x - 6 = (3x + 2)(2x - 3)$$

REVIEW

95. $\sqrt{2x} = 4$ 8

96. $\sqrt{7 + 2x} = -4$
no solution

97. $\sqrt{3t - 9} = \sqrt{t + 1}$ 5

98. $x - 3 = \sqrt{x^2 - 15}$ 4

▶ **6.4**

Special Factorizations and a Factoring Strategy

In this section, you will learn about

Factoring perfect square trinomials ■ Factoring the difference of two squares ■ Multistep factoring ■ Factoring the sum and difference of two cubes ■ A factoring strategy

Introduction In Chapter 4, we considered three special product formulas:

$$(x + y)^2 = x^2 + 2xy + y^2$$
$$(x - y)^2 = x^2 - 2xy + y^2$$
$$(x - y)(x + y) = x^2 - y^2$$

In this section, we will reverse the multiplication process and factor trinomials and binomials like those on the right-hand sides of these equations. We will also develop a technique for factoring two special types of binomials: the sum of two cubes and the difference of two cubes.

Factoring Perfect Square Trinomials

In Section 4.6, we saw that the squares of binomials are trinomials.

$(x + y)^2 \quad = \quad x^2 \quad + \quad 2xy \quad + \quad y^2$

| This is the square of the first term of the binomial. | This is twice the product of the two terms of the binomial. | This is the square of the last term of the binomial. |

$(x - y)^2 \quad = \quad x^2 \quad - \quad 2xy \quad + \quad y^2$

Trinomials that are squares of a binomial are called **perfect square trinomials**. Some examples of perfect square trinomials are

$y^2 + 6y + 9$ Because it is the square of $(y + 3)$: $(y + 3)^2 = y^2 + 6y + 9$.

$t^2 - 14t + 49$ Because it is the square of $(t - 7)$: $(t - 7)^2 = t^2 - 14t + 49$.

$4m^2 - 20m + 25$ Because it is the square of $(2m - 5)$: $(2m - 5)^2 = 4m^2 - 20m + 25$.

EXAMPLE 1

Recognizing perfect square trinomials. Determine whether the following trinomials are perfect square trinomials: **a.** $x^2 + 10x + 25$, **b.** $c^2 - 12c - 36$, and **c.** $25y^2 - 30y + 9$.

Solution **a.** To determine whether $x^2 + 10x + 25$ is a perfect square trinomial, we note that
- ■ the first term is the square of x,
- ■ the last term is the square of 5, and
- ■ the middle term is twice the product of x and 5.

Thus, $x^2 + 10x + 25$ is a perfect square trinomial.

b. To determine whether $c^2 - 12c - 36$ is a perfect square trinomial, we note that

- the first term is the square of c and
- the last term is negative.

Thus, $c^2 - 12c - 36$ is not a perfect square trinomial.

c. To determine whether $25y^2 - 30y + 9$ is a perfect square trinomial, we note that

- the first term is the square of $5y$,
- the last term is the square of 3, and
- the middle term is the negative of twice the product of $5y$ and 3.

Thus, $25y^2 - 30y + 9$ is a perfect square trinomial.

SELF CHECK   Tell which of the following are perfect square
trinomials: **a.** $y^2 + 4y + 4$,   *Answers:*   **a.** yes,   **b.** no,
b. $b^2 - 6b - 9$,   and   **c.** $4z^2 + 4z + 4$.   **c.** no   ■

Although we can factor perfect square trinomials using techniques discussed earlier in the chapter, we can also factor them by inspecting their terms and applying the special product formulas in reverse.

Factoring perfect square trinomials

$$x^2 + 2xy + y^2 = (x + y)^2$$
$$x^2 - 2xy + y^2 = (x - y)^2$$

For example, let's consider $x^2 + 16x + 64$. It is a perfect square trinomial, because:

- The first term x^2 is the square of x: $(x)^2 = x^2$.
- The last term 64 is the square of 8: $8^2 = 64$.
- The middle term $16x$ is twice the product of x and 8: $2(x)(8) = 16x$.

The factored form of the perfect square trinomial $x^2 + 16x + 64$ involves the terms x and 8.

$$x^2 + 16x + 64 = (x + 8)^2$$ The sign in the binomial is the sign of the middle term of the trinomial.

We can check this factorization by verifying that $(x + 8)^2 = x^2 + 16x + 64$.

EXAMPLE 2   **Factoring perfect square trinomials.** Factor $N^2 - 20N + 100$.

Solution $N^2 - 20N + 100$ is a perfect square trinomial, because:

- The first term N^2 is the square of N: $(N)^2 = N^2$.
- The last term 100 is the square of 10: $10^2 = 100$.
- The middle term is the negative of twice the product of N and 10: $-2(N)(10) = -20N$.

The factored form of the trinomial involves the terms N and 10.

$$N^2 - 20N + 100 = (N - 10)^2$$ The sign in the binomial is the sign of the middle term of the trinomial.

Check by multiplication,

SELF CHECK   Factor $x^2 - 18x + 81$.   *Answer:* $(x - 9)^2$ ■

| EXAMPLE 3 | **Perfect square trinomials in two variables.** Factor $9x^2 - 30xy + 25y^2$. |

Solution $9x^2 - 30xy + 25y^2$ is a perfect square trinomial, because:

- The first term $9x^2$ is the square of $3x$: $(3x)^2 = 9x^2$.
- The last term $25y^2$ is the square of $5y$: $(5y)^2 = 25y^2$.
- The middle term is the negative of twice the product of $3x$ and $5y$: $-2(3x)(5y) = -30xy$.

The factored form of the trinomial involves the terms $3x$ and $5y$.

$$9x^2 - 30xy + 25y^2 = (3x - 5y)^2$$ The sign in the binomial is the sign of the middle term of the trinomial.

Check by multiplication.

SELF CHECK Factor $16x^2 + 8xy + y^2$. *Answer:* $(4x + y)^2$ ∎

Factoring the Difference of Two Squares

Whenever we multiply a binomial of the form $x + y$ by a binomial of the form $x - y$, we obtain a binomial of the form $x^2 - y^2$.

$$(x + y)(x - y) = x^2 - xy + xy - y^2$$ Use the FOIL method.
$$= x^2 - y^2$$ Combine like terms: $-xy + xy = 0$.

The binomial $x^2 - y^2$ is called the **difference of two squares,** because x^2 is the square of x and y^2 is the square of y. The difference of the squares of two quantities always factors into the sum of those two quantities multiplied by the difference of those two quantities.

Factoring the difference of two squares

$$x^2 - y^2 = (x + y)(x - y)$$

If we think of the difference of two squares as the square of a **F**irst quantity minus the square of a **L**ast quantity, we have the formula

$$F^2 - L^2 = (F + L)(F - L)$$

and we say: *To factor the square of a First quantity minus the square of a Last quantity, we multiply the First plus the Last by the First minus the Last.*

To factor $x^2 - 9$, we note that it can be written in the form $x^2 - 3^2$ and use the formula for factoring the difference of two squares:

$$F^2 - L^2 = (F + L)(F - L)$$
$$x^2 - 3^2 = (x + 3)(x - 3)$$ Substitute x for F and 3 for L.

We can check by verifying that $(x + 3)(x - 3) = x^2 - 9$. Because of the commutative property of multiplication, we can also write this factorization as $(x - 3)(x + 3)$.

To factor the difference of two squares, it is helpful to know the integers that are perfect squares. The number 400, for example, is a perfect square, because $20^2 = 400$. The perfect integer squares less than 400 are

1, 4, 9, 16, 25, 36, 49, 64, 81, 100, 121, 144, 169, 196, 225, 256, 289, 324, 361

Expressions containing variables such as $25x^2$ are also perfect squares, because they can be written as the square of a quantity:

$$25x^2 = (5x)^2$$

EXAMPLE 4

Factoring the difference of two squares. Factor $25x^2 - 49$.

Solution   We can write $25x^2 - 49$ in the form $(5x)^2 - 7^2$ and use the formula for factoring the difference of two squares:

$$F^2 - L^2 = (F + L)(F - L)$$
$$(5x)^2 - 7^2 = (5x + 7)(5x - 7) \quad \text{Substitute } 5x \text{ for F and 7 for L.}$$

We can check by multiplying $5x + 7$ and $5x - 7$.

$$(5x + 7)(5x - 7) = 25x^2 - 35x + 35x - 49$$
$$= 25x^2 - 49$$

SELF CHECK   Factor $16a^2 - 81$.   *Answer:* $(4a + 9)(4a - 9)$ ∎

EXAMPLE 5

Factoring the difference of two squares. Factor $4y^4 - 121z^2$.

Solution   We can write $4y^4 - 121z^2$ in the form $(2y^2)^2 - (11z)^2$ and use the formula for factoring the difference of two squares:

$$F^2 - L^2 = (F + L)(F - L)$$
$$(2y^2)^2 - (11z)^2 = (2y^2 + 11z)(2y^2 - 11z)$$

Check by multiplication.

SELF CHECK   Factor $9m^2 - 64n^4$.   *Answer:* $(3m + 8n^2)(3m - 8n^2)$ ∎

Multistep Factoring

We can often factor out a greatest common factor before factoring the difference of two squares. To factor $8x^2 - 32$, for example, we factor out the GCF of 8 and then factor the resulting difference of two squares.

$$8x^2 - 32 = 8(x^2 - 4) \qquad \text{Factor out 8.}$$
$$= 8(x^2 - 2^2) \qquad \text{Write 4 as } 2^2.$$
$$= 8(x + 2)(x - 2) \quad \text{Factor the difference of two squares.}$$

We can check by multiplication:

$$8(x + 2)(x - 2) = 8(x^2 - 4) \qquad \text{Multiply the binomials first.}$$
$$= 8x^2 - 32 \qquad \text{Distribute the 8.}$$

EXAMPLE 6

Factoring out the GCF first. Factor $5a^2x^3y - 20b^2xy$.

Solution   We factor out the GCF of $5xy$ and then factor the resulting difference of two squares.

$$5a^2x^3y - 20b^2xy$$
$$= 5xy \cdot a^2x^2 - 5xy \cdot 4b^2 \qquad \text{The GCF is } 5xy.$$
$$= 5xy(a^2x^2 - 4b^2) \qquad \text{Factor out } 5xy.$$
$$= 5xy[(ax)^2 - (2b)^2] \qquad \text{Write } a^2x^2 \text{ as } (ax)^2 \text{ and } 4b^2 \text{ as } (2b)^2.$$
$$= 5xy(ax + 2b)(ax - 2b) \quad \text{Factor the difference of two squares.}$$

We check by multiplication.

SELF CHECK Factor $6p^2q^2s^2 - 54r^2s^2$. *Answer:* $6s^2(pq + 3r)(pq - 3r)$ ∎

Sometimes we must factor a difference of two squares more than once to completely factor a polynomial. For example, the binomial $625a^4 - 81b^4$ can be written in the form $(25a^2)^2 - (9b^2)^2$, which factors as

$$625a^4 - 81b^4 = (25a^2)^2 - (9b^2)^2$$
$$= (25a^2 + 9b^2)(25a^2 - 9b^2)$$

Since the factor $25a^2 - 9b^2$ can be written in the form $(5a)^2 - (3b)^2$, it is the difference of two squares and can be factored as $(5a + 3b)(5a - 3b)$. Thus,

$$625a^4 - 81b^4 = (25a^2 + 9b^2)(5a + 3b)(5a - 3b)$$

WARNING! The binomial $25a^2 + 9b^2$ is the **sum of two squares,** because it can be written in the form $(5a)^2 + (3b)^2$. If we are limited to integer coefficients and there are no common factors, binomials that are the sum of two squares cannot be factored.

EXAMPLE 7 **Multistep factoring.** Factor $2x^4y - 32y$.

Solution

$$2x^4y - 32y = 2y \cdot x^4 - 2y \cdot 16 \qquad \text{The GCF is } 2y.$$
$$= 2y(x^4 - 16) \qquad \text{Factor out the GCF of } 2y.$$
$$= 2y(x^2 + 4)(x^2 - 4) \qquad \text{Factor } x^4 - 16.$$
$$= 2y(x^2 + 4)(x + 2)(x - 2) \qquad \text{Factor } x^2 - 4. \text{ Note that } x^2 + 4 \text{ does not factor.}$$

SELF CHECK Factor $48a^5 - 3ab^4$. *Answer:* $3a(4a^2 + b^2)(2a + b)(2a - b)$ ∎

Factoring the Sum and Difference of Two Cubes

We have seen that the sum of two squares, such as $x^2 + 16$ or $25a^2 + 9b^2$, cannot be factored. However, the sum of two cubes and the difference of two cubes can be factored.

The sum of two cubes   **The difference of two cubes**

$x^3 + 8$ $a^3 - 64b^3$

↑   ↑   ↑   ↑

This term is   This term is   This term is   This term is $4b$
x cubed.   2 cubed: $2^3 = 8.$ a cubed.   cubed: $(4b)^3 = 64b^3.$

To find the formulas for factoring the sum of two cubes and the difference of two cubes, we need to find the following two products:

$$(x + y)(x^2 - xy + y^2) = (x + y)x^2 - (x + y)xy + (x + y)y^2 \quad \text{Use the distributive property.}$$
$$= x^3 + x^2y - x^2y - xy^2 + xy^2 + y^3 \quad \text{Combine like terms.}$$
$$= x^3 + y^3$$

$$(x - y)(x^2 + xy + y^2) = (x - y)x^2 + (x - y)xy + (x - y)y^2 \quad \text{Use the distributive property.}$$
$$= x^3 - x^2y + x^2y - xy^2 + xy^2 - y^3 \quad \text{Combine like terms.}$$
$$= x^3 - y^3$$

These results justify the formulas for factoring the **sum and difference of two cubes.**

Factoring the sum and difference of two cubes

$$x^3 + y^3 = (x + y)(x^2 - xy + y^2)$$
$$x^3 - y^3 = (x - y)(x^2 + xy + y^2)$$

If we think of the sum of two cubes as the cube of a **F**irst quantity plus the cube of a **L**ast quantity, we have the formula

$$F^3 + L^3 = (F + L)(F^2 - FL + L^2)$$

In words, we say, *To factor the cube of a **First** quantity plus the cube of a **Last** quantity, we multiply the **First** plus the **Last** by*

- *the **First** squared*
- *minus the **First** times the **Last***
- *plus the **Last** squared.*

The formula for the difference of two cubes is

$$F^3 - L^3 = (F - L)(F^2 + FL + L^2)$$

In words, we say, *To factor the cube of a **First** quantity minus the cube of a **Last** quantity, we multiply the **First** minus the **Last** by*

- *the **First** squared*
- *plus the **First** times the **Last***
- *plus the **Last** squared.*

To factor the sum or difference of two cubes, it's helpful to know the cubes of the numbers from 1 to 10:

1, 8, 27, 64, 125, 216, 343, 512, 729, 1,000

Expressions containing variables such as $64b^3$ are also perfect cubes, because they can be written as the cube of a quantity:

$$64b^3 = (4b)^3$$

EXAMPLE 8

Factoring the sum of two cubes. Factor $x^3 + 8$.

Solution   We think of $x^3 + 8$ as the cube of a **F**irst quantity, x, plus the cube of a **L**ast quantity, 2.

$$x^3 + 8 = x^3 + 2^3$$

Thus, $x^3 + 8$ factors as the product of the sum of x and 2 and the trinomial $x^2 - 2x + 2^2$.

$$F^3 + L^3 = (F + L)(F^2 - FL + L^2)$$
$$x^3 + 2^3 = (x + 2)(x^2 - x\,2 + 2^2) \quad \text{Substitute } x \text{ for F and 2 for L.}$$
$$= (x + 2)(x^2 - 2x + 4)$$

Check by multiplication.

$$(x + 2)(x^2 - 2x + 4) = (x + 2)x^2 - (x + 2)2x + (x + 2)4$$
$$= x^3 + 2x^2 - 2x^2 - 4x + 4x + 8$$
$$= x^3 + 8 \qquad \text{Combine like terms.}$$

SELF CHECK   Factor $h^3 + 27$.   *Answer:* $(h + 3)(h^2 - 3h + 9)$ ∎

EXAMPLE 9	**Factoring the difference of two cubes.** Factor $a^3 - 64b^3$.

Solution We think of $a^3 - 64b^3$ as the cube of a **F**irst quantity, a, minus the cube of a **L**ast quantity, $4b$.

$$a^3 - 64b^3 = a^3 - (4b)^3$$

Thus, its factors are the difference $a - 4b$ and the trinomial $a^2 + a(4b) + (4b)^2$.

$$\mathbf{F}^3 - \mathbf{L}^3 = (\mathbf{F} - \mathbf{L})(\mathbf{F}^2 + \mathbf{F}\,\mathbf{L} + \mathbf{L}^2)$$
$$a^3 - (4b)^3 = (a - 4b)[a^2 + a(4b) + (4b)^2]$$
$$= (a - 4b)(a^2 + 4ab + 16b^2)$$

Check by multiplication.

SELF CHECK Factor $8c^3 - 1$. *Answer:* $(2c - 1)(4c^2 + 2c + 1)$ ∎

Sometimes we must factor out a greatest common factor before factoring a sum or difference of two cubes.

EXAMPLE 10	**Factoring out the GCF first.** Factor $-2t^5 + 250t^2$.

Solution Each term contains the factor $-2t^2$.

$$-2t^5 + 250t^2 = -2t^2(t^3 - 125) \qquad \text{Factor out } -2t^2.$$
$$= -2t^2(t - 5)(t^2 + 5t + 25) \quad \text{Factor } t^3 - 125.$$

Verify this factorization by multiplication.

SELF CHECK Factor $4c^3 + 4d^3$. *Answer:* $4(c + d)(c^2 - cd + d^2)$ ∎

EXAMPLE 11	**Multistep factoring.** Factor $x^6 - 64$.

Solution The binomial $x^6 - 64$ is both the difference of two squares and the difference of two cubes. Since it's easier to factor the difference of two squares first, the expression factors into the product of a sum and a difference.

$$x^6 - 64 = (x^3)^2 - 8^2$$
$$= (x^3 + 8)(x^3 - 8)$$

Because $x^3 + 8$ is the sum of two cubes and $x^3 - 8$ is the difference of two cubes, each of these binomials can be factored.

$$x^6 - 64 = (x^3 + 8)(x^3 - 8)$$
$$= (x + 2)(x^2 - 2x + 4)(x - 2)(x^2 + 2x + 4)$$

Verify this factorization by multiplication.

A Factoring Strategy

Factoring plays an important role in the remainder of this chapter, as well as in the next chapter. When we solve equations and simplify expressions containing polynomials, we won't be told what type of factoring technique to apply—we will have to determine that ourselves. The following strategy is helpful when factoring a random polynomial. Remember that a polynomial is *factored completely* when it is expressed as the product of prime polynomials.

Steps for factoring a polynomial

1. Factor out all common factors.
2. If a polynomial has two terms, check for the following problem types:
   a. **The difference of two squares:** $x^2 - y^2 = (x + y)(x - y)$
   b. **The sum of two cubes:** $x^3 + y^3 = (x + y)(x^2 - xy + y^2)$
   c. **The difference of two cubes:** $x^3 - y^3 = (x - y)(x^2 + xy + y^2)$
3. If a polynomial has three terms, check for the following problem types:
   a. **A perfect square trinomial:**

   $$x^2 + 2xy + y^2 = (x + y)^2$$
   $$x^2 - 2xy + y^2 = (x - y)^2$$

   b. If the trinomial is not a perfect square, attempt to factor it as a general trinomial using the **trial-and-check method** or **factoring by grouping**.
4. If a polynomial has four or more terms, try **factoring by grouping**.
5. Continue until each individual factor is prime.
6. Check the results by multiplying.

STUDY SET

Section 6.4

VOCABULARY

In Exercises 1–4, fill in the blanks to make the statements true.

1. The binomial $x^2 - 25$ is called a ___difference___ of two squares.

2. $x^2 + 6x + 9$ is a ___perfect___ square trinomial because it is the square of the binomial $(x + 3)$.

3. The binomial $x^3 + 27$ is called a sum of two ___cubes___. The binomial $x^3 - 8$ is called a ___difference___ of two cubes.

4. To ___factor___ $4x^2 - 12x + 9$ means to write it as the product of two binomials.

CONCEPTS

In Exercises 5–10, fill in the blanks to make the statements true.

5. Consider $25x^2 + 30x + 9$.
   a. The first term is the square of ___5x___.
   b. The last term is the square of ___3___.
   c. The middle term is twice the product of ___5x___ and ___3___.

6. Consider $49x^2 - 28xy + 4y^2$.
   a. The first term is the square of ___7x___.
   b. The last term is the square of ___2y___.
   c. The middle term is the negative (opposite) of twice the product of ___7x___ and ___2y___.

7. To factor the square of a First quantity minus the square of a Last quantity, we multiply the ___First___ plus the ___Last___ by the ___First___ minus the ___Last___.

8. If a trinomial is the ___square___ of one quantity, plus the square of a second quantity, plus ___twice___ the product of the quantities, it factors into the square of the ___sum___ of the quantities.

9. a. $36x^2 = (\ 6x\ )^2$ b. $100x^4 = (\ 10x^2\ )^2$
   c. $27m^3 = (\ 3m\ )^3$ d. $a^6 = (\ a^2\ )^3$

10. a. $4x^2 - 9 = (\ 2x\ )^2 - (\ 3\ )^2$
    b. $8x^3 - 27 = (\ 2x\ )^3 - (\ 3\ )^3$
    c. $x^3 - 64y^3 = (\ x\ )^3 - (\ 4y\ )^3$

11. List the first ten perfect integer squares.
    1, 4, 9, 16, 25, 36, 49, 64, 81, 100

12. List the first five perfect integer cubes.
    1, 8, 27, 64, 125

13. Explain why each trinomial is not a perfect square trinomial.
    a. $9h^2 - 6h + 7$ 7 is not a perfect square.
    b. $j^2 - 8j - 16$
       The sign of the last term must be positive.
    c. $25r^2 + 20r + 16$
       The middle term is not twice the product of 5r and 4.

✓ **14. a.** Three incorrect factorizations of $x^2 + 36$ are given below. Use the FOIL method to show why each is wrong.

$$(x + 6)(x - 6) \quad x^2 - 36$$
$$(x + 6)(x + 6) \quad x^2 + 12x + 36$$
$$(x - 6)(x - 6) \quad x^2 - 12x + 36$$

b. Can $x^2 + 36$ be factored using only integers? no

NOTATION

In Exercises 15–18, write each expression as a polynomial in simpler form.

✓ **15.** $(6x)^2 - (5y)^2$ $\quad 36x^2 - 25y^2$

16. $(4x)^2 - (9y)^2$ $\quad 16x^2 - 81y^2$

17. $(3a)^2 - 2(3a)(5b) + (5b)^2$ $\quad 9a^2 - 30ab + 25b^2$

18. $(2s)^2 + 2(2s)(9t) + (9t)^2$ $\quad 4s^2 + 36st + 81t^2$

In Exercises 19–20, use an exponent to write each expression in simpler form.

✓ **19.** $(x + 8)(x + 8)$ $\quad (x + 8)^2$ **20.** $(x - 8)(x - 8)$ $\quad (x - 8)^2$

PRACTICE

In Exercises 21–24, complete each factorization.

21. $a^2 - 6a + 9 = \left(a - \boxed{3}\right)^2$

22. $t^2 + 2t + 1 = \left(t \boxed{+} 1\right)^2$

23. $4x^2 + 4x + 1 = \left(2x \boxed{+} 1\right)^2$

24. $9y^2 - 12y + 4 = \left(3y - \boxed{2}\right)^2$

In Exercises 25–40, factor each polynomial.

25. $x^2 + 6x + 9$ $\quad (x + 3)^2$

26. $x^2 + 10x + 25$ $\quad (x + 5)^2$

27. $y^2 - 8y + 16$ $\quad (y - 4)^2$

28. $z^2 - 2z + 1$ $\quad (z - 1)^2$

29. $t^2 + 20t + 100$ $\quad (t + 10)^2$

30. $r^2 + 24r + 144$ $\quad (r + 12)^2$

31. $u^2 - 18u + 81$ $\quad (u - 9)^2$

32. $v^2 - 14v + 49$ $\quad (v - 7)^2$

33. $4x^2 + 12x + 9$ $\quad (2x + 3)^2$

34. $4x^2 - 4x + 1$ $\quad (2x - 1)^2$

35. $36x^2 + 12x + 1$ $\quad (6x + 1)^2$

36. $4x^2 - 20x + 25$ $\quad (2x - 5)^2$

37. $a^2 + 2ab + b^2$ $\quad (a + b)^2$

38. $a^2 - 2ab + b^2$ $\quad (a - b)^2$

39. $16x^2 - 8xy + y^2$ $\quad (4x - y)^2$

40. $25x^2 + 20xy + 4y^2$ $\quad (5x + 2y)^2$

In Exercises 41–44, complete each factorization.

✓ **41.** $y^2 - 49 = \left(y + \boxed{7}\right)\left(y - \boxed{7}\right)$

42. $p^4 - q^2 = (p^2 + q)\left(\boxed{p^2} - \boxed{q}\right)$

43. $t^2 - w^2 = \left(\boxed{t} + \boxed{w}\right)(t - w)$

44. $49u^2 - 64v^2 = \left(\boxed{7u} + 8v\right)\left(7u \boxed{-} 8v\right)$

In Exercises 45–58, factor each polynomial, if possible.

45. $x^2 - 16$ $\quad (x + 4)(x - 4)$

46. $x^2 - 25$ $\quad (x + 5)(x - 5)$

47. $4y^2 - 1$ $\quad (2y + 1)(2y - 1)$

48. $9z^2 - 1$ $\quad (3z + 1)(3z - 1)$

49. $9x^2 - y^2$ $\quad (3x + y)(3x - y)$

50. $4x^2 - z^2$ $\quad (2x + z)(2x - z)$

51. $16a^2 - 25b^2$ $\quad (4a + 5b)(4a - 5b)$

52. $36a^2 - 121b^2$ $\quad (6a + 11b)(6a - 11b)$

53. $a^2 + b^2$ $\quad$ prime

54. $121a^2 + 144b^2$ $\quad$ prime

55. $a^4 - 144b^2$ $\quad (a^2 + 12b)(a^2 - 12b)$

56. $81y^4 - 100z^2$ $\quad (9y^2 + 10z)(9y^2 - 10z)$

57. $t^2z^2 - 64$ $\quad (tz + 8)(tz - 8)$

58. $900 - B^2C^2$ $\quad (30 + BC)(30 - BC)$

In Exercises 59–70, factor each polynomial.

59. $8x^2 - 32y^2$ $\quad 8(x + 2y)(x - 2y)$

60. $2a^2 - 200b^2$ $\quad 2(a + 10b)(a - 10b)$

61. $7a^2 - 7$ $\quad 7(a + 1)(a - 1)$

62. $20x^2 - 5$ $\quad 5(2x + 1)(2x - 1)$

63. $6x^4 - 6x^2y^2$ $\quad 6x^2(x + y)(x - y)$

64. $4b^2y - 16c^2y$ $\quad 4y(b + 2c)(b - 2c)$

65. $x^4 - 81$ $\quad (x^2 + 9)(x + 3)(x - 3)$

66. $y^4 - 625$ $\quad (y^2 + 25)(y + 5)(y - 5)$

67. $a^4 - 16$ $\quad (a^2 + 4)(a + 2)(a - 2)$

68. $b^4 - 256$ $\quad (b^2 + 16)(b + 4)(b - 4)$

69. $81r^4 - 256s^4$ $\quad (9r^2 + 16s^2)(3r + 4s)(3r - 4s)$

70. $16y^8 - 81z^4$ $\quad (4y^4 + 9z^2)(2y^2 + 3z)(2y^2 - 3z)$

In Exercises 71–74, complete each factorization.

71. $a^3 + 8 = (a + 2)\left(a^2 - \boxed{2a} + 4\right)$

72. $x^3 - 1 = (x - 1)\left(x^2 + \boxed{x} + 1\right)$

73. $b^3 + 27 = \left(\boxed{b + 3}\right)(b^2 - 3b + 9)$

74. $z^3 - 125 = \left(\boxed{z - 5}\right)(z^2 + 5z + 25)$

In Exercises 75–100, factor each polynomial.

75. $y^3 + 1$ $\quad (y + 1)(y^2 - y + 1)$

76. $x^3 - 8$ $\quad (x - 2)(x^2 + 2x + 4)$

77. $a^3 - 27$ $\quad (a - 3)(a^2 + 3a + 9)$

78. $b^3 + 125$ $\quad (b + 5)(b^2 - 5b + 25)$

79. $8 + x^3$ $\quad (2 + x)(4 - 2x + x^2)$

80. $27 - y^3$ $\quad (3 - y)(9 + 3y + y^2)$

81. $s^3 - t^3$ $\quad (s - t)(s^2 + st + t^2)$

82. $8u^3 + w^3$ $(2u + w)(4u^2 - 2uw + w^2)$

83. $a^3 + 8b^3$ $(a + 2b)(a^2 - 2ab + 4b^2)$

84. $27a^3 - b^3$ $(3a - b)(9a^2 + 3ab + b^2)$

85. $64x^3 - 27$ $(4x - 3)(16x^2 + 12x + 9)$

86. $27x^3 + 125$ $(3x + 5)(9x^2 - 15x + 25)$

87. $a^6 - b^3$ $(a^2 - b)(a^4 + a^2b + b^2)$

88. $a^3 + b^6$ $(a + b^2)(a^2 - ab^2 + b^4)$

89. $x^9 + y^6$ $(x^3 + y^2)(x^6 - x^3y^2 + y^4)$

90. $x^3 - y^9$ $(x - y^3)(x^2 + xy^3 + y^6)$

91. $2x^3 + 54$ $2(x + 3)(x^2 - 3x + 9)$

92. $2x^3 - 2$ $2(x - 1)(x^2 + x + 1)$

93. $-x^3 + 216$ $-(x - 6)(x^2 + 6x + 36)$

94. $-x^3 - 125$ $-(x + 5)(x^2 - 5x + 25)$

95. $64m^3x - 8n^3x$ $8x(2m - n)(4m^2 + 2mn + n^2)$

96. $16r^4 + 128rs^3$ $16r(r + 2s)(r^2 - 2rs + 4s^2)$

97. $x^4y + 216xy^4$ $xy(x + 6y)(x^2 - 6xy + 36y^2)$

98. $16a^5 - 54a^2b^3$ $2a^2(2a - 3b)(4a^2 + 6ab + 9b^2)$

99. $81r^4s^2 - 24rs^5$ $3rs^2(3r - 2s)(9r^2 + 6rs + 4s^2)$

100. $4m^5n + 500m^2n^4$ $4m^2n(m + 5n)(m^2 - 5mn + 25n^2)$

In Exercises 101–104, factor each expression completely. Factor a difference of two squares first.

101. $x^6 - 1$ $(x + 1)(x^2 - x + 1)(x - 1)(x^2 + x + 1)$

102. $x^6 - y^6$ $(x - y)(x^2 + xy + y^2)(x + y)(x^2 - xy + y^2)$

103. $x^{12} - y^6$
$(x^2 + y)(x^4 - x^2y + y^2)(x^2 - y)(x^4 + x^2y + y^2)$

104. $a^{12} - 64$ $(a^2 + 2)(a^4 - 2a^2 + 4)(a^2 - 2)(a^4 + 2a^2 + 4)$

APPLICATIONS

105. GENETICS The Hardy–Weinberg equation, one of the fundamental concepts in population genetics, is

$$p^2 + 2pq + q^2 = 1$$

where p represents the frequency of a certain dominant gene and q represents the frequency of a certain recessive gene. Factor the left-hand side of the equation. $(p + q)^2$

106. SPACE TRAVEL The first Soviet manned spacecraft, Vostok, is shown in Illustration 1. The surface area of the spherical part of the craft is given by $(36\pi r^2 - 48\pi r + 16\pi)$ m². Factor the expression.
$4\pi(3r - 2)^2$ m²

ILLUSTRATION 1

107. PHYSICS Illustration 2 shows a time-sequence picture of a falling apple. Factor the expression, which gives the difference in the distance fallen by the apple during the time interval from t_1 to t_2 seconds.
$0.5g(t_1 + t_2)(t_1 - t_2)$

ILLUSTRATION 2

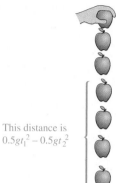

This distance is $0.5gt_1^2 - 0.5gt_2^2$

108. DARTS A circular dart board has a series of rings around a solid center, called the bullseye. (See Illustration 3.) To find the area of the outer white ring, we can use the formula

$$A = \pi R^2 - \pi r^2$$

Factor the expression on the right-hand side of the equation. $\pi(R + r)(R - r)$

ILLUSTRATION 3

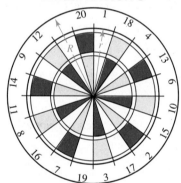

WRITING

109. When asked to factor $x^2 - 25$, one student wrote $(x + 5)(x - 5)$, and another student wrote $(x - 5)(x + 5)$. Are both answers correct? Explain.

110. Write a comment to the student whose work is shown below, explaining the initial error that was made.

Factor $4x^2 - 16y^2$.

$(2x + 4y)(2x - 4y)$

111. Explain why $x^6 - 1$ can be thought of as a difference of two squares or as a difference of two cubes.

112. Why is $a^2 + 2a + 1$ a perfect square trinomial, and why isn't $a^2 + 4a + 1$ a perfect square trinomial?

REVIEW

In Exercises 113–116, do each division.

113. $\dfrac{5x^2 + 10y^2 - 15xy}{5xy}$ $\quad \dfrac{x}{y} + \dfrac{2y}{x} - 3$

114. $\dfrac{-30c^2d^2 - 15c^2d - 10cd^2}{-10cd}$ $\quad 3cd + \dfrac{3c}{2} + d$

115. $2a - 1\,)\overline{a - 2 + 6a^2}$ $\quad 3a + 2$

116. $4b + 3\,)\overline{4b^3 - 5b^2 - 2b + 3}$ $\quad b^2 - 2b + 1$

ADDITIONAL FACTORING PROBLEMS

Exercises 117–150, apply the factoring strategy to factor each polynomial completely. If a polynomial is not factorable, write "prime."

117. $x^2y^2 - 2x^2 - y^2 + 2$ $\quad (y^2 - 2)(x + 1)(x - 1)$

118. $a^2c + a^2d^2 + bc + bd^2$ $\quad (c + d^2)(a^2 + b)$

119. $70p^4q^3 - 35p^4q^2 + 49p^5q^2$ $\quad 7p^4q^2(10q - 5 + 7p)$

120. $a^2b^2 - 144$ $\quad (ab + 12)(ab - 12)$

121. $2ab^2 + 8ab - 24a$ $\quad 2a(b + 6)(b - 2)$

122. $t^4 - 16$ $\quad (t^2 + 4)(t + 2)(t - 2)$

123. $-8p^3q^7 - 4p^2q^3$ $\quad -4p^2q^3(2pq^4 + 1)$

124. $8m^2n^3 - 24mn^4$ $\quad 8mn^3(m - 3n)$

125. $20m^2 + 100m + 125$ $\quad 5(2m + 5)^2$

126. $3rs + 6r^2 - 18s^2$ $\quad 3(2r - 3s)(r + 2s)$

127. $x^2 + 7x + 1$ $\quad$ prime

128. $3a^3 + 24b^3$ $\quad 3(a + 2b)(a^2 - 2ab + 4b^2)$

129. $-2x^5 + 128x^2$ $\quad -2x^2(x - 4)(x^2 + 4x + 16)$

130. $16 - 40z + 25z^2$ $\quad (5z - 4)^2$

131. $14t^3 - 40t^2 + 6t^4$ $\quad 2t^2(3t - 5)(t + 4)$

132. $-9x^2y^2 + 6xy - 1$ $\quad -(3xy - 1)^2$

133. $a^2(x - a) - b^2(x - a)$ $\quad (x - a)(a + b)(a - b)$

134. $5x^3y^3z^4 + 25x^2y^3z^2 - 35x^3y^2z^5$ $\quad 5x^2y^2z^2(xyz^2 + 5y - 7xz^3)$

135. $8p^6 - 27q^6$ $\quad (2p^2 - 3q^2)(4p^4 + 6p^2q^2 + 9q^4)$

136. $2c^2 - 5cd - 3d^2$ $\quad (2c + d)(c - 3d)$

137. $125p^3 - 64y^3$ $\quad (5p - 4y)(25p^2 + 20py + 16y^2)$

138. $8a^2x^3y - 2b^2xy$ $\quad 2xy(2ax + b)(2ax - b)$

139. $-16x^4y^2z + 24x^5y^3z^4 - 15x^2y^3z^7$ $\quad -x^2y^2z(16x^2 - 24x^3yz^3 + 15yz^6)$

140. $2ac + 4ad + bc + 2bd$ $\quad (c + 2d)(2a + b)$

141. $81p^4 - 16q^4$ $\quad (9p^2 + 4q^2)(3p + 2q)(3p - 2q)$

142. $6x^2 - x - 16$ $\quad$ prime

143. $4x^2 + 9y^2$ $\quad$ prime

144. $30a^4 + 5a^3 - 200a^2$ $\quad 5a^2(3a + 8)(2a - 5)$

145. $54x^3 + 250y^6$ $\quad 2(3x + 5y^2)(9x^2 - 15xy^2 + 25y^4)$

146. $6a^3 + 35a^2 - 6a$ $\quad a(6a - 1)(a + 6)$

147. $10r^2 - 13r - 4$ $\quad$ prime

148. $21t^3 - 10t^2 + t$ $\quad t(7t - 1)(3t - 1)$

149. $49p^2 + 28pq + 4q^2$ $\quad (7p + 2q)^2$

150. $16x^2 - 40x^3 + 25x^4$ $\quad x^2(4 - 5x)^2$

▶ **6.5**

Quadratic Equations

In this section, you will learn about

Quadratic equations ■ Solving quadratic equations by factoring ■ Applications

Introduction Equations that involve first-degree polynomials, such as $9x - 6 = 0$, are called *linear equations*. Equations that involve second-degree polynomials, such as $9x^2 - 6x = 0$, are called **quadratic equations.** In this section, we will define quadratic equations and learn how to solve many of them by factoring.

Quadratic Equations

If a polynomial contains one variable with an exponent to the second (but no higher) power, it is called a **second-degree polynomial.** Equations in which a second-degree polynomial is equal to zero are called **quadratic equations.** Some examples are

$$9x^2 - 6x = 0, \qquad x^2 - 2x - 63 = 0, \qquad \text{and} \qquad 2x^2 + 3x - 2 = 0$$

Quadratic equations

A **quadratic equation** is an equation that can be written in the form

$$ax^2 + bx + c = 0 \quad (a \neq 0) \quad \text{This form is called \textbf{quadratic form.}}$$

where a, b, and c are real numbers.

Quadratic equations don't always appear in quadratic form. To write a quadratic equation such as $21x = 10 - 10x^2$ in $ax^2 + bx + c = 0$ form, we use the addition and subtraction properties of equality to get 0 on the right-hand side.

$21x = 10 - 10x^2$ This is a quadratic equation: It contains one variable with an exponent of 2, but none larger.

$10x^2 + 21x = 10 - 10x^2 + 10x^2$ Add $10x^2$ to both sides.

$10x^2 + 21x = 10$ On the right-hand side, combine like terms: $-10x^2 + 10x^2 = 0$.

$10x^2 + 21x - 10 = 0$ Subtract 10 from both sides.

When it is written in quadratic form, we see that the equation $21x = 10 - 10x^2$ is $10x^2 + 21x - 10 = 0$, where $a = 10$, $b = 21$, and $c = -10$.

The techniques we have used to solve linear equations cannot be used to solve a quadratic equation, because those techniques cannot isolate x on one side of the equation. However, we can often solve quadratic equations using factoring and the following property of real numbers.

The zero-factor property of real numbers

Suppose a and b represent two real numbers. Then

If $ab = 0$, then $a = 0$ or $b = 0$.

In words, the zero-factor property states that when the product of two numbers is zero, at least one of them must be zero.

EXAMPLE 1   **Applying the zero-factor property.** Solve $(4y - 1)(y + 6) = 0$.

Solution   The left-hand side of the equation is the product of $4y - 1$ and $y + 6$. By the zero-factor property, one of these factors must be 0.

$$4y - 1 = 0 \quad \text{or} \quad y + 6 = 0$$

We can solve each of the linear equations.

$$4y - 1 = 0 \quad \text{or} \quad y + 6 = 0$$
$$4y = 1 \qquad\qquad y = -6$$
$$y = \frac{1}{4}$$

The equation has two solutions, $\frac{1}{4}$ and -6. To check, we substitute the results for y in the original equation and simplify.

For x = $\frac{1}{4}$	*For y = -6*
$(4y - 1)(y + 6) = 0$	$(4y - 1)(y + 6) = 0$
$\left[4\left(\frac{1}{4}\right) - 1\right]\left(\frac{1}{4} + 6\right) \overset{?}{=} 0$	$[4(-6) - 1](-6 + 6) \overset{?}{=} 0$
$(1 - 1)\left(6\frac{1}{4}\right) \overset{?}{=} 0$	$(-24 - 1)(0) \overset{?}{=} 0$
$0\left(6\frac{1}{4}\right) = 0$	$-25(0) \overset{?}{=} 0$
$0 = 0$	$0 = 0$

Both solutions check.

SELF CHECK   Solve $b(5b - 3) = 0$. *Answer:* $0, \frac{3}{5}$ ▪

Solving Quadratic Equations by Factoring

In Example 1, the left-hand side of the equation was in factored form, so we were able to apply the zero-factor property immediately. This is not always the case. To solve many quadratic equations, we must first do the factoring.

EXAMPLE 2   **Solving quadratic equations by factoring.** Solve $9x^2 - 6x = 0$.

Solution   We begin by factoring the left-hand side of the equation.

$$9x^2 - 6x = 0$$
$$3x(3x - 2) = 0 \quad \text{Factor out the GCF of } 3x.$$

By the zero-factor property, we have

$$3x = 0 \quad \text{or} \quad 3x - 2 = 0$$

We can solve each of the linear equations to get

$$x = 0 \quad \text{or} \quad x = \frac{2}{3}$$

To check, we substitute the results for x in the original equation and simplify.

For x = 0	*For x = $\frac{2}{3}$*
$9x^2 - 6x = 0$	$9x^2 - 6x = 0$
$9(0)^2 - 6(0) \overset{?}{=} 0$	$9\left(\frac{2}{3}\right)^2 - 6\left(\frac{2}{3}\right) \overset{?}{=} 0$
$0 - 0 \overset{?}{=} 0$	$9\left(\frac{4}{9}\right) - 6\left(\frac{2}{3}\right) \overset{?}{=} 0$
$0 = 0$	$4 - 4 \overset{?}{=} 0$
	$0 = 0$

Both solutions check.

SELF CHECK   Solve $5x^2 + 10x = 0$. *Answer:* $0, -2$ ▪

We can use the following steps to solve a quadratic equation by factoring.

Factoring method

1. Write the equation in $ax^2 + bx + c = 0$ form.
2. Factor the left-hand side of the equation.
3. Use the zero-factor property to set each factor equal to zero.
4. Solve each resulting linear equation.
5. Check the solutions in the original equation.

Sometimes we must factor a difference of two squares to solve a quadratic equation.

EXAMPLE 3

Solving quadratic equations by factoring. Solve $x^2 = 9$.

Solution   Before we can use the zero-factor property, we must subtract 9 from both sides to make the right-hand side zero.

$$x^2 = 9$$
$$x^2 - 9 = 0 \qquad \text{Subtract 9 from both sides.}$$
$$(x + 3)(x - 3) = 0 \qquad \text{Factor the difference of two squares.}$$
$$x + 3 = 0 \quad \text{or} \quad x - 3 = 0 \quad \text{Set each factor equal to zero.}$$
$$x = -3 \qquad\qquad x = 3 \quad \text{Solve each linear equation.}$$

Check each possible solution by substituting it into the original equation.

$$
\begin{array}{cc}
\textbf{\textit{For }} x = -3 & \textbf{\textit{For }} x = 3 \\
x^2 = 9 & x^2 = 9 \\
(-3)^2 \overset{?}{=} 9 & (3)^2 \overset{?}{=} 9 \\
9 = 9 & 9 = 9
\end{array}
$$

Both solutions check.

SELF CHECK   Solve $9x^2 - 36 = 0$. *Answer:* 2, −2   ■

EXAMPLE 4

Solving quadratic equations by factoring. Solve $x^2 - 2x - 63 = 0$.

Solution   In this case, we must factor a trinomial to solve the equation.

$$x^2 - 2x - 63 = 0$$
$$(x + 7)(x - 9) = 0 \qquad \text{Factor the trinomial } x^2 - 2x - 63.$$
$$x + 7 = 0 \quad \text{or} \quad x - 9 = 0 \quad \text{Set each factor equal to zero.}$$
$$x = -7 \qquad\qquad x = 9 \quad \text{Solve each linear equation.}$$

The solutions are −7 and 9. Check each one.

SELF CHECK   Solve $x^2 + 5x + 6 = 0$. *Answer:* −2, −3   ■

EXAMPLE 5

Writing an equation in quadratic form. Solve $2x^2 + 3x = 2$.

Solution   We write the equation in the form $ax^2 + bx + c = 0$ and then solve for x.

$$2x^2 + 3x = 2$$

$$2x^2 + 3x - 2 = 0 \qquad \text{Subtract 2 from both sides so that the right-hand side is zero.}$$

$$(2x - 1)(x + 2) = 0 \qquad \text{Factor } 2x^2 + 3x - 2.$$

$$2x - 1 = 0 \quad \text{or} \quad x + 2 = 0 \qquad \text{Set each factor equal to zero.}$$

$$2x = 1 \qquad\qquad x = -2 \qquad \text{Solve each linear equation.}$$

$$x = \frac{1}{2}$$

Check each solution.

SELF CHECK   Solve $3x^2 - 6 = -7x$.   *Answer:* $\frac{2}{3}, -3$ ■

EXAMPLE 6

A repeated solution. Solve $-2 = \frac{1}{2}x(9x - 12)$.

Solution   First, we need to write the equation in the form $ax^2 + bx + c = 0$.

$$-2 = \frac{1}{2}x(9x - 12)$$

$$-4 = x(9x - 12) \qquad \text{Multiply both sides of the equation by 2 to clear the equation of the fraction.}$$

$$-4 = 9x^2 - 12x \qquad \text{Remove parentheses.}$$

$$0 = 9x^2 - 12x + 4 \qquad \text{Add 4 to both sides to make the left-hand side zero.}$$

$$0 = (3x - 2)(3x - 2) \qquad \text{Factor the trinomial.}$$

$$3x - 2 = 0 \quad \text{or} \quad 3x - 2 = 0 \qquad \text{Set each factor equal to zero.}$$

$$3x = 2 \qquad\qquad 3x = 2 \qquad \text{Add 2 to both sides.}$$

$$x = \frac{2}{3} \qquad\qquad x = \frac{2}{3} \qquad \text{Divide both sides by 3.}$$

The equation has two solutions that are the same. We call $\frac{2}{3}$ a *repeated solution*. Check by substituting it into the original equation.

SELF CHECK   Solve $\frac{1}{3}x(4x + 12) = -3$.   *Answer:* $-\frac{3}{2}, -\frac{3}{2}$ ■

EXAMPLE 7

An equation with three solutions. Solve $6x^3 + 12x = 17x^2$.

Solution   This is not a quadratic equation, because it contains the term x^3. However, we can solve it using factoring and an extension of the zero-factor property.

$$6x^3 + 12x = 17x^2$$

$$6x^3 - 17x^2 + 12x = 0 \qquad \text{Add } -17x^2 \text{ to both sides to get 0 on the right-hand side.}$$

$$x(6x^2 - 17x + 12) = 0 \qquad \text{Factor out the GCF of } x.$$

$$x(2x - 3)(3x - 4) = 0 \qquad \text{Factor } 6x^2 - 17x + 12.$$

$$x = 0 \quad \text{or} \quad 2x - 3 = 0 \qquad \text{or} \qquad 3x - 4 = 0 \qquad \text{Set each factor equal to zero.}$$

$$2x = 3 \qquad\qquad\qquad 3x = 4 \qquad \text{Solve the linear equations.}$$

$$x = \frac{3}{2} \qquad\qquad\qquad x = \frac{4}{3}$$

This equation has three solutions. Check all three.

SELF CHECK   Solve $10x^3 + x^2 - 2x = 0$. 　　　　　*Answer:* $0, \frac{2}{5}, -\frac{1}{2}$ ■

Applications

The solutions of many problems involve the use of quadratic equations.

EXAMPLE 8

Softball. A softball pitcher can throw a "fastball" underhand at about 55 mph (80 feet per second). If she throws a ball up into the air with that velocity, as in Figure 6-2, its height h in feet, t seconds after being released, is given by the formula

$$h = 80t - 16t^2$$

After the ball is thrown, in how many seconds will it hit the ground?

Solution   When the ball hits the ground, its height will be zero. Thus, we set h equal to zero and solve for t.

$$h = 80t - 16t^2$$
$$0 = 80t - 16t^2$$

$$0 = 16t(5 - t) \qquad \text{Factor out the GCF of } 16t.$$

$$16t = 0 \quad \text{or} \quad 5 - t = 0 \qquad \text{Set each factor equal to zero.}$$

$$t = 0 \qquad\qquad t = 5 \qquad \text{Solve each linear equation.}$$

When $t = 0$, the ball's height above the ground is 0 feet. When $t = 5$, the height is again 0 feet, and the object has hit the ground. The solution is 5 seconds. ■

FIGURE 6-2

EXAMPLE 9

Perimeter of a rectangle. Assume that the rectangle in Figure 6-3 has an area of 52 square centimeters and that its length is 1 centimeter more than 3 times its width. Find the perimeter of the rectangle.

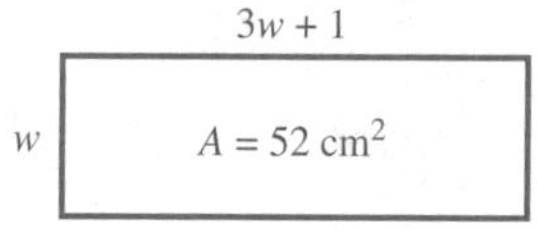

ANALYZE THE PROBLEM   The area of the rectangle is 52 square centimeters. Recall that the formula that gives the area of a rectangle is $A = lw$. To find the perimeter of the rectangle, we need to know its length and width. We are told that its length is related to its width; the length is 1 centimeter more than 3 times the width.

FORM AN EQUATION   Let w represent the width of the rectangle. Then $3w + 1$ represents its length. Because the area is 52 square centimeters, we substitute 52 for A and $3w + 1$ for l in the formula $A = lw$.

$$A = lw$$
$$52 = (3w + 1)w$$

FIGURE 6-3

SOLVE THE EQUATION   Now we solve the equation for w.

$$52 = (3w + 1)w \qquad \text{The equation to solve.}$$

$$52 = 3w^2 + w \qquad \text{Remove parentheses.}$$

$$0 = 3w^2 + w - 52 \qquad \text{Subtract 52 from both sides to make the left-hand side zero.}$$

$$0 = (3w + 13)(w - 4) \qquad \text{Factor the trinomial.}$$

$$3w + 13 = 0 \quad \text{or} \quad w - 4 = 0 \qquad \text{Set each factor equal to zero.}$$

$$3w = -13 \qquad\qquad w = 4 \qquad \text{Solve each linear equation.}$$

$$w = -\frac{13}{3}$$

STATE THE CONCLUSION Since the width cannot be negative, we discard the result $w = -\frac{13}{3}$. Thus, the width of the rectangle is 4, and the length is given by

$$3w + 1 = 3(4) + 1 \quad \text{Substitute 4 for } w.$$
$$= 12 + 1$$
$$= 13$$

The dimensions of the rectangle are 4 centimeters by 13 centimeters. We find the perimeter by substituting 13 for l and 4 for w in the formula for the perimeter of a rectangle.

$$P = 2l + 2w$$
$$= 2(13) + 2(4)$$
$$= 26 + 8$$
$$= 34$$

The perimeter of the rectangle is 34 centimeters.

CHECK THE RESULT A rectangle with dimensions of 13 centimeters by 4 centimeters does have an area of 52 square centimeters, and the length is 1 centimeter more than 3 times the width. A rectangle with these dimensions has a perimeter of 34 centimeters. ∎

EXAMPLE 10

Recording ozone levels. The positions of three pollution monitoring stations are shown in Figure 6-4. The east county station is 3 miles farther from the downtown station than is the west county station. The distance between the east and west county stations is 6 miles longer than the distance between the downtown and west stations. How far apart are the stations?

FIGURE 6-4

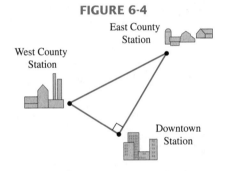

West County Station

East County Station

Downtown Station

ANALYZE THE PROBLEM To find the distances between the stations, we need to find the lengths of the sides of the right triangle formed by connecting their positions. Recall that the Pythagorean theorem gives the relationship between the sides of a right triangle: $a^2 + b^2 = c^2$.

FORM AN EQUATION We let a represent the distance in miles from the downtown station to the west station, because the other two distances can be expressed in terms of it. The distance between the downtown and east stations is $(a + 3)$ miles, and that between the east and west stations is $(a + 6)$ miles. We substitute these distances into the Pythagorean theorem, noting that the length of the hypotenuse c is $(a + 6)$ miles.

$$a^2 + b^2 = c^2 \qquad \text{The Pythagorean theorem.}$$
$$a^2 + (a + 3)^2 = (a + 6)^2 \qquad \text{Substitute } (a + 3) \text{ for } b \text{ and } (a + 6) \text{ for } c.$$
$$a^2 + a^2 + 6a + 9 = a^2 + 12a + 36 \qquad \text{Use the FOIL method to find } (a + 3)^2 \text{ and } (a + 6)^2.$$
$$2a^2 + 6a + 9 = a^2 + 12a + 36 \qquad \text{Combine like terms on the left-hand side.}$$
$$a^2 - 6a - 27 = 0 \qquad \text{Subtract } a^2, 12a, \text{ and } 36 \text{ from both sides to make the right-hand side zero.}$$

SOLVE THE EQUATION Now we solve the equation for a.

$$a^2 - 6a - 27 = 0$$
$$(a - 9)(a + 3) = 0 \qquad \text{Factor.}$$
$$a - 9 = 0 \quad \text{or} \quad a + 3 = 0 \qquad \text{Set each factor to zero.}$$
$$a = 9 \qquad\qquad a = -3 \qquad \text{Solve each linear equation.}$$

STATE THE CONCLUSION   Since a triangle cannot have a negative number for the length of a side, we discard the result $a = -3$. The distance from the downtown station to the west county station is 9 miles. The distance from the downtown station to the east county station is $9 + 3$, or 12 miles. The distance between the east and west stations is $9 + 6$, or 15 miles.

CHECK THE RESULT   The differences in the distances between stations meet the requirements stated in the problem. The distances also satisfy the Pythagorean theorem. So the solutions check.

$$9^2 + 12^2 \stackrel{?}{=} 15^2$$
$$81 + 144 \stackrel{?}{=} 225$$
$$225 = 225$$

STUDY SET
Section 6.5

VOCABULARY

In Exercises 1–2, fill in the blanks to make the statements true.

1. Any equation that can be written in the form $ax^2 + bx + c = 0$ is called a _____quadratic_____ equation.

2. To _____factor_____ a binomial or trinomial means to write it as a product.

CONCEPTS

In Exercises 3–6, fill in the blanks to make the statements true.

3. When the product of two numbers is zero, at least one of them is _____zero_____. Symbolically, we can state this: If $ab = 0$, then $a = \boxed{0}$ or $b = \boxed{0}$.

4. The techniques used to solve linear equations cannot be used to solve quadratic equations, because those techniques cannot _____isolate_____ the variable on one side of the equation.

5. To write a quadratic equation in *quadratic form* means that one side of the equation must be _____zero_____ and the other side must be in the form $ax^2 + bx + c$.

6. In the quadratic equation $ax^2 + bx + c = 0$, $a \neq \boxed{0}$.

7. Classify each equation as quadratic or linear.
   a. $3x^2 + 4x + 2 = 0$ **b.** $3x + 7 = 0$ linear
   quadratic
   c. $2 = -16 - 4x$ **d.** $-6x + 2 = x^2$ quadratic
   linear

8. Check to see whether the given number is a solution of the given quadratic equation.
   a. $x^2 - 4x = 0$; $x = 4$ yes
   b. $x^2 + 2x - 4 = 0$; $x = -2$ no
   c. $4x^2 - x + 3 = 0$; $x = 1$ no

9. **a.** Evaluate $x^2 + 6x - 16$ for $x = 0$. -16
   b. Factor $x^2 + 6x - 16$. $(x-2)(x+8)$
   c. Solve $x^2 + 6x - 16 = 0$. $2, -8$

10. The equation $3x^2 - 4x + 5 = 0$ is written in $ax^2 + bx + c = 0$ form. What are a, b, and c?
    $3, -4, 5$

11. What is the first step that should be performed to solve each equation?
    a. $x^2 + 7x = -6$ Add 6 to both sides.
    b. $\dfrac{1}{2}x(x + 7) = -3$ Multiply both sides by 2.

12. **a.** How many solutions does the linear equation $2a + 3 = 2$ have?   1
    b. How many solutions does the quadratic equation $2a^2 + 3a = 2$ have?   2

NOTATION

In Exercises 13–14, complete each solution.

13. $7y^2 + 14y = 0$
    $\boxed{7y}\,(y + 2) = 0$
    $7y = 0$ or $\boxed{y + 2} = 0$
    $y = 0$ $y = -2$

14. $12p^2 - p - 6 = 0$
    $\left(\boxed{4p} - 3\right)\left(3p + \boxed{2}\right) = 0$
    $\boxed{4p - 3} = 0$ or $3p + 2 = \boxed{0}$
    $4p = \boxed{3}$ $3p = \boxed{-2}$
    $p = \dfrac{3}{4}$ $p = -\dfrac{2}{3}$

PRACTICE

In Exercises 15–72, solve each equation.

15. $(x - 2)(x + 3) = 0$ $2, -3$

16. $(x - 3)(x - 2) = 0$ $3, 2$

17. $(2s - 5)(s + 6) = 0$ $\frac{5}{2}, -6$

18. $(3h - 4)(h + 1) = 0$ $\frac{4}{3}, -1$

19. $(x - 1)(x + 2)(x - 3) = 0$ $1, -2, 3$

20. $(x + 2)(x + 3)(x - 4) = 0$ $-2, -3, 4$

21. $x(x - 3) = 0$ $0, 3$ **22.** $x(x + 5) = 0$ $0, -5$

23. $x(2x - 5) = 0$ $0, \frac{5}{2}$ **24.** $x(5x + 7) = 0$ $0, -\frac{7}{5}$

25. $w^2 - 7w = 0$ $0, 7$ **26.** $p^2 + 5p = 0$ $0, -5$

27. $3x^2 + 8x = 0$ $0, -\frac{8}{3}$ **28.** $5x^2 - x = 0$ $0, \frac{1}{5}$

29. $8s^2 - 16s = 0$ $0, 2$ **30.** $15s^2 - 20s = 0$ $0, \frac{4}{3}$

31. $x^2 - 25 = 0$ $-5, 5$ **32.** $x^2 - 36 = 0$ $-6, 6$

33. $4x^2 - 1 = 0$ $-\frac{1}{2}, \frac{1}{2}$ **34.** $9y^2 - 1 = 0$ $-\frac{1}{3}, \frac{1}{3}$

35. $9y^2 - 4 = 0$ $-\frac{2}{3}, \frac{2}{3}$ **36.** $16z^2 - 25 = 0$ $-\frac{5}{4}, \frac{5}{4}$

37. $x^2 = 100$ $-10, 10$ **38.** $z^2 = 25$ $-5, 5$

39. $4x^2 = 81$ $-\frac{9}{2}, \frac{9}{2}$ **40.** $9y^2 = 64$ $-\frac{8}{3}, \frac{8}{3}$

41. $x^2 - 13x + 12 = 0$ $12, 1$

42. $x^2 + 7x + 6 = 0$ $-1, -6$

43. $x^2 - 4x - 21 = 0$ $-3, 7$

44. $x^2 + 2x - 15 = 0$ $3, -5$

45. $x^2 - 9x + 8 = 0$ $8, 1$

46. $x^2 - 14x + 45 = 0$ $9, 5$

47. $a^2 + 8a = -15$ $-3, -5$

48. $a^2 - a = 56$ $8, -7$

49. $2y - 8 = -y^2$ $-4, 2$

50. $-3y + 18 = y^2$ $3, -6$

51. $x^3 + 3x^2 + 2x = 0$ $0, -1, -2$

52. $x^3 - 7x^2 + 10x = 0$ $0, 5, 2$

53. $k^3 - 27k - 6k^2 = 0$ $0, 9, -3$

54. $j^3 - 22j - 9j^2 = 0$ $0, 11, -2$

55. $(x - 1)(x^2 + 5x + 6) = 0$ $1, -2, -3$

56. $(x - 2)(x^2 - 8x + 7) = 0$ $2, 7, 1$

57. $2x^2 - 5x + 2 = 0$ $\frac{1}{2}, 2$

58. $2x^2 + x - 3 = 0$ $-\frac{3}{2}, 1$

59. $5x^2 - 6x + 1 = 0$ $\frac{1}{5}, 1$

60. $6x^2 - 5x + 1 = 0$ $\frac{1}{3}, \frac{1}{2}$

61. $4r^2 + 4r = -1$ $-\frac{1}{2}, -\frac{1}{2}$

62. $9m^2 + 6m = -1$ $-\frac{1}{3}, -\frac{1}{3}$

63. $-15x^2 + 2 = -7x$ $\frac{2}{3}, -\frac{1}{5}$

64. $-8x^2 - 10x = -3$ $\frac{1}{4}, -\frac{3}{2}$

65. $\frac{1}{2}x(2x - 3) = 10$ $-\frac{5}{2}, 4$

66. $\frac{1}{2}x(2x - 3) = 7$ $\frac{7}{2}, -2$

67. $(d + 1)(8d + 1) = 18d$ $\frac{1}{8}, 1$

68. $4h(3h + 2) = h + 12$ $-\frac{4}{3}, \frac{3}{4}$

69. $2x(3x^2 + 10x) = -6x$ $0, -3, -\frac{1}{3}$

70. $2x^3 = 2x(x + 2)$ $0, -1, 2$

71. $x^3 + 7x^2 = x^2 - 9x$ $0, -3, -3$

72. $x^2(x + 10) = 2x(x - 8)$ $0, -4, -4$

APPLICATIONS

In Exercises 73–74, an object has been thrown straight up into the air. The formula $h = vt - 16t^2$ gives the height h of the object above the ground after t seconds, when it is thrown upward with an initial velocity v.

73. TIME OF FLIGHT After how many seconds will the object hit the ground if it is thrown with a velocity of 144 feet per second? 9 sec

74. TIME OF FLIGHT After how many seconds will the object hit the ground if it is thrown with a velocity of 160 feet per second? 10 sec

75. OFFICIATING Before a football game, a coin toss is used to determine which team will kick off. See Illustration 1. The height h (in feet) of a coin above the ground t seconds after being flipped up into the air is given by

$h = -16t^2 + 22t + 3$

ILLUSTRATION 1

How long does a team captain have to call heads or tails if it must be done while the coin is in the air? $\frac{3}{2} = 1.5$ sec

76. DOLPHINS See Illustration 2. The height h in feet reached by a dolphin t seconds after breaking the surface of the water is given by

$h = -16t^2 + 32t$

How long will it take the dolphin to jump out of the water and touch the trainer's hand? 1 sec

ILLUSTRATION 2

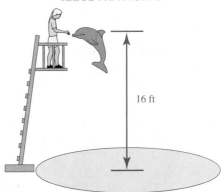

16 ft

77. EXHIBITION DIVING In Acapulco, Mexico, men diving from a cliff to the water 64 feet below are quite a tourist attraction. A diver's height, h, above the water t seconds after diving is given by $h = -16t^2 + 64$. How long does a dive last? 2 sec

78. FORENSIC MEDICINE The kinetic energy E of a moving object is given by $E = \frac{1}{2}mv^2$, where m is the mass of the object (in kilograms) and v is the object's velocity (in meters per second). Kinetic energy is measured in joules. Examining the damage done to a victim, a police pathologist determines that the energy of a 3-kilogram mass at impact was 54 joules. Find the velocity at impact. 6 m/s

79. CHOREOGRAPHY For the finale of a musical, 36 dancers are to assemble in a triangular-shaped series of rows, where each successive row has one more dancer than the previous row. Illustration 3 shows the beginning of such a formation. The relationship between the number of rows r and the number of dancers d is given by

$$d = \frac{1}{2}r(r + 1)$$

Determine the number of rows in the formation. 8

ILLUSTRATION 3

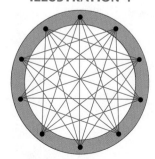

80. CRAFTS Illustration 4 shows how a geometric wall hanging can be created by stretching yarn from peg to peg across a wooden ring. The relationship between the number of pegs p placed evenly around the ring and the number of yarn segments s that crisscross the ring is given by the formula

$$s = \frac{p(p - 3)}{2}$$

How many pegs are needed if the designer wants 27 segments to criss-cross the ring? 9

ILLUSTRATION 4

81. INSULATION The area of the rectangular slab of foam insulation in Illustration 5 is 36 square meters. Find the dimensions of the slab. 4 m by 9 m

ILLUSTRATION 5

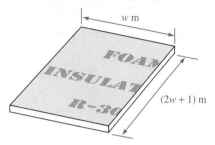
w m

$(2w + 1)$ m

82. SHIPPING PALLETS The length of a rectangular shipping pallet is 2 feet less than 3 times its width. Its area is 21 square feet. Find the dimensions of the pallet. 3 ft by 7 ft

83. BOATING The inclined ramp of the boat launch shown in Illustration 6 is 8 meters longer than the "rise" of the ramp. The "run" is 7 meters longer than the "rise." How long are the three sides of the ramp? 5 m, 12 m, 13 m

ILLUSTRATION 6

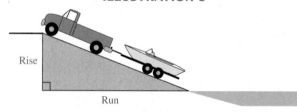
Rise

Run

84. CAR REPAIR To create some space to work under the front end of a car, a mechanic drives it up steel ramps. See Illustration 7. The ramp is 1 foot longer than the back, and the base is 2 feet longer than the back of the ramp. Find the length of each side of the ramp. 3 ft, 4 ft, 5 ft

ILLUSTRATION 7

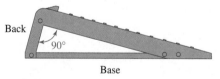
Back

90°

Base

85. GARDENING TOOLS The dimensions (in millimeters) of the teeth of a pruning saw blade are given in Illustration 8. Find each length. 3 mm, 4 mm, 5 mm

ILLUSTRATION 8

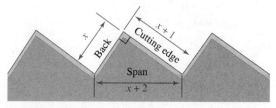
x

Back

Cutting edge

$x + 1$

Span

$x + 2$

86. HARDWARE An aluminum brace used to support a wooden shelf has a length that is 2 inches less than twice the width of the shelf. The brace is anchored to the wall 8 inches below the shelf, as shown in Illustration 9. Find the width of the shelf and the length of the brace.
$w = 6$ in., $l = 10$ in.

ILLUSTRATION 9

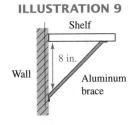

Shelf

Wall

8 in.

Aluminum brace

87. DESIGNING A TENT The length of the base of the triangular sheet of canvas above the door of the tent in Illustration 10 is 2 feet more than twice its height. The area is 30 square feet. Find the height and the length of the base of the triangle. $h = 5$ ft, $b = 12$ ft

ILLUSTRATION 10

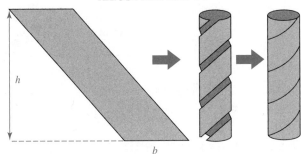

88. DIMENSIONS OF A TRIANGLE The height of a triangle is 2 inches less than 5 times the length of its base. The area is 36 square inches. Find the length of the base and the height of the triangle.
$b = 4$ in., $h = 18$ in.

89. TUBING A piece of cardboard in the shape of a parallelogram is twisted to form the tube for a roll of paper towels. (See Illustration 11.) The parallelogram has an area of 60 square inches. If its height h is 7 inches more than the length of the base b, what is the circumference of the tube? (*Hint:* The formula for the area of a parallellogram is $A = bh$.) 5 in.

ILLUSTRATION 11

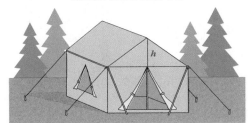

h

b

90. SWIMMING POOL BORDER The owners of the rectangular swimming pool in Illustration 12 want to surround the pool with a crushed-stone border of uniform width. They have enough stone to cover 74

square meters. How wide should they make the border? (*Hint:* The area of the larger rectangle minus the area of the smaller is the area of the border.) 1 m

ILLUSTRATION 12

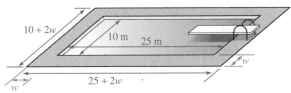

$10 + 2w$

10 m 25 m

$25 + 2w$

w

w

91. HOUSE CONSTRUCTION The formula for the area of a trapezoid is

$$A = \frac{h(B + b)}{2}$$

The area of the trapezoidal truss in Illustration 13 is 24 square meters. Find the height of the trapezoid if one base is 8 meters and the other base is the same as the height. 4 m

ILLUSTRATION 13

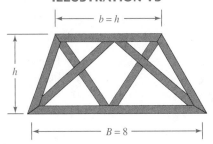

$b = h$

h

$B = 8$

92. VOLUME OF A PYRAMID The volume of a pyramid is given by the formula

$$V = \frac{Bh}{3}$$

where B is the area of its base and h is its height. The volume of the pyramid in Illustration 14 is 192 cubic centimeters. Find the dimensions of its rectangular base if one edge of the base is 2 centimeters longer than the other and the height of the pyramid is 12 centimeters. 6 cm by 8 cm

ILLUSTRATION 14

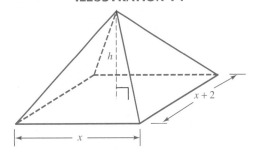

h

$x + 2$

x

WRITING

93. What is wrong with the logic used by a student to "solve" $x^2 + x = 6$?

$$x(x + 1) = 6$$
$$x = 6 \quad \text{or} \quad x + 1 = 6$$
$$x = 5$$
So the solutions are
6 or 5.

94. Suppose that to find the length of the base of a triangle, you write a quadratic equation and solve it to find $b = 6$ or $b = -8$. Explain why one solution should be discarded.

REVIEW

95. A doctor advises one patient to exercise at least 15 minutes but less than 30 minutes per day. Use a compound inequality to express the range of these times in minutes. $15 \text{ min} \leq E < 30 \text{ min}$

96. A bag of peanuts is worth $0.30 less than a bag of cashews. Equal amounts of peanuts and cashews are used to make 40 bags of a mixture that is worth $1.05 per bag. How much is a bag of cashews worth? $1.20

97. A rectangle is 3 times as long as it is wide, and its perimeter is 120 centimeters. Find its area. 675 cm^2

98. A woman invests $15,000, part at 7% annual interest and part at 8% annual interest. If she receives $1,100 interest per year, how much did she invest at 7%? $10,000

▶ **6.6**

Completing the Square

In this section, you will learn about

The square root method ■ Completing the square ■ Solving equations with leading coefficients of 1 ■ Solving equations with leading coefficients other than 1

Introduction The factoring method can't be used to solve all quadratic equations. For example, the trinomial in the equation $x^2 + 5x + 1 = 0$ cannot be factored using integer coefficients. To solve such equations, we need other methods. In this section, we will discuss the method of completing the square.

The Square Root Method

If $x^2 = 9$, x is a number whose square is 9. Since $3^2 = 9$ and $(-3)^2 = 9$, the equation $x^2 = 9$ has two solutions, $x = \sqrt{9} = 3$ and $x = -\sqrt{9} = -3$. In general, any equation of the form $x^2 = c$ $(c > 0)$ has two solutions.

The square root method

If $c > 0$, the equation $x^2 = c$ has two solutions:

$$x = \sqrt{c} \quad \text{or} \quad x = -\sqrt{c}$$

We can write this result with double-sign notation. The equation $x = \pm\sqrt{c}$ (read as "x equals plus or minus $\sqrt{c}$") means that $x = \sqrt{c}$ or $x = -\sqrt{c}$.

EXAMPLE 1

Solving equations using the square root method. Solve $x^2 = 16$.

Solution We use the square root method to find that the equation $x^2 = 16$ has two solutions:

$$x = \sqrt{16} \quad \text{or} \quad x = -\sqrt{16}$$
$$x = 4 \qquad\qquad x = -4$$

Using double-sign notation, we have $x = \pm 4$.

Check: *For x = 4*   *For x = −4*

$$x^2 = 16 \qquad\qquad x^2 = 16$$
$$4^2 \overset{?}{=} 16 \qquad\quad (-4)^2 \overset{?}{=} 16$$
$$16 = 16 \qquad\qquad 16 = 16$$

SELF CHECK   Solve $x^2 = 25$.   *Answer:* ± 5 ∎

The equation in Example 1 can also be solved by factoring.

$$x^2 = 16$$
$$x^2 - 16 = 0 \qquad\qquad \text{Subtract 16 from both sides.}$$
$$(x + 4)(x - 4) = 0 \qquad\quad \text{Factor the difference of two squares.}$$
$$x + 4 = 0 \quad \text{or} \quad x - 4 = 0$$
$$x = -4 \qquad\qquad x = 4$$

E X A M P L E   2   **Solving equations using the square root method.** Solve $3x^2 - 9 = 0$.

Solution   To solve the equation by the square root method, we first isolate x^2.

$$3x^2 - 9 = 0$$
$$3x^2 = 9 \quad \text{Add 9 to both sides.}$$
$$x^2 = 3 \quad \text{Divide both sides by 3.}$$

This equation has two solutions:

$$x = \sqrt{3} \qquad \text{or} \qquad x = -\sqrt{3}$$

The solutions can be written as $x = \pm\sqrt{3}$. Note that these are the exact solutions. Using a calculator, we can *approximate* them to the nearest hundredth: $x \approx \pm 1.73$.

Check: *For $x = \sqrt{3}$*   *For $x = -\sqrt{3}$*

$$3x^2 - 9 = 0 \qquad\qquad 3x^2 - 9 = 0$$
$$3\left(\sqrt{3}\right)^2 - 9 \overset{?}{=} 0 \qquad 3\left(-\sqrt{3}\right)^2 - 9 = 0$$
$$3(3) - 9 \overset{?}{=} 0 \qquad\qquad 3(3) - 9 \overset{?}{=} 0$$
$$9 - 9 \overset{?}{=} 0 \qquad\qquad 9 - 9 \overset{?}{=} 0$$
$$0 = 0 \qquad\qquad\qquad 0 = 0$$

SELF CHECK   Solve $2x^2 - 10 = 0$. Give the exact solutions and
approximations to the nearest hundredth.   *Answer:* $\pm\sqrt{5}$; ± 2.24 ∎

E X A M P L E   3   **Hurricanes.** In 1998, Hurricane Mitch dumped heavy rains on Central America, causing extensive flooding in Honduras, Belize, and Guatemala. Figure 6-5 shows the position of the storm on October 26, at which time the weather service estimated that it covered an area of about 71,000 square miles. What was the diameter of the storm?

Solution   We can use the formula for the area of a circle to find the *radius* of the circular-shaped storm.

$$A = \pi r^2$$
$$71,000 = \pi r^2 \qquad \text{Substitute 71,000 for the area } A.$$
$$\frac{71,000}{\pi} = r^2 \qquad\qquad \text{Divide both sides by } \pi \text{ to isolate } r^2.$$

FIGURE 6-5

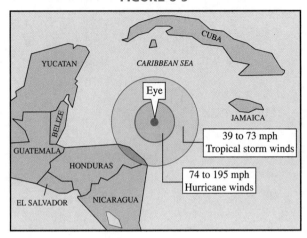

Now we use the square root method to solve for r.

$$r = \sqrt{\frac{71,000}{\pi}} \quad \text{or} \quad r = -\sqrt{\frac{71,000}{\pi}}$$

Using a calculator to approximate the square root, we have

$r \approx 150.3329702$ The second solution is discarded, because the radius cannot be negative.

If we multiply the radius by 2, we find that the diameter of the storm was about 300 miles.   ■

EXAMPLE 4	**Solving equations using the square root method.** Solve $(x + 1)^2 = 9$.

Solution $(x + 1)^2 = 9$

By the square root method, the two solutions are

$$x + 1 = \sqrt{9} \quad \text{or} \quad x + 1 = -\sqrt{9}$$
$$x + 1 = 3 \qquad\qquad x + 1 = -3$$
$$x = 2 \qquad\qquad\quad x = -4 \quad \text{Subtract 1 from both sides.}$$

Check: For x = 2   **For x = −4**
$$(x + 1)^2 = 9 \qquad (x + 1)^2 = 9$$
$$(2 + 1)^2 \overset{?}{=} 9 \qquad (-4 + 1)^2 \overset{?}{=} 9$$
$$3^2 \overset{?}{=} 9 \qquad\quad (-3)^2 \overset{?}{=} 9$$
$$9 = 9 \qquad\qquad 9 = 9$$

SELF CHECK   Solve $(x + 2)^2 = 4$.   *Answer:* 0, −4   ■

EXAMPLE 5	**Solving equations using the square root method.** Solve $(x - 2)^2 - 18 = 0$.

Solution $(x - 2)^2 - 18 = 0$

$\qquad\qquad (x - 2)^2 = 18$ Add 18 to both sides to isolate $(x - 2)^2$.

The two solutions are

$$x - 2 = \sqrt{18} \qquad \text{or} \qquad x - 2 = -\sqrt{18}$$
$$x = 2 + \sqrt{18} \qquad\qquad\qquad x = 2 - \sqrt{18} \quad \text{Add 2 to both sides.}$$
$$= 2 + 3\sqrt{2} \qquad\qquad\qquad = 2 - 3\sqrt{2} \quad \sqrt{18} = \sqrt{9}\sqrt{2} = 3\sqrt{2}.$$

Check: **For** $x = 2 + 3\sqrt{2}$ **For** $x = 2 - 3\sqrt{2}$

$$(x - 2)^2 - 18 = 0 \qquad\qquad (x - 2)^2 - 18 = 0$$
$$\left(2 + 3\sqrt{2} - 2\right)^2 - 18 \stackrel{?}{=} 0 \qquad \left(2 - 3\sqrt{2} - 2\right)^2 - 18 \stackrel{?}{=} 0$$
$$\left(3\sqrt{2}\right)^2 - 18 \stackrel{?}{=} 0 \qquad\qquad \left(-3\sqrt{2}\right)^2 - 18 \stackrel{?}{=} 0$$
$$18 - 18 \stackrel{?}{=} 0 \qquad\qquad 18 - 18 \stackrel{?}{=} 0$$
$$0 = 0 \qquad\qquad\qquad 0 = 0$$

SELF CHECK   Solve $(x + 3)^2 - 12 = 0$.   *Answer:* $-3 \pm 2\sqrt{3}$ ∎

Completing the Square

When the polynomial in a quadratic equation doesn't factor, we can solve the equation by using a method called **completing the square.**

The method of completing the square is based on the following special products:

$$x^2 + 2bx + b^2 = (x + b)^2 \qquad \text{and} \qquad x^2 - 2bx + b^2 = (x - b)^2$$

The trinomials $x^2 + 2bx + b^2$ and $x^2 - 2bx + b^2$ are both perfect square trinomials, since each one factors as the square of a binomial. In each trinomial, if we take one-half of the coefficient of the x and square it, we get the third term.

> In $x^2 + 2bx + b^2$, if we take $\frac{1}{2}(2b)$, which is b, and square it, we get the third term, b^2.
>
> In $x^2 - 2bx + b^2$, if we take $\frac{1}{2}(-2b) = -b$ and square it, we get $(-b)^2 = b^2$, which is the third term.

To change a binomial such as $x^2 + 12x$ into a perfect square trinomial, we take one-half of the coefficient of x (the 12), square it, and add it to $x^2 + 12x$.

$$x^2 + 12x + \left[\frac{1}{2}(12)\right]^2 = x^2 + 12x + (6)^2$$
$$= x^2 + 12x + 36$$

This result is a perfect square trinomial, because $x^2 + 12x + 36 = (x + 6)^2$.

EXAMPLE 6   **Completing the square.** Change each expression into a perfect square trinomial: **a.** $x^2 + 4x$, **b.** $x^2 - 6x$, and **c.** $x^2 - 5x$.

Solution   **a.** $x^2 + 4x + \left[\frac{1}{2}(4)\right]^2 = x^2 + 4x + (2)^2$ The coefficient of x is 4. We add the square of one-half of 4.

$$= x^2 + 4x + 4 \qquad \text{This is } (x + 2)^2.$$

b. $x^2 - 6x + \left[\frac{1}{2}(-6)\right]^2 = x^2 - 6x + (-3)^2$ The coefficient of x is -6. We add the square of one-half of -6.

$$= x^2 - 6x + 9 \qquad \text{This is } (x - 3)^2.$$

c. $x^2 - 5x + \left[\frac{1}{2}(-5)\right]^2 = x^2 - 5x + \left(-\frac{5}{2}\right)^2$ One-half of -5 is $-\frac{5}{2}$. We add its square.

$$= x^2 - 5x + \frac{25}{4} \qquad \text{This is } \left(x - \frac{5}{2}\right)^2.$$

Change each expression into a perfect
square trinomial: **a.** $y^2 + 6y$,   *Answers:*   **a.** $y^2 + 6y + 9$,
b. $y^2 - 8y$, and   **c.** $y^2 + 3y$ **b.** $y^2 - 8y + 16$,   **c.** $y^2 + 3y + \frac{9}{4}$ ■

Solving Equations with Leading Coefficients of 1

If the quadratic equation $ax^2 + bx + c = 0$ has a leading coefficient of 1, it's easy to
solve by completing the square.

EXAMPLE 7   **Solving equations by completing the square.** Solve $x^2 - 4x - 13 = 0$.
Give each answer to the nearest hundredth.

Solution   Since the coefficient of x^2 is 1, we can complete the square as follows:

$$x^2 - 4x - 13 = 0$$
$$x^2 - 4x = 13 \quad \text{Add 13 to both sides so that the constant term is on the right-hand side.}$$

We then find one-half of the coefficient of x, square it, and add the result to both sides
to make the left-hand side a perfect square trinomial.

$$x^2 - 4x + \left[\frac{1}{2}(-4)\right]^2 = 13 + \left[\frac{1}{2}(-4)\right]^2 \quad \text{The coefficient of } x \text{ is } -4.$$
$$x^2 - 4x + 4 = 13 + 4 \quad \text{Simplify.}$$
$$(x - 2)^2 = 17 \quad \text{Factor } x^2 - 4x + 4 \text{ and simplify.}$$
$$x - 2 = \pm\sqrt{17} \quad \text{Use the square root method to solve for } x - 2.$$
$$x = 2 \pm \sqrt{17} \quad \text{Add 2 to both sides to isolate } x.$$

Because of the $\pm$ sign, there are two solutions. We can use a calculator to approximate
each one.

$$x = 2 + \sqrt{17} \qquad \text{or} \quad x = 2 - \sqrt{17}$$
$$\approx 2 + 4.123105626 \qquad \approx 2 - 4.123105626$$
$$x \approx 6.12 \qquad\qquad\qquad x \approx -2.12$$

Solve $x^2 + 10x - 4 = 0$. Give the exact solutions   *Answers:* $-5 + \sqrt{29}$,
and approximations to the nearest hundredth. $-5 - \sqrt{29}$; 0.39, -10.39 ■

Solving Equations with Leading Coefficients Other Than 1

If the quadratic equation $ax^2 + bx + c = 0$ ($a \neq 0$) has a leading coefficient other than
1, we can make the leading coefficient 1 by dividing both sides of the equation by a.
Then we can solve the equation by completing the square.

E X A M P L E   8

Solving equations by completing the square. Solve $4x^2 + 4x - 3 = 0$.

Solution We divide both sides by 4 so that the coefficient of x^2 is 1. We then proceed as follows:

$$4x^2 + 4x - 3 = 0$$

$$x^2 + x - \frac{3}{4} = 0 \qquad \text{Divide both sides by 4: } \frac{4x^2}{4} + \frac{4x}{4} - \frac{3}{4} = \frac{0}{4}.$$

$$x^2 + x = \frac{3}{4} \qquad \text{Add } \tfrac{3}{4} \text{ to both sides so that the constant term is on the right-hand side.}$$

$$x^2 + x + \frac{1}{4} = \frac{3}{4} + \frac{1}{4} \qquad \text{The coefficient of } x \text{ is 1. One-half of 1 is } \tfrac{1}{2}. \\ \text{Add } \left(\tfrac{1}{2}\right)^2 = \tfrac{1}{4} \text{ to both sides to complete the square.}$$

$$\left(x + \frac{1}{2}\right)^2 = 1 \qquad \text{Factor and add.}$$

$$x + \frac{1}{2} = \pm 1 \qquad \text{Solve for } x + \tfrac{1}{2} \text{ using the square root method.}$$

$$x = -\frac{1}{2} \pm 1 \qquad \text{Subtract } \tfrac{1}{2} \text{ from both sides to isolate } x.$$

$$x = -\frac{1}{2} + 1 \quad \text{or} \quad x = -\frac{1}{2} - 1$$

$$= \frac{1}{2} \qquad\qquad\qquad = -\frac{3}{2}$$

Check each solution. Note that this equation can also be solved by factoring.

SELF CHECK Solve $2x^2 - 5x - 3 = 0$. *Answer:* $3, -\tfrac{1}{2}$ ■

The previous examples illustrate that to solve a quadratic equation by completing the square, we follow these steps.

Completing the square to solve a quadratic equation

1. Write the equation in $ax^2 + bx + c = 0$ form. If the coefficient of x^2 is not 1, make it 1 by dividing both sides of the equation by the coefficient of x^2.

2. If necessary, add or subtract a number on both sides of the equation to get the constant term on the right-hand side.

3. Complete the square.
   a. Find half the coefficient of x and square it.
   b. Add that square to both sides of the equation.

4. Factor the perfect square trinomial and combine terms.

5. Solve the resulting quadratic equation using the square root method.

6. Check each solution.

| EXAMPLE 9 | **Solving equations by completing the square.** Solve $2x^2 - 2 = -4x$. |

Solution We write the equation in $ax^2 + bx + c = 0$ form to see if it can be solved by factoring.

$2x^2 + 4x - 2 = 0$ Add $4x$ to both sides to get 0 on the right-hand side.

1. $x^2 + 2x - 1 = 0$ Divide both sides by 2: $\dfrac{2x^2}{2} + \dfrac{4x}{2} - \dfrac{2}{2} = \dfrac{0}{2}$.

Since Equation 1 cannot be solved by factoring, we complete the square.

$$x^2 + 2x = 1$$ Add 1 to both sides.

$$x^2 + 2x + 1 = 1 + 1$$ One-half of 2 is 1. Add $1^2 = 1$ to both sides to complete the square.

$$(x + 1)^2 = 2$$ Factor and simplify.

$$x + 1 = \pm\sqrt{2}$$ Use the square root method to solve for $x + 1$.

$$x = -1 \pm \sqrt{2}$$ Subtract 1 from both sides.

$$x = -1 + \sqrt{2} \quad \text{or} \quad x = -1 - \sqrt{2}$$

Check both solutions.

SELF CHECK   Solve $3x^2 - 18x = -12$.   *Answer:* $3 \pm \sqrt{5}$ ∎

| EXAMPLE 10 | **Solving equations by completing the square.** Solve $x^2 - 7x = 2$. |

Solution The constant term is already on the right-hand side. To complete the square on the left-hand side, we find one-half of the coefficient of x and add its square to both sides.

$$x^2 - 7x = 2$$

$$x^2 - 7x + \frac{49}{4} = 2 + \frac{49}{4}$$ One-half of -7 is $-\frac{7}{2}$. Add $\left(-\frac{7}{2}\right)^2 = \frac{49}{4}$ to both sides.

$$\left(x - \frac{7}{2}\right)^2 = \frac{8}{4} + \frac{49}{4}$$ Factor the left-hand side. Write 2 as $\frac{8}{4}$.

$$\left(x - \frac{7}{2}\right)^2 = \frac{57}{4}$$ The fractions have a common denominator. Add them.

$$x - \frac{7}{2} = \pm\sqrt{\frac{57}{4}}$$ Use the square root method to solve for $x - \frac{7}{2}$.

$$x - \frac{7}{2} = \pm\frac{\sqrt{57}}{2}$$ Simplify: $\sqrt{\dfrac{57}{4}} = \dfrac{\sqrt{57}}{\sqrt{4}} = \dfrac{\sqrt{57}}{2}$.

$$x = \frac{7}{2} \pm \frac{\sqrt{57}}{2}$$ Add $\frac{7}{2}$ to both sides.

$$x = \frac{7 \pm \sqrt{57}}{2}$$ Since the fractions have a common denominator of 2, we can combine them.

If we approximate the solutions to the nearest hundredth, we have

$$\frac{7 + \sqrt{57}}{2} \approx 7.27 \quad \text{and} \quad \frac{7 - \sqrt{57}}{2} \approx -0.27$$

SELF CHECK   Solve $x^2 + 5x = 3$.   *Answer:* $\dfrac{-5 \pm \sqrt{37}}{2}$ ∎

STUDY SET

Section 6.6

VOCABULARY

In Exercises 1–4, fill in the blanks to make the statements true.

1. If the polynomial in the equation $ax^2 + bx + c = 0$ doesn't factor, we can solve the equation by _____completing_____ the square.

2. Since $x^2 + 12x + 36 = (x + 6)^2$, we call the trinomial a perfect _____square_____ trinomial.

3. In the equation $x^2 - 4x + 1 = 0$, the _____coefficient_____ of x is -4.

4. A _____solution_____ of an equation is a value of the variable that makes the equation true.

CONCEPTS

In Exercises 5–8, fill in the blanks to make the statements true.

5. The equation $x^2 = c$ $(c > 0)$ has _____two_____ solutions.

6. The solutions of $x^2 = c$ $(c > 0)$ are $\boxed{\sqrt{c}}$ and $\boxed{-\sqrt{c}}$.

7. To complete the square on $x^2 + 8x$, we add the _____square_____ of one-half of $\boxed{8}$, which is 16.

8. To complete the square on $x^2 - 10x$, we add the square of _____one-half_____ of -10, which is $\boxed{25}$.

9. What is the first step if we solve $x^2 - 2x = 35$
   a. by using the factoring method?
      Subtract 35 from both sides.
   b. by completing the square?   Add 1 to both sides.

10. The equation $n^2 - 4n + 6 = 0$ is written in $ax^2 + bx + c = 0$ form. What is b? -4

11. To solve $x^2 - 2x - 1 = 0$, we must complete the square. Why can't we use the factoring method?
    because $x^2 - 2x - 1 = 0$ doesn't factor

12. Solve $x^2 = 81$ by using the
    a. the square root method. ± 9
    b. the factoring method. ± 9

13. What is one-half of the given number?
    a. 4   2
    b. -8 -4
    c. 5 $\frac{5}{2}$
    d. -7 $-\frac{7}{2}$

14. Find one-half of the given number and then square the result.
    a. 6   9
    b. -12 36
    c. 3 $\frac{9}{4}$
    d. -5 $\frac{25}{4}$

15. What is the result when both sides of $2x^2 + 4x - 8 = 0$ are divided by 2? $x^2 + 2x - 4 = 0$

16. Write $3x^2 = -4x + 8$ in $ax^2 + bx + c = 0$ form.
    $3x^2 + 4x - 8 = 0$

NOTATION

In Exercises 17–18, complete each solution.

17.
$$(y - 1)^2 = 9$$
$$y - 1 = \boxed{\sqrt{9}} \quad \text{or} \quad y - \boxed{1} = -\sqrt{9}$$
$$\boxed{y - 1} = 3 \qquad\qquad y - 1 = \boxed{-3}$$
$$y = 4 \qquad\qquad\qquad y = -2$$

18.
$$y^2 + 2y - 3 = 0$$
$$y^2 + 2y = \boxed{3}$$
$$y^2 + 2y + 1 = 3 + \boxed{1}$$
$$(y + 1)^2 = \boxed{4}$$
$$\boxed{y + 1} = \sqrt{4} \quad \text{or} \quad y + 1 = -\boxed{\sqrt{4}}$$
$$y + 1 = \boxed{2} \qquad\qquad \boxed{y + 1} = -2$$
$$y = 1 \qquad\qquad\qquad y = -3$$

19. In solving a quadratic equation, a student obtains $x = \pm\sqrt{10}$. How many solutions are represented by this notation? List them.   two; $\sqrt{10}, -\sqrt{10}$

20. In solving a quadratic equation, a student obtains $x = 8 \pm \sqrt{3}$. List each solution separately. Then round each one to the nearest hundredth.
    $8 + \sqrt{3}, 8 - \sqrt{3}; 9.73, 6.27.$

PRACTICE

In Exercises 21–38, use the square root method to solve each equation.

21. $x^2 = 1$ ± 1
22. $r^2 = 4$ ± 2 Look at give answer
23. $x^2 = 9$ ± 3
24. $x^2 = 32$ $\pm 4\sqrt{2}$
25. $t^2 = 20$ $\pm 2\sqrt{5}$
26. $x^2 = 0$ $0, 0$
27. $3m^2 = 27$ ± 3
28. $4x^2 = 64$ ± 4
29. $4x^2 = 16$ ± 2
30. $5x^2 = 125$ ± 5
31. $x^2 = \dfrac{9}{16}$ $\pm\frac{3}{4}$
32. $x^2 = \dfrac{81}{25}$ $\pm\frac{9}{5}$
33. $(x + 1)^2 = 25$ $-6, 4$
34. $(x - 1)^2 = 49$ $-6, 8$
35. $(x + 2)^2 = 81$ $7, -11$
36. $(x + 3)^2 = 16$ $1, -7$
37. $(x - 2)^2 = 8$ $2 \pm 2\sqrt{2}$
38. $(x + 2)^2 = 50$ $-2 \pm 5\sqrt{2}$

In Exercises 39–42, use the square root method to solve each equation. Use a calculator to approximate the solutions. Round to the nearest hundredth.

39. $x^2 = 45.82$ ± 6.77

40. $x^2 = 6.05$ ± 2.46

41. $(x + 2)^2 = 90.04$
   $7.49, -11.49$

42. $(x - 5)^2 = 33.31$
   $10.77, -0.77$

In Exercises 43–46, factor the trinomial square and use the square root method to solve each equation.

43. $y^2 + 4y + 4 = 4$ $0, -4$

44. $y^2 - 6y + 9 = 9$ $0, 6$

45. $9x^2 - 12x + 4 = 16$ $2, -\frac{2}{3}$

46. $4x^2 - 20x + 25 = 36$ $\frac{11}{2}, -\frac{1}{2}$

In Exercises 47–56, complete the square to make a perfect square trinomial.

47. $x^2 + 2x$ $x^2 + 2x + 1$

48. $x^2 + 12x$
   $x^2 + 12x + 36$

49. $x^2 - 4x$ $x^2 - 4x + 4$

50. $x^2 - 14x$
   $x^2 - 14x + 49$

51. $x^2 + 7x$ $x^2 + 7x + \frac{49}{4}$

52. $x^2 + 21x$
   $x^2 + 21x + \frac{441}{4}$

53. $a^2 - 3a$ $a^2 - 3a + \frac{9}{4}$

54. $b^2 - 13b$
   $b^2 - 13b + \frac{169}{4}$

55. $b^2 + \frac{2}{3}b$ $b^2 + \frac{2}{3}b + \frac{1}{9}$

56. $c^2 - \frac{5}{2}c$ $c^2 - \frac{5}{2}c + \frac{25}{16}$

In Exercises 57–72, solve each equation by completing the square.

57. $x^2 + 6x + 8 = 0$
   $-2, -4$

58. $x^2 + 8x + 12 = 0$
   $-2, -6$

59. $k^2 - 8k + 12 = 0$ $2, 6$

60. $p^2 - 4p + 3 = 0$ $3, 1$

61. $x^2 - 2x = 15$ $5, -3$

62. $x^2 - 2x = 8$ $4, -2$

63. $g^2 + 5g - 6 = 0$
   $1, -6$

64. $s^2 = 14 - 5s$ $2, -7$

65. $2x^2 = 4 - 2x$ $1, -2$

66. $3q^2 = 3q + 6$ $2, -1$

67. $3x^2 + 9x + 6 = 0$
   $-1, -2$

68. $3d^2 + 48 = -24d$
   $-4, -4$

69. $2x^2 = 3x + 2$ $2, -\frac{1}{2}$

70. $3x^2 = 2 - 5x$ $-2, \frac{1}{3}$

71. $4x^2 = 2 - 7x$ $-2, \frac{1}{4}$

72. $2x^2 = 5x + 3$ $3, -\frac{1}{2}$

check by factoring.

In Exercises 73–80, solve each equation. Give the exact solutions, and then give the solutions rounded to the nearest hundredth.

73. $x^2 + 4x + 1 = 0$
   $-2 \pm \sqrt{3}$; $-0.27, -3.73$

74. $x^2 + 6x + 2 = 0$
   $-3 \pm \sqrt{7}$; $-0.35, -5.65$

75. $x^2 - 2x - 4 = 0$
   $1 \pm \sqrt{5}$; $3.24, -1.24$

76. $x^2 - 4x = 2$
   $2 \pm \sqrt{6}$; $4.45, -0.45$

77. $x^2 = 4x + 3$
   $2 \pm \sqrt{7}$; $4.65, -0.65$

78. $x^2 = 6x - 3$
   $3 \pm \sqrt{6}$; $5.45, 0.55$

79. $4x^2 + 4x + 1 = 20$
   $\frac{-1 \pm 2\sqrt{5}}{2}$; $-2.74, 1.74$

80. $9x^2 = 8 - 12x$
   $\frac{-2 \pm 2\sqrt{3}}{3}$; $0.49, -1.82$

In Exercises 81–86, write each equation in the form $ax^2 + bx + c = 0$ and solve it by completing the square.

81. $2x(x + 3) = 8$ $1, -4$

82. $3x(x - 2) = 9$ $3, -1$

83. $6(x^2 - 1) = 5x$ $\frac{3}{2}, -\frac{2}{3}$

84. $2(3x^2 - 2) = 5x$ $\frac{4}{3}, -\frac{1}{2}$

85. $x(x + 3) - \frac{1}{2} = -2$ $\frac{-3 \pm \sqrt{3}}{2}$

86. $x[(x - 2) + 3] = 3\left(x - \frac{2}{9}\right)$ $\frac{3 \pm \sqrt{3}}{3}$

APPLICATIONS

87. CAROUSELS In 1999, the city of Lancaster, Pennsylvania considered installing a classic Dentzel carousel in an abandoned downtown building. After learning that the circular-shaped carousel (like that shown in Illustration 1) would occupy 2,376 square feet of floor space and that it was 26 feet high, the proposal was determined to be impractical because of the large remodeling costs. Find the diameter of the carousel to the nearest foot.   55 ft

ILLUSTRATION 1

88. ESCAPE VELOCITY The speed at which a rocket must be fired for it to leave the earth's gravitational attraction is called the *escape velocity*. See Illustration 2. If the escape velocity, v_e in miles per hour, is given by

$$\frac{v_e^2}{2g} = R$$

where $g = 78,545$ and $R = 3,960$, find v_e. Round to the nearest mi/hr.   24,941 mi/hr

ILLUSTRATION 2

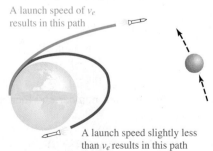

A launch speed of v_e results in this path

A launch speed slightly less than v_e results in this path

89. BICYCLE SAFETY A bicycle training program for children uses a figure-8 course to help them improve their balance and steering. The course is laid out over a paved area covering 800 square feet, as shown in Illustration 3. Find its dimensions.   20 ft by 40 ft

ILLUSTRATION 3

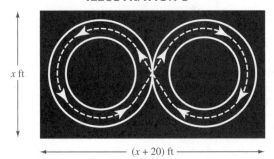

x ft

$(x + 20)$ ft

90. BADMINTON The badminton court shown in Illustration 4 occupies 880 square feet of the floor space of a gymnasium. If its length is 4 feet more than twice its width, find its dimensions.   20 ft by 44 ft

WRITING

91. Explain how to complete the square on $x^2 - 5x$.

92. Explain how to solve $x^2 = 81$ with the square root method.

ILLUSTRATION 4

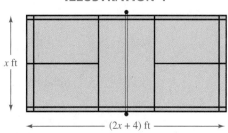

x ft

$(2x + 4)$ ft

93. Rounded to the nearest hundredth, one solution of the equation $x^2 + 4x + 1 = 0$ is -0.27. Use your calculator to check it. How could it be a solution if it doesn't make the left-hand side zero? Explain.

94. Give an example of a perfect square trinomial. Why do you think the word "perfect" is used to describe it?

REVIEW

In Exercises 95–100, do each operation and simplify.

95. $(y - 1)^2$ $y^2 - 2y + 1$ **96.** $(z + 2)^2$ $z^2 + 4z + 4$

97. $(x + y)^2$ $x^2 + 2xy + y^2$ **98.** $(a - b)^2$
$a^2 - 2ab + b^2$

99. $(2z)^2$ $4z^2$ **100.** $(xy)^2$ x^2y^2

▶ **6.7**

The Quadratic Formula

In this section, you will learn about

   The quadratic formula ■ Quadratic equations with no real solutions
  ■ Applications

Introduction We can solve any quadratic equation by completing the square, but the work is often tedious. Fortunately, there is an easier way. In this section, we will develop a formula, called the *quadratic formula,* that will enable us to solve quadratic equations with much less effort.

The Quadratic Formula

We can solve the **general quadratic equation** $ax^2 + bx + c = 0$ $(a \neq 0)$ by completing the square.

$$ax^2 + bx + c = 0$$

$$\frac{ax^2}{a} + \frac{bx}{a} + \frac{c}{a} = \frac{0}{a} \qquad \text{Divide both sides by } a.$$

$$x^2 + \frac{b}{a}x + \frac{c}{a} = 0 \qquad \text{Simplify: } \frac{a}{a} = 1. \text{ Write } \frac{bx}{a} \text{ as } \frac{b}{a}x.$$

$$x^2 + \frac{b}{a}x = -\frac{c}{a} \qquad \text{Subtract } \frac{c}{a} \text{ from both sides.}$$

Since the coefficient of x is $\frac{b}{a}$, we can complete the square on x by adding

$$\left(\frac{1}{2} \cdot \frac{b}{a}\right)^2 \qquad \text{or} \qquad \frac{b^2}{4a^2}$$

to both sides:

$$x^2 + \frac{b}{a}x + \frac{b^2}{4a^2} = \frac{b^2}{4a^2} - \frac{c}{a}$$

After factoring the perfect square trinomial on the left-hand side, we have

$$\left(x + \frac{b}{2a}\right)\left(x + \frac{b}{2a}\right) = \frac{b^2}{4a^2} - \frac{4ac}{4aa}$$ The lowest common denominator on the right-hand side is $4a^2$.

1. $$\left(x + \frac{b}{2a}\right)^2 = \frac{b^2 - 4ac}{4a^2}$$ Subtract the numerators and write the difference over the common denominator.

Equation 1 can be solved by the square root method to obtain

$$x + \frac{b}{2a} = \sqrt{\frac{b^2 - 4ac}{4a^2}} \qquad \text{or} \qquad x + \frac{b}{2a} = -\sqrt{\frac{b^2 - 4ac}{4a^2}}$$

$$x + \frac{b}{2a} = \frac{\sqrt{b^2 - 4ac}}{\sqrt{4a^2}} \qquad\qquad x + \frac{b}{2a} = -\frac{\sqrt{b^2 - 4ac}}{\sqrt{4a^2}}$$

$$x = -\frac{b}{2a} + \frac{\sqrt{b^2 - 4ac}}{2a} \qquad\qquad x = -\frac{b}{2a} - \frac{\sqrt{b^2 - 4ac}}{2a}$$

$$x = \frac{-b + \sqrt{b^2 - 4ac}}{2a} \qquad\qquad x = \frac{-b - \sqrt{b^2 - 4ac}}{2a}$$

These solutions are usually written in one formula called the **quadratic formula.**

Quadratic formula

The solutions of the quadratic equation $ax^2 + bx + c = 0$ are

$$x = \frac{-b \pm \sqrt{b^2 - 4ac}}{2a} \qquad (a \neq 0)$$

WARNING! When you write the quadratic formula, be careful to draw the fraction bar so that it includes the complete numerator. Do not write

$$x = -b \pm \frac{\sqrt{b^2 - 4ac}}{2a}$$

E X A M P L E   1

Solving equations using the quadratic formula. Solve $x^2 + 5x + 6 = 0$.

Solution   The equation is written in $ax^2 + bx + c = 0$ form with $a = 1$, $b = 5$, and $c = 6$. We substitute these values into the quadratic formula and simplify using the rules for the order of operations.

$$x = \frac{-b \pm \sqrt{b^2 - 4ac}}{2a}$$ The quadratic formula.

$$= \frac{-5 \pm \sqrt{5^2 - 4(1)(6)}}{2(1)}$$ Substitute 1 for a, 5 for b, and 6 for c.

$$= \frac{-5 \pm \sqrt{25 - 24}}{2}$$ Evaluate the power and do the multiplication within the radical.

$$= \frac{-5 \pm \sqrt{1}}{2}$$ Do the subtraction within the radical.

$$x = \frac{-5 \pm 1}{2}$$ Simplify: $\sqrt{1} = 1$.

This notation represents two solutions. We simplify them separately, first using the $+$ sign and then using the $-$ sign.

$$x = \frac{-5 + 1}{2} \quad \text{or} \quad x = \frac{-5 - 1}{2}$$

$$x = \frac{-4}{2} \qquad\qquad x = \frac{-6}{2}$$

$$x = -2 \qquad\qquad\quad x = -3$$

Check both solutions, $x = -2$ and $x = -3$.

SELF CHECK   Solve $x^2 + 6x + 5 = 0$.   *Answer:* $-1, -5$ ■

WARNING! Be sure to write a quadratic equation in quadratic form ($ax^2 + bx + c = 0$) before identifying the values of a, b, and c.

EXAMPLE 2

Writing equations in quadratic form.   Solve $2x^2 = 5x + 3$.

Solution   We begin by writing the equation in quadratic form.

$$2x^2 = 5x + 3$$

$$2x^2 - 5x - 3 = 0$$ Subtract $5x$ and 3 from both sides.

In this equation, $a = 2$, $b = -5$, and $c = -3$. We substitute these values into the quadratic formula and simplify.

$$x = \frac{-b \pm \sqrt{b^2 - 4ac}}{2a}$$ The quadratic formula.

$$= \frac{-(-5) \pm \sqrt{(-5)^2 - 4(2)(-3)}}{2(2)}$$ Substitute 2 for a, -5 for b, and -3 for c.

$$= \frac{5 \pm \sqrt{25 - (-24)}}{4}$$ $-(-5) = 5$. Evaluate the power and do the multiplication within the radical.

$$= \frac{5 \pm \sqrt{49}}{4}$$ Do the subtraction within the radical; $25 - (-24) = 25 + 24 = 49$.

$$= \frac{5 \pm 7}{4}$$ Simplify: $\sqrt{49} = 7$.

Thus,

$$x = \frac{5 + 7}{4} \quad \text{or} \quad x = \frac{5 - 7}{4} \qquad \text{Evaluate each solution separately.}$$

$$x = \frac{12}{4} \qquad\qquad x = \frac{-2}{4}$$

$$x = 3 \qquad\qquad x = -\frac{1}{2}$$

Check both solutions in the original equation.

SELF CHECK Solve $4x^2 - 11x = 3$. *Answer:* 3, $-\frac{1}{4}$ ■

EXAMPLE 3

Approximating solutions. Solve $3x^2 = 2x + 4$. Round each solution to the nearest hundredth.

Solution We begin by writing the given equation in $ax^2 + bx + c = 0$ form.

$$3x^2 = 2x + 4$$
$$3x^2 - 2x - 4 = 0 \qquad \text{Subtract } 2x \text{ and 4 from both sides.}$$

In this equation, $a = 3$, $b = -2$, and $c = -4$. We substitute these values into the quadratic formula and simplify.

$$x = \frac{-b \pm \sqrt{b^2 - 4ac}}{2a} \qquad \text{The quadratic formula.}$$

$$= \frac{-(-2) \pm \sqrt{(-2)^2 - 4(3)(-4)}}{2(3)} \qquad \text{Substitute 3 for } a, -2 \text{ for } b, \text{ and } -4 \text{ for } c.$$

$$= \frac{2 \pm \sqrt{4 + 48}}{6} \qquad -(-2) = 2. \text{ Simplify within the radical.}$$

$$= \frac{2 \pm \sqrt{52}}{6} \qquad \text{Do the addition within the radical.}$$

$$= \frac{2 \pm 2\sqrt{13}}{6} \qquad \text{Simplify: } \sqrt{52} = \sqrt{4 \cdot 13} = \sqrt{4}\sqrt{13} = 2\sqrt{13}.$$

$$= \frac{2(1 \pm \sqrt{13})}{2 \cdot 3} \qquad \text{Factor out 2 from } 2 \pm 2\sqrt{13}:$$
$$2 \pm 2\sqrt{13} = 2(1 \pm \sqrt{13}). \text{ Write 6 as } 2 \cdot 3.$$

$$\qquad\qquad\qquad\qquad \text{Divide out the common factor of 2:}$$
$$x = \frac{1 \pm \sqrt{13}}{3} \qquad \frac{\overset{1}{\cancel{2}}(1 \pm \sqrt{13})}{\underset{1}{\cancel{2} \cdot 3}}.$$

Thus,

$$x = \frac{1 + \sqrt{13}}{3} \quad \text{or} \quad x = \frac{1 - \sqrt{13}}{3} \qquad \text{Express the solutions separately.}$$

$$x = \frac{1}{3} + \frac{\sqrt{13}}{3} \quad \text{or} \quad x = \frac{1}{3} - \frac{\sqrt{13}}{3} \qquad \text{This is an alternative form for expressing the solutions.}$$

Check these solutions in the original equation.

We can use a calculator to approximate each of these solutions. To the nearest hundredth,

$$\frac{1 + \sqrt{13}}{3} \approx 1.54 \quad \text{and} \quad \frac{1 - \sqrt{13}}{3} \approx -0.87$$

SELF CHECK   Solve $2x^2 - 1 = 2x$. Round to the nearest hundredth.

Answers: $\dfrac{1 + \sqrt{3}}{2} \approx 1.37, \dfrac{1 - \sqrt{3}}{2} \approx -0.37$ ∎

Quadratic Equations with No Real Solutions

The next example shows that some quadratic equations have no real-number solutions.

E X A M P L E   4   **An equation with no real-number solutions.** Solve $x^2 + 2x + 5 = 0$.

Solution   In this equation, $a = 1$, $b = 2$, and $c = 5$. We substitute these values into the quadratic formula.

$$x = \frac{-b \pm \sqrt{b^2 - 4ac}}{2a} \qquad \text{The quadratic formula.}$$

$$= \frac{-2 \pm \sqrt{2^2 - 4(1)(5)}}{2(1)} \qquad \text{Substitute 1 for } a, 2 \text{ for } b, \text{ and } 5 \text{ for } c.$$

$$= \frac{-2 \pm \sqrt{4 - 20}}{2} \qquad \text{Evaluate the power and do the multiplication within the radical.}$$

$$x = \frac{-2 \pm \sqrt{-16}}{2} \qquad \text{Do the subtraction within the radical. The result is a negative number, } -16.$$

Since $\sqrt{-16}$ is not a real number, there are no real-number solutions.

SELF CHECK   Does the equation

$$2x^2 + x + 1 = 0$$

have any real-number solutions?

Answer: no ∎

Applications

Quadratic equations are used to solve many types of problems. In the following examples, when we get to the "solve the equation" step in the solution, we will need to determine the most efficient method to use to solve the equation. To do this, we can use the following strategy.

Strategy for solving quadratic equations

1. First, see whether the equation is in a form such that the **square root method** is easily applied.
2. If the square root method can't be used, write the equation in $ax^2 + bx + c = 0$ form.
3. Then see whether the equation can be solved using the **factoring method.**
4. If you can't factor the quadratic, solve the equation by **completing the square** or by the **quadratic formula.**

E X A M P L E   5

Nutrition. The poster in Figure 6-6 shows the six basic food groups, as established by the U.S. Department of Agriculture. If the area of the poster is 90 square inches and the base is 3 inches longer than the height, find the length of the base and the height of the poster.

FIGURE 6-6

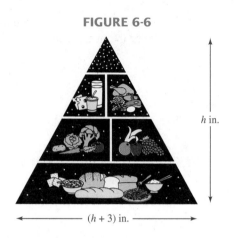

h in.

⟵ $(h + 3)$ in. ⟶

ANALYZE THE PROBLEM   We are given the area of the triangular-shaped poster and asked to find the length of its base and its height.

FORM AN EQUATION   Since the length of the base is related to the height, we will let h represent the height of the triangle. Then $h + 3$ represents the length of the base. The area of a triangle is given by the formula $A = \frac{1}{2}bh$, which gives the equation

$\frac{1}{2}$	times	the length of the base	times	the height	equals	the area of the triangle.
$\frac{1}{2}$	·	$(h + 3)$	·	h	=	90

SOLVE THE EQUATION   To solve the equation $\frac{1}{2}(h + 3)h = 90$, we first write it in quadratic form.

$$\frac{1}{2}(h + 3)h = 90$$

$(h + 3)h = 180$ Multiply both sides by 2.

$h^2 + 3h = 180$ Use the distributive property to remove parentheses.

$h^2 + 3h - 180 = 0$ Subtract 180 from both sides. The equation is now in quadratic form.

By inspection, we see that -180 has factors -12 and 15 and that their sum is 3. Therefore, we can use the factoring method to solve the equation.

$(h - 12)(h + 15) = 0$ Factor $h^2 + 3h - 180$.

$h - 12 = 0$ or $h + 15 = 0$ Set each factor equal to 0.

$h = 12$ $h = -15$ Solve each linear equation.

STATE THE CONCLUSION   When $h = 12$, the length of the base, $h + 3$, is 15. We discard the solution $h = -15$, because the triangle cannot have a negative height. So the length of the base is 15 inches, and the height is 12 inches.

CHECK THE RESULT   With a base of 15 inches and a height of 12 inches, the base of the triangle is 3 inches longer than its height. Its area is $\frac{1}{2}(15)(12) = 90$ square inches. The solution checks. ■

EXAMPLE 6

Action movies. As part of an action scene in a movie, a stuntman is to fall from the top of a 95-foot-tall building into a large airbag directly below him on the ground, as shown in Figure 6-7. If an object falls s feet in t seconds, where $s = 16t^2$, and if the bag is inflated to a height of 10 feet, how long will the stuntman fall before making contact with the airbag?

FIGURE 6-7

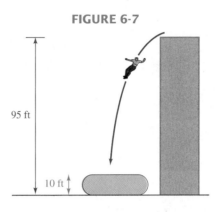

95 ft

10 ft

Solution   If we subtract the height of the airbag from the height of the building, we find that the stuntman will fall $95 - 10 = 85$ feet. We substitute 85 for s in the formula and find that the equation is in a form that allows us to use the square root method.

$$s = 16t^2 \qquad \text{The given formula.}$$

$$85 = 16t^2 \qquad \text{Substitute 85 for } s.$$

$$\frac{85}{16} = t^2 \qquad \text{Divide both sides by 16.}$$

$$\pm\sqrt{\frac{85}{16}} = t \qquad \text{Use the square root method to solve the equation.}$$

$$\pm\frac{\sqrt{85}}{\sqrt{16}} = t \qquad \text{The square root of a quotient is the quotient of the square roots.}$$

$$\pm\frac{\sqrt{85}}{4} = t \qquad \sqrt{16} = 4.$$

The stuntman will fall for $\frac{\sqrt{85}}{4}$ seconds before making contact with the airbag. To the nearest tenth, this is 2.3 seconds. We discard the other solution, $-\frac{\sqrt{85}}{4}$, because a negative time does not make sense in this context.   ■

EXAMPLE 7

Manufacturing. A manufacturer of television parts receives an order for 52-inch picture tubes, measured along the diagonal, as shown in Figure 6-8. The tubes are to be rectangular in shape and 4 inches wider than they are high. Find the dimensions of the tube.

FIGURE 6-8

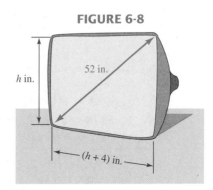

h in.

52 in.

$(h + 4)$ in.

ANALYZE THE PROBLEM   We need to find the height and width of the rectangular picture tube. We note that two adjacent sides of the picture tube and a diagonal form a right triangle.

FORM AN EQUATION   We can let h represent the height of the picture tube. Then $h + 4$ will represent the width. Since two adjacent sides and a diagonal of the tube form a right triangle, we can use the Pythagorean theorem to form the equation.

$a^2 + b^2 = c^2$	The Pythagorean theorem.
$h^2 + (h + 4)^2 = 52^2$	Substitute h for a, $(h + 4)$ for b, and 52 for c.
$h^2 + h^2 + 8h + 16 = 2,704$	Use the FOIL method: $(h + 4)^2 = h^2 + 8h + 16$.
$2h^2 + 8h - 2,688 = 0$	Subtract 2,704 from both sides and combine like terms.
$h^2 + 4h - 1,344 = 0$	Divide both sides by 2.

SOLVE THE EQUATION   To solve $h^2 + 4h - 1,344 = 0$, we cannot use the square root method, and the factoring method looks difficult because of the cumbersome last term $(-1,344)$. We will use the quadratic formula.

$h = \dfrac{-b \pm \sqrt{b^2 - 4ac}}{2a}$	The quadratic formula.
$= \dfrac{-4 \pm \sqrt{(4)^2 - 4(1)(-1,344)}}{2(1)}$	Substitute 1 for a, 4 for b, and $-1,344$ for c.
$= \dfrac{-4 \pm \sqrt{16 + 5,376}}{2}$	Use a calculator to find $4(1)(-1,344)$.
$= \dfrac{-4 \pm \sqrt{5,392}}{2}$	Do the addition within the radical.
$\approx \dfrac{-4 \pm 73.430239}{2}$	Use a calculator to approximate $\sqrt{5,392}$.

$$h \approx \frac{-4 + 73.430239}{2} \quad \text{or} \quad h \approx \frac{-4 - 73.430239}{2}$$ Use a calculator to evaluate each solution.

$$\approx \frac{69.430239}{2} \qquad\qquad \approx \frac{-77.430239}{2}$$

$$h \approx 34.7151195 \qquad\qquad h \approx -38.7151195$$

STATE THE CONCLUSION   The width of each tube will be approximately 34.7 inches, and the length will be approximately $34.7 + 4 = 38.7$ inches. We discard the second solution, because the diagonal measure of a TV picture tube cannot be negative.

CHECK THE RESULT   Check the solution by substituting 34.7, 38.7, and 52 into the Pythagorean theorem.   ∎

EXAMPLE 8

Finance. If $\$P$ is invested at an annual rate r, it will grow to an amount of $\$A$ in n years according to the formula $A = P(1 + r)^n$. What interest rate is needed so that a \$5,000 investment will grow to \$5,618 after 2 years?

Solution   We can substitute 5,000 for P, 5,618 for A, and 2 for n in the formula and solve for r.

$A = P(1 + r)^n$	
$5,618 = 5,000(1 + r)^2$	
$5,618 = 5,000(1 + 2r + r^2)$	Use FOIL: $(1 + r)^2 = 1 + 2r + r^2$.
$5,618 = 5,000 + 10,000r + 5,000r^2$	Remove parentheses.
$0 = 5,000r^2 + 10,000r - 618$	Subtract 5,618 from both sides.

We can use a calculator and solve this equation with the quadratic formula.

$$r = \frac{-b \pm \sqrt{b^2 - 4ac}}{2a}$$

$$= \frac{-10{,}000 \pm \sqrt{10{,}000^2 - 4(5{,}000)(-618)}}{2(5{,}000)}$$

$$= \frac{-10{,}000 \pm \sqrt{100{,}000{,}000 + 12{,}360{,}000}}{10{,}000}$$

$$= \frac{-10{,}000 \pm \sqrt{112{,}360{,}000}}{10{,}000}$$

$$= \frac{-10{,}000 \pm 10{,}600}{10{,}000}$$

$$r = \frac{-10{,}000 + 10{,}600}{10{,}000} \quad \text{or} \quad r = \frac{-10{,}000 - 10{,}600}{10{,}000}$$

$$= \frac{600}{10{,}000} \qquad\qquad\qquad = \frac{-20{,}600}{10{,}000}$$

$$= 0.06 \qquad\qquad\qquad\quad = -2.06$$

$$r = 6\% \qquad\qquad\qquad\quad r = -206\%$$

The required rate is 6%. The rate of −206% has no meaning in this problem. ∎

STUDY SET

Section 6.7

VOCABULARY

In Exercises 1–4, fill in the blanks to make the statements true.

1. The general _____quadratic_____ equation is $ax^2 + bx + c = 0$.

2. The formula
$$x = \frac{-b \pm \sqrt{b^2 - 4ac}}{2a}$$
is called the _____quadratic_____ formula.

3. To _____solve_____ a quadratic equation means to find all the values of the variable that make the equation true.

4. $\sqrt{-16}$ is not a _____real_____ number.

CONCEPTS

In Exercises 5–10, fill in the blanks to make the statements true.

5. In the quadratic equation $ax^2 + bx + c = 0$, a cannot equal 0 .

6. Before we can determine a, b, and c for $x = 3x^2 - 1$, we must write the equation in _____quadratic_____ form.

7. In the quadratic equation $3x^2 - 5 = 0$, $a = $ 3 , $b = $ 0 , and $c = $ −5 .

8. In the quadratic equation $-4x^2 + 8x = 0$, $a = $ −4 , $b = $ 8 , and $c = $ 0 .

9. The formula for the area of a rectangle is $A = $ lw and the formula for the area of a triangle is $A = \frac{1}{2}bh$.

10. If a, b, and c are three sides of a right triangle and c is the hypotenuse, then $c^2 = $ $a^2 + b^2$.

11. In evaluating the numerator of
$$\frac{-5 \pm \sqrt{5^2 - 4(2)(1)}}{2(2)}$$
what operation should be performed first?
Evaluate 5^2.

12. Consider the expression
$$\frac{3 \pm 6\sqrt{2}}{3}$$
   a. How many terms does the numerator contain? 2
   b. What common factor do the terms have? 3
   c. Simplify the expression. $1 \pm 2\sqrt{2}$

13. A student used the quadratic formula to solve an equation and obtained

$$x = \frac{-3 \pm \sqrt{15}}{2}$$

a. How many solutions does the equation have?
2

b. What are they *exactly?*
$$\frac{-3 + \sqrt{15}}{2}, \frac{-3 - \sqrt{15}}{2}$$

c. Approximate them to the nearest hundredth.
$0.44, -3.44$

14. Write the steps of the strategy for solving quadratic equations in the proper order.

■ Use the quadratic formula.
■ Write the equation in $ax^2 + bx + c = 0$ form.
■ Use the factoring method.
■ Use the square root method.

square root, write, factor, quadratic formula

15. The solutions of a quadratic equation are

$$x = \frac{-1 \pm \sqrt{5}}{2}$$

Graph them on a number line.

16. The solutions of a quadratic equation are

$$x = 2 \pm \sqrt{3}$$

Graph them on a number line.

NOTATION

In Exercises 17–18, complete each solution.

17. Solve $x^2 - 5x - 6 = 0$.

$$x = \frac{-b \pm \sqrt{b^2 - 4ac}}{2a}$$

$$= \frac{-\left(\boxed{-5} \right) \pm \sqrt{(-5)^2 - 4(1)(-6)}}{2(1)}$$

$$= \frac{\boxed{5} \pm \sqrt{25 + \boxed{24}}}{2}$$

$$= \frac{5 \pm \sqrt{\boxed{49}}}{2}$$

$$x = \frac{\boxed{5} \pm 7}{2}$$

$$x = \frac{5 \boxed{+} 7}{2} = 6 \quad \text{or} \quad x = \frac{5 \boxed{-} 7}{2} = -1$$

18. Solve $3x^2 + 2x - 2 = 0$.

$$x = \frac{-b \pm \sqrt{b^2 - 4ac}}{2a}$$

$$= \frac{-2 \pm \sqrt{\left(\boxed{2} \right)^2 - 4(3)\left(\boxed{-2} \right)}}{2\left(\boxed{3} \right)}$$

$$= \frac{-2 \pm \sqrt{4 \boxed{+} 24}}{6}$$

$$= \frac{-2 \pm \sqrt{\boxed{28}}}{6}$$

$$= \frac{-2 \pm \boxed{2} \sqrt{7}}{6}$$

$$= \frac{\boxed{2} \left(-1 \pm \sqrt{7} \right)}{2 \cdot 3}$$

$$x = \frac{-1 \pm \sqrt{7}}{3}$$

19. What is **wrong** with this student's work?
Solve $x^2 + 4x - 5 = 0$.

$$x = -4 \pm \frac{\sqrt{16 - 4(1)(-5)}}{2}$$

The student didn't extend the fraction bar so that it underlines the complete numerator.

20. In reading

$$\frac{-b \pm \sqrt{b^2 - 4ac}}{2a}$$

we say, "The ___opposite (negative)___ of b, plus or
___minus___ the ___square___ root of b
___squared___ minus 4 ___times___ a times
c, all ___over___ $2a$."

PRACTICE

In Exercises 21–30, change each equation into quadratic form, if necessary, and find the values of a, b, and c. **Do not solve the equation.**

21. $x^2 + 4x + 3 = 0$ $a = 1, b = 4, c = 3$

22. $x^2 - x - 4 = 0$ $a = 1, b = -1, c = -4$

23. $3x^2 - 2x + 7 = 0$ $a = 3, b = -2, c = 7$

24. $4x^2 + 7x - 3 = 0$ $a = 4, b = 7, c = -3$

25. $4y^2 = 2y - 1$ $a = 4, b = -2, c = 1$

26. $2x = 3x^2 + 4$ $a = 3, b = -2, c = 4$

27. $x(3x - 5) = 2$ $a = 3, b = -5, c = -2$

28. $y(5y + 10) = 8$ $a = 5, b = 10, c = -8$

29. $7(x^2 + 3) = -14x$ $a = 7, b = 14, c = 21$

30. $(2a + 3)(a - 2) = (a + 1)(a - 1)$
$a = 1, b = -1, c = -5$

In Exercises 31–50, use the quadratic formula to find all real solutions.

31. $x^2 - 5x + 6 = 0$ 2, 3

32. $x^2 + 5x + 4 = 0$ $-1, -4$

33. $x^2 + 7x + 12 = 0$ $-3, -4$

34. $x^2 - x - 12 = 0$ $-3, 4$ *Could factor*

35. $2x^2 - x - 1 = 0$ $1, -\frac{1}{2}$

36. $2x^2 + 3x - 2 = 0$ $-2, \frac{1}{2}$

37. $3x^2 + 5x + 2 = 0$ $-1, -\frac{2}{3}$

38. $3x^2 - 4x + 1 = 0$ $1, \frac{1}{3}$

39. $4x^2 + 4x - 3 = 0$ $\frac{1}{2}, -\frac{3}{2}$

40. $4x^2 + 3x - 1 = 0$ $\frac{1}{4}, -1$

41. $x^2 + 3x + 1 = 0$ $\dfrac{-3 \pm \sqrt{5}}{2}$

42. $x^2 + 3x - 2 = 0$ $\dfrac{-3 \pm \sqrt{17}}{2}$

43. $3x^2 - x = 3$ $\dfrac{1 \pm \sqrt{37}}{6}$

44. $5x^2 = 3x + 1$ $\dfrac{3 \pm \sqrt{29}}{10}$

45. $x^2 + 5 = 2x$ no real solutions

46. $2x^2 + 3x = -3$ no real solutions

47. $x^2 = 1 - 2x$ $-1 \pm \sqrt{2}$

48. $x^2 = 4 + 2x$ $1 \pm \sqrt{5}$

49. $3x^2 = 6x + 2$ $\dfrac{3 \pm \sqrt{15}}{3}$

50. $3x^2 = -8x - 2$ $\dfrac{-4 \pm \sqrt{10}}{3}$

In Exercises 51–64, use the most convenient method to find all real solutions. If a solution contains a radical, give the exact solution and then approximate it to the nearest hundredth.

discuss how to choose method.

51. $(2y - 1)^2 = 25$ $-2, 3$

52. $m^2 + 14m + 49 = 0$ $-7, -7$

53. $2x^2 + x = 5$ $\dfrac{-1 \pm \sqrt{41}}{4}$; $-1.85, 1.35$

54. $2x^2 - x + 2 = 0$ no real solutions

55. $x^2 - 2x - 1 = 0$ $1 \pm \sqrt{2}$; $-0.41; 2.41$

56. $b^2 = 18$ $\pm 3\sqrt{2}$; ± 4.24

57. $x^2 - 2x - 35 = 0$ $-5, 7$

58. $x^2 + 5x + 3 = 0$ $\dfrac{-5 \pm \sqrt{13}}{2}$; $-4.30, -0.70$

59. $x^2 + 2x + 7 = 0$ no real solutions

60. $3x^2 - x = 1$ $\dfrac{1 \pm \sqrt{13}}{6}$; $-0.43, 0.77$

61. $4c^2 + 16c = 0$ $-4, 0$

62. $t^2 - 1 = 0$ ± 1

63. $18 = 3y^2$ $\pm \sqrt{6}$; ± 2.45

64. $25x - 50x^2 = 0$ $0, \frac{1}{2}$

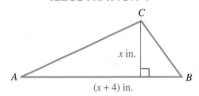

In Exercises 65–66, use a calculator to solve each equation. Round to the nearest tenth.

65. $2.4x^2 - 9.5x + 6.2 = 0$ 0.8, 3.1

66. $-1.7x^2 + 0.5x + 0.9 = 0$ $-0.6, 0.9$

APPLICATIONS

67. HEIGHT OF A TRIANGLE The triangle shown in Illustration 1 has an area of 30 square inches. Find its height.   6 in.

ILLUSTRATION 1

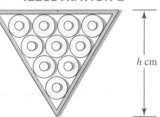

68. BOWLING When the pins for a children's bowling game are set up, they occupy 418 cm² of floor space. See Illustration 2. If the base of the triangular-shaped region is 6 cm longer than twice the height, how wide is the last row of pins?   44 cm

See appendix to find that form. for area of △ on p141

ILLUSTRATION 2

h cm

69. FLAGS According to the *Guinness Book of World Records 1998*, the largest flag flown from a flagpole was a Brazilian national flag, having an area of 3,102 ft². If the flag is 19 feet longer than it is wide, find its width and length.   47 ft by 66 ft

70. COMICS See Illustration 3. A comic strip occupies 96 square centimeters of space in a newspaper. The length of the rectangular space is 4 centimeters more than twice its width. Find its dimensions.
6 cm by 16 cm

ILLUSTRATION 3

71. COMMUNITY GARDENS See Illustration 4. Residents of a community can work their own 16 ft × 24 ft plot of city-owned land if they agree to the following stipulations:

- The area of the garden cannot exceed 180 square feet.

- A path of uniform width must be maintained around the garden.

Find the dimensions of the largest possible garden.
10 ft by 18 ft

ILLUSTRATION 4

24 ft

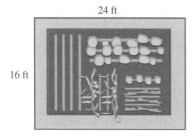

16 ft

72. DECKING The owner of the pool in Illustration 5 wants to surround it with a concrete deck of uniform width (shown in gray). If he can afford 368 square feet of decking, how wide can he make the deck?
4 ft

ILLUSTRATION 5

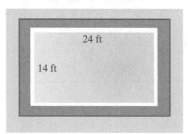

24 ft

14 ft

73. DAREDEVIL In 1873, Henry Bellini combined a tightrope walk over the Niagara River with a leap into the churning river below, where he was picked up by a boat. If the rope was 200 feet above the water, for how many seconds did he fall before hitting the water? Round to the nearest tenth.   3.5 sec

74. FALLING OBJECTS A tourist drops a penny from the observation deck of the World Trade Center, 1,377 feet above the ground. How long will it take for the penny to hit the ground?   about 9.3 sec

75. ABACUS The Chinese abacus shown in Illustration 6 consists of a frame, parallel wires, and beads that are moved to perform arithmetic computations. The frame is 21 centimeters wider than it is high. Find its dimensions.   15 cm by 36 cm

ILLUSTRATION 6

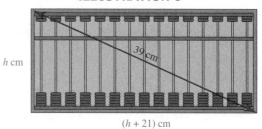

h cm

39 cm

$(h + 21)$ cm

76. INSTALLING A SIDEWALK A 170-meter-long sidewalk from the mathematics building M to the student center C is shown in Illustration 7. However, students prefer to walk directly from M to C. How long are the two segments of the existing sidewalk?
50 m and 120 m

ILLUSTRATION 7

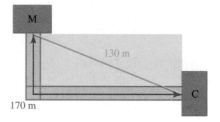

M

130 m

C

170 m

77. NAVIGATION Two boats leave port at the same time, one sailing east and one sailing south. If one boat sails 10 nautical miles more than the other and they are then 50 nautical miles apart, how far does each boat sail?   30 and 40 nautical miles

78. NAVIGATION One plane heads west from an airport, flying at 200 mph. One hour later, a second plane heads north from the same airport, flying at the same speed. When will the planes be 1,000 miles apart?
3 hr after the second plane takes off.

79. INVESTING We can use the formula $A = P(1 + r)^2$ to find the amount A that P will become when invested at an annual rate of r% for 2 years. What interest rate is needed for $5,000 to grow to $5,724.50 in 2 years?   7%

80. INVESTING What interest rate is needed for $7,000 to grow to $8,470 in 2 years? See Exercise 79.   10%

81. MANUFACTURING An electronics firm has found that its revenue for manufacturing and selling x television sets is given by the formula $R = -\frac{1}{6}x^2 + 450x$. How much revenue will be earned by manufacturing 600 television sets? (*Hint:* Multiply both sides of the equation by −6.) $210,000

82. RETAILING When a wholesaler sells n CD players, his revenue R is given by the formula $R = 150n - \frac{1}{2}n^2$. How many players would he have to sell to receive $11,250? (*Hint:* Multiply both sides of the equation by −2.)   150

83. METAL FABRICATION A piece of tin, 12 inches on a side, is to have four equal squares cut from its corners, as shown in Illustration 8. If the edges are then to be folded up to make a box with a floor area of 64 square inches, find the depth of the box. 2 in.

ILLUSTRATION 8

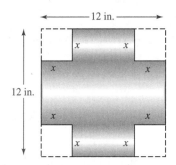

84. MAKING GUTTERS A piece of sheet metal, 18 inches wide, is bent to form the gutter shown in Illustration 9. If the cross-sectional area is 36 square inches, find the depth of the gutter. 3 in. or 6 in.

ILLUSTRATION 9

WRITING

85. Why is the quadratic formula useful in solving some types of quadratic equations?

86. Explain the meaning of the $\pm$ symbol.

87. Use the quadratic formula to solve $x^2 - 2x - 4 = 0$. What is an exact solution, and what is an approximate solution of this equation? Explain the difference.

88. The binomial $b^2 - 4ac$ is called the **discriminant.** From its value (be it positive, negative, or zero), you can predict whether the solutions of a given quadratic equation are real or nonreal numbers. Explain.

REVIEW

In Exercises 89–90, solve each equation for the indicated variable.

89. $A = p + prt$; for r $r = \dfrac{A - p}{pt}$

90. $F = \dfrac{GMm}{d^2}$, for M $M = \dfrac{Fd^2}{Gm}$

In Exercises 91–92, write the equation of the line with the given properties in general form.

91. Slope of $\frac{3}{5}$ and passing through $(0, 12)$
  $3x - 5y = -60$

92. Passes through $(6, 8)$ and the origin $4x - 3y = 0$

In Exercises 93–96, simplify each expression.

93. $\sqrt{80}$ $4\sqrt{5}$

94. $2\sqrt{x^3 y^2}$ $2xy\sqrt{x}$

95. $\dfrac{x}{\sqrt{7x}}$ $\dfrac{\sqrt{7x}}{7}$

96. $\dfrac{\sqrt{x} + 2}{\sqrt{x} - 2}$ $\dfrac{x + 4\sqrt{x} + 4}{x - 4}$

▶ 6.8   Graphing Quadratic Functions

In this section, you will learn about

> Quadratic functions ■ Finding the vertex and the intercepts of a parabola ■ A strategy for graphing quadratic functions ■ Finding a maximum value ■ Solving quadratic equations graphically

Introduction In Chapter 4, we discussed polynomial functions. We now consider a special type of polynomial function called a *quadratic function*. When graphing polynomial functions, we constructed a table of values and plotted points. In this section, we will develop a more general strategy for graphing quadratic functions by analyzing the given function and determining the important characteristics of its graph.

FIGURE 6-9

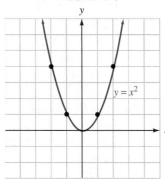

$y = x^2$

Quadratic Functions

Quadratic functions are defined by equations of the form $y = ax^2 + bx + c$ $(a \neq 0)$ where the right-hand side is a second-degree polynomial in the variable x. Three examples of quadratic functions are

$$y = x^2 - 3 \qquad y = x^2 - 2x - 3 \qquad y = -2x^2 - 4x + 2$$

We can replace y with the function notation $f(x)$ to express the defining equation in the form $f(x) = ax^2 + bx + c$. For the functions above, we can write

$$f(x) = x^2 - 3 \qquad f(x) = x^2 - 2x - 3 \qquad f(x) = -2x^2 - 4x + 2$$

In Section 3.2, we constructed the graph of $y = x^2$ by plotting points. The result was the **parabola** shown in Figure 6-9.

EXAMPLE 1

Graphing quadratic functions by plotting points. Graph $y = x^2 - 3$. Compare the graph to that of $y = x^2$.

Solution The function is written in $y = ax^2 + bx + c$ form, where $a = 1$, $b = 0$, and $c = -3$. To find ordered pairs (x, y) that satisfy the equation, we pick several numbers x and find the corresponding values of y. If we let $x = 3$, we have

$$y = x^2 - 3$$
$$= 3^2 - 3 \quad \text{Substitute 3 for } x.$$
$$= 6$$

The ordered pair $(3, 6)$ and six others satisfying the equation appear in the table shown in Figure 6-10. To graph the equation, we plot each point and draw a smooth curve passing through them. The resulting parabola is the graph of $y = x^2 - 3$. The parabola opens upward, and the lowest point on the graph, called the **vertex of the parabola,** is the point $(0, -3)$.

Note that the graph of $y = x^2 - 3$ looks just like the graph of $y = x^2$, except that it is 3 units lower.

FIGURE 6-10

$$y = x^2 - 3$$

x	y	(x, y)
3	6	$(3, 6)$
2	1	$(2, 1)$
1	-2	$(1, -2)$
0	-3	$(0, -3)$
-1	-2	$(-1, -2)$
-2	1	$(-2, 1)$
-3	6	$(-3, 6)$

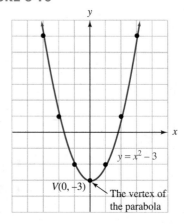

$y = x^2 - 3$

$V(0, -3)$
The vertex of the parabola

SELF CHECK Graph $y = x^2 + 2$. Compare the graph to that of $y = x^2$.

Answer: The graph has the same shape as the graph of $y = x^2$, but it is 2 units higher.

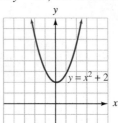

$y = x^2 + 2$

If we draw a vertical line through the vertex of a parabola and fold the graph on this line, the two sides of the graph will match. We call the vertical line the **axis of symmetry.**

EXAMPLE 2 **Graphing quadratic functions.** Graph $f(x) = -2x^2 - 4x + 2$, find its vertex, and draw its axis of symmetry.

Solution The function is written in $f(x) = ax^2 + bx + c$ form, where $a = -2$, $b = -4$, and $c = 2$. We construct the table shown in Figure 6-11, plot the points, and draw the graph.

FIGURE 6-11

$f(x) = -2x^2 - 4x + 2$

x	$f(x)$	$(x, f(x))$
-3	-4	$(-3, -4)$
-2	2	$(-2, 2)$
-1	4	$(-1, 4)$
0	2	$(0, 2)$
1	-4	$(1, -4)$

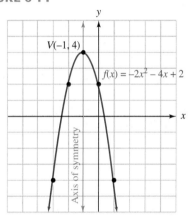

The parabola opens downward, so its vertex is its highest point, the point $(-1, 4)$.

SELF CHECK Graph $f(x) = -x^2 - 4x - 4$, find its vertex, and draw its axis of symmetry.

Answer: The vertex is at $(-2, 0)$.

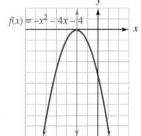

In Example 1, the coefficient of the x^2 term in $y = x^2 - 3$ is positive ($a = 1$). In Example 2, the coefficient of the x^2 term in $f(x) = -2x^2 - 4x + 2$ is negative ($a = -2$). The results of these first two examples suggest the following fact.

Graphs of quadratic functions

The graph of the function $y = ax^2 + bx + c$ or $f(x) = ax^2 + bx + c$ ($a \neq 0$) is a parabola. It opens upward when $a > 0$ and downward when $a < 0$.

The cup-like shape of a parabola can be seen in a wide variety of real-world settings. Some examples are shown in Figure 6-12.

FIGURE 6-12

The path of a thrown object

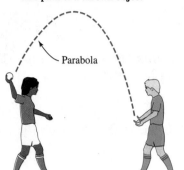

The pursuit path of a shark seeking its prey

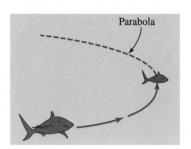

The shape of a satellite antenna dish

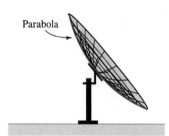

The path of a stream of water

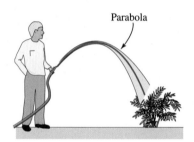

Finding the Vertex and the Intercepts of a Parabola

It is easier to graph a quadratic function when we know the coordinates of the vertex of its parabolic graph. For a parabola defined by $y = ax^2 + bx + c$ or $f(x) = ax^2 + bx + c$, it can be shown that the x-coordinate of the vertex is given by $-\frac{b}{2a}$. This fact enables us to find the coordinates of its vertex.

Finding the vertex of a parabola

> The graph of the quadratic function $y = ax^2 + bx + c$ or $f(x) = ax^2 + bx + c$ is a parabola whose vertex has an x-coordinate of $-\frac{b}{2a}$. To find the y-coordinate of the vertex, substitute $-\frac{b}{2a}$ into the defining equation and find y.

EXAMPLE 3

Finding the vertex of a parabola. Find the vertex of the parabola defined by $y = x^2 - 2x - 3$.

Solution   For $y = x^2 - 2x - 3$, we have $a = 1$, $b = -2$, and $c = -3$. To find the x-coordinate of the vertex, we substitute the values for a and b into the formula $x = -\frac{b}{2a}$.

$$x = -\frac{b}{2a}$$

$$x = -\frac{-2}{2(1)}$$

$$= 1$$

The x-coordinate of the vertex is $x = 1$. To find the y-coordinate, we substitute 1 for x:

$$y = x^2 - 2x - 3$$
$$y = 1^2 - 2(1) - 3$$
$$= 1 - 2 - 3$$
$$= -4$$

The vertex of the parabola is the point $(1, -4)$. See Figure 6-13(a) on the next page.

SELF CHECK Find the vertex of the parabola defined by $y = -x^2 + 6x - 8$. *Answer:* $(3, 1)$ ∎

When graphing quadratic functions, it is often helpful to find the x- and y-intercepts of the parabola.

EXAMPLE 4

Finding the intercepts of a parabola. Find the x- and y-intercepts of the parabola defined by $y = x^2 - 2x - 3$.

Solution To find the y-intercept of the parabola, we let $x = 0$ and solve for y.

$$y = x^2 - 2x - 3$$
$$y = 0^2 - 2(0) - 3$$
$$y = -3$$

The parabola passes through the point $(0, -3)$. We note that the y-coordinate of the y-intercept is the same as the value of the constant term c on the right-hand side of $y = x^2 - 2x - 3$.

To find the x-intercepts of the graph, we set y equal to 0 and solve the resulting quadratic equation.

$$y = x^2 - 2x - 3$$
$$0 = x^2 - 2x - 3 \qquad \text{Substitute 0 for } y.$$
$$0 = (x - 3)(x + 1) \qquad \text{Factor the trinomial.}$$
$$x - 3 = 0 \quad \text{or} \quad x + 1 = 0 \qquad \text{Set each factor equal to 0.}$$
$$x = 3 \qquad\qquad x = -1$$

Since there are two solutions, the graph has two x-intercepts: $(3, 0)$ and $(-1, 0)$. See Figure 6-13(a) on the next page.

SELF CHECK Find the x- and y-intercepts of the parabola defined by $y = -x^2 + 6x - 8$. *Answers:* y-intercept: $(0, -8)$ x-intercepts: $(2, 0)$, $(4, 0)$ ∎

A Strategy for Graphing Quadratic Functions

There is another way to graph quadratic functions. Rather than plotting random points, we can determine the important characteristics of a parabola. For example, to graph $y = x^2 - 2x - 3$, we note that the coefficient of the x^2 term is positive ($a = 1$). Therefore, the parabola defined by this function opens upward. In Examples 3 and 4, we found that the vertex of the graph of $y = x^2 - 2x - 3$ is at $(1, -4)$ and that it has a y-intercept of $(0, -3)$ and the x-intercepts of $(3, 0)$ and $(-1, 0)$. See Figure 6-13(a).

We can locate other points on the parabola by noting that the graph has the axis of symmetry shown in Figure 6-13(a). If the point $(0, -3)$, which is 1 unit to the left of the axis of symmetry, is on the graph, the point $(2, -3)$, which is 1 unit to the right of the axis of symmetry, is also on the graph.

FIGURE 6-13

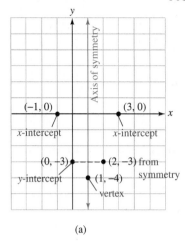

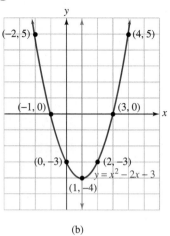

(a)   (b)

$y = x^2 - 2x - 3$

x	y	(x, y)
-2	5	$(-2, 5)$

We can complete the graph by plotting two more points. If $x = -2$, then $y = 5$, and the parabola passes through $(-2, 5)$. Again using symmetry, the parabola must also pass through $(4, 5)$. The completed graph of $y = x^2 - 2x - 3$ is shown in Figure 6-13(b).

Much can be determined about the graph of $y = ax^2 + bx + c$ from the coefficients a, b, and c. This information is summarized below.

Graphing a quadratic function
$y = ax^2 + bx + c \ (a \neq 0)$

Determine whether the parabola opens upward or downward by examining a.

The x-coordinate of the vertex of the parabola is $x = -\frac{b}{2a}$.

To find the y-coordinate of the vertex, substitute $-\frac{b}{2a}$ for x into the equation and find y.

The axis of symmetry is the vertical line passing throught the vertex.

The y-intercept $(0, y)$ is determined by the value of y attained when $x = 0$: the y-intercept is $(0, c)$.

The x-intercepts (if any) are determined by the numbers x that make $y = 0$. To find them, solve the quadratic equation $ax^2 + bx + c = 0$.

EXAMPLE 5

Graphing quadratic functions. Graph $f(x) = -2x^2 - 8x - 8$.

Solution   **Step 1:** *Determine whether the parabola opens upward or downward.* The equation is in the form $f(x) = ax^2 + bx + c$, with $a = -2$, $b = -8$, and $c = -8$. Since $a < 0$, the parabola opens downward.

Step 2: *Find the vertex and draw the axis of symmetry.* To find the x-coordinate of the vertex, we substitute the values for a and b into the formula $x = -\frac{b}{2a}$.

$$x = -\frac{b}{2a}$$

$$x = -\frac{-8}{2(-2)}$$

$$= -2$$

The x-coordinate of the vertex is -2. To find the y-coordinate, we substitute -2 for x in the equation and find $f(x)$.

$$f(x) = -2x^2 - 8x - 8$$
$$f(-2) = -2(-2)^2 - 8(-2) - 8$$
$$= -8 + 16 - 8$$
$$= 0 \qquad \text{If } f(-2) = 0, \text{ then } y = 0 \text{ for } x = -2.$$

The vertex of the parabola is the point $(-2, 0)$. This point is the blue dot in Figure 6-14.

Step 3: *Find the x- and y-intercepts.* Since $c = -8$, the y-intercept of the parabola is $(0, -8)$. The point $(-4, -8)$, two units to the left of the axis of symmetry, must also be on the graph. We plot both points in black in Figure 6-14.

To find the x-intercepts, we set $f(x)$ equal to 0 and solve the resulting quadratic equation.

$$f(x) = -2x^2 - 8x - 8$$
$$0 = -2x^2 - 8x - 8 \quad \text{Set } f(x) = 0.$$
$$0 = x^2 + 4x + 4 \qquad \text{Divide both sides by } -2.$$
$$= (x + 2)(x + 2) \quad \text{Factor the trinomial.}$$
$$x + 2 = 0 \quad \text{or} \quad x + 2 = 0 \qquad \text{Set each factor equal to 0.}$$
$$x = -2 \qquad\qquad x = -2$$

Since the solutions are the same, the graph has only one x-intercept: $(-2, 0)$. This point is the vertex of the parabola and has already been plotted.

Step 4: *Plot another point.* Finally, we find another point on the parabola. If $x = -3$, then $y = -2$. We plot $(-3, -2)$ in Figure 6-14 and use symmetry to determine that $(-1, -2)$ is also on the graph. Both points are in green.

FIGURE 6-14

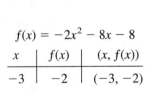

$$f(x) = -2x^2 - 8x - 8$$

x	$f(x)$	$(x, f(x))$
-3	-2	$(-3, -2)$

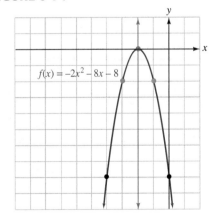

SELF CHECK   Graph the function $y = -x^2 + 6x - 8$.   *Answer:*

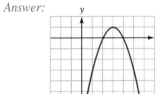

$y = -x^2 + 6x - 8$ ∎

Finding a Maximum Value

E X A M P L E   6

Finding maximum revenue. An electronics firm manufactures radios. Over the past 10 years, the firm has learned that it can sell x radios at a price of $\left(200 - \frac{1}{5}x\right)$ dollars. How many radios should the firm manufacture and sell to maximize its revenue? Find the maximum revenue.

Solution   The revenue obtained is the product of the number of radios sold (x) and the price of each radio $\left(200 - \frac{1}{5}x\right)$. Thus, the revenue R is given by the function

$$R = x\left(200 - \frac{1}{5}x\right) \qquad \text{or} \qquad R = -\frac{1}{5}x^2 + 200x$$

Since the graph of this function is a parabola that opens downward, the *maximum* value of R will be the value of R determined by the vertex of the parabola. Because the x-coordinate of the vertex is at $x = -\frac{b}{2a}$, we have

$$x = -\frac{b}{2a}$$

$$= -\frac{200}{2\left(-\frac{1}{5}\right)} \qquad \text{Substitute 200 for } b \text{ and } -\frac{1}{5} \text{ for } a.$$

$$= -\frac{200}{-\frac{2}{5}} \qquad \text{Do the multiplication in the denominator.}$$

$$= (-200)\left(-\frac{5}{2}\right) \qquad \text{Division by } -\frac{2}{5} \text{ is the same as multiplication by its reciprocal, which is } -\frac{5}{2}.$$

$$= 500$$

If the firm manufactures 500 radios, the maximum revenue will be

$$R = -\frac{1}{5}x^2 + 200x \qquad \text{The revenue formula.}$$

$$= -\frac{1}{5}(500)^2 + 200(500) \qquad \text{Substitute 500 for } x, \text{ the number of radios.}$$

$$= 50,000$$

The firm should manufacture 500 radios to get a maximum revenue of $50,000. This fact is verified by examining the graph of $R = -\frac{1}{5}x^2 + 200x$, which appears in Figure 6-15.

FIGURE 6-15

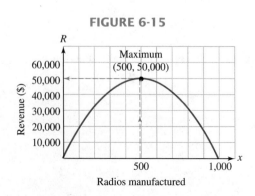

Radios manufactured

Solving Quadratic Equations Graphically

We can also solve quadratic equations graphically. For example, the solutions of $x^2 - 6x + 8 = 0$ are the numbers x that will make y equal to 0 in the quadratic function $y = x^2 - 6x + 8$. To find these numbers, we inspect the graph of $y = x^2 - 6x + 8$ and locate the points on the graph with a y-coordinate of 0. In Figure 6-16, these points are $(2, 0)$ and $(4, 0)$, the x-intercepts of the graph. We can conclude that the x-coordinates of the x-intercepts, $x = 2$ and $x = 4$, are the solutions of $x^2 - 6x + 8 = 0$.

FIGURE 6-16

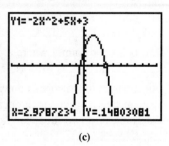

ACCENT ON TECHNOLOGY *Solving Quadratic Equations Graphically*

The solutions of $-2x^2 + 5x + 3 = 0$ are the numbers x that will make y equal to 0 in the quadratic function $y = -2x^2 + 5x + 3$. To find these numbers, we can graph the function and read the x-intercepts of the graph.

Using the standard window settings, we press the $\boxed{Y=}$ key and enter $-2x^2 + 5x + 3$. The screen should read

$$Y_1 = -2x\verb|^|2 + 5x + 3$$

The resulting graph is shown in Figure 6-17(a). We then trace and move the cursor to identify each x-intercept. From the calculator, we can then find the x-coordinate of the x-intercept. See Figure 6-17(b) and 6-17(c), where the approximate solutions to the quadratic equation $-2x^2 + 5x + 3 = 0$ are the x-values shown on the screen.

FIGURE 6-17

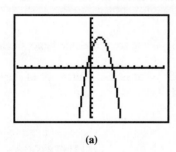

(a)

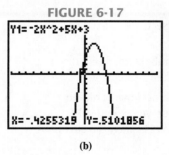

(b)

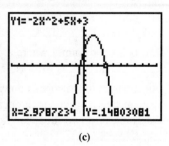

(c)

For better results, we can use the ZOOM key. We would then conclude that the two solutions of $-2x^2 + 5x + 3 = 0$ are

$$x = -0.5 \qquad \text{and} \qquad x = 3$$

Another way to find the solutions of a quadratic equation is by using the zero feature of the calculator. With this feature, the x-coordinate of the x-intercepts of the graph of $y = -2x^2 + 5x + 3$ can be found. The *zeros*, in this case -0.5 and 3, are solutions of $-2x^2 + 5x + 3 = 0$. Consult your owner's manual for the specific instructions to use this feature.

When solving quadratic equations graphically, there are three possibilities to consider. If the graph of the associated quadratic function has two x-intercepts, the quadratic equation has two real-number solutions. Figure 6-18(a) shows an example of this. If the graph has one x-intercept, as shown in Figure 6-18(b), the equation has one real-number solution. Finally, if the graph does not have an x-intercept, as shown in Figure 6-18(c), the equation does not have any real-number solutions.

FIGURE 6-18

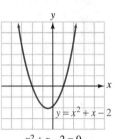

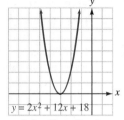

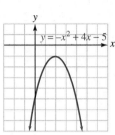

$x^2 + x - 2 = 0$ has two solutions, $x = -2$ and $x = 1$.	$2x^2 + 12x + 18 = 0$ has one solution, $x = -3$.	$-x^2 + 4x - 5 = 0$ has no real-number solutions.
(a)	(b)	(c)

STUDY SET

Section 6.8

VOCABULARY

In Exercises 1–6, fill in the blanks to make the statements true.

1. A function defined by the equation $y = ax^2 + bx + c$ ($a \neq 0$) is called a ___quadratic___ function.

2. The lowest (or highest) point on a parabola is called the ___vertex___ of the parabola.

3. The point where a parabola intersects the y-axis is called the ___y-intercept___.

4. The point (or points) where a parabola intersects the ___x-axis___ is (are) called the x-intercept(s).

5. For a parabola that opens upward or downward, the vertical line that passes through its vertex and splits the graph into two identical pieces is called the axis of ___symmetry___.

6. For the graph of $y = ax^2 + bx + c$, the ___coefficient___ of the x^2 term indicates whether the parabola opens upward or downward.

CONCEPTS

In Exercises 7–10, fill in the blanks to make the statements true.

7. The graph of $y = ax^2 + bx + c$ ($a \neq 0$) opens upward when $a > 0$.

8. The graph of $f(x) = ax^2 + bx + c$ ($a \neq 0$) opens downward when $a < 0$.

9. The y-intercept of the graph of $f(x) = ax^2 + bx + c$ is the point $(0, c)$.

10. The x-coordinate of the vertex of the parabola that results when we graph $y = ax^2 + bx + c$ is $x = -\frac{b}{2a}$.

11. Refer to the graph in Illustration 1.
    a. What do we call the curve shown there?
       a parabola
    b. What are the x-intercepts of the graph?
       $(1, 0), (3, 0)$
    c. What is the y-intercept of the graph? $(0, -3)$
    d. What is the vertex? $(2, 1)$

ILLUSTRATION 1

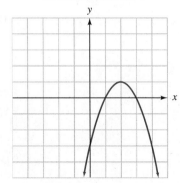

12. The vertex of a parabola is at $(1, -3)$, its y-intercept is $(0, -2)$, and it passes through the point $(3, 1)$, as shown in Illustration 2. Draw the axis of symmetry and use it to help determine two other points on the parabola.

$(2, -2), (-1, 1)$

ILLUSTRATION 2

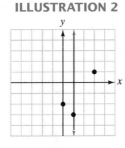

13. Sketch the graphs of parabolas with zero, one, and two x-intercepts.

14. Sketch the graph of a parabola that doesn't have a y-intercept, if possible.   not possible

15. HEALTH DEPARTMENT The number of cases of flu seen by doctors at a county health clinic each week during a 10-week period is described by the quadratic function graphed in Illustration 3. Write a brief summary report about the flu outbreak. What important piece of information does the vertex give?

The most cases of flu (25) were reported the fifth week.

ILLUSTRATION 3

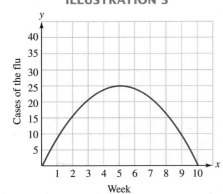

16. COST ANALYSIS A company has found that when it assembles x carburetors in a production run, the manufacturing cost \$$y$ per carburetor is given by the quadratic function graphed in Illustration 4. What important piece of information does the vertex give?

The cost to manufacture a carburetor is lowest (\$100) for a production run of 30 units.

ILLUSTRATION 4

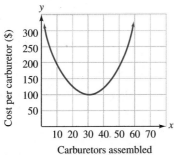

17. The graph of $f(x) = -x^2 + 4x - 3$ is shown in Illustration 5. What are the solutions of $-x^2 + 4x - 3 = 0$?   1, 3

ILLUSTRATION 5

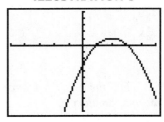

18. The graph of $y = 2x^2 - 4x - 6$ is shown in Illustration 6. What are the solutions of $2x^2 - 4x - 6 = 0$? $-1, 3$

ILLUSTRATION 6

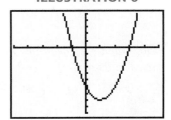

NOTATION

19. Tell whether this statement is true or false: The equations $y = 2x^2 - x - 2$ and $f(x) = 2x^2 - x - 2$ are the same.   true

20. The function $y = -x^2 + 3x - 5$ is written in $y = ax^2 + bx + c$ form. What are a, b, and c?

$-1, 3, -5$

PRACTICE

In Exercises 21–24, graph each quadratic function and compare the graph to the graph of $y = x^2$.

21. $y = x^2 + 1$
moved up 1

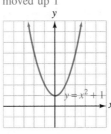

22. $y = x^2 - 4$
moved down 4

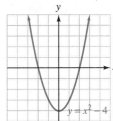

23. $f(x) = -x^2$
opens the opposite direction

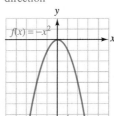

24. $f(x) = (x - 1)^2$
moved 1 to the right

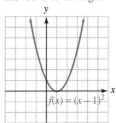

In Exercises 25–28, find the vertex of the graph of each quadratic function.

25. $y = -x^2 + 6x - 8$ $(3, 1)$

26. $y = -x^2 - 2x - 1$ $(-1, 0)$

27. $f(x) = 2x^2 - 4x + 1$ $(1, -1)$

28. $f(x) = 2x^2 + 8x - 4$ $(-2, -12)$

In Exercises 29–32, find the x- and y-intercepts of the graph of the quadratic function.

29. $f(x) = x^2 - 2x + 1$ $(1, 0); (0, 1)$

30. $f(x) = 2x^2 - 4x$ $(0, 0), (2, 0); (0, 0)$

31. $y = -x^2 - 10x - 21$ $(-3, 0), (-7, 0); (0, -21)$

32. $y = 3x^2 + 6x - 9$ $(-3, 0), (1, 0); (0, -9)$

In Exercises 33–50, graph each quadratic function. Use the method discussed in Example 5.

33. $y = x^2 - 2x$

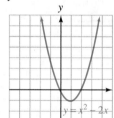

34. $f(x) = -x^2 - 4x$

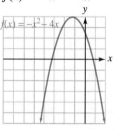

35. $f(x) = -x^2 + 2x$

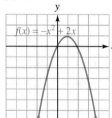

36. $y = x^2 + x$ solve by factoring also

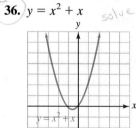

37. $f(x) = x^2 + 4x + 4$

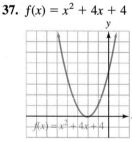

38. $f(x) = x^2 - 6x + 9$

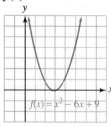

39. $y = -x^2 - 2x - 1$

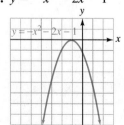

40. $y = -x^2 + 2x - 1$ solve by factoring also

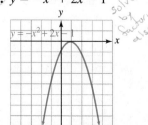

41. $y = x^2 + 2x - 3$

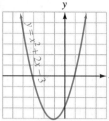

42. $y = x^2 + 6x + 5$

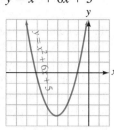

43. $f(x) = 2x^2 + 8x + 6$

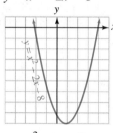

44. $f(x) = 3x^2 - 12x + 9$

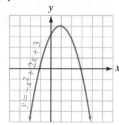

45. $y = x^2 - 2x - 8$

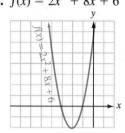

46. $y = -x^2 + 2x + 3$

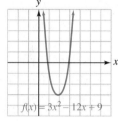

47. $y = x^2 - x - 2$

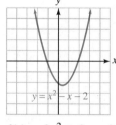

48. $y = -x^2 + 5x - 4$

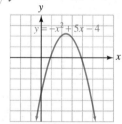

49. $f(x) = 2x^2 + 3x - 2$

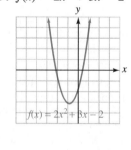

50. $f(x) = 3x^2 - 7x + 2$

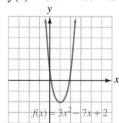

In Exercises 51–60, solve each quadratic equation using a graphing calculator.

51. $x^2 - 3x + 2 = 0$ $1, 2$

52. $x^2 + x - 6 = 0$ $-3, 2$

53. $-4x^2 - 8x - 3 = 0$ $-1.5, -0.5$

54. $16x^2 + 8x - 3 = 0$ $-0.75, 0.25$

55. $-6x^2 - 11x + 10 = 0$ $-2.5, 0.67$

56. $5x^2 + 11x = 0$ $-2.2, 0$

57. $0.5x^2 - 0.7x - 3 = 0$ $\quad -1.85, 3.25$

58. $x^2 - 2x + 1 = 0$ $\quad 1, 1$

59. $x^2 - x + 0.25 = 0$ $\quad 0.5, 0.5$

60. $-16x^2 + 56x - 49 = 0$ $\quad 1.75, 1.75$

APPLICATIONS

61. REFLECTIVE PROPERTY OF PARABOLAS In
Illustration 7, plot (0, 0), (1, 0.2), (2, 0.8), (3, 1.8),
(4, 3.2), (4.5, 4.1), (−1, 0.2), (−2, 0.8), (−3, 1.8),
(−4, 3.2), and (−4.5, 4.1) and connect the points with
a smooth curve. This is a side view of a mirrored re-
flector of a flashlight. Draw
the parabola's axis of symmetry. Plot the point (0, 1.25),
which locates the filament of
the lightbulb. Finally, draw a
line representing a "ray" of
light coming from the bulb,
striking the mirrored surface
at (2, 0.8) and then reflecting
outward, parallel to the axis
of symmetry.

ILLUSTRATION 7

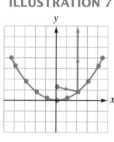

62. PARABOLIC DISH In Illustration 8, plot (0, 0),
(0.1, 1), (0.4, 2), (0.9, 3), (1.6, 4), (2.5, 5), (0.1, −1),
(0.4, −2), (0.9, −3), (1.6, −4), and (2.5, −5) and
connect the points with a smooth curve. This is a side
view of a parabolic dish used to pick up and amplify
sound waves. In this case, the parabola opens to the
right. Draw the axis of
symmetry. Plot the point
(2.5, 0), which locates the
microphone that picks up
the reflected sound waves.
Finally, draw a line representing a single wave of
sound, coming *into* the dish,
parallel to the axis of symmetry. Show it striking the
dish at (0.9, −3) and re-
flecting into the micro-
phone.

ILLUSTRATION 8

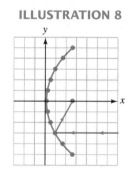

63. BRIDGES The shapes of the suspension cables in
certain types of bridges are parabolic. The suspension
cable for the bridge shown in Illustration 9 is described by $y = 0.005x^2$. Finish the mathematical
model of the bridge by completing the table of values
using a calculator, plotting the points, and drawing a
smooth curve through them to represent the cable.
Finally, from each plotted point, draw a vertical support cable attached to the roadway.

x	−80	−60	−40	−20	0	20	40	60	80
y	32	18	8	2	0	2	8	18	32

ILLUSTRATION 9

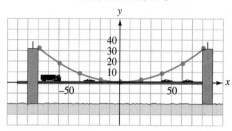

64. PROJECTILES If we disregard air resistance and
other outside factors, the path of a projectile, such as
a kicked soccer ball, is parabolic. Suppose the path
of the soccer ball after it is kicked is given by the
quadratic function $y = -0.5x^2 + 2x$. Use a calculator
to complete the table of values in Illustration 10 and
then plot the points and draw a smooth curve through
them to depict the ball's path.

x	0	0.5	1	1.5	2	2.5	3	3.5	4
y	0	0.875	1.5	1.875	2	1.875	1.5	0.875	0

ILLUSTRATION 10

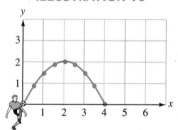

65. SELLING TVS A company has found that it can sell
x TVs at a price of $\$\left(450 - \frac{1}{6}x\right)$.
  a. How many TVs must the company sell to maxi-
     mize its revenue? $\quad 1,350$
  b. Find the maximum revenue. $\quad \$303,750$

66. SELLING CD PLAYERS A wholesaler sells CD
players for $150 each. However, she gives volume
discounts on purchases of 500 to 1,000 units accord-
ing to the formula $\left(150 - \frac{1}{10}n\right)$, where n represents
the number of units purchased.
  a. How many units would a retailer have to buy
     for the wholesaler to obtain maximum revenue?
     $\quad 750$
  b. Find the maximum revenue. $\quad \$56,250$

WRITING

67. Explain why the y-intercept of the graph of
$y = ax^2 + bx + c$ $(a \neq 0)$ is $(0, c)$.

68. Use the example of a stream of water from a drinking fountain to explain the concept of the vertex of a
parabola.

69. Explain why parabolas that open left or right aren't the graphs of functions.

70. Is it possible for the graph of a parabola not to have a *y*-intercept? Explain.

REVIEW

In Exercises 71–74, simplify each expression.

71. $\sqrt{12} + \sqrt{27}$ $5\sqrt{3}$

72. $3\sqrt{6y}\left(-4\sqrt{3y}\right)$
$-36y\sqrt{2}$

73. $\left(\sqrt{3} + 1\right)\left(\sqrt{3} - 1\right)$ 2

74. $\left(\sqrt{x} + 2\right)^2$
$x + 4\sqrt{x} + 4$

Quadratic Equations

In this chapter, we have studied several ways to solve quadratic equations. We have also graphed quadratic functions and seen that their graphs are parabolas.

What Is a Quadratic Equation?

A quadratic equation can be written in the form $ax^2 + bx + c = 0$ where $a \neq 0$. In Exercises 1–12, which are quadratic equations?

1. $y = 3x + 7$ no

2. $4(x + 5) = 2x$ no

3. $2x^2 - 3x + 4 = 0$ yes

4. $y(y - 6) = 0$ yes

5. $a^2 + 7a - 1 > 0$ no

6. $3y^2 - y + 4$ no

7. $5 = y - y^2$ yes

8. $|x - 8|$ no

9. $\sqrt{x + 7} = 4$ no

10. $x^2 = 16$ yes

11. $\dfrac{m}{2} - \dfrac{1}{3} = \dfrac{1}{4}$ no

12. $C = \dfrac{5}{9}(F - 32)$ no

Solving Quadratic Equations

The techniques we have used to solve linear equations cannot be used to solve a quadratic equation, because those techniques cannot isolate the variable on one side of the equation. Exercises 13 and 14 show examples of student work to solve quadratic equations. In each case, what did the student do **wrong** and why is it incorrect?

13. Solve $x^2 = 6$.

$$\frac{x^2}{2} = \frac{6}{2}$$

$$x = 3$$

The student divided both sides by 2 and incorrectly thought that $\frac{x^2}{2}$ equals x.

14. Solve $x^2 - x = 10$.

$$x(x - 1) = 10$$
$$x = 10 \quad \text{or} \quad x - 1 = 10$$
$$x = 11$$

The student factored the left-hand side instead of first subtracting 10 from both sides and then factoring.

In Exercises 15–20, solve each quadratic equation using the method listed.

15. $4x^2 - x = 0$; factoring method $0, \frac{1}{4}$

16. $x^2 + 3x + 1 = 0$; quadratic formula $\dfrac{-3 \pm \sqrt{5}}{2}$

17. $x^2 = 36$; square root method ± 6

18. $a^2 - a - 56 = 0$; factoring method $8, -7$

19. $x^2 + 4x + 1 = 0$; complete the square $-2 \pm \sqrt{3}$

20. $(x + 3)^2 = 16$; square root method $1, -7$

Quadratic Functions

The graph of the quadratic function $y = ax^2 + bx + c \ (a \neq 0)$ is a parabola. It opens upward when $a > 0$ and downward when $a < 0$.

21. The formula $R = 4x - x^2$ gives the revenue R (in tens of thousands of dollars) that a business obtains from the manufacture and sale of x patio chairs (in hundreds). Graph $R = 4x - x^2$ in Illustration 1.

22. Find the vertex of the parabola. What is its significance concerning the revenue the business brings in?

Vertex (2, 4); the company should manufacture 200 chairs to get a maximum revenue of $40,000.

ILLUSTRATION 1

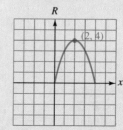

Accent on Teamwork

Section 6.1

Prime numbers We can use a procedure called the *sieve of Eratosthenes* to find all the prime numbers in the set of the first 100 whole numbers. Give each member in your group a copy of the table shown in Illustration 1. Cross out 1, since it is not a prime number by definition. Cross out any numbers divisible by 2, 3, 5, 7, or 11, because they have a factor of 2, 3, 5, 7, or 11 and thus would not be prime. Don't cross out 2, 3, 5, 7, or 11, because they are prime numbers. At the end of this process, you should end up with the first 25 prime numbers.

ILLUSTRATION 1

1	2	3	4	5	6	7	8	9	10
11	12	13	14	15	16	17	18	19	20
21	22	23	24	25	26	27	28	29	30
31	32	33	34	35	36	37	38	39	40
41	42	43	44	45	46	47	48	49	50
51	52	53	54	55	56	57	58	59	60
61	62	63	64	65	66	67	68	69	70
71	72	73	74	75	76	77	78	79	80
81	82	83	84	85	86	87	88	89	90
91	92	93	94	95	96	97	98	99	100

Section 6.2

Factoring trinomials Use colored paper to create the model shown in Illustration 2. Note that the total area of the figure is $(x + 1)(x + 1)$ square units. Disassemble the model and show that it is composed of four pieces with areas of x^2, x, x, and 1 square units. Label each piece with its area. What is the relationship between $x^2 + 2x + 1$ and $(x + 1)(x + 1)$? Next, create a model with an area of $(x + 2)(x + 2)$ square units. What observations do you have about this example?

ILLUSTRATION 2

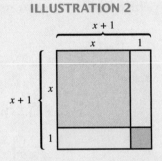

Section 6.3

Comparing methods Factor $18x^2 + 3x - 10$ by using the trial-and-check method and by using the grouping method. Which method do you think is better? Explain why.

Section 6.4

Comparing methods Show how $x^6 - 1$ can be factored in two ways: first as a difference of two squares and then as a difference of two cubes.

Section 6.5

Solving quadratic equations The techniques used to solve linear equations cannot be used to solve a quadratic equation. Write a brief note to the student whose work in "solving" $x^2 + 3x - 1 = 0$ is shown here and explain the misunderstanding.

$$x^2 + 3x - 1 = 0$$
$$x^2 + 3x = 1 \qquad \text{Add 1.}$$
$$3x = 1 - x^2 \qquad \text{Subtract } x^2.$$
$$\frac{3x}{3} = \frac{1 - x^2}{3} \qquad \text{Divide by 3.}$$
$$\text{So } x = \frac{1 - x^2}{3}$$

Section 6.6

Completing the square Construct the model in Illustration 3. Label each piece of the model with its respective area. Show that the total area of the model is $(x^2 + 4x)$ square units. Next, add enough 1×1 squares to make the model a square. How many does it take to do this? Show that the area of the new figure can be expressed as $(x^2 + 4x + 4)$ or $(x + 2)(x + 2)$ square units. Explain how this model demonstrates the process of completing the square on $x^2 + 4x$.

ILLUSTRATION 3

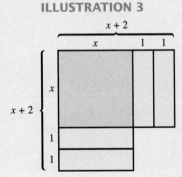

Section 6.7

Solving quadratic equations Solve the quadratic equation $2x^2 - x - 1 = 0$ using these methods: factoring, completing the square, and the quadratic formula. Write each solution on a separate piece of poster board. Under each solution, in two columns, list the advantages and the drawbacks of each method.

Section 6.8

Parabolas Use a home video camera to make a "documentary" showing examples of parabolic shapes you see in everyday life. Write a script for your video and have a narrator explain the setting, point out the vertex, and tell whether the parabola opens upward or downward in each case.

Section 6.1

Factoring Out the Greatest Common Factor and Factoring by Grouping

CONCEPTS

A *prime number* is a natural number greater than 1 whose only factors are 1 and itself. A natural number is in *prime-factored form* when it is written as the product of prime numbers.

To find the *greatest common factor* (GCF) of several monomials:

1. Prime-factor each monomial.
2. List each common factor the least number of times it appears in any one monomial.
3. Find the product of the factors in the list to obtain the GCF.

To *factor by grouping,* arrange the polynomial so that the first two terms have a common factor and the last two terms have a common factor. Factor out the common factor from both groups. Then factor out the resulting common binomial factor.

REVIEW EXERCISES

1. Find the prime factorization of each number.
   a. 35 $5 \cdot 7$ **b.** 45 $3^2 \cdot 5$

   c. 96 $2^5 \cdot 3$ **d.** 99 $3^2 \cdot 11$

   e. 2,050 $2 \cdot 5^2 \cdot 41$ **f.** 4,096 2^{12}

2. Factor out the GCF.
   a. $3x + 9y$ $3(x + 3y)$ **b.** $5ax^2 + 15a$ $5a(x^2 + 3)$

   c. $7s^2 + 14s$ $7s(s + 2)$ **d.** $\pi ab - \pi ac$ $\pi a(b - c)$

   e. $2x^3 + 4x^2 - 8x$ $2x(x^2 + 2x - 4)$ **f.** $x^2yz + xy^2z + xyz$ $xyz(x + y + 1)$

   g. $-5ab^2 + 10a^2b - 15ab$ $-5ab(b - 2a + 3)$
   h. $4(x - 2) - x(x - 2)$ $(x - 2)(4 - x)$

3. Factor out -1 from each polynomial.
   a. $-a - 7$ $-(a + 7)$ **b.** $-4t^2 + 3t - 1$ $-(4t^2 - 3t + 1)$

4. Factor by grouping:
   a. $2c + 2d + ac + ad$
   $(c + d)(2 + a)$ **b.** $3xy + 9x - 2y - 6$ $(y + 3)(3x - 2)$

   c. $2a^3 - a + 2a^2 - 1$ **d.** $4m^2n + 12m^2 - 8mn - 24m$
   $(2a^2 - 1)(a + 1)$ $4m(n + 3)(m - 2)$

Section 6.2

Factoring Trinomials of the Form $x^2 + bx + c$

To *factor a trinomial* of the form $x^2 + bx + c$ means to write it as the product of two binomials.

To factor $x^2 + bx + c$, find two integers whose product is c and whose sum is b.

$(x +)(x +)$

Write the trinomial in descending powers of the variable and factor out -1 when applicable.

5. Complete the table.

Product of the factors of 6	Sum of the factors of 6
1(6)	7
2(3)	5
-1 (-6)	-7
$-2(-3)$	-5

6. Factor each trinomial, if possible.

a. $x^2 + 2x - 24$ $(x + 6)(x - 4)$ **b.** $x^2 - 4x - 12$ $(x - 6)(x + 2)$

c. $n^2 - 7x + 10$ $(n - 5)(n - 2)$ **d.** $t^2 + 10t + 15$ prime

e. $-y^2 + 9y - 20$
$-(y - 5)(y - 4)$

f. $10y + 9 + y^2$ $(y + 9)(y + 1)$

g. $c^2 + 3cd - 10d^2$
$(c + 5d)(c - 2d)$

h. $-3mn + m^2 + 2n^2$
$(m - 2n)(m - n)$

If a trinomial cannot be factored using only integers, it is called a *prime polynomial.*

The *GCF* should always be factored out first. A trinomial is *factored completely* when it is expressed as a product of prime polynomials.

7. Explain how we can check to see if $(x - 4)(x + 5)$ is the factorization of $x^2 + x - 20$. Use the FOIL method to see if $(x - 4)(x + 5) = x^2 + x - 20$.

8. Completely factor each trinomial.

a. $5a^2 + 45a - 50$
$5(a + 10)(a - 1)$

b. $-4x^2y - 4x^3 + 24xy^2$
$-4x(x + 3y)(x - 2y)$

Section 6.3

Factoring Trinomials of the Form $ax^2 + bx + c$

To factor $ax^2 + bx + c$ using the *trial-and-check* factoring method, we must determine four integers. Use the FOIL method to check your work.

Factors
of a

$$\Big(\boxed{}\, x + \boxed{} \Big)\Big(\boxed{}\, x + \boxed{} \Big)$$

Factors
of c

To factor $ax^2 + bx + c$ using the *grouping method,* we write it as

$$ax^2 + \boxed{}\, x + \boxed{}\, x + c$$

9. Factor each trinomial completely, if possible.

a. $2x^2 - 5x - 3$ $(2x + 1)(x - 3)$ **b.** $10y^2 + 21y - 10$ $(2y + 5)(5y - 2)$

c. $-3x^2 + 14x + 5$
$-(3x + 1)(x - 5)$

d. $8a^2 + 16a + 6$ $2(2a + 3)(2a + 1)$

e. $-9p^2 - 6p + 6p^3$
$3p(2p + 1)(p - 2)$

f. $4b^2 + 15bc - 4c^2$ $(4b - c)(b + 4c)$

g. $3y^2 + 7y - 11$ prime

h. $7r^4 + 31r^2 + 12$ $(7r^2 + 3)(r^2 + 4)$

10. ENTERTAINING The rectangular-shaped area occupied by a table setting shown in Illustration 1 is $(12x^2 - x - 1)$ square inches. Factor the expression to find the binomials that represent the length and width of the table setting.

$(4x + 1)$ in., $(3x - 1)$ in.

ILLUSTRATION 1

Section 6.4

Special Factorizations and a Factoring Strategy

Special product formulas are used to factor *perfect square trinomials.*

$$x^2 + 2xy + y^2 = (x + y)^2$$
$$x^2 - 2xy + y^2 = (x - y)^2$$

To factor the *difference of two squares,* use the formula

$$F^2 - L^2 = (F + L)(F - L)$$

11. Factor each polynomial completely.

a. $x^2 + 10x + 25$ $(x + 5)^2$ **b.** $9y^2 - 24y + 16$ $(3y - 4)^2$

c. $-z^2 + 2z - 1$ $-(z - 1)^2$ **d.** $25a^2 + 20ab + 4b^2$ $(5a + 2b)^2$

12. Factor each polynomial completely, if possible.

a. $x^2 - 9$ $(x + 3)(x - 3)$ **b.** $49t^2 - 25y^2$ $(7t + 5y)(7t - 5y)$

c. $x^2y^2 - 400$ $(xy + 20)(xy - 20)$ **d.** $8at^2 - 32a$ $8a(t + 2)(t - 2)$

e. $c^4 - 64$ $(c^2 + 16)(c + 4)(c - 4)$ **f.** $h^2 + 36$ prime

To factor the *sum* and *difference of two cubes,* use the formulas

$$F^3 + L^3$$
$$= (F + L)(F^2 - FL + L^2)$$
$$F^3 - L^3$$
$$= (F - L)(F^2 + FL + L^2)$$

To factor a random polynomial, use the *factoring strategy* shown on page 462.

13. Factor each polynomial completely, if possible.
   a. $h^3 + 1$ $(h + 1)(h^2 - h + 1)$
   b. $125p^3 + q^3$
        $(5p + q)(25p^2 - 5pq + q^2)$
   c. $x^3 - 27$ $(x - 3)(x^2 + 3x + 9)$
   d. $16x^5 - 54x^2y^3$
        $2x^2(2x - 3y)(4x^2 + 6xy + 9y^2)$

14. Factor each polynomial completely, if possible.
   a. $14y^3 + 6y^4 - 40y^2$
        $2y^2(3y - 5)(y + 4)$
   b. $s^2t + s^2u^2 + vt + vu^2$
        $(t + u^2)(s^2 + v)$
   c. $j^4 - 16$ $(j^2 + 4)(j + 2)(j - 2)$
   d. $3j^3 - 24k^3$
        $-3(j + 2k)(j^2 - 2jk + 4k^2)$
   e. $12w^2 - 36w + 27$ $3(2w - 3)^2$
   f. $121p^2 + 36q^2$ prime

Section 6.5 Quadratic Equations

A *quadratic equation* is an equation of the form
$ax^2 + bx + c = 0$ $(a \neq 0)$,
where a, b, and c are real numbers.

To use the *factoring method* to solve a quadratic equation:
1. Write the equation in $ax^2 + bx + c = 0$ form.
2. Factor the left-hand side.
3. Use the *zero-factor property* (if $ab = 0$, then $a = 0$ or $b = 0$) and set each factor equal to zero.
4. Solve each resulting linear equation.

15. Solve each quadratic equation by factoring.
   a. $x^2 + 2x = 0$ $0, -2$
   b. $2x^2 - 6x = 0$ $0, 3$
   c. $x^2 - 9 = 0$ $-3, 3$
   d. $36p^2 = 25$ $-\frac{5}{6}, \frac{5}{6}$
   e. $a^2 - 7a + 12 = 0$ $3, 4$
   f. $t^2 + 4t + 4 = 0$ $-2, -2$
   g. $2x - x^2 + 24 = 0$ $6, -4$
   h. $(t + 1)(8t + 1) = 18t$ $\frac{1}{8}, 1$
   i. $x^2 - 11x = 12$ $-1, 12$
   j. $2p^3 = 2p(p + 2)$ $0, -1, 2$

16. CONSTRUCTION The face of the triangular preformed concrete panel shown in Illustration 2 has an area of 45 square meters, and its base is 3 meters longer than twice its height. How long is its base?   15 m

ILLUSTRATION 2

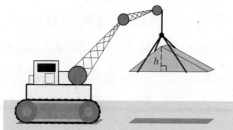

17. GARDENING A rectangular flower bed occupies 27 square feet and is 3 feet longer than twice its width. Find its dimensions.   3 ft by 9 ft

---

Section 6.6 Completing the Square

We can use the *square root method* to solve $x^2 = c$, where $c > 0$. The two solutions are $x = \sqrt{c}$ and $x = -\sqrt{c}$ $\left(\text{or } x = \pm\sqrt{c}\right)$.

18. Use the square root method to solve each quadratic equation.
   a. $x^2 = 25$ ± 5
   b. $x^2 = 400$ ± 20
   c. $2x^2 = 18$ ± 3
   d. $4y^2 = 9$ $\pm\frac{3}{2}$
   e. $t^2 = 8$ $\pm 2\sqrt{2}$
   f. $2x^2 - 1 = 149$ $\pm 5\sqrt{3}$

19. Use the square root method to solve each equation.
   a. $(x - 1)^2 = 25$ $-4, 6$
   b. $4(x - 2)^2 = 9$ $\frac{7}{2}, \frac{1}{2}$
   c. $(x - 8)^2 = 8$ $8 \pm 2\sqrt{2}$
   d. $(x + 5)^2 = 75$ $-5 \pm 5\sqrt{3}$

20. Use the square root method to solve each equation. Round each solution to the nearest hundredth.
   a. $x^2 = 12$ ± 3.46
   b. $(x - 1)^2 = 55$ $-6.42, 8.42$

To make $x^2 + bx$ a trinomial square, add the square of one-half of the coefficient of x.

The factoring method doesn't always work in solving many quadratic equations. In these cases, we can use a method called *completing the square*.

To solve a quadratic equation by completing the square:
1. If necessary, divide both sides of the equation by the coefficient of x^2 to make its coefficient 1.
2. If necessary, get the constant on the right-hand side of the equation.
3. Complete the square and factor the resulting trinomial square.
4. Solve the quadratic equation using the square root method.
5. Check each solution.

21. Complete the square to make each expression a trinomial square.
  a. $x^2 + 4x$ $x^2 + 4x + 4$
  b. $z^2 - 10z$ $z^2 - 10z + 25$
  c. $t^2 - 5t$ $t^2 - 5t + \frac{25}{4}$
  d. $a^2 + \frac{3}{4}a$ $a^2 + \frac{3}{4}a + \frac{9}{64}$

22. Explain why the quadratic equation $x^2 + 4x + 1 = 0$ can't be solved by the factoring method. $x^2 + 4x + 1$ doesn't factor.

23. Solve each quadratic equation by completing the square.
  a. $x^2 - 8x + 15 = 0$ $3, 5$
  b. $x^2 + 5x - 14 = 0$ $2, -7$
  c. $2x^2 + 5x - 3 = 0$ $\frac{1}{2}, -3$
  d. $2x^2 - 2x - 1 = 0$ $\dfrac{1 \pm \sqrt{3}}{2}$

24. Solve $x^2 + 4x + 1 = 0$ by completing the square. Round each solution to the nearest hundredth. $-0.27, -3.73$

25. PLAYGROUND EQUIPMENT
The large tractor tire shown in Illustration 3 makes a good container for sand. If the circular area that the "sandbox" covers is 28.3 square feet, what is the radius of the tire? Round to the nearest tenth of a foot.
3.0 ft

ILLUSTRATION 3

Section 6.7

For the *general quadratic equation* $ax^2 + bx + c = 0$, where $a \neq 0$,

$$x = \frac{-b \pm \sqrt{b^2 - 4ac}}{2a}$$

This is called the *quadratic formula*.

The Quadratic Formula

26. Use the quadratic formula to solve each quadratic equation.
  a. $x^2 - 2x - 15 = 0$ $5, -3$
  b. $x^2 - 6x - 7 = 0$ $7, -1$
  c. $6x^2 - 7x - 3 = 0$ $\frac{3}{2}, -\frac{1}{3}$
  d. $x^2 - 6x + 7 = 0$ $3 \pm \sqrt{2}$

27. Use the quadratic formula to solve $3x^2 + 2x - 2 = 0$. Give the solutions in exact form and then rounded to the nearest hundredth.
$\dfrac{-1 \pm \sqrt{7}}{3}$, $-1.22, 0.55$

28. Use the quadratic formula to solve $10x^2 + 2x + 1 = 0$.   no real solutions

29. SECURITY GATE The length of the frame for the iron gate in Illustration 4 is 14 feet longer than the width. A diagonal crossbrace is 26 feet long. Find the width and length of the gate frame.
10 ft, 24 ft

ILLUSTRATION 4

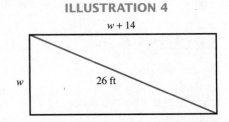

30. MILITARY A pilot releases a bomb from an altitude of 3,000 feet. The bomb's height h above the target t seconds after its release is given by the formula

$$h = 3,000 + 40t - 16t^2$$

How long will it be until the bomb hits its target?   15 sec

Strategy for solving quadratic equations:

1. Try the square root method.
2. If it doesn't apply, write the equation in $ax^2 + bx + c = 0$ form.
3. Try the factoring method.
4. If it doesn't work, complete the square or use the quadratic formula.

31. Use the most convenient method to find all real solutions of each equation.

   a. $x^2 + 6x + 2 = 0$ $-3 \pm \sqrt{7}$ **b.** $(y + 3)^2 = 16$ $1, -7$

   c. $x^2 + 5x = 0$ $0, -5$ **d.** $2x^2 + x = 5$ $\dfrac{-1 \pm \sqrt{41}}{4}$

   e. $g^2 - 20 = 0$ $\pm 2\sqrt{5}$ **f.** $a^2 = 4a - 4$ $2, 2$

Section 6.8

Graphing Quadratic Functions

The *vertex* of a parabola is the lowest (or highest) point on the parabola.

32. See the graph in Illustration 5.

   a. What are the x-intercepts of the parabola? $(-3, 0), (1, 0)$

   b. What is the y-intercept of the parabola? $(0, -3)$

   c. What is the vertex of the parabola? $(-1, -4)$

   d. Draw the axis of symmetry of the parabola on the graph.

A vertical line through the vertex of a parabola that opens upward or downward is its *axis of symmetry*.

33. What important information can be obtained from the vertex of the parabola in Illustration 6?

The maximum profit of \$16,000 is obtained from the sale of 400 units.

ILLUSTRATION 5   ILLUSTRATION 6

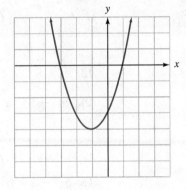 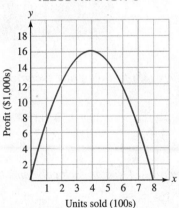

The graph of the quadratic function $y = ax^2 + bx + c$ is a parabola. It opens upward when $a > 0$ and downward when $a < 0$.

The x-coordinate of the vertex of the parabola $y = ax^2 + bx + c$ is $x = -\frac{b}{2a}$. To find the y-coordinate of the vertex, substitute $-\frac{b}{2a}$ for x in the equation of the parabola and find y.

34. Find the vertex of the graph of each quadratic function and tell which direction the parabola opens. **Do not draw the graph.**

   a. $y = 2x^2 - 4x + 7$ **b.** $y = -3x^2 + 18x - 11$

     $(1, 5)$; upward $(3, 16)$; downward

35. Find the x- and y-intercepts of the graph of $y = x^2 + 6x + 5$.

   $(-5, 0), (-1, 0); (0, 5)$

The x-intercepts of a parabola are determined by solving $ax^2 + bx + c = 0$. The y-intercept is $(0, c)$.

36. Graph each quadratic function by finding the vertex, x- and y-intercepts, and axis of symmetry of its graph.

a. $y = x^2 + 2x - 3$

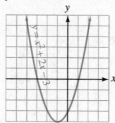

b. $f(x) = -2x^2 + 4x - 2$

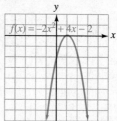

37. Use the graph in Illustration 7 to solve the quadratic equation $x^2 + 2x - 3 = 0$.

$-3, 1$

ILLUSTRATION 7

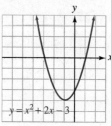

CHAPTER 6

Test

In Problems 1–2, find the prime factorization of each number.

1. 196 $2^2 \cdot 7^2$

2. 111 $3 \cdot 37$

In Problems 3–14, factor each polynomial, if possible.

3. $4x + 16$ $4(x + 4)$

4. $30a^2b^3 - 20a^3b^2 + 5abc$ $5ab(6ab^2 - 4a^2b + c)$

5. $q^2 - 81$ $(q + 9)(q - 9)$

6. $x^2 + 9$ prime

7. $16x^4 - 81$ $(4x^2 + 9)(2x + 3)(2x - 3)$

8. $x^2 + 4x + 3$ $(x + 3)(x + 1)$

9. $-x^2 + 9x + 22$ $-(x - 11)(x + 2)$

10. $9a - 9b + ax - bx$ $(a - b)(9 + x)$

11. $2a^2 + 5a - 12$ $(2a - 3)(a + 4)$

12. $18x^2 - 60xy + 50y^2$ $2(3x - 5y)^2$

13. $x^3 + 8$ $(x + 2)(x^2 - 2x + 4)$

14. $2a^3 - 54$ $2(a - 3)(a^2 + 3a + 9)$

In Problems 15–18, solve each equation by factoring.

15. $6x^2 - x = 0$ $0, \frac{1}{6}$

16. $x^2 + 6x + 9 = 0$ $-3, -3$

17. $6x^2 + x - 1 = 0$ $\frac{1}{3}, -\frac{1}{2}$

18. $10x^2 + 43x = 9$ $\frac{1}{5}, -\frac{9}{2}$

In Problems 19–20, solve each equation by the square root method.

19. $x^2 = 16$ $4, -4$

20. $(x - 2)^2 = 3$ $2 \pm \sqrt{3}$

21. Find the number required to complete the square on $x^2 - 14x$.   49

22. Complete the square to solve $a^2 + 2a - 4 = 0$. Give the exact solutions and then round them to the nearest hundredth. $-1 \pm \sqrt{5}$; $-3.24, 1.24$

In Problems 23–25, use the quadratic formula to solve each equation.

23. $x^2 + 3x - 10 = 0$ 2, -5

24. $2x^2 - 5x = 12$ $-\frac{3}{2}$, 4

25. $x^2 = 4x - 2$ $2 \pm \sqrt{2}$

26. FLAGS According to the *Guinness Book of World Records 1998,* the largest flag in the world is the American "Superflag" with an area of 128,775 ft². If its length is 5 feet less than twice its width, find its width and length.   255 ft, 505 ft.

27. ADVERTISING When a business runs *x* advertisements per week on television, the number *y* of air conditioners it sells is given by the quadratic function graphed in Illustration 1. What important information can be obtained from the vertex?

The most air conditioners sold in a week (18) occurred when 3 ads were run.

28. Graph the function $y = x^2 + x - 2$ by finding the vertex, *x*- and *y*-intercepts, and axis of symmetry.

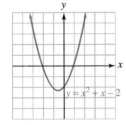

ILLUSTRATION 1

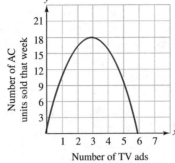

Number of TV ads
run during the week

29. Solve the equation $-x^2 - 2x - 1 = 0$, given the graph of $f(x) = -x^2 - 2x - 1$ in Illustration 2.   -1

ILLUSTRATION 2

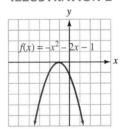

30. Explain how the FOIL method can be used to check the factorization of $x^2 - 3x - 54$.

CHAPTERS 1–6

Cumulative Review Exercises

1. Evaluate $9^2 - 3[45 - 3(6 + 4)]$.   36

2. PAIN RELIEVERS For the 12-month period ending August 16, 1998, Tylenol had sales of $567,600,000. Use the information in Illustration 1 to determine the total amount of money spent on pain-relieving tablets for that 12-month period. $2,580,000,000

3. Find the average (mean) test score of a student in a history class with scores of 80, 73, 61, 73, and 98.   77

4. What is the value in cents of *x* 33¢ stamps?   33x¢

5. Solve $\frac{3}{4} = \frac{1}{2} + \frac{x}{5}$. $\frac{5}{4}$

6. Change 40°C to degrees Fahrenheit.   104°F

7. Find the volume of a pyramid that has a square base, measuring 6 feet on a side, and whose height is 20 feet.   240 ft³

ILLUSTRATION 1

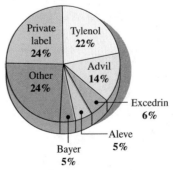

Based on information from *Los Angeles Times* (Sept. 24, 1998)

8. Solve $\frac{4}{5}d = -4$. -5

9. BLENDING TEA One grade of tea, worth $3.20 per pound, is to be mixed with another grade worth $2 per pound to make 20 pounds of a mixture that will be worth $2.72 per pound. How much of each grade of tea must be used?

12 lb of the $3.20 tea and 8 lb of the $2 tea

10. SPEED OF A PLANE Two planes are 6,000 miles apart, and their speeds differ by 200 mph. If they travel toward each other and meet in 5 hours, find the speed of the slower plane. 500 mph

11. Find the slope of the line passing through $(-1, 3)$ and $(3, -1)$. -1

12. Write the equation of a line that has slope 3 and passes through the point $(1, 5)$. $y = 3x + 2$

13. Graph $3x - 2y = 6$.

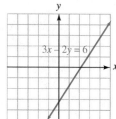

14. Is this the graph of a function? no

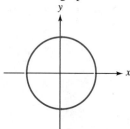

15. CUTTING STEEL The graph in Illustration 2 shows the amount of wear (in mm) on a cutting blade for a given length of a cut (in m). Find the rate of change in the length of the cutting blade. 0.008 mm/m

ILLUSTRATION 2

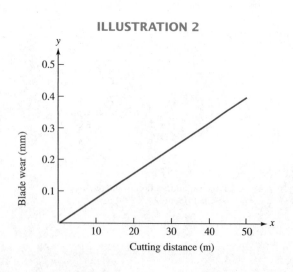

16. Graph $y = 2x - 3$.

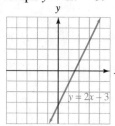

In Exercises 17–24, write each expression using one positive exponent.

17. $(y^3y^5)y^6$ y^{14}

18. $(x^3x^4)^2$ x^{14}

19. $\dfrac{x^3x^4}{x^2x^3}$ x^2

20. $\dfrac{a^4b^0}{a^{-3}}$ a^7

21. x^{-5} $\dfrac{1}{x^5}$

22. $(-2y)^{-4}$ $\dfrac{1}{16y^4}$

23. $(x^{-4})^2$ $\dfrac{1}{x^8}$

24. $\left(-\dfrac{x^3}{x^{-2}}\right)^3$ $-x^{15}$

In Exercises 25–26, write each number in scientific notation.

25. 615,000 6.15×10^5

26. 0.0000013 1.3×10^{-6}

In Exercises 27–28, write each number in standard notation.

27. 5.25×10^{-4} 0.000525

28. 2.77×10^3 2,770

In Exercises 29–30, give the degree of each polynomial.

29. $3x^2 + 2x - 5$ 2

30. $-3x^3y^2 + 3x^2y^2 - xy$ 5

31. Find $f(-4)$ if $f(x) = \dfrac{x^2 - 2x}{2}$. 12

32. Complete the table and then graph the function.

$f(x) = x^3 + 3x^2$

x	$f(x)$
-3	0
-2	4
-1	2
0	0
1	4

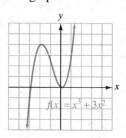

In Exercises 33–42, do the operations.

33. $(3x^2 + 2x - 7) - (2x^2 - 2x + 7)$ $x^2 + 4x - 14$

34. $(2x^2 - 3x + 4) + (2x^2 + 2x - 5)$ $4x^2 - x - 1$

35. $-5x^2(7x^3 - 2x^2 - 2)$ $-35x^5 + 10x^4 + 10x^2$

36. $(3x^3y^2)(-4x^2y^3)$ $-12x^5y^5$

37. $(3x - 7)(2x + 8)$ $6x^2 + 10x - 56$

38. $(5x - 4y)(3x + 2y)$ $15x^2 - 2xy - 8y^2$

39. $(3x + 1)^2$ $9x^2 + 6x + 1$

40. $(x - 2)(x^2 + 2x + 4)$ $x^3 - 8$

41. $\dfrac{6x^2 - 8x}{2x}$ $3x - 4$

42. $x - 3\overline{)2x^2 - 5x - 3}$ $2x + 1$

In Exercises 43–46, find each square root, if one exists.

43. $\sqrt{169}$ 13

44. $\sqrt{\dfrac{1}{4}}$ $\frac{1}{2}$

45. $\sqrt{0.36}$ 0.6

46. $\sqrt{-9}$ None exists.

47. Find the missing side length. 12

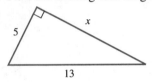

48. Complete the table and then graph the function.

$$f(x) = \sqrt{x}$$

x	f(x)
0	0
1	1
2	1.4
4	2
6	2.4
8	2.8
9	3

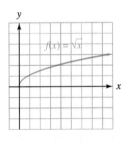

In Exercises 49–52, find each cube root.

49. $\sqrt[3]{27}$ 3

50. $\sqrt[3]{-64}$ -4

51. $\sqrt[3]{\dfrac{1}{8}}$ $\frac{1}{2}$

52. $\sqrt[3]{\dfrac{8}{125}}$ $\frac{2}{5}$

In Exercises 53–56, find each root.

53. $\sqrt{9x^2}$ $3x$

54. $\sqrt{x^8y^4}$ x^4y^2

55. $\sqrt[4]{x^8y^{12}}$ x^2y^3

56. $\sqrt[5]{x^{10}y^5}$ x^2y

In Exercises 57–58, solve each equation.

57. $\sqrt{6x + 1} + 3 = 8$ 4

58. $2\sqrt{x + 2} + 6 = 3$ no solution

In Exercises 59–60, find the distance between each pair of points. If an answer is not exact, round to the nearest hundredth.

59. $(-2, 3)$ and $(4, -5)$ 10

60. $(-2, -8)$ and $(8, 4)$ 15.62

In Exercises 61–64, find each root.

61. $\sqrt{27x^3}$ $3x\sqrt{3x}$

62. $\sqrt{175x^2y^3}$ $5xy\sqrt{7y}$

63. $\sqrt[3]{-80}$ $-2\sqrt[3]{10}$

64. $\sqrt[3]{x^{10}y^5}$ $x^3y\sqrt[3]{xy^2}$

In Exercises 65–70, do the operations and simplify when possible.

65. $\sqrt{27} - \sqrt{12}$ $\sqrt{3}$

66. $\sqrt{25x^2} - \sqrt{16x^2}$ x

67. $\left(3\sqrt{6x}\right)\left(2\sqrt{3x}\right)$ $18x\sqrt{2}$

68. $\sqrt{2}\left(\sqrt{2} + 2\right)$ $2 + 2\sqrt{2}$

69. $\left(2\sqrt{3} + 1\right)\left(\sqrt{3} - 1\right)$ $5 - \sqrt{3}$

70. $\left(\sqrt{x} + 2\right)\left(\sqrt{x} - 2\right)$ $x - 4$

In Exercises 71–72, rationalize each denominator.

71. $\dfrac{2}{\sqrt{3}}$ $\frac{2\sqrt{3}}{3}$

72. $\dfrac{x - 16}{\sqrt{x} - 4}$ $\sqrt{x} + 4$

In Exercises 73–76, simplify each expression.

73. $4^{1/2}$ 2

74. $\left(\dfrac{8}{27}\right)^{-2/3}$ $\frac{9}{4}$

75. $x^{1/2}x^{1/2}$ x

76. $(x^{1/2})^4$ x^2

77. 🖩 CONCENTRIC CIRCLES The area of the ring between the two concentric circles of radius r and R (see Illustration 3) is given by the formula

$$A = \pi(R + r)(R - r)$$

If $r = 3$ inches and $R = 17$ inches, find A to the nearest tenth. 879.6 in.^{2}

ILLUSTRATION 3

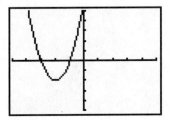

78. The graph of $f(x) = 2x^2 + 8x + 6$ is shown in Illustration 4. Use it to solve the quadratic equation $2x^2 + 8x + 6 = 0$. $-3, -1$

ILLUSTRATION 4

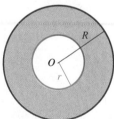

In Exercises 79–82, factor each polynomial completely.

79. $k^3t - 3k^2t$ $k^2t(k - 3)$

80. $2ab + 2ac + 3b + 3c$ $(b + c)(2a + 3)$

81. $2a^2 - 200b^2$ $2(a + 10b)(a - 10b)$

82. $b^3 + 125$ $(b + 5)(b^2 - 5b + 25)$

In Exercises 83–84, solve each equation by factoring.

83. $5x^2 + x = 0$ $0, -\frac{1}{5}$ **84.** $6x^2 - 5x = -1$ $\frac{1}{3}, \frac{1}{2}$

85. QUILT According to the *Guinness Book of World Records 1998*, the world's largest quilt was made by the Seniors' Association of Saskatchewan, Canada, in 1994. If its length is 11 feet less than twice its width and it has an area of 12,865 ft^2, find its width and length.   83 ft $\times$ 155 ft

86. Sketch the graph of a parabola with vertex at $(-1, -4)$, x-intercepts of $(1, 0)$ and $(-3, 0)$, and a y-intercept of $(0, -3)$.

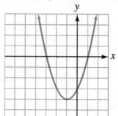

7

Proportion and Rational Expressions

CAMPUS CONNECTION

The Public Works Department

In Public Works classes, students learn about water systems and water management. As their instruction progresses, they see that algebra is used by people who work in the field as well as those in the office. For instance, if a manager knows the time it takes each of two inlet pipes to fill a water storage tank working alone, he can use algebra to find the time it would take the two pipes working together to fill the empty tank. This is done by solving an *equation that contains algebraic fractions*, known as rational expressions. In this chapter, we will use such equations to solve problems concerning finances, travel, and public works projects.

Ratios can be used to solve problems involving pricing, travel, mixtures, and manufacturing. They are examples of a broader set of algebraic expressions called rational expressions.

▶ 7.1   Ratios

In this section, you will learn about

Ratios ■ Unit costs ■ Rates

Introduction The concept of *ratio* is often applicable in real-life situations. For example,

■ To prepare fuel for a Lawnboy lawnmower, gasoline must be mixed with oil in the ratio of 50 to 1.

■ To make 14-karat jewelry, gold is mixed with other metals in the ratio of 14 to 10.

■ In the stock market, winning stocks might outnumber losing stocks in the ratio of 7 to 4.

■ At Citrus College, the ratio of students to faculty is 21 to 1.

In this section, we will discuss ratios and use them to solve problems. Since ratios are fractions, you may want to read Appendix I, which reviews the basic properties of fractions.

Ratios

Ratios give us a way to compare numerical quantities.

Ratios

A **ratio** is the quotient of two numbers or two quantities.

There are three common ways to write a ratio: as a fraction, with the word *to*, or with a colon.
Some examples of ratios are

$$\frac{7}{9}, \quad 21 \text{ to } 27, \quad \text{and} \quad 2{,}290 : 1{,}317$$

■ The fraction $\frac{7}{9}$ can be read as "the ratio of 7 to 9."

■ 21 to 27 can be read as "the ratio of 21 to 27" and can be written as $\frac{21}{27}$.

■ 2,290 : 1,317 can be read as "the ratio of 2,290 to 1,317" and can be written as $\frac{2{,}290}{1{,}137}$.

Because the fractions $\frac{7}{9}$ and $\frac{21}{27}$ represent equal numbers, they are **equal ratios.**

 WARNING! Unlike the fraction $\frac{a}{b}$, b can be zero in the ratio $\frac{a}{b}$. For example, the ratio of women to men on a women's softball team could be 25 to 0. However, these applications are rare.

Writing ratios. Express each phrase as a ratio in lowest terms: **a.** the ratio of 15 to 12 and **b.** the ratio of 0.3 to 1.2.

Solution **a.** The ratio of 15 to 12 can be written as the fraction $\frac{15}{12}$. After simplifying, the ratio is $\frac{5}{4}$.

$$\frac{15}{12} = \frac{\overset{1}{\cancel{3}} \cdot 5}{\underset{1}{\cancel{3}} \cdot 4}$$ Factor 15 and 12. Then divide out the common factor of 3.

$$= \frac{5}{4}$$

b. The ratio of 0.3 to 1.2 can be written as the fraction $\frac{0.3}{1.2}$. We can simplify this fraction as follows:

$$\frac{0.3}{1.2} = \frac{0.3 \cdot \mathbf{10}}{1.2 \cdot \mathbf{10}}$$ To clear the fraction of decimals, multiply both the numerator and the denominator by 10.

$$= \frac{3}{12}$$ Multiply: $0.3 \cdot 10 = 3$ and $1.2 \cdot 10 = 12$.

$$= \frac{1}{4}$$ Simplify the fraction: $\frac{3}{12} = \frac{\overset{1}{\cancel{3}} \cdot 1}{\underset{1}{\cancel{3}} \cdot 4} = \frac{1}{4}$.

SELF CHECK Express each ratio in lowest terms: **a.** The ratio of 8 to 12 and **b.** The ratio of 3.2 to 16. *Answers:* **a.** $\frac{2}{3}$, **b.** $\frac{1}{5}$ ■

Writing ratios. Express each phrase as a ratio in lowest terms: **a.** the ratio of 3 meters to 8 meters and **b.** the ratio of 4 ounces to 1 pound.

Solution **a.** The ratio of 3 meters to 8 meters can be written as the fraction $\frac{3 \text{ meters}}{8 \text{ meters}}$, or simply $\frac{3}{8}$.

b. When possible, we should express ratios in the same units. Since there are 16 ounces in 1 pound, the proper ratio is $\frac{4 \text{ ounces}}{16 \text{ ounces}}$, which simplifies to $\frac{1}{4}$.

SELF CHECK Express each ratio in lowest terms: **a.** the ratio of 8 ounces to 2 pounds and **b.** the ratio of 1 foot to 2 yards. (*Hint:* 3 feet = 1 yard.) *Answers:* **a.** $\frac{1}{4}$, **b.** $\frac{1}{6}$ ■

Student-to-faculty ratios. At a college, there are 2,772 students and 154 faculty members. Write a fraction in simplified form to express the ratio of students per faculty member.

Solution The ratio of students to faculty is 2,772 to 154. We can write this ratio as the fraction $\frac{2,772}{154}$ and simplify it.

$$\frac{2,772}{154} = \frac{\overset{1}{\cancel{2}} \cdot 2 \cdot 3 \cdot 3 \cdot \overset{1}{\cancel{7}} \cdot \overset{1}{\cancel{11}}}{\underset{1}{\cancel{2}} \cdot \underset{1}{\cancel{7}} \cdot \underset{1}{\cancel{11}}}$$ Prime factor 2,772 and 154. Divide out common factors.

$$= \frac{18}{1}$$

The ratio of students to faculty is 18 to 1.

SELF CHECK In a college graduating class, 224 students out of 632 went on to graduate school. Write a fraction in simplified form to express the ratio of the number of students continuing their education to the number in the graduating class. *Answer:* $\frac{28}{79}$ ∎

Unit Costs

The *unit cost* of an item is the ratio of its cost to its quantity. For example, the unit cost (the cost per pound) of 5 pounds of coffee priced at $20.75 is given by the ratio

$$\frac{\$20.75}{5 \text{ pounds}} = \frac{\$2,075}{500 \text{ pounds}} \qquad \text{To eliminate the decimal, multiply the numerator and the denominator by 100.}$$

$$= \$4.15 \text{ per pound} \quad \text{Do the division.}$$

The unit cost is $4.15 per pound.

EXAMPLE 4

Comparison shopping. Olives come packaged in a 12-ounce jar, which sells for $3.09, or in a 6-ounce jar, which sells for $1.53. Which is the better buy?

Solution To find the better buy, we must compare the unit costs. The unit cost of the 12-ounce jar is

$$\frac{\$3.09}{12 \text{ ounces}} = \frac{309¢}{12 \text{ ounces}} \qquad \text{Change \$3.09 to 309¢.}$$

$$= 25.75¢ \text{ per ounce.} \quad \text{Simplify by dividing.}$$

The unit cost of the 6-ounce jar is

$$\frac{\$1.53}{6 \text{ ounces}} = \frac{153¢}{6 \text{ ounces}} \qquad \text{Change \$1.53 to 153¢.}$$

$$= 25.5¢ \text{ per ounce.} \quad \text{Do the division.}$$

Since the unit cost is less when olives are packaged in 6-ounce jars, that is the better buy.

SELF CHECK A fast-food restaurant sells a 12-ounce Coke for 79¢ and a 16-ounce Coke for 99¢. Which is the better buy? *Answer:* the 16-oz Coke ∎

Rates

When ratios are used to compare quantities with different units, they are called *rates*. For example, if the 495-mile drive from New Orleans to Dallas takes 9 hours, the average rate of speed is the ratio of the miles driven to the length of time of the trip.

$$\text{Average rate of speed} = \frac{495 \text{ miles}}{9 \text{ hours}} = \frac{55 \text{ miles}}{1 \text{ hour}} \qquad \frac{495}{9} = \frac{\overset{1}{\cancel{9} \cdot 55}}{\underset{1}{\cancel{9} \cdot 1}} = \frac{55}{1}.$$

The ratio $\frac{55 \text{ miles}}{1 \text{ hour}}$ can be expressed in any of the following forms:

$$55 \frac{\text{miles}}{\text{hour}}, \qquad 55 \text{ miles per hour}, \qquad 55 \text{ miles/hour}, \qquad \text{or} \qquad 55 \text{ mph}$$

EXAMPLE 5

Finding hourly rates of pay. Find the hourly rate of pay for a student who earns $370 for working 40 hours.

Solution   We can write the rate of pay as the ratio

$$\text{Rate of pay} = \frac{\$370}{40 \text{ hours}}$$

and simplify by dividing 370 by 40.

$$\text{Rate of pay} = 9.25 \frac{\$}{\text{hour}}$$

The rate is $9.25 per hour.

SELF CHECK

In 1997, the average weekly pay of a member of a union working full-time was $640. Based on a 40-hour work week, what was the average rate of pay per hour?   *Answer:* $16 per hour   ∎

EXAMPLE 6

Electric bill. The amount of electricity a household uses is measured in kilowatt hours by a meter like that shown in Figure 7-1. Determine the rate of energy consumption in kilowatt hours per day from the readings listed in the table.

FIGURE 7-1

Meter Reading

Previous From 4/1/98	Present To 4/30/98
4589	5384

Solution   To find the number of kilowatt hours of electricity used, we subtract the meter reading on the first day of the month from the reading on the last day of the month.

$$5,384 - 4,589 = 795$$

We can write the rate of energy consumption as the ratio

$$\text{Rate of energy consumption} = \frac{795 \text{ kilowatt hours}}{30 \text{ days}}$$

From 4/1/98 to 4/30/98 is 30 days.

and simplify by dividing 795 by 30.

$$\text{Rate of energy consumption} = 26.5 \frac{\text{kilowatt hours}}{\text{day}}$$

The rate of consumption in April was 26.5 kilowatt hours per day.

EXAMPLE 7

Tax rates. A book cost $49.22, including $3.22 sales tax. Find the sales tax rate.

Solution   Since the tax was $3.22, the cost of the book alone was

$$\$49.22 - \$3.22 = \$46.00$$

We can write the sales tax rate as the ratio

$$\text{Sales tax rate} = \frac{\text{amount of sales tax}}{\text{cost of the book, without tax}}$$

$$= \frac{\$3.22}{\$46}$$

and simplify by dividing 3.22 by 46.

$$\text{Sales tax rate} = 0.07$$

The tax rate is 0.07, or 7%.

SELF CHECK   A married couple, filing jointly, had a taxable income of $22,420. They paid $3,363 in federal income tax. What tax bracket were they in? That is, what income tax rate did they have to pay?   *Answer:* 15%

ACCENT ON TECHNOLOGY *Computing Gas Mileage*

A man drove a total of 775 miles. Along the way, he stopped for gas three times, pumping 10.5, 11.3, and 8.75 gallons of gas. He started with the tank half full and ended with the tank half full. To find how many miles he got per gallon, we need to divide the total distance by the total number of gallons of gas consumed.

$$\frac{775}{10.5 + 11.3 + 8.75}$$ ← Total distance

← Total number of gallons consumed

Using a scientific calculator, we enter these numbers and press these keys:

Keystrokes   775 $\div$ (10.5 + 11.3 + 8.75) = $\boxed{25.36824877}$

Using a graphing calculator, we enter these numbers and press these keys:

Keystrokes   775 $\div$ (10.5 + 11.3 + 8.75) ENTER

$\boxed{\begin{array}{l}775/(10.5+11.3+8 \\ .75) \\ \qquad\qquad 25.36824877\end{array}}$

To the nearest hundredth, he got 25.37 mpg.

STUDY SET
Section 7.1

VOCABULARY

In Exercises 1–4, fill in the blanks to make the statements true.

1. A ratio is a ___comparison___ of two numbers by their indicated ___quotient___.
2. The ___unit___ cost of an item is the ratio of its cost to its quantity.
3. The ratios $\frac{2}{3}$ and $\frac{4}{6}$ are ___equal___ ratios.
4. The ratio $\frac{500 \text{ miles}}{15 \text{ hours}}$ is called a ___rate___.

CONCEPTS

5. Give three examples of ratios that you have encountered this past week.
6. Suppose that a basketball player made 8 free throws out of 12 tries. The ratio of $\frac{8}{12}$ can be simplified as $\frac{2}{3}$. Interpret this result.
   The player made 2 of every 3 attempts.

NOTATION

7. **a.** Write the ratio 3 to 11 as a fraction. $\frac{3}{11}$
   b. Write the ratio 27:32 as a fraction. $\frac{27}{32}$
8. Express the ratio $\frac{468 \text{ miles}}{9 \text{ hours}}$ in two different ways.
   52 miles per hour, 52 mph, etc.

PRACTICE

In Exercises 9–24, express each phrase as a ratio in lowest terms.

9. 5 to 7 $\frac{5}{7}$
10. 3 to 5 $\frac{3}{5}$
11. 17:34 $\frac{1}{2}$
12. 19:38 $\frac{1}{2}$
13. 22 to 33 $\frac{2}{3}$
14. 14 to 21 $\frac{2}{3}$
15. 7 to 24.5 $\frac{2}{7}$
16. 0.65 to 0.15 $\frac{13}{3}$
17. 4 ounces to 12 ounces $\frac{1}{3}$
18. 3 inches to 15 inches $\frac{1}{5}$
19. 12 minutes to 1 hour $\frac{1}{5}$
20. 8 ounces to 1 pound $\frac{1}{2}$
21. 3 days to 1 week $\frac{3}{7}$
22. 4 inches to 2 yards $\frac{1}{18}$
23. 18 months to 2 years $\frac{3}{4}$
24. 8 feet to 4 yards $\frac{2}{3}$

In Exercises 25–28, refer to the monthly family budget shown in Illustration 1. Give each ratio in lowest terms.

25. Find the total amount of the budget. $1,825
26. Find the ratio of the amount budgeted for rent to the total budget. $\frac{30}{73}$

27. Find the ratio of the amount budgeted for entertainment to the total budget. $\frac{22}{365}$
28. Find the ratio of the amount budgeted for phone to the amount budgeted for entertainment. Then complete this statement: For every $25 spent on the phone, $ 22 is spent on entertainment. $\frac{25}{22}$

ILLUSTRATION 1

Item	Amount
Rent	$750
Food	$652
Gas and Electric	$188
Phone	$125
Entertainment	$110

In Exercises 29–32, refer to the tax deductions listed in Illustration 2. Give each ratio in lowest terms.

29. Find the total amount of deductions. $8,725
30. Find the ratio of the real estate tax deduction to the total deductions. $\frac{249}{1,745}$
31. Find the ratio of the contributions to the total deductions. $\frac{336}{1,745}$
32. Find the ratio of the mortgage deduction to the union dues deduction. $\frac{916}{45}$

ILLUSTRATION 2

Item	Amount
Medical	$ 995
Real estate tax	$1,245
Contributions	$1,680
Mortgage	$4,580
Union dues	$ 225

APPLICATIONS

In Exercises 33–52, find each ratio and express it in lowest terms. You may use a calculator.

33. **FACULTY-TO-STUDENT RATIO** At a college, there are 125 faculty members and 2,000 students. Find the faculty-to-student ratio. $\frac{1}{16}$

34. U.S. SENATE In the 105th Congress, which convened January 7, 1997, there were 9 women in the 100-member Senate. Find the ratio of women to men senators. $\frac{9}{91}$

35. BLOOD PRESSURE A person's blood pressure is expressed as the ratio of the systolic pressure (when the heart contracts to empty its blood) to the diastolic pressure (when the heart relaxes to fill with blood). This ratio, written as a fraction, is not simplified. Complete the table in Illustration 3, which lists the "normal" readings for three age groups. Do you know your blood pressure ratio?

ILLUSTRATION 3

	Systolic	Diastolic	Ratio
Infant	80	45	$\frac{80}{45}$
Age 30	120	80	$\frac{120}{80}$
Age 40	140	85	$\frac{140}{85}$

36. BICYCLING Illustration 4 shows a "high wheeler" bicycle, which was popular in the 1870s. Find the ratio of the circumference of the front wheel to the circumference of the rear wheel. $\frac{16}{5}$

ILLUSTRATION 4

48 in.

15 in.

37. UNIT COST OF GASOLINE A driver pumped 17 gallons of gasoline into her tank at a cost of $21.59. Write a ratio of dollars to gallons, and give the unit cost of gasoline. $\frac{\$21.59}{17 \text{ gallons}}$; $1.27 per gallon

38. UNIT COST OF GRASS SEED A 50-pound bag of grass seed costs $222.50. Write a ratio of dollars to pounds, and give the unit cost of grass seed. $\frac{\$222.50}{50 \text{ pounds}}$; $4.45 per pound

39. ENTERTAINMENT Disneyland has 63 rides and attractions. If an adult admission ticket costs $36, what is the unit cost of each ride and attraction if a person rides each of them once? Round to the nearest cent. 57¢ per ride or attraction

40. UNIT COST OF BEANS A 24-ounce package of green beans sells for $1.29. Give the unit cost in cents per ounce. 5.375¢ per ounce

41. COMPARISON SHOPPING A 6-ounce can of orange juice sells for 89¢, and an 8-ounce can sells for $1.19. Which is the better buy? the 6-ounce can

42. COMPARING SPEEDS A car travels 345 miles in 6 hours, and a truck travels 376 miles in 6.2 hours. Which vehicle is going faster? the truck

43. COMPARING READING SPEEDS One seventh-grader read a 54-page book in 40 minutes, and another read an 80-page book in 62 minutes. If the books were equally difficult, which student reads faster? the first student

44. COMPARATIVE SHOPPING A 30-pound bag of fertilizer costs $12.25, and an 80-pound bag costs $30.25. Which is the better buy? the 80-pound bag

45. EMPTYING A TANK An 11,880-gallon tank can be emptied in 27 minutes. Write a ratio of gallons to minutes, and give the rate of flow in gallons per minute. $\frac{11,880 \text{ gallons}}{27 \text{ minutes}}$; 440 gallons per minute

46. PRESIDENTIAL PAY The president of the United States receives a salary of $200,000 per year. Write a ratio of dollars to days, and give the daily rate of presidential pay, rounded to the nearest cent. $\frac{\$200,000}{365 \text{ days}}$; $547.95 per day

47. SALES TAX A sweater cost $36.75 after sales tax had been added. Find the tax rate as a percent if the sweater retailed for $35. 5%

48. REAL ESTATE TAXES The real estate taxes on a summer home assessed at $75,000 were $1,500. Find the tax rate as a percent. 2%

49. RATE OF SPEED A car travels 325 miles in 5 hours. Find its rate of speed in miles per hour. 65 mph

50. RATE OF SPEED An airplane travels from Chicago to San Francisco, a distance of 1,883 miles, in 3.5 hours. Find the rate of speed of the plane. 538 mph

51. COMPARING GAS MILEAGE One car went 1,235 miles on 51.3 gallons of gasoline, and another went 1,456 miles on 55.78 gallons. Which car has the better mpg rating? the second car

52. COMPARING ELECTRIC RATES In one community, a bill for 575 kilowatt hours (kwh) of electricity was $38.81. In a second community, a bill for 831 kwh was $58.10. In which community is electricity cheaper? the first community

53. DIVORCE RATE See the graph in Illustration 5.
   a. Express the 1920 U.S. marriage/divorce ratio as a fraction. Explain what it means.
   $\frac{7.5 \text{ marriages}}{1 \text{ divorce}}$; for every 7.5 marriages, there was 1 divorce
   b. According to the graph, in 1996 for every 2 marriages, there was (were) 1 divorce(s).
   c. How has the marriage/divorce ratio changed from 1920 to 1996?
   The number of divorces for a given number of marriages is increasing.

ILLUSTRATION 5

United States Marriage/Divorce Ratio

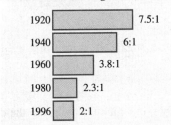

Year	Ratio
1920	7.5:1
1940	6:1
1960	3.8:1
1980	2.3:1
1996	2:1

54. SPREADING TECHNOLOGY For each of the regions listed in Illustration 6, write the ratio of the number of homes with TV to the number of homes without TV. U.S.: $\frac{95}{2}$; W.E.: $\frac{142}{18} = \frac{71}{9}$; A.P.: $\frac{390}{197}$; L.A.: $\frac{83}{13}$

ILLUSTRATION 6

	United States	Western Europe	Asia-Pacific	Latin America
Homes (millions)	97	160	587	96
Homes with TV (millions)	95	142	390	83

Based on data from *Los Angeles Times* (June 6, 1997), page D4

55. SCALE DRAWINGS The map of Tokyo in Illustration 7 provides a reduced representation of a very large area. To express the relationship between a length on the map and the actual length of the geographic area it portrays, a scale is used. The scale for this map is written as a ratio. Explain its meaning. The units used are inches.

One inch on the map represents 1,000,000 inches on the earth.

ILLUSTRATION 7

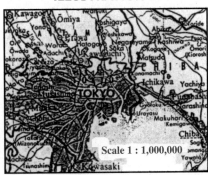

Scale 1 : 1,000,000

56. HAWAII'S AGRICULTURE The graph in Illustration 8 shows sales of sugar and pineapple and all other agricultural products of Hawaii.

   a. What was the ratio of sugar and pineapple sales to the sales of all other agricultural products in 1986 and in 1995? $\frac{340}{240} = \frac{17}{12}$; $\frac{215}{275} = \frac{43}{55}$

   b. Approximately when was the ratio 1:1? mid-1991

   c. Explain what we can learn about Hawaii's agriculture from the change in the ratios.

   Hawaii has steadily moved away from dependence on the two traditional crops to other agricultural products.

ILLUSTRATION 8

Hawaiian Agriculture

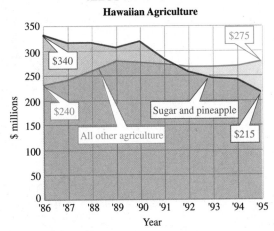

Based on data from the Hawaii Agricultural Statistics Service

WRITING

57. Some people think that the word *ratio* comes from the words *rational number*. Explain why this may be true.

58. In the fraction $\frac{a}{b}$, b cannot be zero. Explain why. In the ratio $\frac{a}{b}$, b can be zero. Explain why.

REVIEW

In Exercises 59–62, solve each equation.

59. $0.2x + 4 = 3.8$ -1

60. $\frac{x}{2} - 4 = 38$ 84

61. $3(x + 2) = 24$ 6

62. $\frac{x - 6}{3} = 20$ 66

In Exercises 63–66, find each square root. All variables represent positive numbers.

63. $\sqrt{16}$ 4

64. $\sqrt{25x^4}$ $5x^2$

65. $\sqrt{81x^4y^8}$ $9x^2y^4$

66. $\sqrt{\dfrac{x^{12}}{25y^4}}$ $\dfrac{x^6}{5y^2}$

▶ 7.2   Proportions and Similar Triangles

In this section, you will learn about

     Proportions ■ Solving proportions ■ Problem solving ■ Similar triangles

Introduction   Consider the following table, in which we are given the costs of various numbers of gallons of gasoline.

Gallons of gas	Cost
2	$ 2.72
5	$ 6.80
8	$10.88
12	$16.32
20	$27.20

If we find the ratios of the costs to the number of gallons purchased, we will see that they are equal. In this example, each ratio represents the cost of 1 gallon of gasoline, which is $1.36 per gallon.

$$\frac{\$2.72}{2} = \$1.36, \qquad \frac{\$6.80}{5} = \$1.36, \qquad \frac{\$10.88}{8} = \$1.36,$$

$$\frac{\$16.32}{12} = \$1.36, \qquad \text{and} \qquad \frac{\$27.20}{20} = \$1.36$$

When two ratios such as $\frac{\$2.72}{2}$ and $\frac{\$6.80}{5}$ are equal, they form a *proportion*. In this section, we will discuss proportions and use them to solve problems.

Proportions

Proportions

A **proportion** is a statement that two ratios are equal.

Some examples of proportions are

$$\frac{1}{2} = \frac{3}{6}, \qquad \frac{7}{3} = \frac{21}{9}, \qquad \frac{8x}{1} = \frac{40x}{5}, \qquad \text{and} \qquad \frac{a}{b} = \frac{c}{d}$$

■   The proportion $\frac{1}{2} = \frac{3}{6}$ can be read as "1 is to 2 as 3 is to 6."

■   The proportion $\frac{7}{3} = \frac{21}{9}$ can be read as "7 is to 3 as 21 is to 9."

■   The proportion $\frac{8x}{1} = \frac{40x}{5}$ can be read as "8x is to 1 as 40x is to 5."

■   The proportion $\frac{a}{b} = \frac{c}{d}$ can be read as "a is to b as c is to d."

The terms of the proportion $\frac{a}{b} = \frac{c}{d}$ are numbered as follows:

First term ⟶ a c ⟵ Third term
$$\frac{a}{b} = \frac{c}{d}$$
Second term ⟶ b d ⟵ Fourth term

In the proportion $\frac{1}{2} = \frac{3}{6}$, the numbers 1 and 6 are called the **extremes,** and the numbers 2 and 3 are called the **means.**

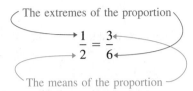

The extremes of the proportion

$$\frac{1}{2} = \frac{3}{6}$$

The means of the proportion

In this proportion, the product of the extremes is equal to the product of the means.

$$1 \cdot 6 = 6 \quad \text{and} \quad 2 \cdot 3 = 6$$

This illustrates a fundamental property of proportions.

Fundamental property of proportions

In any proportion, the product of the extremes is equal to the product of the means.

In the proportion $\frac{a}{b} = \frac{c}{d}$, a and d are the extremes, and b and c are the means. We can show that the product of the extremes (ad) is equal to the product of the means (bc) by multiplying both sides of the proportion by bd and observing that $ad = bc$.

$$\frac{a}{b} = \frac{c}{d}$$

$$\frac{bd}{1} \cdot \frac{a}{b} = \frac{bd}{1} \cdot \frac{c}{d} \qquad \text{To eliminate the fractions, multiply both sides by } \frac{bd}{1}.$$

$$\frac{abd}{b} = \frac{bcd}{d} \qquad \text{Multiply the numerators and multiply the denominators.}$$

$$ad = bc \qquad \text{Divide out the common factors: } \frac{b}{b} = 1 \text{ and } \frac{d}{d} = 1.$$

Since $ad = bc$, the product of the extremes equals the product of the means.

To determine whether an equation is a proportion, we can check to see whether the product of the extremes is equal to the product of the means.

EXAMPLE 1

Determining whether an equation is a proportion. Determine whether each equation is a proportion: **a.** $\frac{3}{7} = \frac{9}{21}$ and **b.** $\frac{8}{3} = \frac{13}{5}$.

Solution

In each case, we check to see whether the product of the extremes is equal to the product of the means.

a. The product of the extremes is $3 \cdot 21 = 63$. The product of the means is $7 \cdot 9 = 63$. Since the products are equal, the equation is a proportion: $\frac{3}{7} = \frac{9}{21}$.

b. The product of the extremes is $8 \cdot 5 = 40$. The product of the means is $3 \cdot 13 = 39$. Since the products are not equal, the equation is not a proportion: $\frac{8}{3} \neq \frac{13}{5}$.

SELF CHECK

Determine whether this equation is a proportion: $\frac{6}{13} = \frac{24}{53}$. *Answer:* **no** ∎

When two pairs of numbers such as 2, 3 and 8, 12 form a proportion, we say that they are **proportional.** To show that 2, 3 and 8, 12 are proportional, we check to see whether the equation

$$\frac{2}{3} = \frac{8}{12}$$

is a proportion. To do so, we find the product of the extremes and the product of the means:

$$2 \cdot 12 = 24 \qquad 3 \cdot 8 = 24$$

Since the products are equal, the equation is a proportion, and the numbers are proportional.

EXAMPLE 2

Determining whether numbers are proportional. Determine whether 3, 7 and 36, 91 are proportional.

Solution   We check to see whether $\frac{3}{7} = \frac{36}{91}$ is a proportion by finding two products:

$$3 \cdot 91 = 273 \quad \text{The product of the extremes.}$$
$$7 \cdot 36 = 252 \quad \text{The product of the means.}$$

Since the products are not equal, the numbers are not proportional.

SELF CHECK   Determine whether 6, 11 and 54, 99 are proportional.   *Answer:* yes   ■

Solving Proportions

Suppose that we know three terms in the proportion

$$\frac{x}{5} = \frac{24}{20}$$

To find the unknown term, we multiply the extremes and multiply the means, set them equal, and solve for x:

$$\frac{x}{5} = \frac{24}{20}$$

$$20 \cdot x = 5 \cdot 24 \quad \text{In a proportion, the product of the extremes is equal to the product of the means.}$$

$$20x = 120 \quad \text{Multiply: } 5 \cdot 24 = 120.$$

$$\frac{20x}{20} = \frac{120}{20} \quad \text{To undo the multiplication by 20, divide both sides by 20.}$$

$$x = 6 \quad \text{Simplify: } \frac{120}{20} = 6.$$

The first term is 6.

EXAMPLE 3

Solving proportions. Solve $\dfrac{12}{18} = \dfrac{3}{x}$.

Solution
$$\frac{12}{18} = \frac{3}{x}$$

$$12 \cdot x = 18 \cdot 3 \quad \text{In a proportion, the product of the extremes equals the product of the means.}$$

$$12x = 54 \quad \text{Multiply: } 18 \cdot 3 = 54.$$

$$\frac{12x}{12} = \frac{54}{12} \quad \text{To undo the multiplication by 12, divide both sides by 12.}$$

$$x = \frac{9}{2} \quad \text{Simplify: } \frac{54}{12} = \frac{9}{2}.$$

Thus, $x = \frac{9}{2}$.

SELF CHECK   Solve $\frac{15}{x} = \frac{25}{40}$ *Answer:* 24   ■

EXAMPLE 4

Solving proportions. Find the third term of the proportion $\dfrac{3.5}{7.2} = \dfrac{x}{15.84}$.

Solution

$$\frac{3.5}{7.2} = \frac{x}{15.84}$$

$3.5(15.84) = 7.2x$ In a proportion, the product of the extremes equals the product of the means.

$55.44 = 7.2x$ Multiply: $3.5(15.84) = 55.44$.

$\dfrac{55.44}{7.2} = \dfrac{7.2x}{7.2}$ To undo the multiplication by 7.2, divide both sides by 7.2.

$7.7 = x$ Simplify: $\frac{55.44}{7.2} = 7.7$.

The third term is 7.7

SELF CHECK Find the second term of the proportion $\frac{6.7}{x} = \frac{33.5}{38}$ *Answer:* **7.6** ∎

ACCENT ON TECHNOLOGY *Solving Proportions with a Calculator*

To solve the proportion in Example 4 with a calculator, we can proceed as follows.

$$\frac{3.5}{7.2} = \frac{x}{15.84}$$

$\dfrac{3.5(15.84)}{7.2} = x$ Multiply both sides by 15.84.

We can find x by entering these numbers into a scientific calculator:

Keystrokes 3.5 $\boxed{\times}$ 15.84 $\boxed{\div}$ 7.2 $\boxed{=}$ $\boxed{\quad\quad\quad\quad 7.7\ }$

Using a graphing calculator, we enter these numbers and press these keys:

Keystrokes 3.5 $\boxed{\times}$ 15.84 $\boxed{\div}$ 7.2 $\boxed{\text{ENTER}}$ $\boxed{\begin{array}{l}\text{3.5*15.84/7.2} \\ \qquad\qquad\qquad 7.7\end{array}}$

Thus, $x = 7.7$.

EXAMPLE 5

Solving proportions. Solve $\dfrac{2x+1}{4} = \dfrac{10}{8}$.

Solution

$$\frac{2x+1}{4} = \frac{10}{8}$$

$8(2x+1) = 40$ In a proportion, the product of the extremes equals the product of the means.

$16x + 8 = 40$ Use the distributive property to remove parentheses.

$16x + 8 - 8 = 40 - 8$ To undo the addition of 8, subtract 8 from both sides.

$16x = 32$ Combine like terms.

$\dfrac{16x}{16} = \dfrac{32}{16}$ To undo the multiplication by 16, divide both sides by 16.

$x = 2$ Simplify: $\frac{32}{16} = 2$.

Thus, $x = 2$.

SELF CHECK   Solve $\frac{3x-1}{2} = \frac{12.5}{5}$.   *Answer:* 2   ■

Problem Solving

We can use proportions to solve problems.

EXAMPLE 6	**Grocery shopping.** If 6 apples cost $1.38, how much will 16 apples cost?

Solution

ANALYZE THE PROBLEM   We know the cost of 6 apples; we are to find the cost of 16 apples.

FORM AN EQUATION   Let c represent the cost of 16 apples. The ratios of the numbers of apples to their costs are equal.

6 apples is to $1.38 as 16 apples is to $c.

$$\begin{array}{c} 6 \text{ apples} \longrightarrow \\ \text{Cost of 6 apples} \longrightarrow \end{array} \frac{6}{1.38} = \frac{16}{c} \begin{array}{c} \longleftarrow 16 \text{ apples} \\ \longleftarrow \text{Cost of 16 apples} \end{array}$$

SOLVE THE EQUATION

$6 \cdot c = 1.38(16)$ In a proportion, the product of the extremes is equal to the product of the means.

$6c = 22.08$ Do the multiplication: $1.38(16) = 22.08$.

$\dfrac{6c}{6} = \dfrac{22.08}{6}$ To undo the multiplication by 6, divide both sides by 6.

$c = 3.68$ Simplify: $\frac{22.08}{6} = 3.68$.

STATE THE CONCLUSION   Sixteen apples will cost $3.68.

CHECK THE RESULT   If 16 apples are bought, this is *about* 3 times as many as a purchase of 6 apples, which cost $1.38. If we multiply $1.38 by 3, we get an estimate of the cost of 16 apples. $1.38 \cdot 3 = \$4.14$. The result of $3.68 seems reasonable.

SELF CHECK   If 9 tickets to a concert cost $112.50, how much will 15 tickets cost?   *Answer:* $187.50   ■

EXAMPLE 7	**Mixing solutions.** A solution contains 2 quarts of antifreeze and 5 quarts of water. How many quarts of antifreeze must be mixed with 18 quarts of water to have the same concentration?

Solution

ANALYZE THE PROBLEM   We want to maintain the same concentration of antifreeze to water. This problem can be solved using a proportion.

FORM AN EQUATION   Let q represent the number of quarts of antifreeze to be mixed with the water. The ratios of the quarts of antifreeze to the quarts of water are equal.

2 quarts antifreeze is to 5 quarts water as q quarts antifreeze is to 18 quarts water.

$$\begin{array}{c} 2 \text{ quarts antifreeze} \longrightarrow \\ 5 \text{ quarts water} \longrightarrow \end{array} \frac{2}{5} = \frac{q}{18} \begin{array}{c} \longleftarrow \text{quarts of antifreeze} \\ \longleftarrow 18 \text{ quarts of water} \end{array}$$

SOLVE THE EQUATION

$2 \cdot 18 = 5q$ In a proportion, the product of the extremes is equal to the product of the means.

$36 = 5q$ Do the multiplication: $2 \cdot 18 = 36$.

$\dfrac{36}{5} = \dfrac{5q}{5}$ To undo the multiplication by 5, divide both sides by 5.

$\dfrac{36}{5} = q$

STATE THE CONCLUSION   The mixture should contain $\frac{36}{5}$ or 7.2 quarts of antifreeze.

CHECK THE RESULT   See if the result seems reasonable.

SELF CHECK   A solution should contain 2 ounces of alcohol for every 7 ounces of water. How much alcohol should be added to 20 ounces of water to get the proper concentration?   *Answer:* $\frac{40}{7}$ oz   ∎

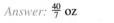

EXAMPLE 8

Baking. A recipe for rhubarb cake calls for $1\frac{1}{4}$ cups of sugar for every $2\frac{1}{2}$ cups of flour. How many cups of flour are needed if the baker intends to use 3 cups of sugar?

Solution

ANALYZE THE PROBLEM   The baker needs to maintain the same ratio between the sugar and flour as is called for in the original recipe.

FORM AN EQUATION   Let f represent the number of cups of flour to be mixed with the 3 cups of sugar. The ratios of the cups of sugar to the cups of flour are equal.

$1\frac{1}{4}$ cups sugar is to $2\frac{1}{2}$ cups flour as 3 cups sugar is to f cups flour.

$1\frac{1}{4}$ cups sugar ⟶ $\dfrac{1\frac{1}{4}}{2\frac{1}{2}} = \dfrac{3}{f}$ ⟵ 3 cups sugar
$2\frac{1}{2}$ cups flour ⟶   ⟵ f cups flour

SOLVE THE EQUATION

$\dfrac{1.25}{2.5} = \dfrac{3}{f}$ Change the fractions to decimals.

$1.25f = 2.5 \cdot 3$ In a proportion, the product of the extremes is equal to the product of the means.

$1.25f = 7.5$ Do the multiplication: $2.5 \cdot 3 = 7.5$.

$\dfrac{1.25f}{1.25} = \dfrac{7.5}{1.25}$ To undo the multiplication by 1.25, divide both sides by 1.25.

$f = 6$ Divide: $\frac{7.5}{1.25} = 6$.

STATE THE CONCLUSION   The baker should use 6 cups of flour.

CHECK THE RESULT   The recipe calls for about 2 cups of flour for about 1 cup of sugar. If 3 cups of sugar are used, 6 cups of flour seems reasonable.

SELF CHECK   How many cups of sugar will be needed to make several cakes that will require a total of 25 cups of flour?   *Answer:* $12\frac{1}{2}$ ∎

Similar Triangles

If two angles of one triangle have the same measures as two angles of a second triangle, the triangles will have the same shape. Triangles with the same shape are called **similar triangles.** In Figure 7-2, $\triangle ABC \sim \triangle DEF$. (Read the symbol $\sim$ as "is similar to.")

FIGURE 7-2

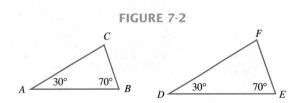

Property of similar triangles

If two triangles are **similar,** all pairs of corresponding sides are in proportion.

In the similar triangles shown in Figure 7-2, the following proportions are true.

$$\frac{AB}{DE} = \frac{BC}{EF}, \qquad \frac{BC}{EF} = \frac{CA}{FD}, \qquad \text{and} \qquad \frac{CA}{FD} = \frac{AB}{DE}$$ Read AB as "the length of segment AB."

EXAMPLE 9

Finding the height of a tree. A tree casts a shadow 18 feet long at the same time as a woman 5 feet tall casts a shadow 1.5 feet long. Find the height of the tree.

Solution

ANALYZE THE PROBLEM   Figure 7-3 shows the triangles determined by the tree and its shadow and the woman and her shadow. Since the triangles have the same shape, they are similar, and the lengths of their corresponding sides are in proportion.

FIGURE 7-3

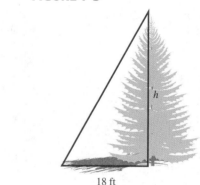

5 ft

h

1.5 ft

18 ft

FORM AN EQUATION   If we let h represent the height of the tree, we can find h by solving the following proportion.

$$\frac{h}{5} = \frac{18}{1.5} \qquad \frac{\text{Height of the tree}}{\text{Height of the woman}} = \frac{\text{Length of shadow of the tree}}{\text{Length of shadow of the woman}}$$

SOLVE THE EQUATION $1.5h = 5(18)$ In a proportion, the product of the extremes is equal to the product of the means.

$1.5h = 90$ Do the multiplication.

$h = 60$ To undo the multiplication by 1.5, divide both sides by 1.5 and simplify.

STATE THE CONCLUSION   The tree is 60 feet tall.

CHECK THE RESULT $\frac{18}{1.5} = 12$ and $\frac{60}{5} = 12$. The ratios are the same. The solution checks.

SELF CHECK

Find the height of the tree in Example 9 if the woman is 5 feet, 6 inches tall.   *Answer:* **66 ft**   ∎

STUDY SET

Section 7.2

VOCABULARY

In Exercises 1–6, fill in the blanks to make the statements true.

1. A ___proportion___ is a statement that two ratios are equal.

2. In the proportion $\frac{a}{b} = \frac{c}{d}$, a and d are called the ___extremes___ of the proportion.

3. The second and third terms of a proportion are called the ___means___ of the proportion.

4. When two pairs of numbers form a proportion, we say that the numbers are ___proportional___.

5. If two triangles have the same ___shape___, they are said to be similar.

6. If two triangles are similar, their corresponding sides are in ___proportion___.

CONCEPTS

In Exercises 7–8, fill in the blanks to make the statements true.

7. The equation $\frac{a}{b} = \frac{c}{d}$ is a proportion if the product ___ad___ is equal to the product ___bc___.

8. If $3 \cdot 10 = x \cdot 17$, then ___$\frac{3}{x} = \frac{17}{10}$___ is a proportion.

9. DEFENSE SPENDING See Illustration 1.
  a. Find the ratio of defense spending to the gross domestic product for the United States and then for China. U.S.: $\frac{278}{7,316} = \frac{139}{3,658}$; China: $\frac{32}{561}$
  b. Set the two ratios from part a equal. Is this equation a proportion? Explain what this means in terms of defense spending by the two countries.
  No; their ratios of defense spending to gross domestic product are different. China's is larger.

ILLUSTRATION 1

Country	Defense spending (billions)	Gross domestic product (billions)
United States	$278	$7,316
China	$ 32	$ 561

Based on data from the International Institute for Strategic Studies

10. IMPORT/EXPORT RATIO Examine the graph in Illustration 2 to determine the year when the ratio of imports to exports was the closest to 1:1. Explain in words what this 1:1 ratio means.
  1994; it means the export and import totals were the same.

ILLUSTRATION 2

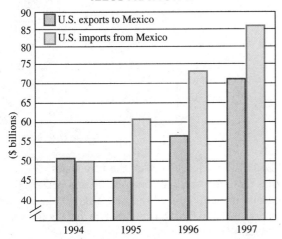

Based on data from the Census Bureau, Foreign Trade Division

In Exercises 11–12, complete each solution.

11. Solve for x: $\frac{12}{18} = \frac{x}{24}$.

$$12 \cdot 24 = 18 \cdot \boxed{x}$$
$$\boxed{288} = 18x$$
$$\frac{288}{\boxed{18}} = \frac{18x}{\boxed{18}}$$
$$16 = x$$

12. Solve for x: $\frac{14}{x} = \frac{49}{17.5}$.

$$14 \cdot \boxed{17.5} = 49x$$
$$\boxed{245} = 49x$$
$$\frac{245}{\boxed{49}} = \frac{49x}{\boxed{49}}$$
$$5 = x$$

13. Read $\triangle ABC$ as ___triangle___ ABC.

14. The symbol $\sim$ is read as ___is similar to___.

PRACTICE

In Exercises 15–22, tell whether each statement is a proportion.

15. $\frac{9}{7} = \frac{81}{70}$ no

16. $\frac{5}{2} = \frac{20}{8}$ yes

17. $\frac{7}{3} = \frac{14}{6}$ yes

18. $\frac{13}{19} = \frac{65}{95}$ yes

19. $\frac{9}{19} = \frac{38}{80}$ no

20. $\frac{40}{29} = \frac{29}{22}$ no

21. $\frac{10.4}{3.6} = \frac{41.6}{14.4}$ yes

22. $\frac{13.23}{3.45} = \frac{39.96}{11.35}$ no

In Exercises 23–38, solve for the variable in each proportion.

23. $\frac{2}{3} = \frac{x}{6}$ 4

24. $\frac{3}{6} = \frac{x}{8}$ 4

25. $\frac{5}{10} = \frac{3}{c}$ 6

26. $\frac{7}{14} = \frac{2}{x}$ 4

27. $\frac{6}{x} = \frac{8}{4}$ 3

28. $\frac{4}{x} = \frac{2}{8}$ 16

29. $\frac{x}{3} = \frac{9}{3}$ 9

30. $\frac{x}{2} = \frac{18}{6}$ 6

31. $\frac{x+1}{5} = \frac{3}{15}$ 0

32. $\frac{x-1}{7} = \frac{2}{21}$ $\frac{5}{3}$

33. $\frac{x+3}{12} = \frac{-7}{6}$ −17

34. $\frac{x+7}{-4} = \frac{1}{4}$ −8

✓ **35.** $\dfrac{4-x}{13} = \dfrac{11}{26}$ $-\dfrac{3}{2}$ **36.** $\dfrac{5-x}{17} = \dfrac{13}{34}$ $-\dfrac{3}{2}$

37. $\dfrac{2x+1}{18} = \dfrac{14}{3}$ $\dfrac{83}{2}$ **38.** $\dfrac{2x-1}{18} = \dfrac{9}{54}$ 2

▦ APPLICATIONS

In Exercises 39–58, set up and solve a proportion. Use a calculator when it is helpful.

✓ **39.** GROCERY SHOPPING If 3 pints of yogurt cost $1, how much will 51 pints cost? $17

40. SHOPPING FOR CLOTHES If shirts are on sale at two for $25, how much will five shirts cost? $62.50

41. ADVERTISING In 1997, a 30-second TV ad during the Super Bowl telecast cost $1.2 million. At this rate, what was the cost of a 45-second ad? $1.8 million

42. COOKING A recipe for spaghetti sauce requires four 16-ounce bottles of ketchup to make two gallons of sauce. How many bottles of ketchup are needed to make 10 gallons of sauce? 20

✓ **43.** MIXING PERFUME A perfume is to be mixed in the ratio of 3 drops of pure essence to 7 drops of alcohol. How many drops of pure essence should be mixed with 56 drops of alcohol? 24

44. CPR A first aid handbook states that when performing cardiopulmonary resuscitation on an adult, the ratio of chest compressions to breaths should be 5:2. If 210 compressions were administered to an adult patient, how many breaths should have been given? 84

45. COOKING A recipe for wild rice soup is shown in Illustration 3. Find the amount of each ingredient needed to make 15 servings. $7\frac{1}{2}, 1\frac{2}{3}, \frac{5}{8}, 1\frac{1}{4}, 2\frac{1}{2}, 5, \frac{5}{16}$

ILLUSTRATION 3

Wild Rice Soup

A sumptuous side dish with a nutty flavor

3 cups chicken broth   1 cup light cream

$\frac{2}{3}$ cup uncooked rice   2 tablespoons flour

$\frac{1}{4}$ cup sliced onions $\frac{1}{8}$ teaspoon pepper

$\frac{1}{2}$ cup shredded carrots

Serves: 6

46. PHOTO ENLARGEMENTS In Illustration 4, the 3-by-5 photo is to be blown up to the larger size. Find *x*. $3\frac{3}{4}$ in.

ILLUSTRATION 4

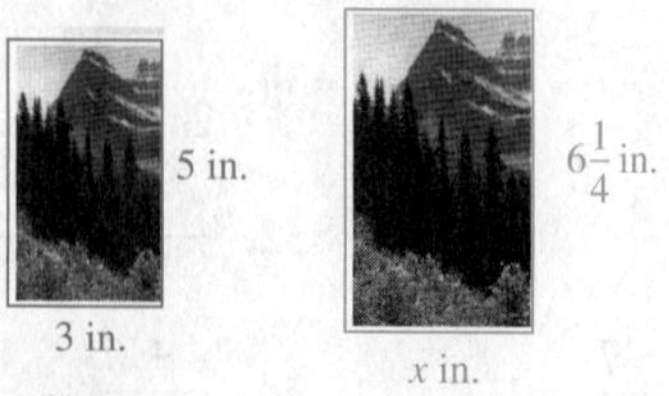

5 in. $6\frac{1}{4}$ in.

3 in.   *x* in.

✓ **47.** QUALITY CONTROL In a manufacturing process, 95% of the parts made are to be within specifications. How many defective parts would be expected in a run of 940 pieces? 47

48. QUALITY CONTROL Out of a sample of 500 men's shirts, 17 were rejected because of crooked collars. How many crooked collars would you expect to find in a run of 15,000 shirts? 510

49. GAS CONSUMPTION If a car can travel 42 miles on 1 gallon of gas, how much gas is needed to travel 315 miles? $7\frac{1}{2}$ gal

50. RAPPERS According to the *Guinness Book of World Records 1998*, Rebel X.D. of Chicago, IL rapped 674 syllables in 54.9 seconds. At this rate, how many syllables could he rap in 1 minute? Round to the nearest syllable. 737

✓ **51.** BANKRUPTCY After filing for bankruptcy, a company was only able to pay its creditors 15 cents on the dollar. If the company owed a lumberyard $9,712, how much could the lumberyard expect to be paid? $1,456.80

52. COMPUTING PAYCHECKS Billie earns $412 for a 40-hour week. If she missed 10 hours of work last week, how much did she get paid? $309

53. MODEL RAILROADING A model railroad engine is 9 inches long. If the scale is 87 feet to 1 foot, how long is a real engine? 65 ft, 3 in.

54. MODEL RAILROADING A model railroad caboose is 3.5 inches long. If the scale is 169 feet to 1 foot, how long is a real caboose? 49 ft, $3\frac{1}{2}$ in.

✓ **55.** NUTRITION Illustration 5 shows the nutritional facts about a 10-oz chocolate milkshake sold by a fast-food restaurant. Use the information to complete the table for the 16-oz shake. Round to the nearest unit when an answer is not exact.

ILLUSTRATION 5

	Calories	Fat (gm)	Protein (gm)
10-oz chocolate milkshake	355	8	9
16-oz chocolate milkshake	568	13	14

56. DRIVER'S LICENSES Of the fifty states, Oregon has the largest ratio of licensed drivers per 1,000 residents. If the ratio is 824 to 1,000 and Oregon's population is 3,140,000, how many Oregonians have a driver's license? 2,587,360

57. MIXING FUEL The instructions on a can of oil intended to be added to lawnmower gasoline read as follows:

Recommended	Gasoline	Oil
50 to 1	6 gal	16 oz

Are these instructions correct? (*Hint:* There are 128 ounces in 1 gallon.)   not exactly, but close

58. MIXING FUEL In Exercise 57, how much oil should be mixed with 28 gallons of gas?   0.56 gal (71.68 oz)

In Exercises 59–64, use similar triangles to solve each problem.

59. HEIGHT OF A TREE A tree casts a shadow of 26 feet at the same time as a 6-foot man casts a shadow of 4 feet. (See Illustration 6.) Find the height of the tree.   39 ft

ILLUSTRATION 6

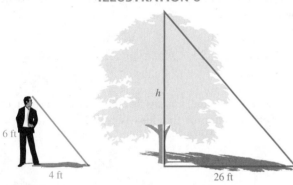

6 ft

4 ft   26 ft

60. HEIGHT OF A BUILDING A man places a mirror on the ground and sees the reflection of the top of a building, as shown in Illustration 7. The two triangles in the illustration are similar. Find the height, h, of the building.   25 ft

ILLUSTRATION 7

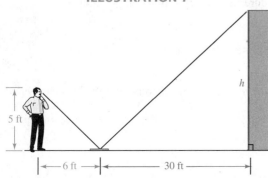

5 ft

h

← 6 ft → ← 30 ft →

61. WIDTH OF A RIVER Use the dimensions in Illustration 8 to find w, the width of the river. The two triangles in the illustration are similar. $46\frac{7}{8}$ ft

ILLUSTRATION 8

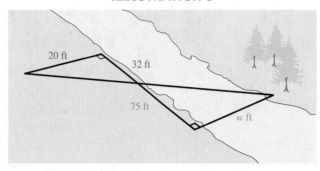

20 ft   32 ft

75 ft

w ft

62. FLIGHT PATH An airplane ascends 100 feet as it flies a horizontal distance of 1,000 feet. How much altitude will it gain as it flies a horizontal distance of 1 mile? See Illustration 9. (*Hint:* 5,280 feet = 1 mile.)   528 ft

ILLUSTRATION 9

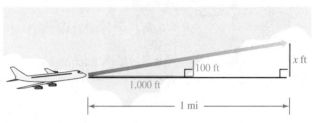

100 ft x ft

1,000 ft

1 mi

63. FLIGHT PATH An airplane descends 1,350 feet as it flies a horizontal distance of 1 mile. How much altitude is lost as it flies a horizontal distance of 5 miles?   6,750 ft

64. SKI RUNS A ski course falls 100 feet in every 300 feet of horizontal run. If the total horizontal run is $\frac{1}{2}$ mile, find the height of the hill.   880 ft

WRITING

65. Explain the difference between a ratio and a proportion.

66. Explain how to tell whether the equation $\frac{3.2}{3.7} = \frac{5.44}{6.29}$ is a proportion.

REVIEW

67. Change $\frac{9}{10}$ to a percent.   90%

68. Change $\frac{7}{8}$ to a percent. $87\frac{1}{2}$%

69. Change $33\frac{1}{3}$% to a fraction. $\frac{1}{3}$

70. Change 75% to a fraction. $\frac{3}{4}$

71. Find 30% of 1,600.   480

72. Find $\frac{1}{2}$% of 520.   2.6

73. SHOPPING Maria bought a dress for 25% off the original price of $98. How much did the dress cost? $73.50

74. SHOPPING Liz purchased a shirt on sale for $17.50. Find the original cost of the shirt if it was marked down 30%. $25

▶ 7.3   Rational Expressions and Rational Functions

In this section, you will learn about

Simplifying rational expressions ■ Division by 1 ■ Dividing polynomials that are negatives ■ Rational functions ■ Graphing rational functions

Introduction   Fractions such as $\frac{1}{2}$ and $\frac{3}{4}$ that are the quotient of two integers are *rational numbers*. Fractions such as

$$\frac{3}{2y}, \qquad \frac{x}{x+2}, \qquad \text{and} \qquad \frac{5a^2 + b^2}{3a - b}$$

where the numerators and denominators are polynomials, are called **algebraic fractions** or **rational expressions.**

Simplifying Rational Expressions

A fraction can be simplified by dividing out common factors shared by its numerator and denominator. For example,

$$\frac{18}{30} = \frac{3 \cdot 6}{5 \cdot 6} = \frac{3 \cdot \overset{1}{\cancel{6}}}{5 \cdot \cancel{6}} = \frac{3}{5} \qquad \text{and} \qquad -\frac{6}{15} = -\frac{3 \cdot 2}{3 \cdot 5} = -\frac{\overset{1}{\cancel{3}} \cdot 2}{\cancel{3} \cdot 5} = -\frac{2}{5}$$

When all common factors have been divided out, we say that the fraction has been **expressed in lowest terms**. The generalization of this idea is called the *fundamental property of fractions*.

Fundamental property of fractions

If a is a real number and b and c are nonzero real numbers, then

$$\frac{ac}{bc} = \frac{a}{b}$$

We can use the same procedure to simplify rational expressions.

EXAMPLE 1

Simplifying rational expressions. Simplify $\dfrac{21x^2 y}{14xy^2}$.

Solution   We look for common factors in the numerator and denominator that can be divided out.

$$\frac{21x^2 y}{14xy^2} = \frac{3 \cdot 7 \cdot x \cdot x \cdot y}{2 \cdot 7 \cdot x \cdot y \cdot y}$$

   Factor the numerator and denominator.

$$= \frac{3 \cdot \overset{1}{\cancel{7}} \cdot \overset{1}{\cancel{x}} \cdot x \cdot \overset{1}{\cancel{y}}}{2 \cdot \underset{1}{\cancel{7}} \cdot \underset{1}{\cancel{x}} \cdot y \cdot \underset{1}{\cancel{y}}}$$

   Divide out the common factors of 7, x, and y.

$$= \frac{3x}{2y}$$

   Do the multiplications in the numerator and in the denominator.

We can also simplify by using the rules of exponents:

$$\frac{21x^2y}{14xy^2} = \frac{3 \cdot 7}{2 \cdot 7}x^{2-1}y^{1-2} \qquad \frac{x^2}{x} = x^{2-1} \text{ and } \frac{y}{y^2} = y^{1-2}.$$

$$= \frac{3}{2}xy^{-1} \qquad 2 - 1 = 1 \text{ and } 1 - 2 = -1.$$

$$= \frac{3}{2} \cdot \frac{x}{y} \qquad y^{-1} = \frac{1}{y}.$$

$$= \frac{3x}{2y} \qquad \text{Multiply.}$$

SELF CHECK   Simplify $\dfrac{32a^3b^2}{24ab^4}$.   *Answer:* $\dfrac{4a^2}{3b^2}$ ∎

EXAMPLE 2   **Factoring to simplify rational expressions.** Write $\dfrac{x^2 + 3x}{3x + 9}$ in lowest terms.

Solution   We note that the terms of the numerator have a common factor of x and the terms of the denominator have a common factor of 3.

$$\frac{x^2 + 3x}{3x + 9} = \frac{x(x + 3)}{3(x + 3)} \qquad \text{Factor the numerator and the denominator.}$$

$$= \frac{x\cancel{(x + 3)}^{\ 1}}{3\cancel{(x + 3)}_{\ 1}} \qquad \text{Divide out the common factor of } x + 3.$$

$$= \frac{x}{3} \qquad \text{Simplify the numerator and the denominator.}$$

SELF CHECK   Simplify $\dfrac{x^2 - 5x}{5x - 25}$.   *Answer:* $\dfrac{x}{5}$ ∎

EXAMPLE 3   **Factoring to simplify rational expressions.** Simplify $\dfrac{x^2 + 13x + 12}{x^2 - 144}$.

Solution   The numerator is a trinomial, and the denominator is a difference of two squares.

$$\frac{x^2 + 13x + 12}{x^2 - 144} = \frac{(x + 1)(x + 12)}{(x + 12)(x - 12)} \qquad \text{Factor the numerator and the denominator.}$$

$$= \frac{(x + 1)\cancel{(x + 12)}^{\ 1}}{\cancel{(x + 12)}_{\ 1}(x - 12)} \qquad \text{Divide out the common factor of } x + 12.$$

$$= \frac{x + 1}{x - 12}$$

SELF CHECK   Simplify $\dfrac{3x^2 - 8x - 3}{x^2 - 9}$.   *Answer:* $\dfrac{3x + 1}{x + 3}$ ∎

Division by 1

Any number or algebraic expression divided by 1 remains unchanged. For example,

$$\frac{37}{1} = 37, \qquad \frac{5x}{1} = 5x, \qquad \text{and} \qquad \frac{3x + y}{1} = 3x + y$$

In general, we have the following.

Division by 1

For any real number a, $\dfrac{a}{1} = a$

E X A M P L E   4 **Simplifying rational expressions.** Simplify $\dfrac{x^3 + x^2}{x + 1}$.

Solution
$$\frac{x^3 + x^2}{x + 1} = \frac{x^2(x + 1)}{x + 1} \qquad \text{Factor the numerator.}$$

$$= \frac{x^2 \overset{1}{\cancel{(x + 1)}}}{\underset{1}{\cancel{x + 1}}} \qquad \text{Divide out the common factor of } x + 1.$$

$$= \frac{x^2}{1} \qquad \text{Simplify the numerator.}$$

$$= x^2 \qquad \text{Denominators of 1 need not be written.}$$

SELF CHECK   Simplify $\dfrac{a^2 + a - 2}{a - 1}$.   *Answer: $a + 2$*

 WARNING! Remember that only *factors* that are common to the *entire numerator* and the *entire denominator* can be divided out. For example, consider the correct simplification

$$\frac{5 + 8}{5} = \frac{13}{5}$$

It would be incorrect to divide out the common *term* of 5 in this simplification. Doing so gives an incorrect answer:

$$\frac{5 + 8}{5} = \frac{\overset{1}{\cancel{5}} + 8}{\underset{1}{\cancel{5}}} = \frac{1 + 8}{1} = 9$$

When simplifying algebraic fractions, it is also incorrect to divide out terms common to both the numerator and denominator.

$$\frac{\overset{1}{\cancel{x}} + 5}{\underset{1}{\cancel{x}} + 6} \qquad \frac{a^2 - 3\overset{1}{\cancel{a}} + \overset{1}{\cancel{2}}}{\underset{1}{\cancel{a}} + \underset{1}{\cancel{2}}} \qquad \frac{\overset{1}{\cancel{y^2}} - 36}{\underset{1}{\cancel{y^2}} - y - 7}$$

EXAMPLE 5

Dividing out common factors. Write $\dfrac{5(x+3)-5}{7(x+3)-7}$ in lowest terms.

Solution We cannot divide out $x+3$, because it is not a factor of the entire numerator, nor is it a factor of the entire denominator. Instead, we simplify the numerator and denominator, factor them, and then divide out any common factors.

$$\frac{5(x+3)-5}{7(x+3)-7}=\frac{5x+15-5}{7x+21-7}\qquad \text{Remove parentheses.}$$

$$=\frac{5x+10}{7x+14}\qquad \text{Combine like terms.}$$

$$=\frac{5(x+2)}{7(x+2)}\qquad \text{Factor the numerator and the denominator.}$$

$$=\frac{\overset{1}{5\cancel{(x+2)}}}{\underset{1}{7\cancel{(x+2)}}}\qquad \text{Divide out the common factor of } x+2.$$

$$=\frac{5}{7}$$

SELF CHECK Simplify $\frac{4(x-2)+4}{3(x-2)+3}$. *Answer:* $\frac{4}{3}$ ∎

EXAMPLE 6

Combining like terms. Simplify $\dfrac{x(x+3)-3(x-1)}{x^2+3}$.

Solution $$\frac{x(x+3)-3(x-1)}{x^2+3}=\frac{x^2+3x-3x+3}{x^2+3}\qquad \text{Remove parentheses in the numerator.}$$

$$=\frac{x^2+3}{x^2+3}\qquad \begin{array}{l}\text{Combine like terms in the numerator:}\\ 3x-3x=0.\end{array}$$

$$=\frac{\overset{1}{\cancel{x^2+3}}}{\underset{1}{\cancel{x^2+3}}}\qquad \text{Divide out the common factor of } x^2+3.$$

$$=1$$

SELF CHECK Simplify $\dfrac{a(a+2)-2(a-1)}{a^2+2}$. *Answer:* **1** ∎

Sometimes a fraction does not simplify. For example, to attempt to simplify

$$\frac{x^2+x-2}{x^2+x}$$

we factor the numerator and the denominator.

$$\frac{x^2+x-2}{x^2+x}=\frac{(x+2)(x-1)}{x(x+1)}$$

Because there are no factors common to the numerator and denominator, this fraction is already in lowest terms.

Dividing Polynomials That Are Negatives

If the terms of two polynomials are the same, except for sign, the polynomials are called **negatives** (**opposites**) of each other. For example, the following pairs are negatives of each other:

$$x - y \quad \text{and} \quad -x + y$$
$$2a - 1 \quad \text{and} \quad -2a + 1$$
$$-3x^2 - 2x + 5 \quad \text{and} \quad 3x^2 + 2x - 5$$

Example 7 shows why the quotient of two binomials that are negatives is always -1.

EXAMPLE 7

Quotients of negatives. Simplify **a.** $\dfrac{x - y}{y - x}$ and **b.** $\dfrac{2a - 1}{1 - 2a}$.

Solution   We can rearrange terms in each numerator, factor out -1, and proceed as follows:

a. $\dfrac{x - y}{y - x} = \dfrac{-y + x}{y - x}$

$\qquad = \dfrac{-(y - x)}{y - x}$

$\qquad = \dfrac{-\overset{1}{\cancel{(y - x)}}}{\underset{1}{\cancel{y - x}}}$

$\qquad = -1$

b. $\dfrac{2a - 1}{1 - 2a} = \dfrac{-1 + 2a}{1 - 2a}$

$\qquad = \dfrac{-(1 - 2a)}{1 - 2a}$

$\qquad = \dfrac{-\overset{1}{\cancel{(1 - 2a)}}}{\underset{1}{\cancel{1 - 2a}}}$

$\qquad = -1$

SELF CHECK   Simplify $\dfrac{3p - 2q}{2q - 3p}$.

Answer: -1 ∎

In general, we have this important fact.

Division of negatives

The quotient of any nonzero expression and its negative is -1.

Rational Functions

Rational expressions can be used to model many situations. In Figure 7-4, for example, the time t (in minutes) that it takes the cardiac rehabilitation patient to complete his $\frac{1}{4}$-mile treadmill workout is given by

$$t = \frac{15}{r}$$

FIGURE 7-4

where r is the rate that he walks (in miles per hour). The expression on the right-hand side of this equation is a rational expression.

The *rational function* that gives the time it takes for the patient to complete the workout, when walking at a rate of r mph, can be written

$$f(r) = \frac{15}{r}$$ Assume that the patient walks at a constant rate r and that $r > 0$.

Rational functions

> A **rational function** is a function whose equation is defined by a rational expression in one variable. The value of the polynomial in the denominator of the expression cannot be zero.

EXAMPLE 8

Evaluating rational functions. Use the function $f(r) = \frac{15}{r}$ to find the time it will take the patient to complete the treadmill workout if he walks at a rate of 3 mph.

Solution

To find the workout time if the patient walks at a rate of 3 mph, we find $f(3)$:

$$f(3) = \frac{15}{3}$$ Input 3 for r.

$$= 5$$ Do the division. The units of the output are minutes.

It will take the patient 5 minutes to complete the workout.

SELF CHECK

Find the time it will take the patient to complete the workout if he walks at a rate of 4 mph.   *Answer:* $3\frac{3}{4}$ min   ∎

Graphing Rational Functions

To graph the rational function $f(r) = \frac{15}{r}$, we substitute values for r (the inputs) in the equation, compute the corresponding values of $f(r)$ (the outputs), and express the results as ordered pairs. From the evaluation in Example 8 and its self check, we know two ordered pairs that satisfy the equation: (3, 5) and $\left(4, 3\frac{3}{4}\right)$. These pairs and others are listed in the table shown in Figure 7-5. (To show the entire graph of the rational function, we have chosen rates of 6, 10, and 15 mph, although they are unrealistic rates for a rehabilitation patient.) We plot each point and draw a smooth curve through them to get the graph.

FIGURE 7-5

$f(r) = \frac{15}{r}$

r	$f(r)$	$(r, f(r))$
1	15	(1, 15)
2	7.5	(2, 7.5)
3	5	(3, 5)
4	$3\frac{3}{4}$	$(4, 3\frac{3}{4})$
5	3	(5, 3)
6	$2\frac{1}{2}$	$(6, 2\frac{1}{2})$
10	$1\frac{1}{2}$	$(10, 1\frac{1}{2})$
15	1	(15, 1)

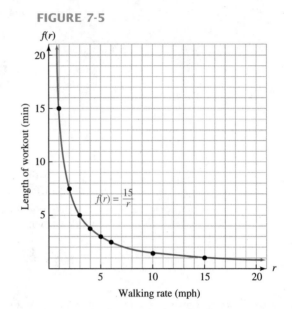

From the graph, we can see that the time to complete the treadmill workout decreases as the patient's rate of walking increases. We note that the graph approaches the x-axis as r increases without bound. When a graph approaches a line, we call the line an **asymptote**. The x-axis is a **horizontal asymptote** of the graph.

As r gets smaller and approaches 0, the graph approaches the y-axis. The y-axis is a **vertical asymptote** of the graph.

ACCENT ON TECHNOLOGY *Graphing Rational Functions*

We can use a graphing calculator to generate tables and graphs for rational functions. For example, Figure 7-6 (a) shows how to enter the function $f(r) = \frac{15}{r}$, using x as the independent variable instead of r. Figures 7-6(b) and (c) show the calculator display of a table and the graph of the function.

FIGURE 7-6

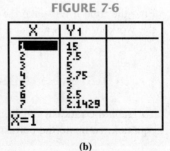

(a)

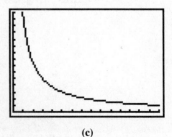

(b)

(c)

STUDY SET

Section 7.3

VOCABULARY

In Exercises 1–6, fill in the blanks to make the statements true.

1. In a fraction, the part above the fraction bar is called the _____numerator_____. In a fraction, the part below the fraction bar is called the _____denominator_____.

2. A fraction that has polynomials in its numerator and denominator, such as $\dfrac{x+2}{x-3}$, is called a _____rational_____ expression.

3. The denominator of a fraction cannot be _0_.

4. $x - 2$ and $2 - x$ are called _____negatives_____ of each other.

5. "To simplify a fraction" means to write it in _____lowest_____ terms.

6. $f(x) = \dfrac{3}{x}$ is a _____rational_____ function, because its equation is defined by a rational expression in one variable.

CONCEPTS

In Exercises 7–10, fill in the blanks to make the statements true.

7. The fundamental property of fractions states that $\dfrac{ac}{bc} = \dfrac{a}{b}$.

8. Any number x divided by 1 is _x_.

9. To simplify a rational expression, we _____factor_____ the numerator and denominator and divide out _____common_____ factors.

10. A rational expression cannot be simplified when it is written in _____lowest_____ terms.

11. Use the graph of a rational function shown below to complete the table.

x	$f(x)$
1	6
2	3
6	1
12	0.5

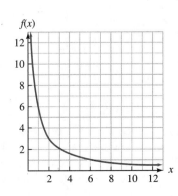

12. Illustration 1 shows a table of values for the rational function $f(x) = \dfrac{5}{x}$. Explain why the word ERROR appears in the second column.

For an input of $x = 0$, the output is $f(0) = \dfrac{5}{0}$, which is undefined.

ILLUSTRATION 1

X	Y1
0	ERROR
1	5
2	2.5
3	1.6667
4	1.25
5	1
6	.83333

X=0

NOTATION

In Exercises 13–14, complete each solution.

13. $\dfrac{x^2 + 5x - 6}{x^2 - 1} = \dfrac{(x + \boxed{6})(x - 1)}{(x + 1)(x - \boxed{1})}$

$= \dfrac{x + 6}{x + 1}$

14. $\dfrac{5(x + 2) - 5}{4(x + 2) - 4} = \dfrac{5x + \boxed{10} - 5}{4x + \boxed{8} - 4}$

$= \dfrac{5x + \boxed{5}}{4x + \boxed{4}}$

$= \dfrac{5\boxed{(x+1)}}{4(x + 1)}$

$= \dfrac{5}{4}$

PRACTICE

In Exercises 15–82 simplify each rational expression. If it is already in lowest terms, so indicate. Assume that no denominators are zero.

✓ 15. $\dfrac{8}{10}$ $\frac{4}{5}$

16. $\dfrac{16}{28}$ $\frac{4}{7}$

17. $\dfrac{28}{35}$ $\frac{4}{5}$

18. $\dfrac{14}{20}$ $\frac{7}{10}$

✓ 19. $\dfrac{8}{52}$ $\frac{2}{13}$

20. $\dfrac{15}{21}$ $\frac{5}{7}$

✓ 21. $\dfrac{10}{45}$ $\frac{2}{9}$

22. $\dfrac{21}{35}$ $\frac{3}{5}$

23. $\dfrac{-18}{54}$ $-\frac{1}{3}$

24. $\dfrac{-16}{40}$ $-\frac{2}{5}$

25. $\dfrac{4x}{2}$ $2x$

26. $\dfrac{2x}{4}$ $\frac{x}{2}$

27. $-\dfrac{6x}{18}$ $-\frac{x}{3}$

28. $-\dfrac{25y}{5}$ $-5y$

29. $\dfrac{45}{9a}$ $\frac{5}{a}$

30. $\dfrac{48}{16y}$ $\frac{3}{y}$

31. $\dfrac{5+5}{5z}$ $\frac{2}{z}$

32. $\dfrac{(3-18)k}{25}$ $-\frac{3k}{5}$

33. $\dfrac{(3+4)a}{24-3}$ $\frac{a}{3}$

34. $\dfrac{x+x}{2}$ x

35. $\dfrac{2x}{3x}$ $\frac{2}{3}$

36. $\dfrac{5y}{7y}$ $\frac{5}{7}$

37. $\dfrac{6x^2}{4x^2}$ $\frac{3}{2}$

38. $\dfrac{9xy}{6xy}$ $\frac{3}{2}$

39. $\dfrac{2x^2}{3y}$ in lowest terms

40. $\dfrac{5y^2}{2y^2}$ $\frac{5}{2}$

41. $\dfrac{15x^2y}{5xy^2}$ $\frac{3x}{y}$

42. $\dfrac{12xz}{4xz^2}$ $\frac{3}{z}$

43. $\dfrac{28x}{32y}$ $\frac{7x}{8y}$

44. $\dfrac{14xz^2}{7x^2z^2}$ $\frac{2}{x}$

45. $\dfrac{x+3}{3(x+3)}$ $\frac{1}{3}$

46. $\dfrac{2(x+7)}{x+7}$ 2

47. $\dfrac{5x+35}{x+7}$ 5

48. $\dfrac{x-9}{3x-27}$ $\frac{1}{3}$

49. $\dfrac{x^2+3x}{2x+6}$ $\frac{x}{2}$

50. $\dfrac{xz-2x}{yz-2y}$ $\frac{x}{y}$

51. $\dfrac{15x-3x^2}{25y-5xy}$ $\frac{3x}{5y}$

52. $\dfrac{3y+xy}{3x+xy}$ in lowest terms

53. $\dfrac{6a-6b+6c}{9a-9b+9c}$ $\frac{2}{3}$

54. $\dfrac{3a-3b-6}{2a-2b-4}$ $\frac{3}{2}$

55. $\dfrac{x-7}{7-x}$ -1

56. $\dfrac{d-c}{c-d}$ -1

57. $\dfrac{6x-3y}{3y-6x}$ -1

58. $\dfrac{3c-4d}{4c-3d}$ in lowest terms

59. $\dfrac{a+b-c}{c-a-b}$ -1

60. $\dfrac{x-y-z}{z+y-x}$ -1

61. $\dfrac{x^2+3x+2}{x^2+x-2}$ $\frac{x+1}{x-1}$

62. $\dfrac{x^2+x-6}{x^2-x-2}$ $\frac{x+3}{x+1}$

63. $\dfrac{x^2-8x+15}{x^2-x-6}$ $\frac{x-5}{x+2}$

64. $\dfrac{x^2-6x-7}{x^2+8x+7}$ $\frac{x-7}{x+7}$

65. $\dfrac{2x^2-8x}{x^2-6x+8}$ $\frac{2x}{x-2}$

66. $\dfrac{3y^2-15y}{y^2-3y-10}$ $\frac{3y}{y+2}$

67. $\dfrac{xy+2x^2}{2xy+y^2}$ $\frac{x}{y}$

68. $\dfrac{3x+3y}{x^2+xy}$ $\frac{3}{x}$

69. $\dfrac{x^2+3x+2}{x^3+x^2}$ $\frac{x+2}{x^2}$

70. $\dfrac{6x^2-13x+6}{3x^2+x-2}$ $\frac{2x-3}{x+1}$

71. $\dfrac{x^2-8x+16}{x^2-16}$ $\frac{x-4}{x+4}$

72. $\dfrac{3x+15}{x^2-25}$ $\frac{3}{x-5}$

73. $\dfrac{2x^2-8}{x^2-3x+2}$ $\frac{2(x+2)}{x-1}$

74. $\dfrac{3x^2-27}{x^2+3x-18}$ $\frac{3(x+3)}{x+6}$

75. $\dfrac{x^2-2x-15}{x^2+2x-15}$ in lowest terms

76. $\dfrac{x^2+4x-77}{x^2-4x-21}$ $\frac{x+11}{x+3}$

77. $\dfrac{x^2-3(2x-3)}{9-x^2}$ $\frac{3-x}{3+x}$ or $-\frac{x-3}{x+3}$

78. $\dfrac{x(x-8)+16}{16-x^2}$ $\frac{4-x}{4+x}$ or $-\frac{x-4}{x+4}$

79. $\dfrac{4(x+3)+4}{3(x+2)+6}$ $\frac{4}{3}$

80. $\dfrac{4+2(x-5)}{3x-5(x-2)}$ $\frac{x-3}{5-x}$

81. $\dfrac{x^2-9}{(2x+3)-(x+6)}$ $x+3$

82. $\dfrac{x^2+5x+4}{2(x+3)-(x+2)}$ $x+1$

In Exercises 83–84, complete the table of values for each rational function (round to the nearest tenth when applicable). Then graph it. Each function is defined for x > 0.

83. $f(x) = \dfrac{4}{x}$

x	$f(x)$
1	4
2	2
3	1.3
4	1
5	0.8
6	0.7
8	0.5

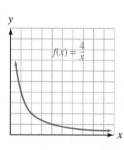

84. $f(x) = \dfrac{5}{2x}$

x	$f(x)$
0.5	5
1	2.5
2	1.3
3	0.8
4	0.6
5	0.5

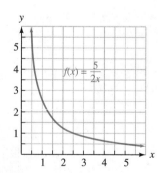

APPLICATIONS

85. ROOFING The *pitch* of a roof is a measure of how steep or how flat the roof is. If pitch $= \frac{\text{rise}}{\text{run}}$, find the pitch of the roof of the cabin shown in Illustration 2. Express the result in lowest terms. $\frac{x+2}{x-2}$

ILLUSTRATION 2

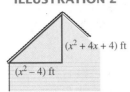

$(x^2 + 4x + 4)$ ft

$(x^2 - 4)$ ft

86. GRAPHIC DESIGN A chart of the basic food groups, in the shape of an equilateral triangle, is to be enlarged and distributed to schools for display in their health classes. (See Illustration 3.) What is the ratio of the length of a side of the original design to a length of a side of the enlargement? Express the result in lowest terms. $\frac{2}{x+1}$

ILLUSTRATION 3

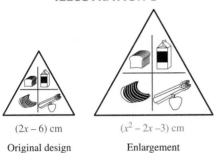

$(2x - 6)$ cm $(x^2 - 2x - 3)$ cm

Original design   Enlargement

✓ **87.** LIGHTING See Illustration 4.
  a. As you move away from a light bulb, the intensity of light reaching you decreases. Explain how the shape of the graph shows this.

  As the x-values (distance from the light bulb) get larger, the y-values (intensity of the light) get smaller.

  b. When you stand far away from a light bulb, the intensity of light reaching you is almost zero. Explain how the shape of the graph shows this.

  As the x-values (distance from the light bulb) get very large, the y-values (intensity of the light) are very small, almost zero.

ILLUSTRATION 4

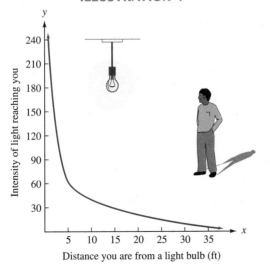

Distance you are from a light bulb (ft)

88. WORD PROCESSOR For the word processor shown in Illustration 5, the number of words that can be typed on a piece of paper is given by

$$f(x) = \frac{8,000}{x}$$

where x is the font size used. Find the number of words that can be typed on a page for each font size choice shown.

1,000, 800, about 667, 500, about 333, about 222.

ILLUSTRATION 5

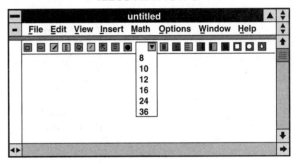

WRITING

89. Explain why $\frac{x-7}{7-x} = -1$.

90. Explain the difference between a factor and a term. Give several examples.

91. Explain the error.

$$\frac{\overset{1}{\cancel{x}} + 6}{\underset{1}{\cancel{x}} + 1} = \frac{7}{2}$$

92. Explain the error.

$$\frac{3(\cancel{x+1}) - x}{\underset{1}{\cancel{x+1}}} = 3 - x$$

REVIEW

93. State the associative property of addition using the variables a, b, and c. $(a+b)+c = a+(b+c)$

94. State the distributive property using the variables x, y, and z. $x(y+z) = xy + xz$

95. If $ab = 0$, what must be true about a or b?
  one of them is zero

96. What is the product of a number and 1? the number

97. What is the opposite of $-\frac{5}{3}$? $\frac{5}{3}$

98. What is the cube of 2 squared? 64

99. What is the sum of a number and zero? the number

100. What is the quotient of a nonzero number and itself? 1

▶ 7.4 Multiplying and Dividing Rational Expressions

In this section, you will learn about

Multiplying rational expressions ▪ Multiplying a rational expression by a polynomial ▪ Dividing rational expressions ▪ Dividing a rational expression by a polynomial ▪ Combined operations

Introduction   In this section, we will extend the rules for multiplying and dividing numerical fractions to problems involving multiplication and division of rational expressions.

Multiplying Rational Expressions

To multiply fractions, we multiply their numerators and multiply their denominators. For example,

$$\frac{4}{7} \cdot \frac{3}{5} = \frac{4 \cdot 3}{7 \cdot 5} \quad \text{Multiply the numerators and multiply the denominators.}$$

$$= \frac{12}{35} \quad \begin{array}{l}\text{Do the multiplication in the numerator: } 4 \cdot 3 = 12.\\ \text{Do the multiplication in the denominator: } 7 \cdot 5 = 35.\end{array}$$

In general, we have the following rule.

Rule for multiplying fractions

If a, b, c, and d are real numbers and $b \neq 0$ and $d \neq 0$,

$$\frac{a}{b} \cdot \frac{c}{d} = \frac{ac}{bd}$$

We use the same procedure to multiply rational expressions.

EXAMPLE 1

Multiplying rational expressions. Multiply   **a.** $\dfrac{x}{3} \cdot \dfrac{2}{5}$,   **b.** $\dfrac{7}{9} \cdot \dfrac{-5}{3x}$, **c.** $\dfrac{x^2}{2} \cdot \dfrac{3}{y^2}$,   and   **d.** $\dfrac{t+1}{t} \cdot \dfrac{t-1}{t-2}$.

Solution   **a.** $\dfrac{x}{3} \cdot \dfrac{2}{5} = \dfrac{x \cdot 2}{3 \cdot 5}$

$$= \frac{2x}{15}$$

b. $\dfrac{7}{9} \cdot \dfrac{-5}{3x} = \dfrac{7(-5)}{9 \cdot 3x}$

$$= \frac{-35}{27x}$$

$$= -\frac{35}{27x}$$

c. $\dfrac{x^2}{2} \cdot \dfrac{3}{y^2} = \dfrac{x^2 \cdot 3}{2 \cdot y^2}$

$$= \frac{3x^2}{2y^2}$$

d. $\dfrac{t+1}{t} \cdot \dfrac{t-1}{t-2} = \dfrac{(t+1)(t-1)}{t(t-2)}$

SELF CHECK Multiply $\dfrac{3x}{4} \cdot \dfrac{x-3}{5}$.

Answer: $\dfrac{3x(x-3)}{20}$ ∎

EXAMPLE 2 **Multiplying rational expressions.** Multiply $\dfrac{35x^2y}{7y^2z} \cdot \dfrac{z}{5xy}$.

Solution

$$\frac{35x^2y}{7y^2z} \cdot \frac{z}{5xy} = \frac{35x^2y \cdot z}{7y^2z \cdot 5xy}$$ Multiply the numerators and multiply the denominators.

$$= \frac{5 \cdot 7 \cdot x \cdot x \cdot y \cdot z}{7 \cdot y \cdot y \cdot z \cdot 5 \cdot x \cdot y}$$ Factor: $35x^2 = 5 \cdot 7 \cdot x \cdot x$. Factor: $y^2 = y \cdot y$.

$$= \frac{\overset{1}{\cancel{5}} \cdot \overset{1}{\cancel{7}} \cdot \overset{1}{\cancel{x}} \cdot x \cdot \overset{1}{\cancel{y}} \cdot \overset{1}{\cancel{z}}}{\underset{1}{\cancel{7}} \cdot \underset{1}{\cancel{y}} \cdot y \cdot \underset{1}{\cancel{z}} \cdot \underset{1}{\cancel{5}} \cdot \underset{1}{\cancel{x}} \cdot y}$$ Divide out the common factors of 5, 7, x, y, and z.

$$= \frac{x}{y^2}$$ Do the multiplications in the numerator and the denominator.

SELF CHECK Multiply $\dfrac{a^2b^2}{2a} \cdot \dfrac{9a^3}{3b^3}$.

Answer: $\dfrac{3a^4}{2b}$ ∎

EXAMPLE 3 **Factoring to simplify a product.** Multiply $\dfrac{x^2-x}{2x+4} \cdot \dfrac{x+2}{x}$.

Solution

$$\frac{x^2-x}{2x+4} \cdot \frac{x+2}{x} = \frac{(x^2-x)(x+2)}{(2x+4)(x)}$$ Multiply the numerators and multiply the denominators.

We now factor the numerator and denominator to see if this product can be simplified.

$$\frac{x^2-x}{2x+y} \cdot \frac{x+2}{x} = \frac{x(x-1)(x+2)}{2(x+2)x}$$ Factor the numerator: $(x^2-x) = x(x-1)$. Factor the denominator: $(2x+4) = 2(x+2)$.

$$= \frac{\overset{1}{\cancel{x}}(x-1)\overset{1}{\cancel{(x+2)}}}{2\underset{1}{\cancel{(x+2)}}\underset{1}{\cancel{x}}}$$ Divide out common factors.

$$= \frac{x-1}{2}$$

SELF CHECK Multiply $\dfrac{x^2+x}{3x+6} \cdot \dfrac{x+2}{x+1}$.

Answer: $\dfrac{x}{3}$ ∎

EXAMPLE 4 **Factoring to simplify a product.** Multiply $\dfrac{x^2-3x}{x^2-x-6} \cdot \dfrac{x^2+x-2}{x^2-x}$.

Solution

$$\frac{x^2-3x}{x^2-x-6} \cdot \frac{x^2+x-2}{x^2-x}$$

$$= \frac{(x^2-3x)(x^2+x-2)}{(x^2-x-6)(x^2-x)}$$ Multiply the numerators and multiply the denominators.

$$= \frac{x(x-3)(x+2)(x-1)}{(x+2)(x-3)x(x-1)}$$ Factor the numerator and denominator to see if the result can be simplified.

$$= \frac{\overset{1}{\cancel{x}}\overset{1}{\cancel{(x-3)}}\overset{1}{\cancel{(x+2)}}\overset{1}{\cancel{(x-1)}}}{\underset{1}{\cancel{(x+2)}}\underset{1}{\cancel{(x-3)}}\underset{1}{\cancel{x}}\underset{1}{\cancel{(x-1)}}}$$ Divide out common factors.

$$= 1$$

SELF CHECK Multiply $\dfrac{a^2 + a}{a^2 - 4} \cdot \dfrac{a^2 - a - 2}{a^2 + 2a + 1}$.

Answer: $\dfrac{a}{a + 2}$ ∎

Multiplying a Rational Expression by a Polynomial

Since any number divided by 1 remains unchanged, we can write any polynomial as a fraction by inserting a denominator of 1.

E X A M P L E   5 **Multiplying a rational expression by a binomial.** Multiply

$\dfrac{x^2 + x}{x^2 + 8x + 7} \cdot (x + 7)$.

Solution
$$\dfrac{x^2 + x}{x^2 + 8x + 7} \cdot (x + 7)$$

$$= \dfrac{x^2 + x}{x^2 + 8x + 7} \cdot \dfrac{x + 7}{1} \qquad \text{Write } x + 7 \text{ as a fraction with a denominator of 1.}$$

$$= \dfrac{x(x + 1)(x + 7)}{(x + 1)(x + 7)1} \qquad \begin{array}{l}\text{Multiply the numerators and multiply the denominators.}\\\text{Factor where possible.}\end{array}$$

$$= \dfrac{x\overset{1}{\cancel{(x + 1)}}\overset{1}{\cancel{(x + 7)}}}{\underset{1}{\cancel{(x + 1)}}\underset{1}{\cancel{(x + 7)}}1} \qquad \text{Divide out common factors.}$$

$$= x$$

SELF CHECK Multiply $(a - 7) \cdot \dfrac{a^2 - a}{a^2 - 8a + 7}$.

Answer: a ∎

Dividing Rational Expressions

Division by a nonzero number is equivalent to multiplying by its reciprocal. Thus, to divide two fractions, we can invert the divisor (the fraction following the ÷ sign) and multiply. For example,

$$\dfrac{4}{7} \div \dfrac{3}{5} = \dfrac{4}{7} \cdot \dfrac{5}{3} \qquad \text{Invert } \tfrac{3}{5} \text{ and change the division to a multiplication.}$$

$$= \dfrac{20}{21} \qquad \text{Multiply the numerators and multiply the denominators.}$$

In general, we have the following rule.

Division of fractions

If a is a real number and b, c, and d are nonzero real numbers, then

$$\dfrac{a}{b} \div \dfrac{c}{d} = \dfrac{a}{b} \cdot \dfrac{d}{c}$$

We use the same procedures to divide rational expressions.

EXAMPLE 6

Dividing rational expressions. Divide **a.** $\dfrac{a}{13} \div \dfrac{17}{26}$ and

b. $\dfrac{-9x}{35y} \div \dfrac{15x^2}{14}$.

Solution **a.** $\dfrac{a}{13} \div \dfrac{17}{26} = \dfrac{a}{13} \cdot \dfrac{26}{17}$ Invert the divisor, which is $\frac{17}{26}$, and change the division to a multiplication.

$$= \dfrac{a \cdot 2 \cdot 13}{13 \cdot 17}$$ Multiply. Then factor where possible.

$$= \dfrac{a \cdot 2 \cdot \overset{1}{\cancel{13}}}{\underset{1}{\cancel{13}} \cdot 17}$$ Divide out common factors.

$$= \dfrac{2a}{17}$$

b. $\dfrac{-9x}{35y} \div \dfrac{15x^2}{14} = \dfrac{-9x}{35y} \cdot \dfrac{14}{15x^2}$ Multiply by the reciprocal of $\dfrac{15x^2}{14}$.

$$= \dfrac{-3 \cdot 3 \cdot x \cdot 2 \cdot 7}{5 \cdot 7 \cdot y \cdot 3 \cdot 5 \cdot x \cdot x}$$ Multiply. Then factor where possible.

$$= \dfrac{-3 \cdot \overset{1}{\cancel{3}} \cdot \overset{1}{\cancel{x}} \cdot 2 \cdot \overset{1}{\cancel{7}}}{5 \cdot \underset{1}{\cancel{7}} \cdot y \cdot \underset{1}{\cancel{3}} \cdot 5 \cdot \underset{1}{\cancel{x}} \cdot x}$$ Divide out common factors.

$$= -\dfrac{6}{25xy}$$ Multiply the remaining factors.

SELF CHECK   Divide $\dfrac{-8a}{3b} \div \dfrac{16a^2}{9b^2}$.   *Answer:* $-\dfrac{3b}{2a}$ ∎

EXAMPLE 7

Dividing rational expressions. Divide $\dfrac{x^2 + x}{3x - 15} \div \dfrac{x^2 + 2x + 1}{6x - 30}$.

Solution $\dfrac{x^2 + x}{3x - 15} \div \dfrac{x^2 + 2x + 1}{6x - 30}$

$$= \dfrac{x^2 + x}{3x - 15} \cdot \dfrac{6x - 30}{x^2 + 2x + 1}$$ Invert the divisor and multiply.

$$= \dfrac{x(x + 1) \cdot 2 \cdot 3(x - 5)}{3(x - 5)(x + 1)(x + 1)}$$ Multiply. Then factor.

$$= \dfrac{x\overset{1}{\cancel{(x + 1)}} \cdot 2 \cdot \overset{1}{\cancel{3}}\overset{1}{\cancel{(x - 5)}}}{\underset{1}{\cancel{3}}\underset{1}{\cancel{(x - 5)}}\underset{1}{\cancel{(x + 1)}}(x + 1)}$$ Divide out common factors.

$$= \dfrac{2x}{x + 1}$$

SELF CHECK   Divide $\dfrac{z^2 - 1}{z^2 + 4z + 3} \div \dfrac{z - 1}{z^2 + 2z - 3}$.   *Answer:* $z - 1$ ∎

Dividing a Rational Expression by a Polynomial

To divide a rational expression by a polynomial, we write the polynomial as a fraction by inserting a denominator of 1, and then we divide the fractions.

EXAMPLE 8

Dividing by a polynomial. Divide $\dfrac{2x^2 - 3x - 2}{2x + 1} \div (4 - x^2)$.

Solution

$$\dfrac{2x^2 - 3x - 2}{2x + 1} \div (4 - x^2)$$

$$= \dfrac{2x^2 - 3x - 2}{2x + 1} \div \dfrac{4 - x^2}{1}$$ Write $4 - x^2$ as a fraction with a denominator of 1.

$$= \dfrac{2x^2 - 3x - 2}{2x + 1} \cdot \dfrac{1}{4 - x^2}$$ Invert the divisor and multiply.

$$= \dfrac{(2x + 1)(x - 2) \cdot 1}{(2x + 1)(2 + x)(2 - x)}$$ Multiply. Then factor where possible.

$$= \dfrac{\overset{1}{\cancel{(2x + 1)}}\overset{-1}{\cancel{(x - 2)}} \cdot 1}{\underset{1}{\cancel{(2x + 1)}}(2 + x)\underset{1}{\cancel{(2 - x)}}}$$ Divide out common factors. The binomials $x - 2$ and $2 - x$ are negatives: $\frac{x - 2}{2 - x} = -1$.

$$= \dfrac{-1}{2 + x}$$

$$= -\dfrac{1}{2 + x}$$

SELF CHECK Divide $(b - a) \div \dfrac{a^2 - b^2}{a^2 + ab}$.

Answer: $-a$ ∎

Combined Operations

Unless parentheses indicate otherwise, we do multiplications and divisions in order from left to right.

EXAMPLE 9

Multiplying and dividing rational expressions Simplify
$\dfrac{x^2 - x - 6}{x - 2} \div \dfrac{x^2 - 4x}{x^2 - x - 2} \cdot \dfrac{x - 4}{x^2 + x}$.

Solution Since there are no parentheses to indicate otherwise, we do the division first.

$$\dfrac{x^2 - x - 6}{x - 2} \div \dfrac{x^2 - 4x}{x^2 - x - 2} \cdot \dfrac{x - 4}{x^2 + x}$$

$$= \dfrac{x^2 - x - 6}{x - 2} \cdot \dfrac{x^2 - x - 2}{x^2 - 4x} \cdot \dfrac{x - 4}{x^2 + x}$$ Invert the divisor, which is $\dfrac{x^2 - 4x}{x^2 - x - 2}$, and change the division to a multiplication.

$$= \dfrac{(x + 2)(x - 3)(x + 1)(x - 2)(x - 4)}{(x - 2)x(x - 4)x(x + 1)}$$ Multiply. Then factor.

$$= \dfrac{(x + 2)(x - 3)\overset{1}{\cancel{(x + 1)}}\cancel{(x - 2)}\overset{1}{\cancel{(x - 4)}}}{\cancel{(x - 2)}x\cancel{(x - 4)}x\underset{1}{\cancel{(x + 1)}}}$$ Divide out common factors.

$$= \dfrac{(x + 2)(x - 3)}{x^2}$$

SELF CHECK Simplify $\dfrac{a^2 + ab}{ab - b^2} \cdot \dfrac{a^2 - b^2}{a^2 + ab} \div \dfrac{a + b}{b}$.

Answer: 1 ∎

E X A M P L E 1 0

Multiplying and dividing rational expressions. Simplify
$$\frac{x^2 + 6x + 9}{x^2 - 2x}\left(\frac{x^2 - 4}{x^2 + 3x} \div \frac{x + 2}{x}\right).$$

Solution We do the division within the parentheses first.

$$\frac{x^2 + 6x + 9}{x^2 - 2x}\left(\frac{x^2 - 4}{x^2 + 3x} \div \frac{x + 2}{x}\right)$$

$$= \frac{x^2 + 6x + 9}{x^2 - 2x}\left(\frac{x^2 - 4}{x^2 + 3x} \cdot \frac{x}{x + 2}\right) \quad \text{Invert the divisor and change the division to a multiplication.}$$

$$= \frac{(x + 3)(x + 3)(x - 2)(x + 2)x}{x(x - 2)x(x + 3)(x + 2)} \quad \text{Multiply and factor where possible.}$$

$$= \frac{\overset{1}{\cancel{(x + 3)}}(x + 3)\overset{1}{\cancel{(x - 2)}}\overset{1}{\cancel{(x + 2)}}\overset{1}{\cancel{x}}}{\underset{1}{\cancel{x}}\underset{1}{\cancel{(x - 2)}}x\underset{1}{\cancel{(x + 3)}}\underset{1}{\cancel{(x + 2)}}} \quad \text{Divide out common factors.}$$

$$= \frac{x + 3}{x}$$

SELF CHECK Simplify $\dfrac{x^2 - 2x}{x^2 + 6x + 9} \div \left(\dfrac{x^2 - 4}{x^2 + 3x} \cdot \dfrac{x}{x + 2}\right)$ *Answer:* $\dfrac{x}{x + 3}$ ∎

S T U D Y S E T

Section 7.4

VOCABULARY

In Exercises 1–2, fill in the blanks to make the statements true.

1. In a fraction, the part above the fraction bar is called the ___numerator___.

2. In a fraction, the part below the fraction bar is called the ___denominator___.

CONCEPTS

In Exercises 3–8, fill in the blanks to make the statements true.

3. To multiply fractions, we multiply their ___numerators___ and multiply their ___denominators___.

4. $\dfrac{a}{b} \cdot \dfrac{c}{d} = \dfrac{ac}{bd}$

5. To write a polynomial in fractional form, we insert a denominator of 1.

6. $\dfrac{a}{b} \div \dfrac{c}{d} = \dfrac{a}{b} \cdot \dfrac{d}{c}$

7. To divide fractions, we invert the ___divisor___ and ___multiply___.

8. The ___reciprocal___ of $\dfrac{x}{x + 2}$ is $\dfrac{x + 2}{x}$.

NOTATION

In Exercises 9–10, complete each solution.

9. $\dfrac{x^2 + x}{3x - 6} \cdot \dfrac{x - 2}{x + 1} = \dfrac{(x^2 + x)\,(x - 2)}{(3x - 6)\,(x + 1)}$

$$= \dfrac{x(x + 1)\,(x - 2)}{3(x - 2)\,(x + 1)}$$

$$= \dfrac{x}{3}$$

10. $\dfrac{x^2 - x}{4x + 12} \div \dfrac{x - 1}{x + 3} = \dfrac{x^2 - x}{4x + 12} \cdot \boxed{\dfrac{x + 3}{x - 1}}$

$$= \dfrac{(x^2 - x)\,(x + 3)}{(4x + 12)\,(x - 1)}$$

$$= \dfrac{x(x - 1)\,(x + 3)}{4(x + 3)\,(x - 1)}$$

$$= \dfrac{x}{4}$$

PRACTICE

In Exercises 11–54, do the multiplications. Simplify answers if possible.

✓ **11.** $\dfrac{5}{7} \cdot \dfrac{9}{13}$ $\dfrac{45}{91}$

12. $\dfrac{2}{7} \cdot \dfrac{5}{11}$ $\dfrac{10}{77}$

13. $\dfrac{25}{35} \cdot \dfrac{-21}{55}$ $-\dfrac{3}{11}$

14. $-\dfrac{27}{24} \cdot \left(-\dfrac{56}{35}\right)$ $\dfrac{9}{5}$

15. $\dfrac{2}{3} \cdot \dfrac{15}{2} \cdot \dfrac{1}{7}$ $\dfrac{5}{7}$

16. $\dfrac{2}{5} \cdot \dfrac{10}{9} \cdot \dfrac{3}{2}$ $\dfrac{2}{3}$

✓ **17.** $\dfrac{3x}{y} \cdot \dfrac{y}{2}$ $\dfrac{3x}{2}$

18. $\dfrac{2y}{z} \cdot \dfrac{z}{3}$ $\dfrac{2y}{3}$

19. $\dfrac{5y}{7} \cdot \dfrac{7x}{5z}$ $\dfrac{xy}{z}$

20. $\dfrac{4x}{3y} \cdot \dfrac{3y}{7x}$ $\dfrac{4}{7}$

21. $\dfrac{7z}{9z} \cdot \dfrac{4z}{2z}$ $\dfrac{14}{9}$

22. $\dfrac{8z}{2x} \cdot \dfrac{16x}{3x}$ $\dfrac{64z}{3x}$

✓ **23.** $\dfrac{2x^2y}{3xy} \cdot \dfrac{3xy^2}{2}$ x^2y^2

24. $\dfrac{2x^2z}{z} \cdot \dfrac{5x}{z}$ $\dfrac{10x^3}{z}$

25. $\dfrac{8x^2y^2}{4x^2} \cdot \dfrac{2xy}{2y}$ $2xy^2$

26. $\dfrac{9x^2y}{3x} \cdot \dfrac{3xy}{3y}$ $3x^2y$

27. $\dfrac{-2xy}{x^2} \cdot \dfrac{3xy}{2}$ $-3y^2$

28. $\dfrac{-3x}{x^2} \cdot \dfrac{2xz}{3}$ $-2z$

✓ **29.** $\dfrac{ab^2}{a^2b} \cdot \dfrac{b^2c^2}{abc} \cdot \dfrac{abc^2}{a^3c^2}$ $\dfrac{b^3c}{a^4}$

30. $\dfrac{x^3y}{z} \cdot \dfrac{xz^3}{x^2y^2} \cdot \dfrac{yz}{xyz}$ $\dfrac{xz^2}{y}$

31. $\dfrac{10r^2st^3}{6rs^2} \cdot \dfrac{3r^3t}{2rst} \cdot \dfrac{2s^3t^4}{5s^2t^3}$ $\dfrac{r^3t^4}{s}$

32. $\dfrac{3a^3b}{25cd^3} \cdot \dfrac{-5cd^2}{6ab} \cdot \dfrac{10abc^2}{2bc^2d}$ $-\dfrac{a^3}{2d^2}$

33. $\dfrac{z+7}{7} \cdot \dfrac{z+2}{z}$ $\dfrac{(z+7)(z+2)}{7z}$

34. $\dfrac{a-3}{a} \cdot \dfrac{a+3}{5}$ $\dfrac{(a-3)(a+3)}{5a}$

✓ **35.** $\dfrac{x-2}{2} \cdot \dfrac{2x}{x-2}$ x

36. $\dfrac{y+3}{y} \cdot \dfrac{3y}{y+3}$ 3

37. $\dfrac{x+5}{5} \cdot \dfrac{x}{x+5}$ $\dfrac{x}{5}$

38. $\dfrac{y-9}{y+9} \cdot \dfrac{y}{9}$ $\dfrac{y(y-9)}{9(y+9)}$

39. $\dfrac{(x+1)^2}{x+1} \cdot \dfrac{x+2}{x+1}$ $x+2$

40. $\dfrac{(y-3)^2}{y-3} \cdot \dfrac{y-3}{y-3}$ $y-3$

✓ **41.** $\dfrac{2x+6}{x+3} \cdot \dfrac{3}{4x}$ $\dfrac{3}{2x}$

42. $\dfrac{3y-9}{y-3} \cdot \dfrac{y}{3y^2}$ $\dfrac{1}{y}$

43. $\dfrac{x^2-x}{x} \cdot \dfrac{3x-6}{3x-3}$ $x-2$

44. $\dfrac{5z-10}{z+2} \cdot \dfrac{3}{3z-6}$ $\dfrac{5}{z+2}$

45. $\dfrac{7y-14}{y-2} \cdot \dfrac{x^2}{7x}$ x

46. $\dfrac{y^2+3y}{9} \cdot \dfrac{3x}{y+3}$ $\dfrac{xy}{3}$

✓ **47.** $\dfrac{x^2+x-6}{5x} \cdot \dfrac{5x-10}{x+3}$ $\dfrac{(x-2)^2}{x}$

48. $\dfrac{z^2+4z-5}{5z-5} \cdot \dfrac{5z}{z+5}$ z

49. $\dfrac{m^2-2m-3}{2m+4} \cdot \dfrac{m^2-4}{m^2+3m+2}$ $\dfrac{(m-2)(m-3)}{2(m+2)}$

50. $\dfrac{p^2-p-6}{3p-9} \cdot \dfrac{p^2-9}{p^2+6p+9}$ $\dfrac{(p+2)(p-3)}{3(p+3)}$

51. $\dfrac{abc^2}{a+1} \cdot \dfrac{c}{a^2b^2} \cdot \dfrac{a^2+a}{ac}$ $\dfrac{c^2}{ab}$

52. $\dfrac{x^3yz^2}{4x+8} \cdot \dfrac{x^2-4}{2x^2y^2z^2} \cdot \dfrac{8yz}{x-2}$ xz

✓ **53.** $\dfrac{3x^2+5x+2}{x^2-9} \cdot \dfrac{x-3}{x^2-4} \cdot \dfrac{x^2+5x+6}{6x+4}$ $\dfrac{x+1}{2(x-2)}$

54. $\dfrac{x^2-25}{3x+6} \cdot \dfrac{x^2+x-2}{2x+10} \cdot \dfrac{6x}{3x^2-18x+15}$ $\dfrac{x}{3}$

In Exercises 55–84, do each division. Simplify answers when possible.

✓ **55.** $\dfrac{1}{3} \div \dfrac{1}{2}$ $\dfrac{2}{3}$

56. $\dfrac{3}{4} \div \dfrac{1}{3}$ $\dfrac{9}{4}$

57. $\dfrac{21}{14} \div \dfrac{5}{2}$ $\dfrac{3}{5}$

58. $\dfrac{14}{3} \div \dfrac{10}{3}$ $\dfrac{7}{5}$

✓ **59.** $\dfrac{2}{y} \div \dfrac{4}{3}$ $\dfrac{3}{2y}$

60. $\dfrac{3}{a} \div \dfrac{a}{9}$ $\dfrac{27}{a^2}$

61. $\dfrac{3x}{2} \div \dfrac{x}{2}$ 3

62. $\dfrac{y}{6} \div \dfrac{2}{3y}$ $\dfrac{y^2}{4}$

63. $\dfrac{3x}{y} \div \dfrac{2x}{4}$ $\dfrac{6}{y}$

64. $\dfrac{3y}{8} \div \dfrac{2y}{4y}$ $\dfrac{3y}{4}$

✓ **65.** $\dfrac{4x}{3x} \div \dfrac{2y}{9y}$ 6

66. $\dfrac{14}{7y} \div \dfrac{10}{5z}$ $\dfrac{z}{y}$

67. $\dfrac{x^2}{3} \div \dfrac{2x}{4}$ $\dfrac{2x}{3}$

68. $\dfrac{z^2}{z} \div \dfrac{z}{3z}$ $3z$

69. $\dfrac{x^2y}{3xy} \div \dfrac{xy^2}{6y}$ $\dfrac{2}{y}$

70. $\dfrac{2xz}{z} \div \dfrac{4x^2}{z^2}$ $\dfrac{z^2}{2x}$

✓ **71.** $\dfrac{x+2}{3x} \div \dfrac{x+2}{2}$ $\dfrac{2}{3x}$

72. $\dfrac{z-3}{3z} \div \dfrac{z+3}{z}$ $\dfrac{z-3}{3(z+3)}$

73. $\dfrac{(z-2)^2}{3z^2} \div \dfrac{z-2}{6z}$

$\dfrac{2(z-2)}{z}$

74. $\dfrac{(x+7)^2}{x+7} \div \dfrac{(x-3)^2}{x+7}$

$\dfrac{(x+7)^2}{(x-3)^2}$

75. $\dfrac{(z-7)^2}{z+2} \div \dfrac{z(z-7)}{5z^2}$

$\dfrac{5z(z-7)}{z+2}$

76. $\dfrac{y(y+2)}{y^2(y-3)} \div \dfrac{y^2(y+2)}{(y-3)^2}$

$\dfrac{y-3}{y^3}$

77. $\dfrac{x^2-4}{3x+6} \div \dfrac{x-2}{x+2}$

$\dfrac{x+2}{3}$

78. $\dfrac{x^2-9}{5x+15} \div \dfrac{x-3}{x+3}$

$\dfrac{x+3}{5}$

79. $\dfrac{x^2-1}{3x-3} \div \dfrac{x+1}{3}$ 1

80. $\dfrac{x^2-16}{x-4} \div \dfrac{3x+12}{x}$

$\dfrac{x}{3}$

81. $\dfrac{x^2-2x-35}{3x^2+27x} \div \dfrac{x^2+7x+10}{6x^2+12x}$ $\dfrac{2(x-7)}{x+9}$

82. $\dfrac{x^2-x-6}{2x^2+9x+10} \div \dfrac{x^2-25}{2x^2+15x+25}$ $\dfrac{x-3}{x-5}$

83. $\dfrac{2d^2+8d-42}{d-3} \div \dfrac{2d^2+14d}{d^2+5d}$ $d+5$

84. $\dfrac{5x^2+13x-6}{x+3} \div \dfrac{5x^2-17x+6}{x-2}$ $\dfrac{x-2}{x-3}$

In Exercises 85–96, do the operations.

85. $\dfrac{x}{3} \cdot \dfrac{9}{4} \div \dfrac{x^2}{6}$ $\dfrac{9}{2x}$

86. $\dfrac{y^2}{2} \div \dfrac{4}{y} \cdot \dfrac{y^2}{8}$ $\dfrac{y^5}{64}$

87. $\dfrac{x^2}{18} \div \dfrac{x^3}{6} \div \dfrac{12}{x^2}$ $\dfrac{x}{36}$

88. $\dfrac{y^3}{3y} \cdot \dfrac{3y^2}{4} \div \dfrac{15}{20}$ $\dfrac{y^4}{3}$

89. $\dfrac{z^2-4}{2z+6} \div \dfrac{z+2}{4} \cdot \dfrac{z+3}{z-2}$ 2

90. $\dfrac{2}{3x-3} \div \dfrac{2x+2}{x-1} \cdot \dfrac{5}{x+1}$ $\dfrac{5}{3(x+1)^2}$

91. $\dfrac{x-x^2}{x^2-4}\left(\dfrac{2x+4}{x+2} \div \dfrac{5}{x+2}\right)$ $\dfrac{2x(1-x)}{5(x-2)}$

92. $\dfrac{2}{3x-3} \div \left(\dfrac{2x+2}{x-1} \cdot \dfrac{5}{x+1}\right)$ $\dfrac{1}{15}$

93. $\dfrac{y^2}{x+1} \cdot \dfrac{x^2+2x+1}{x^2-1} \div \dfrac{3y}{xy-y}$ $\dfrac{y^2}{3}$

94. $\dfrac{x^2-y^2}{x^4-x^3} \div \dfrac{x-y}{x^2} \div \dfrac{x^2+2xy+y^2}{x+y}$ $\dfrac{1}{x(x-1)}$

95. $\dfrac{x^2+x-6}{x^2-4} \cdot \dfrac{x^2+2x}{x-2} \div \dfrac{x^2+3x}{x+2}$ $\dfrac{x+2}{x-2}$

96. $\dfrac{x^2-x-6}{x^2+6x-7} \cdot \dfrac{x^2+x-2}{x^2+2x} \div \dfrac{x^2+7x}{x^2-3x}$

$\dfrac{(x-3)^2(x+2)}{x(x+7)^2}$

APPLICATIONS

97. INTERNATIONAL ALPHABET The symbols representing the letters A, B, C, D, E, and F of an international code used at sea are printed six to a sheet and then cut into separate cards. If each card is a square, find the area of the large printed sheet shown in Illustration 1. $\dfrac{12x^2+12x+3}{2}$ in.²

ILLUSTRATION 1

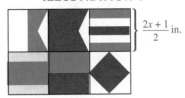

$\dfrac{2x+1}{2}$ in.

98. PHYSICS EXPERIMENT The table in Illustration 2 contains algebraic expressions for the rate an object travels, and the time traveled at that rate, in terms of a constant k. Complete the table.

ILLUSTRATION 2

Rate (mph)	Time (hr)	Distance (mi)
$\dfrac{k^2+k-6}{k-3}$	$\dfrac{k^2-9}{k^2-4}$	$\dfrac{(k+3)^2}{k+2}$

WRITING

99. Explain how to multiply two fractions and how to simplify the result.

100. Explain why any mathematical expression can be written as a fraction.

101. To divide fractions, you must first know how to multiply fractions. Explain.

102. Explain how to do the division $\dfrac{a}{b} \div \dfrac{c}{d} \div \dfrac{e}{f}$.

REVIEW

In Exercises 103–106, simplify each expression. Write all answers without using negative exponents.

103. $2x^3y^2(-3x^2y^4)$ $-6x^5y^6$

104. $\dfrac{8x^4y^5}{-2x^3y^2}$ $-4xy^3$

105. $(3y)^{-4}$ $\dfrac{1}{81y^4}$

106. $\dfrac{x^{3m}}{x^{4m}}$ $\dfrac{1}{x^m}$

In Exercises 107–108, do the operations and simplify.

107. $-4(y^3-4y^2+3y-2)-4(-2y^3-y)$

$4y^3+16y^2-8y+8$

108. $y-5\overline{)5y^3-3y^2+4y-1}$ $5y^2+22y+114+\dfrac{569}{y-5}$

▶ 7.5   Adding and Subtracting Rational Expressions

In this section, you will learn about

Adding rational expressions with like denominators ■ Subtracting rational expressions with like denominators ■ Combined operations ■ The LCD ■ Adding rational expressions with unlike denominators ■ Subtracting rational expressions with unlike denominators ■ Combined operations

Introduction   In this section, we will extend the rules for adding and subtracting numerical fractions to problems involving addition and subtraction of rational expressions.

Adding Rational Expressions with Like Denominators

To add fractions with a common denominator, we add their numerators and keep the common denominator. For example,

$$\frac{2}{7} + \frac{3}{7} = \frac{2+3}{7} \qquad \text{Add the numerators and keep the common denominator.}$$

$$= \frac{5}{7}$$

In general, we have the following rule.

Adding fractions with like denominators

If a, b, and d represent real numbers, then

$$\frac{a}{d} + \frac{b}{d} = \frac{a+b}{d} \qquad (d \neq 0)$$

We use the same procedure to add rational expressions with like denominators.

EXAMPLE 1

Adding rational expressions. Do each addition.

a. $\dfrac{x}{8} + \dfrac{3x}{8} = \dfrac{x + 3x}{8}$ Add the numerators and keep the common denominator.

$= \dfrac{4x}{8}$ Combine like terms: $x + 3x = 4x$.

$= \dfrac{\overset{1}{\cancel{4}} \cdot x}{\underset{1}{\cancel{4}} \cdot 2}$ Factor and divide out the common factor of 4.

$= \dfrac{x}{2}$ Simplify.

b. $\dfrac{3x + y}{5x} + \dfrac{x + y}{5x} = \dfrac{3x + y + x + y}{5x}$ Add the numerators and keep the common denominator.

$\qquad\qquad\qquad = \dfrac{4x + 2y}{5x}$ Combine like terms.

SELF CHECK   Add: **a.** $\dfrac{x}{7} + \dfrac{4x}{7}$ and   **b.** $\dfrac{3x}{7y} + \dfrac{4x}{7y}$.   *Answers:*   **a.** $\frac{5x}{7}$,   **b.** $\frac{x}{y}$ ■

EXAMPLE 2   **Adding rational expressions.**   Add $\dfrac{3x + 21}{5x + 10} + \dfrac{8x + 1}{5x + 10}$.

Solution   Because the fractions have the same denominator, we add their numerators and keep the common denominator.

$\qquad \dfrac{3x + 21}{5x + 10} + \dfrac{8x + 1}{5x + 10} = \dfrac{3x + 21 + 8x + 1}{5x + 10}$ Add.

$\qquad\qquad\qquad\qquad = \dfrac{11x + 22}{5x + 10}$ Combine like terms.

$\qquad\qquad\qquad\qquad = \dfrac{\overset{1}{11\cancel{(x + 2)}}}{\underset{1}{5\cancel{(x + 2)}}}$ Simplify the result by factoring the numerator and denominator. Divide out the common factor of $x + 2$.

$\qquad\qquad\qquad\qquad = \dfrac{11}{5}$

SELF CHECK   Add $\frac{x+4}{6x-12} + \frac{x-8}{6x-12}$.   *Answer:* $\frac{1}{3}$ ■

Subtracting Rational Expressions with Like Denominators

To subtract fractions with a common denominator, we subtract their numerators and keep the common denominator.

Subtracting fractions with like denominators

If a, b, and d represent real numbers, then

$$\frac{a}{d} - \frac{b}{d} = \frac{a - b}{d} \qquad (d \neq 0)$$

We use the same procedure to subtract rational expressions.

EXAMPLE 3   **Subtracting rational expressions.**   Subtract   **a.** $\dfrac{5x}{3} - \dfrac{2x}{3}$ and

b. $\dfrac{5x + 1}{x - 3} - \dfrac{4x - 2}{x - 3}$.

Solution In each part, the fractions have the same denominator. To subtract them, we subtract their numerators and keep the common denominator.

a. $\dfrac{5x}{3} - \dfrac{2x}{3} = \dfrac{5x - 2x}{3}$ Subtract.

$\qquad\qquad = \dfrac{3x}{3}$ Combine like terms: $5x - 2x = 3x$.

$\qquad\qquad = \dfrac{x}{1}$ Divide out the common factor of 3.

$\qquad\qquad = x$ Denominators of 1 need not be written.

b. $\dfrac{5x + 1}{x - 3} - \dfrac{4x - 2}{x - 3} = \dfrac{(5x + 1) - (4x - 2)}{x - 3}$ Subtract. Write each numerator in parentheses.

$\qquad\qquad = \dfrac{5x + 1 - 4x + 2}{x - 3}$ Remove parentheses: $-(4x - 2) = -4x + 2$.

$\qquad\qquad = \dfrac{x + 3}{x - 3}$ Combine like terms.

SELF CHECK   Subtract $\dfrac{2y + 1}{y + 5} - \dfrac{y - 4}{y + 5}$.   *Answer:* 1   ■

Combined Operations

To add and/or subtract three or more rational expressions, we follow the rules for the order of operations.

EXAMPLE 4   **Combined operations.** Simplify $\dfrac{3x + 1}{x^2 + x + 1} - \dfrac{5x + 2}{x^2 + x + 1} + \dfrac{2x + 1}{x^2 + x + 1}$.

Solution This example combines addition and subtraction. Unless parentheses indicate otherwise, we do additions and subtractions from left to right.

$\dfrac{3x + 1}{x^2 + x + 1} - \dfrac{5x + 2}{x^2 + x + 1} + \dfrac{2x + 1}{x^2 + x + 1}$

$\qquad = \dfrac{(3x + 1) - (5x + 2) + (2x + 1)}{x^2 + x + 1}$ Combine the numerators and keep the common denominator.

$\qquad = \dfrac{3x + 1 - 5x - 2 + 2x + 1}{x^2 + x + 1}$ Remove parentheses: $-(5x + 2) = -5x - 2$.

$\qquad = \dfrac{0}{x^2 + x + 1}$ Combine like terms.

$\qquad = 0$ If the numerator of a fraction is zero and the denominator is not zero, the fraction's value is zero.

SELF CHECK   Simplify $\dfrac{2a^2 - 3}{a - 5} + \dfrac{3a^2 + 2}{a - 5} - \dfrac{5a^2}{a - 5}$.   *Answer:* $-\dfrac{1}{a - 5}$ ■

The LCD

Since the denominators of the fractions in the addition $\frac{4}{7} + \frac{3}{5}$ are different, we cannot add the fractions in their present form.

four-sevenths   +   three-fifths

└── Different denominators ──┘

To add these fractions, we need to find a common denominator. The smallest common denominator (called the **least** or **lowest common denominator**) is usually the easiest one to work with.

Least common denominator

The **least common denominator (LCD)** for a set of fractions is the smallest number that each denominator will divide exactly.

In the addition $\frac{4}{7} + \frac{3}{5}$, the denominators are 7 and 5. The smallest number that 7 and 5 will divide exactly is 35. This is the LCD. We now **build** each fraction into a fraction with a denominator of 35. To do so, we use the fundamental property of fractions to multiply both the numerator and the denominator of each fraction by some appropriate number.

$$\frac{4}{7} + \frac{3}{5} = \frac{4 \cdot 5}{7 \cdot 5} + \frac{3 \cdot 7}{5 \cdot 7}$$ Multiply numerator and denominator of $\frac{4}{7}$ by 5, and multiply numerator and denominator of $\frac{3}{5}$ by 7.

$$= \frac{20}{35} + \frac{21}{35}$$ Do the multiplications.

Now that the fractions have a common denominator, we can add them.

$$\frac{20}{35} + \frac{21}{35} = \frac{20 + 21}{35} = \frac{41}{35}$$

EXAMPLE 5

Building fractions. Change each fraction into one with a denominator of $30y$:
a. $\frac{1}{2y}$, **b.** $\frac{3y}{5}$, and **c.** $\frac{7 + x}{10y}$.

Solution

To build each fraction, we multiply the numerator and denominator by the factor that makes the denominator $30y$.

a. $\dfrac{1}{2y} = \dfrac{1 \cdot 15}{2y \cdot 15} = \dfrac{15}{30y}$ Multiply numerator and denominator by 15, because $2y \cdot 15 = 30y$.

b. $\dfrac{3y}{5} = \dfrac{3y \cdot 6y}{5 \cdot 6y} = \dfrac{18y^2}{30y}$ Multiply numerator and denominator by $6y$, because $5 \cdot 6y = 30y$.

c. $\dfrac{7 + x}{10y} = \dfrac{(7 + x)3}{(10y)3} = \dfrac{21 + 3x}{30y}$ Multiply numerator and denominator by 3, because $10y \cdot 3 = 30y$.

SELF CHECK

Change $\frac{5}{6b}$ into a fraction with a denominator of $30ab$.   *Answer:* $\frac{25a}{30ab}$ ■

There is a process that we can use to find the least common denominator of several fractions.

Finding the least common denominator (LCD)

1. List the different denominators that appear in the fraction.
2. Completely factor each denominator.
3. Form a product using each different factor obtained in step 2. Use each different factor the *greatest* number of times it appears in any one factorization. The product formed by multiplying these factors is the LCD.

EXAMPLE 6

Finding the LCD. Find the LCD of $\dfrac{5}{24b}$ and $\dfrac{11}{18b}$.

Solution   We list and factor each denominator into the product of prime numbers.

$$24b = 2 \cdot 2 \cdot 2 \cdot 3 \cdot b$$
$$18b = 2 \cdot 3 \cdot 3 \cdot b$$

To find the LCD, we use each of these factors the greatest number of times it appears in any one factorization. We use 2 three times, because it appears three times as a factor of 24. We use 3 twice, because it occurs twice as a factor of 18. We use b once.

$$\begin{aligned}
\text{LCD} &= 2 \cdot 2 \cdot 2 \cdot 3 \cdot 3 \cdot b \\
&= 8 \cdot 9 \cdot b \\
&= 72b
\end{aligned}$$

SELF CHECK   Find the LCD of $\frac{3}{28z}$ and $\frac{5}{21z}$.   *Answer:* $84z$ ■

Adding Rational Expressions with Unlike Denominators

The following steps summarize how to add fractions that have unlike denominators.

Adding fractions with unlike denominators

To add fractions with different denominators,

1. Find the LCD.
2. Write each fraction as a fraction whose denominator is the LCD.
3. Add the resulting fractions and simplify the result, if possible.

EXAMPLE 7

Adding rational expressions. Add $\dfrac{4x}{7} + \dfrac{3x}{5}$.

Solution   The LCD is 35. We build each fraction so that it has a denominator of 35 and then add the resulting fractions.

$$\dfrac{4x}{7} + \dfrac{3x}{5} = \dfrac{4x \cdot 5}{7 \cdot 5} + \dfrac{3x \cdot 7}{5 \cdot 7} \qquad \text{Multiply the numerator and the denominator of } \tfrac{4x}{7} \text{ by 5 and the numerator and denominator of } \tfrac{3x}{5} \text{ by 7.}$$

$$= \dfrac{20x}{35} + \dfrac{21x}{35} \qquad \text{Do the multiplication.}$$

$$= \dfrac{41x}{35} \qquad \text{Add the numerators and keep the common denominator.}$$

SELF CHECK   Add $\frac{y}{2} + \frac{6y}{7}$.   *Answer:* $\frac{19y}{14}$ ■

EXAMPLE 8

Adding rational expressions. Add $\dfrac{5}{24b} + \dfrac{11}{18b}$.

Solution   In Example 6, we saw that the LCD of these fractions is $2 \cdot 2 \cdot 2 \cdot 3 \cdot 3 \cdot b = 72b$. To add them, we first factor each denominator:

$$\dfrac{5}{24b} + \dfrac{11}{18b} = \dfrac{5}{2 \cdot 2 \cdot 2 \cdot 3 \cdot b} + \dfrac{11}{2 \cdot 3 \cdot 3 \cdot b}$$

In each resulting fraction, we multiply the numerator and the denominator by whatever it takes to build the denominator to the LCD of $2 \cdot 2 \cdot 2 \cdot 3 \cdot 3 \cdot b$.

$$= \frac{5 \cdot 3}{2 \cdot 2 \cdot 2 \cdot 3 \cdot b \cdot 3} + \frac{11 \cdot 2 \cdot 2}{2 \cdot 3 \cdot 3 \cdot b \cdot 2 \cdot 2}$$

$$= \frac{15}{72b} + \frac{44}{72b} \qquad \text{Do the multiplications.}$$

$$= \frac{59}{72b} \qquad \text{Add the numerators and keep the common denominator.}$$

·SELF CHECK   Add $\frac{3}{28z} + \frac{5}{21z}$.   *Answer:* $\frac{29}{84z}$ ■

EXAMPLE 9	**Adding rational expressions.** Add $\dfrac{x+4}{x^2} + \dfrac{x-5}{4x}$.

Solution   First we find the LCD.

$$\left.\begin{array}{l} x^2 = x \cdot x \\ 4x = 2 \cdot 2 \cdot x \end{array}\right\} \qquad \text{LCD} = x \cdot x \cdot 2 \cdot 2 = 4x^2$$

$$\frac{x+4}{x^2} + \frac{x-5}{4x} = \frac{(x+4)4}{(x^2)4} + \frac{(x-5)x}{(4x)x} \qquad \begin{array}{l}\text{Build the fractions to get the common} \\ \text{denominator of } 4x^2.\end{array}$$

$$= \frac{4x+16}{4x^2} + \frac{x^2-5x}{4x^2} \qquad \text{Do the multiplications.}$$

$$= \frac{4x+16+x^2-5x}{4x^2} \qquad \begin{array}{l}\text{Add the numerators and keep the common} \\ \text{denominator.}\end{array}$$

$$= \frac{x^2-x+16}{4x^2} \qquad \text{Combine like terms.}$$

SELF CHECK   Add $\dfrac{a-1}{9a} + \dfrac{2-a}{a^2}$.   *Answer:* $\dfrac{a^2-10a+18}{9a^2}$ ■

Subtracting Rational Expressions with Unlike Denominators

To subtract fractions with unlike denominators, we first change them into fractions with the same denominators. We use the same procedure to subtract rational expressions.

EXAMPLE 10	**Subtracting rational expressions.** Subtract $\dfrac{x}{x+1} - \dfrac{3}{x}$.

Solution   By inspection, the least common denominator is $(x+1)x$.

$$\frac{x}{x+1} - \frac{3}{x} = \frac{x(x)}{(x+1)x} - \frac{3(x+1)}{x(x+1)} \qquad \begin{array}{l}\text{Build the fractions to get the common denom-} \\ \text{inator.}\end{array}$$

$$= \frac{x(x)-3(x+1)}{x(x+1)} \qquad \begin{array}{l}\text{Subtract the numerators and keep the common} \\ \text{denominator.}\end{array}$$

$$= \frac{x^2-3x-3}{x(x+1)} \qquad \text{Do the multiplications in the numerator.}$$

SELF CHECK   Subtract $\dfrac{a}{a-1} - \dfrac{5}{a}$.   *Answer:* $\dfrac{a^2-5a+5}{a(a-1)}$ ■

EXAMPLE 11

Simplifying the result. Subtract $\dfrac{a}{a-1} - \dfrac{2}{a^2-1}$.

Solution We factor $a^2 - 1$ to see that the LCD $= (a+1)(a-1)$.

$$\dfrac{a}{a-1} - \dfrac{2}{a^2-1}$$

$$= \dfrac{a(a+1)}{(a-1)(a+1)} - \dfrac{2}{(a+1)(a-1)} \qquad \text{Build the first fraction to get the LCD.}$$

$$= \dfrac{a(a+1)-2}{(a-1)(a+1)} \qquad \begin{array}{l}\text{Subtract the numerators and keep the}\\ \text{common denominator.}\end{array}$$

$$= \dfrac{a^2+a-2}{(a-1)(a+1)} \qquad \text{Remove parentheses.}$$

$$= \dfrac{(a+2)\overset{1}{\cancel{(a-1)}}}{\underset{1}{\cancel{(a-1)}}(a+1)} \qquad \begin{array}{l}\text{Simplify the result by factoring}\\ a^2+a-2.\text{ Divide out the common}\\ \text{factor of }a-1.\end{array}$$

$$= \dfrac{a+2}{a+1}.$$

SELF CHECK Subtract $\dfrac{b}{b-2} - \dfrac{8}{b^2-4}$.   *Answer:* $\dfrac{b+4}{b+2}$ ∎

EXAMPLE 12

Factoring to find the LCD. Subtract $\dfrac{2a}{a^2+4a+4} - \dfrac{1}{2a+4}$.

Solution Find the least common denominator by factoring each denominator.

$$\left.\begin{array}{l}a^2+4a+4 = (a+2)(a+2)\\ 2a+4 = 2(a+2)\end{array}\right\} \qquad \text{LCD} = (a+2)(a+2)2$$

We build each fraction into a new fraction with a denominator of $2(a+2)(a+2)$.

$$\dfrac{2a}{a^2+4a+4} - \dfrac{1}{2a+4}$$

$$= \dfrac{2a}{(a+2)(a+2)} - \dfrac{1}{2(a+2)} \qquad \begin{array}{l}\text{Write the denominators in factored}\\ \text{form.}\end{array}$$

$$= \dfrac{2a \cdot 2}{(a+2)(a+2)2} - \dfrac{1(a+2)}{2(a+2)(a+2)} \qquad \begin{array}{l}\text{Build each fraction to get a common}\\ \text{denominator.}\end{array}$$

$$= \dfrac{4a-1(a+2)}{2(a+2)^2} \qquad \begin{array}{l}\text{Subtract the numerators and keep the}\\ \text{common denominator. Write}\\ (a+2)(a+2)\text{ as }(a+2)^2.\end{array}$$

$$= \dfrac{4a-a-2}{2(a+2)^2} \qquad \text{Remove parentheses.}$$

$$= \dfrac{3a-2}{2(a+2)^2} \qquad \text{Combine like terms.}$$

SELF CHECK Subtract $\dfrac{a}{a^2-2a+1} - \dfrac{1}{6a-6}$.   *Answer:* $\dfrac{5a+1}{6(a-1)^2}$. ∎

EXAMPLE 13

Denominators that are opposites. Subtract $\dfrac{3}{x-y} - \dfrac{x}{y-x}$.

Solution We note that the second denominator is the negative of the first. So we can multiply the numerator and denominator of the second fraction by -1 to get

$$\frac{3}{x-y} - \frac{x}{y-x} = \frac{3}{x-y} - \frac{-1x}{-1(y-x)} \qquad \text{Multiply numerator and denominator by } -1.$$

$$= \frac{3}{x-y} - \frac{-x}{-y+x} \qquad \text{Remove parentheses: } -1(y-x) = -y + x.$$

$$= \frac{3}{x-y} - \frac{-x}{x-y} \qquad \begin{array}{l} -y + x = x - y. \text{ The fractions have a} \\ \text{common denominator of } x - y. \end{array}$$

$$= \frac{3 - (-x)}{x-y} \qquad \begin{array}{l} \text{Subtract the numerators and keep the com-} \\ \text{mon denominator.} \end{array}$$

$$= \frac{3+x}{x-y} \qquad -(-x) = x.$$

SELF CHECK Subtract $\dfrac{5}{a-b} - \dfrac{2}{b-a}$.   *Answer:* $\dfrac{7}{a-b}$ ∎

Combined Operations

To add and/or subtract three or more rational expressions, we follow the rules for the order of operations.

EXAMPLE 14

Addition and subtraction. Do the operations: $\dfrac{3}{x^2y} + \dfrac{2}{xy} - \dfrac{1}{xy^2}$.

Solution Find the least common denominator.

$$\left. \begin{array}{l} x^2y = x \cdot x \cdot y \\ xy = x \cdot y \\ xy^2 = x \cdot y \cdot y \end{array} \right\} \quad \text{Factor each denominator.}$$

In any one of these denominators, the factor x occurs at most twice, and the factor y occurs at most twice. Thus,

$$\begin{aligned} \text{LCD} &= x \cdot x \cdot y \cdot y \\ &= x^2y^2 \end{aligned}$$

We build each fraction into one with a denominator of x^2y^2.

$$\frac{3}{x^2y} + \frac{2}{xy} - \frac{1}{xy^2}$$

$$= \frac{3 \cdot y}{x \cdot x \cdot y \cdot y} + \frac{2 \cdot x \cdot y}{x \cdot y \cdot x \cdot y} - \frac{1 \cdot x}{x \cdot y \cdot y \cdot x} \qquad \begin{array}{l} \text{Factor each denominator and build} \\ \text{each fraction.} \end{array}$$

$$= \frac{3y + 2xy - x}{x^2y^2} \qquad \begin{array}{l} \text{Do the multiplications and combine} \\ \text{the numerators.} \end{array}$$

SELF CHECK Combine $\dfrac{5}{ab^2} - \dfrac{b}{a} + \dfrac{a}{b}$.   *Answer:* $\dfrac{5 - b^3 + a^2b}{ab^2}$ ∎

VOCABULARY

In Exercises 1–2, fill in the blanks to make the statements true.

1. The _____LCD_____ for a set of fractions is the smallest number that each denominator divides exactly.

2. When we multiply the numerator and denominator of a fraction by some number to get a common denominator, we say that we are _____building_____ the fraction.

CONCEPTS

In Exercises 3–4, fill in the blanks to make the statements true.

3. To add two fractions with like denominators, we add their _____numerators_____ and keep the _____common denominator_____.

4. To subtract two fractions with _____unlike_____ denominators, we need to find a common denominator.

NOTATION

In Exercises 5–6, complete each solution.

5. $\dfrac{6a - 1}{4a + 1} + \dfrac{2a + 3}{4a + 1} = \dfrac{6a - 1 + \boxed{2a + 3}}{4a + 1}$

$= \dfrac{8a + \boxed{2}}{4a + 1}$

$= \dfrac{2\,\boxed{(4a + 1)}}{4a + 1}$

$= 2$

6. $\dfrac{x}{2x + 1} - \dfrac{1}{3x} = \dfrac{x\,\boxed{(3x)}}{(2x + 1)(3x)} - \dfrac{1(2x + 1)}{3x\,\boxed{(2x + 1)}}$

$= \dfrac{x(3x) - 1\,\boxed{(2x + 1)}}{3x(2x + 1)}$

$= \dfrac{3x^2 - \boxed{2x} - \boxed{1}}{3x(2x + 1)}$

$= \dfrac{(3x + 1)(x - 1)}{3x(2x + 1)}$

PRACTICE

In Exercises 7–18, do each addition. Simplify answers, if possible.

7. $\dfrac{x}{9} + \dfrac{2x}{9}$ $\quad \frac{x}{3}$

8. $\dfrac{5x}{7} + \dfrac{9x}{7}$ $\quad 2x$

9. $\dfrac{2x}{y} + \dfrac{2x}{y}$ $\quad \frac{4x}{y}$

10. $\dfrac{4y}{3x} + \dfrac{2y}{3x}$ $\quad \frac{2y}{x}$

11. $\dfrac{4}{7y} + \dfrac{10}{7y}$ $\quad \frac{2}{y}$

12. $\dfrac{x^2}{4y} + \dfrac{x^2}{4y}$ $\quad \frac{x^2}{2y}$

13. $\dfrac{y + 2}{10z} + \dfrac{y + 4}{10z}$ $\quad \frac{y + 3}{5z}$

14. $\dfrac{x + 3}{2x^2} + \dfrac{x + 5}{2x^2}$ $\quad \frac{x + 4}{x^2}$

15. $\dfrac{3x - 5}{x - 2} + \dfrac{6x - 13}{x - 2}$ $\quad 9$

16. $\dfrac{8x - 7}{x + 3} + \dfrac{2x + 37}{x + 3}$ $\quad 10$

17. $\dfrac{a}{a^2 + 5a + 6} + \dfrac{3}{a^2 + 5a + 6}$ $\quad \frac{1}{a + 2}$

18. $\dfrac{b}{b^2 - 4} + \dfrac{2}{b^2 - 4}$ $\quad \frac{1}{b - 2}$

In Exercises 19–30, do each subtraction. Simplify answers, if possible.

19. $\dfrac{35y}{72} - \dfrac{44y}{72}$ $\quad -\frac{y}{8}$

20. $\dfrac{13t}{99} - \dfrac{35t}{99}$ $\quad -\frac{2t}{9}$

21. $\dfrac{2x}{y} - \dfrac{x}{y}$ $\quad \frac{x}{y}$

22. $\dfrac{7y}{5} - \dfrac{4y}{5}$ $\quad \frac{3y}{5}$

23. $\dfrac{9y}{3x} - \dfrac{6y}{3x}$ $\quad \frac{y}{x}$

24. $\dfrac{5r^2}{2r} - \dfrac{r^2}{2r}$ $\quad 2r$

25. $\dfrac{6x - 5}{3xy} - \dfrac{3x - 5}{3xy}$ $\quad \frac{1}{y}$

26. $\dfrac{7x + 7}{5y} - \dfrac{2x + 7}{5y}$ $\quad \frac{x}{y}$

27. $\dfrac{3y - 2}{2y + 6} - \dfrac{2y - 5}{2y + 6}$ $\quad \frac{1}{2}$

28. $\dfrac{5x + 8}{3x + 15} - \dfrac{3x - 2}{3x + 15}$ $\quad \frac{2}{3}$

29. $\dfrac{2c}{c^2 - d^2} - \dfrac{2d}{c^2 - d^2}$ $\quad \frac{2}{c + d}$

30. $\dfrac{3t}{t^2 - 8t + 7} - \dfrac{3}{t^2 - 8t + 7}$ $\quad \frac{3}{t - 7}$

In Exercises 31–38, do the operations. Simplify answers if possible.

31. $\dfrac{13x}{15} + \dfrac{12x}{15} - \dfrac{5x}{15}$ $\quad \frac{4x}{3}$

32. $\dfrac{13y}{32} + \dfrac{13y}{32} - \dfrac{10y}{32}$ $\quad \frac{y}{2}$

33. $-\dfrac{x}{y} + \dfrac{2x}{y} - \dfrac{x}{y}$ $\quad 0$

34. $\dfrac{5y}{8x} + \dfrac{4y}{8x} - \dfrac{9y}{8x}$ $\quad 0$

35. $\dfrac{3x}{y + 2} - \dfrac{3y}{y + 2} + \dfrac{x + y}{y + 2}$ $\quad \frac{4x - 2y}{y + 2}$

36. $\dfrac{3y}{x-5} + \dfrac{x}{x-5} - \dfrac{y-x}{x-5}$ $\dfrac{2(y+x)}{x-5}$

✓ **37.** $\dfrac{x+1}{x-2} - \dfrac{2(x-3)}{x-2} + \dfrac{3(x+1)}{x-2}$ $\dfrac{2x+10}{x-2}$

38. $\dfrac{3xy}{x-y} - \dfrac{x(3y-x)}{x-y} - \dfrac{x(x-y)}{x-y}$ $\dfrac{xy}{x-y}$

In Exercises 39–50, build each fraction into an equivalent fraction with the indicated denominator.

39. $\dfrac{25}{4}$; $20x$ $\dfrac{125x}{20x}$

40. $\dfrac{5}{y}$; y^2 $\dfrac{5y}{y^2}$

41. $\dfrac{8}{x}$; x^2y $\dfrac{8xy}{x^2y}$

42. $\dfrac{7}{y}$; xy^2 $\dfrac{7xy}{xy^2}$

✓ **43.** $\dfrac{3x}{x+1}$; $(x+1)^2$

$\dfrac{3x(x+1)}{(x+1)^2}$

44. $\dfrac{5y}{y-2}$; $(y-2)^2$

$\dfrac{5y(y-2)}{(y-2)^2}$

45. $\dfrac{2y}{x}$; x^2+x $\dfrac{2y(x+1)}{x^2+x}$

46. $\dfrac{3x}{y}$; y^2-y $\dfrac{3x(y-1)}{y^2-y}$

47. $\dfrac{z}{z-1}$; z^2-1

$\dfrac{z(z+1)}{z^2-1}$

48. $\dfrac{y}{y+2}$; y^2-4

$\dfrac{y(y-2)}{y^2-4}$

✓ **49.** $\dfrac{2}{x+1}$; x^2+3x+2

$\dfrac{2(x+2)}{x^2+3x+2}$

50. $\dfrac{3}{x-1}$; x^2+x-2

$\dfrac{3(x+2)}{x^2+x-2}$

In Exercises 51–60, several denominators are given. Find the LCD.

51. $2x$, $6x$ $6x$

52. $3y$, $9y$ $9y$

53. $6y$, $9xy^2$ $18xy^2$

54. $6y$, $3x^2y$ $6x^2y$

✓ **55.** x^2-1, $x+1$ x^2-1

56. y^2-9, $y-3$ y^2-9

57. x^2+6x, $x+6$, x x^2+6x

58. xy^2-xy, xy, $y-1$ xy^2-xy

59. x^2-4x-5, x^2-25 $(x+1)(x+5)(x-5)$

60. x^2-x-6, x^2-9 $(x-3)(x+2)(x+3)$

In Exercises 61–96, do the operations. Simplify answers, if possible.

✓ **61.** $\dfrac{2y}{9} + \dfrac{y}{3}$ $\dfrac{5y}{9}$

62. $\dfrac{8a}{15} - \dfrac{5a}{12}$ $\dfrac{7a}{60}$

63. $\dfrac{21x}{14} - \dfrac{5x}{21}$ $\dfrac{53x}{42}$

64. $\dfrac{7y}{6} + \dfrac{10y}{9}$ $\dfrac{41y}{18}$

65. $\dfrac{4x}{3} + \dfrac{2x}{y}$ $\dfrac{4xy+6x}{3y}$

66. $\dfrac{2y}{5x} - \dfrac{y}{2}$ $\dfrac{4y-5xy}{10x}$

✓ **67.** $\dfrac{2}{x} - 3x$ $\left(Hint:\ 3x=\tfrac{3x}{1}\right)$ $\dfrac{2-3x^2}{x}$

68. $14 + \dfrac{10}{y^2}$ $\left(Hint:\ 14=\tfrac{14}{1}\right)$ $\dfrac{14y^2+10}{y^2}$

69. $\dfrac{y+2}{5y^2} + \dfrac{y+4}{15y}$ $\dfrac{y^2+7y+6}{15y^2}$

70. $\dfrac{x+3}{x^2} + \dfrac{x+5}{2x}$ $\dfrac{x^2+7x+6}{2x^2}$

71. $\dfrac{x+5}{xy} - \dfrac{x-1}{x^2y}$ $\dfrac{x^2+4x+1}{x^2y}$

72. $\dfrac{y-7}{y^2} - \dfrac{y+7}{2y}$ $\dfrac{-y^2+5y+14}{2y^2}$

✓ **73.** $\dfrac{x}{x+1} + \dfrac{x-1}{x}$ $\dfrac{2x^2-1}{x(x+1)}$

74. $\dfrac{3x}{xy} + \dfrac{x+1}{y-1}$ $\dfrac{4y-3+xy}{y(y-1)}$

75. $\dfrac{x-1}{x} + \dfrac{y+1}{y}$ $\dfrac{2xy+x-y}{xy}$

76. $\dfrac{a+2}{b} + \dfrac{b-2}{a}$ $\dfrac{a^2+2a+b^2-2b}{ab}$

77. $\dfrac{x}{x-2} + \dfrac{4+2x}{x^2-4}$ $\dfrac{x+2}{x-2}$

78. $\dfrac{y}{y+3} - \dfrac{2y-6}{y^2-9}$ $\dfrac{y-2}{y+3}$

✓ **79.** $\dfrac{x+1}{x-1} + \dfrac{x-1}{x+1}$ $\dfrac{2x^2+2}{(x-1)(x+1)}$

80. $\dfrac{2x}{x+2} + \dfrac{x+1}{x-3}$ $\dfrac{3x^2-3x+2}{(x-3)(x+2)}$

81. $\dfrac{5}{a-4} + \dfrac{7}{4-a}$ $-\dfrac{2}{a-4}$

82. $\dfrac{4}{b-6} - \dfrac{b}{6-b}$ $\dfrac{b+4}{b-6}$

83. $\dfrac{t+1}{t-7} - \dfrac{t+1}{7-t}$ $\dfrac{2t+2}{t-7}$

84. $\dfrac{r+2}{r^2-4} + \dfrac{4}{4-r^2}$ $\dfrac{1}{r+2}$

✓ **85.** $\dfrac{2x+2}{x-2} - \dfrac{2x}{2-x}$ $\dfrac{4x+2}{x-2}$

86. $\dfrac{y+3}{y-1} - \dfrac{y+4}{1-y}$ $\dfrac{2y+7}{y-1}$

87. $\dfrac{b}{b+1} - \dfrac{b+1}{2b+2}$ $\dfrac{b-1}{2(b+1)}$

88. $\dfrac{4x+1}{8x-12} + \dfrac{x-3}{2x-3}$ $\dfrac{8x-11}{4(2x-3)}$

89. $\dfrac{2}{a^2+4a+3} + \dfrac{1}{a+3}$ $\dfrac{1}{a+1}$

90. $\dfrac{1}{c+6} - \dfrac{-4}{c^2+8a+12}$ $\quad \dfrac{1}{c+2}$

91. $\dfrac{x+1}{2x+4} - \dfrac{x^2}{2x^2-8}$ $\quad -\dfrac{1}{2(x-2)}$

92. $\dfrac{x+1}{x+2} - \dfrac{x^2+1}{x^2-x-6}$ $\quad -\dfrac{2}{x-3}$

93. $\dfrac{2x}{x^2-3x+2} + \dfrac{2x}{x-1} - \dfrac{x}{x-2}$ $\quad \dfrac{x}{x-2}$

94. $\dfrac{4a}{a-2} - \dfrac{3a}{a-3} + \dfrac{4a}{a^2-5a+6}$ $\quad \dfrac{a}{a-3}$

95. $\dfrac{2x}{x-1} + \dfrac{3x}{x+1} - \dfrac{x+3}{x^2-1}$ $\quad \dfrac{5x+3}{x+1}$

96. $\dfrac{a}{a-1} - \dfrac{2}{a+2} + \dfrac{3(a-2)}{a^2+a-2}$ $\quad \dfrac{a+4}{a+2}$

APPLICATIONS

In Exercises 97–98, refer to Illustration 1.

97. Find the total height of the funnel. $\dfrac{20x+9}{6x^2}$ cm

98. What is the difference between the diameter of the opening at the top of the funnel and the diameter of its spout? $\dfrac{16x^2-3}{6x^3}$ cm

ILLUSTRATION 1

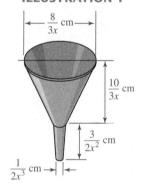

WRITING

99. Explain how to add fractions with the same denominator.

100. Explain how to find a lowest common denominator.

101. Explain what is **wrong** with the following solution:

$$\frac{2x+3}{x+5} - \frac{x+2}{x+5} = \frac{2x+3-x+2}{x+5}$$

$$= \frac{x+5}{x+5}$$

$$= 1$$

102. Explain what is **wrong** with the following solution:

$$\frac{5x-4}{y} + \frac{x}{y} = \frac{5x-4+x}{y+y}$$

$$= \frac{6x-4}{2y}$$

$$= \frac{2(x-2)}{2y}$$

$$= \frac{3x-2}{y}$$

REVIEW

In Exercises 103–106, write each number in prime factored form.

103. 49 $\quad 7^2$

104. 64 $\quad 2^6$

105. 136 $\quad 2^3 \cdot 17$

106. 315 $\quad 3^2 \cdot 5 \cdot 7$

▶ **7.6**

Complex Fractions

In this section, you will learn about

Simplifying complex fractions ■ Simplifying fractions with terms containing negative exponents

Introduction   Rational expressions such as

$$\frac{\dfrac{5x}{3}}{\dfrac{2y}{9}}, \qquad \frac{x+\dfrac{1}{2}}{3-x}, \qquad \text{and} \qquad \frac{\dfrac{x+1}{2}}{x+\dfrac{1}{x}}$$

which contain fractions in their numerators or denominators, are called **complex fractions.** In this section, we will show how to use the properties of fractions to simplify complex fractions.

Simplifying Complex Fractions

Complex fractions can often be simplified.

$$\dfrac{\dfrac{5x}{3}}{\dfrac{2y}{9}} \quad \leftarrow \text{The main fraction bar indicates division.}$$

We can simplify the complex fraction by doing the division:

$$\dfrac{\dfrac{5x}{3}}{\dfrac{2y}{9}} = \dfrac{5x}{3} \div \dfrac{2y}{9} = \dfrac{5x}{3} \cdot \dfrac{9}{2y} = \dfrac{5x \cdot 3 \cdot \overset{1}{\cancel{3}}}{\underset{1}{\cancel{3}} \cdot 2y} = \dfrac{15x}{2y}$$

There are two ways to simplify complex fractions.

Methods for simplifying complex fractions

Method 1: Write the numerator and denominator of the complex fraction as single fractions. Then divide the fractions and simplify.

Method 2: Multiply the numerator and denominator of the complex fraction by the LCD of the fractions in its numerator and denominator. Then simplify the results, if possible.

To simplify the complex fraction

$$\dfrac{\dfrac{3x}{5} + 1}{2 - \dfrac{x}{5}}$$

using method 1, we proceed as follows:

$$\dfrac{\dfrac{3x}{5} + 1}{2 - \dfrac{x}{5}} = \dfrac{\dfrac{3x}{5} + \dfrac{5}{5}}{\dfrac{10}{5} - \dfrac{x}{5}} \qquad$$ Change 1 to $\frac{5}{5}$ and 2 to $\frac{10}{5}$ so that we can write the numerator and denominator as single fractions.

$$= \dfrac{\dfrac{3x + 5}{5}}{\dfrac{10 - x}{5}} \qquad$$ Add the fractions in the numerator and subtract the fractions in the denominator.

$$= \dfrac{3x + 5}{5} \div \dfrac{10 - x}{5} \qquad$$ Write the complex fraction as an equivalent division problem.

$$= \dfrac{3x + 5}{5} \cdot \dfrac{5}{10 - x} \qquad$$ Invert the divisor and multiply.

$$= \dfrac{(3x + 5)5}{5(10 - x)} \qquad$$ Multiply the fractions.

$$= \dfrac{3x + 5}{10 - x} \qquad$$ Divide out the common factor of 5.

To use method 2, we proceed as follows:

$$\dfrac{\dfrac{3x}{5} + 1}{2 - \dfrac{x}{5}} = \dfrac{5\left(\dfrac{3x}{5} + 1\right)}{5\left(2 - \dfrac{x}{5}\right)}$$ Multiply both the numerator and denominator of the complex fraction by 5, the LCD of $\frac{3x}{5}$ and $\frac{x}{5}$.

$$= \dfrac{5 \cdot \dfrac{3x}{5} + 5 \cdot 1}{5 \cdot 2 - 5 \cdot \dfrac{x}{5}}$$ Remove parentheses.

$$= \dfrac{3x + 5}{10 - x}$$ Do the multiplications.

In this example, method 2 is easier than method 1. Either method can be used to simplify complex fractions. With practice, you will be able to see which method is best in a given situation.

EXAMPLE 1

Simplifying complex fractions. Simplify $\dfrac{\dfrac{x}{3}}{\dfrac{y}{3}}$.

Solution

Method 1

$$\dfrac{\dfrac{x}{3}}{\dfrac{y}{3}} = \dfrac{x}{3} \div \dfrac{y}{3}$$

$$= \dfrac{x}{3} \cdot \dfrac{3}{y}$$

$$= \dfrac{3x}{3y}$$

$$= \dfrac{x}{y}$$

Method 2

$$\dfrac{\dfrac{x}{3}}{\dfrac{y}{3}} = \dfrac{3\left(\dfrac{x}{3}\right)}{3\left(\dfrac{y}{3}\right)}$$

$$= \dfrac{x}{y}$$

SELF CHECK Simplify $\dfrac{\dfrac{a}{4}}{\dfrac{5}{b}}$.

Answer: $\frac{ab}{20}$

EXAMPLE 2

Simplifying complex fractions. Simplify $\dfrac{\dfrac{x}{x + 1}}{\dfrac{y}{x}}$.

Solution

Method 1

$$\dfrac{\dfrac{x}{x + 1}}{\dfrac{y}{x}} = \dfrac{x}{x + 1} \div \dfrac{y}{x}$$

$$= \dfrac{x}{x + 1} \cdot \dfrac{x}{y}$$

$$= \dfrac{x^2}{y(x + 1)}$$

Method 2

$$\dfrac{\dfrac{x}{x + 1}}{\dfrac{y}{x}} = \dfrac{x(x + 1)\left(\dfrac{x}{x + 1}\right)}{x(x + 1)\left(\dfrac{y}{x}\right)}$$

$$= \dfrac{x^2}{y(x + 1)}$$

SELF CHECK Simplify $\dfrac{\dfrac{x}{y}}{\dfrac{x}{y+1}}$.

Answer: $\frac{y+1}{y}$ ■

E X A M P L E   3

Simplifying complex fractions. Simplify $\dfrac{1+\dfrac{1}{x}}{1-\dfrac{1}{x}}$.

Solution

Method 1	*Method 2*

Method 1

$$\dfrac{1+\dfrac{1}{x}}{1-\dfrac{1}{x}} = \dfrac{\dfrac{x}{x}+\dfrac{1}{x}}{\dfrac{x}{x}-\dfrac{1}{x}}$$

$$= \dfrac{\dfrac{x+1}{x}}{\dfrac{x-1}{x}}$$

$$= \dfrac{x+1}{x} \div \dfrac{x-1}{x}$$

$$= \dfrac{x+1}{x} \cdot \dfrac{x}{x-1}$$

$$= \dfrac{(x+1)\cancel{x}}{\cancel{x}(x-1)}$$

$$= \dfrac{x+1}{x-1}$$

Method 2

$$\dfrac{1+\dfrac{1}{x}}{1-\dfrac{1}{x}} = \dfrac{x\left(1+\dfrac{1}{x}\right)}{x\left(1-\dfrac{1}{x}\right)}$$

$$= \dfrac{x+1}{x-1}$$

SELF CHECK Simplify $\dfrac{\dfrac{1}{x}+1}{\dfrac{1}{x}-1}$.

Answer: $\frac{1+x}{1-x}$ ■

E X A M P L E   4

Simplifying complex fractions. Simplify $\dfrac{1}{1+\dfrac{1}{x+1}}$.

Solution We use method 2.

$$\dfrac{1}{1+\dfrac{1}{x+1}} = \dfrac{(x+1)\cdot 1}{(x+1)\left(1+\dfrac{1}{x+1}\right)}$$ Multiply the numerator and the denominator of the complex fraction by $x+1$.

$$= \dfrac{x+1}{(x+1)1+1}$$ In the denominator, distribute $x+1$.

$$= \dfrac{x+1}{x+2}$$ Simplify.

SELF CHECK Simplify $\dfrac{2}{\dfrac{1}{x+2}-2}$.

Answer: $\frac{2(x+2)}{-2x-3}$ ■

Simplifying Fractions with Terms Containing Negative Exponents

Many fractions with terms containing negative exponents are complex fractions in disguise.

EXAMPLE 5

Simplifying complex fractions. Simplify $\dfrac{x^{-1} + y^{-2}}{x^{-2} - y^{-1}}$.

Solution   Write the fraction in complex fraction form and simplify using method 2.

$$\frac{x^{-1} + y^{-2}}{x^{-2} - y^{-1}} = \frac{\dfrac{1}{x} + \dfrac{1}{y^2}}{\dfrac{1}{x^2} - \dfrac{1}{y}}$$

$$= \frac{x^2 y^2 \left(\dfrac{1}{x} + \dfrac{1}{y^2} \right)}{x^2 y^2 \left(\dfrac{1}{x^2} - \dfrac{1}{y} \right)}$$ Multiply the numerator and denominator by $x^2 y^2$, which is the LCD of the fractions in the numerator and the denominator.

$$= \frac{xy^2 + x^2}{y^2 - x^2 y}$$ Remove parentheses.

$$= \frac{x(y^2 + x)}{y(y - x^2)}$$ Attempt to simplify the fraction by factoring the numerator and the denominator. The result cannot be simplified.

SELF CHECK   Simplify $\dfrac{x^{-2} - y^{-1}}{x^{-1} + y^{-2}}$.   *Answer:* $\dfrac{y(y - x^2)}{x(y^2 + x)}$ ∎

STUDY SET

Section 7.6

VOCABULARY

In Exercises 1–2, fill in the blanks to make the statements true.

1. If a fraction has a fraction in its numerator or denominator, it is called a ___complex fraction___.

2. The denominator of the complex fraction $\dfrac{\frac{3}{x} + \frac{x}{y}}{\frac{1}{x} + 2}$ is

   $\dfrac{1}{x} + 2$

CONCEPTS

In Exercises 3–4, fill in the blanks to make the statements true.

3. In method 1, we write the numerator and denominator of a complex fraction as ___single___ fractions and then ___divide___.

4. In method 2, we multiply the numerator and denominator of the complex fraction by the ___LCD___ of the fractions in its numerator and denominator.

NOTATION

In Exercises 5–6, complete each solution.

5.
$$\frac{\dfrac{2}{a} - \dfrac{1}{b}}{\dfrac{1}{a} + \dfrac{2}{b}} = \frac{\dfrac{2b - a}{ab}}{\dfrac{b + 2a}{ab}}$$

$$= \frac{2b - a}{ab} \div \frac{b + 2a}{ab}$$

$$= \frac{2b - a}{ab} \cdot \frac{ab}{b + 2a}$$

$$= \frac{(2b - a)\,ab}{ab\,(b + 2a)}$$

$$= \frac{2b - a}{b + 2a}$$

$$6. \ \ \frac{\dfrac{2}{a}-\dfrac{1}{b}}{\dfrac{1}{a}+\dfrac{2}{b}} = \frac{ab\left(\dfrac{2}{a}-\dfrac{1}{b}\right)}{ab\left(\dfrac{1}{a}+\dfrac{2}{b}\right)}$$

$$= \frac{2b-a}{b+2a}$$

PRACTICE

In Exercises 7–40, simplify each complex fraction.

7. $\dfrac{\dfrac{2}{3}}{\dfrac{3}{4}}$ $\quad \frac{8}{9}$

8. $\dfrac{\dfrac{3}{5}}{\dfrac{2}{7}}$ $\quad \frac{21}{10}$

9. $\dfrac{\dfrac{4}{5}}{\dfrac{32}{15}}$ $\quad \frac{3}{8}$

10. $\dfrac{\dfrac{7}{8}}{\dfrac{49}{4}}$ $\quad \frac{1}{14}$

11. $\dfrac{\dfrac{2}{3}+1}{\dfrac{1}{3}+1}$ $\quad \frac{5}{4}$

12. $\dfrac{\dfrac{3}{5}-2}{\dfrac{2}{5}-2}$ $\quad \frac{7}{8}$

13. $\dfrac{\dfrac{1}{2}+\dfrac{3}{4}}{\dfrac{3}{2}+\dfrac{1}{4}}$ $\quad \frac{5}{7}$

14. $\dfrac{\dfrac{2}{3}-\dfrac{5}{2}}{\dfrac{2}{3}-\dfrac{3}{2}}$ $\quad \frac{11}{5}$

15. $\dfrac{\dfrac{x}{y}}{\dfrac{1}{x}}$ $\quad \frac{x^2}{y}$

16. $\dfrac{\dfrac{y}{x}}{\dfrac{x}{xy}}$ $\quad \frac{y^2}{x}$

17. $\dfrac{\dfrac{5t^2}{9x^2}}{\dfrac{3t}{x^2t}}$ $\quad \frac{5t^2}{27}$

18. $\dfrac{\dfrac{5w^2}{4tz}}{\dfrac{15wt}{z^2}}$ $\quad \frac{wz}{12t^2}$

19. $\dfrac{\dfrac{1}{x}-3}{\dfrac{5}{x}+2}$ $\quad \frac{1-3x}{5+2x}$

20. $\dfrac{\dfrac{1}{y}+3}{\dfrac{3}{y}-2}$ $\quad \frac{1+3y}{3-2y}$

21. $\dfrac{\dfrac{2}{x}+2}{\dfrac{4}{x}+2}$ $\quad \frac{1+x}{2+x}$

22. $\dfrac{\dfrac{3}{x}-3}{\dfrac{9}{x}-3}$ $\quad \frac{1-x}{3-x}$

23. $\dfrac{\dfrac{3y}{x}-y}{y-\dfrac{y}{x}}$ $\quad \frac{3-x}{x-1}$

24. $\dfrac{\dfrac{y}{x}+3y}{y+\dfrac{2y}{x}}$ $\quad \frac{3x+1}{x+2}$

25. $\dfrac{\dfrac{1}{x+1}}{1+\dfrac{1}{x+1}}$ $\quad \frac{1}{x+2}$

26. $\dfrac{\dfrac{1}{x-1}}{1-\dfrac{1}{x-1}}$ $\quad \frac{1}{x-2}$

27. $\dfrac{\dfrac{x}{x+2}}{\dfrac{x}{x+2}+x}$ $\quad \frac{1}{x+3}$

28. $\dfrac{\dfrac{2}{x-2}}{\dfrac{2}{x-2}-1}$ $\quad \frac{2}{4-x}$

29. $\dfrac{\dfrac{1}{1}}{\dfrac{1}{x}+\dfrac{1}{y}}$ $\quad \frac{xy}{y+x}$

30. $\dfrac{\dfrac{1}{1}}{\dfrac{b}{a}-\dfrac{a}{b}}$ $\quad \frac{ab}{b^2-a^2}$

31. $\dfrac{\dfrac{2}{x}}{\dfrac{2}{y}-\dfrac{4}{x}}$ $\quad \frac{y}{x-2y}$

32. $\dfrac{\dfrac{2y}{3}}{\dfrac{2y}{3}-\dfrac{8}{y}}$ $\quad \frac{y^2}{y^2-12}$

33. $\dfrac{3+\dfrac{3}{x-1}}{3-\dfrac{3}{x}}$ $\quad \frac{x^2}{(x-1)^2}$

34. $\dfrac{2-\dfrac{2}{x+1}}{2+\dfrac{2}{x}}$ $\quad \frac{x^2}{(x+1)^2}$

35. $\dfrac{\dfrac{3}{x}+\dfrac{4}{x+1}}{\dfrac{2}{x+1}-\dfrac{3}{x}}$ $\quad \frac{7x+3}{-x-3}$

36. $\dfrac{\dfrac{5}{y-3}-\dfrac{2}{y}}{\dfrac{1}{y}+\dfrac{2}{y-3}}$ $\quad \frac{y+2}{y-1}$

37. $\dfrac{\dfrac{2}{x}-\dfrac{3}{x+1}}{\dfrac{2}{x+1}-\dfrac{3}{x}}$ $\quad \frac{x-2}{x+3}$

38. $\dfrac{\dfrac{5}{y}+\dfrac{4}{y+1}}{\dfrac{4}{y}-\dfrac{5}{y+1}}$ $\quad \frac{9y+5}{4-y}$

39. $\dfrac{\dfrac{1}{y^2+y}-\dfrac{1}{xy+x}}{\dfrac{1}{xy+x}-\dfrac{1}{y^2+y}}$ $\quad -1$

40. $\dfrac{\dfrac{2}{b^2-1}-\dfrac{3}{ab-a}}{\dfrac{3}{ab-a}-\dfrac{2}{b^2-1}}$ $\quad -1$

In Exercises 41–50, simplify each complex fraction.

41. $\dfrac{x^{-2}}{y^{-1}}$ $\quad \frac{y}{x^2}$

42. $\dfrac{a^{-4}}{b^{-2}}$ $\quad \frac{b^2}{a^4}$

43. $\dfrac{1+x^{-1}}{x^{-1}-1}$ $\quad \frac{x+1}{1-x}$

44. $\dfrac{y^{-2}+1}{y^{-2}-1}$ $\quad \frac{1+y^2}{1-y^2}$

45. $\dfrac{a^{-2}+a}{a}$ $\quad \frac{1+a^3}{a^3}$

46. $\dfrac{t-t^{-2}}{t^{-1}}$ $\quad \frac{t^3-1}{t}$

47. $\dfrac{2x^{-1}+4x^{-2}}{2x^{-2}+x^{-1}}$ $\quad 2$

48. $\dfrac{x^{-2}-3x^{-3}}{3x^{-2}-9x^{-3}}$ $\quad \frac{1}{3}$

49. $\dfrac{\dfrac{1-25y^{-2}}{1+10y^{-1}+25y^{-2}}}{\dfrac{y-5}{y+5}}$

50. $\dfrac{\dfrac{1-9x^{-2}}{1-6x^{-1}+9x^{-2}}}{\dfrac{x+3}{x-3}}$

APPLICATIONS

51. GARDENING TOOLS In Illustration 1, what is the ratio of the opening of the cutting blades to the opening of the handles for the lopping shears? Express the result in simplest form. $\frac{3}{14}$

ILLUSTRATION 1

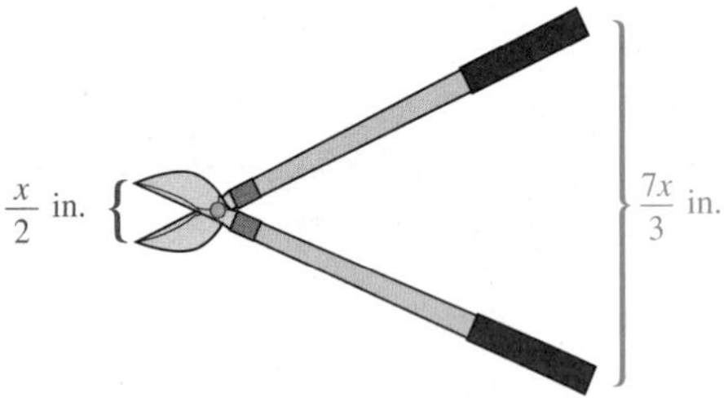

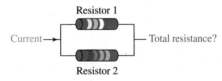

$\frac{x}{2}$ in.

$\frac{7x}{3}$ in.

52. EARNED RUN AVERAGE The earned run average (ERA) is a statistic that gives the average number of earned runs a pitcher allows. For a softball pitcher, this is based on a six-inning game. The formula for ERA is

$$\text{ERA} = \frac{\dfrac{\text{earned runs}}{\text{innings pitched}}}{6}$$

Simplify the complex fraction on the right-hand side of the equation. $\text{ERA} = \dfrac{6 \cdot \text{earned runs}}{\text{innings pitched}}$

53. ELECTRONICS In electronic circuits, resistors oppose the flow of an electric current. To find the total resistance of a parallel combination of two resistors (see Illustration 2), we can use the formula

$$\text{Total resistance} = \frac{1}{\dfrac{1}{R_1} + \dfrac{1}{R_2}}$$

ILLUSTRATION 2

Resistor 1

Current → Total resistance?

Resistor 2

where R_1 is the resistance of the first resistor and R_2 is the resistance of the second. Simplify the complex fraction on the right-hand side of the formula. $\dfrac{R_1 R_2}{R_2 + R_1}$

54. DATA ANALYSIS Use the data in Illustration 3 to find the average measurement for the three-trial experiment. $\frac{4k}{9}$

ILLUSTRATION 3

	Trial 1	Trial 2	Trial 3
Measurement	$\frac{k}{2}$	$\frac{k}{3}$	$\frac{k}{2}$

WRITING

55. Explain how to use method 1 to simplify

$$\frac{1 + \dfrac{1}{x}}{3 - \dfrac{1}{x}}$$

56. Explain how to use method 2 to simplify the expression in Exercise 55.

REVIEW

In Exercises 57–60, write each expression as an expression involving only one exponent.

57. $t^3 t^4 t^2$ t^9

58. $(a^0 a^2)^3$ a^6

59. $-2r(r^3)^2$ $-2r^7$

60. $(s^3)^2 (s^4)^0$ s^6

In Exercises 61–64, write each expression without using parentheses or negative exponents.

61. $\left(\dfrac{3r}{4r^3}\right)^4$ $\dfrac{81}{256 r^8}$

62. $\left(\dfrac{12y^{-3}}{3y^2}\right)^{-2}$ $\dfrac{y^{10}}{16}$

63. $\left(\dfrac{6r^{-2}}{2r^3}\right)^{-2}$ $\dfrac{r^{10}}{9}$

64. $\left(\dfrac{4x^3}{5x^{-3}}\right)^{-2}$ $\dfrac{25}{16x^{12}}$

▶ 7.7   Solving Equations That Contain Rational Expressions; Inverse Variation

In this section, you will learn about

Solving equations that contain rational expressions ■ Extraneous solutions ■ Literal equations ■ Inverse variation

---

Introduction   In this section, we will discuss how to solve equations containing rational expressions. We will make use of the properties of equality and several concepts from this chapter, one of which is LCD.

Solving Equations That Contain Rational Expressions

To solve equations containing fractions, it's usually best to eliminate those fractions. To do so, we multiply both sides of the equation by the LCD. For example, to solve $\frac{x}{3} + 1 = \frac{x}{6}$, we multiply both sides of the equation by 6:

$$\frac{x}{3} + 1 = \frac{x}{6}$$

$$6\left(\frac{x}{3} + 1\right) = 6\left(\frac{x}{6}\right)$$

We then use the distributive property to remove parentheses, simplify, and solve the resulting equation for x.

$$6 \cdot \frac{x}{3} + 6 \cdot 1 = 6 \cdot \frac{x}{6}$$

$$2x + 6 = x$$

$$x + 6 = 0 \qquad \text{Subtract } x \text{ from both sides.}$$

$$x = -6 \qquad \text{Subtract 6 from both sides.}$$

Check: $\qquad \dfrac{x}{3} + 1 = \dfrac{x}{6}$

$$\frac{-6}{3} + 1 \stackrel{?}{=} \frac{-6}{6} \qquad \text{Substitute } -6 \text{ for } x.$$

$$-2 + 1 \stackrel{?}{=} -1 \qquad \text{Simplify.}$$

$$-1 = -1$$

EXAMPLE 1

Solving equations containing rational expressions. Solve $\dfrac{4}{x} + 1 = \dfrac{6}{x}$.

Solution To clear the equation of fractions, we multiply both sides by the LCD of $\frac{4}{x}$ and $\frac{6}{x}$, which is x.

$$\frac{4}{x} + 1 = \frac{6}{x}$$

$$x\left(\frac{4}{x} + 1\right) = x\left(\frac{6}{x}\right) \qquad \text{Multiply both sides by } x.$$

$$x \cdot \frac{4}{x} + x \cdot 1 = x \cdot \frac{6}{x} \qquad \text{Remove parentheses.}$$

$$4 + x = 6 \qquad \text{Simplify.}$$

$$x = 2 \qquad \text{Subtract 4 from both sides.}$$

Check: $\qquad \dfrac{4}{x} + 1 = \dfrac{6}{x}$

$$\frac{4}{2} + 1 \stackrel{?}{=} \frac{6}{2} \qquad \text{Substitute 2 for } x.$$

$$2 + 1 \stackrel{?}{=} 3 \qquad \text{Simplify.}$$

$$3 = 3$$

SELF CHECK Solve $\frac{6}{x} - 1 = \frac{3}{x}$. *Answer:* 3 ∎

Extraneous Solutions

If we multiply both sides of an equation by an expression that involves a variable, as we did in Example 1, we must check the apparent solutions. The next example shows why.

EXAMPLE 2

Checking apparent solutions. Solve $\dfrac{x+3}{x-1} = \dfrac{4}{x-1}$.

Solution   To clear the equation of fractions, we multiply both sides by the LCD, which is $x-1$.

$$\frac{x+3}{x-1} = \frac{4}{x-1}$$

$$(x-1)\frac{x+3}{x-1} = (x-1)\frac{4}{x-1} \qquad \text{Multiply both sides by } x-1.$$

$$x+3 = 4 \qquad\qquad \text{Simplify each side of the equation.}$$

$$x = 1 \qquad\qquad \text{Subtract 3 from both sides.}$$

Because both sides were multiplied by an expression containing a variable, we must check the apparent solution.

$$\frac{x+3}{x-1} = \frac{4}{x-1}$$

$$\frac{1+3}{1-1} \overset{?}{=} \frac{4}{1-1} \qquad \text{Substitute 1 for } x.$$

$$\frac{4}{0} \overset{?}{=} \frac{4}{0} \qquad \text{Simplify.}$$

Since zeros appear in the denominators, the fractions are undefined. Thus, 1 is a false solution, and the equation has no solutions. Such false solutions are often called **extraneous solutions.**

SELF CHECK

Solve $\frac{x+5}{x-2} = \frac{7}{x-2}$.

Answer: 2 is extraneous.   ∎

EXAMPLE 3

Equations containing rational expressions. Solve
$\dfrac{3x+1}{x+1} - 2 = \dfrac{3(x-3)}{x+1}$.

Solution   To clear the equation of fractions, we multiply both sides by the LCD, which is $x+1$.

$$\frac{3x+1}{x+1} - 2 = \frac{3(x-3)}{x+1}$$

$$(x+1)\left(\frac{3x+1}{x+1} - 2\right) = (x+1)\left[\frac{3(x-3)}{x+1}\right]$$

$$3x+1 - 2(x+1) = 3(x-3) \qquad\qquad \begin{array}{l}\text{Use the distributive property to re-}\\ \text{move parentheses.}\end{array}$$

$$3x+1-2x-2 = 3x-9 \qquad\qquad \text{Remove parentheses.}$$

$$x - 1 = 3x - 9 \qquad\qquad \text{Combine like terms.}$$

$$-2x = -8 \qquad\qquad \text{Subtract } 3x \text{ and add 1 to both sides.}$$

$$x = 4 \qquad\qquad \text{Divide both sides by } -2.$$

Check:
$$\frac{3x+1}{x+1} - 2 = \frac{3(x-3)}{x+1}$$

$$\frac{3(4)+1}{4+1} - 2 \overset{?}{=} \frac{3(4-3)}{4+1} \qquad \text{Substitute 4 for } x.$$

$$\frac{13}{5} - \frac{10}{5} \overset{?}{=} \frac{3(1)}{5}$$

$$\frac{3}{5} = \frac{3}{5}$$

SELF CHECK Solve $\frac{12}{x+1} - 5 = \frac{2}{x+1}$.   *Answer:* 1   ■

Many times, we will have to factor a denominator to find the LCD.

E X A M P L E   4 **Factoring to find the LCD.** Solve $\dfrac{x+2}{x+3} + \dfrac{1}{x^2 + 2x - 3} = 1$.

Solution To find the LCD, we must factor the second denominator.

$$\frac{x+2}{x+3} + \frac{1}{x^2 + 2x - 3} = 1$$

$$\frac{x+2}{x+3} + \frac{1}{(x+3)(x-1)} = 1 \quad \text{Factor } x^2 + 2x - 3.$$

To clear the equation of fractions, we multiply both sides by the LCD, which is $(x+3)(x-1)$.

$$(x+3)(x-1)\left[\frac{x+2}{x+3} + \frac{1}{(x+3)(x-1)}\right] = (x+3)(x-1)1 \quad \text{Multiply both sides by } (x+3)(x-1).$$

$$(x+3)(x-1)\frac{x+2}{x+3} + (x+3)(x-1)\frac{1}{(x+3)(x-1)} = (x+3)(x-1)1 \quad \text{Remove brackets.}$$

$$(x-1)(x+2) + 1 = (x+3)(x-1) \quad \text{Simplify.}$$

$$x^2 + x - 2 + 1 = x^2 + 2x - 3 \quad \text{Use the FOIL method to remove parentheses.}$$

$$x - 2 + 1 = 2x - 3 \quad \text{Subtract } x^2 \text{ from both sides.}$$

$$x - 1 = 2x - 3 \quad \text{Combine like terms.}$$

$$-x - 1 = -3 \quad \text{Add } -2x \text{ to both sides.}$$

$$-x = -2 \quad \text{Add 1 to both sides.}$$

$$x = 2 \quad \text{Divide both sides by } -1.$$

Verify that 2 is a solution of the given equation.   ■

E X A M P L E   5 **Factoring to find the LCD.** Solve $\dfrac{4}{5} + y = \dfrac{4y - 50}{5y - 25}$.

Solution

$$\frac{4}{5} + y = \frac{4y - 50}{5y - 25} \qquad \text{To clear the equation of fractions, we must find the LCD.}$$

$$\frac{4}{5} + y = \frac{4y - 50}{5(y - 5)} \qquad \text{Factor } 5y - 25 \text{ to help determine the LCD.}$$

$$5(y - 5)\left[\frac{4}{5} + y\right] = 5(y - 5)\left[\frac{4y - 50}{5(y - 5)}\right] \qquad \text{Multiply both sides by the LCD, which is } 5(y - 5).$$

$$4(y - 5) + 5y(y - 5) = 4y - 50 \qquad \text{Remove brackets and simplify.}$$

$$4y - 20 + 5y^2 - 25y = 4y - 50 \qquad \text{Remove parentheses.}$$

$$5y^2 - 25y - 20 = -50 \qquad \text{Subtract } 4y \text{ from both sides and rearrange terms.}$$

$$5y^2 - 25y + 30 = 0 \qquad \text{Add 50 to both sides.}$$

$$y^2 - 5y + 6 = 0 \qquad \text{Divide both sides by 5.}$$

$$(y - 3)(y - 2) = 0 \qquad \text{Factor } y^2 - 5y + 6.$$

$$y - 3 = 0 \quad \text{or} \quad y - 2 = 0 \qquad \text{Set each factor equal to zero.}$$

$$y = 3 \qquad\qquad y = 2 \qquad \text{Solve each equation.}$$

Verify that 3 and 2 satisfy the original equation.

SELF CHECK Solve $\frac{x-6}{3x-9} - \frac{1}{3} = \frac{x}{2}$.   *Answer:* 1, 2   ■

Literal Equations

Many formulas are equations that contain rational expressions.

EXAMPLE 6

Solving formulas. The formula

$$\frac{1}{r} = \frac{1}{r_1} + \frac{1}{r_2}$$

is used in electronics to calculate parallel resistances. Solve the equation for r.

Solution   Clear the equation of fractions by multiplying both sides by the LCD, which is rr_1r_2.

$$\frac{1}{r} = \frac{1}{r_1} + \frac{1}{r_2}$$

$$rr_1r_2\left(\frac{1}{r}\right) = rr_1r_2\left(\frac{1}{r_1} + \frac{1}{r_2}\right) \qquad \text{Multiply both sides by } rr_1r_2.$$

$$\frac{rr_1r_2}{r} = \frac{rr_1r_2}{r_1} + \frac{rr_1r_2}{r_2} \qquad \text{Remove parentheses.}$$

$$r_1r_2 = rr_2 + rr_1 \qquad \text{Simplify each fraction.}$$

$$r_1r_2 = r(r_2 + r_1) \qquad \text{Factor out } r.$$

$$\frac{r_1r_2}{r_2 + r_1} = r \qquad \text{To isolate } r, \text{ divide both sides by } r_2 + r_1.$$

or

$$r = \frac{r_1r_2}{r_2 + r_1}$$

SELF CHECK   Solve the formula in Example 6 for r_1.   *Answer:* $r_1 = \dfrac{rr_2}{r_2 - r}$ ∎

Inverse Variation

In Section 3.7, we discussed direct variation.

Direct variation

The words y *varies directly with* x mean that

$$y = kx$$

for some constant k, called the **constant of variation.**

We now discuss a type of variation that involves rational expressions.

Inverse variation

The words y *varies inversely with* x mean that

$$y = \frac{k}{x}$$

for some constant k, called the **constant of variation.**

Scientists have found that under constant temperature, the volume occupied by a gas varies inversely with its pressure. That is, the volume occupied by the gas increases as the pressure placed on it decreases. If V represents volume and p represents pressure, this relationship is expressed by the equation $V = \frac{k}{p}$.

EXAMPLE 7

A gas occupies a volume of 15 cubic inches when placed under 4 pounds per square inch (psi) of pressure. How much pressure is needed to compress the gas into a volume of 10 cubic inches?

Solution   To find the constant of variation, we substitute 15 for V and 4 for p in the formula $V = \frac{k}{p}$ and solve for k.

$$V = \frac{k}{p}$$

$$15 = \frac{k}{4}$$

$60 = k$ Multiply both sides by 4.

To find the pressure needed to compress the gas into a volume of 10 cubic inches, we substitute 60 for k and 10 for V in the formula and solve for p.

$$V = \frac{k}{p}$$

$$10 = \frac{60}{p}$$

$10p = 60$ Multiply both sides by p.

$p = 6$ Divide both sides by 10.

It will take 6 pounds per square inch (psi) of pressure to compress the gas into a volume of 10 cubic inches.

SELF CHECK   How much pressure is needed to compress the gas in Example 7 into a volume of 8 cubic inches?   *Answer:* 7.5 psi ■

STUDY SET

Section 7.7

VOCABULARY

In Exercises 1–2, fill in the blanks to make the statements true.

1. False solutions that result from multiplying both sides of an equation by a variable are called _____extraneous_____ solutions.

2. "y varies ____inversely____ with x" means that $y = \frac{k}{x}$ for some constant k.

CONCEPTS

In Exercises 3–6, fill in the blanks to make the statements true.

3. To clear an equation of fractions, we multiply both sides by the ____LCD____ of the fractions of the equation.

4. If you multiply both sides of an equation by an expression that involves a variable, you must ____check____ the solution.

5. To clear the equation

$$\frac{1}{x} + \frac{2}{x} = 5$$

of fractions, we multiply both sides by x .

6. To clear the equation

$$\frac{x}{x - 2} - \frac{x}{x - 1} = 5$$

of fractions, we multiply both sides by $(x - 2)(x - 1)$.

7. a. Simplify the expressions.

$$\frac{3}{2} - \frac{3}{r} + \frac{1}{3} \quad \frac{11r - 18}{6r}$$

b. Solve the equation.

$$\frac{3}{2} - \frac{3}{r} = \frac{1}{3} \quad \frac{18}{7}$$

8. Is $x = 5$ a solution of the following equations?

a. $\dfrac{1}{x - 1} = 1 - \dfrac{3}{x - 1}$ yes

b. $\dfrac{x}{x - 5} = 3 + \dfrac{5}{x - 5}$ no

NOTATION

In Exercises 9–10, complete each solution.

9.
$$\frac{2}{a} + \frac{1}{2} = \frac{7}{2a}$$

$$2a\left(\frac{2}{a} + \frac{1}{2}\right) = 2a\left(\frac{7}{2a}\right)$$

$$\frac{4a}{a} + \frac{2a}{2} = \frac{14a}{2a}$$

$$\boxed{4} + a = 7$$

$$4 + a - \boxed{4} = 7 - \boxed{4}$$

$$a = 3$$

10.
$$\frac{3}{5} + \frac{7}{a + 2} = 2$$

$$\boxed{5(a + 2)}\left(\frac{3}{5} + \frac{7}{a + 2}\right) = \boxed{5(a + 2)}\,2$$

$$\frac{5(a + 2)3}{5} + \frac{\boxed{5(a + 2)}\,7}{a + 2} = \boxed{10}\,(a + 2)$$

$$3(a + 2) + \boxed{35} = 10(a + 2)$$

$$3a + \boxed{6} + 35 = 10a + \boxed{20}$$

$$3a + \boxed{41} = 10a + 20$$

$$-7a = \boxed{-21}$$

$$a = 3$$

PRACTICE

In Exercises 11–62, solve each equation and check the solution. If an equation has no solution, so indicate.

11. $\dfrac{x}{2} + 4 = \dfrac{3x}{2}$ 4

12. $\dfrac{y}{3} + 6 = \dfrac{4y}{3}$ 6

13. $\dfrac{2y}{5} - 8 = \dfrac{4y}{5}$ −20

14. $\dfrac{3x}{4} - 6 = \dfrac{x}{4}$ 12

15. $\dfrac{3a}{2} + \dfrac{a}{3} + 22 = 0$ −12

16. $\dfrac{x}{2} + x - \dfrac{9}{2} = 0$ 3

17. $\dfrac{5(x + 1)}{8} = x + 1$ −1

18. $\dfrac{3(x - 1)}{2} + 2 = x$ −1

19. $\dfrac{x + 1}{3} + \dfrac{x - 1}{5} = \dfrac{2}{15}$ 0

20. $\dfrac{y - 5}{7} + \dfrac{y - 7}{5} = \dfrac{-2}{5}$ 5

21. $\dfrac{3x - 1}{6} - \dfrac{x + 3}{2} = \dfrac{3x + 4}{3}$ −3

22. $\dfrac{2x + 3}{3} + \dfrac{3x - 4}{6} = \dfrac{x - 2}{2}$ −2

23. $\dfrac{3}{x} + 2 = 3$ 3

24. $\dfrac{2}{x} + 9 = 11$ 1

25. $\dfrac{5}{a} - \dfrac{4}{a} = 8 + \dfrac{1}{a}$

No solution; 0 is extraneous.

26. $\dfrac{11}{b} + \dfrac{13}{b} = 12$ 2

27. $\dfrac{3}{4h} + \dfrac{2}{h} = 5 - 4$ $\frac{11}{4}$

28. $\dfrac{5}{3k} + \dfrac{1}{k} = 2 - 4$ $-\frac{4}{3}$

29. $\dfrac{a}{4} - \dfrac{4}{a} = 0$ −4, 4

30. $0 = \dfrac{t}{3} - \dfrac{12}{t}$ −6, 6

31. $\dfrac{2}{y + 1} + 5 = \dfrac{12}{y + 1}$ 1

32. $\dfrac{1}{t - 3} = \dfrac{-2}{t - 3} + 1$ 6

33. $\dfrac{1}{x - 1} + \dfrac{3}{x - 1} = 1$ 5

34. $\dfrac{3}{p + 6} - 2 = \dfrac{7}{p + 6}$ −8

35. $\dfrac{a^2}{a + 2} - \dfrac{4}{a + 2} = a$ No solution; −2 is extraneous.

36. $\dfrac{z^2}{z + 1} + 2 = \dfrac{1}{z + 1}$ No solution; −1 is extraneous.

37. $\dfrac{x}{x - 5} - \dfrac{5}{x - 5} = 3$ No solution; 5 is extraneous.

38. $\dfrac{3}{y - 2} + 1 = \dfrac{3}{y - 2}$ No solution; 2 is extraneous.

39. $\dfrac{3r}{2} - \dfrac{3}{r} = \dfrac{3r}{2} + 3$ −1

40. $\dfrac{2p}{3} - \dfrac{1}{p} = \dfrac{2p - 1}{3}$ 3

41. $\dfrac{1}{3} + \dfrac{2}{x - 3} = 1$ 6

42. $\dfrac{3}{5} + \dfrac{7}{x + 2} = 2$ 3

43. $\dfrac{u}{u - 1} + \dfrac{1}{u} = \dfrac{u^2 + 1}{u^2 - u}$ 2

44. $\dfrac{v}{v + 2} + \dfrac{1}{v - 1} = 1$ 4

45. $\dfrac{3}{x - 2} + \dfrac{1}{x} = \dfrac{2(3x + 2)}{x^2 - 2x}$ −3

46. $\dfrac{5}{x} + \dfrac{3}{x + 2} = \dfrac{-6}{x(x + 2)}$ No solution; −2 is extraneous.

47. $\dfrac{7}{q^2 - q - 2} + \dfrac{1}{q + 1} = \dfrac{3}{q - 2}$ 1

48. $\dfrac{-5}{s^2 + s - 2} + \dfrac{3}{s + 2} = \dfrac{1}{s - 1}$ 5

49. $\dfrac{3y}{3y - 6} + \dfrac{8}{y^2 - 4} = \dfrac{2y}{2y + 4}$

No solution; −2 is extraneous

50. $\dfrac{x - 3}{4x - 4} + \dfrac{1}{9} = \dfrac{x - 5}{6x - 6}$ $\frac{1}{7}$

51. $y + \dfrac{2}{3} = \dfrac{2y - 12}{3y - 9}$ 1, 2   **52.** $y + \dfrac{3}{4} = \dfrac{3y - 50}{4y - 24}$ 2, 4

53. $\dfrac{5}{4y + 12} - \dfrac{3}{4} = \dfrac{5}{4y + 12} - \dfrac{y}{4}$ 3; -3 is extraneous.

54. $\dfrac{3}{5x - 20} + \dfrac{4}{5} = \dfrac{3}{5x - 20} - \dfrac{x}{5}$ -4; 4 is extraneous.

55. $\dfrac{x}{x - 1} - \dfrac{12}{x^2 - x} = \dfrac{-1}{x - 1}$ 3, -4

56. $1 - \dfrac{3}{b} = \dfrac{-8b}{b^2 + 3b}$ 1, -9

57. $\dfrac{z - 4}{z - 3} = \dfrac{z + 2}{z + 1}$ 1

58. $\dfrac{a + 2}{a + 8} = \dfrac{a - 3}{a - 2}$ 4

59. $\dfrac{n}{n^2 - 9} + \dfrac{n + 8}{n + 3} = \dfrac{n - 8}{n - 3}$ 0

60. $\dfrac{x - 3}{x - 2} - \dfrac{1}{x} = \dfrac{x - 3}{x}$ 4

61. $\dfrac{b + 2}{b + 3} + 1 = \dfrac{-7}{b - 5}$ -2, 1

62. $\dfrac{x - 4}{x - 3} + \dfrac{x - 2}{x - 3} = x - 3$ 5; 3 is extraneous.

63. Solve the formula $\dfrac{1}{a} + \dfrac{1}{b} = 1$ for a. $a = \frac{b}{b - 1}$

64. Solve the formula $\dfrac{1}{a} - \dfrac{1}{b} = 1$ for b. $b = \frac{a}{1 - a}$

APPLICATIONS

65. OPTICS The focal length f of a lens is given by the formula

$$\frac{1}{f} = \frac{1}{d_1} + \frac{1}{d_2}$$

where d_1 is the distance from the object to the lens and d_2 is the distance from the lens to the image.

Solve the formula for f. $f = \frac{d_1 d_2}{d_1 + d_2}$

66. OPTICS Solve the formula in Exercise 65 for d_1.

$d_1 = \frac{f d_2}{d_2 - f}$

67. MEDICINE Radioactive tracers are used for diagnostic work in nuclear medicine. The **effective half-life,** H, of a radioactive material in a biological organism is given by the formula

$$H = \frac{RB}{R + B}$$

where R is the radioactive half-life and B is the biological half-life of the tracer. Solve the formula for R.

$R = \frac{HB}{B - H}$

68. CHEMISTRY Charles's Law describes the relationship between the volume and the temperature of a gas that is kept at a constant pressure. It states that as the temperature of the gas increases, the volume of the gas will increase:

$$\frac{V_1}{V_2} = \frac{T_1}{T_2}$$

Solve the equation for V_2. $V_2 = \frac{V_1 T_2}{T_1}$

69. COMMUTING TIME The time it takes a car to travel a certain distance varies inversely with its rate of speed. If a certain trip takes 3 hours when the driver travels at 50 miles per hour, how long will the trip take if the driver travels at 60 miles per hour? $2\frac{1}{2}$ hr

70. GEOMETRY For a fixed area, the length of a rectangle is inversely proportional to its width. A rectangle has a width of 12 feet and a length of 20 feet. If the length is increased to 24 feet, find the width of the rectangle. 10 ft

71. COMPUTING PRESSURES If the temperature of a gas is constant, the volume occupied varies inversely with the pressure. If a gas occupies a volume of 40 cubic meters under a pressure of 8 atmospheres, find the volume when the pressure is changed to 6 atmospheres. $53\frac{1}{3}$ m^3

72. COMPUTING DEPRECIATION Assume that the value of a machine varies inversely with its age. If a drill press is worth \$300 when it is 2 years old, find its value when it is 6 years old. How much has the machine depreciated over that 4-year period? \$100; \$200

WRITING

73. Explain how you would decide what to do first to solve an equation that involves fractions.

74. Explain why it is important to check your solutions to an equation that contains fractions with variables in the denominator.

REVIEW

In Exercises 75–80, factor each expression.

75. $x^2 + 4x$ $x(x + 4)$

76. $x^2 - 16y^2$ $(x + 4y)(x - 4y)$

77. $2x^2 + x - 3$ $(2x + 3)(x - 1)$

78. $6a^2 - 5a - 6$ $(3a + 2)(2a - 3)$

79. $x^4 - 16$ $(x^2 + 4)(x + 2)(x - 2)$

80. $4x^2 + 10x - 6$ $2(x + 3)(2x - 1)$

▶ 7.8 Applications of Equations That Contain Rational Expressions

In this section, you will learn about

Solving number problems ■ Solving shared-work problems
■ Solving uniform motion problems ■ Solving investment problems

Introduction Many problems involve equations that contain rational expressions. In this section, we will consider several of these applications.

Solving Number Problems

EXAMPLE 1

Number problem. If the same number is added to both the numerator and the denominator of the fraction $\frac{3}{5}$, the result is $\frac{4}{5}$. Find the number.

ANALYZE THE PROBLEM We are asked to find a number. If we add it to both the numerator and the denominator of a fraction, we will get $\frac{4}{5}$.

FORM AN EQUATION Let n represent the unknown number and add n to both the numerator and the denominator of $\frac{3}{5}$. Then set the result equal to $\frac{4}{5}$ to get the equation

$$\frac{3+n}{5+n} = \frac{4}{5}$$

SOLVE THE EQUATION To solve the equation, we proceed as follows:

$$\frac{3+n}{5+n} = \frac{4}{5}$$

$$5(5+n)\frac{3+n}{5+n} = 5(5+n)\frac{4}{5} \qquad \text{Multiply both sides by } 5(5+n), \text{ which is the LCD of the fractions appearing in the equation.}$$

$$5(3+n) = (5+n)4 \qquad \text{Simplify.}$$

$$15 + 5n = 20 + 4n \qquad \text{Use the distributive property to remove parentheses.}$$

$$15 + n = 20 \qquad \text{Subtract } 4n \text{ from both sides.}$$

$$n = 5 \qquad \text{Subtract 15 from both sides.}$$

STATE THE CONCLUSION The number is 5.

CHECK THE RESULT When we add 5 to both the numerator and denominator of $\frac{3}{5}$, we get

$$\frac{3+5}{5+5} = \frac{8}{10} = \frac{4}{5}$$

The result checks. ■

Solving Shared-Work Problems

EXAMPLE 2

Filling an oil tank. An inlet pipe can fill an oil tank in 7 days, and a second inlet pipe can fill the same tank in 9 days. If both pipes are used, how long will it take to fill the tank?

ANALYZE THE PROBLEM The key is to note what each pipe can do in 1 day. If we add what the first pipe can do in 1 day to what the second pipe can do in 1 day, the sum is what they can do together in 1 day. Since the first pipe can fill the tank in 7 days, it can do $\frac{1}{7}$ of the job in 1 day.

Since the second pipe can fill the tank in 9 days, it can do $\frac{1}{9}$ of the job in 1 day. If it takes x days for both pipes to fill the tank, together they can do $\frac{1}{x}$ of the job in 1 day.

FORM AN EQUATION Let x represent the number of days it will take to fill the tank if both inlet pipes are used. Then form the equation.

What the first inlet pipe can do in 1 day	plus	what the second inlet pipe can do in 1 day	equals	what they can do together in 1 day.
$\frac{1}{7}$	$+$	$\frac{1}{9}$	$=$	$\frac{1}{x}$

SOLVE THE EQUATION To solve the equation, we proceed as follows:

$$\frac{1}{7} + \frac{1}{9} = \frac{1}{x}$$

$$63x\left(\frac{1}{7} + \frac{1}{9}\right) = 63x\left(\frac{1}{x}\right) \qquad \text{Multiply both sides by } 63x \text{ to clear the equation of fractions.}$$

$$9x + 7x = 63 \qquad \text{Use the distributive property to remove parentheses and simplify.}$$

$$16x = 63 \qquad \text{Combine like terms.}$$

$$x = \frac{63}{16} \qquad \text{Divide both sides by 16.}$$

STATE THE CONCLUSION It will take $\frac{63}{16}$ or $3\frac{15}{16}$ days for both inlet pipes to fill the tank.

CHECK THE RESULT In $\frac{63}{16}$ days, the first pipe fills $\frac{1}{7}\left(\frac{63}{16}\right)$ of the tank and the second pipe fills $\frac{1}{9}\left(\frac{63}{16}\right)$ of the tank. The sum of these efforts, $\frac{9}{16} + \frac{7}{16}$, is equal to one full tank. ■

Solving Uniform Motion Problems

EXAMPLE 3

Track and field. A coach can run 10 miles in the same amount of time that his best student-athlete can run 12 miles. If the student can run 1 mile per hour faster than the coach, how fast can the student run?

ANALYZE THE PROBLEM This is a uniform motion problem. We use the formula $d = rt$, where d is the distance traveled, r is the rate, and t is the time. If we solve this formula for t, we obtain

$$t = \frac{d}{r}$$

FORM AN EQUATION It will take $\frac{10}{r}$ hours for the coach to run 10 miles at some unknown rate of r mph. It will take $\frac{12}{r+1}$ hours for the student to run 12 miles at some unknown rate of $(r + 1)$ mph. We can organize the information of the problem in a table, as shown in Figure 7-7.

FIGURE 7-7

	r	$\cdot$	t	$=$	d
Student	$r + 1$		$\frac{12}{r+1}$		12
Coach	r		$\frac{10}{r}$		10

The time it takes the student to run 12 miles	equals	the time it takes the coach to run 10 miles.

$$\frac{12}{r+1} \qquad = \qquad \frac{10}{r}$$

SOLVE THE EQUATION   We can solve the equation as follows:

$$\frac{12}{r+1} = \frac{10}{r}$$

$$r(r+1)\frac{12}{r+1} = r(r+1)\frac{10}{r} \qquad \text{Multiply both sides by } r(r+1).$$

$$12r = 10(r+1) \qquad \text{Simplify.}$$

$$12r = 10r + 10 \qquad \text{Use the distributive property to remove parentheses.}$$

$$2r = 10 \qquad \text{Subtract } 10r \text{ from both sides.}$$

$$r = 5 \qquad \text{Divide both sides by 2.}$$

STATE THE CONCLUSION   The coach can run 5 mph. The student, running 1 mph faster, can run 6 mph.

CHECK THE RESULT   Verify that these results check.   ∎

Solving Investment Problems

EXAMPLE 4

Comparing investments. At one bank, a sum of money invested for one year will earn $96 interest. If invested in bonds, that money would earn $108, because the interest rate paid by the bonds is 1% greater than that paid by the bank. Find the bank's rate.

ANALYZE THE PROBLEM   This interest problem is based on the formula $I = Pr$, where I is the interest earned in 1 year, P is the principal (the amount invested), and r is the annual rate of interest. If we solve this formula for P, we obtain

$$P = \frac{I}{r}$$

FORM AN EQUATION   If we let r represent the bank's rate of interest, then $r + 0.01$ represents the rate paid by the bonds. If a person earns $96 interest at a bank at some unknown rate r, the principal invested was $\frac{96}{r}$. If a person earns $108 interest in bonds at some unknown rate $(r + 0.01)$, the principal invested was $\frac{108}{r+0.01}$. We can organize the information of the problem in a table, as shown in Figure 7-8.

FIGURE 7-8

	Principal	·	Rate	=	Interest
Bank	$\frac{96}{r}$		r		96
Bonds	$\frac{108}{r+0.01}$		$r + 0.01$		108

Because the same principal would be invested in either account, we can set up the following equation:

$$\frac{96}{r} = \frac{108}{r+0.01}$$

SOLVE THE EQUATION   We can solve the equation as follows:

$$\frac{96}{r} = \frac{108}{r + 0.01}$$

$$r(r + 0.01) \cdot \frac{96}{r} = r(r + 0.01) \cdot \frac{108}{r + 0.01}$$ Multiply both sides by $r(r + 0.01)$.

$$96(r + 0.01) = 108r$$

$$96r + 0.96 = 108r$$ Remove parentheses.

$$0.96 = 12r$$ Subtract $96r$ from both sides.

$$0.08 = r$$ Divide both sides by 12.

STATE THE CONCLUSION   The bank's interest rate is 0.08, or 8%. The bonds pay 9% interest, a rate 1% greater than that paid by the bank.

CHECK THE RESULT   Verify that these rates check.   ∎

STUDY SET

Section 7.8

VOCABULARY

In Exercises 1–4, fill in the blanks to make the statements true.

1. In the formula $I = Pr$, I stands for the amount of _____interest_____ earned in one year, P stands for the _____principal_____, and r stands for the interest _____rate_____.

2. In the formula $d = rt$, d stands for the _____distance_____ traveled, r is the _____rate_____, and t is the _____time_____.

3. We can clear an equation of fractions by multiplying both sides by the _____LCD_____ of the fractions of the equation.

4. The distributive property is used to _____remove_____ parentheses.

CONCEPTS

5. List the five steps used in problem solving.
   analyze, form, solve, state, check

6. Write 6% as a decimal.   0.06

7. Solve $d = rt$
   a. for r $r = \frac{d}{t}$ **b.** for t $t = \frac{d}{r}$

8. Solve $I = Pr$
   a. for r $r = \frac{I}{P}$ **b.** for P $P = \frac{I}{r}$

9. Illustration 1 shows the length of time it takes each of two hardware store employees to assemble a metal storage shed working alone.

a. Complete the table.

ILLUSTRATION 1

	Time to assemble the shed (hr)	Amount of the shed assembled in 1 hr
Marvin	6	$\frac{1}{6}$
Kyla	5	$\frac{1}{5}$

b. If we assume that working together would not change their individual rates, how much of the shed could they assemble in one hour if they worked together? $\frac{11}{30}$

10. **a.** Complete the table in Illustration 2.

ILLUSTRATION 2

	r	$\cdot$	t	$=$	d
Snowmobile	r		$\frac{4}{r}$		4
4×4 truck	$r - 5$		$\frac{3}{r-5}$		3

b. Complete the table in Illustration 3.

ILLUSTRATION 3

	P	$\cdot$	r	$=$	I
City Savings	$\frac{50}{r}$		r		50
Credit Union	$\frac{75}{r-0.02}$		$r - 0.02$		75

11. When two ice machines are both running, they can fill a supermarket's order in x hours. At this rate, how much of the order do they fill in 1 hour? $\frac{1}{x}$

12. If the exits at the front of a theater are opened, a full theater can be emptied of all occupants in 6 minutes. How much of the theater is emptied in 1 minute? $\frac{1}{6}$

PRACTICE

In Exercises 13–18, use the statement to find the number or numbers.

13. If the denominator of $\frac{3}{4}$ is increased by a number and the numerator of the fraction is doubled, the result is 1.
    2

14. If a number is added to the numerator of $\frac{7}{8}$ and the same number is subtracted from the denominator, the result is 2. 3

15. If a number is added to the numerator of $\frac{3}{4}$ and twice as much is added to the denominator, the result is $\frac{4}{7}$.
    5

16. If a number is added to the numerator of $\frac{5}{7}$ and twice as much is subtracted from the denominator, the result is 8. 3

17. The sum of a number and its reciprocal is $\frac{13}{6}$. $\frac{2}{3}, \frac{3}{2}$

18. The sum of the reciprocals of two consecutive even integers is $\frac{7}{24}$. 6 and 8   1ˢᵗ nbr x, then next is x+2
    $\frac{1}{x} + \frac{1}{x+2} = \frac{7}{24} \Rightarrow x = -\frac{8}{7}, x = 6$
    (N/A)

APPLICATIONS

19. **FILLING A POOL** An inlet pipe can fill an empty swimming pool in 5 hours, and another inlet pipe can fill the pool in 4 hours. How long will it take both pipes to fill the pool? $2\frac{2}{9}$ hr

20. **FILLING A POOL** One inlet pipe can fill an empty pool in 4 hours, and a drain can empty the pool in 8 hours. How long will it take the pipe to fill the pool if the drain is left open? 8 hr

21. **ROOFING A HOUSE** A homeowner estimates that it will take her 7 days to roof her house. A professional roofer estimates that he could roof the house in 4 days. How long will it take if the homeowner helps the roofer? $2\frac{6}{11}$ days

22. **SEWAGE TREATMENT** A sludge pool is filled by two inlet pipes. One pipe can fill the pool in 15 days, and the other can fill it in 21 days. However, if no sewage is added, continuous waste removal will empty the pool in 36 days. How long will it take the two inlet pipes to fill an empty sludge pool? $11\frac{61}{109}$ days

23. **TOURING** A woman can bicycle 28 miles in the same time as it takes her to walk 8 miles. If she can ride 10 mph faster than she can walk, how much time should she allow to walk a 30-mile trail? See Illustration 4. (*Hint:* How fast can she walk?) $7\frac{1}{2}$ hr

13-27
odds

ILLUSTRATION 4

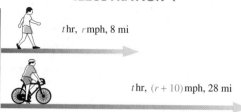

t hr, r mph, 8 mi

t hr, $(r + 10)$ mph, 28 mi

24. **COMPARING TRAVEL** A plane can fly 300 miles in the same time as it takes a car to go 120 miles. If the car travels 90 mph slower than the plane, find the speed of the plane. 150 mph

25. **BOATING** A boat that travels 18 mph in still water can travel 22 miles downstream in the same time as it takes to travel 14 miles upstream. Find the speed of the current in the river. (See Illustration 5.) 4 mph

ILLUSTRATION 5

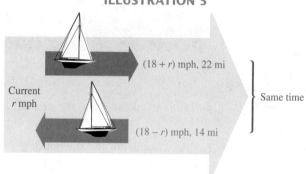

$(18 + r)$ mph, 22 mi

Current
r mph

$(18 - r)$ mph, 14 mi

Same time

26. **WIND SPEED** A plane can fly 300 miles downwind in the same time as it can travel 210 miles upwind. Find the velocity of the wind if the plane can fly 255 mph in still air. 45 mph

27. **COMPARING INVESTMENTS** Two certificates of deposit (CDs) pay interest at rates that differ by 1%. Money invested for one year in the first CD earns $175 interest. The same principal invested in the second CD earns $200. Find the two rates of interest.
    7% and 8%

28. **COMPARING INTEREST RATES** Two bond funds pay interest at rates that differ by 2%. Money invested for one year in the first fund earns $315 interest. The same amount invested in the second fund earns $385. Find the lower rate of interest. 9%

29. **SHARING COSTS** Several office workers bought a $35 gift for their boss. If there had been two more employees to contribute, everyone's cost would have been $2 less. How many workers contributed to the gift? 5

30. **SALES** A dealer bought some radios for a total of $1,200. She gave away 6 radios as gifts, sold the rest for $10 more than she paid for each radio, and broke even. How many radios did she buy? 30

31. SALES A bookstore can purchase several calculators for a total cost of $120. If each calculator cost $1 less, the bookstore could purchase 10 additional calculators at the same total cost. How many calculators can be purchased at the regular price? 30

32. FURNACE REPAIR A repairman purchased several furnace-blower motors for a total cost of $210. If his cost per motor had been $5 less, he could have purchased one additional motor. How many motors did he buy at the regular rate? 6

33. RIVER TOURS A river boat tour begins by going 60 miles upstream against a 5-mph current. There, the boat turns around and returns with the current. What still-water speed should the captain use to complete the tour in 5 hours? 25 mph

34. TRAVEL TIME A company president flew 680 miles one way in the corporate jet but returned in a smaller plane that could fly only half as fast. If the total travel time was 6 hours, find the speeds of the planes. 340 mph and 170 mph

WRITING

35. The key to solving shared-work problems is to ask, "How much of the job can be done in 1 hour?" Explain.

36. It's difficult to check the solution of a shared-work problem. Explain how you could decide whether an answer is at least reasonable.

REVIEW

In Exercises 37–44, solve each equation.

37. $x^2 - 5x - 6 = 0$ $-1, 6$

38. $x^2 - 25 = 0$ $5, -5$

39. $(t + 2)(t^2 + 7t + 12) = 0$ $-2, -3, -4$

40. $2(y - 4) = -y^2$ $2, -4$

41. $y^3 - y^2 = 0$ $0, 0, 1$

42. $5a^3 - 125a = 0$ $0, 5, -5$

43. $(x^2 - 1)(x^2 - 4) = 0$ $1, -1, 2, -2$

44. $6t^3 + 35t^2 = 6t$ $0, -6, \frac{1}{6}$

The Language of Algebra

Algebra is a language in its own right. One of the keys to becoming a good algebra student is to know the vocabulary of algebra.

In Exercises 1–21, match each instruction in column I with the most appropriate problem in column II. Each letter in column II is used only once.

Column I	*Column II*
1. Use the FOIL method. g	**a.** $R = cd + 2t$
2. Graph the function. o	**b.** 5^3
3. Simplify the expression. i	**c.** 144
4. Rationalize the denominator. l	**d.** $2(3x - 4)$
5. Factor completely. u	**e.** $4x - 7 > -3$
6. Evaluate the expression for $a = -1$ and $b = -6$. r	**f.** $\dfrac{x}{3} + \dfrac{1}{2} = \dfrac{1}{6}$
7. Express in lowest terms. p	**g.** $(x - 1)(x + 8)$
8. Solve for t. a	**h.** $a + (b + c) = (a + b) + c$
9. Combine like terms. m	**i.** $\dfrac{1}{2x^2} + \dfrac{5}{4x}$
10. Remove parentheses. d	**j.** $\sqrt{4x^2}$
11. Solve the equation by clearing it of fractions. f	**k.** the sum of four cubed and y
12. Prime factor. c	**l.** $\dfrac{4}{\sqrt{6}}$
13. Solve using the quadratic formula. q	**m.** $2x - 8 + 6y - 14$
14. Identify the base and the exponent. b	**n.** $(3, -2)$ and $(0, -5)$
15. Write without a radical sign. j	**o.** $f(x) = x^3 + 1$
16. Write the equation of the line. s	**p.** $\dfrac{4x^2}{16x}$
17. Solve the inequality. e	**q.** $x^2 - 3x - 4 = 0$
18. Complete the square to make a trinomial square. t	**r.** $2a^2 - 3b$
19. Find the slope of the line passing through the given points. n	**s.** with $m = \frac{2}{3}$ and $b = 2$
20. Translate into mathematical symbols. k	**t.** $x^2 + 4x$
21. Name the property shown. h	**u.** $y^4 - 81$

Accent on Teamwork

Section 7.1

Pi The Greek letter pi (π) represents the ratio of the circumference C of any circle to its diameter d. That is, $\pi = \frac{C}{d}$. Use a tape measure to find the circumference and the diameter of various objects that are circular in shape. You can measure anything round: for example, a swimming pool spa, the top of a can, or a ring. Enter your results in a table like that in Illustration 1. Convert each measurement to a decimal and use a calculator to compute the ratio of C to d. Make some observations about your results.

ILLUSTRATION 1

Object	Circumference	Diameter	$\frac{C}{d}$
Quarter-dollar	$2\frac{15}{16}$ in. 2.9375 in.	$\frac{15}{16}$ in. 0.9375 in.	3.13333. . .

Section 7.2

Cooking Find a simple recipe for a treat that you can make for your class. Use a proportion to determine the amount of each ingredient needed to make enough for the exact number of people in your class. See Exercise 45 in Study Set 7.2 for an example. Write the old recipe and the new recipe on separate pieces of poster board. Did the recipe serve the correct number of people? Share with the class how you made the calculations, as well as any difficulties you encountered.

Section 7.3

Warning In Section 7.3, the following two-part warning was given: "Remember that only *factors* that are common to the entire numerator and the entire denominator can be divided out. *Terms* that are common to both the numerator and denominator cannot be divided out." Explain how each part of the warning applies to the rational expression $\frac{2x+8}{8}$.

Section 7.4

The rule of four The trend in mathematics education today is to present a topic in four ways:

- symbolically (using variables)
- geometrically (using a diagram)
- numerically (using numbers)
- verbally (using words)

Use the rule of four as a guide in developing a presentation explaining the concept of multiplying fractions, as shown in the beginning of Section 7.4.

Section 7.5

First, add

$$\frac{1}{2x^2} + \frac{1}{8x}$$

by expressing each fraction in terms of a common denominator $16x^3$. (This is the *product* of their denominators.) Then add the fractions again by expressing each of them in terms of their lowest common denominator. What is one advantage and one drawback of each method?

Section 7.6

Unit analysis Simplify each complex fraction. The units can be divided out just as in the case of common factors.

$$\frac{\dfrac{36 \text{ inches}}{3 \text{ feet}}}{\dfrac{1 \text{ yard}}{3 \text{ feet}}} \qquad \frac{\dfrac{60 \text{ minutes}}{1 \text{ hour}}}{\dfrac{3{,}600 \text{ seconds}}{1 \text{ hour}}} \qquad \frac{\dfrac{4 \text{ quarts}}{1 \text{ gallon}}}{\dfrac{8 \text{ pints}}{1 \text{ gallon}}}$$

Section 7.7

Simplify and solve The two problems below look similar. Explain why their one-word instructions can't be switched. Write a solution for each problem and then identify the major similarity and the major difference in the solution methods.

Simplify

$$\frac{2}{4x-4} + \frac{3}{x-1}$$

Solve

$$\frac{2}{4x-4} + \frac{3}{x-1} = \frac{7}{4}$$

Section 7.8

Problem solving Problems such as the filling of a water tank (Section 7.8) are often called *shared-work problems*. For each of the following equations, write a shared-work problem that could be solved using it.

$$\frac{1}{3} + \frac{1}{8} = \frac{1}{x}$$

$$\frac{1}{3} - \frac{1}{8} = \frac{1}{x}$$

$$\frac{1}{3} + \frac{1}{8} - \frac{1}{16} = \frac{1}{x}$$

Section 7.1

Ratios

CONCEPTS

A *ratio* is the comparison of two numbers by their indicated quotient. The ratio of a to b is written $\frac{a}{b}$ or $a:b$.

Where possible, express ratios using the same units.

REVIEW EXERCISES

1. Write each ratio as a fraction in lowest terms.
   a. 3 to 6 $\frac{1}{2}$
   b. $12x$ to $15x$ $\frac{4}{5}$
   c. 2 feet to 1 yard $\frac{2}{3}$
   d. 5 pints to 3 quarts $\frac{5}{6}$

2. BICYCLE GEARS See Illustration 1. Find the ratio of the number of teeth on the front sprocket of the bicycle to the number of teeth on the smallest rear sprocket. Express the ratio in lowest terms using $a:b$ form. Then write ratios for the following sprockets: the front to the middle rear and the front to the large rear. $20:7$; $40:17$; $2:1$

ILLUSTRATION 1

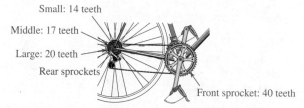

Small: 14 teeth
Middle: 17 teeth
Large: 20 teeth
Rear sprockets
Front sprocket: 40 teeth

The *unit cost* of an item is the ratio of its cost to its quantity.

3. If three pounds of coffee cost \$8.79, find its unit cost (the cost per pound).   \$2.93

When ratios are used to compare quantities with different units, they are called *rates*.

4. If a factory used 2,275 kilowatt hours of electricity in February, what was the rate of energy consumption in kilowatt hours per week?   568.75 kwh per week

5. SUPREME COURT The annual salary of a U.S. Supreme Court Justice is \$164,100. What is their weekly rate of pay? Round to the nearest dollar.   \$3,156

Section 7.2

Proportions and Similar Triangles

A *proportion* is a statement that two ratios are equal.

In the proportion $\frac{a}{b} = \frac{c}{d}$, a and d are the *extremes*, and b and c are the *means*.

In any proportion, the product of the extremes is equal to the product of the means.

6. Determine whether the following equations are proportions.
   a. $\frac{4}{7} = \frac{20}{34}$ no
   b. $\frac{5}{7} = \frac{30}{42}$ yes

7. Solve each proportion.
   a. $\frac{3}{x} = \frac{6}{9}$ $\frac{9}{2}$
   b. $\frac{x}{3} = \frac{x}{5}$ 0
   c. $\frac{x-2}{5} = \frac{x}{7}$ 7
   d. $\frac{4x-1}{18} = \frac{x}{6}$ 1

The measures of corresponding sides of *similar triangles* are in proportion.

8. A telephone pole casts a shadow 12 feet long at the same time that a man 6 feet tall casts a shadow of 3.6 feet. How tall is the pole?   20 ft

9. DENTISTRY The diagram in Illustration 2 was displayed in a dentist's office. According to the diagram, if the dentist has 340 adult patients, how many will develop gum disese? 255

ILLUSTRATION 2

3 out of 4 adults will develop gum disease.

Section 7.3

Rational Expressions and Rational Functions

A *rational expression* is a ratio of two polynomials.

The fundamental property of fractions:
If b and c are not zero, then

$$\frac{ac}{bc} = \frac{a}{b}$$

When all common factors have been divided out, a fraction is in *lowest terms*.

$$\frac{a}{1} = a$$

$\dfrac{a}{0}$ is not defined

If the terms of two polynomials are the same, except for sign, the polynomials are called *negatives* of each other.

The quotient of any nonzero expression and its negative is −1.

10. Write each fraction in lowest terms. If it is already in lowest terms, so indicate.

a. $\dfrac{10}{25}$ $\frac{2}{5}$

b. $-\dfrac{12}{18}$ $-\frac{2}{3}$

c. $\dfrac{3x^2}{6x^3}$ $\frac{1}{2x}$

d. $\dfrac{5xy^2}{2x^2y^2}$ $\frac{5}{2x}$

e. $\dfrac{x^2}{x^2 + x}$ $\frac{x}{x+1}$

f. $\dfrac{x+2}{x^2-4}$ $\frac{1}{x-2}$

g. $\dfrac{3p-2}{2-3p}$ -1

h. $\dfrac{8-x}{x^2-5x-24}$ $-\frac{1}{x+3}$

i. $\dfrac{2x^2-16x}{2x^2-18x+16}$ $\frac{x}{x-1}$

j. $\dfrac{x^2+x-2}{x^2-x-2}$ in lowest terms

11. Explain why it would be incorrect to divide out the common x's in $\frac{x+1}{x}$. x is not a common factor of the numerator and the denominator.

12. Complete the table and graph the rational function for $x > 0$. Round to the nearest tenth when necessary.

$$f(x) = \frac{8}{x}$$

x	$f(x)$
1	8
2	4
3	2.7
4	2
5	1.6
6	1.3
7	1.1
8	1

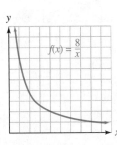

Section 7.4

Multiplying and Dividing Rational Expressions

Rule for multiplying fractions:

$$\frac{a}{b} \cdot \frac{c}{d} = \frac{ac}{bd} \quad (b, d \neq 0)$$

13. Do each multiplication and simplify.

    a. $\dfrac{3xy}{2x} \cdot \dfrac{4x}{2y^2}$ $\frac{3x}{y}$

    b. $\dfrac{3x}{x^2 - x} \cdot \dfrac{2x - 2}{x^2}$ $\frac{6}{x^2}$

    c. $\dfrac{x^2 - 1}{x^2 + 2x} \cdot \dfrac{x}{x + 1}$ $\frac{x-1}{x+2}$

    d. $\dfrac{x^2 + x}{3x - 15} \cdot \dfrac{6x - 30}{x^2 + 2x + 1}$ $\frac{2x}{x+1}$

Rule for dividing fractions:

$$\frac{a}{b} \div \frac{c}{d} = \frac{a}{b} \cdot \frac{d}{c} \quad (b, c, d \neq 0)$$

To write the *reciprocal* of a fraction, we invert the fraction.

14. Do each division and simplify.

    a. $\dfrac{3x^2}{5x^2y} \div \dfrac{6x}{15xy^2}$ $\frac{3y}{2}$

    b. $\dfrac{x^2 + 5x}{x^2 + 4x - 5} \div \dfrac{x^2}{x - 1}$ $\frac{1}{x}$

    c. $\dfrac{x^2 - x - 6}{2x - 1} \div \dfrac{x^2 - 2x - 3}{2x^2 + x - 1}$ $x + 2$

    d. $\dfrac{x^2 - 3x}{x^2 - x - 6} \div \dfrac{x^2 - x}{4 - x^2}$ $\frac{2-x}{x-1}$

    e. $\dfrac{b^2 + 4b + 4}{b^2 + b - 6}\left(\dfrac{b - 2}{b - 1} \div \dfrac{b + 2}{b^2 + 2b - 3}\right)$ $b + 2$

Section 7.5

Adding and Subtracting Rational Expressions

Adding and subtracting fractions with like denominators:

$$\frac{a}{d} + \frac{b}{d} = \frac{a + b}{d} \quad (d \neq 0)$$

$$\frac{a}{d} - \frac{b}{d} = \frac{a - b}{d} \quad (d \neq 0)$$

15. Do each operation. Simplify all answers.

    a. $\dfrac{x}{x + y} + \dfrac{y}{x + y}$ 1

    b. $\dfrac{3x}{x - 7} - \dfrac{x - 2}{x - 7}$ $\frac{2x + 2}{x - 7}$

    c. $\dfrac{a}{a^2 - 2a - 8} + \dfrac{2}{a^2 - 2a - 8}$ $\frac{1}{a - 4}$

To find the *LCD,* completely factor each denominator. Form a product using each different factor the greatest number of times it appears in any one factorization.

16. Several denominators are given. Find the lowest common denominator (LCD).

    a. $2x^2, 4x$ $4x^2$

    b. $3y^2, 9x, 6x$ $18xy^2$

    c. $x + 1, x + 2$ $(x + 1)(x + 2)$

    d. $y^2 - 25, y - 5$ $y^2 - 25$

To add or subtract fractions with unlike denominators, first find the LCD of the fractions. Then express each fraction in equivalent form with a common denominator. Finally, add or subtract the fractions.

17. Do each operation. Simplify all answers.

    a. $\dfrac{x}{x - 1} + \dfrac{1}{x}$ $\frac{x^2 + x - 1}{x(x - 1)}$

    b. $\dfrac{1}{7} - \dfrac{1}{c}$ $\frac{c - 7}{7c}$

    c. $\dfrac{x + 2}{2x} - \dfrac{2 - x}{x^2}$ $\frac{x^2 + 4x - 4}{2x^2}$

    d. $\dfrac{2t + 2}{t^2 + 2t + 1} - \dfrac{1}{t + 1}$ $\frac{1}{t + 1}$

    e. $\dfrac{x}{x + 2} + \dfrac{3}{x} - \dfrac{4}{x^2 + 2x}$ $\frac{x + 1}{x}$

    f. $\dfrac{6}{b - 1} - \dfrac{b}{1 - b}$ $\frac{b + 6}{b - 1}$

Section 7.6

Complex fractions contain fractions in their numerators or denominators.

To simplify a complex fraction, use either of these methods:

1. Write the numerator and denominator of the complex fraction as single fractions, do the division of the fractions, and simplify.

2. Multiply both the numerator and the denominator of the complex fraction by the LCD of the fractions that appear in the numerator and denominator, then simplify.

Complex Fractions

18. Simplify each complex fraction.

a. $\dfrac{\dfrac{3}{2}}{\dfrac{2}{3}}$ $\dfrac{9}{4}$

b. $\dfrac{\dfrac{3}{2}+1}{\dfrac{2}{3}+1}$ $\dfrac{3}{2}$

c. $\dfrac{\dfrac{1}{y}+1}{\dfrac{1}{y}-1}$ $\dfrac{1+y}{1-y}$

d. $\dfrac{1+\dfrac{3}{x}}{2-\dfrac{1}{x^2}}$ $\dfrac{x(x+3)}{2x^2-1}$

e. $\dfrac{\dfrac{2}{x-1}+\dfrac{x-1}{x+1}}{\dfrac{1}{x^2-1}}$ x^2+3

f. $\dfrac{x^{-2}+1}{x^{-2}-1}$ $\dfrac{1+x^2}{1-x^2}$

Section 7.7

To solve an equation that contains fractions, change it to another equation without fractions. Do so by multiplying both sides by the LCD of the fractions. Check all solutions.

An apparent solution that does not satisfy the original equation is called an *extraneous* solution.

Solving Equations That Contain Rational Expressions; Inverse Variation

19. Solve each equation and check all answers.

a. $\dfrac{3}{x}=\dfrac{2}{x-1}$ 3

b. $\dfrac{5}{r+4}=\dfrac{3}{r+2}$ 1

c. $\dfrac{2}{3t}+\dfrac{1}{t}=\dfrac{5}{9}$ 3

d. $\dfrac{2x}{x+4}=\dfrac{3}{x-1}$ $4,-\dfrac{3}{2}$

e. $a=\dfrac{3a-50}{4a-24}-\dfrac{3}{4}$ $2,4$

f. $\dfrac{4}{x+2}-\dfrac{3}{x+3}=\dfrac{6}{x^2+5x+6}$ 0

20. The efficiency E of a Carnot engine is given by the formula

$$E = 1 - \dfrac{T_2}{T_1}$$

Solve the formula for T_1. $T_1 = \dfrac{T_2}{1-E}$

21. Solve for r_1: $\dfrac{1}{r}=\dfrac{1}{r_1}+\dfrac{1}{r_2}$. $r_1 = \dfrac{rr_2}{r_2-r}$

The words "y varies *inversely* with x" mean that $y=\frac{k}{x}$ for some constant k.

22. l varies inversely with w. Find the constant of variation if $l=30$ when $w=20$. 600

Section 7.8

To solve a problem, follow these steps:

1. Analyze the problem.
2. Form an equation.
3. Solve the equation.
4. State the conclusion.
5. Check the result.

Interest = principal · rate · time

Distance = rate · time

Applications of Equations That Contain Rational Expressions

23. NUMBER PROBLEM If a number is subtracted from the denominator of $\frac{4}{5}$ and twice as much is added to the numerator, the result is 5. Find the number. **3**

24. If a maid can clean a house in 4 hours, how much of the house does she clean in 1 hour? $\frac{1}{4}$

25. PAINTING HOUSES If a homeowner can paint a house in 14 days and a professional painter can paint it in 10 days, how long will it take if they work together? $5\frac{5}{6}$ days

26. INVESTMENTS In one year, a student earned $100 interest on money she deposited at a savings and loan. She later learned that the money would have earned $120 if she had deposited it at a credit union, because the credit union paid 1% more interest at the time. Find the rate she received from the savings and loan. **5%**

27. EXERCISE A jogger can bicycle 30 miles in the same time that it takes her to jog 10 miles. If she can ride 10 mph faster than she can jog, how fast can she jog? **5 mph**

28. WIND SPEED A plane flies 400 miles downwind in the same amount of time as it takes to travel 320 miles upwind. If the plane can fly at 360 mph in still air, find the velocity of the wind. **40 mph**

CHAPTER 7

Test

1. Express as a ratio in lowest terms: 6 feet to 3 yards. $\frac{2}{3}$

2. Is the equation $\dfrac{3}{5} = \dfrac{6xt}{10xt}$ a proportion? yes

3. Solve the proportion for y: $\dfrac{y}{y-1} = \dfrac{y-2}{y}$. $\frac{2}{3}$

4. HEALTH RISKS A medical newsletter states that a "healthy" waist-to-hip ratio for men is $19:20$ or less. Does the patient shown in Illustration 1 fall within the "healthy" range? yes

5. Simplify $\dfrac{48x^2y}{54xy^2}$. $\frac{8x}{9y}$

6. Simplify $\dfrac{2x^2 - x - 3}{4x^2 - 9}$. $\frac{x+1}{2x+3}$

ILLUSTRATION 1

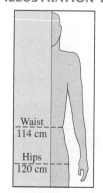

Waist
114 cm

Hips
120 cm

7. Simplify $\dfrac{3(x+2) - 3}{2x - 4 - (x-5)}$. 3

8. Multiply and simplify $-\dfrac{12x^2y}{15xy} \cdot \dfrac{25y^2}{16x}$. $-\frac{5y^2}{4}$

9. Multiply and simplify $\dfrac{x^2 + 3x + 2}{3x + 9} \cdot \dfrac{x+3}{x^2 - 4}$. $\frac{x+1}{3(x-2)}$

10. Divide and simplify $\dfrac{8x^2}{25x} \div \dfrac{16x^2}{30x}$. $\frac{3}{5}$

11. Divide and simplify $\dfrac{x - x^2}{3x^2 + 6x} \div \dfrac{3x - 3}{3x^3 + 6x^2}$. $-\frac{x^2}{3}$

12. Simplify $\dfrac{x^2 + x}{x - 1} \cdot \dfrac{x^2 - 1}{x^2 - 2x} \div \dfrac{x^2 + 2x + 1}{x^2 - 4}$. $x + 2$

13. Add $\dfrac{5x - 4}{x - 1} + \dfrac{5x + 3}{x - 1}$. $\frac{10x - 1}{x - 1}$

14. Subtract $\dfrac{3y + 7}{2y + 3} - \dfrac{3(y - 2)}{2y + 3}$. $\dfrac{13}{2y + 3}$

15. Add $\dfrac{x + 1}{x} + \dfrac{x - 1}{x + 1}$. $\dfrac{2x^2 + x + 1}{x(x + 1)}$

16. Subtract $\dfrac{a + 3}{a - 1} - \dfrac{a + 4}{1 - a}$. $\dfrac{2a + 7}{a - 1}$

17. Subtract $\dfrac{2n}{5m} - \dfrac{n}{2}$. $\dfrac{4n - 5mn}{10m}$

18. Simplify $\dfrac{1 + \dfrac{y}{x}}{\dfrac{y}{x} - 1}$. $\dfrac{x + y}{y - x}$

19. Solve for c: $\dfrac{2}{3} = \dfrac{2c - 12}{3c - 9} - c$.   1, 2

20. Solve for x: $3x - \dfrac{2(x + 3)}{3} = 16 - \dfrac{x + 2}{2}$.   6

21. Solve for x: $\dfrac{7}{x + 4} - \dfrac{1}{2} = \dfrac{3}{x + 4}$.   4

22. Solve for B: $H = \dfrac{RB}{R + B}$. $B = \dfrac{HR}{R - H}$

23. If i varies inversely with d, find the constant of variation if $i = 100$ when $d = 2$.   200

24. CLEANING HIGHWAYS One highway worker can pick up all the trash on a strip of highway in 7 hours, and his helper can pick up the trash in 9 hours. How long will it take them if they work together? $3\frac{15}{16}$ hr

25. BOATING A boat can motor 28 miles downstream in the same amount of time as it can motor 18 miles upstream. Find the speed of the current if the boat can motor at 23 mph in still water.   5 mph

26. FLIGHT PATH A plane drops 575 feet as it flies a horizontal distance of $\frac{1}{2}$ mile, as shown in Illustration 2. How much altitude will it lose as it flies a horizontal distance of 7 miles?   8,050 ft

ILLUSTRATION 2

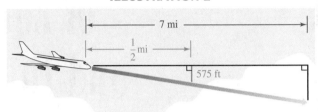

27. Explain why we can divide out the 5's in $\frac{5x}{5}$ and why we can't divide them out in $\frac{5 + x}{5}$.

We can only divide out common factors, as in the first expression. We can't divide out common terms, as in the second expression.

28. Complete the table and graph the rational function for $x > 0$. Round to the nearest tenth when necessary.

$f(x) = \dfrac{2}{x}$

x	$f(x)$
$\frac{1}{2}$	4
1	2
2	1
3	0.7
4	0.5
5	0.4
6	0.3

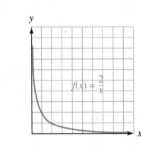

8

Solving Systems of Equations and Inequalities

CAMPUS CONNECTION

The Theater Arts Department

Theater arts students often use mathematics to help them plan and produce stage plays. Using algebra, the ticket sales department can determine the number of regular admissions and the number of senior citizen tickets sold for a performance if the total attendance and the total dollar receipts are known. This is done by solving a *system of equations* in which each equation contains two variables. In this chapter, we will introduce systems of equations. You will see that they can be used to solve a wide variety of problems, including some from theater arts.

To solve many problems, we must use two variables. This requires that we solve a system of equations.

▶ 8.1

Solving Systems of Equations by Graphing

In this section, you will learn about

> Systems of equations ■ The graphing method ■ Inconsistent systems ■ Dependent equations

Introduction   The lines graphed in Figure 8-1 approximate the per-person consumption of chicken and beef in the United States for the years 1990–1997. We can see that consumption of chicken increased, while that of beef decreased.

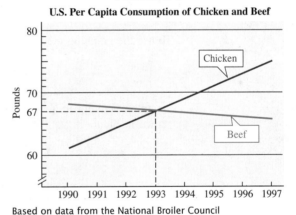

FIGURE 8-1

U.S. Per Capita Consumption of Chicken and Beef

Based on data from the National Broiler Council

By graphing this *pair* of lines on the same coordinate system, it is apparent that Americans consumed equal amounts of chicken and beef in 1993—about 67 pounds of each. In this section, we will work with pairs of linear equations. We call such a pair of equations a *system of equations.*

Systems of Equations

We have considered equations that contain two variables, such as $x + y = 3$. Because there are infinitely many pairs of numbers whose sum is 3, there are infinitely many pairs (x, y) that satisfy this equation. Some of these pairs are

$$x + y = 3$$

x	y
0	3
1	**2**
2	1
3	0

Likewise, there are infinitely many pairs (x, y) that satisfy the equation $3x - y = 1$. Some of these pairs are

$$3x - y = 1$$

x	y
0	-1
1	2
2	5
3	8

Although there are infinitely many pairs that satisfy each of these equations, only the pair $(1, 2)$ satisfies both equations at the same time. The pair of equations

$$\begin{cases} x + y = 3 \\ 3x - y = 1 \end{cases}$$

is called a **system of equations.** Because the ordered pair $(1, 2)$ satisfies both equations simultaneously, it is called a **simultaneous solution,** or a **solution of the system of equations.** In this chapter, we will discuss three methods for finding the simultaneous solution of a system of two equations that each have two variables. First, we consider the graphing method.

The Graphing Method

To use the graphing method to solve the system

$$\begin{cases} x + y = 3 \\ 3x - y = 1 \end{cases}$$

we graph both equations on one set of coordinate axes using the intercept method, as shown in Figure 8-2.

FIGURE 8-2

$x + y = 3$

x	y	(x, y)
0	3	$(0, 3)$
3	0	$(3, 0)$
2	1	$(2, 1)$

$3x - y = 1$

x	y	(x, y)
0	-1	$(0, -1)$
$\frac{1}{3}$	0	$\left(\frac{1}{3}, 0\right)$
2	5	$(2, 5)$

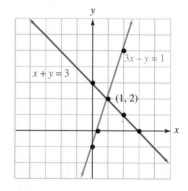

Although there are infinitely many pairs (x, y) that satisfy $x + y = 3$, and infinitely many pairs (x, y) that satisfy $3x - y = 1$, only the coordinates of the point where their graphs intersect satisfy both equations simultaneously. Thus, the solution of the system is $x = 1$ and $y = 2$, or $(1, 2)$.

To check this solution, we substitute 1 for x and 2 for y in each equation and verify that the pair $(1, 2)$ satisfies each equation.

First equation	*Second equation*
$x + y = 3$	$3x - y = 1$
$1 + 2 \overset{?}{=} 3$	$3(1) - 2 \overset{?}{=} 1$
$3 = 3$	$3 - 2 \overset{?}{=} 1$
	$1 = 1$

When the graphs of two equations in a system are different lines, the equations are called **independent equations.** When a system of equations has a solution, the system is called a **consistent system.**

To solve a system of equations in two variables by graphing, we follow these steps.

The graphing method

1. Carefully graph each equation.

2. When possible, find the coordinates of the point where the graphs intersect.

3. Check the solution in the equations of the original system.

E X A M P L E   1

Solving systems by graphing. Use graphing to solve $\begin{cases} 2x + 3y = 2 \\ 3x = 2y + 16 \end{cases}$.

Solution   Using the intercept method, we graph both equations on one set of coordinate axes, as shown in Figure 8-3.

FIGURE 8-3

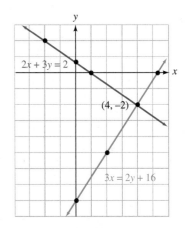

	$2x + 3y = 2$			$3x = 2y + 16$	
x	y	(x, y)	x	y	(x, y)
0	$\frac{2}{3}$	$\left(0, \frac{2}{3}\right)$	0	-8	$(0, -8)$
1	0	$(1, 0)$	$\frac{16}{3}$	0	$\left(\frac{16}{3}, 0\right)$
-2	2	$(-2, 2)$	2	-5	$(2, -5)$

Although there are infinitely many pairs (x, y) that satisfy $2x + 3y = 2$, and infinitely many pairs (x, y) that satisfy $3x = 2y + 16$, only the coordinates of the point where the graphs intersect satisfy both equations at the same time. The solution is $x = 4$ and $y = -2$, or $(4, -2)$.

To check, we substitute 4 for x and -2 for y in each equation and verify that the pair $(4, -2)$ satisfies each equation.

$$2x + 3y = 2 \qquad\qquad 3x = 2y + 16$$
$$2(4) + 3(-2) \overset{?}{=} 2 \qquad\qquad 3(4) \overset{?}{=} 2(-2) + 16$$
$$8 - 6 \overset{?}{=} 2 \qquad\qquad 12 \overset{?}{=} -4 + 16$$
$$2 = 2 \qquad\qquad 12 = 12$$

The equations in this system are independent equations, and the system is a consistent system of equations.

SELF CHECK   Use graphing to solve $\begin{cases} 2x = y - 5 \\ x + y = -1 \end{cases}$.   *Answer:* $(-2, 1)$

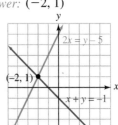

Inconsistent Systems

Sometimes a system of equations has no solution. Such systems are called **inconsistent systems.**

EXAMPLE 2

A system having no solution. Solve $\begin{cases} y = -2x - 6 \\ 4x + 2y = 8 \end{cases}$.

Solution Since $y = -2x - 6$ is written in slope-intercept form, we can graph it by plotting the y-intercept $(0, -6)$ and then drawing a slope of -2. (The run is 1, and the rise is -2.) We graph $4x + 2y = 8$ using the intercept method.

$y = -2x - 6$

so

$m = -2 = -\frac{2}{1}$

and

$b = -6$

$4x + 2y = 8$

x	y	(x, y)
0	4	$(0, 4)$
2	0	$(2, 0)$
1	2	$(1, 2)$

The system is graphed in Figure 8-4. Since the lines in the figure are parallel, they have the same slope. We can verify this by writing the second equation in slope–intercept form and observing that the coefficients of x in each equation are equal.

$$y = -2x - 6 \qquad 4x + 2y = 8$$
$$2y = -4x + 8$$
$$y = -2x + 4$$

Because parallel lines do not intersect, this system has no solution and is inconsistent. Since the graphs are different lines, the equations of the system are independent.

FIGURE 8-4

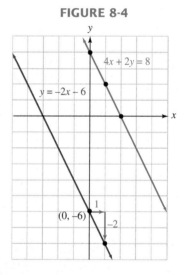

SELF CHECK Solve $\begin{cases} y = \frac{3}{2}x \\ 3x - 2y = 6 \end{cases}$.

Answer: The lines are parallel; therefore, the system has no solution.

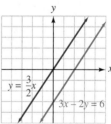

Dependent Equations

Sometimes a system has an infinite number of solutions. In this case, we say that the equations of the system are **dependent equations.**

EXAMPLE 3

Infinitely many solutions. Solve $\begin{cases} y - 4 = 2x \\ 4x + 8 = 2y \end{cases}$.

Solution We graph both equations on one set of axes, as shown in Figure 8-5.

FIGURE 8-5

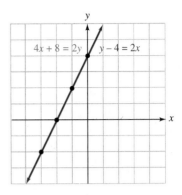

$y - 4 = 2x$

x	y	(x, y)
0	4	$(0, 4)$
-2	0	$(-2, 0)$
-1	2	$(-1, 2)$

$4x + 8 = 2y$

x	y	(x, y)
0	4	$(0, 4)$
-2	0	$(-2, 0)$
-3	-2	$(-3, -2)$

The lines in Figure 8-5 coincide (they are the same line). Because the lines intersect at infinitely many points, there is an infinite number of solutions. Any pair (x, y) that satisfies one of the equations also satisfies the other.

From the graph, we can see that some possible solutions are $(0, 4)$, $(-1, 2)$, and $(-3, -2)$, since each of these points lies on the one line that is the graph of both equations.

SELF CHECK   Solve $\begin{cases} 6x - 2y = 4 \\ y + 2 = 3x \end{cases}$.

Answer: The graphs are the same line. There is an infinite number of solutions.

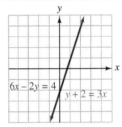

The possibilities that can occur when graphing two linear equations, each with two variables, are summarized as follows.

Possible graph	If the	then
(graph: lines intersecting)	lines are different and intersect,	the equations are independent and the system is consistent. One solution exists.
(graph: parallel lines)	lines are different and parallel,	the equations are independent and the system is inconsistent. No solutions exist.
(graph: coinciding lines)	lines coincide,	the equations are dependent and the system is consistent. Infinitely many solutions exist.

EXAMPLE 4

Solving an equivalent system.   Solve $\begin{cases} -\frac{x}{2} - 1 = \frac{y}{2} \\ \frac{1}{3}x - \frac{1}{2}y = -4 \end{cases}$.

Solution   We can multiply both sides of the first equation by 2 to clear it of fractions.

$$-\frac{x}{2} - 1 = \frac{y}{2}$$

$$2\left(-\frac{x}{2} - 1\right) = 2\left(\frac{y}{2}\right)$$

1. $-x - 2 = y$ We will call this Equation 1.

We then multiply both sides of the second equation by 6 to clear it of fractions.

$$\frac{1}{3}x - \frac{1}{2}y = -4$$

$$6\left(\frac{1}{3}x - \frac{1}{2}y\right) = 6(-4)$$

2. $2x - 3y = -24$ We will call this Equation 2.

Equations 1 and 2 form the following **equivalent system**, which has the same solutions as the original system:

$$\begin{cases} -x - 2 = y \\ 2x - 3y = -24 \end{cases}$$

In Figure 8-6, we graph $-x - 2 = y$ by plotting the y-intercept $(0, -2)$ and then drawing a slope of -1. We graph $2x - 3y = -24$ using the intercept method. We find that $(-6, 4)$ is the point of intersection. The solution is $x = -6$ and $y = 4$, or $(-6, 4)$.

$$y = -x - 2$$

so

$$m = -1 = \frac{-1}{1}$$

and

$$b = -2$$

$$2x - 3y = -24$$

x	y	(x, y)
0	8	$(0, 8)$
-12	0	$(-12, 0)$
-3	6	$(-3, 6)$

FIGURE 8-6

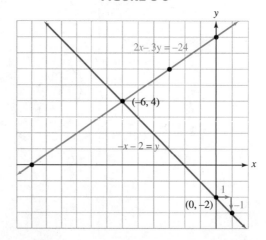

A check will show that when the coordinates of $(-6, 4)$ are substituted into the two original equations, true statements result. Therefore, the equations are independent and the system is consistent.

SELF CHECK   Solve $\begin{cases} -\frac{x}{2} = \frac{y}{4} \\ \frac{1}{4}x - \frac{3}{8}y = -2 \end{cases}.$

Answer: $(-2, 4)$

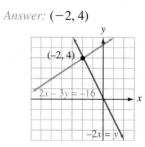

ACCENT ON TECHNOLOGY *Solving Systems with a Graphing Calculator*

We can use a graphing calculator to solve the system

$$\begin{cases} 2x + y = 12 \\ 2x - y = -2 \end{cases}$$

However, before we can enter the equations into the calculator, we must solve them for y.

$$2x + y = 12 \qquad\qquad 2x - y = -2$$
$$y = -2x + 12 \qquad\qquad -y = -2x - 2$$
$$\qquad\qquad\qquad\qquad\qquad y = 2x + 2$$

We enter the resulting equations and graph them on the same coordinate axes. If we use the standard window settings, their graphs will look like Figure 8-7(a). We can use the $\boxed{\text{TRACE}}$ key to find that the coordinates of the intersection point are

$$x = 2.5531915 \qquad \text{and} \qquad y = 6.893617$$

See Figure 8-7(b). For better results, we can zoom in on the intersection point, use the $\boxed{\text{TRACE}}$ key again, and find that

$$x = 2.5 \qquad \text{and} \qquad y = 7$$

See Figure 8-7(c). Check each solution.

FIGURE 8-7

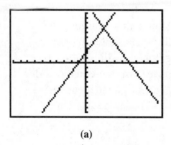

(a)

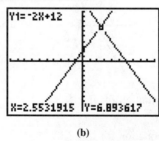

(b)

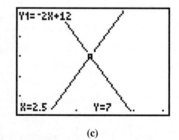

(c)

An easier way to find the coordinates of the point of intersection of two graphs is using the INTERSECT feature that is found on most graphing calculators. With this option, the cursor automatically moves to the point of intersection shown on the screen. Then the coordinates of the point are displayed. See Figure 8-8. Consult your owner's manual for the specific keystrokes to use INTERSECT.

FIGURE 8-8

STUDY SET

Section 8.1

VOCABULARY

In Exercises 1–6, fill in the blanks to make the statements true.

1. The pair of equations $\begin{cases} x - y = -1 \\ 2x - y = 1 \end{cases}$ is called a _____system_____ of equations.

2. Because the ordered pair (2, 3) satisfies both equations in Exercise 1, it is called a _____solution_____ of the system of equations.

3. When the graphs of two equations in a system are different lines, the equations are called _____independent_____ equations.

4. When a system of equations has a solution, the system is called a _____consistent system_____.

5. Systems of equations that have no solution are called _____inconsistent_____ systems.

6. When a system has infinitely many solutions, the equations of the system are said to be _____dependent_____ equations.

CONCEPTS

In Exercises 7–12, refer to Illustration 1. Tell whether a true or false statement would be obtained when the coordinates of

7. point A are substituted into the equation for line l_1.   true

8. point B are substituted into the equation for line l_2.   true

9. point A are substituted into the equation for line l_2.   false

10. point B are substituted into the equation for line l_1.   false

11. point C are substituted into the equation for line l_1.
    true

12. point C are substituted into the equation for line l_2.
    true

ILLUSTRATION 1

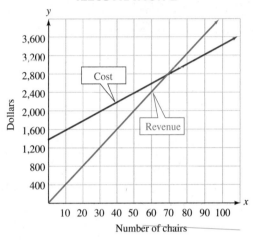

In Exercises 13–14, a furniture company is considering manufacturing a new line of oak chairs. Graphs showing the cost to make the chairs and the revenue the company will receive from their sale are given in Illustration 2.

13. **a.** What will it cost to make 30 chairs? $2,000
    b. What revenue will their sale bring? $1,200
    c. How much money will the company make or lose in this case?   lose $800

14. **a.** How many chairs must be built and sold so that the costs and the revenue are the same?   70
    b. Why do you think (70, 2,800) is called the "break-even point"?   costs = revenue

ILLUSTRATION 2

15. How many solutions does the system of equations graphed in Illustration 3 have? Is the system consistent or inconsistent?
    1 solution; consistent

ILLUSTRATION 3

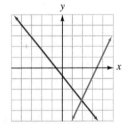

16. How many solutions does the system of equations graphed in Illustration 4 have? Are the equations dependent or independent?
    no solution; independent

ILLUSTRATION 4

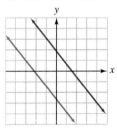

17. The solution of the system of equations graphed in Illustration 5 is $\left(\frac{2}{5}, -\frac{1}{3}\right)$. Knowing this, can you see any disadvantages to the graphing method?
    The method is not accurate enough to find a solution such as $\left(\frac{2}{5}, -\frac{1}{3}\right)$.

ILLUSTRATION 5

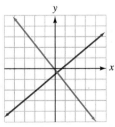

18. Draw the graphs of two linear equations so that the system has
   a. one solution $(-3, -2)$.
   Answers may vary.

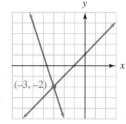

   b. infinitely many solutions, three of which are $(-2, 0)$, $(1, 2)$, and $(4, 4)$.

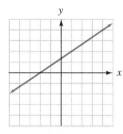

28. $\left(2, \frac{1}{3}\right)$, $\begin{cases} x - 3y = 1 \\ -2x + 6 = -6y \end{cases}$ no

29. $\left(-\frac{2}{5}, \frac{1}{4}\right)$, $\begin{cases} x - 4y = -6 \\ 8y = 10x + 12 \end{cases}$ no

30. $\left(-\frac{1}{3}, \frac{3}{4}\right)$, $\begin{cases} 3x + 4y = 2 \\ 12y = 3(2 - 3x) \end{cases}$ yes

31. $(0.2, 0.3)$, $\begin{cases} 20x + 10y = 7 \\ 20y = 15x + 3 \end{cases}$ yes

32. $(2.5, 3.5)$, $\begin{cases} 4x - 3 = 2y \\ 4y + 1 = 6x \end{cases}$ yes

NOTATION

In Exercises 19–20, clear the equation of the fractions.

19.
$$\frac{1}{6}x - \frac{1}{3}y = \frac{11}{2}$$

$$6\left(\frac{1}{6}x - \frac{1}{3}y\right) = 6\left(\frac{11}{2}\right)$$

$$6\left(\frac{1}{6}x\right) - 6\,\tfrac{1}{3y} = 6\left(\frac{11}{2}\right)$$

$$x - 2y = 33$$

20.
$$\frac{3x}{5} - \frac{4y}{5} = -1$$

$$5\left(\frac{3x}{5} - \frac{4y}{5}\right) = 5\,(-1)$$

$$5\left(\frac{3x}{5}\right) - 5\left(\tfrac{4y}{5}\right) = 5(-1)$$

$$3x - 4y = -5$$

PRACTICE

In Exercises 21–32, tell whether the ordered pair is a solution of the given system.

21. $(1, 1)$, $\begin{cases} x + y = 2 \\ 2x - y = 1 \end{cases}$ yes

22. $(1, 3)$, $\begin{cases} 2x + y = 5 \\ 3x - y = 0 \end{cases}$ yes

23. $(3, -2)$, $\begin{cases} 2x + y = 4 \\ y = 1 - x \end{cases}$ yes

24. $(-2, 4)$, $\begin{cases} 2x + 2y = 4 \\ 3y = 10 - x \end{cases}$ yes

25. $(-2, -4)$, $\begin{cases} 4x + 5y = -23 \\ -3x + 2y = 0 \end{cases}$ no

26. $(-5, 2)$, $\begin{cases} -2x + 7y = 17 \\ 3x - 4y = -19 \end{cases}$ no

27. $\left(\frac{1}{2}, 3\right)$, $\begin{cases} 2x + y = 4 \\ 4x - 11 = 3y \end{cases}$ no

In Exercises 33–52, solve each system by the graphing method. If the equations of a system are dependent or if a system is inconsistent, so indicate.

33. $\begin{cases} x + y = 2 \\ x - y = 0 \end{cases}$

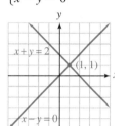

34. $\begin{cases} x + y = 4 \\ x - y = 0 \end{cases}$

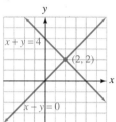

35. $\begin{cases} x + y = 2 \\ y = x - 4 \end{cases}$

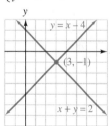

36. $\begin{cases} x + y = 1 \\ y = x + 5 \end{cases}$

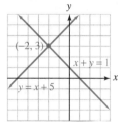

37. $\begin{cases} 3x + 2y = -8 \\ 2x - 3y = -1 \end{cases}$

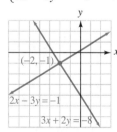

38. $\begin{cases} x + 4y = -2 \\ y = -x - 5 \end{cases}$

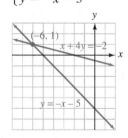

39. $\begin{cases} 4x - 2y = 8 \\ y = 2x - 4 \end{cases}$

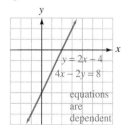
equations are dependent

40. $\begin{cases} 3x - 6y = 18 \\ x = 2y + 3 \end{cases}$

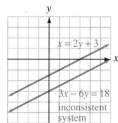
inconsistent system

41. $\begin{cases} 2x - 3y = -18 \\ 3x + 2y = -1 \end{cases}$

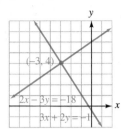

42. $\begin{cases} -x + 3y = -11 \\ 3x - y = 17 \end{cases}$

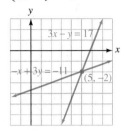

51. $\begin{cases} \frac{1}{3}x - \frac{1}{2}y = \frac{1}{6} \\ \frac{2x}{5} + \frac{y}{2} = \frac{13}{10} \end{cases}$

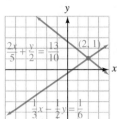

52. $\begin{cases} \frac{3x}{4} + \frac{2y}{3} = -\frac{19}{6} \\ 3y = -x \end{cases}$

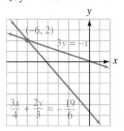

43. $\begin{cases} x = 4 \\ 2y = 12 - 4x \end{cases}$

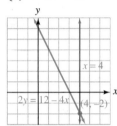

44. $\begin{cases} x = 3 \\ 3y = 6 - 2x \end{cases}$

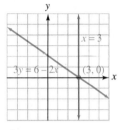

In Exercises 53–56, use a graphing calculator to solve each psystem, if possible.

53. $\begin{cases} y = 4 - x \\ y = 2 + x \end{cases}$ (1, 3)

54. $\begin{cases} 3x - 6y = 4 \\ 2x + y = 1 \end{cases}$ $\left(\frac{2}{3}, -\frac{1}{3}\right)$

55. $\begin{cases} 6x - 2y = 5 \\ 3x = y + 10 \end{cases}$
no solution

56. $\begin{cases} x - 3y = -2 \\ 5x + y = 10 \end{cases}$
(1.75, 1.25)

45. $\begin{cases} x + 2y = -4 \\ x - \frac{1}{2}y = 6 \end{cases}$

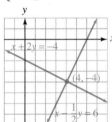

46. $\begin{cases} \frac{2}{3}x - y = -3 \\ 3x + y = 3 \end{cases}$

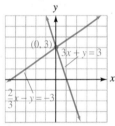

APPLICATIONS odd

57. TRANSPLANTS See Illustration 6.
a. What was the relationship between the number of donors and those awaiting a transplant in 1989?
Donors outnumbered those needing a transplant.
b. In what year were the number of donors and the number waiting for a transplant the same? Estimate the number. 1994; 4,100
c. Explain the most recent trend.
People needing a transplant outnumber the donors.

ILLUSTRATION 6

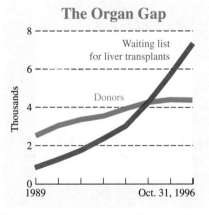

Based on data from *Business Week* (Nov. 25, 1996)

47. $\begin{cases} -\frac{3}{4}x + y = 3 \\ \frac{1}{4}x + y = -1 \end{cases}$

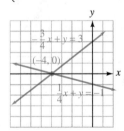

48. $\begin{cases} \frac{1}{3}x + y = 7 \\ \frac{2x}{3} - y = -4 \end{cases}$

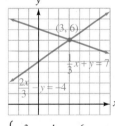

49. $\begin{cases} 2y = 3x + 2 \\ \frac{3}{2}x - y = 3 \end{cases}$

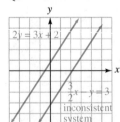
inconsistent system

50. $\begin{cases} -\frac{3}{5}x - \frac{1}{5}y = \frac{6}{5} \\ x + \frac{y}{3} = -2 \end{cases}$

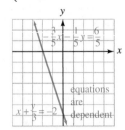
equations are dependent

58. DAILY TRACKING POLL See Illustration 7.
a. Which political candidate was ahead on October 28 and by how much? the incumbent; 7%
b. On what day did the challenger pull even with the incumbent? November 2
c. If the election was held November 4, who did the poll predict would win, and by how many percentage points? the challenger; 3

ILLUSTRATION 7

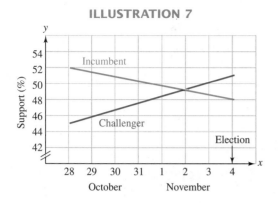

59. LATITUDE AND LONGITUDE See Illustration 8.
a. Name three American cities that lie on a latitude line of 30° north.
    Houston, New Orleans, St. Augustine
b. Name three American cities that lie on a longitude line of 90° west.   St. Louis, Memphis, New Orleans
c. What city lies on both lines?   New Orleans

ILLUSTRATION 8

60. ECONOMICS The graph in Illustration 9 illustrates the law of supply and demand.
a. Complete this sentence: As the price of an item increases, the *supply* of the item ____increases____.
b. Complete this sentence: As the price of an item increases, the *demand* for the item ____decreases____.
c. For what price will the supply equal the demand? How many items will be supplied for this price?
    $6; 30,000

ILLUSTRATION 9

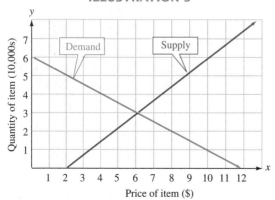

61. AIR TRAFFIC CONTROL The equations describing the paths of two airplanes are $y = -\frac{1}{2}x + 3$ and $3y = 2x + 2$. Graph each equation on the radar screen shown in Illustration 10. Is there a possibility of a mid-air collision? If so, where?   yes, $(2, 2)$

ILLUSTRATION 10

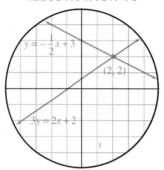

62. TV COVERAGE A television camera is located at $(-2, 0)$ and will follow the launch of a space shuttle, as shown in Illustration 11. (Each unit in the illustration is 1 mile.) As the shuttle rises vertically on a path described by $x = 2$, the farthest the camera can tilt back is a line of sight given by $y = \frac{5}{2}x + 5$. For how many miles of the shuttle's flight will it be in view of the camera?   10 mi

ILLUSTRATION 11

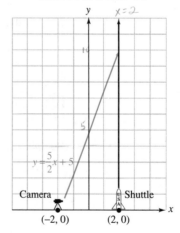

WRITING

63. Look up the word *simultaneous* in a dictionary and give its definition. In mathematics, what is meant by a simultaneous solution of a system of equations?

64. Suppose the solution of a system is $\left(\frac{1}{3}, -\frac{3}{5}\right)$. Do you think you would be able to find the solution using the graphing method? Explain.

REVIEW

65. What is the slope and the y-intercept of the graph of the line $y = -3x + 4$? -3; $(0, 4)$

66. Are the graphs of the lines $y = 5x$ and $y = -\frac{1}{5}x$ parallel, perpendicular, or neither?   perpendicular

67. If $f(x) = -4x - x^2$, find $f(3)$. -21

68. In what quadrant does $(-12, 15)$ lie?   quadrant II

69. Write the equation for the y-axis. $x = 0$

70. Does $(1, 2)$ lie on the line $2x + y = 4$?   yes

71. What point does the line with equation $y - 2 = 7(x - 5)$ pass through? $(5, 2)$

72. Is the word *domain* associated with the inputs or the outputs of a function?   inputs

---

▶ **8.2**

Solving Systems of Equations by Substitution

In this section, you will learn about

> The substitution method ■ Inconsistent systems ■ Dependent equations

---

Introduction   When solving a system of equations by the graphing method, it is often difficult to determine the exact coordinates of the point of intersection. For example, it would be virtually impossible to distinguish that two lines intersect at the point $\left(\frac{1}{16}, -\frac{3}{5}\right)$. In this section, we introduce an algebraic method that finds *exact* solutions. It is called the **substitution method.** This method is based on the **substitution principle,** which states: If $a = b$, then a may replace b or b may replace a in any statement.

The Substitution Method

To solve the system

$$\begin{cases} y = 3x - 2 \\ 2x + y = 8 \end{cases}$$

by the substitution method, we note that the first equation, $y = 3x - 2$, is *solved for y* (or *y is expressed in terms of x*). Because $y = 3x - 2$, we can substitute $3x - 2$ for y in the equation $2x + y = 8$ to get

$$2x + y = 8 \quad \text{The second equation.}$$
$$2x + (\mathbf{3x - 2}) = 8 \quad \text{Substitute } 3x - 2 \text{ for } y. \text{ Write parentheses around the expression that was substituted for } y.$$

The resulting equation has only one variable and can be solved for x.

$$2x + (3x - 2) = 8$$
$$2x + 3x - 2 = 8 \quad \text{Remove the parentheses.}$$
$$5x - 2 = 8 \quad \text{Combine like terms: } 2x + 3x = 5x.$$
$$5x = 10 \quad \text{Add 2 to both sides.}$$
$$x = 2 \quad \text{Divide both sides by 5.}$$

We can find y by substituting 2 for x in either equation of the given system. Because $y = 3x - 2$ is already solved for y, it is easier to substitute into this equation.

$$y = 3x - 2 \quad \text{The first equation.}$$
$$= 3(\mathbf{2}) - 2 \quad \text{Substitute 2 for } x.$$
$$= 6 - 2$$
$$y = 4$$

The solution to the given system is $x = 2$ and $y = 4$, or $(2, 4)$.

Check: First equation **Second equation**

$$y = 3x - 2 \qquad\qquad 2x + y = 8$$

$$4 \stackrel{?}{=} 3(2) - 2 \qquad 2(2) + 4 \stackrel{?}{=} 8$$

$$4 \stackrel{?}{=} 6 - 2 \qquad\qquad 4 + 4 \stackrel{?}{=} 8$$

$$4 = 4 \qquad\qquad\qquad 8 = 8$$

If we graphed the lines represented by the equations of the given system, they would intersect at the point $(2, 4)$. The equations of this system are independent, and the system is consistent.

To solve a system of equations in x and y by the substitution method, we follow these steps.

The substitution method

1. Solve one of the equations for either x or y. (This step will not be necessary if an equation is already solved for x or y.)

2. Substitute the resulting expression for the variable obtained in step 1 into the remaining equation and solve that equation.

3. Find the value of the other variable by substituting the solution found in step 2 into any equation containing both variables.

4. Check the solution in the equations of the original system.

EXAMPLE 1

Solving systems by substitution. Solve $\begin{cases} 2x + y = -10 \\ x = -3y \end{cases}$.

Solution The second equation, $x = -3y$, tells us that x and $-3y$ have the same value. Therefore, we may substitute $-3y$ for x in the first equation.

$$2x + y = -10 \qquad \text{The first equation.}$$

$$2(-3y) + y = -10 \qquad \text{Replace } x \text{ with } -3y.$$

$$-6y + y = -10 \qquad \text{Do the multiplication.}$$

$$-5y = -10 \qquad \text{Combine like terms.}$$

$$y = 2 \qquad \text{Divide both sides by } -5.$$

We can find x by substituting 2 for y in the equation $x = -3y$.

$$x = -3y \qquad \text{The second equation.}$$

$$= -3(2) \qquad \text{Substitute 2 for } y.$$

$$= -6$$

The solution is $x = -6$ and $y = 2$, or $(-6, 2)$.

Check: First equation **Second equation**

$$2x + y = -10 \qquad\qquad x = -3y$$

$$2(-6) + 2 \stackrel{?}{=} -10 \qquad -6 \stackrel{?}{=} -3(2)$$

$$-12 + 2 \stackrel{?}{=} -10 \qquad -6 = -6$$

$$-10 = -10$$

SELF CHECK Solve $\begin{cases} y = -2x \\ 3x - 2y = -7 \end{cases}$.

Answer: $(-1, 2)$ ∎

E X A M P L E   2

Solving for a variable first. Solve $\begin{cases} 2x + y = -5 \\ 3x + 5y = -4 \end{cases}$.

Solution   We solve one of the equations for one of the variables. Since the term y in the first equation has a coefficient of 1, we solve the first equation for y.

$$2x + y = -5 \qquad \text{The first equation.}$$
$$y = -5 - 2x \qquad \text{Subtract } 2x \text{ from both sides to isolate } y.$$

We then substitute $-5 - 2x$ for y in the second equation and solve for x.

$$3x + 5y = -4 \qquad \text{The second equation.}$$
$$3x + 5(-5 - 2x) = -4 \qquad \text{Substitute } -5 - 2x \text{ for } y.$$
$$3x - 25 - 10x = -4 \qquad \text{Remove parentheses.}$$
$$-7x - 25 = -4 \qquad \text{Combine like terms: } 3x - 10x = -7x.$$
$$-7x = 21 \qquad \text{Add 25 to both sides.}$$
$$x = -3 \qquad \text{Divide both sides by } -7.$$

We can find y by substituting -3 for x in the equation $y = -5 - 2x$.

$$y = -5 - 2x$$
$$= -5 - 2(-3) \qquad \text{Substitute } -3 \text{ for } x.$$
$$= -5 + 6$$
$$= 1$$

The solution is $(-3, 1)$. Check it in the original equations.

SELF CHECK   Solve $\begin{cases} 2x - 3y = 13 \\ 3x + y = 3 \end{cases}$.

Answer: $(2, -3)$ ∎

Systems of equations are sometimes written in variables other than x and y. For example, the system

$$\begin{cases} 3a - 3b = 5 \\ 3 - a = -2b \end{cases}$$

is written in a and b. Regardless of the variables used, the procedures used to solve the system remain the same. The solution should be expressed in the form (a, b).

E X A M P L E   3

Solving for a variable first. Solve $\begin{cases} 3a - 3b = 5 \\ 3 - a = -2b \end{cases}$.

Solution   Since the coefficient of a in the second equation is -1, we will solve that equation for a.

$$3 - a = -2b \qquad \text{The second equation.}$$
$$-a = -2b - 3 \qquad \text{Subtract 3 from both sides.}$$

To obtain a on the left-hand side, we can multiply (or divide) both sides of the equation by -1.

$$-1(-a) = -1(-2b - 3) \qquad \text{Multiply both sides by } -1.$$
$$a = 2b + 3 \qquad \text{Do the multiplications.}$$

We then substitute $2b + 3$ for a in the first equation and proceed as follows:

$$3a - 3b = 5$$
$$3(2b + 3) - 3b = 5 \qquad \text{Substitute.}$$
$$6b + 9 - 3b = 5 \qquad \text{Remove parentheses.}$$
$$3b + 9 = 5 \qquad \text{Combine like terms.}$$
$$3b = -4 \qquad \text{Subtract 9 from both sides: } 5 - 9 = -4.$$
$$b = -\frac{4}{3} \qquad \text{Divide both sides by 3.}$$

To find a, we substitute $-\frac{4}{3}$ for b in $a = 2b + 3$ and simplify.

$$a = 2b + 3$$
$$= 2\left(-\frac{4}{3}\right) + 3 \qquad \text{Substitute.}$$
$$= -\frac{8}{3} + \frac{9}{3} \qquad \text{Do the multiplication: } 2\left(-\frac{4}{3}\right) = -\frac{8}{3}. \text{ Write 3 as } \frac{9}{3}.$$
$$= \frac{1}{3} \qquad \text{Add the numerators and keep the common denominator.}$$

The solution is $\left(\frac{1}{3}, -\frac{4}{3}\right)$. Check it in the original equations.

SELF CHECK   Solve $\begin{cases} 2s - t = 4 \\ 3s - 5t = 2 \end{cases}$.   *Answer:* $\left(\frac{18}{7}, \frac{8}{7}\right)$ ■

EXAMPLE 4

Solving an equivalent system. Solve $\begin{cases} \frac{x}{2} + \frac{y}{4} = -\frac{1}{4} \\ 2x - y = 2 + y - x \end{cases}$.

Solution   It is helpful to rewrite each equation in simpler form before performing a substitution. We begin by clearing the first equation of fractions.

$$\frac{x}{2} + \frac{y}{4} = -\frac{1}{4}$$
$$4\left(\frac{x}{2} + \frac{y}{4}\right) = 4\left(-\frac{1}{4}\right) \qquad \text{Multiply both sides by the LCD, 4.}$$
$$2x + y = -1$$

We can write the second equation in general form ($Ax + By = C$) by adding x and subtracting y from both sides.

$$2x - y = 2 + y - x$$
$$2x - y + x - y = 2 + y - x + x - y$$
$$3x - 2y = 2 \qquad \text{Combine like terms.}$$

The two results form the following equivalent system, which has the same solution as the original one.

1. $\begin{cases} 2x + y = -1 \\ 3x - 2y = 2 \end{cases}$
2.

To solve this system, we solve Equation 1 for y.

$$2x + y = -1$$
$$2x + y - 2x = -1 - 2x \qquad \text{Subtract } 2x \text{ from both sides.}$$
3. $\qquad\qquad\quad y = -1 - 2x \qquad \text{Combine like terms.}$

To find x, we substitute $-1 - 2x$ for y in Equation 2 and proceed as follows:

$$3x - 2y = 2$$
$$3x - 2(-1 - 2x) = 2 \quad \text{Substitute.}$$
$$3x + 2 + 4x = 2 \quad \text{Remove parentheses.}$$
$$7x + 2 = 2 \quad \text{Combine like terms.}$$
$$7x = 0 \quad \text{Subtract 2 from both sides.}$$
$$x = 0 \quad \text{Divide both sides by 7.}$$

To find y, we substitute 0 for x in Equation 3.

$$y = -1 - 2x$$
$$y = -1 - 2(0)$$
$$y = -1$$

The solution is $(0, -1)$. Check it in the original equations.

SELF CHECK   Solve $\begin{cases} \frac{1}{3}x - \frac{1}{6}y = -\frac{1}{3} \\ x + y = -3 - 2x - y \end{cases}$.   *Answer:* $(-1, 0)$ ■

Inconsistent Systems

EXAMPLE 5

A system with no solution.   Solve $\begin{cases} 0.01x = 0.12 - 0.04y \\ 2x = 4(3 - 2y) \end{cases}$.

Solution   The first equation contains decimal coefficients. We can clear the equation of decimals by multiplying both sides by 100.

$$\begin{cases} x = 12 - 4y \\ 2x = 4(3 - 2y) \end{cases}$$

FIGURE 8-9

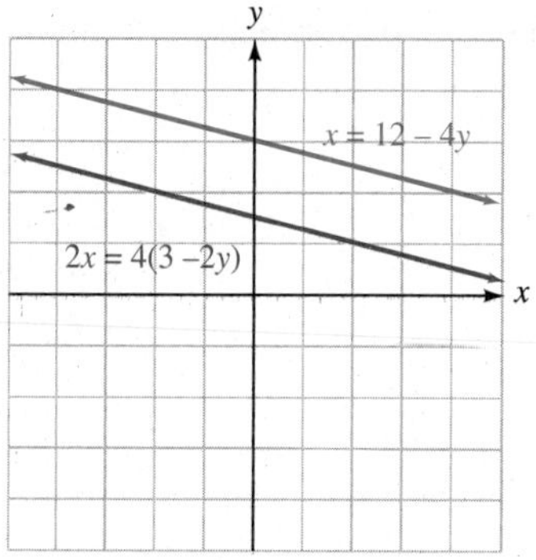

Since $x = 12 - 4y$, we can substitute $12 - 4y$ for x in the second equation and solve for y.

$$2x = 4(3 - 2y) \quad \text{The second equation.}$$
$$2(12 - 4y) = 4(3 - 2y) \quad \text{Substitute.}$$
$$24 - 8y = 12 - 8y \quad \text{Remove parentheses.}$$
$$24 \neq 12 \quad \text{Add } 8y \text{ to both sides.}$$

This result indicates that the equations are independent and also that the system is inconsistent. As we see in Figure 8-9, when the equations are graphed, the graphs are parallel lines. This system has no solution.

SELF CHECK   Solve $\begin{cases} 0.1x - 0.4 = 0.1y \\ -2y = 2(2 - x) \end{cases}$.   *Answer:* no solution   ■

Dependent Equations

EXAMPLE 6

Infinitely many solutions.   Solve $\begin{cases} x = -3y + 6 \\ 2x + 6y = 12 \end{cases}$.

Solution   We can substitute $-3y + 6$ for x in the second equation and proceed as follows:

$$2x + 6y = 12 \quad \text{The second equation.}$$
$$2(-3y + 6) + 6y = 12 \quad \text{Substitute.}$$
$$-6y + 12 + 6y = 12 \quad \text{Remove parentheses.}$$
$$12 = 12 \quad \text{Combine like terms.}$$

Although $12 = 12$ is true, we did not find y. This indicates that the equations are dependent. As we see in Figure 8-10, when these equations are graphed, their graphs are identical.

Because any ordered pair that satisfies one equation of the system also satisfies the other, the system has infinitely many solutions. To find some, we substitute 0, 3, and 6 for x in either equation and solve for y. The pairs $(0, 2)$, $(3, 1)$, and $(6, 0)$ are some of the solutions.

FIGURE 8-10

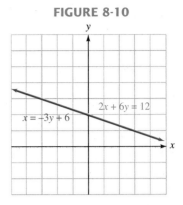

SELF CHECK Solve $\begin{cases} y = 2 - x \\ 3x + 3y = 6 \end{cases}$.

Answer: infinitely many solutions ∎

STUDY SET

Section 8.2

VOCABULARY

In Exercises 1–6, fill in the blanks to make the statements true.

1. We say that the equation $y = 2x + 4$ is solved for _y_ or that y is expressed in _____terms_____ of x.

2. "To _____check_____ a solution of a system" means to see whether the coordinates of the ordered pair satisfy both equations.

3. Consider $2(x - 6) = 2x - 12$. The distributive property was applied to _____remove_____ parentheses.

4. In mathematics, "to _____substitute_____" means to replace an expression with one that is equivalent to it.

5. A dependent system has _____infinitely_____ many solutions.

6. In the term y, the _____coefficient_____ is understood to be 1.

CONCEPTS

7. Consider the system $\begin{cases} 2x + 3y = 12 \\ y = 2x + 4 \end{cases}$.

  a. How many variables does each equation of the system contain? 2

  b. Substitute $2x + 4$ for y in the first equation. How many variables does the resulting equation contain?

  1

8. For each equation, solve for y.

  a. $y + 2 = x$ $y = x - 2$

  b. $2 - y = x$ $y = 2 - x$

  c. $2 + x + y = 0$ $y = -x - 2$

9. Given the equation $x - 2y = -10$,

  a. solve it for x. $x = 2y - 10$

  b. solve it for y. $y = \frac{x}{2} + 5$

  c. which variable was easier to solve for, x or y? Explain. x; it involved only one step.

10. Which variable in which equation should be solved for in step 1 of the substitution method?

  a. $\begin{cases} x - 2y = 2 \\ 2x + 3y = 11 \end{cases}$ Solve for x in the first equation.

  b. $\begin{cases} 2x - 3y = 2 \\ 2x - y = 11 \end{cases}$ Solve for y in the second equation.

  c. $\begin{cases} 7x - 3y = 2 \\ 2x - 8y = 0 \end{cases}$ Solve for x in the second equation.

11. a. Find the **error** in the following work when $x - 4$ is substituted for y.

$$x + 2y = 5 \quad \text{The first equation of the system.}$$
$$x + 2x - 4 = 5 \quad \text{Substitute for } y: y = x - 4.$$
$$3x - 4 = 5 \quad \text{Combine like terms.}$$
$$3x = 9 \quad \text{Add 4 to both sides.}$$
$$x = 3 \quad \text{Do the divisions.}$$

Parentheses must be written around $x - 4$ in line 2.

  b. Rework the problem to find the correct value of x. $\frac{13}{3}$

12. A student uses the substitution method to solve the system $\begin{cases} 4a + 5b = 2 \\ b = 3a - 11 \end{cases}$. She finds that $a = 3$. What is the easiest way for her to determine the value of b?

Substitute 3 for a in the second equation.

13. Consider the system $\begin{cases} x - 2y = 0 \\ 6x + 3y = 5 \end{cases}$.

    a. Graph the equations on the same coordinate system. Why is it difficult to determine the solution of the system?

    The coordinates of the intersection point are not integers.

    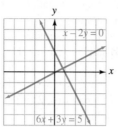

    b. Solve the system by the substitution method.
    $\left(\frac{2}{3}, \frac{1}{3}\right)$

14. The equation $-2 = 1$ is the result when a system is solved by the substitution method. Which graph in Illustration 1 is a possible graph of the system?
    the second graph

ILLUSTRATION 1

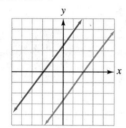

NOTATION

In Exercises 15–16, complete the solution of each system.

15. Solve $\begin{cases} y = 3x \\ x - y = 4 \end{cases}$.

    $x - y = 4$

    $x - \left(\boxed{3x} \right) = 4$

    $-2x = \boxed{4}$

    $x = -2$

    $y = 3x$

    $y = 3\left(\boxed{-2} \right)$

    $y = -6$

    The solution of the system is $(-2, -6)$.

16. Solve $\begin{cases} 2x + y = -5 \\ 2 - 2y = x \end{cases}$.

    $2x + y = -5$

    $2\left(\boxed{2 - 2y} \right) + y = -5$

    $4 - \boxed{4y} + y = -5$

    $\boxed{4} - 3y = -5$

    $-3y = \boxed{-9}$

    $y = 3$

    $2 - 2y = x$

    $2 - 2\left(\boxed{3} \right) = x$

    $2 - 6 = x$

    $-4 = x$

    The solution of the system is $(-4, 3)$.

PRACTICE

In Exercises 17–52, use the substitution method to solve each system. If the equations of a system are dependent or if a system is inconsistent, so indicate.

17. $\begin{cases} y = 2x \\ x + y = 6 \end{cases}$ $(2, 4)$

18. $\begin{cases} y = 3x \\ x + y = 4 \end{cases}$ $(1, 3)$

19. $\begin{cases} y = 2x - 6 \\ 2x + y = 6 \end{cases}$ $(3, 0)$

20. $\begin{cases} y = 2x - 9 \\ x + 3y = 8 \end{cases}$ $(5, 1)$

21. $\begin{cases} y = 2x + 5 \\ x + 2y = -5 \end{cases}$
    $(-3, -1)$

22. $\begin{cases} y = -2x \\ 3x + 2y = -1 \end{cases}$
    $(1, -2)$

23. $\begin{cases} 2a + 4b = -24 \\ a = 20 - 2b \end{cases}$
    inconsistent system

24. $\begin{cases} 3a + 6b = -15 \\ a = -2b - 5 \end{cases}$
    dependent equations

25. $\begin{cases} 2a = 3b - 13 \\ -b = -2a - 7 \end{cases}$
    $(-2, 3)$

26. $\begin{cases} a = 3b - 1 \\ -b = -2a - 2 \end{cases}$
    $(-1, 0)$

27. $\begin{cases} r + 3s = 9 \\ 3r + 2s = 13 \end{cases}$ $(3, 2)$

28. $\begin{cases} x - 2y = 2 \\ 2x + 3y = 11 \end{cases}$ $(4, 1)$

29. $\begin{cases} 0.4x + 0.5y = 0.2 \\ 3x - y = 11 \end{cases}$
    $(3, -2)$

30. $\begin{cases} 0.5u + 0.3v = 0.5 \\ 4u - v = 4 \end{cases}$
    $(1, 0)$

31. $\begin{cases} 6x - 3y = 5 \\ 2y + x = 0 \end{cases}$ $\left(\frac{2}{3}, -\frac{1}{3}\right)$

32. $\begin{cases} 5s + 10t = 3 \\ 2s + t = 0 \end{cases}$ $\left(-\frac{1}{5}, \frac{2}{5}\right)$

33. $\begin{cases} 3x + 4y = -7 \\ 2y - x = -1 \end{cases}$
    $(-1, -1)$

34. $\begin{cases} 4x + 5y = -2 \\ x + 2y = -2 \end{cases}$
    $(2, -2)$

35. $\begin{cases} 9x = 3y + 12 \\ 4 = 3x - y \end{cases}$
    dependent equations

36. $\begin{cases} 8y = 15 - 4x \\ x + 2y = 4 \end{cases}$
    inconsistent system

37. $\begin{cases} 0.02x + 0.05y = -0.02 \\ -\frac{x}{2} = y \end{cases}$ $(4, -2)$

38. $\begin{cases} y = -\frac{x}{2} \\ 0.02x - 0.03y = -0.07 \end{cases}$ $(-2, 1)$

39. $\begin{cases} b = \frac{2}{3}a \\ 8a - 3b = 3 \end{cases}$ $\left(\frac{1}{2}, \frac{1}{3}\right)$

40. $\begin{cases} a = \frac{2}{3}b \\ 9a + 4b = 5 \end{cases}$ $\left(\frac{1}{3}, \frac{1}{2}\right)$

41. $\begin{cases} y - x = 3x \\ 2x + 2y = 14 - y \end{cases}$
    $(1, 4)$

42. $\begin{cases} y + x = 2x + 2 \\ 6x - 4y = 21 - y \end{cases}$
    $(9, 11)$

43. $\begin{cases} 2x - y = x + y \\ -2x + 4y = 6 \end{cases}$
    inconsistent system

44. $\begin{cases} x = -3y + 6 \\ 2x + 4y = 6 + x + y \end{cases}$
    dependent equations

45. $\begin{cases} 3(x - 1) + 3 = 8 + 2y \\ 2(x + 1) = 8 + y \end{cases}$ $(4, 2)$

46. $\begin{cases} 4(x - 2) = 19 - 5y \\ 3(x - 2) - 2y = -y \end{cases}$ $(3, 3)$

47. $\begin{cases} \frac{1}{2}x + \frac{1}{2}y = -1 \\ \frac{1}{3}x - \frac{1}{2}y = -4 \end{cases}$
    $(-6, 4)$

48. $\begin{cases} \frac{2}{3}y + \frac{1}{5}z = 1 \\ \frac{1}{3}y - \frac{2}{5}z = 3 \end{cases}$ $(3, -5)$

49. $\begin{cases} 5x = \frac{1}{2}y - 1 \\ \frac{1}{4}y = 10x - 1 \end{cases}$ $\left(\frac{1}{5}, 4\right)$

50. $\begin{cases} \frac{2}{3}x = 1 - 2y \\ 2(5y - x) + 11 = 0 \end{cases}$
    $\left(3, -\frac{1}{2}\right)$

51. $\begin{cases} \frac{6x-1}{3} - \frac{5}{3} = \frac{3y+1}{2} \\ \frac{1+5y}{4} + \frac{x+3}{4} = \frac{17}{2} \end{cases}$
(5, 5)

52. $\begin{cases} \frac{5x-2}{4} + \frac{1}{2} = \frac{3y+2}{2} \\ \frac{7y+3}{3} = \frac{x}{2} + \frac{7}{3} \end{cases}$
(2, 1)

APPLICATIONS

53. DINING See the breakfast menu in Illustration 2. What substitution from the a la carte menu will the restaurant owner allow customers to make if they don't want hash browns with their country breakfast? Why?   melon, because it's the same price as hash browns

ILLUSTRATION 2

Village Vault Restaurant			
Country Breakfast $5.95			
Includes 2 eggs, 3 pancakes, sausage, bacon			
hash browns, and coffee			
A la Carte Menu–Single Servings			
Strawberries	$1.25	Melon	$0.95
Croissant	$1.70	Orange juice	$1.65
Hash browns	$0.95	Oatmeal	$1.95
Muffin	$1.30	Ham	$1.80

54. DISCOUNT COUPON In mathematics, the substitution property states:

If a = b, then a may replace b or b may replace a in any statement.

Where on the coupon in Illustration 3 is there an application of the substitution property? Explain.

An entree of equal value may be substituted for the second roast beef dinner.

ILLUSTRATION 3

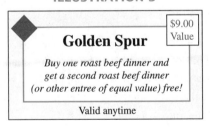

	$9.00 Value
Golden Spur	
Buy one roast beef dinner and get a second roast beef dinner (or other entree of equal value) free!	
Valid anytime	

WRITING

55. Explain how to use substitution to solve a system of equations.

56. If the equations of a system are written in general form, why is it to your advantage to solve for a variable whose coefficient is 1 when using the substitution method?

57. When solving a system, what advantages and disadvantages are there with the graphing method? With the substitution method?

58. In this section, the substitution method for solving a system of two equations was discussed. List some other uses of the word *substitution*, or *substitute*, that you encounter in everyday life.

REVIEW

59. What is the slope of the line $y = -\frac{5}{8}x - 12$? $-\frac{5}{8}$

60. If $g(x) = -3x + 9$, find $g(-3)$.   18

61. Find the y-intercept of $2x - 3y = 18$.   (0, −6)

62. Write the equation of the line passing through $(-1, 5)$ with a slope of -3. $y = -3x + 2$

63. Can a circle represent the graph of a function?   no

64. What is the range of the function $f(x) = |x|$?   all real numbers greater than or equal to 0

65. A bowler has found that his score s varies directly with the time t in minutes he practices. Write an equation describing this relationship. $s = kt$

66. On what axis does $(0, -2)$ lie? y-axis

▶ **8.3**

Solving Systems of Equations by Addition

In this section, you will learn about

 The addition method ■ Inconsistent systems ■ Dependent equations

---

Introduction In step 1 of the substitution method for solving a system of equations, we solve one equation for one of the variables. At times, this can be difficult—especially if neither variable has a coefficient of 1 or −1. In cases such as these, we can use another algebraic method called the **addition** or **elimination method** to find the exact solution of the system. This method is based on the addition property of equality: *When equal quantities are added to both sides of an equation, the results are equal.*

The Addition Method

To solve the system

$$\begin{cases} x + y = 8 \\ x - y = -2 \end{cases}$$

by the addition method, we see that the coefficients of y are opposites and then add the left- and right-hand sides of the equations to eliminate the variable y.

$$\begin{aligned} x + y &= 8 \\ \underline{x - y} &= \underline{-2} \end{aligned}$$ Equal quantities, $x - y$ and -2, are added to both sides of the equation $x + y = 8$. By the addition property of equality, the results will be equal.

Now, column by column, we add like terms. The terms y and $-y$ are eliminated.

$$\begin{aligned} x + y &= 8 \\ \underline{x - y} &= \underline{-2} \\ 2x &= 6 \end{aligned}$$ ↓ ↓ ↓ Combine like terms: $x + x = 2x$, $y + (-y) = 0$, and $8 + (-2) = 6$.

← Write each result here.

We can then solve the resulting equation for x.

$$2x = 6$$
$$x = 3 \quad \text{Divide both sides by 2.}$$

To find y, we substitute 3 for x in either equation and solve it for y.

$$x + y = 8 \quad \text{The first equation of the system.}$$
$$3 + y = 8 \quad \text{Substitute 3 for } x.$$
$$y = 5 \quad \text{Subtract 3 from both sides.}$$

We check the solution by verifying that $(3, 5)$ satisfies each equation of the system. To solve an equation in x and y by the addition method, we follow these steps.

The addition method

1. Write both equations in general form: $Ax + By = C$.

2. If necessary, multiply one or both of the equations by nonzero quantities to make the coefficients of x (or the coefficients of y) opposites.

3. Add the equations to eliminate the term involving x (or y).

4. Solve the equation resulting from step 3.

5. Find the value of the other variable by substituting the solution found in step 4 into any equation containing both variables.

6. Check the solution in the equations of the original system.

EXAMPLE 1

Solving systems by addition. Solve $\begin{cases} 5x + y = -4 \\ -5x + 2y = 7 \end{cases}$.

Solution   When the equations are added, the terms $5x$ and $-5x$ drop out. We can then solve the resulting equation for y.

$$\begin{aligned} 5x + y &= -4 \\ \underline{-5x + 2y} &= \underline{7} \\ 3y &= 3 \end{aligned}$$ Combine like terms: $5x + (-5x) = 0$, $y + 2y = 3y$, and $-4 + 7 = 3$.

$$y = 1 \quad \text{Divide both sides by 3.}$$

To find x, we substitute 1 for y in either equation. If we use $5x + y = -4$, we have

$$5x + y = -4 \quad \text{The first equation of the system.}$$
$$5x + (1) = -4 \quad \text{Substitute 1 for } y.$$
$$5x = -5 \quad \text{Subtract 1 from both sides.}$$
$$x = -1 \quad \text{Divide both sides by 5.}$$

Verify that $(-1, 1)$ satisfies each original equation.

SELF CHECK   Solve $\begin{cases} x + 3y = 7 \\ 2x - 3y = -22 \end{cases}$.   *Answer:* $(-5, 4)$ ■

EXAMPLE 2   **Solving systems by addition.** Solve $\begin{cases} 3x + y = 7 \\ x + 2y = 4 \end{cases}$.

Solution   If we add the equations as they are, neither variable will be eliminated. We must write the equations so that the coefficients of one of the variables are opposites. To eliminate x, we can multiply both sides of the second equation by -3 to get

$$\begin{cases} 3x + y = 7 \\ -3(x + 2y) = -3(4) \end{cases} \longrightarrow \begin{cases} 3x + y = 7 \\ -3x - 6y = -12 \end{cases}$$

The coefficients of the terms $3x$ and $-3x$ are now opposites. When the equations are added, x is eliminated.

$$\begin{array}{r} 3x + y = 7 \\ -3x - 6y = -12 \\ \hline -5y = -5 \end{array}$$
$$y = 1 \quad \text{Divide both sides by } -5.$$

To find x, we substitute 1 for y in the equation $x + 2y = 4$.

$$x + 2y = 4 \quad \text{The second equation of the original system.}$$
$$x + 2(1) = 4 \quad \text{Substitute 1 for } y.$$
$$x + 2 = 4 \quad \text{Do the multiplication.}$$
$$x = 2 \quad \text{Subtract 2 from both sides.}$$

Check the solution $(2, 1)$ in the original system of equations.

SELF CHECK   Solve $\begin{cases} 3x + 4y = 25 \\ 2x + y = 10 \end{cases}$.   *Answer:* $(3, 4)$ ■

EXAMPLE 3   **Solving systems by addition.** Solve $\begin{cases} 2a - 5b = 10 \\ 3a - 2b = -7 \end{cases}$.

Solution   The equations in the system must be written so that one of the variables will be eliminated when the equations are added.

To eliminate a, we can multiply the first equation by 3 and the second equation by -2 to get

$$\begin{cases} 3(2a - 5b) = 3(10) \\ -2(3a - 2b) = -2(-7) \end{cases} \longrightarrow \begin{cases} 6a - 15b = 30 \\ -6a + 4b = 14 \end{cases}$$

When these equations are added, the terms $6a$ and $-6a$ are eliminated.

$$
\begin{array}{r}
6a - 15b = 30 \\
\underline{-6a + 4b = 14} \\
-11b = 44 \\
\end{array}
$$

$$b = -4 \qquad \text{Divide both sides by } -11.$$

To find a, we substitute -4 for b in the equation $2a - 5b = 10$.

$$
\begin{aligned}
2a - 5b &= 10 && \text{The first equation of the original system.} \\
2a - 5(-4) &= 10 && \text{Substitute } -4 \text{ for } b. \\
2a + 20 &= 10 && \text{Simplify.} \\
2a &= -10 && \text{Subtract 20 from both sides.} \\
a &= -5 && \text{Divide both sides by 2.}
\end{aligned}
$$

Check the solution $(-5, -4)$ in the original equations.

SELF CHECK   Solve $\begin{cases} 2a + 3b = 7 \\ 5a + 2b = 1 \end{cases}$.   *Answer:* $(-1, 3)$ ■

E X A M P L E   4

Equations containing fractions. Solve $\begin{cases} \frac{5}{6}x + \frac{2}{3}y = \frac{7}{6} \\ \frac{10}{7}x - \frac{4}{9}y = \frac{17}{21} \end{cases}$.

Solution   To clear the equations of fractions, we multiply both sides of the first equation by 6 and both sides of the second equation by 63. This gives the equivalent system

1. $\begin{cases} 5x + 4y = 7 \\ 90x - 28y = 51 \end{cases}$
2.

We can solve for x by eliminating the terms involving y. To do so, we multiply Equation 1 by 7 and add the result to Equation 2.

$$
\begin{array}{r}
35x + 28y = 49 \\
\underline{90x - 28y = 51} \\
125x = 100 \\
\end{array}
$$

$$x = \frac{100}{125} \qquad \text{Divide both sides by 125.}$$

$$x = \frac{4}{5} \qquad \text{Simplify. Divide out the common factor of 25.}$$

To solve for y, we substitute $\frac{4}{5}$ for x in Equation 1 and simplify.

$$
\begin{aligned}
5x + 4y &= 7 \\
5\left(\frac{4}{5}\right) + 4y &= 7 \\
4 + 4y &= 7 && \text{Simplify.} \\
4y &= 3 && \text{Subtract 4 from both sides.} \\
y &= \frac{3}{4} && \text{Divide both sides by 4.}
\end{aligned}
$$

Check the solution of $\left(\frac{4}{5}, \frac{3}{4}\right)$ in the original equations.

SELF CHECK   Solve $\begin{cases} \frac{1}{3}x + \frac{1}{6}y = 1 \\ \frac{1}{2}x - \frac{1}{4}y = 0 \end{cases}$.   *Answer:* $\left(\frac{3}{2}, 3\right)$ ■

EXAMPLE 5

Writing equations in general form. Solve $\begin{cases} 2(2x + y) = 13 \\ 8x = 2y - 16 \end{cases}$.

Solution

We begin by writing each equation in $Ax + By = C$ form. For the first equation, we need only remove the parentheses. To write the second equation in general form, we subtract $2y$ from both sides.

$$2(2x + y) = 13 \qquad\qquad 8x = 2y - 16$$
$$4x + 2y = 13 \qquad\qquad 8x - 2y = 2y - 16 - 2y$$
$$\qquad\qquad\qquad\qquad 8x - 2y = -16$$

The two resulting equations form the following system.

1. $\begin{cases} 4x + 2y = 13 \\ 8x - 2y = -16 \end{cases}$
2.

When the equations are added, the terms involving y are eliminated.

$$\begin{array}{r} 4x + 2y = 13 \\ 8x - 2y = -16 \\ \hline 12x \quad\;\; = -3 \end{array}$$

$$x = -\frac{1}{4} \quad \text{Divide both sides by 12 and simplify the fraction: } -\frac{3}{12} = -\frac{1}{4}.$$

We can use Equation 1 to find y.

$$4x + 2y = 13$$
$$4\left(-\tfrac{1}{4}\right) + 2y = 13 \quad \text{Substitute } -\tfrac{1}{4} \text{ for } x.$$
$$-1 + 2y = 13 \quad \text{Do the multiplication.}$$
$$2y = 14 \quad \text{Add 1 to both sides.}$$
$$y = 7 \quad \text{Divide both sides by 2.}$$

Verify that $\left(-\tfrac{1}{4}, 7\right)$ satisfies each original equation.

SELF CHECK Solve $\begin{cases} -3y = -5 - x \\ 3(x - y) = -11 \end{cases}$. *Answer:* $\left(-3, \tfrac{2}{3}\right)$ ∎

Inconsistent Systems

EXAMPLE 6

A system with no solutions. Solve $\begin{cases} 3x - 2y = 8 \\ -3x + 2y = -12 \end{cases}$.

Solution

We can add the equations to eliminate the term involving x.

$$\begin{array}{r} 3x - 2y = 8 \\ -3x + 2y = -12 \\ \hline 0 = -4 \end{array}$$

Here the terms involving both x and y drop out, and a false result of $0 = -4$ is obtained. This shows that the equations of the system are independent and that the system is inconsistent. This system has no solution.

SELF CHECK Solve $\begin{cases} 2t - 7v = 5 \\ -2t + 7v = 3 \end{cases}$. *Answer:* no solution ∎

Dependent Equations

EXAMPLE 7

Infinitely many solutions. Solve $\begin{cases} \frac{2x-5y}{2} = \frac{19}{2} \\ -0.2x + 0.5y = -1.9 \end{cases}$.

Solution   We can multiply both sides of the first equation by 2 to clear it of the fractions and both sides of the second equation by 10 to clear it of the decimals.

$$\begin{cases} 2\left(\dfrac{2x-5y}{2}\right) = 2\left(\dfrac{19}{2}\right) \\ 10(-0.2x + 0.5y) = 10(-1.9) \end{cases} \longrightarrow \begin{cases} 2x - 5y = 19 \\ -2x + 5y = -19 \end{cases}$$

We add the resulting equations to get

$$\begin{array}{rcr} 2x - 5y = & 19 \\ -2x + 5y = & -19 \\ \hline 0 = & 0 \end{array}$$

As in Example 6, both x and y drop out. However, this time a true result is obtained. This shows that the equations are dependent and that the system has infinitely many solutions.

Any ordered pair that satisfies one equation also satisfies the other equation. Some solutions are $(2, -3)$, $(12, 1)$, and $\left(0, -\frac{19}{5}\right)$.

SELF CHECK   Solve $\begin{cases} \frac{3x+y}{6} = \frac{1}{3} \\ -0.3x - 0.1y = -0.2 \end{cases}$.

Answer: infinitely many solutions ∎

STUDY SET

Section 8.3

VOCABULARY

In Exercises 1–4, fill in the blanks to make the statements true.

1. The ___coefficient___ of the term $-3x$ is -3.

2. The ___opposite___ of 4 is -4.

3. $Ax + By = C$ is the ___general___ form of the equation of a line.

4. When adding the equations
$$\begin{array}{r} 5x - 6y = 10 \\ -3x + 6y = 24 \\ \hline \end{array}$$
the variable y will be ___eliminated___.

CONCEPTS

5. If the addition method is to be used to solve this system, what is wrong with the form in which it is written?
$$\begin{cases} 2x - 5y = -3 \\ -2y + 3x = 10 \end{cases}$$
The second equation should be written in general form: $3x - 2y = 10$.

6. Can the system
$$\begin{cases} 2x + 5y = -13 \\ -2x - 3y = -5 \end{cases}$$
be solved more easily using the addition method or the substitution method? Explain.
addition; the coefficients of x are opposites

7. What algebraic step should be performed to clear this equation of the fractions?
$$\frac{2}{3}x + 4y = -\frac{4}{5} \quad \text{Multiply both sides by 15.}$$

8. If the addition method is used to solve
$$\begin{cases} 3x + 12y = 4 \\ 6x - 4y = 8 \end{cases}$$
a. By what would we multiply the first equation to eliminate x? -2
b. By what would we multiply the second equation to eliminate y?   3

9. Solve $\begin{cases} 4x + 2y = 2 \\ 3x - 2y = 12 \end{cases}$.

   a. Use the graphing method.
   $(2, -3)$

   b. Use the substitution method. $(2, -3)$

   c. Use the addition method. $(2, -3)$

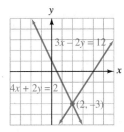

10. The addition method was used to solve three different systems. The results after x was eliminated in each case are listed here. Match each result with a possible graph of the system.

   Result when solving

   a. System 1:   **b.** System 2:   **c.** System 3:
   $-1 = -1$ $y = -1$ $-1 = -2$
   ii   iii   i

   Possible graph of the system

   i   ii   iii

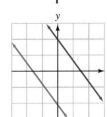

NOTATION

In Exercises 11–12, complete the solution of each system.

11. Solve $\begin{cases} x + y = 5 \\ x - y = -3 \end{cases}$.

$$\begin{array}{r} x + y = 5 \\ x - y = -3 \\ \hline \boxed{2x} = 2 \\ x = \boxed{1} \end{array}$$

$$x + y = 5$$
$$\left(\boxed{1}\right) + y = 5$$
$$y = 4$$

The solution of the system is $\boxed{(1, 4)}$.

12. Solve $\begin{cases} x - 2y = 8 \\ -x + 5y = -17 \end{cases}$.

$$\begin{array}{r} x - 2y = 8 \\ -x + 5y = -17 \\ \hline \boxed{3y} = -9 \\ y = \boxed{-3} \end{array}$$

$$x - 2y = 8$$
$$x - 2\left(\boxed{-3}\right) = 8$$
$$x + 6 = 8$$
$$x = 2$$

The solution of the system is $\boxed{(2, -3)}$.

PRACTICE

In Exercises 13–20, use the addition method to solve each system.

13. $\begin{cases} x - y = -5 \\ x + y = 1 \end{cases}$ $(-2, 3)$

14. $\begin{cases} x + y = 1 \\ x - y = 5 \end{cases}$ $(3, -2)$

15. $\begin{cases} 2r + s = -1 \\ -2r + s = 3 \end{cases}$ $(-1, 1)$

16. $\begin{cases} 3m + n = -6 \\ m - n = -2 \end{cases}$
$(-2, 0)$

17. $\begin{cases} 2x + y = -2 \\ -2x - 3y = -6 \end{cases}$
$(-3, 4)$

18. $\begin{cases} 3x + 4y = 8 \\ 5x - 4y = 24 \end{cases}$
$(4, -1)$

19. $\begin{cases} 4x + 3y = 24 \\ 4x - 3y = -24 \end{cases}$
$(0, 8)$

20. $\begin{cases} 5x - 4y = 8 \\ -5x - 4y = 8 \end{cases}$
$(0, -2)$

In Exercises 21–48, use the addition method to solve each system of equations. If the equations of a system are dependent or if a system is inconsistent, so indicate.

21. $\begin{cases} x + y = 5 \\ x + 2y = 8 \end{cases}$ $(2, 3)$

22. $\begin{cases} x + 2y = 0 \\ x - y = -3 \end{cases}$ $(-2, 1)$

23. $\begin{cases} 2x + y = 4 \\ 2x + 3y = 0 \end{cases}$ $(3, -2)$

24. $\begin{cases} 2x + 5y = -13 \\ 2x - 3y = -5 \end{cases}$
$(-4, -1)$

25. $\begin{cases} 3x - 5y = -29 \\ 3x + 4y = 34 \end{cases}$
$(2, 7)$

26. $\begin{cases} 3x - 5y = 16 \\ 4x + 5y = 33 \end{cases}$ $(7, 1)$

27. $\begin{cases} 2a - 3b = -6 \\ 2a - 3b = 8 \end{cases}$
inconsistent system

28. $\begin{cases} 3a - 4b = 6 \\ 2(2b + 3) = 3a \end{cases}$
dependent equations

29. $\begin{cases} 8x - 4y = 18 \\ 3x - 2y = 8 \end{cases}$ $\left(1, -\tfrac{5}{2}\right)$

30. $\begin{cases} 4x + 6y = 5 \\ 8x - 9y = 3 \end{cases}$ $\left(\tfrac{3}{4}, \tfrac{1}{3}\right)$

31. $\begin{cases} 2x + y = 10 \\ 0.1x + 0.2y = 1.0 \end{cases}$
$\left(\tfrac{10}{3}, \tfrac{10}{3}\right)$

32. $\begin{cases} 0.3x + 0.2y = 0 \\ 2x - 3y = -13 \end{cases}$
$(-2, 3)$

33. $\begin{cases} 2x - y = 16 \\ 0.03x + 0.02y = 0.03 \end{cases}$ $(5, -6)$

34. $\begin{cases} -5y + 2x = 4 \\ -0.02y + 0.03x = 0.04 \end{cases}$ $\left(\tfrac{12}{11}, -\tfrac{4}{11}\right)$

35. $\begin{cases} 6x + 3y = 0 \\ 5y = 2x + 12 \end{cases}$ $(-1, 2)$

36. $\begin{cases} 0 = 4x - 3y \\ 5x = 4y - 2 \end{cases}$ $(6, 8)$

37. $\begin{cases} -2(x + 1) = 3y - 6 \\ 3(y + 2) = 10 - 2x \end{cases}$ dependent equations

38. $\begin{cases} 3x + 2y + 1 = 5 \\ 3(x - 1) = -2y - 4 \end{cases}$ inconsistent system

39. $\begin{cases} 4(x + 1) = 17 - 3(y - 1) \\ 2(x + 2) + 3(y - 1) = 9 \end{cases}$ $(4, 0)$

40. $\begin{cases} 5(x - 1) = 8 - 3(y + 2) \\ 4(x + 2) - 7 = 3(2 - y) \end{cases}$ $(2, -1)$

41. $\begin{cases} \frac{3}{5}s + \frac{4}{5}t = 1 \\ -\frac{1}{4}s + \frac{3}{8}t = 1 \end{cases}$ $(-1, 2)$

42. $\begin{cases} \frac{1}{2}s - \frac{1}{4}t = 1 \\ \frac{1}{3}s + t = 3 \end{cases}$ $(3, 2)$

43. $\begin{cases} \frac{3}{5}x + y = 1 \\ \frac{4}{5}x - y = -1 \end{cases}$ $(0, 1)$

44. $\begin{cases} \frac{1}{2}x + \frac{4}{7}y = -1 \\ 5x - \frac{4}{5}y = -10 \end{cases}$
$(-2, 0)$

45. $\begin{cases} \frac{x}{2} - \frac{y}{3} = -2 \\ \frac{2x-3}{2} + \frac{6y+1}{3} = \frac{17}{6} \end{cases}$
$(-2, 3)$

46. $\begin{cases} \frac{x+2}{4} + \frac{y-1}{3} = \frac{1}{12} \\ \frac{x+4}{5} - \frac{y-2}{2} = \frac{5}{2} \end{cases}$
$(1, -1)$

47. $\begin{cases} \frac{x-3}{2} + \frac{y+5}{3} = \frac{11}{6} \\ \frac{x+3}{3} - \frac{5}{12} = \frac{y+3}{4} \end{cases}$
$(2, 2)$

48. $\begin{cases} \frac{x+2}{3} = \frac{3-y}{2} \\ \frac{x+3}{2} = \frac{2-y}{3} \end{cases}$ $(-5, 5)$

WRITING

49. Why is it usually to your advantage to write the equations of a system in general form before using the addition method to solve it?

50. How would you decide whether to use substitution or addition to solve a system of equations?

51. In this section, we discussed the addition method for solving a system of two equations. Some instructors call it the *elimination method*. Why do you think it would be known by this name?

52. Write a note to the student whose work is shown below, explaining his error.

Solve $\begin{cases} x + y = 1 \\ x - y = 5 \end{cases}$.

$\begin{array}{r} + \quad x + y = 1 \\ x - y = 5 \\ \hline 2x \quad\quad = 6 \end{array}$

$\dfrac{2x}{2} = \dfrac{6}{2}$

$\boxed{x = 3}$ —Done—

REVIEW

53. Solve $8(3x - 5) - 12 = 4(2x + 3)$. 4

54. Solve $3y + \dfrac{y + 2}{2} = \dfrac{2(y + 3)}{3} + 16$. 6

55. Simplify $x - x$. 0

56. Simplify $3.2m - 4.4 + 2.1m + 16$. $5.3m - 11.6$

57. Find the area of a triangular-shaped sign with a base of 4 feet and a height of 3.75 feet. 7.5 ft^2

58. Translate to mathematical symbols: *the product of the sum of x and y and the difference of x and y.*
$(x + y)(x - y)$

59. What is 10 less than x? $x - 10$

60. Factor $6x^2 + 7x - 20$. $(3x - 4)(2x + 5)$

▶ 8.4   Applications of Systems of Equations

In this section, you will learn about

Solving problems using two variables

Introduction   We have previously formed equations involving one variable to solve problems. In this section, we consider ways to solve problems by using two variables.

Solving Problems Using Two Variables

The following steps are helpful when solving problems involving two unknown quantities.

Problem-solving strategy

1. Read the problem several times and *analyze* the facts. Occasionally, a sketch, table, or diagram will help you visualize the facts of the problem.

2. Pick different variables to represent two unknown quantities. *Form* two equations involving each of the two variables. This will give a system of two equations in two variables.

3. *Solve* the system of equations using the most convenient method: graphing, substitution, or addition.

4. *State* the conclusion.

5. *Check* the result in the words of the problem.

EXAMPLE 1

Farming. A farmer raises wheat and soybeans on 215 acres. If he wants to plant 31 more acres in wheat than in soybeans, how many acres of each should he plant?

ANALYZE THE PROBLEM

A farmer plants two fields, one in wheat and one in soybeans. We know that the number of acres of wheat planted plus the number of acres of soybeans planted will equal a total of 215 acres.

FORM TWO EQUATIONS

If w represents the number of acres of wheat and s the number of acres of soybeans to be planted, we can form the two equations.

The number of acres planted in wheat	plus	the number of acres planted in soybeans	is	215 acres.
w	$+$	s	$=$	215

Since the farmer wants to plant 31 more acres in wheat than in soybeans, we have

The number of acres planted in wheat	less	the number of acres planted in soybeans	is	31 acres.
w	$-$	s	$=$	31

SOLVE THE SYSTEM

We can now solve the system

$$\textbf{1.} \quad \begin{cases} w + s = 215 \\ w - s = 31 \end{cases} \textbf{2.}$$

using the addition method.

$$\begin{array}{r} w + s = 215 \\ \underline{w - s = 31} \\ 2w = 246 \end{array}$$

$$w = 123 \quad \text{Divide both sides by 2.}$$

To find s, we substitute 123 for w in Equation 1.

$$w + s = 215$$
$$123 + s = 215 \quad \text{Substitute.}$$
$$s = 92 \quad \text{Subtract 123 from both sides.}$$

STATE THE CONCLUSION

The farmer should plant 123 acres of wheat and 92 acres of soybeans.

CHECK THE RESULT

The total acreage planted is $123 + 92$, or 215 acres. The area planted in wheat is 31 acres greater than that planted in soybeans, because $123 - 92 = 31$. The answers check. ∎

EXAMPLE 2

Lawn care. An installer of underground irrigation systems wants to cut a 20-foot length of plastic tubing into two pieces. The longer piece is to be 2 feet longer than twice the shorter piece. Find the length of each piece.

FIGURE 8-11

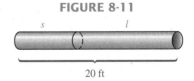

20 ft

ANALYZE THE PROBLEM

Refer to Figure 8-11, which shows the pipe.

FORM TWO EQUATIONS

We can let s represent the length of the shorter piece and l the length of the longer piece. Then we can form the two equations.

The length of the shorter piece	plus	the length of the longer piece	is	20 feet.
s	$+$	l	$=$	20

Since the longer piece is 2 feet longer than twice the shorter piece, we have

The length of the longer piece	is	2	times	the length of the shorter piece	plus	2 feet.
l	$=$	2	$\cdot$	s	$+$	2

SOLVE THE SYSTEM   We can use the substitution method to solve the system.

1. $\begin{cases} s + l = 20 \\ l = 2s + 2 \end{cases}$
2.

$$s + (2s + 2) = 20 \quad \text{Substitute } 2s + 2 \text{ for } l \text{ in Equation 1.}$$
$$3s + 2 = 20 \quad \text{Combine like terms.}$$
$$3s = 18 \quad \text{Subtract 2 from both sides.}$$
$$s = 6 \quad \text{Divide both sides by 3.}$$

The shorter piece should be 6 feet long. To find the length of the longer piece, we substitute 6 for s in Equation 2 and find l.

$$l = 2s + 2$$
$$= 2(6) + 2 \quad \text{Substitute.}$$
$$= 12 + 2 \quad \text{Simplify.}$$
$$l = 14$$

STATE THE CONCLUSION   The longer piece should be 14 feet long, and the shorter piece 6 feet long.

CHECK THE RESULT   The sum of 6 and 14 is 20, and 14 is 2 more than twice 6. The answers check.   ■

EXAMPLE 3

Gardening. Tom has 150 feet of fencing to enclose a rectangular garden. If the garden's length is to be 5 feet less than 3 times its width, find the area of the garden.

FIGURE 8-12

ANALYZE THE PROBLEM   To find the area of a rectangle, we need to know its length and width.

FORM TWO EQUATIONS   We can let l represent the length of the garden and w its width, as shown in Figure 8-12. Since the perimeter of a rectangle is two lengths plus two widths, we can form the two equations.

2	times	the length of the garden	plus	2	times	the width of the garden	is	150 feet.
2	$\cdot$	l	$+$	2	$\cdot$	w	$=$	150

Since the length is 5 feet less than 3 times the width,

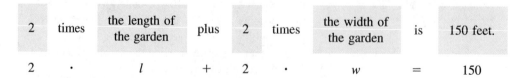

The length of the garden	is	3	times	the width of the garden	minus	5 feet.
l	$=$	3	$\cdot$	w	$-$	5

SOLVE THE SYSTEM   We can use the substitution method to solve this system.

1. $\begin{cases} 2l + 2w = 150 \\ l = 3w - 5 \end{cases}$
2.

$2(3w - 5) + 2w = 150$ Substitute $3w - 5$ for l in Equation 1.

$6w - 10 + 2w = 150$ Remove parentheses.

$8w - 10 = 150$ Combine like terms.

$8w = 160$ Add 10 to both sides.

$w = 20$ Divide both sides by 8.

The width of the garden is 20 feet. To find the length, we substitute 20 for w in Equation 2 and simplify.

$l = 3w - 5$

$= 3(20) - 5$ Substitute.

$= 60 - 5$

$l = 55$

Now we find the area of the rectangle with dimensions 55 feet by 20 feet.

$A = l \cdot w$ The formula for the area of a rectangle.

$= 55 \cdot 20$ Substitute 55 for l and 20 for w.

$A = 1,100$

STATE THE CONCLUSION   The garden covers an area of 1,100 square feet.

CHECK THE RESULT   Because the dimensions of the garden are 55 feet by 20 feet, the perimeter is

$P = 2l + 2w$

$= 2(55) + 2(20)$ Substitute for l and w.

$= 110 + 40$

$P = 150$

It is also true that 55 feet is 5 feet less than 3 times 20 feet. The answers check. ∎

EXAMPLE 4

Manufacturing. The setup cost of a machine that mills brass plates is $750. After setup, it costs $0.25 to mill each plate. Management is considering the purchase of a larger machine that can produce the same plate at a cost of $0.20 per plate. If the setup cost of the larger machine is $1,200, how many plates would the company have to produce to make the purchase worthwhile?

ANALYZE THE PROBLEM   We need to find the number of plates (called the **break point**) that will cost equal amounts to produce on either machine.

FORM TWO EQUATIONS   We can let c represent the cost of milling p plates. If we call the machine currently being used machine 1, and the new, larger one machine 2, we can form the two equations.

The cost of making p plates on machine 1	is	the setup cost of machine 1	plus	the cost per plate on machine 1	times	the number of plates p to be made.
c	=	750	+	0.25	·	p

The cost of making p plates on machine 2	is	the setup cost of machine 2	plus	the cost per plate on machine 2	times	the number of plates p to be made.
c	=	1,200	+	0.20	·	p

SOLVE THE SYSTEM   Since the costs are equal, we can use the substitution method to solve the system

$$
\textbf{1.} \quad \begin{cases} c = 750 + 0.25p \\ c = 1,200 + 0.20p \end{cases}
$$
$$\textbf{2.}$$

$750 + 0.25p = 1,200 + 0.20p$	Substitute $750 + 0.25p$ for c in the second equation.
$0.25p = 450 + 0.20p$	Subtract 750 from both sides.
$0.05p = 450$	Subtract $0.20p$ from both sides.
$p = 9,000$	Divide both sides by 0.05.

STATE THE CONCLUSION   If 9,000 plates are milled, the cost will be the same on either machine. If more than 9,000 plates are milled, the cost will be cheaper on the larger machine, because it mills the plates less expensively than the smaller machine.

CHECK THE RESULT   We check the solution by substituting 9,000 for p in Equations 1 and 2 and verifying that 3,000 is the value of c in both cases.

If we graph the two equations, we can illustrate the break point. (See Figure 8-13.)

FIGURE 8-13

Machine 1
$$c = 750 + 0.25p$$

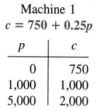

p	c
0	750
1,000	1,000
5,000	2,000

Machine 2
$$c = 1,200 + 0.20p$$

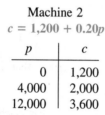

p	c
0	1,200
4,000	2,000
12,000	3,600

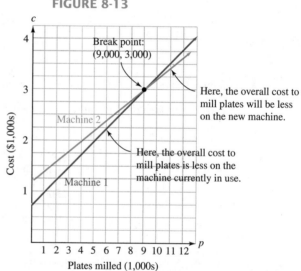

EXAMPLE 5

White-collar crime. Federal investigators discovered that a company secretly moved $150,000 out of the country to avoid paying corporate income tax on it. Some of the money was invested in a Swiss bank account that paid 8% interest annually. The remainder was deposited in a Cayman Islands account, paying 7% annual interest. The investigation also revealed that the combined interest earned the first year was $11,500. How much money was invested in each account?

ANALYZE THE PROBLEM   We are told that an unknown part of the $150,000 was invested at an annual rate of 8% and the rest at 7%. Together, the accounts earned $11,500 in interest.

FORM TWO EQUATIONS   We can let x represent the amount invested in the Swiss bank account and y represent the amount invested in the Cayman Islands account. Because the total investment was $150,000, we have

The amount invested in the Swiss account	+	the amount invested in the Cayman Is. account	is	$150,000.
x	+	y	=	150,000

Since the annual income on x dollars invested at 8% is $0.08x$, the income on y dollars invested at 7% is $0.07y$, and the combined income is $11,500, we have

The income on the 8% investment	+	the income on the 7% investment	is	$11,500.
$0.08x$	+	$0.07y$	=	$11,500$

The resulting system is

1. $\begin{cases} x + y = 150{,}000 \\ 0.08x + 0.07y = 11{,}500 \end{cases}$
2.

SOLVE THE SYSTEM   To solve the system, we use the addition method to eliminate x.

$$
\begin{aligned}
-8x - 8y &= -1{,}200{,}000 \quad &\text{Multiply both sides of Equation 1 by } -8.\\
\underline{8x + 7y} &= \underline{1{,}150{,}000} \quad &\text{Multiply both sides of Equation 2 by 100.}\\
-y &= -50{,}000\\
y &= 50{,}000 \quad &\text{Multiply both sides by } -1.
\end{aligned}
$$

To find x, we substitute 50,000 for y in Equation 1 and simplify.

$$
\begin{aligned}
x + y &= 150{,}000\\
x + 50{,}000 &= 150{,}000 \quad &\text{Substitute.}\\
x &= 100{,}000 \quad &\text{Subtract 50,000 from both sides.}
\end{aligned}
$$

STATE THE CONCLUSION $100,000 was invested in the Swiss bank account, and $50,000 was invested in the Cayman Islands account.

CHECK THE RESULT

$$
\begin{aligned}
\$100{,}000 + \$50{,}000 &= \$150{,}000 \quad &\text{The two investments total } \$150{,}000.\\
0.08(\$100{,}000) &= \$8{,}000 \quad &\text{The Swiss bank account earned } \$8{,}000.\\
0.07(\$50{,}000) &= \$3{,}500 \quad &\text{The Cayman Islands account earned } \$3{,}500.
\end{aligned}
$$

The combined interest is $8,000 + $3,500 = $11,500. The answers check. ∎

EXAMPLE 6

Boating. A boat traveled 30 kilometers downstream in 3 hours and made the return trip in 5 hours. Find the speed of the boat in still water.

ANALYZE THE PROBLEM   Traveling downstream, the speed of the boat will be faster than it would be in still water. Traveling upstream, the speed of the boat will be less than it would be in still water.

FORM TWO EQUATIONS   We can let s represent the speed of the boat in still water and c the speed of the current. Then the rate of the boat going downstream is $s + c$, and its rate going upstream is $s - c$. We can organize the information as shown in Figure 8-14.

FIGURE 8-14

	Rate	·	Time	=	Distance
Downstream	$s + c$		3		30
Upstream	$s - c$		5		30

Because $d = r \cdot t$, the information in the table gives two equations in two variables.

$$
\begin{cases} 3(s + c) = 30 \\ 5(s - c) = 30 \end{cases}
$$

After removing parentheses, we have

1. $\begin{cases} 3s + 3c = 30 \\ 5s - 5c = 30 \end{cases}$
2.

SOLVE THE SYSTEM   To solve this system by addition, we multiply Equation 1 by 5, multiply Equation 2 by 3, add the equations, and solve for s.

$$15s + 15c = 150$$
$$\underline{15s - 15c = 90}$$
$$30s = 240$$

$$s = 8 \qquad \text{Divide both sides by 30.}$$

STATE THE CONCLUSION   The speed of the boat in still water is 8 kilometers per hour.

CHECK THE RESULT   We leave the check to the reader.   ∎

EXAMPLE 7

Medical technology. A laboratory technician has one batch of antiseptic that is 40% alcohol and a second batch that is 60% alcohol. She would like to make 8 liters of solution that is 55% alcohol. How many liters of each batch should she use?

ANALYZE THE PROBLEM   Some 60% alcohol solution must be added to some 40% alcohol solution to make a 55% alcohol solution.

FORM TWO EQUATIONS   We can let x represent the number of liters to be used from batch 1 and y the number of liters to be used from batch 2. We then organize the information as shown in Figure 8-15.

FIGURE 8-15

	Fractional part that is alcohol	Number of liters of solution	Number of liters of alcohol
Batch 1	0.40	x	0.40x
Batch 2	0.60	y	0.60y
Mixture	0.55	8	0.55(8)

One equation comes from information in this column. | Another equation comes from information in this column.

The information in Figure 8-15 provides two equations.

1. $\begin{cases} x + y = 8 & \text{The number of liters of batch 1 plus the number of liters} \\ & \text{of batch 2 equals the total number of liters in the mixture.} \\ 0.40x + 0.60y = 0.55(8) & \text{The amount of alcohol in batch 1 plus the amount of alcohol} \\ & \text{in batch 2 equals the amount of alcohol in the mixture.} \end{cases}$
2.

SOLVE THE SYSTEM   We can use addition to solve this system.

$$-40x - 40y = -320 \qquad \text{Multiply both sides of Equation 1 by } -40.$$
$$\underline{40x + 60y = 440} \qquad \text{Multiply both sides of Equation 2 by 100.}$$
$$20y = 120$$

$$y = 6 \qquad \text{Divide both sides by 20.}$$

To find x, we substitute 6 for y in Equation 1 and simplify.

$$x + y = 8$$
$$x + 6 = 8 \quad \text{Substitute.}$$
$$x = 2 \quad \text{Subtract 6 from both sides.}$$

STATE THE CONCLUSION   The technician should use 2 liters of the 40% solution and 6 liters of the 60% solution.

CHECK THE RESULT   The check is left to the reader.   ∎

STUDY SET

Section 8.4

VOCABULARY

In Exercises 1–4, fill in the blanks to make the statements true.

1. A ___variable___ is a letter that stands for a number.
2. An ___equation___ is a statement indicating that two quantities are equal.
3. $\begin{cases} a + b = 20 \\ a = 2b + 4 \end{cases}$ is a ___system___ of linear equations.
4. A ___solution___ of a system of linear equations satisfies both equations simultaneously.

CONCEPTS

5. For each case in Illustration 1, write an algebraic expression that represents the speed of the canoe in miles per hour if its speed in still water is x miles per hour. $x - c, x + c$

ILLUSTRATION 1

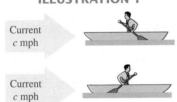

Current
c mph

Current
c mph

6. See Illustration 2.
   a. If the contents of the two test tubes are poured into a third test tube, how much solution will the third test tube contain? $(x + y)$ mL
   b. Which is the best estimate of the concentration of the solution in the third test tube— 25%, 35%, or 45% acid solution? 35%

ILLUSTRATION 2

x mL

y mL

30% acid solution   40% acid solution

7. Use the information in the table to answer the questions about two investments.

	Principal ·	Rate ·	Time =	Interest
City Bank	x	5%	1 yr	$0.05x$
USA Savings	y	11%	1 yr	$0.11y$

   a. How much money was deposited in the USA Savings account? $\$y$

   b. What interest rate did the City Bank account earn? 5%
   c. Complete the table.

8. Use the information in the table to answer the questions about a plane flying in windy conditions.

	Rate ·	Time =	Distance
With	$x + y$	3 hr	450 mi
Against	$x - y$	5 hr	450 mi

   a. For how long did the plane fly against the wind? 5 hr
   b. At what rate did the plane travel when flying with the wind? $(x + y)$ mph
   c. Write two equations that could be used to solve for x and y. $3(x + y) = 450, 5(x - y) = 450$

9. a. If a problem contains two unknowns, and if two variables are used to represent them, how many equations must be written to find the unknowns? two
   b. Name three methods that can be used to solve a system of linear equations. graphing, substitution, addition

10. Put the steps of the five-step problem-solving strategy listed below in the correct order.

    State the conclusion   Form two equations
    Analyze the problem   Check the result
    Solve the system

    analyze, form, solve, state, check

NOTATION

In Exercises 11–14, write a formula that relates the given quantities.

11. length, width, area of a rectangle $A = lw$
12. length, width, perimeter of a rectangle $P = 2l + 2w$
13. rate, time, distance traveled $d = rt$
14. principal, rate, time, interest earned $I = Prt$

In Exercises 15–16, translate each verbal model into mathematical symbols. Use variables to represent any unknowns.

15. 2 · length of pool $+$ 2 · width of pool is 90 yards.

    $2l + 2w = 90$

16. $\boxed{\$6}$ $\cdot$ $\boxed{\begin{array}{c}\text{number}\\\text{of adults}\end{array}}$ $+$ $\boxed{\$2}$ $\cdot$ $\boxed{\begin{array}{c}\text{number}\\\text{of children}\end{array}}$ is $\boxed{\$26}$.

$6a + 2c = 26$

PRACTICE

In Exercises 17–20, use two equations in two variables to find the integers.

17. One integer is twice another. Their sum is 96.   32, 64

18. The sum of two integers is 38. Their difference is 12.   13, 25

19. Three times one integer plus another integer is 29. The first integer plus twice the second is 18.   8, 5

20. Twice one integer plus another integer is 21. The first integer plus 3 times the second is 33.   6, 9

APPLICATIONS

In Exercises 21–46, use two equations in two variables to solve each problem.

21. TREE TRIMMING
When fully extended, the arm on the tree service truck shown in Illustration 3 is 51 feet long. If the upper part of the arm is 7 feet shorter than the lower part, how long is each part of the arm?
22 ft, 29 ft

ILLUSTRATION 3

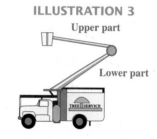

Upper part

Lower part

TREE SERVICE

22. TV PROGRAMMING The producer of a 30-minute TV documentary about World War I divided it into two parts. Four times as much program time was devoted to the causes of the war as to the outcome. How long is each part of the documentary?
causes: 24 min; outcome: 6 min

23. EXECUTIVE BRANCH The salaries of the president and vice president of the United States total $371,500 a year. If the president makes $28,500 more than the vice president, find each of their salaries.
president: $200,000; vice president: $171,500

24. CAUSES OF DEATH In 1993, the number of Americans dying from cancer was 6 times the number who died from accidents. If the number of deaths from these two causes totaled 630,000, how many Americans died from each cause?
90,000 in accidents; 540,000 from cancer

25. BUYING PAINTING SUPPLIES Two partial receipts for paint supplies are shown in Illustration 4. How much does each gallon of paint and each brush cost? $15, $5

ILLUSTRATION 4

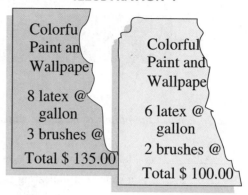

Colorful Paint and Wallpaper
8 latex @ gallon
3 brushes @
Total $ 135.00

Colorful Paint and Wallpaper
6 latex @ gallon
2 brushes @
Total $ 100.00

26. WEDDING PICTURES A photographer sells the two wedding picture packages shown in Illustration 5.

ILLUSTRATION 5

Package #1	Package #2
1 10 x 14	1 10 x 14
10 8 x 10	5 8 x 10
color photos	color photos
Cost $239.50	Cost $134.50

How much does a 10×14 photo cost? An 8×10 photo? $29.50, $21

27. BUYING TICKETS If receipts for the movie advertised in Illustration 6 were $720 for an audience of 190 people, how many senior citizens attended?   40

ILLUSTRATION 6

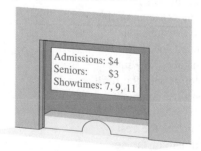

Admissions: $4
Seniors: $3
Showtimes: 7, 9, 11

28. SELLING ICE CREAM At a store, ice cream cones cost $0.90 and sundaes cost $1.65. One day the receipts for a total of 148 cones and sundaes were $180.45. How many cones were sold?   85

29. MARINE CORPS The Marine Corps War Memorial in Arlington, Virginia, portrays the raising of the U.S. flag on Iwo Jima during World War II. Find the two angles shown in Illustration 7 if the measure of one of the angles is 15° less than twice the other.
65°, 115°

ILLUSTRATION 7

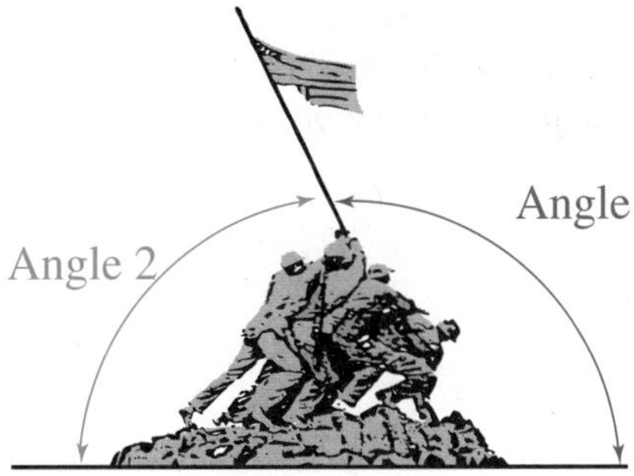

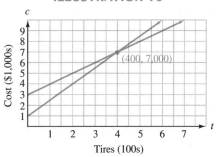

Angle 1

Angle 2

30. PHYSICAL THERAPY To rehabilitate her knee, an athlete does leg extensions. Her goal is to regain a full 90° range of motion in this exercise. Use the information in Illustration 8 to determine her current range of motion in degrees. 72°

ILLUSTRATION 8

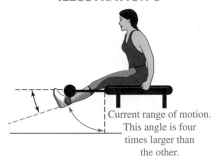

Current range of motion.
This angle is four
times larger than
the other.

31. THEATER SCREEN At an IMAX theater, the giant rectangular movie screen has a width 26 feet less than its length. If its perimeter is 332 feet, find the area of the screen. 6,720 ft²

32. GEOMETRY A 50-meter path surrounds the rectangular garden shown in Illustration 9. The width of the garden is two-thirds its length. Find its area. 150 m²

ILLUSTRATION 9

33. MAKING TIRES A company has two molds to form tires. One mold has a setup cost of $1,000, and the other has a setup cost of $3,000. The cost to make each tire with the first mold is $15, and the cost to make each tire with the second mold is $10.
a. Find the break point. 400 tires
b. Check your result by graphing both equations on the coordinate system in Illustration 10.

ILLUSTRATION 10

Cost ($1,000s)

(400, 7,000)

Tires (100s)

c. If a production run of 500 tires is planned, determine which mold should be used.
the second mold

34. CHOOSING A FURNACE A high-efficiency 90+ furnace can be purchased for $2,250 and costs an average of $412 per year to operate in Rockford, Illinois. An 80+ furnace can be purchased for only $1,715, but it costs $466 per year to operate.
a. Find the break point. about 9.9 yr
b. If you intended to live in a Rockford house for 7 years, which furnace would you choose? 80+

35. STUDENT LOANS A college used a $5,000 gift from an alumnus to make two student loans. The first was at 5% annual interest to a nursing student. The second was at 7% to a business major. If the college collected $310 in interest the first year, how much was loaned to each student?
nursing: $2,000; business: $3,000

36. FINANCIAL PLANNING In investing $6,000 of a couple's money, a financial planner put some of it into a savings account paying 6% annual interest. The rest was invested in a riskier mini-mall development plan paying 12% annually. The combined interest earned for the first year was $540. How much money was invested at each rate? 6%: $3,000; 12%: $3,000

37. GULF STREAM The Gulf Stream is a warm ocean current of the North Atlantic Ocean that flows northward, as shown in Illustration 11. Heading north with the Gulf Stream, a cruise ship traveled 300 miles in 10 hours. Against the current, it took 15 hours to make the return trip. Find the speed of the current.
5 mph

ILLUSTRATION 11

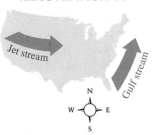

38. JET STREAM The jet stream is a strong wind current that flows across the United States, as shown in Illustration 11. Flying with the jet stream, a plane flew 3,000 miles in 5 hours. Against the same wind, the trip took 6 hours. Find the airspeed of the plane (the speed in still air).   550 mph

39. AVIATION An airplane can fly downwind a distance of 600 miles in 2 hours. However, the return trip against the same wind takes 3 hours. Find the speed of the wind.   50 mph

40. BOATING A boat can travel 24 miles downstream in 2 hours and can make the return trip in 3 hours. Find the speed of the boat in still water.   10 mph

Example 7 ✓
41. MARINE BIOLOGY A marine biologist wants to set up an aquarium containing 3% salt water. He has two tanks on hand that contain 6% and 2% salt water. How much water from each tank must he use to fill a 16-liter aquarium with a 3% saltwater mixture?
4 L 6% salt water, 12 L 2% salt water

Example 7.
42. COMMEMORATIVE COINS A foundry has been commissioned to make souvenir coins. The coins are to be made from an alloy that is 40% silver. The foundry has on hand two alloys, one with 50% silver content and one with a 25% silver content. How many kilograms of each alloy should be used to make 20 kilograms of the 40% silver alloy?
12 kg 50% alloy, 8 kg 25% alloy

43. MIXING NUTS A merchant wants to mix peanuts with cashews, as shown in Illustration 12, to get 48 pounds of mixed nuts that will be sold at $4 per pound. How many pounds of each should the merchant use?   32 lb peanuts, 16 lb cashews

ILLUSTRATION 12

44. COFFEE SALES A coffee supply store waits until the orders for its special coffee blend reach 100 pounds before making up a batch. Coffee selling for $8.75 a pound is blended with coffee selling for $3.75 a pound to make a product that sells for $6.35 a pound. How much of each type of coffee should be used to make the blend that will fill the orders?
52 lb $8.75, 48 lb $3.75

45. MARKDOWN A set of golf clubs has been marked down 40% to a sale price of $384. Let r represent the retail price and d the discount. Then use the following equations to find the original retail price. $640

Retail price	−	discount	=	sale price

Discount	=	discount rate	·	retail price

46. MARKUP A stereo system retailing at $565.50 has been marked up 45% from wholesale. Let w represent the wholesale cost and m the markup. Then use the following equations to find the wholesale cost. $390

w m 565.50

Wholesale cost	+	markup	=	retail price

Markup	=	markup rate	·	wholesale cost

m $.45$ w

WRITING

47. When solving a problem using two variables, why isn't one equation sufficient to find the two unknown quantities?

48. Describe an everyday situation in which you might need to make a mixture.

REVIEW

In Exercises 49–52, graph each inequality.

49. $x < 4$

50. $x \geq -3$

51. $-1 < x \leq 2$

52. $-2 \leq x \leq 0$

In Exercises 53–56, solve each equation.

53. $x^2 - 4 = 0$ $-2, 2$

54. $x^2 - 4x = 0$ $0, 4$

55. $x^2 - 4x + 4 = 0$ $2, 2$

56. $x^2 - 4x - 2 = 0$ $2 \pm \sqrt{6}$

▶ 8.5

Graphing Linear Inequalities

In this section, you will learn about

> Solutions of linear inequalities ■ Graphing linear inequalities
> ■ An application of linear inequalities

Introduction We have seen that the solutions of a linear *equation* in x and y can be expressed as ordered pairs (x, y) and that when graphed, the ordered pairs form a line. In this section, we consider linear *inequalities*. Solutions of linear inequalities can also be expressed as ordered pairs and graphed.

Solutions of Linear Inequalities

A linear equation in x and y is an equation that can be written in the form $Ax + By = C$. A **linear inequality** in x and y is an inequality that can be written in one of four forms:

$$Ax + By > C, \qquad Ax + By < C, \qquad Ax + By \geq C, \qquad \text{or} \qquad Ax + By \leq C$$

where A, B, and C represent real numbers and A and B are not both zero. Some examples of linear inequalities are

$$2x - y > -3, \qquad y < 3, \qquad x + 4y \geq 6, \qquad \text{and} \qquad x \leq -2$$

As with linear equations, an ordered pair (x, y) is a solution of an inequality in x and y if a true statement results when the variables in the inequality are replaced by the coordinates of the ordered pair.

EXAMPLE 1

Verifying a solution. Determine whether each ordered pair is a solution of $x - y \leq 5$. Then graph each solution: **a.** $(4, 2)$, **b.** $(0, -6)$, and **c.** $(1, -4)$.

Solution In each case, we substitute the x-coordinate for x and the y-coordinate for y in the inequality $x - y \leq 5$. If the ordered pair is a solution, a true statement will be obtained.

a. For $(4, 2)$:

$$x - y \leq 5 \qquad \text{The original inequality.}$$
$$4 - 2 \leq 5 \qquad \text{Replace } x \text{ with 4 and } y \text{ with 2.}$$
$$2 \leq 5 \qquad \text{True.}$$

Because $2 \leq 5$ is true, $(4, 2)$ is a solution of the inequality, and we graph it in Figure 8-16.

b. For $(0, -6)$:

$$x - y \leq 5 \qquad \text{The original inequality.}$$
$$0 - (-6) \leq 5 \qquad \text{Replace } x \text{ with 0 and } y \text{ with } -6.$$
$$6 \leq 5 \qquad \text{False.}$$

Because $6 \leq 5$ is false, $(0, -6)$ is not a solution.

c. For $(1, -4)$:

$$x - y \leq 5 \qquad \text{The original inequality.}$$
$$1 - (-4) \leq 5 \qquad \text{Replace } x \text{ with 1 and } y \text{ with } -4.$$
$$5 \leq 5 \qquad \text{True.}$$

Because $5 \leq 5$ is true, $(1, -4)$ is a solution, and we graph it in Figure 8-16.

FIGURE 8-16

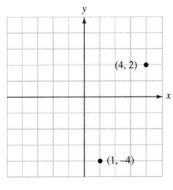

SELF CHECK   Using the inequality in Example 1, determine
whether each ordered pair is a solution. If it is,
graph that solution on the coordinate system in   *Answers:* **a.** not a solution,
Figure 8-16. **a.** $(8, 2)$, **b.** $(4, -1)$,   **b.** solution, **c.** solution,
c. $(-2, 4)$, and **d.** $(-3, -5)$.   **d.** solution ■

The graph in Figure 8-16 contains some solutions of the inequality $x - y \leq 5$. Intuition tells us that there are many more ordered pairs (x, y) such that $x - y$ is less than
or equal to 5. How then do we get a complete graph of the solutions of $x - y \leq 5$? We
address this question in the following discussion.

Graphing Linear Inequalities

The graph of $x - y = 5$ is a line consisting of the points whose coordinates satisfy the
equation. The graph of the inequality $x - y \leq 5$ is not a line, but an area bounded by
a line, called a **half-plane.** The half-plane consists of the points whose coordinates satisfy the inequality.

E X A M P L E   2   **Graphing a linear inequality.** Graph $x - y \leq 5$.

Solution   Since the inequality symbol $\leq$ includes an equals sign, the graph of $x - y \leq 5$ includes
the graph of $x - y = 5$. So we begin by graphing the equation $x - y = 5$, as shown in
Figure 8-17(a).

FIGURE 8-17

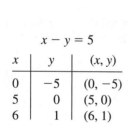

$x - y = 5$

x	y	(x, y)
0	-5	$(0, -5)$
5	0	$(5, 0)$
6	1	$(6, 1)$

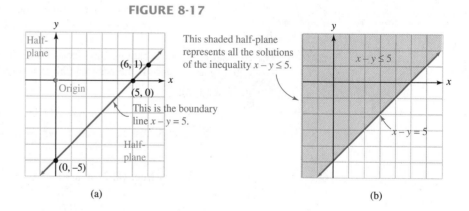

(a)   (b)

Since the inequality $x - y \leq 5$ allows $x - y$ to be less than 5, the coordinates of
points other than those shown on the line in Figure 8-17(a) satisfy the inequality. For
example, the coordinates of the origin $(0, 0)$ satisfy the inequality. We can verify this
by letting x and y be zero in the given inequality:

$x - y \leq 5$

$\mathbf{0} - \mathbf{0} \leq 5$ Substitute 0 for x and 0 for y.

$0 \leq 5$ True.

Because $0 \leq 5$, the coordinates of the origin satisfy the original inequality. In fact,
the coordinates of every point on the *same side* of the line as the origin satisfy the inequality. The graph of $x - y \leq 5$ is the half-plane that is shaded in Figure 8-17(b).
Since the **boundary line** $x - y = 5$ is included, we draw it with a solid line. ■

EXAMPLE 3	**Graphing a linear inequality.** Graph $2(x - 3) - (x - y) \geq -3$.

Solution   We begin by simplifying the inequality as follows:

$$2(x - 3) - (x - y) \geq -3$$
$$2x - 6 - x + y \geq -3 \quad \text{Remove the parentheses.}$$
$$x - 6 + y \geq -3 \quad \text{Combine like terms.}$$
$$x + y \geq 3 \quad \text{Add 6 to both sides.}$$

To graph the inequality $x + y \geq 3$, we graph the boundary line whose equation is $x + y = 3$. Since the graph of $x + y \geq 3$ includes the line $x + y = 3$, we draw the boundary with a solid line. Note that it divides the coordinate plane into two half-planes. See Figure 8-18(a).

To decide which half-plane to shade, we substitute the coordinates of some point that lies on one side of the boundary line into the inequality. If we use the origin $(0, 0)$ for the **test point,** we have

$$x + y \geq 3$$
$$\mathbf{0} + \mathbf{0} \geq 3 \quad \text{Substitute 0 for } x \text{ and 0 for } y.$$
$$0 \geq 3 \quad \text{False.}$$

Since $0 \geq 3$ is a false statement, the origin is not in the graph. In fact, the coordinates of *every* point on the origin's side of the boundary line will not satisfy the inequality. However, every point on the other side of the boundary line will satisfy the inequality. We shade that half-plane. The graph of $x + y \geq 3$ is the half-plane that appears in color in Figure 8-18(b).

FIGURE 8-18

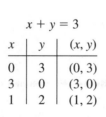

$x + y = 3$

x	y	(x, y)
0	3	$(0, 3)$
3	0	$(3, 0)$
1	2	$(1, 2)$

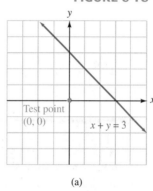

(a)

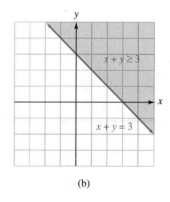

(b)

SELF CHECK	Graph $3(x - 1) - (2x + y) \leq -5$. *Answer:*

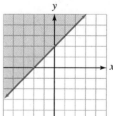

EXAMPLE 4	**Graphing a linear inequality.** Graph $y > 2x$.

Solution   To find the boundary line, we graph $y = 2x$. Since the symbol $>$ does not include an equals sign, the points on the graph of $y = 2x$ are not part of the graph of $y > 2x$. We draw the boundary line as a broken line to show this, as in Figure 8-19(a).

To determine which half-plane to shade, we substitute the coordinates of some point that lies on one side of the boundary line into $y > 2x$. Since the origin is on the boundary, we cannot use it as a test point. Point $T(2, 0)$, for example, is below the boundary line. See Figure 8-19(a). To see whether point $T(2, 0)$ satisfies $y > 2x$, we substitute 2 for x and 0 for y in the inequality.

$$y > 2x$$
$$0 > 2(2) \quad \text{Substitute 2 for } x \text{ and 0 for } y.$$
$$0 > 4 \quad \text{False.}$$

Since $0 > 4$ is a false statement, the coordinates of point T do not satisfy the inequality, and point T is not on the side of the line we wish to shade. Instead, we shade the other side of the boundary line. The graph of the solution set of $y > 2x$ is shown in Figure 8-19(b).

FIGURE 8-19

$y = 2x$

x	y	(x, y)
0	0	$(0, 0)$
-1	-2	$(-1, -2)$
1	2	$(1, 2)$

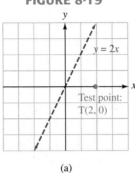

(a)

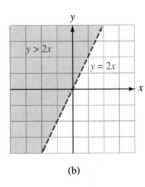

(b)

SELF CHECK   Graph $y < 3x$.

Answer:

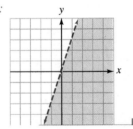

EXAMPLE 5   **Graphing a linear inequality.** Graph $x + 2y < 6$.

Solution   We find the boundary by graphing the equation $x + 2y = 6$. We draw the boundary as a broken line to show that it is not part of the solution. We then choose a test point not on the boundary and see whether its coordinates satisfy $x + 2y < 6$. The origin is a convenient choice.

$$x + 2y < 6$$
$$0 + 2(0) < 6 \quad \text{Substitute 0 for } x \text{ and 0 for } y.$$
$$0 < 6 \quad \text{True.}$$

Since $0 < 6$ is a true statement, we shade the side of the line that includes the origin. The graph is shown in Figure 8-20.

SELF CHECK   Graph $2x - y < 4$.

Answer:

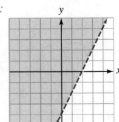

FIGURE 8-20

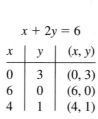

$$x + 2y = 6$$

x	y	(x, y)
0	3	(0, 3)
6	0	(6, 0)
4	1	(4, 1)

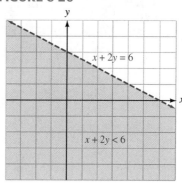

E X A M P L E   6

Graphing a linear inequality. Graph $y \geq 0$.

Solution   We find the boundary by graphing the equation $y = 0$. We draw the boundary as a solid line to show that it is part of the solution. We then choose a test point not on the boundary and see whether its coordinates satisfy $y \geq 0$. The point $T(0, 1)$ is a convenient choice.

$$y \geq 0$$
$$\mathbf{1 \geq 0} \quad \text{Substitute 1 for } y.$$

Since $1 \geq 0$ is a true statement, we shade the side of the line that includes point T. The graph is shown in Figure 8-21.

FIGURE 8-21

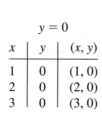

$$y = 0$$

x	y	(x, y)
1	0	(1, 0)
2	0	(2, 0)
3	0	(3, 0)

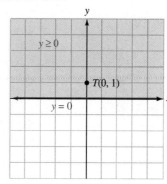

SELF CHECK   Graph $x \geq 2$.

Answer:

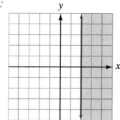

The following is a summary of the procedure for graphing linear inequalities.

Graphing linear inequalities in two variables

1. Graph the boundary line of the region. If the inequality allows the possibility of equality (the symbol is either ≤ or ≥), draw the boundary line as a solid line. If equality is not allowed (< or >), draw the boundary line as a broken line.

2. Pick a test point that is on one side of the boundary line. (Use the origin if possible.) Replace x and y in the inequality with the coordinates of that point. If the inequality is satisfied, shade the side that contains that point. If the inequality is not satisfied, shade the other side of the boundary.

An Application of Linear Inequalities

EXAMPLE 7

Earning money. Carlos has two part-time jobs, one paying $5 per hour and another paying $6 per hour. He must earn at least $120 per week to pay his expenses while attending college. Write an inequality that shows the various ways he can schedule his time to achieve his goal.

Solution   If we let x represent the number of hours Carlos works on the first job and y the number of hours he works on the second job, we have

The hourly rate on the first job	times	the hours worked on the first job	plus	the hourly rate on the second job	times	the hours worked on the second job	is at least	$120.
$5	·	x	+	6	·	y	≥	$120

The graph of the inequality $5x + 6y \geq 120$ is shown in Figure 8-22. Any point in the shaded region indicates a possible way Carlos can schedule his time and earn $120 or more per week. For example, if he works 20 hours on the first job and 10 hours on the second job, he will earn

$$\$5(20) + \$6(10) = \$100 + \$60$$
$$= \$160$$

Since Carlos cannot work a negative number of hours, a graph showing negative values of x or y would have no meaning.   ∎

FIGURE 8-22

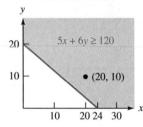

$5x + 6y \geq 120$

• (20, 10)

STUDY SET

Section 8.5

VOCABULARY

In Exercises 1–4, fill in the blanks to make the statements true.

1. $2x - y \leq 4$ is a linear ___inequality___ in x and y.

2. The symbol ≤ means ___is less than___ or ___equal to___.

3. In the graph in Illustration 1, the line $2x - y = 4$ is the ___boundary___.

4. In Illustration 1, the line $2x - y = 4$ divides the rectangular coordinate system into two ___half-planes___.

ILLUSTRATION 1

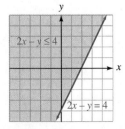

$2x - y \leq 4$

$2x - y = 4$

CONCEPTS

5. Tell whether each ordered pair is a solution of
$5x - 3y \geq 0$.

   a. $(1, 1)$ yes   **b.** $(-2, -3)$ no

   c. $(0, 0)$ yes   **d.** $\left(\dfrac{1}{5}, \dfrac{4}{3}\right)$ no

6. Tell whether each ordered pair is a solution of
$x + 4y < -1$.

   a. $(3, 1)$ no   **b.** $(-2, 0)$ yes

   c. $(-0.5, 0.2)$ no   **d.** $\left(-2, \dfrac{1}{4}\right)$ no

7. Tell whether the graph of each linear inequality includes the boundary line.

   a. $y > -x$ no   **b.** $5x - 3y \leq -2$ yes

8. If a false statement results when the coordinates of a test point are substituted into a linear inequality, which half-plane should be shaded to represent the solution of the inequality?

   the half-plane opposite that in which the test point lies

9. A linear inequality has been graphed in Illustration 2. Tell whether each point satisfies the inequality.

   ILLUSTRATION 2

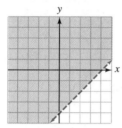

   a. $(1, -3)$ no
   b. $(-2, -1)$ yes
   c. $(2, 3)$ yes
   d. $(3, -4)$ no

10. A linear inequality has been graphed in Illustration 3. Tell whether each point satisfies the inequality.

   ILLUSTRATION 3

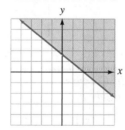

   a. $(2, 1)$ yes
   b. $(-2, -4)$ no
   c. $(4, -2)$ no
   d. $(-3, 4)$ yes

11. The boundary for the graph of a linear inequality is shown in Illustration 4. Why can't the origin be used as a test point to decide which side to shade?

   ILLUSTRATION 4

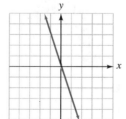

   The test point must be on one side of the boundary.

12. To decide how many pallets (x) and barrels (y) a delivery truck can hold, a dispatcher refers to the loading sheet in Illustration 5. Can a truck make a delivery of 4 pallets and 10 barrels in one trip?

   no

ILLUSTRATION 5

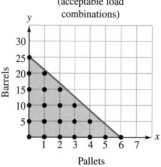

Truck Loading Sheet
(acceptable load combinations)

PRACTICE

In Exercises 13–20, complete the graph by shading the correct side of the boundary.

13. $y \leq x + 2$

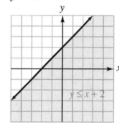

14. $y > x - 3$

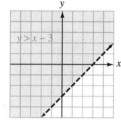

15. $y > 2x - 4$

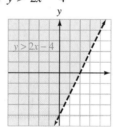

16. $y \leq -x + 1$

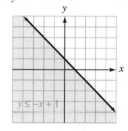

17. $x - 2y \geq 4$

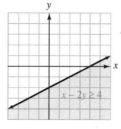

18. $3x + 2y > 12$

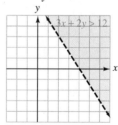

19. $y \leq 4x$

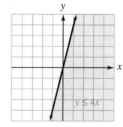

20. $y + 2x < 0$

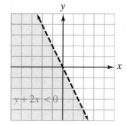

In Exercises 21–40, graph each inequality.

21. $y \geq 3 - x$

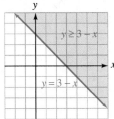

22. $y < 2 - x$

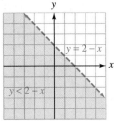

23. $y < 2 - 3x$

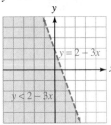

24. $y \geq 5 - 2x$ $y = -2x + 5$

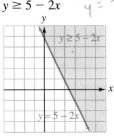

25. $y \geq 2x$

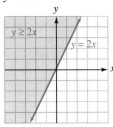

26. $y < 3x$

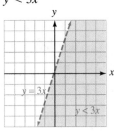

27. $2y - x < 8$

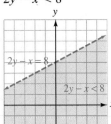

28. $y + 9x \geq 3$

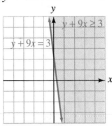

29. $y - x \geq 0$

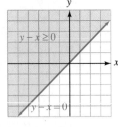

30. $y + x < 0$

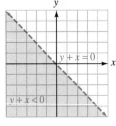

31. $2x + y > 2$

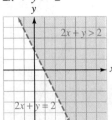

32. $3x - 2y > 6$

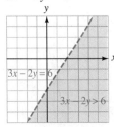

33. $3x - 4y > 12$

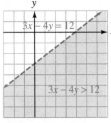

34. $4x + 3y \leq 12$

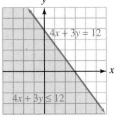

35. $5x + 4y \geq 20$

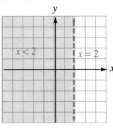

36. $7x - 2y < 21$

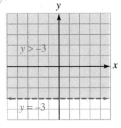

37. $x < 2$

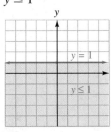

38. $y > -3$

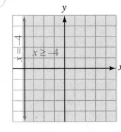

39. $y \leq 1$

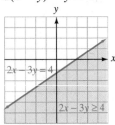

40. $x \geq -4$

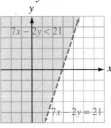

In Exercises 41–44, simplify each inequality and then graph it.

41. $3(x + y) + x < 6$

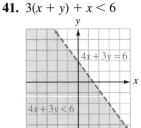

42. $2(x - y) - y \geq 4$

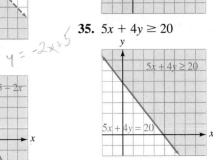

43. $4x - 3(x + 2y) \geq -6y$

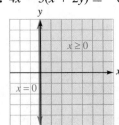

44. $3y + 2(x + y) < 5y$

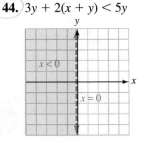

APPLICATIONS

45. NATO In March of 1999, NATO aircraft and cruise missiles targeted Serbian military forces that were south of the 44th parallel of north latitude in Yugoslavia, Montenegro, and Kosovo. See Illustration 6. Shade the geographic area that NATO was trying to rid of Serbian forces.

ILLUSTRATION 6

Based on data from *Los Angeles Times* (March 24, 1999)

46. U.S. HISTORY When he ran for president in 1844, the campaign slogan of James K. Polk was "54-40 or fight!" It meant that Polk was willing to fight Great Britain for the possession of the Oregon Territory north to the 54°40′ parallel, as shown in Illustration 7. In 1846, Polk accepted a compromise to establish the 49th parallel as the permanent boundary of the United States. Shade the area of land that Polk conceded to the British.

ILLUSTRATION 7

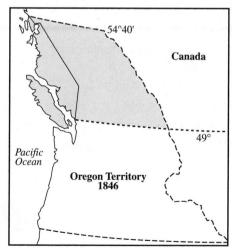

In Exercises 47–52, write an inequality and graph it for nonnegative values of x and y. Then give three ordered pairs that satisfy the inequality.

47. PRODUCTION PLANNING It costs a bakery $3 to make a cake and $4 to make a pie. Production costs cannot exceed $120 per day. Use Illustration 8 to graph an inequality that shows the possible combinations of cakes (x) and pies (y) that can be made. $(10, 10)$, $(20, 10)$, $(10, 20)$

ILLUSTRATION 8

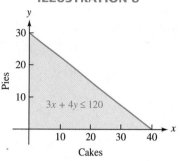

48. HIRING BABYSITTERS Mary has a choice of two babysitters. Sitter 1 charges $6 per hour, and sitter 2 charges $7 per hour. If Mary can afford no more than $42 per week for sitters, use Illustration 9 to graph an inequality that shows the possible ways that she can hire sitter 1 (x) and sitter 2 (y). $(2, 2)$, $(4, 2)$, $(3, 3)$

ILLUSTRATION 9

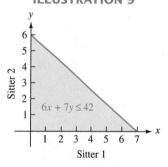

49. INVENTORY A clothing store advertises that it maintains an inventory of at least $4,400 worth of men's jackets. If a leather jacket costs $100 and a nylon jacket costs $88, use Illustration 10 to graph an inequality that shows the possible ways that leather jackets (x) and nylon jackets (y) can be stocked. $(50, 50)$, $(30, 40)$, $(40, 40)$

ILLUSTRATION 10

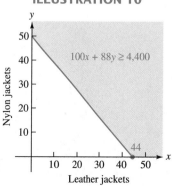

50. MAKING SPORTING GOODS To keep up with demand, a sporting goods manufacturer allocates at least 2,400 units of production time per day to make baseballs and footballs. If it takes 20 units of time to make a baseball and 30 units of time to make a football, use Illustration 11 to graph an inequality that shows the possible ways to schedule the production time to make baseballs (x) and footballs (y).
(60, 80), (100, 60), (120, 40)

ILLUSTRATION 11

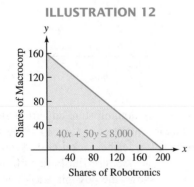

51. INVESTING IN STOCKS Robert has up to $8,000 to invest in two companies. If stock in Robotronics sells for $40 per share and stock in Macrocorp sells for $50 per share, use Illustration 12 to graph an inequality that shows the possible ways that he can buy shares of Robotronics (x) and Macrocorp (y).
(80, 40), (80, 80), (120, 40)

ILLUSTRATION 12

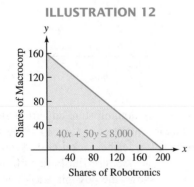

52. BUYING BASEBALL TICKETS Tickets to the Rockford Rox baseball games cost $6 for reserved

seats and $4 for general admission. If nightly receipts must average at least $10,200 to meet expenses, use Illustration 13 to graph an inequality that shows the possible ways that the Rox can sell reserved seats (x) and general admission tickets (y).
(1,200, 2,000), (1,600, 1,200), (800, 2,400)

ILLUSTRATION 13

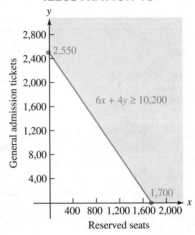

WRITING

53. Explain how to find the boundary for the graph of an inequality.

54. Explain how to decide which side of the boundary line to shade when graphing an inequality.

REVIEW

55. Let $g(x) = 3x^2 - 4x + 3$. Find $g(2)$.   7

56. Solve $2(x - 4) \leq -12$. $x \leq -2$

57. Factor $x^3 + 27$. $(x + 3)(x^2 - 3x + 9)$

58. Factor $9p - 9q + mp - mq$. $(p - q)(9 + m)$

59. Write a formula relating distance, rate, and time. $d = rt$

60. What is the slope of the line $2x - 3y = 2$? $\frac{2}{3}$

61. Solve $A = P + Prt$ for t. $t = \frac{A - P}{Pr}$

62. What is the sum of the measures of the three angles of any triangle?   180°

▶**8.6**   **Solving Systems of Linear Inequalities**

In this section, you will learn about

    Systems of linear inequalities ■ An application of systems of linear inequalities

---

Introduction   We have previously solved systems of linear *equations* by the graphing method. The solution of such a system is the point of intersection of the straight lines.

We now consider how to solve systems of linear *inequalities* graphically. When the solution of a linear inequality is graphed, the result is a half-plane. Therefore, we would expect to find the graphical solution of a system of inequalities by looking for the intersection, or "overlap," of shaded half-planes.

Systems of Linear Inequalities

To solve the system

$$\begin{cases} x + y \geq 1 \\ x - y \geq 1 \end{cases}$$

we first graph each inequality. For instructional purposes, we will initially graph each inequality on a separate set of axes, although in practice we will draw them on the same axes.

The graph of $x + y \geq 1$ includes the graph of the equation $x + y = 1$ and all points above it. Because the boundary line is included, we draw it with a solid line, as shown in Figure 8-23(a).

The graph of $x - y \geq 1$ includes the graph of the equation $x - y = 1$ and all points below it. Because the boundary line is included here also, it is drawn with a solid line, as shown in Figure 8-23(b).

FIGURE 8-23

$x + y = 1$

x	y	(x, y)
0	1	$(0, 1)$
1	0	$(1, 0)$
2	-1	$(2, -1)$

$x - y = 1$

x	y	(x, y)
0	-1	$(0, -1)$
1	0	$(1, 0)$
2	1	$(2, 1)$

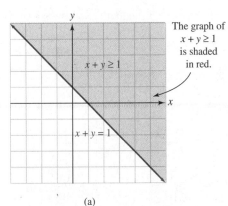

The graph of $x + y \geq 1$ is shaded in red.

(a)

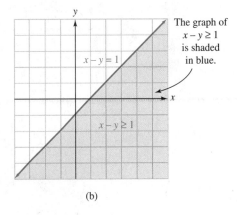

The graph of $x - y \geq 1$ is shaded in blue.

(b)

In Figure 8-24 on the next page, we show the result when the inequalities $x + y \geq 1$ and $x - y \geq 1$ are graphed one at a time on the same coordinate axes. The area that is shaded twice represents the set of simultaneous solutions of the given system of inequalities. Any point in the doubly shaded region has coordinates that satisfy both inequalities of the system.

To see whether this is true, we can pick a point, such as point A, that lies in the doubly shaded region and show that its coordinates satisfy both inequalities. Because point A has coordinates $(4, 1)$, we have

$$x + y \geq 1 \quad \text{and} \quad x - y \geq 1$$
$$4 + 1 \geq 1 \qquad\qquad 4 - 1 \geq 1$$
$$5 \geq 1 \qquad\qquad\quad 3 \geq 1$$

FIGURE 8-24

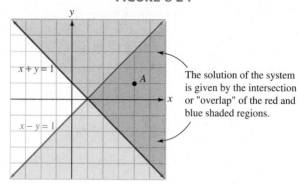

Since the coordinates of point A satisfy each inequality, point A is a solution of the system. If we pick a point that is not in the doubly shaded region, its coordinates will fail to satisfy at least one of the inequalities.

In general, to solve systems of linear inequalities, we will follow these steps.

Solving systems of inequalities

1. Graph each inequality in the system on the same coordinate axes.
2. Find the region that is common to every graph.
3. Pick a test point from the region to verify the solution.

E X A M P L E   1   **Solving systems of inequalities.** Graph the solution of $\begin{cases} 2x + y < 4 \\ -2x + y > 2 \end{cases}$.

Solution   First, we graph each inequality on one set of axes, as shown in Figure 8-25.

FIGURE 8-25

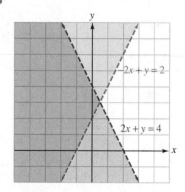

$2x + y = 4$			$-2x + y = 2$		
x	y	(x, y)	x	y	(x, y)
0	4	$(0, 4)$	-1	0	$(-1, 0)$
2	0	$(2, 0)$	0	2	$(0, 2)$
1	2	$(1, 2)$	2	6	$(2, 6)$

We note that

- The graph of $2x + y < 4$ includes all points below the line $2x + y = 4$. Since the boundary is not included, we draw it as a broken line.

- The graph of $-2x + y > 2$ includes all points above the line $-2x + y = 2$. Since the boundary is not included, we also draw it as a broken line.

The area that is shaded twice (the region in purple) is the solution of the given system of inequalities. Any point in the doubly shaded region has coordinates that will satisfy both inequalities of the system.

Pick a point in the doubly shaded region and show that it satisfies both inequalities.

SELF CHECK   Graph the solution of $\begin{cases} x + 3y < 3 \\ -x + 3y > 3 \end{cases}$.   *Answer:*

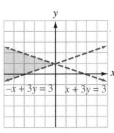

EXAMPLE 2   **Solving systems of inequalities.** Graph the solution of $\begin{cases} x \le 2 \\ y > 3 \end{cases}$.

Solution   We graph each inequality on one set of axes, as shown in Figure 8-26.

FIGURE 8-26

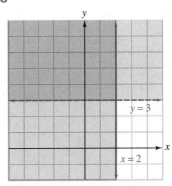

$x = 2$				$y = 3$		
x	y	(x, y)		x	y	(x, y)
2	0	$(2, 0)$		0	3	$(0, 3)$
2	2	$(2, 2)$		1	3	$(1, 3)$
2	4	$(2, 4)$		4	3	$(4, 3)$

We note that

- The graph of $x \le 2$ includes all points to the left of the line $x = 2$. Since the boundary line is included, we draw it as a solid line.

- The graph $y > 3$ includes all points above the line $y = 3$. Since the boundary is not included, we draw it as a broken line.

The area that is shaded twice is the solution of the given system of inequalities. Any point in the doubly shaded region (purple) has coordinates that will satisfy both inequalities of the system. Pick a point in the doubly shaded region and show that this is true.

SELF CHECK   Graph the solution of $\begin{cases} y \le 1 \\ x > 2 \end{cases}$.   *Answer:*

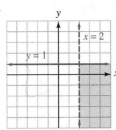

EXAMPLE 3   **Solving systems of inequalities.** Graph the solution of $\begin{cases} y < 3x - 1 \\ y \ge 3x + 1 \end{cases}$.

Solution   We graph each inequality as shown in Figure 8-27 and make the following observations:

- The graph of $y < 3x - 1$ includes all points below the broken line $y = 3x - 1$.

- The graph of $y \ge 3x + 1$ includes all points on and above the solid line $y = 3x + 1$.

FIGURE 8-27

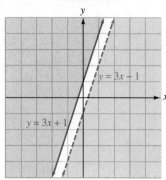

Because the graphs of these inequalities do not intersect, the solution set is empty. There are no solutions.

SELF CHECK   Graph the solution of $\begin{cases} y \geq -\frac{1}{2}x + 1 \\ y < -\frac{1}{2}x - 1 \end{cases}$.   *Answer:* no solutions

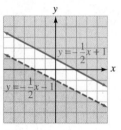

EXAMPLE 4   **Solving systems of inequalities.** Graph the solution of $\begin{cases} x \geq 0 \\ y \geq 0 \\ x + 2y \leq 6 \end{cases}$.

Solution   We graph each inequality as shown in Figure 8-28 and make the following observations:

- The graph of $x \geq 0$ includes all points on the y-axis and to the right.
- The graph of $y \geq 0$ includes all points on the x-axis and above.
- The graph of $x + 2y \leq 6$ includes all points on the line $x + 2y = 6$ and below.

The solution is the region that is shaded three times. This includes triangle OPQ and the triangular region it encloses.

FIGURE 8-28

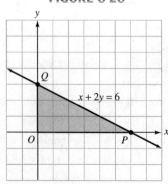

SELF CHECK Graph the solution of $\begin{cases} x \leq 1 \\ y \leq 2 \\ 2x - y \leq 4 \end{cases}$. *Answer:*

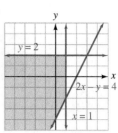

An Application of Systems of Linear Inequalities

EXAMPLE 5

Landscaping. A homeowner budgets from $300 to $600 for trees and bushes to landscape his yard. After shopping around, he finds that good trees cost $150 and mature bushes cost $75. What combinations of trees and bushes can he afford to buy?

ANALYZE THE PROBLEM The homeowner wants to spend *at least* $300 but *not more than* $600 for trees and bushes.

FORM TWO INEQUALITIES We can let x represent the number of trees purchased and y the number of bushes purchased. We then form the following system of inequalities:

The cost of a tree	times	the number of trees purchased	plus	the cost of a bush	times	the number of bushes purchased	should at least be	$300.
$150	·	x	+	$75	·	y	≥	$300

The cost of a tree	times	the number of trees purchased	plus	the cost of a bush	times	the number of bushes purchased	should not be more than	$600.
$150	·	x	+	$75	·	y	≤	$600

SOLVE THE SYSTEM We graph the system

$$\begin{cases} 150x + 75y \geq 300 \\ 150x + 75y \leq 600 \end{cases}$$

FIGURE 8-29

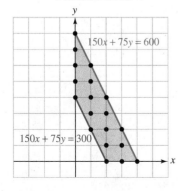

as shown in Figure 8-29. The coordinates of each point shown in the graph give a possible combination of the number of trees (x) and the number of bushes (y) that can be purchased. These possibilities are

$(0, 4), (0, 5), (0, 6), (0, 7), (0, 8)$

$(1, 2), (1, 3), (1, 4), (1, 5), (1, 6)$

$(2, 0), (2, 1), (2, 2), (2, 3), (2, 4)$

$(3, 0), (3, 1), (3, 2), (4, 0)$

Only these points can be used, because the homeowner cannot buy a portion of a tree or a bush.

STUDY SET

Section 8.6

VOCABULARY

In Exercises 1–4, fill in the blanks to make the statements true.

1. $\begin{cases} x + y > 2 \\ x + y < 4 \end{cases}$ This is a system of linear __inequalities__.

2. The ___solution___ of a system of linear inequalities is all the ordered pairs that make all inequalities of the system true at the same time.

3. Any point in the ___doubly shaded___ region of the graph of the solution of a system of two linear inequalities has coordinates that satisfy both inequalities of the system.

4. To graph a linear inequality such as $x + y > 2$, first graph the boundary. Then pick a test ___point___ to determine which half-plane to shade.

CONCEPTS

5. In Illustration 1, the solution of linear inequality 1 is shaded in red, and the solution of linear inequality 2 is shaded in blue. Tell whether a true or a false statement results when the coordinates of the given point are substituted into the given inequality.

   ILLUSTRATION 1

   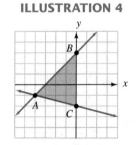

   a. *A*, inequality 1   true
   b. *A*, inequality 2   false
   c. *B*, inequality 1   false
   d. *B*, inequality 2   true
   e. *C*, inequality 1   true
   f. *C*, inequality 2   true

6. Match each equation, inequality, or system with the graph of its solution.

   a. $x + y = 2$ ii
   b. $x + y \geq 2$ iii
   c. $\begin{cases} x + y = 2 \\ x - y = 2 \end{cases}$ iv
   d. $\begin{cases} x + y \geq 2 \\ x - y \leq 2 \end{cases}$ i

   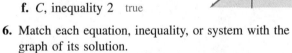

   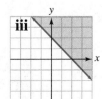 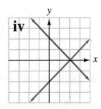

7. The graph of the solution of a system of linear inequalities is shown in Illustration 2. Tell whether each point is a part of the solution set.

   a. $(4, -2)$ yes
   b. $(1, 3)$ no
   c. the origin   no

   ILLUSTRATION 2

   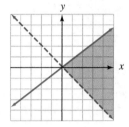

8. Use a system of inequalities to describe the shaded region in Illustration 3.

   $\begin{cases} x \leq -2 \\ y > 2 \end{cases}$

   ILLUSTRATION 3

   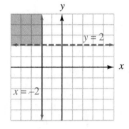

NOTATION

9. Fill in the blank to make the statement true: The graph of the solution of a system of linear inequalities shown in Illustration 4 can be described as the triangle ___ABC___ and the triangular region it encloses.

   ILLUSTRATION 4

10. Represent each phrase using either $>$, $<$, $\geq$, or $\leq$.

   a. is not more than $\leq$
   b. must be at least $\geq$
   c. should not surpass $\leq$
   d. cannot go below $\geq$

PRACTICE

In Exercises 11–34, graph the solution set of each system of inequalities, when possible.

11. $\begin{cases} x + 2y \leq 3 \\ 2x - y \geq 1 \end{cases}$

   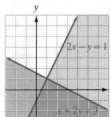

12. $\begin{cases} 2x + y \geq 3 \\ x - 2y \leq -1 \end{cases}$

   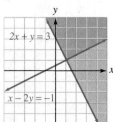

13. $\begin{cases} x + y < -1 \\ x - y > -1 \end{cases}$

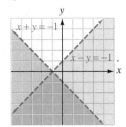

14. $\begin{cases} x + y > 2 \\ x - y < -2 \end{cases}$

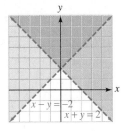

23. $\begin{cases} 3x + 4y \geq -7 \\ 2x - 3y \geq 1 \end{cases}$

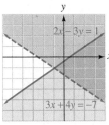

24. $\begin{cases} 3x + y \leq 1 \\ 4x - y \geq -8 \end{cases}$

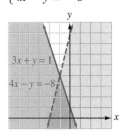

15. $\begin{cases} x \geq 2 \\ y \leq 3 \end{cases}$

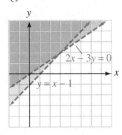

16. $\begin{cases} x \geq -1 \\ y > -2 \end{cases}$

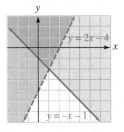

25. $\begin{cases} 2x + y < 7 \\ y > 2(1 - x) \end{cases}$

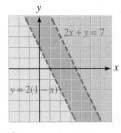

26. $\begin{cases} 2x + y \geq 6 \\ y \leq 2(2x - 3) \end{cases}$

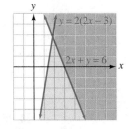

17. $\begin{cases} 2x - 3y \leq 0 \\ y \geq x - 1 \end{cases}$

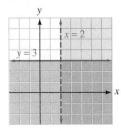

18. $\begin{cases} y > 2x - 4 \\ y \geq -x - 1 \end{cases}$

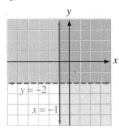

27. $\begin{cases} 2x - 4y > -6 \\ 3x + y \geq 5 \end{cases}$

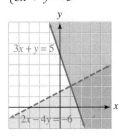

28. $\begin{cases} 2x - 3y < 0 \\ 2x + 3y \geq 12 \end{cases}$

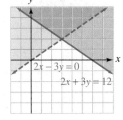

19. $\begin{cases} y < -x + 1 \\ y > -x + 3 \end{cases}$

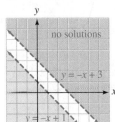

20. $\begin{cases} y > -x + 2 \\ y < -x + 4 \end{cases}$

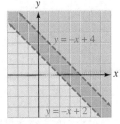

29. $\begin{cases} 3x - y \leq -4 \\ 3y > -2(x + 5) \end{cases}$

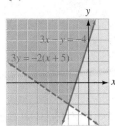

30. $\begin{cases} 3x + y < -2 \\ y > 3(1 - x) \end{cases}$

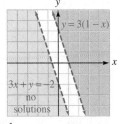

21. $\begin{cases} x > 0 \\ y > 0 \end{cases}$

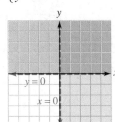

22. $\begin{cases} x \leq 0 \\ y < 0 \end{cases}$

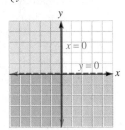

31. $\begin{cases} \dfrac{x}{2} + \dfrac{y}{3} \geq 2 \\ \dfrac{x}{2} - \dfrac{y}{2} < -1 \end{cases}$

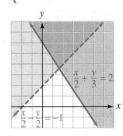

32. $\begin{cases} \dfrac{x}{3} - \dfrac{y}{2} < -3 \\ \dfrac{x}{3} + \dfrac{y}{2} > -1 \end{cases}$

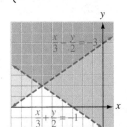

33. $\begin{cases} x \geq 0 \\ y \geq 0 \\ x + y \leq 3 \end{cases}$

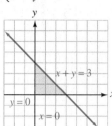

34. $\begin{cases} x - y \leq 6 \\ x + 2y \leq 6 \\ x \geq 0 \end{cases}$

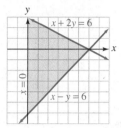

APPLICATIONS

35. BIRDS OF PREY Parts a and b of Illustration 5 show the individual fields of vision for each eye of an owl. In part c, shade the area where the fields of vision overlap—that is, the area that is seen by both eyes.

ILLUSTRATION 5

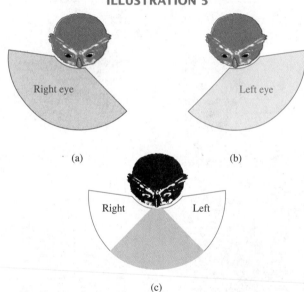

36. EARTH SCIENCE
In Illustration 6, shade the area of the earth's surface that is north of the Tropic of Capricorn and south of the Tropic of Cancer.

ILLUSTRATION 6

In Exercises 37–40, graph each system of inequalities and give two possible solutions.

37. BUYING COMPACT DISCS Melodic Music has compact discs on sale for either $10 or $15. If a customer wants to spend at least $30 but no more than $60 on CDs, use Illustration 7 to graph a system of inequalities showing the possible combinations of $10 CDs (x) and $15 CDs (y) that the customer can buy.
1 $10 CD and 2 $15 CDs; 4 $10 CDs and 1 $15 CD

ILLUSTRATION 7

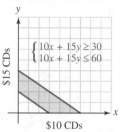

$\begin{cases} 10x + 15y \geq 30 \\ 10x + 15y \leq 60 \end{cases}$

38. BUYING BOATS Dry Boatworks wholesales aluminum boats for $800 and fiberglass boats for $600. Northland Marina wants to make a purchase totaling at least $2,400, but no more than $4,800. Use Illustration 8 to graph a system of inequalities showing the possible combinations of aluminum boats (x) and fiberglass boats (y) that can be ordered.

4 alum. and 1 glass; 1 alum. and 4 glass

ILLUSTRATION 8

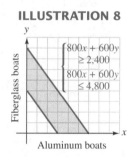

$\begin{cases} 800x + 600y \geq 2,400 \\ 800x + 600y \leq 4,800 \end{cases}$

39. BUYING FURNITURE A distributor wholesales desk chairs for $150 and side chairs for $100. Best Furniture wants its order to total no more than $900; Best also wants to order more side chairs than desk chairs. Use Illustration 9 to graph a system of inequalities showing the possible combinations of desk chairs (x) and side chairs (y) that can be ordered.

2 desk chairs and 4 side chairs; 1 desk chair and 5 side chairs

ILLUSTRATION 9

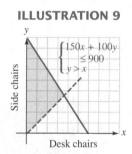

$\begin{cases} 150x + 100y \leq 900 \\ y > x \end{cases}$

40. ORDERING FURNACE EQUIPMENT J. Bolden Heating Company wants to order no more than $2,000 worth of electronic air cleaners and humidifiers from a wholesaler that charges $500 for air cleaners and $200 for humidifiers. If Bolden wants more humidifiers than air cleaners, use Illustration 10 to graph a system of inequalities showing the possible combinations of air cleaners (x) and humidifiers (y) that can be ordered.

1 air cleaner and 2 humidifiers; 2 air cleaners and 3 humidifiers

ILLUSTRATION 10

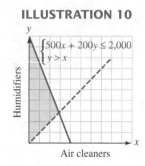

$\begin{cases} 500x + 200y \leq 2,000 \\ y > x \end{cases}$

41. PESTICIDE To eradicate a fruit fly infestation, helicopters sprayed an area of a city that can be described by $y \geq -2x + 1$ (within the city limits). In another two weeks, more spraying was ordered over the area described by $y \geq \frac{1}{4}x - 4$ (within the city limits). In Illustration 11, show the part of the city that was sprayed twice.

ILLUSTRATION 11

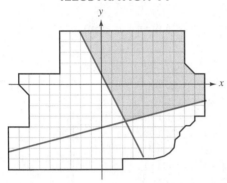

42. REDEVELOPMENT A government agency has declared an area of a city east of First Street, north of Second Avenue, south of Sixth Avenue, and west of Fifth Street as eligible for federal redevelopment funds. See Illustration 12. Describe this area of the city mathematically using a system of four inequalities, if the corner of Central Avenue and Main Street is considered the origin.
$$\begin{cases} x > 1 \\ x < 5 \\ y > 2 \\ y < 6 \end{cases}$$

ILLUSTRATION 12

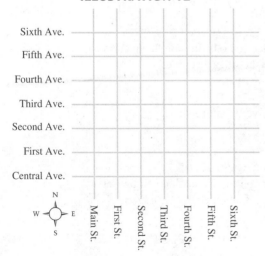

WRITING

43. Explain how to use graphing to solve a system of inequalities.

44. Explain when a system of inequalities will have no solutions.

45. Describe how the graphs of the solutions of these systems are similar and how they differ.
$$\begin{cases} x + y = 4 \\ x - y = 4 \end{cases} \quad \text{and} \quad \begin{cases} x + y \geq 4 \\ x - y \geq 4 \end{cases}$$

46. When a solution of a system of linear inequalities is graphed, what does the shading represent?

REVIEW

In Exercises 47–50, complete each table of values.

47. $y = 2x^2$

x	y
8	128
−2	8

48. $t = -|s + 2|$

s	t
−3	−1
−10	−8

49. $f(x) = 4 + x^3$

Input	Output
0	4
−3	−23

50. $g(x) = 2x - x^2$

x	$g(x)$
5	−15
−5	−35

Systems of Equations and Inequalities

In Chapter 8, we have solved problems that required the use of two variables to represent two unknown quantities. To find the unknowns, we write a pair of equations (or inequalities) called a **system.**

Solving Systems of Equations by Graphing

A system of linear equations can be solved by graphing both equations and locating the point of intersection of the two lines.

1. FOOD SERVICE The two equations in the table give the fees two different catering companies charge a Hollywood studio for on-location meal service.

Caterer	Setup fee	Cost per meal	Equation
Sunshine	$1,000	$4	$y = 4x + 1,000$
Lucy's	$500	$5	$y = 5x + 500$

Complete Illustration 1, using the graphing method to find the break point. That is, find the number of meals and the corresponding fee for which the two caterers will charge the studio the same amount.
(500, 3,000)

ILLUSTRATION 1

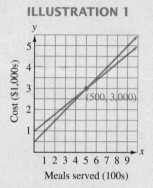

Solving Systems of Equations by Substitution

The substitution method for solving a system of equations works well when a variable in either equation has a coefficient of 1 or -1.

2. Solve by substitution: $\begin{cases} y = 2x - 9 \\ x + 3y = 8 \end{cases}$. (5, 1)

3. Solve by substitution: $\begin{cases} 3x + 4y = -7 \\ 2y - x = -1 \end{cases}$. $(-1, -1)$

Solving Systems of Equations by Addition

With the addition method, equal quantities are added to both sides of an equation to eliminate one of the variables. Then we solve for the other variable.

4. Solve by addition: $\begin{cases} x + y = 1 \\ x - y = 5 \end{cases}$. (3, -2)

5. Solve by addition: $\begin{cases} 2x - 3y = -18 \\ 3x + 2y = -1 \end{cases}$. $(-3, 4)$

Solving Systems of Inequalities

To solve a system of two linear inequalities, we graph the inequalities on the same coordinate axes. The area that is shaded twice represents the set of solutions.

6. This system of inequalities describes the number of $20 shirts ($x$) and $40 pants ($y$) a person can buy if he or she plans to spend not less than $80 but not more than $120. Using Illustration 2, graph the system. Then give three solutions.
   (1, 2), (2, 2), (3, 1) (answers may vary)

$$\begin{cases} 20x + 40y \geq 80 \\ 20x + 40y \leq 120 \end{cases}$$

ILLUSTRATION 2

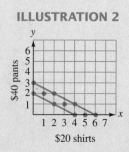

Accent on Teamwork

Section 8.1

TV ratings Illustration 1 shows the Nielsen ratings of the four major television networks from 1995 to 1997. The ratings indicate the number of viewers a network had. Write a report that describes the ratings of each network over this period. Use descriptive phrases such as *sharp decrease* and *moderate gain* to tell how the ratings changed. Explain the significance of the points of intersection of the graphs, and estimate the dates when they occurred. As of 1997, which network do you think had reason to be the most optimistic about its future?

ILLUSTRATION 1

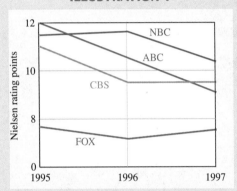

Based on data from Nielsen Media Research, Trade-Line

Section 8.2

Substitutions The word *substitution* is used in several ways. Explain how it is used in the context of a sporting event such as a basketball game. Explain how it is sometimes used when ordering food at a restaurant. Explain what is meant by a *substitute* teacher. Finally, explain how the substitution method is used to solve a system such as

$$\begin{cases} y = -2x - 5 \\ 3x + 5y = -4 \end{cases}$$

What is the difference in the mathematical meaning of the word *substitution* as opposed to the everyday usage of the word?

Section 8.3

Solving systems Solve the system

$$\begin{cases} 2x + y = 4 \\ 2x + 3y = 0 \end{cases}$$

using the graphing method, the substitution method, and the addition method. Which method do you think is the best to use in this case? Why?

Section 8.4

Two unknowns Consider the following problem: A man paid $89 for two white shirts and four pairs of black socks. Find the cost of a white shirt.

If we let x represent the cost of a white shirt and y represent the cost of a pair of black socks, an equation describing the situation is $2x + 4y = 89$. Explain why there is not enough information to solve the problem.

Section 8.5

Matching game Have a student in your group write ten linear inequalities on 3×5 note cards, one inequality per card. Then have him or her graph each of the inequalities on separate cards. Mix up the cards and put all the inequality cards on one side of a table and all the cards with graphs on the other side. Work together to match each inequality with its proper graph.

Section 8.6

Systems of inequalities The points A, B, C, D, E, F, and G, are labeled in the graph of the solution of the system of inequalities in Illustration 2. Tell whether the coordinates of each point make the first inequality (whose solution is shown in red) and the second inequality (whose solution is shown in blue) true or false. Use a table of the following form to keep track of your results.

Point	Coordinates	1st inequality	2nd inequality
A			

ILLUSTRATION 2

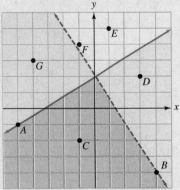

Section 8.1

Solving Systems of Equations by Graphing

CONCEPTS

An ordered pair that satisfies both equations simultaneously is a *solution* of the system.

REVIEW EXERCISES

1. Tell whether the ordered pair is a solution of the system.

a. $(2, -3)$, $\begin{cases} 3x - 2y = 12 \\ 2x + 3y = -5 \end{cases}$ yes

b. $\left(\dfrac{7}{2}, -\dfrac{2}{3}\right)$, $\begin{cases} 4x - 6y = 18 \\ \dfrac{x}{3} + \dfrac{y}{2} = \dfrac{5}{6} \end{cases}$ yes

2. INJURY COMPARISON Illustration 1 shows the number of skiing and snowboarding injuries nationally for the years 1993–1997. If the number of injuries continued to occur at the 1996–1997 rates, estimate when they would be the same for skiing and snowboarding. About how many injuries would that be?
2000; 60,000 per year

ILLUSTRATION 1

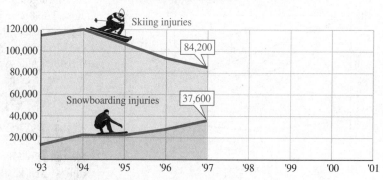

Based on data from *Los Angeles Times* (March 23, 1999)

To *solve a system graphically:*
1. Carefully graph each equation.
2. If the lines intersect, the coordinates of the point of intersection give the solution of the system.
3. Check the solution in the original equations.

When a system of equations has a solution, it is a *consistent* system. Systems with no solutions are *inconsistent*.

When the graphs of two equations in a system are different lines, the equations are *independent* equations.

The equations of a system with infinitely many solutions are *dependent*.

3. Use the graphing method to solve each system.

a. $\begin{cases} x + y = 7 \\ 2x - y = 5 \end{cases}$

b. $\begin{cases} y = -\dfrac{x}{3} \\ 2x + y = 5 \end{cases}$

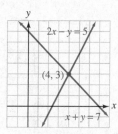

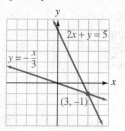

c. $\begin{cases} 3x + 6y = 6 \\ x + 2y = 2 \end{cases}$

d. $\begin{cases} 6x + 3y = 12 \\ 2x + y = 2 \end{cases}$

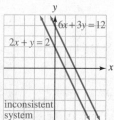

dependent equations

inconsistent system

Section 8.2

Solving Systems of Equations by Substitution

To solve a system of equations in x and y by the *substitution method*:

1. Solve one of the equations for either x or y.
2. Substitute the resulting expression for the variable in step 1 into the other equation, and solve the equation.
3. Find the value of the other variable by substituting the solution found in step 2 in any equation containing x and y.
4. Check the solution in the original equations.

4. Use the substitution method to solve each system.

a. $\begin{cases} x = y \\ 5x - 4y = 3 \end{cases}$ (3, 3)

b. $\begin{cases} y = 15 - 3x \\ 7y + 3x = 15 \end{cases}$ (5, 0)

c. $\begin{cases} 0.2x + 0.2y = 0.6 \\ 3x = 2 - y \end{cases}$ $\left(-\frac{1}{2}, \frac{7}{2}\right)$

d. $\begin{cases} 6(r + 2) = s - 1 \\ r - 5s = -7 \end{cases}$ (−2, 1)

e. $\begin{cases} 9x + 3y = 5 \\ 3x + y = \dfrac{5}{3} \end{cases}$ dependent equations

f. $\begin{cases} \dfrac{x}{6} + \dfrac{y}{10} = 3 \\ \dfrac{5x}{16} - \dfrac{3y}{16} = \dfrac{15}{8} \end{cases}$ (12, 10)

5. In solving a system using the substitution method, suppose you obtain the result of $8 = 9$.

a. How many solutions does the system have? no solutions

b. Describe the graph of the system. two parallel lines

c. What term is used to describe the system? inconsistent system

Section 8.3

Solving Systems of Equations by Addition

To solve a system of equations using the *addition method*:

1. Write each equation in $Ax + By = C$ form.
2. Multiply one or both equations by nonzero quantities to make the coefficients of x (or y) opposites.
3. Add the equations to eliminate the term involving x (or y).
4. Solve the equation resulting from step 3.
5. Find the value of the other variable by substituting the value of the variable found in step 4 into any equation containing both variables.
6. Check the solution in the original equations.

6. Solve each system using the addition method.

a. $\begin{cases} 2x + y = 1 \\ 5x - y = 20 \end{cases}$ (3, −5)

b. $\begin{cases} x + 8y = 7 \\ x - 4y = 1 \end{cases}$ $\left(3, \frac{1}{2}\right)$

c. $\begin{cases} 5a + b = 2 \\ 3a + 2b = 11 \end{cases}$ (−1, 7)

d. $\begin{cases} 11x + 3y = 27 \\ 8x + 4y = 36 \end{cases}$ (0, 9)

e. $\begin{cases} 9x + 3y = 15 \\ 3x = 5 - y \end{cases}$ dependent equations

f. $\begin{cases} \dfrac{x}{3} + \dfrac{y + 2}{2} = 1 \\ \dfrac{x + 8}{8} + \dfrac{y - 3}{3} = 0 \end{cases}$ (0, 0)

g. $\begin{cases} 0.02x + 0.05y = 0 \\ 0.3x - 0.2y = -1.9 \end{cases}$ (−5, 2)

h. $\begin{cases} -\dfrac{1}{4}x = 1 - \dfrac{2}{3}y \\ 6(x - 3y) + 2y = 5 \end{cases}$ inconsistent system

7. For each system, tell which method, substitution or addition, would be easier to use to solve the system and why.

a. $\begin{cases} 6x + 2y = 5 \\ 3x - 3y = -4 \end{cases}$

Addition; no variables have a coefficient of 1 or −1.

b. $\begin{cases} x = 5 - 7y \\ 3x - 3y = -4 \end{cases}$

Substitution; equation 1 is solved for x.

Section 8.4

In this section, we considered ways to solve problems by using *two* variables.

To solve problems involving two unknown quantities:

1. *Analyze* the facts of the problem. Make a table or diagram if necessary.
2. Pick different variables to represent two unknown quantities. *Form* two equations involving the variables.
3. *Solve* the system of equations.
4. *State* the conclusion.
5. *Check* the results.

The *break point* of a linear system is the point of intersection of the graph.

Applications of Systems of Equations

In Exercises 8–16, use two equations in two variables to solve each problem.

8. CAUSES OF DEATH In 1993, the number of Americans dying from heart disease was 5 times more than the number dying from a stroke. If the total number of deaths from these causes was 900,000, how many deaths were attributed to each? stroke: 150,000; heart disease: 750,000

9. PAINTING EQUIPMENT When fully extended, the ladder shown in Illustration 2 is 35 feet in length. If the extension is 7 feet shorter than the base, how long is each part of the ladder? base: 21 ft; extension: 14 ft

ILLUSTRATION 2

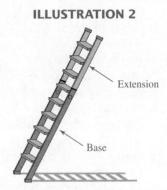

Extension

Base

10. CRASH INVESTIGATION In an effort to protect evidence, investigators used 420 yards of yellow "Police Line—Do Not Cross" tape to seal off a large rectangular-shaped area around an airplane crash site. How much area will the investigators have to search if the width of the rectangle is three-fourths of the length? 10,800 yd^2

11. CELEBRITY ENDORSEMENT A company selling a home juicing machine is contemplating hiring either an athlete or an actor to serve as a spokesperson for a product. The terms of each contract would be as follows:

Celebrity	Base pay	Commission per item sold
Athlete	$30,000	$5
Actor	$20,000	$10

a. For each celebrity, write an equation giving the money (y) the celebrity would earn if x juicers were sold. $y = 5x + 30,000$, $y = 10x + 20,000$

b. For what number of juicers would the athlete and the actor earn the same amount? 2,000

c. Using Illustration 3, graph the equations from part a. The company expects to sell over 3,000 juicers. Which celebrity would cost the company the least money to serve as a spokesperson? the athlete

ILLUSTRATION 3

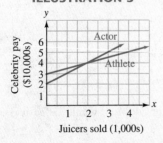

Juicers sold (1,000s)

12. CANDY OUTLET STORE A merchant wants to mix gummy worms worth $3 per pound and gummy bears worth $1.50 per pound to make 30 pounds of a mixture worth $2.10 per pound. How many pounds of each type of candy should he use? 12 lb worms, 18 lb bears

13. BOATING It takes a motorboat 4 hours to travel 56 miles down a river, and 3 hours longer to make the return trip. Find the speed of the current. 3 mph

14. SHOPPING Packages containing two bottles of contact lens cleaner and three bottles of soaking solution cost $29.40, and packages containing three bottles of cleaner and two bottles of soaking solution cost $28.60. Find the cost of a bottle of cleaner and a bottle of soaking solution. $5.40, $6.20

15. INVESTING Carlos invested part of $3,000 in a 10% certificate account and the rest in a 6% passbook account. The total annual interest from both accounts is $270. How much did he invest at 6%? $750

16. ANTIFREEZE How much of a 40% antifreeze solution must a mechanic mix with a 70% antifreeze solution if he needs 20 gallons of a 50% antifreeze solution? $13\frac{1}{3}$ gal 40%, $6\frac{2}{3}$ gal 70%

Section 8.5 — Graphing Linear Inequalities

An ordered pair (x, y) is a *solution* of an inequality in x and y if a true statement results when the variables are replaced by the coordinates of the ordered pair.

To graph a linear inequality:
1. Graph the *boundary line*. Draw a solid line if the inequality contains $\leq$ or $\geq$ and a broken line if it contains $<$ or $>$.

2. Pick a *test point* on one side of the boundary. Use the origin if possible. Replace x and y with the coordinates of that point. If the inequality is satisfied, shade the side that contains the point. If the inequality is not satisfied, shade the other side.

17. Determine whether each ordered pair is a solution of $2x - y \leq -4$.
 a. $(0, 5)$ yes
 b. $(2, 8)$ yes
 c. $(-3, -2)$ yes
 d. $\left(\frac{1}{2}, -5\right)$ no

18. Graph each inequality.
 a. $x - y < 5$

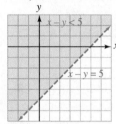

 b. $2x - 3y \geq 6$

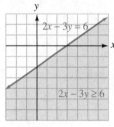

 c. $y \leq -2x$

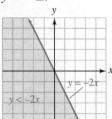

 d. $y < -4$

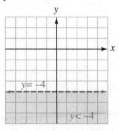

19. In Illustration 4, the graph of a linear inequality is shown. Would a true or a false statement result if the coordinates of
 a. point A were substituted into the inequality? true
 b. point B were substituted into the inequality? false
 c. point C were substituted into the inequality? false

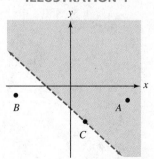

ILLUSTRATION 4

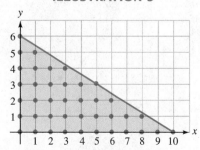

ILLUSTRATION 5

20. WORK SCHEDULE A student told her employer that during the school year, she would be available for up to 30 hours a week, working either 3- or 5-hour shifts. Find an inequality that shows the possible ways to schedule the number of 3-hour (x) and 5-hour shifts (y) she can work, and graph it in Illustration 5 on the previous page. Give three ordered pairs that satisfy the inequality.

$3x + 5y \le 30$; (2, 4), (5, 3), (6, 2) (answers may vary)

Section 8.6

To graph a system of linear inequalities:
1. Graph the individual inequalities of the system on the same coordinate axes.
2. The final solution, if one exists, is that region where all individual graphs intersect.

Systems of linear inequalities can be used to solve application problems.

Solving Systems of Linear Inequalities

21. Solve each system of inequalities.

a. $\begin{cases} 5x + 3y < 15 \\ 3x - y > 3 \end{cases}$

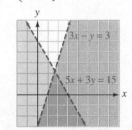

b. $\begin{cases} x \ge 3y \\ y < 3x \end{cases}$

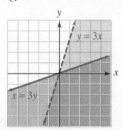

22. GIFT SHOPPING A grandmother wants to spend at least \$40 but no more than \$60 on school clothes for her grandson. If T-shirts sell for \$10 and pants sell for \$20, write a system of inequalities that describes the possible combinations of T-shirts (x) and pants (y) she can buy. Graph the system in Illustration 6. Give two possible solutions.

$10x + 20y \ge 40$, $10x + 20y \le 60$; (3, 1): 3 shirts and 1 pair of pants; (1, 2): 1 shirt and 2 pairs of pants (answers may vary)

ILLUSTRATION 6

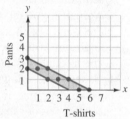

T-shirts

CHAPTER 8

Test

In Problems 1–2, tell whether the given ordered pair is a solution of the given system.

1. (5, 3), $\begin{cases} 3x + 2y = 21 \\ x + y = 8 \end{cases}$ yes

2. (−2, −1), $\begin{cases} 4x + y = -9 \\ 2x - 3y = -7 \end{cases}$ no

3. Solve the system by graphing: $\begin{cases} 3x + y = 7 \\ x - 2y = 0 \end{cases}$ (2, 1)

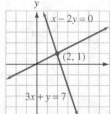

4. To solve a system of two linear equations in x and y, a student used a graphing calculator. From the calculator display in Illustration 1, determine whether the system has a solution. Explain your answer.

The lines appear to be parallel. Since the lines do not intersect, the system does not have a solution.

ILLUSTRATION 1

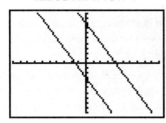

In Problems 5–6, solve each system by substitution.

5. $\begin{cases} y = x - 1 \\ 2x + y = -7 \end{cases}$
$(-2, -3)$

6. $\begin{cases} 3a + 4b = -7 \\ 2b - a = -1 \end{cases}$
$(-1, -1)$

In Problems 7–8, solve each system by addition.

7. $\begin{cases} 3x - y = 2 \\ 2x + y = 8 \end{cases}$ $(2, 4)$

8. $\begin{cases} 4x + 3y = -3 \\ -3x = -4y + 21 \end{cases}$
$(-3, 3)$

In Problems 9–10, classify each system as consistent or inconsistent.

9. $\begin{cases} x + y = 4 \\ x + y = 6 \end{cases}$
inconsistent

10. $\begin{cases} \dfrac{x}{3} + y = 4 \\ x + 3y = 12 \end{cases}$
consistent

11. Which method would be most efficient to solve the following system?

$$\begin{cases} 5x - 3y = 5 \\ 3x + 3y = 3 \end{cases}$$

Explain your answer. (You do not need to solve the system.)

Addition method; the terms involving y can be eliminated easily.

12. FINANCIAL PLANNING A woman invested some money at 8% and some at 9%. The interest for 1 year on the combined investment of $10,000 was $840. How much was invested at 9%? Use a system of equations in two variables to solve this problem. $4,000

In Problems 13–14, tell whether the given ordered pair is a solution of $2x - 4y > 8$.

13. $(7, 1)$ yes

14. $(0, -2)$ no

15. Graph the inequality $x - y > -2$.

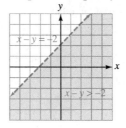

16. Solve the system by graphing.

$$\begin{cases} 2x + 3y \leq 6 \\ x \geq 2 \end{cases}$$

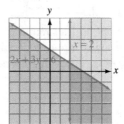

For Problems 17–18, see the graph in Illustration 2, which shows two different ways in which a salesperson can be paid according to the number of items he or she sells.

17. What is the point of intersection of the graphs? Explain its significance.

$(30, 3)$; if 30 items are sold, the salesperson gets paid the same by both plans, $3,000.

18. Which plan do you think is better for the salesperson? Explain why.

If sales of less than 30 items are anticipated, Plan 1 is better. Otherwise, Plan 2 is more profitable.

ILLUSTRATION 2

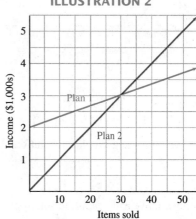

1. Tell whether each statement is true or false.
   a. All whole numbers are integers.   true
   b. π is a rational number.   false
   c. A real number is either rational or irrational.   true

2. Find the value of the expression
$$\frac{-3(3 + 2)^2 - (-5)}{17 - 3|-4|} \quad -14$$

3. BACKPACKS Pediatricians advise that children should not carry more than 20% of their own body weight in a backpack. According to this warning, how much weight can a fifth-grade girl who weighs 85 pounds safely carry in her backpack?   17 lb

4. POWER OUTPUT The graph in Illustration 1 shows the power output (in horsepower, hp) of a certain engine for various engine speeds (in revolutions per minute, rpm).
   a. At an engine speed of 3,000 rpm, what is the power output?   150 hp
   b. For what engine speed(s) is the power output 125 hp?
      2,000 rpm and 5,000 rpm
   c. For what engine speed does the power output reach a maximum?   4,000 rpm

ILLUSTRATION 1

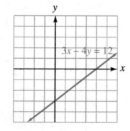

5. Simplify $3p - 6(p + z) + p$. $-2p - 6z$
6. Solve $2 - (4x + 7) = 3 + 2(x + 2)$. -2
7. Solve the inequality and graph the solution:
   $3 - 3x \geq 6 + x$. $x \leq -\frac{3}{4}$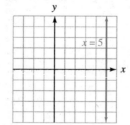

8. Solve the inequality and graph the solution:
   $0 \leq \frac{4 - x}{3} \leq 2$. $-2 \leq x \leq 4$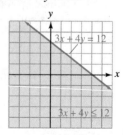

9. SEARCH AND RESCUE Two search and rescue teams leave base at the same time, looking for a lost boy. The first team, on foot, heads north at 2 mph and the other, on horseback, south at 4 mph. How long will it take them to search a distance of 21 miles between them?   3.5 hr

10. BLENDING COFFEE A store sells regular coffee for $4 a pound and gourmet coffee for $7 a pound. Using 40 pounds of the gourmet coffee, the owner makes a blend to put on sale for $5 a pound. How many pounds of regular coffee should he use?   80

11. SURFACE AREA The total surface area A of a box with dimensions l, w, and h (see Illustration 2) is given by the formula
$$A = 2lw + 2wh + 2lh$$
If $A = 202$ square inches, $l = 9$ inches, and $w = 5$ inches, find h.   4 in.

ILLUSTRATION 2

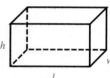

In Exercises 12–15, graph each equation or inequality.

12. $3x - 4y = 12$

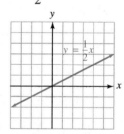

13. $y = \frac{1}{2}x$.

14. $x = 5$

15. $3x + 4y \leq 12$

16. SHOPPING SURGE On the graph in Illustration 3, draw a line through the points (1991, 724) and (1996, 974). The line approximates the total annual sales at U.S. shopping centers for the years 1991–1996. Find the rate of increase in sales over this period by finding the slope of the line. $50 billion/yr

ILLUSTRATION 3

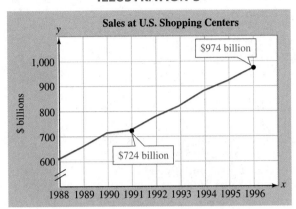

Sales at U.S. Shopping Centers

$974 billion

$724 billion

Based on data from International Council of Shopping Centers

17. Write the equation of the line passing through $(-2, 5)$ and $(4, 8)$. $x - 2y = -12$

18. What is the slope of the line defined by each equation?
  a. $y = 3x - 7$ 3 **b.** $2x + 3y = -10$ $-\frac{2}{3}$

19. What is true about the slopes of two
  a. parallel lines? they are the same
  b. perpendicular lines? they are negative reciprocals

20. If $f(x) = x^3 - x + 5$, find $f(-2)$. -1

21. Complete the table and graph the function. Then give the domain and range of the function.

$f(x) = |1 - x|$

x	$f(x)$
0	1
1	0
2	1
3	2
-1	2
-2	3

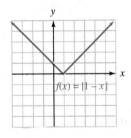

$f(x) = |1 - x|$

D: all reals; R: all real numbers greater than or equal to 0

22. BOATING The graph in Illustration 4 shows the vertical distance from a point on the tip of a propeller to the centerline as the propeller spins. Is this the graph of a function? yes

ILLUSTRATION 4

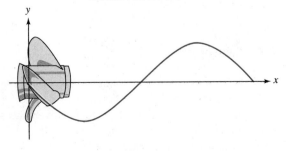

In Exercises 23–26, simplify each expression. Write each answer without using parentheses or negative exponents.

23. $(x^5)^2(x^7)^3$ x^{31} **24.** $\left(\dfrac{a^3b}{c^4}\right)^5$ $\dfrac{a^{15}b^5}{c^{20}}$

25. $4^{-3} \cdot 4^{-2} \cdot 4^5$ 1 **26.** $(a^{-2}b^3)^{-4}$ $\dfrac{a^8}{b^{12}}$

27. ASTRONOMY The **parsec**, a unit of distance used in astronomy, is 3×10^{16} meters. The distance to Betelgeuse, a star in the constellation Orion, is 1.6×10^2 parsecs. Use scientific notation to express this distance in meters. 4.8×10^{18} m

28. NCAA MEN'S BASKETBALL The graph in Illustration 5 shows the University of Connecticut's lead or deficit during the second half of the 1999 championship game with Duke University.
  a. How many x-intercepts does the graph have? Explain their importance.
    3; they indicate that the game was tied 3 times in the second half.
  b. Give the coordinates of the highest point on the graph. What is its importance?
    (11, 6); In the second half, UConn had its largest lead (6 points) after 11 minutes had elapsed.
  c. Give the coordinates of the lowest point on the graph. What is its importance?
    (4, −5); in the second half, UConn faced its largest deficit (5 points) after 4 minutes had elapsed.

ILLUSTRATION 5

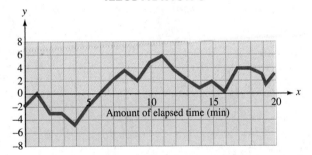

Amount of elapsed time (min)

In Exercises 29–34, do the indicated operations.

29. $(-r^4st^2)(2r^2st)(rst)$ $-2r^7s^3t^4$

30. $(-3t + 2s)(2t - 3s)$ $-6t^2 + 13st - 6s^2$

31. $(3a^2 - 2a + 4) - (a^2 - 3a + 7)$ $2a^2 + a - 3$

32. $(y - 6)^2$ $y^2 - 12y + 36$

33. $\dfrac{4x - 3y + 8z}{4xy}$ $\frac{1}{y} - \frac{3}{4x} + \frac{2z}{xy}$

34. $2 + x\overline{)3x + 2x^2 - 2}$ $2x - 1$

In Exercises 35–43, simplify each expression. All variables represent positive numbers.

35. $\sqrt{\frac{49}{225}}$　$\frac{7}{15}$

36. $-\sqrt[3]{-27}$　3

37. $\sqrt[4]{x^4}$　x

38. $-12x\sqrt{16x^2y^3}$　$-48x^2y\sqrt{y}$

39. $\sqrt{48} - \sqrt{8} + \sqrt{27} - \sqrt{32}$　$7\sqrt{3} - 6\sqrt{2}$

40. $\left(\sqrt{y} - 4\right)\left(\sqrt{y} - 5\right)$　$y - 9\sqrt{y} + 20$

41. $\left(-5\sqrt{6}\right)\left(4\sqrt{3}\right)$　$-60\sqrt{2}$

42. $\dfrac{4}{\sqrt{20}}$　$\dfrac{2\sqrt{5}}{5}$

43. $\dfrac{\sqrt{x} - 3}{\sqrt{x} + 3}$　$\dfrac{x - 6\sqrt{x} + 9}{x - 9}$

44. CUBICLES The diagonal distance across the face of each of the stacking cubes shown in Illustration 6 is 15 inches. What is the height of the storage arrangement? Round to the nearest tenth of an inch.　21.2 in.

ILLUSTRATION 6

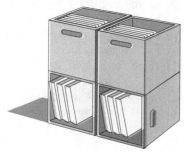

In Exercises 45–52, factor each expression completely.

45. $3x^2y - 6xy^2$　$3xy(x - 2y)$

46. $2x^2 + 2xy - 3x - 3y$　$(x + y)(2x - 3)$

47. $25p^4 - 16q^2$　$(5p^2 + 4q)(5p^2 - 4q)$

48. $3x^3 - 243x$　$3x(x + 9)(x - 9)$

49. $x^2 - 11x - 12$　$(x - 12)(x + 1)$

50. $a^3 + 8b^3$　$(a + 2b)(a^2 - 2ab + 4b^2)$

51. $6a^2 - 7a - 20$　$(3a + 4)(2a - 5)$

52. $16m^2 - 20m - 6$　$2(4m + 1)(2m - 3)$

In Exercises 53–56, solve each equation.

53. $x^2 + 3x + 2 = 0$　$-1, -2$

54. $5x^2 = 10x$　$0, 2$

55. $6x^2 - x - 2 = 0$　$\frac{2}{3}, -\frac{1}{2}$

56. $2y^2 = 12 - 5y$　$\frac{3}{2}, -4$

57. Find the number that must be added to the binomial $y^2 - 5y$ to complete the square.　$\frac{25}{4}$

58. Solve $p^2 - 4p + 3 = 0$ by completing the square.　1, 3

In Exercises 59–60, use the quadratic formula to solve each equation.

59. $x^2 + 2x - 8 = 0$　$2, -4$

60. $-6x = x^2 + 4$　$-3 \pm \sqrt{5}$ (-0.8 and -5.2)

61. DAREDEVILS On October 7, 1829, 22-year-old Sam Patch became the first person to challenge the Niagara River by diving head first into the churning waters from an 85-foot-tall platform. How long did the dive take him? Round to the nearest hundredth. (The distance d in feet that an object will fall in t seconds is given by the formula $d = 16t^2$.)　2.30 sec

62. FLAGS According to the *Guinness Book of World Records 1998*, the largest flag in the world is the American "super flag" owned by Ski Demski of Long Beach, California. Its length is 5 feet less than twice its width, and its area is 128,775 square feet. Find the width and length of the flag.　255 ft, 505 ft

63. Graph the quadratic function $f(x) = x^2 + x - 6$ by finding the x- and y-intercepts and the vertex of the parabola. Draw the axis of symmetry.
$(-3, 0)$, $(2, 0)$; $(0, -6)$; $\left(-\frac{1}{2}, -6\frac{1}{4}\right)$

64. Illustration 7 shows the graph of $f(x) = -2x^2 - 3x + 2$. Use it to estimate the solutions of the quadratic equation $-2x^2 - 3x + 2 = 0$.　$-2, \frac{1}{2}$

ILLUSTRATION 7

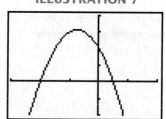

65. Solve the proportion $\dfrac{4 - a}{13} = \dfrac{11}{26}$.　$-\frac{3}{2}$

66. ONLINE SALES A company found that, on average, it made 9 online sales transactions for every 500 hits on its Internet web site. If the company's web site had 360,000 hits in one year, how many sales transactions did it have that year?　6,480

In Exercises 67–68, simplify each fraction.

67. $\dfrac{x^2 + 2x + 1}{x^2 - 1}$ $\dfrac{x + 1}{x - 1}$ **68.** $\dfrac{-15a^2}{25a^3}$ $-\dfrac{3}{5a}$

In Exercises 69–76, do the operation(s) and simplify when possible.

69. $\dfrac{x^2 + x - 6}{5x - 5} \cdot \dfrac{10 - 5x}{x + 3}$ $-\dfrac{(x - 2)^2}{x - 1}$

70. $\dfrac{p^2 - p - 6}{3p - 9} \div \dfrac{p^2 + 6p + 9}{p^2 - 9}$ $\dfrac{(p + 2)(p - 3)}{3(p + 3)}$

71. $\dfrac{x^2 y^2}{cd} \cdot \dfrac{d^2}{c^2 x}$ $\dfrac{xy^2 d}{c^3}$

72. $\dfrac{x + 2}{x + 5} - \dfrac{x - 3}{x + 7}$ $\dfrac{7x + 29}{(x + 5)(x + 7)}$

73. $\dfrac{3x}{x + 2} + \dfrac{5x}{x + 2} - \dfrac{7x - 2}{x + 2}$ 1

74. $\dfrac{3a}{2b} - \dfrac{2b}{3a}$ $\dfrac{9a^2 - 4b^2}{6ab}$

75. $\dfrac{a + 1}{2a + 4} - \dfrac{a^2}{2a^2 - 8}$ $-\dfrac{1}{2(a - 2)}$

76. $\dfrac{\dfrac{1}{x} + \dfrac{1}{y}}{\dfrac{1}{x} - \dfrac{1}{y}}$ $\dfrac{y + x}{y - x}$

In Exercises 77–78 solve each equation.

77. $\dfrac{4}{a} = \dfrac{6}{a} - 1$ 2

78. $\dfrac{a + 2}{a + 3} - 1 = \dfrac{-1}{a^2 + 2a - 3}$ 2

79. Assume that y varies inversely with x. If $y = 8$ when $x = 2$, find y when $x = 8$. 2

80. FILLING A POOL An inlet pipe can fill an empty swimming pool in 5 hours, and another inlet pipe can fill the pool in 4 hours. How long will it take both pipes to fill the pool? $2\frac{2}{9}$ hr

81. Solve the formula $\dfrac{1}{r} = \dfrac{1}{r_1} + \dfrac{1}{r_2}$ for r. $r = \dfrac{r_1 r_2}{r_2 + r_1}$

82. FINANCIAL PLANNING In investing \$6,000 of a couple's money, a financial planner put some of it into a savings account paying 6% annual interest. The rest was invested in a riskier mini-mall development plan paying 12% annually. The combined interest earned for the first year was \$540. How much money was invested at each rate? Use two variables to solve this problem.   6%: \$3,000; 12%: \$3,000

83. The graphing calculator display in Illustration 8 shows the graphs of $x + y = 1$ and $y = x + 5$. What is the solution of the system $\begin{cases} x + y = 1 \\ y = x + 5 \end{cases}$? Check the solution. $(-2, 3)$

ILLUSTRATION 8

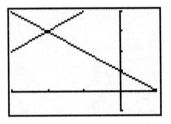

84. Graph the solution of $\begin{cases} 3x + 2y \geq 6 \\ x + 3y \leq 6 \end{cases}$

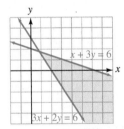

In Exercises 85–86, solve each system of equations. If the equations of a system are dependent or if a system is inconsistent, so indicate.

85. $\begin{cases} x = y + 4 \\ 2x + y = 5 \end{cases}$ $(3, -1)$ **86.** $\begin{cases} \frac{3}{5}s + \frac{4}{5}t = 1 \\ -\frac{1}{4}s + \frac{3}{8}t = 1 \end{cases}$ $(-1, 2)$

Appendix I

Arithmetic Fractions

In this appendix, you will learn about

Simplifying fractions ■ Multiplying fractions ■ Dividing fractions
■ Adding fractions ■ Subtracting fractions ■ Mixed numbers
■ Decimal fractions ■ Rounding decimals ■ Applications

Introduction In this appendix, we will review the properties of arithmetic fractions. In the **fractions**

$$\frac{1}{2}, \quad \frac{3}{5}, \quad \frac{2}{17}, \quad \text{and} \quad \frac{37}{7}$$

the number above the fraction bar is called the **numerator,** and the number below is called the **denominator.**

Fractions are used to indicate parts of a whole. In Figure I-1(a), a rectangle has been divided into 5 equal parts with 3 of the parts shaded. The fraction $\frac{3}{5}$ indicates how much of the figure is shaded. In Figure I-1(b), $\frac{5}{7}$ of the rectangle is shaded. In either example, the denominator of the fraction shows the total number of equal parts into which the whole is divided, and the numerator shows the number of these equal parts that are being considered.

FIGURE I-1

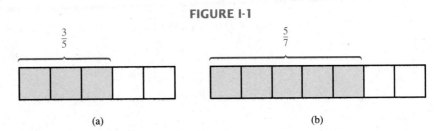

(a)   (b)

Fractions are also used to indicate division. For example, the fraction $\frac{8}{2}$ indicates that 8 is to be divided by 2:

$$\frac{8}{2} = 8 \div 2 = 4$$

We note that $\frac{8}{2} = 4$, because $4 \cdot 2 = 8$, and that $\frac{0}{7} = 0$, because $0 \cdot 7 = 0$. However, the fraction $\frac{6}{0}$ is undefined, because no number multiplied by 0 gives 6. The fraction $\frac{0}{0}$ is indeterminate, because every number multiplied by 0 gives 0.

 WARNING! Remember that the denominator of a fraction cannot be zero.

Simplifying Fractions

A fraction is in **lowest terms** when no natural number greater than 1 will divide both its numerator and denominator exactly. The fraction $\frac{6}{11}$ is in lowest terms, because only 1 divides both 6 and 11 exactly. The fraction $\frac{6}{8}$ is not in lowest terms, because 2 divides both 6 and 8 exactly.

We can **simplify** (or **reduce**) a fraction that is not in lowest terms by dividing both its numerator and denominator by the same number. For example, to simplify the fraction $\frac{6}{8}$, we divide both numerator and denominator by 2:

$$\frac{6}{8} = \frac{6 \div 2}{8 \div 2} = \frac{3}{4}$$

From Figure I-2, we see that $\frac{6}{8}$ and $\frac{3}{4}$ are equal fractions, because each one represents the same part of the rectangle.

FIGURE I-2

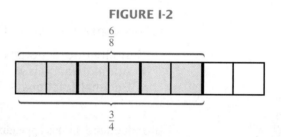

We can use the fact that nonprime numbers can be written as the product of other natural numbers to simplify fractions. For example, 12 can be written as the product of 4 and 3:

$$12 = 4 \cdot 3$$

When a number has been written as the product of other natural numbers, we say that it has been **factored.** The numbers 3 and 4 are called **factors** of 12. When a number is written as the product of prime numbers, we say that the number is written in **prime-factored form.**

EXAMPLE 1

Write 210 in prime-factored form.

Solution   We can write 210 as the product of 21 and 10 and proceed as follows:

$$210 = \mathbf{21 \cdot 10}$$
$$210 = \mathbf{3 \cdot 7 \cdot 2 \cdot 5} \quad \text{Factor 21 as } 3 \cdot 7 \text{ and factor 10 as } 2 \cdot 5.$$

Since 210 is now written as the product of prime numbers, its prime-factored form is $3 \cdot 7 \cdot 2 \cdot 5$. ∎

To simplify a fraction, we factor its numerator and denominator, and then divide out all common factors that appear in both the numerator and denominator. To simplify the fractions $\frac{6}{8}$ and $\frac{15}{18}$, for example, we proceed as follows:

$$\frac{6}{8} = \frac{3 \cdot 2}{4 \cdot 2} = \frac{3 \cdot \overset{1}{\cancel{2}}}{4 \cdot \cancel{2}} = \frac{3}{4} \quad \text{and} \quad \frac{15}{18} = \frac{5 \cdot 3}{6 \cdot 3} = \frac{5 \cdot \overset{1}{\cancel{3}}}{6 \cdot \cancel{3}} = \frac{5}{6}$$

 WARNING! Remember that a fraction is in lowest terms only when its numerator and denominator have no common factors.

EXAMPLE 2

a. To simplify $\frac{6}{30}$, we factor the numerator and denominator and divide out the common factor of 6.

$$\frac{6}{30} = \frac{6 \cdot 1}{6 \cdot 5} = \frac{\overset{1}{\cancel{6}} \cdot 1}{\underset{1}{\cancel{6}} \cdot 5} = \frac{1}{5}$$

b. To show that $\frac{33}{40}$ is in lowest terms, we must show that the numerator and the denominator share no common factors, by writing both the numerator and denominator in prime-factored form.

$$\frac{33}{40} = \frac{3 \cdot 11}{2 \cdot 2 \cdot 2 \cdot 5}$$

Since the numerator and denominator have no common factors, $\frac{33}{40}$ is in lowest terms. ■

The previous examples illustrate the **fundamental property of fractions.**

The fundamental property of fractions

If a, b, and c are real numbers, then

$$\frac{a \cdot c}{b \cdot c} = \frac{a}{b} \quad (b \neq 0, c \neq 0)$$

Multiplying Fractions

Multiplying fractions

To multiply two fractions, we multiply the numerators and multiply the denominators. In symbols, if a, b, c, and d are real numbers, then

$$\frac{a}{b} \cdot \frac{c}{d} = \frac{a \cdot c}{b \cdot d} \quad (b \neq 0, d \neq 0)$$

For example,

$$\frac{4}{7} \cdot \frac{2}{3} = \frac{4 \cdot 2}{7 \cdot 3} \qquad \text{and} \qquad \frac{4}{5} \cdot \frac{13}{9} = \frac{4 \cdot 13}{5 \cdot 9}$$

$$= \frac{8}{21} \qquad\qquad\qquad\qquad = \frac{52}{45}$$

FIGURE I-3

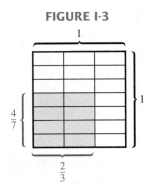

To justify the rule for multiplying fractions, we consider the square in Figure I-3. Since the length of each side of the square is 1 unit and the area is the product of the lengths of two sides, the area is 1 square unit.

If this square is divided into 3 equal parts vertically and 7 equal parts horizontally, it is divided into 21 equal parts, and each represents $\frac{1}{21}$ of the total area. The area of the shaded rectangle in the square is $\frac{8}{21}$, because it contains 8 of the 21 parts. The width w of the shaded rectangle is $\frac{4}{7}$, its length l is $\frac{2}{3}$, and its area A is the product of l and w:

$$A = l \cdot w$$

$$\frac{8}{21} = \frac{4}{7} \cdot \frac{2}{3}$$

This suggests that we can find the product of

$$\frac{4}{7} \quad \text{and} \quad \frac{2}{3}$$

by multiplying their numerators and multiplying their denominators.

Fractions with a numerator that is smaller than the denominator, such as $\frac{8}{21}$, are called **proper fractions**. Fractions with a numerator that is larger than the denominator, such as $\frac{52}{45}$, are called **improper fractions**.

EXAMPLE 3

a. $\dfrac{3}{7} \cdot \dfrac{13}{5} = \dfrac{3 \cdot 13}{7 \cdot 5}$ Multiply the numerators and multiply the denominators.

$= \dfrac{39}{35}$ Do the multiplication in the numerator.
Do the multiplication in the denominator.

b. $5 \cdot \dfrac{3}{15} = \dfrac{5}{1} \cdot \dfrac{3}{15}$ Write 5 as the improper fraction $\frac{5}{1}$.

$= \dfrac{5 \cdot 3}{1 \cdot 15}$ Multiply the numerators and multiply the denominators.

$= \dfrac{5 \cdot 3}{1 \cdot 5 \cdot 3}$ To attempt to simplify the fraction, factor the denominator.

$= \dfrac{\overset{1}{\cancel{5}} \cdot \overset{1}{\cancel{3}}}{1 \cdot \underset{1}{\cancel{5}} \cdot \underset{1}{\cancel{3}}}$ Divide out the common factors of 3 and 5.

$= \dfrac{1}{1}$

$= 1$ ∎

EXAMPLE 4

European travel. Out of 36 students in a history class, three-fourths have signed up for a trip to Europe. If there are 30 places available on the chartered flight, will there be room for one more student?

Solution   We first find three-fourths of 36 by multiplying 36 by $\frac{3}{4}$.

$\dfrac{3}{4} \cdot 36 = \dfrac{3}{4} \cdot \dfrac{36}{1}$ Write 36 as $\frac{36}{1}$.

$= \dfrac{3 \cdot 36}{4 \cdot 1}$ Multiply the numerators and multiply the denominators.

$= \dfrac{3 \cdot 4 \cdot 9}{4 \cdot 1}$ To simplify the fractions, factor the numerator.

$= \dfrac{3 \cdot \overset{1}{\cancel{4}} \cdot 9}{\underset{1}{\cancel{4}} \cdot 1}$ Divide out the common factor of 4.

$= \dfrac{27}{1}$

$= 27$

Twenty-seven students plan to go on the trip. Since there is room for 30 passengers, there is plenty of room for one more.   ∎

Dividing Fractions

One number is called the **reciprocal** of another if their product is 1. For example, $\frac{3}{5}$ is the reciprocal of $\frac{5}{3}$, because

$$\frac{3}{5} \cdot \frac{5}{3} = \frac{15}{15} = 1$$

Dividing fractions

To divide two fractions, we multiply the first fraction by the reciprocal of the second fraction. In symbols, if a, b, c, and d are real numbers, then

$$\frac{a}{b} \div \frac{c}{d} = \frac{a}{b} \cdot \frac{d}{c} \quad (b \neq 0,\ c \neq 0,\ \text{and}\ d \neq 0)$$

EXAMPLE 5

a. $\dfrac{3}{5} \div \dfrac{6}{5} = \dfrac{3}{5} \cdot \dfrac{5}{6}$ Multiply $\frac{3}{5}$ by the reciprocal of $\frac{6}{5}$.

$\qquad = \dfrac{3 \cdot 5}{5 \cdot 6}$ Multiply the numerators and multiply the denominators.

$\qquad = \dfrac{3 \cdot 5}{5 \cdot 2 \cdot 3}$ Factor the denominator.

$\qquad = \dfrac{\overset{1}{\cancel{3}} \cdot \overset{1}{\cancel{5}}}{\underset{1}{\cancel{5}} \cdot 2 \cdot \underset{1}{\cancel{3}}}$ Divide out the common factors of 3 and 5.

$\qquad = \dfrac{1}{2}$

b. $\dfrac{15}{7} \div 10 = \dfrac{15}{7} \div \dfrac{10}{1}$ Write 10 as the improper fraction $\frac{10}{1}$.

$\qquad = \dfrac{15}{7} \cdot \dfrac{1}{10}$ Multiply $\frac{15}{7}$ by the reciprocal of $\frac{10}{1}$.

$\qquad = \dfrac{15 \cdot 1}{7 \cdot 10}$ Multiply the numerators and multiply the denominators.

$\qquad = \dfrac{3 \cdot \overset{1}{\cancel{5}} \cdot 1}{7 \cdot 2 \cdot \underset{1}{\cancel{5}}}$ Factor the numerator and denominator.

$\qquad = \dfrac{3}{14}$ Divide out the common factor of 5 and simplify.   ∎

Adding Fractions

Adding fractions with like denominators

To add two fractions with the same denominator, we add the numerators and keep the common denominator. In symbols, if a, b, and d are real numbers, then

$$\frac{a}{d} + \frac{b}{d} = \frac{a + b}{d} \quad (d \neq 0)$$

For example,

$$\frac{3}{7} + \frac{2}{7} = \frac{3+2}{7}$$

$$= \frac{5}{7}$$

Figure I-4 shows why $\frac{3}{7} + \frac{2}{7} = \frac{5}{7}$.

FIGURE I-4

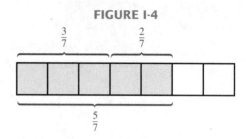

To add fractions with unlike denominators, we rewrite the fractions so they have the same denominator. For example, we can multiply both the numerator and denominator of the fraction $\frac{1}{3}$ by 5, to obtain an equivalent fraction with a denominator of 15:

$$\frac{1}{3} = \frac{1 \cdot 5}{3 \cdot 5}$$

$$= \frac{5}{15}$$

Similarly, we can rewrite the fraction $\frac{1}{5}$ as an equal fraction with a denominator of 15. We multiply both the numerator and the denominator by 3:

$$\frac{1}{5} = \frac{1 \cdot 3}{5 \cdot 3}$$

$$= \frac{3}{15}$$

Because 15 is the smallest number that can be used as a denominator for $\frac{1}{3}$ and $\frac{1}{5}$, it is called the **least** or **lowest common denominator (LCD).**

To add the fractions $\frac{1}{3}$ and $\frac{1}{5}$, we rewrite each fraction as an equivalent fraction having a denominator of 15 and add the results:

$$\frac{1}{3} + \frac{1}{5} = \frac{1 \cdot 5}{3 \cdot 5} + \frac{1 \cdot 3}{5 \cdot 3}$$

$$= \frac{5}{15} + \frac{3}{15}$$

$$= \frac{5+3}{15} \qquad \text{Add the numerators and keep the common denominator.}$$

$$= \frac{8}{15}$$

E X A M P L E   6   Add $\dfrac{3}{10} + \dfrac{5}{28}$.

Solution   To find the least common denominator (LCD), we find the prime factorization of both denominators and use each prime factor the greatest number of times it appears in either factorization:

$$\left. \begin{array}{l} 10 = 2 \cdot 5 \\ 28 = 2 \cdot 2 \cdot 7 \end{array} \right\} \qquad \text{LCD} = 2 \cdot 2 \cdot 5 \cdot 7 = 140$$

Since 140 is the smallest number that 10 and 28 divide exactly, we write both fractions as fractions with the LCD of 140.

$$\frac{3}{10} + \frac{5}{28} = \frac{3 \cdot 14}{10 \cdot 14} + \frac{5 \cdot 5}{28 \cdot 5}$$ Write each fraction as a fraction with a denominator of 140.

$$= \frac{42}{140} + \frac{25}{140}$$ Do the multiplications.

$$= \frac{42 + 25}{140}$$ Add the numerators and keep the common denominator.

$$= \frac{67}{140}$$ Do the addition.

Since 67 is a prime number, it has no common factor with 140. Thus, $\frac{67}{140}$ is in lowest terms and cannot be simplified.   ∎

Subtracting Fractions

Subtracting fractions with like denominators

To subtract two fractions with the same denominator, we subtract their numerators and keep their common denominator. In symbols, if a, b, and d are real numbers, then

$$\frac{a}{d} - \frac{b}{d} = \frac{a - b}{d} \quad (d \neq 0)$$

For example,

$$\frac{7}{9} - \frac{2}{9} = \frac{7 - 2}{9}$$ Subtract the numerators and keep the common denominator.

$$= \frac{5}{9}$$

To subtract fractions with unlike denominators, we write them as equivalent fractions with a common denominator. For example, to subtract $\frac{2}{5}$ from $\frac{3}{4}$, we write $\frac{3}{4} - \frac{2}{5}$, find the LCD of 20, and proceed as follows:

$$\frac{3}{4} - \frac{2}{5} = \frac{3 \cdot 5}{4 \cdot 5} - \frac{2 \cdot 4}{5 \cdot 4}$$

$$= \frac{15}{20} - \frac{8}{20}$$

$$= \frac{15 - 8}{20}$$ Subtract the numerators and keep the common denominator.

$$= \frac{7}{20}$$

EXAMPLE 7	Subtract 5 from $\dfrac{23}{3}$.

Solution

$$\dfrac{23}{3} - 5 = \dfrac{23}{3} - \dfrac{5}{1} \qquad \text{Write 5 as the improper fraction } \tfrac{5}{1}.$$

$$= \dfrac{23}{3} - \dfrac{5 \cdot 3}{1 \cdot 3} \qquad \text{Write } \tfrac{5}{1} \text{ as a fraction with a denominator of 3.}$$

$$= \dfrac{23}{3} - \dfrac{15}{3} \qquad \text{Do the multiplication.}$$

$$= \dfrac{23 - 15}{3} \qquad \text{Subtract the numerators and keep the common denominator.}$$

$$= \dfrac{8}{3}$$

∎

Mixed Numbers

The **mixed number** $3\frac{1}{2}$ represents the sum of 3 and $\frac{1}{2}$. We can write $3\frac{1}{2}$ as an improper fraction as follows:

$$3\dfrac{1}{2} = 3 + \dfrac{1}{2}$$

$$= \dfrac{6}{2} + \dfrac{1}{2} \qquad 3 = \tfrac{6}{2}.$$

$$= \dfrac{6 + 1}{2} \qquad \text{Add the numerators and keep the common denominator.}$$

$$= \dfrac{7}{2}$$

To write the fraction $\frac{19}{5}$ as a mixed number, we divide 19 by 5 to get 3, with a remainder of 4. Thus,

$$\dfrac{19}{5} = 3 + \dfrac{4}{5}$$

$$= 3\dfrac{4}{5}$$

EXAMPLE 8	Add $2\dfrac{1}{4}$ and $1\dfrac{1}{3}$.

Solution We first change each mixed number to an improper fraction:

$$2\dfrac{1}{4} = 2 + \dfrac{1}{4} \qquad\qquad 1\dfrac{1}{3} = 1 + \dfrac{1}{3}$$

$$= \dfrac{8}{4} + \dfrac{1}{4} \qquad\qquad = \dfrac{3}{3} + \dfrac{1}{3}$$

$$= \dfrac{9}{4} \qquad\qquad\qquad = \dfrac{4}{3}$$

and then add the fractions.

$$2\frac{1}{4} + 1\frac{1}{3} = \frac{9}{4} + \frac{4}{3}$$

$$= \frac{9 \cdot 3}{4 \cdot 3} + \frac{4 \cdot 4}{3 \cdot 4} \qquad \text{The LCD is 12.}$$

$$= \frac{27}{12} + \frac{16}{12}$$

$$= \frac{43}{12}$$

Finally, we change $\frac{43}{12}$ to a mixed number.

$$\frac{43}{12} = 3 + \frac{7}{12} = 3\frac{7}{12}$$

■

EXAMPLE 9

Fencing land. The three sides of a triangular piece of land measure $33\frac{1}{4}$, $57\frac{3}{4}$, and $72\frac{1}{2}$ meters. How much fencing will be needed to enclose the area?

Solution   We can find the sum of the lengths by adding the whole-number parts and the fractional parts of the dimensions separately:

$$33\frac{1}{4} + 57\frac{3}{4} + 72\frac{1}{2} = 33 + 57 + 72 + \frac{1}{4} + \frac{3}{4} + \frac{1}{2}$$

$$= 162 + \frac{1}{4} + \frac{3}{4} + \frac{2}{4} \qquad \begin{array}{l}\text{Change } \frac{1}{2} \text{ to } \frac{2}{4} \text{ to obtain a}\\ \text{common denominator.}\end{array}$$

$$= 162 + \frac{6}{4} \qquad \begin{array}{l}\text{Add the fractions by adding}\\ \text{the numerators and keeping}\\ \text{the common denominator.}\end{array}$$

$$= 162 + \frac{3}{2} \qquad \frac{6}{4} = \frac{2 \cdot 3}{2 \cdot 2} = \frac{\overset{1}{\cancel{2}} \cdot 3}{\underset{1}{\cancel{2}} \cdot 2} = \frac{3}{2}.$$

$$= 162 + 1\frac{1}{2} \qquad \text{Change } \frac{3}{2} \text{ to a mixed number.}$$

$$= 163\frac{1}{2}$$

It will require $163\frac{1}{2}$ meters of fencing to enclose the triangular area.   ■

Decimal Fractions

Rational numbers can always be changed into decimal fractions. For example, to change $\frac{1}{4}$ and $\frac{5}{22}$ to decimal fractions, we use long division:

```
      0.25                  0.22727. . .
   4)1.00               22)5.00000
      8                      44
     ──                      ──
     20                      60
     20                      44
     ──                      ──
      0                     160
                            154
                            ───
                             60
                             44
                            ───
                            160
```

The decimal fraction 0.25 is called a **terminating decimal,** and the decimal fraction 0.2272727. . . (often written as $0.2\overline{27}$) is called a **repeating decimal,** because it repeats the block of digits 27. Every rational number can be changed into either a **terminating** or a **repeating decimal.**

Terminating decimals	*Repeating decimals*
$\dfrac{1}{2} = 0.5$	$\dfrac{1}{3} = 0.3333. . .$ or $0.\overline{3}$
$\dfrac{3}{4} = 0.75$	$\dfrac{1}{6} = 0.16666. . .$ or $0.1\overline{6}$
$\dfrac{5}{8} = 0.625$	$\dfrac{5}{22} = 0.2272727. . .$ or $0.2\overline{27}$

The decimal 0.5 has one **decimal place,** because it has one digit to the right of the decimal point. The decimal 0.75 has two decimal places, and 0.625 has three.

To *add* or *subtract* decimal fractions, we first align their decimal points and then add or subtract.

$$\begin{array}{r} 25.568 \\ 2.74 \\ \hline 28.308 \end{array} \qquad \begin{array}{r} 25.568 \\ -\ 2.74 \\ \hline 22.828 \end{array}$$

To do the previous operations with a calculator, we would press these keys:

25.568 $+$ 2.74 $=$ and 25.568 $-$ 2.74 $=$

To *multiply* decimal fractions, we multiply the numbers and then place the decimal point so that the number of decimal places in the answer is equal to the sum of the decimal places in the factors.

$$\begin{array}{r} 3.453 \\ 9.25 \\ \hline 17265 \\ 6906 \\ 31077 \\ \hline 31.94025 \end{array}$$

← Here there are three decimal places.
← Here there are two decimal places.

← The product has 3 + 2 = 5 decimal places.

To do the previous multiplication with a calculator, we would press these keys:

3.453 $\times$ 9.25 $=$

To *divide* decimal fractions, we move the decimal point in the divisor to the right to make the divisor a whole number. We then move the decimal point in the dividend the same number of places to the right.

$1.23\overline{)30.258}$ Move the decimal point in both the divisor and the dividend two places to the right.

We align the decimal point in the quotient with the repositioned decimal point in the dividend and then use long division.

$$\begin{array}{r} 24.6 \\ 123\overline{)3025.8} \\ \underline{246} \\ 565 \\ \underline{492} \\ 73\ 8 \\ \underline{73\ 8} \\ 0 \end{array}$$

To do the previous division with a calculator, we would press these keys:

30.258 $\div$ 1.23 $=$

Rounding Decimals

When decimal fractions are long, we often **round** them to a specific number of decimal places. For example, the decimal fraction 25.36124 rounded to one place (or to the nearest tenth) is 25.4. Rounded to two places (or to the nearest hundredth), the decimal is 25.36. To round decimals, we use the following rules.

Rounding decimals

1. Determine to how many decimal places you wish to round.
2. Look at the first digit to the right of that decimal place.
3. If that digit is 4 or less, drop it and all digits that follow. If it is 5 or greater, add 1 to the digit in the position to which you wish to round, and drop all digits that follow.

Applications

A **percent** is the numerator of a fraction with a denominator of 100. For example, $6\frac{1}{4}$ percent, written $6\frac{1}{4}\%$, is the fraction $\frac{6.25}{100}$, or the decimal 0.0625. In problems involving percent, the word *of* often indicates multiplication. For example, $6\frac{1}{4}\%$ of 8,500 is the product 0.0625(8,500).

EXAMPLE 10

Auto loans. Juan signs a one-year note to borrow $8,500 to buy a car. If the interest rate is $6\frac{1}{4}\%$, how much interest will he pay?

Solution   For the privilege of using the bank's money for one year, Juan must pay $6\frac{1}{4}\%$ of $8,500. We calculate the interest I as follows:

$$I = 6\tfrac{1}{4}\% \text{ of } 8,500$$
$$= 0.0625 \cdot 8,500 \quad \text{Here, } of \text{ means } times.$$
$$= 531.25$$

He will pay $531.25 interest. ∎

STUDY SET

Appendix I

PRACTICE

In Exercises 1–8, write each fraction in lowest terms. If the fraction is already in lowest terms, so indicate.

In Exercises 9–20, do each multiplication. Simplify each result when possible.

1. $\dfrac{6}{12}$ $\frac{1}{2}$

2. $\dfrac{3}{9}$ $\frac{1}{3}$

3. $\dfrac{15}{20}$ $\frac{3}{4}$

4. $\dfrac{22}{77}$ $\frac{2}{7}$

5. $\dfrac{24}{18}$ $\frac{4}{3}$

6. $\dfrac{35}{14}$ $\frac{5}{2}$

7. $\dfrac{72}{64}$ $\frac{9}{8}$

8. $\dfrac{26}{21}$ in lowest terms

9. $\dfrac{1}{2} \cdot \dfrac{3}{5}$ $\frac{3}{10}$

10. $\dfrac{3}{4} \cdot \dfrac{5}{7}$ $\frac{15}{28}$

11. $\dfrac{4}{3} \cdot \dfrac{6}{5}$ $\frac{8}{5}$

12. $\dfrac{7}{8} \cdot \dfrac{6}{15}$ $\frac{7}{20}$

13. $\dfrac{5}{12} \cdot \dfrac{18}{5}$ $\frac{3}{2}$

14. $\dfrac{5}{4} \cdot \dfrac{12}{10}$ $\frac{3}{2}$

15. $\dfrac{17}{34} \cdot \dfrac{3}{6}$ $\frac{1}{4}$

16. $\dfrac{21}{14} \cdot \dfrac{3}{6}$ $\frac{3}{4}$

17. $12 \cdot \dfrac{5}{6}$ 10

18. $9 \cdot \dfrac{7}{12}$ $\frac{21}{4}$

19. $\dfrac{10}{21} \cdot 14$ $\frac{20}{3}$

20. $\dfrac{5}{24} \cdot 16$ $\frac{10}{3}$

57. $3\dfrac{3}{4} - 2\dfrac{1}{2}$ $1\frac{1}{4}$

58. $15\dfrac{5}{6} + 11\dfrac{5}{8}$ $27\frac{11}{24}$

59. $8\dfrac{2}{9} - 7\dfrac{2}{3}$ $\frac{5}{9}$

60. $3\dfrac{4}{5} - 3\dfrac{1}{10}$ $\frac{7}{10}$

In Exercises 21–32, do each division. Simplify each result when possible.

21. $\dfrac{3}{5} \div \dfrac{2}{3}$ $\frac{9}{10}$

22. $\dfrac{4}{5} \div \dfrac{3}{7}$ $\frac{28}{15}$

23. $\dfrac{3}{4} \div \dfrac{6}{5}$ $\frac{5}{8}$

24. $\dfrac{3}{8} \div \dfrac{15}{28}$ $\frac{7}{10}$

25. $\dfrac{2}{13} \div \dfrac{8}{13}$ $\frac{1}{4}$

26. $\dfrac{4}{7} \div \dfrac{20}{21}$ $\frac{3}{5}$

27. $\dfrac{21}{35} \div \dfrac{3}{14}$ $\frac{14}{5}$

28. $\dfrac{23}{25} \div \dfrac{46}{5}$ $\frac{1}{10}$

29. $6 \div \dfrac{3}{14}$ 28

30. $23 \div \dfrac{46}{5}$ $\frac{5}{2}$

31. $\dfrac{42}{30} \div 7$ $\frac{1}{5}$

32. $\dfrac{34}{8} \div 17$ $\frac{1}{4}$

In Exercises 33–60, do each addition or subtraction. Simplify each result when possible.

33. $\dfrac{3}{5} + \dfrac{3}{5}$ $\frac{6}{5}$

34. $\dfrac{4}{7} - \dfrac{2}{7}$ $\frac{2}{7}$

35. $\dfrac{4}{13} - \dfrac{3}{13}$ $\frac{1}{13}$

36. $\dfrac{2}{11} + \dfrac{9}{11}$ 1

37. $\dfrac{1}{6} + \dfrac{1}{24}$ $\frac{5}{24}$

38. $\dfrac{17}{25} - \dfrac{2}{5}$ $\frac{7}{25}$

39. $\dfrac{3}{5} + \dfrac{2}{3}$ $\frac{19}{15}$

40. $\dfrac{4}{3} + \dfrac{7}{2}$ $\frac{29}{6}$

41. $\dfrac{9}{4} - \dfrac{5}{6}$ $\frac{17}{12}$

42. $\dfrac{2}{15} + \dfrac{7}{9}$ $\frac{41}{45}$

43. $\dfrac{7}{10} - \dfrac{1}{14}$ $\frac{22}{35}$

44. $\dfrac{7}{25} + \dfrac{3}{10}$ $\frac{29}{50}$

45. $\dfrac{5}{14} - \dfrac{4}{21}$ $\frac{1}{6}$

46. $\dfrac{2}{33} + \dfrac{3}{22}$ $\frac{13}{66}$

47. $3 - \dfrac{3}{4}$ $\frac{9}{4}$

48. $5 + \dfrac{21}{5}$ $\frac{46}{5}$

49. $\dfrac{17}{3} + 4$ $\frac{29}{3}$

50. $\dfrac{13}{9} - 1$ $\frac{4}{9}$

51. $\dfrac{3}{15} + \dfrac{6}{10}$ $\frac{4}{5}$

52. $\dfrac{7}{5} - \dfrac{2}{15}$ $\frac{19}{15}$

53. $4\dfrac{3}{5} + \dfrac{3}{5}$ $5\frac{1}{5}$

54. $2\dfrac{1}{8} + \dfrac{3}{8}$ $2\frac{1}{2}$

55. $3\dfrac{1}{3} - 1\dfrac{2}{3}$ $1\frac{2}{3}$

56. $5\dfrac{1}{7} - 3\dfrac{2}{7}$ $1\frac{6}{7}$

APPLICATIONS

61. PERIMETER OF A TRIANGLE Each side of a triangle measures $2\dfrac{3}{7}$ centimeters. Find its perimeter (the sum of the lengths of its three sides). $7\frac{2}{7}$ cm

62. BUYING FENCING Each side of a square field measures $30\dfrac{2}{5}$ meters. How many meters of fencing is needed to enclose the field? $121\frac{3}{5}$ m

63. RUNNING A RACE Jim has run $6\dfrac{3}{10}$ kilometers of a 10-kilometer race. How far is the finish line? $3\frac{7}{10}$ km

64. SPRING PLOWING A farmer has plowed $12\dfrac{1}{3}$ acres of a $43\dfrac{1}{2}$-acre field. How much more needs to be plowed? $31\frac{1}{6}$ acres

65. PERIMETER OF A GARDEN The four sides of a garden measure $7\dfrac{2}{3}$, $15\dfrac{1}{4}$, $19\dfrac{1}{2}$, and $10\dfrac{3}{4}$ feet. Find the length of the fence needed to enclose the garden. $53\frac{1}{6}$ feet

66. MAKING CLOTHES A clothing designer requires $3\dfrac{1}{4}$ yards of material for each dress he makes. How much material will be used to make 14 dresses? $45\frac{1}{2}$ yards

67. MINORITY POPULATION In Illinois, 22% of the 11,431,000 citizens are nonwhite. How many are nonwhite?   2,514,820

68. QUALITY CONTROL Reject rates are high in the manufacture of active-matrix color liquid crystal computer displays. If 23% of a production run of 17,500 units are defective, how many units are acceptable?   13,475

69. FREEZE DRYING Almost all of the water must be removed when food is preserved by freeze drying. Find the weight of the water removed from 750 pounds of a food that is 36% water.   270 lb

70. PLANNING FOR GROWTH This year, sales at Positronics Corporation totaled $18.7 million. If the company's prediction of 12% annual growth is true, what will be next year's sales? $20.944 million

PRACTICE

In Exercises 71–78, do each operation.

71. $23.45 + 135.2$ 158.65

72. $345.213 - 27.35$ 317.863

73. $67.235 - 22.45$ 44.785

74. $12.17 + 3.457$ 15.627

75. $3.4 \cdot 13.2$ 44.88

76. $4.21 \cdot 2.73$ 11.4933

77. $0.23\overline{)1.0465}$ 4.55

78. $4.7\overline{)10.857}$ 2.31

In Exercises 79–86, use a calculator to do each operation, and round the answer to two decimal places.

79. $323.24 + 27.2543$
350.49

80. $843.45213 - 712.765$
130.69

81. $55.77443 - 0.568245$
55.21

82. $0.62317 + 1.3316$
1.95

83. $25.25 \cdot 132.179$
3,337.52

84. $234.874 \cdot 242.46473$
56,948.66

85. $0.456\overline{)4.5694323}$
10.02

86. $43.225\overline{)32.465758}$
0.75

APPLICATIONS

In Exercises 87–100, use a calculator to solve each problem. Round numeric answers to two decimal places.

87. FINDING AREAS Find the area of a square with a side that is 62.17 feet long. (*Hint:* $A = s \cdot s$.)
3,865.11 ft²

88. SPEED SKATING In tryouts for the Olympics, a speed skater had times of 44.47, 43.24, 42.77, and 42.05 seconds. Find the average time. (*Hint:* Add the numbers and divide by 4.) 43.13 sec

89. COST OF GASOLINE Juan drove his car 15,675.2 miles last year, averaging 25.5 miles per gallon of gasoline. The average cost of gasoline was $1.27 per gallon. Find the fuel cost to drive the car. $780.69

90. PAYING TAXES A woman earns $48,712.32 in taxable income. She must pay 15% tax on the first $23,000 and 28% on the rest. In addition, she must pay a Social Security tax of 15.4% on the total amount. How much tax will she need to pay?
$18,151.15

91. SEALING ASPHALT A rectangular parking lot is 253.5 feet long and 178.5 feet wide. A 55-gallon drum of asphalt sealer covers 4,000 square feet and costs $97.50. Find the cost to seal the parking lot. Sealer can be purchased only in full drums. $1,170

92. INSTALLING CARPET What will it cost to carpet a 23-by-17.5-foot living room and a 17.5-by-14-foot dining room with carpet that costs $29.79 per square yard? One square yard is 9 square feet.
$2,143.23

93. INVENTORY COSTS Each television costs $3.25 per day for warehouse storage. What are the warehousing costs to store 37 television sets for three weeks? $2,525.25

94. MANUFACTURING PROFITS A manufacturer of computer memory boards has a profit of $37.50 on each standard-capacity memory board and $57.35 for each high-capacity board. The sales department has orders for 2,530 standard boards and 1,670 high-capacity boards. To receive the greater profit, which order should be filled first?
the high-capacity boards

95. DAIRY PRODUCTION A Holstein cow will produce 7,600 pounds of milk each year, with a $3\frac{1}{2}\%$ butterfat content. Each year, a Guernsey cow will produce about 6,500 pounds of milk that is 5% butterfat. Which cow produces more butterfat?
the Guernsey

96. FEEDING COWS Each year, a typical dairy cow will eat 12,000 pounds of food that is 57% silage. To feed a herd of 30 cows, how many pounds of silage will a farmer use in a year? 205,200 lb

97. COMPARING BIDS Two contractors bid on a home remodeling project. The first bids $9,350 for the entire job. The second contractor will work for $27.50 per hour, plus $4,500 for materials. He estimates that the job will take 150 hours. Which contractor has the lower bid? the second contractor

98. CHOOSING A FURNACE A high-efficiency home heating system can be installed for $4,170, with an average monthly heating bill of $57.50. A regular furnace can be installed for $1,730, but monthly heating bills will average $107.75. After three years, which system is more expensive?
the high-efficiency furnace

99. CHOOSING A FURNACE Refer to Exercise 98. Which furnace system is the more expensive after five years? the regular furnace

100. MORTGAGE INTEREST A mortgage carries an annual interest rate of 9.75%. Each month, the bank's interest charge is one-twelfth of 9.75% of the outstanding balance, currently $72,363. Find the amount of interest that will be paid this month.
$587.95

WRITING

101. Describe how you would find the common denominator of two fractions.

102. Explain how to convert an improper fraction into a mixed number.

103. Explain how to convert a mixed number into an improper fraction.

104. Explain how you would decide which of two decimal fractions is the larger.

Appendix II

Statistics

In this appendix, you will learn about

The mean ■ The median ■ The mode

Introduction Statistics is a branch of mathematics that deals with the analysis of numerical data. In statistics, three types of averages are commonly used as measures of central tendency of a distribution of numbers: the *mean*, the *median*, and the *mode*.

The Mean

We have previously discussed the mean of a distribution.

The mean

The **mean** of several values is the sum of those values divided by the number of values.

$$\text{Mean} = \frac{\text{sum of the values}}{\text{number of values}}$$

EXAMPLE 1

Physiology. As part of a class project, a student measured ten people's reaction time to a visual stimulus. Their reaction times (in hundredths of a second) were

0.36, 0.24, 0.23, 0.41, 0.28, 0.25, 0.20, 0.28, 0.39, 0.26

Find the mean reaction time.

Solution To find the mean, we add the values and divide by the number of values.

$$\text{Mean} = \frac{0.36 + 0.24 + 0.23 + 0.41 + 0.28 + 0.25 + 0.20 + 0.28 + 0.39 + 0.26}{10}$$

$$= \frac{2.9}{10}$$

$$= 0.29$$

The mean reaction time is 0.29 second.

EXAMPLE 2

Banking When the mean (average) daily balance of a checking account falls below $500 in any week, the customer must pay a $20 service charge. What minimum balance must a customer have on Friday to avoid a service charge? See Figure II-1.

FIGURE II-1

Security Savings		
Day	**Date**	**Daily balance**
Mon	5/09	$670.70
Tues	5/10	$540.19
Wed	5/11	−$60.39
Thurs	5/12	$475.65
Fri	5/13	

ANALYZE THE PROBLEM We can find the mean (average) daily balance for the week by adding the daily balances and dividing by 5. If the mean is $500 or more, there will be no service charge.

FORM AN EQUATION We can let x = the minimum balance needed on Friday and translate the words into mathematical symbols.

| The sum of the five daily balances | divided by | 5 | is | $500 |

$$\frac{670.70 + 540.19 + (-60.39) + 475.65 + x}{5} = 500$$

SOLVE THE EQUATION

$$\frac{670.70 + 540.19 + (-60.39) + 475.65 + x}{5} = 500$$

$$\frac{1{,}626.15 + x}{5} = 500 \qquad \text{Simplify the numerator.}$$

$$5\left(\frac{1{,}626.15 + x}{5}\right) = 5(500) \qquad \text{Multiply both sides by 5.}$$

$$1{,}626.15 + x = 2{,}500$$

$$x = 873.85 \qquad \text{Subtract 1,626.15 from both sides.}$$

STATE THE CONCLUSION On Friday, the account balance must be at least $873.85 to avoid a service charge.

CHECK THE RESULT Check the result by adding the five daily balances and dividing by 5. ∎

The Median

The median

The **median** of several values is the middle value. To find the median of several values,

1. Arrange the values in increasing order.
2. If there is an odd number of values, choose the middle value.
3. If there is an even number of values, add the middle two values and divide by 2.

| EXAMPLE 3 | **Finding the median.** In Example 1, the following values were the reaction times of ten people to a visual stimulus. |

$$0.36, 0.24, 0.23, 0.41, 0.28, 0.25, 0.20, 0.28, 0.39, 0.26$$

Find the median of these values.

Solution   To find the median, we first arrange the values in increasing order:

$$0.20, 0.23, 0.24, 0.25, \boxed{0.26}, \boxed{0.28}, 0.28, 0.36, 0.39, 0.41$$

Because there is an even number of values, the median will be the sum of the middle two values, 0.26 and 0.28, divided by 2. Thus, the median is

$$\text{Median} = \frac{0.26 + 0.28}{2} = 0.27$$

The median reaction time is 0.27 second.   ■

The Mode

The mode

The **mode** of several values is the value that occurs most often.

| EXAMPLE 4 | **Finding the mode.** Find the mode of the following values. |

$$0.36, 0.24, 0.23, 0.41, 0.28, 0.25, 0.20, 0.28, 0.39, 0.26$$

Solution   Since the value 0.28 occurs most often, it is the mode.   ■

If two different numbers in a distribution tie for occuring most often, there are two modes, and the distribution is called **bimodal**.

Although the mean is probably the most common measure of average, the median and the mode are frequently used. For example, workers' salaries are usually compared to the median (average) salary. To say the modal (average) shoe size is 10 means that a shoe size of 10 occurs more often than any other shoe size.

STUDY SET

Appendix II

PRACTICE

In Exercises 1–3, use the following distribution of values:
7, 5, 9, 10, 8, 6, 6, 7, 9, 12, 9.

1. Find the mean.   8

2. Find the median.   8

3. Find the mode.   9

In Exercises 4–6, use the following distribution of values:
8, 12, 23, 12, 10, 16, 26, 12, 14, 8, 16, 23.

4. Find the median.   13

5. Find the mode.   12

6. Find the mean.   15

7. Find the mean, median, and mode of the following values: 24, 27, 30, 27, 31, 30, and 27.   28, 27, 27

8. Find the mean, median, and mode of the following golf scores: 85, 87, 88, 82, 85, 91, 88, and 88. 86.75, 87.5, 88

APPLICATIONS

9. FOOTBALL The gains and losses made by a running back on seven plays were −8 yd, 2 yd, −6 yd, 6 yd, 4 yd, −7 yd, and −5 yd. Find his average (mean) yards per carry.   −2 yd

10. SALES If a clerk had the sales shown in Illustration 1 for one week, find the mean of her daily sales. $1,211

ILLUSTRATION 1

Monday	$1,525
Tuesday	$ 785
Wednesday	$1,628
Thursday	$1,214
Friday	$ 917
Saturday	$1,197

11. VIRUSES Illustration 2 gives the approximate lengths (in centimicrons) of the viruses that cause five common diseases. Find the mean length of the viruses.   74.5 centimicrons

ILLUSTRATION 2

Polio	2.5
Influenza	105.1
Pharyngitis	74.9
Chicken pox	137.4
Yellow fever	52.6

12. SALARIES Ten workers in a small business have monthly salaries of $2,500, $1,750, $2,415, $3,240, $2,790, $3,240, $2,650, $2,415, $2,415, and $2,650. Find the average (mean) salary. $2,606.50

13. JOB TESTING To be accepted into a police training program, a recruit must have an average (mean) score of 85 on a battery of four tests. If a candidate scored 78 on the oral test, 91 on the physical test, and 87 on the psychological test, what is the lowest score she can obtain on the written test and be accepted into the program?   84

14. GAS MILEAGE Mileage estimates for four cars owned by a small business are shown in Illustration 3.

If the business buys a fifth car, what must its mileage average be so that the five-car fleet averages 20.8 mpg? 24.5 mpg

ILLUSTRATION 3

Model	City mileage (mpg)
Chevrolet Lumina	20.3
Jeep Cherokee	14.1
Ford Contour	28.2
Dodge Caravan	16.9

15. SPORT FISHING The weights (in pounds) of the trophy fish caught one week in Catfish Lake were 4, 7, 4, 3, 3, 5, 6, 9, 4, 5, 8, 13, 4, 5, 4, 6, and 9. Find the median and modal averages of the fish caught.   5 lb, 4 lb

16. SALARIES Find the median and mode of the ten salaries given in Exercises 12. $2,575, $2,415

17. FUEL EFFICIENCY The ten most fuel-efficient cars in 1997, based on manufacturer's estimates, are shown in Illustration 4. Find the median and mode of the city mileage estimates.   median: 29.5, mode: 29

18. FUEL EFFICIENCY Use the data in Illustration 4 to find the median and mode of the highway mileage estimates.   median 38; mode: 38

ILLUSTRATION 4

Model	mpg city/hwy
Geo Metro LSi	39/43
Honda Civic HX coupe	35/41
Honda Civic LX sedan	33/38
Mazda Protégé	31/35
Nissan Sentra GXE	30/40
Toyota Paseo	29/37
Saturn SL1	29/40
Dodge Neon Sport Coupe	29/38
Hyundai Accent	29/38
Toyota Tercel DX	28/38

WRITING

19. Explain why the mean of two numbers is halfway between the numbers.

20. Can the mean, median, and mode of a distribution be the same number? Explain.

21. Must the mean, median, and mode of a distribution be the same number? Explain.

22. Can the mode of a distribution be greater than the mean? Explain.

Appendix III

Roots and Powers

n	n^2	$\sqrt{n}$	n^3	$\sqrt[3]{n}$	n	n^2	$\sqrt{n}$	n^3	$\sqrt[3]{n}$
1	1	1.000	1	1.000	51	2,601	7.141	132,651	3.708
2	4	1.414	8	1.260	52	2,704	7.211	140,608	3.733
3	9	1.732	27	1.442	53	2,809	7.280	148,877	3.756
4	16	2.000	64	1.587	54	2,916	7.348	157,464	3.780
5	25	2.236	125	1.710	55	3,025	7.416	166,375	3.803
6	36	2.449	216	1.817	56	3,136	7.483	175,616	3.826
7	49	2.646	343	1.913	57	3,249	7.550	185,193	3.849
8	64	2.828	512	2.000	58	3,364	7.616	195,112	3.871
9	81	3.000	729	2.080	59	3,481	7.681	205,379	3.893
10	100	3.162	1,000	2.154	60	3,600	7.746	216,000	3.915
11	121	3.317	1,331	2.224	61	3,721	7.810	226,981	3.936
12	144	3.464	1,728	2.289	62	3,844	7.874	238,328	3.958
13	169	3.606	2,197	2.351	63	3,969	7.937	250,047	3.979
14	196	3.742	2,744	2.410	64	4,096	8.000	262,144	4.000
15	225	3.873	3,375	2.466	65	4,225	8.062	274,625	4.021
16	256	4.000	4,096	2.520	66	4,356	8.124	287,496	4.041
17	289	4.123	4,913	2.571	67	4,489	8.185	300,763	4.062
18	324	4.243	5,832	2.621	68	4,624	8.246	314,432	4.082
19	361	4.359	6,859	2.668	69	4,761	8.307	328,509	4.102
20	400	4.472	8,000	2.714	70	4,900	8.367	343,000	4.121
21	441	4.583	9,261	2.759	71	5,041	8.426	357,911	4.141
22	484	4.690	10,648	2.802	72	5,184	8.485	373,248	4.160
23	529	4.796	12,167	2.844	73	5,329	8.544	389,017	4.179
24	576	4.899	13,824	2.884	74	5,476	8.602	405,224	4.198
25	625	5.000	15,625	2.924	75	5,625	8.660	421,875	4.217
26	676	5.099	17,576	2.962	76	5,776	8.718	438,976	4.236
27	729	5.196	19,683	3.000	77	5,929	8.775	456,533	4.254
28	784	5.292	21,952	3.037	78	6,084	8.832	474,552	4.273
29	841	5.385	24,389	3.072	79	6,241	8.888	493,039	4.291
30	900	5.477	27,000	3.107	80	6,400	8.944	512,000	4.309
31	961	5.568	29,791	3.141	81	6,561	9.000	531,441	4.327
32	1,024	5.657	32,768	3.175	82	6,724	9.055	551,368	4.344
33	1,089	5.745	35,937	3.208	83	6,889	9.110	571,787	4.362
34	1,156	5.831	39,304	3.240	84	7,056	9.165	592,704	4.380
35	1,225	5.916	42,875	3.271	85	7,225	9.220	614,125	4.397
36	1,296	6.000	46,656	3.302	86	7,396	9.274	636,056	4.414
37	1,369	6.083	50,653	3.332	87	7,569	9.327	658,503	4.431
38	1,444	6.164	54,872	3.362	88	7,744	9.381	681,472	4.448
39	1,521	6.245	59,319	3.391	89	7,921	9.434	704,969	4.465
40	1,600	6.325	64,000	3.420	90	8,100	9.487	729,000	4.481
41	1,681	6.403	68,921	3.448	91	8,281	9.539	753,571	4.498
42	1,764	6.481	74,088	3.476	92	8,464	9.592	778,688	4.514
43	1,849	6.557	79,507	3.503	93	8,649	9.644	804,357	4.531
44	1,936	6.633	85,184	3.530	94	8,836	9.695	830,584	4.547
45	2,025	6.708	91,125	3.557	95	9,025	9.747	857,375	4.563
46	2,116	6.782	97,336	3.583	96	9,216	9.798	884,736	4.579
47	2,209	6.856	103,823	3.609	97	9,409	9.849	912,673	4.595
48	2,304	6.928	110,592	3.634	98	9,604	9.899	941,192	4.610
49	2,401	7.000	117,649	3.659	99	9,801	9.950	970,299	4.626
50	2,500	7.071	125,000	3.684	100	10,000	10.000	1,000,000	4.642

Index